29th Ethylene Producers Conference 2017

Topical Conference at the 2017 AIChE Spring Meeting and 13th Global Congress on Process Safety

San Antonio, Texas, USA
26 - 30 March 2017

ISBN: 978-1-5108-4141-3

Printed from e-media with permission by:

Curran Associates, Inc.
57 Morehouse Lane
Red Hook, NY 12571

Some format issues inherent in the e-media version may also appear in this print version.

Printed by Curran Associates, Inc. (2017)

For permission requests, please contact AIChE
at the address below.

AIChE
120 Wall Street, FL 23
New York, NY 10005-4020

Phone: (800) 242-4363
Fax: (203) 775-5177

www.aiche.org

Additional copies of this publication are available from:

Curran Associates, Inc.
57 Morehouse Lane
Red Hook, NY 12571 USA
Phone: 845-758-0400
Fax: 845-758-2633
Email: curran@proceedings.com
Web: www.proceedings.com

TABLE OF CONTENTS

482030 AGILE Award Keynote Address

Abstract:

Factors in Selecting AGILE Honoree

1. Overall sustained excellent industry leadership in businesses that employ chemical engineers

2. Recognized for efforts in innovation and creating strategies that ensure success

3. Value of the keynote to the meeting attendees and the members of the Institute Schedule

Event	Time	Location
Continental Breakfast	7:00AM - 8:00AM	Henry Gonzalez Convention Center, Lila Cockrell Theatre
AGILE Keynote Address	8:00AM - 9:15AM	Henry Gonzalez Convention Center, Lila Cockrell Theatre

The 2017 AIChE Spring Meeting and 13th GCPS AGILE Keynote Address will be presented by AGILE award recipient John Y. Televantos, Partner, Arsenal Capital Partners.

John Televantos is a Partner at Arsenal Capital Partners, a middle market private equity firm focused on the specialty industrial, healthcare and specialty financial service sectors. Before joining Arsenal in 2006, Dr. Televantos was President of Aqualon Company, owned by Hercules, Inc.; President and Founder of Helios Chemical Company; and Chief Executive Officer for Foamex International.

Dr. Televantos began his career with Union Carbide in 1977 in silicones research. There he developed two key processes for the manufacture of Methyl Chloride and Siloxane polymers that were commercialized. These processes not only improved efficiency and reduced cost but also dramatically reduced waste. He moved through a variety of management positions there and later at ARCO Chemical, where he became vice president of strategy and development.

At ARCO John co-invented and led a team that developed and commercialized a breakthrough technology in polyol manufacture (a key raw material for polyurethanes) eliminating a whole process step along with waste and emissions, while at the same time opening the door to the creation of new higher-performance polyurethanes. The technology has since been licensed by most of the major global polyol manufacturers with significant waste reduction worldwide.

At Arsenal Capital, Dr. Televantos serves as co-head of the specialty industrial practice. During his tenure, he has been part of several acquisitions and successful exits in the chemicals and materials sectors. In addition to serving on the boards of several Arsenal-owned companies, he is lead director on the board of Balchem Corp.

AIChE Paper Number 481064

Artificial Intelligence and Machine Learning for Ethylene Plant Operations

Ryan G. McClarren

Assistant Professor

Texas A&M University

John W. Adams, P.E.

Owner

John W. Adams, P.E.

Prepared for Presentation at the 2017 Spring National Meeting

San Antonio, Texas March 27, 2017

Artificial Intelligence and Machine Learning for Ethylene Plant Operations

Ryan G. McClarren

Assistant Professor

Texas A&M University

John W. Adams, P.E.

Owner

John W. Adams, P.E.

Abstract: Ethylene plants represent a large capital investments with large economic returns. These plants have many operational factors that can be controlled and or accounted for. In this talk we present how advances in machine learning, including so-called deep learning, can be used to give real-time guidance and control. We cover the basics of these methods and give concrete examples of techniques such as long-term, short-term memory neural nets and other artificial intelligence (AI) techniques. We then will present details of how this can be applied to ethylene production and other industrial processes. The overall goal is to build a virtual foreman that takes all of the knowledge from operational history to make predictions that optimize operation based on production goals while minimizing downtime. Some possibilities for the system include predicting component failure and scheduling maintenance, as well as controlling the system to optimize output products as a function of feedstock, ambient conditions, and system state. Such as system is now possible without an investment in a large supercomputer, as would have been the case even 5 years ago. With the right system in place, AI can be a competitive advantage for an ethylene producer. Moreover, because machine earning *evolves* and improves over time, the investment in AI will appreciate rather than depreciate.

Introduction

In a typical process system there are production decisions that need to be made based on a host of control variables, constraints, and current conditions [1,2]. There may be a variety of feedstocks, catalysts, ambient conditions etc. to consider. At the same time, operators may wish to optimize for certain output products, and determine when to perform routine maintenance on the system. Moreover, process decisions typically involve nonlinear feedback in the system, making control variables difficult or impossible to optimize.

Situations such as these are typically handled by expert operators assisted by simulations that have decades of operating experience upon which to make decisions. Most of the knowledge of these operators is a combination of empirical knowledge based on experience and knowledge of how a system should behave based on theoretical knowledge. The operator generalizes from previous experience to predict what will happen in the system.

This generalization from past experience to make predictions is exactly what modern machine learning and artificial intelligence (AI) systems do to make decisions. These systems have shown remarkable breakthroughs in a variety of fields based on an ability to mine large datasets to uncover relations that are not obvious between system inputs and outputs.

The widely-publicized successes of AI have been primarily in the retail and tech industries. The Amazon and Netflix recommendation systems are two examples many people have encountered and take a large set of user information to derive commonalities in purchasing or watching behavior. In the realm of control, a team from Google developed a machine learning algorithm that could play Atari video games (such as Space Invaders) better than a human expert [4]. The system took as inputs the value of all the pixels on the screen as inputs and predicted what the number of points gained from a possible manipulation of the controls. The system would then choose the action that maximizes the points gained.

While these examples may not seem to directly apply to process systems, they do have similar hallmarks to the problems in the chemical processing industries. They also have high dimensional inputs and a large range of potential choices. Moreover, these applications require near real time response from the system. These, and other successes of machine learning indicate how process industries can benefit from this technology.

What is Artificial Intelligence and Machine Learning?

Artificial intelligence is the construction of computer systems that perform tasks that usually require human intervention: such as visual recognition, translation, or decision making. Machine learning is a subfield of AI that seeks to develop software the determines how to solve a problem based on data, and can improve and evolve itself as it is exposed to new data.

In artificial intelligence a common tool is the neural network. A neural network works by defining artificial neurons, an analog of the neurons in the human brain. These neurons take inputs, possibly from other neurons. A neuron takes a linear combination of the inputs ("x"), multiplied by a set of weights ("w"), and gives an output that is a function of that linear combination. If the signal to a neuron is strong enough, it produces output to the next layer of neurons in an analogous way to biological neurons. In the figure below, the neuron (represented as a circle), has three inputs and a single output. The output will be a function of the inputs as $y = f(w_1x_1 + w_2x_2 + w_3 x_3)$.

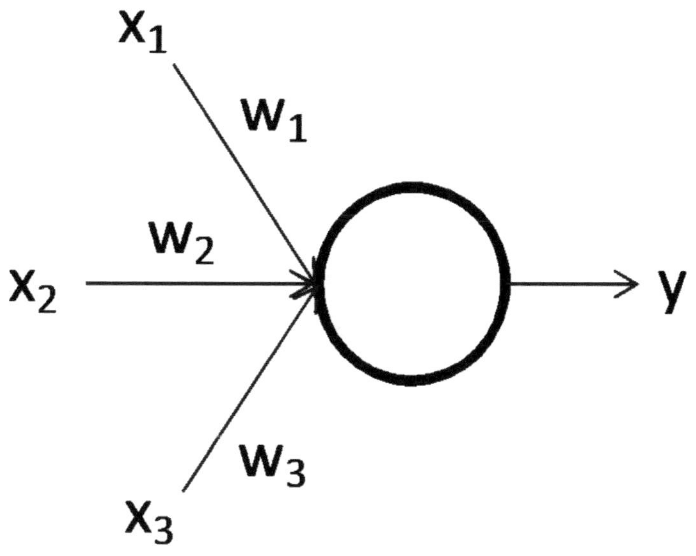

Figure 1: An artificial neuron takes inputs (x's) modified by weights (w's) to produce an output y. The weights are determined by an optimization procedure to fit the known data.

The power of neural networks arises from coupling many layers of neurons, from the original inputs to an output. In the example below there are three inputs to a layer of 4 neurons that then maps to a single output.

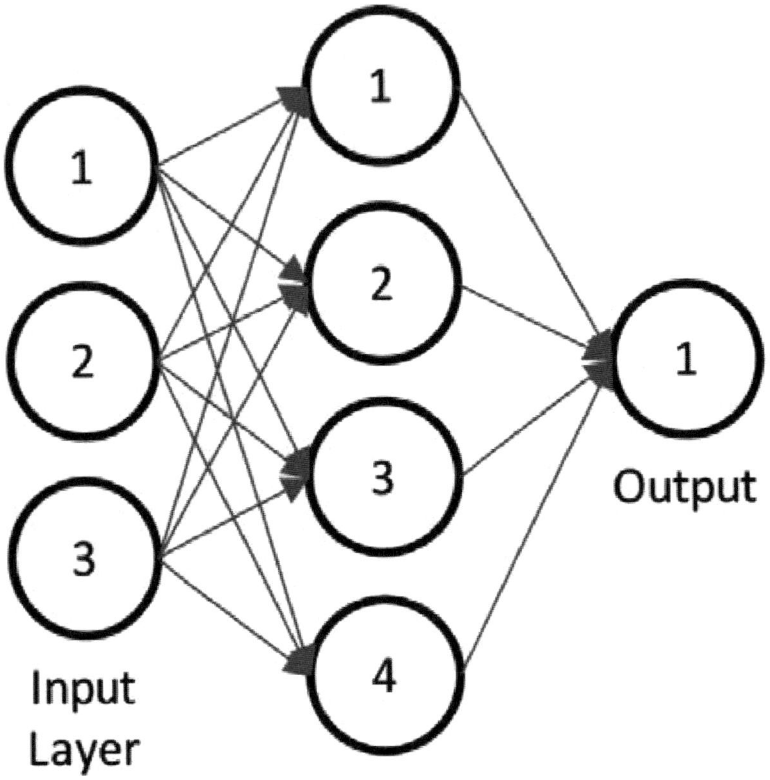

Figure 2: There can be many layers between the input and output, the more layers that exist the more flexible and powerful the network can be.

Machine learning systems take examples of relationships between inputs and outputs to create generalized relationships. These can take several forms from simple linear regression to sophisticated deep learning neural networks. Then, using these relationships, the system determines what is the best action to take given the current state of the system.

A fundamental characteristic of machine learning systems is that the more data they have been exposed to, the better they will perform. That is because the current generation of machine learning systems are flexible in the way they can model reality and adapt and change themselves to deal with newly discovered phenomenon. A common saying in the machine learning community is that more data will beat out the best algorithms and simulations---that is why Google wants to give you free email and searches.

Why is this technology not being used already?

Given all the benefits of machine learning and AI, it seems that these systems should widely be applied to chemical process systems, yet their adoption has been slow. A large part of the lack of adoption in the industry is a human capital issue. Most of the people trained in machine learning and AI are trained for, and employed at technology firms in Silicon Valley and at other places in the tech world. Secondly, there has been a lack of technological communication between the AI community and the "analog" world in general and process systems in particular.

Without a knowledgeable liaison between the AI and chemical process system fields, there is not likely to be an opportunity to leverage the benefits that come with AI in terms of efficiency, cost savings, and system optimization. Everyone from executives down to the process technicians needs some level of understanding of how the AI systems work; what are the limitations, and what the benefits are. This requires a team that can speak the languages of both chemical engineering and machine learning to work with the operations team at an operating level.

The emergence of cheap computing platforms to implement AI has only recently become cost effective for production applications. Many of the algorithms and techniques for machine learning have been around for decades, but there were not fast (and inexpensive) enough computers to implement the algorithms. Given that the technologies had yet to be proven, it was not prudent to invest dollars in computing hardware to deploy an AI system.

We are now at an inflection point in the technology, the computing hardware needed to implement sophisticated AI is about a factor of 100 less expensive thanks to the breakthroughs in general purpose graphics processing units, and other inexpensive number-crunching hardware. Furthermore, the tools to apply and deploy AI have been democratized in the sense that they no longer exist only in the hands of a few academic labs and specialized companies. There are now several open-source machine learning libraries, as well as proprietary solutions that can be used to implement machine learning algorithms on a variety of systems. We believe that AI will revolutionize the chemical industry, the question is who will be the firms that take advantage of its promise, and who will be playing catch up.

Case Study: Olefin Production in Ethylene Steam Cracking Plants

In an olefins steam cracker there are many inputs to the process. These include (but are certainly not limited to) the amount of heat applied, the steam flow, the composition of the input feedstock, and the quality and condition of furnace tube coking, to name a few. Furthermore, there are several outputs that need to be optimized: the output ethylene, propylene, and other less desirable products; co-generation system optimization, and the output steam and energy. There are other considerations outside the outputs themselves such as when to perform decoking maintenance.

For each possible condition the AI system can predict the outputs as a function of the severity and inputs. These predictions are then used to optimize the plant efficiency, and make decisions to balance the competing interests.

One of the outputs of the AI system would be a recommendation as to how to operate the plant given the desired outputs. Additionally, managers in the organization can use the system to perform analyses of different cases such as what is the effect of changing input feedstock to a lower quality in terms of output qualities. This can be used to make decisions in the financial and planning groups about feedstock and other long and short term considerations.

How to implement AI in an existing plant

Each ethylene plant is a unique system, this means that the AI system will have to be trained based on the operations of the specific plant. In the initial phases of operation the AI system will be in learning mode where it observes the effects of different inputs, operator decisions, etc. It is in this phase where the system learns how to make predictions based on the historical plant situation.

The training process for the AI system is where the weights in the neural network are determined. This is done through a training procedure such as backpropagation where the weights are updated to improve the overall accuracy of the neural network on the available data. The learning phase is said to be complete when the prediction accuracy of the system is considered to be satisfactory.

When the learning phase is complete, the network can then be used to make predictions about plant operations. The operation of the system will then go into a mode where it operates alongside a human operator, much like an autonomous vehicle with an engineer in the driver's seat. The AI system will suggest operating decisions based on constraints and goals as determined by plant engineers and management. During this phase, the system can also be used by other members of the organization for scenario planning and profit optimization studies.

Even when the AI system is actively predicting system response and suggesting operator action, it will continue to be trained using actual plant data. This guarantees that the system improves and evolves over time: it will be an asset that appreciates rather than depreciates.

The role of the AI system could potentially transition to a phase where it operates more autonomously, but this is not a required end goal. After a long period of successful performance, the operation of several plants could be centralized so that a smaller staff could manage operations or the operation team at a single plant could be modified to have a smaller number of experienced engineers in favor of a team that has more technicians with a few engineers. The decision to make such a transition will, of course, depend on many factors outside of the actual AI performance, such as perception of risk or other safety features that are built into the plant.

The Symbiosis of AI and Simulators

There has been much work in the industry to develop plant simulators that predict a variety of system responses given plant operations. In many regards the goals of these simulators overlap with some of the goals for AI. In fact an AI system can be used to make a simulator more effective.

One way that machine learning and AI can enhance a simulator is by using the simulator to train AI to predict the simulator output. Such an AI system is commonly called a surrogate model or emulator. The emulator is trained by running the simulator in many different scenarios. The resulting outputs from the simulator are used to train the emulator. When the emulator is trained to an adequate precision, it is then used instead of the simulator to predict plant output. Because it is computationally inexpensive to get predictions from the emulator, it is the emulator that used in scenario planning, and other optimizations.

The state of the art AI systems that underlie the emulator will provide estimates for the uncertainty in the emulator result. This uncertainty estimate can be used to decide when additional simulations are needed to improve the emulator. In this sense an AI system can be bolted onto an existing simulator system for a significant value add to the end client.

The combination of an AI system with a simulator can extend beyond the emulator concept. AI can be used to assimilate data from actual plant operations to correct the simulator or emulator. As originally developed in the statistics community such a system for correcting simulation with experimental data is called a predictive model. The most well-known predictive model is the Kennedy-O'Hagen model. Consider an output of a simulator Y(x) and the actual plant realized output $Y_a(x)$ with x representing a vector of plant conditions. There will be a difference between the simulator response and the actual plant behavior that we can write as a function d(x). Therefore, the plant output is written as $Y_a(x) = Y(x) + d(x)$. The function d(x) can be learned in the AI system based on plant data and simulator/emulator results. In this manner, the AI system improves the simulator over time because more results for the AI system to learn d(x) on, creates a more accurate correction.

References

1. Gómez-Prado, J., Hutton, D. (2017) . Use the right model to unlock utility system potential. *Hydrocarbon Processing,* February.
2. Hay, G., Rasouli, G., Carbognani-Arambarri, L., Suzuki, R., Urata, K., Inoue, M. (2017). Reduce coke formation and save operating costs with optimization of DMDS into ethane cracking furnaces. *Hydrocarbon Processing,* February.

3. Kennedy, M. C., & O'Hagan, A. (2001). Bayesian calibration of computer models. Journal of the Royal Statistical Society: Series B (Statistical Methodology), 63(3), 425-464.
4. Mnih, V., Kavukcuoglu, K., Silver, D., Rusu, A. A., Veness, J., Bellemare, M. G., ... & Petersen, S. (2015). Human-level control through deep reinforcement learning. *Nature*, *518*(7540), 529-533.
5. Wang, H., Wang, Z., Mei, H., Qian, F.,and Tang, Z.,(2011), Intelligent severity control stabilizes ethylene crackers. Oil and Gas Journal, 109 (6):104-109.
6. Zhao, G., Teng, J., Xie, Z., Jin, W., Yang, W., Chen, Q. and Tang, Y., (2007). Effect of phosphorus on HZSM-5 catalyst for C 4-olefin cracking reactions to produce propylene. *Journal of Catalysis*, *248*(1), pp.29-37.

AIChE Paper Number: 2B

Tailoring Big Data Analytics for Optimizing Ethylene Plant Asset Performance

Dr. Pratap Nair

President & CEO

Ingenero

Prepared for Presentation at the 2017 Spring National Meeting

San Antonio, Texas, March, 27, 2017

Tailoring Big Data Analytics for Optimizing Ethylene Plant Asset Performance

Dr. Pratap Nair

Getting the best out of the existing ethylene assets is a key result area for manufacturers. Enhancing the capability to run the plant reliably and safely while producing the required quality product and understanding the gaps between the best production that can be achieved and current operation and finding ways to close this gap are always operational goals.

The information to make this happen resides in the data that is typically collected, partially used, and mostly archived in manufacturing and business databases. The volume of available data, the velocity at which this data is collected and the sheer variety of data available from the plant and the supply chain (albeit often of questionable veracity in raw form) makes it impossible for unaided operating personnel to assimilate and get his/her mind around this data overload. However this "Big data" analysis problem has been getting significant press recently due to successful application in diverse applications.

Applying the "Big Data Analytics" concept to ethylene manufacturing requires tailoring to take advantage of past knowledge and best practice. Additionally combining it with techniques including Machine learning and the Industrial Internet of Things is necessary to achieve the goal. When combined with a methodology to ensure implementation and results, delivery of 5% - 20% increased throughput, improved reliability of operation and a substantial improvement in furnace run lengths and hence uptime, while proactively preventing safety incidents is achieved at ethylene sites where it has been implemented.

This methodology allows operating personnel to drive operations based on data-enhanced insights made readily available to them. It enables operations personnel and managers with the ability to uncover truths and insights that aren't readily obvious or don't align with conventional wisdom or intuition. The analysis from this methodology allows predictions and provides better prescriptive insights to better determine the best course of action, while continuously growing the experience "database" of managers and operating personnel at a faster pace over time. The methodology allows the improvements to be tracked, quantified and controlled, so that it can be sustained and improved continuously.

This paper describes how this Big Data Analytics methodology has been successfully applied to several ethylene plants and case examples are discussed.

Big Data Analytics: Process Manufacturing Perspective

Big data refers to immense and fast-growing volumes of different types of data that can be quickly collected and analyzed. Data collected from diverse sources of connected industrial devices and databases, social media commentary, customer service inquiries and other Web sources such as video, chat and online forums, is providing a tsunami of information. The driving interest in the big data movement is the combination of this data and the new analytic tools that enable incremental revenue gains. Big data analytics has been very effectively used in Media and Customer analysis.

Big data characterized by the four Vs, three of which (Volume, Velocity and Variety) were established early in Big Data driven journey (1), while the fourth V (Veracity) was added later.

In the context of Process Manufacturing, there is a data overload resulting from large Volumes of a wide Variety of data from diverse sources (Historian giving raw measurement data for diverse parameters, Laboratory Information Management systems (LIMS), Enterprise Resource Planning (ERP), Operating logs, Engineering data sheets). The raw measurements are collected at a rapid Velocity (in order of seconds). The data collected is fraught with gross and random errors, typical of process measurements, requiring the Veracity of the data to be checked before analysis. The volume of data being collected and accumulated, the types of data collected and the rate at which it is being collected (the original 3 Vs), have all increased substantially over time, while the Veracity of the data remains a challenge.

The objective is to combine Big Data with Analytics to enable brilliant decisions quickly based on solid evidence and insights creating the elusive "single version of the truth". Analytics has always been a part of managing process operations. However processing limitations and high system costs often meant working with limited data sets, which could compromise analytics accuracy. The phenomenal growth in data storage and data processing capacities has now enabled working with more complete data sets ("Big Data") in process manufacturing, based on which we can look deeper and wider and get better about making more accurate and timely predictions. Big Data analytics provides a new set of eyes upon which prescriptive decisions can be made, reaching into places imperceptible to the unaided human mind and enables a journey towards more informed business and operational decisions.

The basis of most of the Customer analytics and Media analytics has been Machine Learning, which is combination of Data mining, Statistics, Artificial Intelligence and Linear and non-linear Programming (NLP/LP). In Machine Learning the models built and utilized are data driven, where we do not care which parameters drive the prediction, as long as the prediction is accurate on new data (Opaque modeling algorithms). This is adequately utilized to predict buying patterns or market trends. The interest is to generally how to respond to a market trend, more than understanding why a product's sales pattern is behaving in a particular way, partly because of the fickleness of trends in the business. The "what" aspect of data analysis has been the focus in the traditional Big Data applications, where the problem identification is the key that signals a possible action to take (2).

In Process Manufacturing, understanding the reason for a particular trend is equally important, for a problem resolution that does not lead to operational (Reliability and productivity) and safety issues. So the objective of Big Data Analytics in Process Manufacturing would be to:

- Find and interpret patterns in data
- Use for predictive purposes
- Extract meaningful relationships to use to improve and optimize the process

To achieve this objective in Process Manufacturing, Machine Learning techniques (commonly considered when Big Data Analytics is being discussed "traditionally") and domain knowledge captured in fundamental chemical engineering models and techniques can be used synergistically. Machine Learning models are suitable for process monitoring and fault detection activities, which rely on a good description of Normal Operating Conditions as a reference, to detect any assignable or significant deviation in behavior. The Machine Learning models could be used for some amount of prediction, but then the fundamental models are then used as surrogate models to explain the results from the opaque Machine Learning Models and for predictions outside the Normal Operating range.

The Need in Ethylene Plants

Productive, professional analytics initiatives should start with a defined business need and should yield quantifiable results.

Starting up and sustaining production levels consistently from an ethylene plant does not always match expectations. Typical challenges faced when operating an ethylene plant:

- Furnace run lengths below expectation
- Unexpected tube failures and lower than expected tube life
- Rapid coking
- Inefficient furnace decokes
- Equipment reliability and lower than expected uptime
- Below capacity operation
- Recurring fouling in TLEs, chillers, CGC circuit
- Utility limitations
- Refrigerant and caustic losses
- Lower than expected product recovery
- Higher than expected specific energy consumption
- Off spec product
- Excessive flaring

The resulting exposure and risk:

- Safety risks and compliance issues
- Decreased efficiencies and reduced reliability

- ➢ Lost opportunity for strategic change/modification
- ➢ Operational difficulties - lack of control resulting in fire-fighting mode
- ➢ Unplanned shutdowns, extended turnarounds, poor maintenance planning, and problematic start-ups

There is sufficient information and more, in the data that is typically collected and available from the Ethylene plant, analysis of which can address the above needs.

Typical Challenges to Addressing the Need Through Analytics

What is the source of these operating challenges and why is the required data analytics not being done?

Traditional plant operating paradigms contribute to some of the plant operating challenges observed:

- ➢ Plants usually operated based on personal experience of the operator/engineer, guided by SOPs provided by the licensor
- ➢ Early detection of anomalies often missed due to inability to uncover truths and insights that aren't readily obvious or don't align with conventional process wisdom or intuition
- ➢ Each process has its own idiosyncrasies often varying by 30% or more from a plant employing the same process technology. Differences from:
  - ✓ Design execution discrepancies
  - ✓ Raw material variations from that expected in the design
  - ✓ Outdated SOPs not changed to reflect the specific plant
  - ✓ Operators running using experience from another similar process that may not work for the given plant
- ➢ Relying on intuition based on prior experience alone, to make plant operating decisions, can pose challenges on account of the plant idiosyncrasies
  - ✓ It is important for people to have intuition and ideas, but they need to prove them out with data, before taking action to implement or taking no action on account of not having the conviction to back the idea
- ➢ While there is sufficient data to make informed decisions, it is not readily usable
- ➢ Available skills and or time to do the analysis

In a survey of business intelligence, analytics and data management professionals conducted by TDWI in the second half of 2015, the top two challenges that organizations face in trying to enable analytics processes into their system:

#1 – Poor data quality – 40%
#2 – Skills issue – 38%

The analysis of this data requires Data Scientists with sufficient analytic talent and an organization culture that encourages this activity. The team of data scientists need to have a background in Computer Science, Mathematics and have Domain expertise (familiarity with process and engineering modeling calculations and analysis). Unfortunately there are not enough skilled people to develop and execute such analytics projects that involves complex techniques (3).

This is further exacerbated by an activity paradigm for the domain experts, that does not allow them to spend the time required on such analytics work (Figure 1).

Figure 1

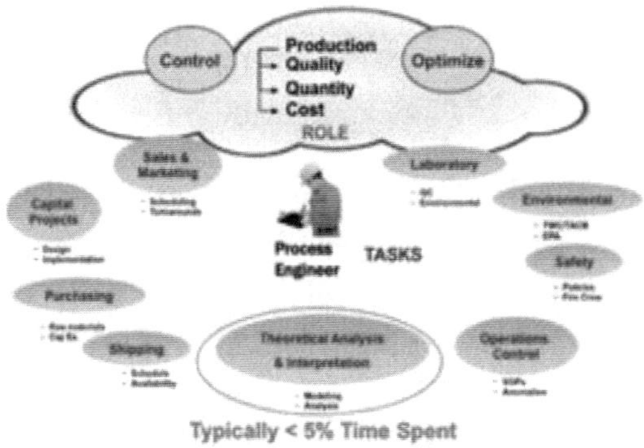

Activity paradigm for domain experts

As a result, 90% or more of the data is typically archived and not even looked at or analyzed.

Figure 2

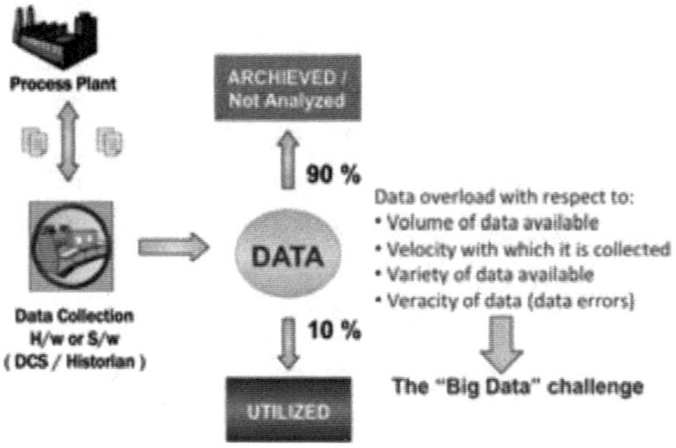

The Process Data Paradigm

Analytics is still a manpower and expertise intensive activity, particularly in the diagnostic, predictive and prescriptive analytics areas. Also, at the onset of setting up any analytics framework, specialists have to spend time building, testing, and refining models (data driven and/or fundamental models). Also, because plant assets age, business processes evolve and given the availability of new types of contextual data, model adjustments become a periodic activity.

Any form of analytics automation can improve data specialists' efficiencies and lower the enterprise professional services costs. However, this is not an easy proposition because it requires a platform that can easily ingest new data, make automatic adjustments to analytics models and even automate prescriptive analytics.

Analytics automation is in its infancy today, but its impact will be profound. The efficiencies it provides to analysts effectively "democratizes" analytics, allowing more problems to be solved and analytics use by more employees and business units (4).

Integrated Big Data Analytics – Application to Ethylene Plants

Figure 3 shows the synergistic utilization of Machine Learning and fundamental process and thermodynamic models, to synthesize and analyze data.

Figure 3

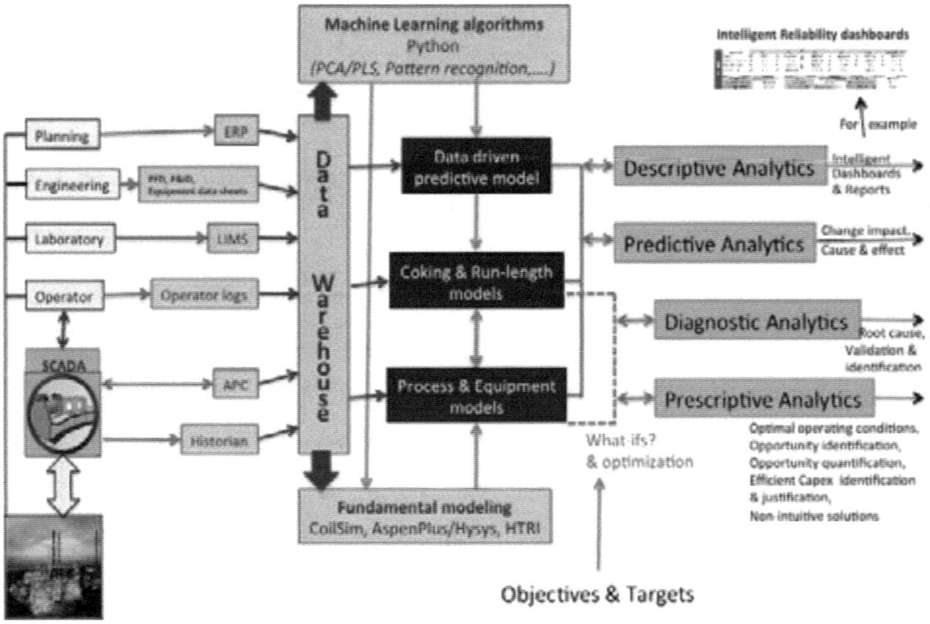

Big Data Analytics Process for Ethylene Plants

Data Inputs

The data comes in from various plants systems, like the Historian, Laboratory Information system, Operator logs, Engineering information, Control systems and ERP.

The data collected has all the characteristics attributed to big data in terms of the Volume, Velocity, Variety and Veracity. The data collected entails volumes of the order of several GB's of data for a typical ethylene plant every day. The data consists of minute averages across the time periods normally considered in a given day of a variety of data like temperatures, pressures, flows and composition measurements, descriptive notes, engineering data sheets and economic information. Typically about 25% of the data will be erroneous and identifying which ones to rely on is important. Analysis of current as well as data history over the past few relevant years of operation are important to gain more accurate insights, which runs into a few TB's of data.

The Analytics engine

Machine Learning does the primary heavy lifting in parsing and doing primary analysis of the data. Table 1 lists the techniques utilized as part of this step. Pattern recognition, PCA and PLS are however the most utilized it the ethylene plant application. Data is clustered, there is a primary identification of outliers and a model is developed based on the data, which is able to predict plant behavior, however only in the range in which the data inputs that were used. This model is effectively used to detect deviations in key process parameters from expected values and hence serving as an early alert for possible anomalies with the process, equipment and instrumentation.

Table 1

Basic data analytics techniques used as part of Machine Learning

Category	Technique	Applications
Classification	Discriminant Analysis Correspondence Analysis Principal Component Analysis	Grouping Similar Equipments/ Furnaces Variables that lead to different groups Outlier Detection
Clustering	k-means Clustering	Root Cause Analysis, Early Fault diagnosis
Multivariate Prediction	Partial Least Squares Principal Component Regression Multiple Linear Regression	Steady State Predictions
Neural Networks	Fuzzy Network	Reactor Modeling
Time Series Modeling	ARIMA, Box Jenkins	Furnace Run Length Prediction
Degradation Analysis	Curve Fitting	KPI monitoring
Hypothesis Testing	Distribution Fitting	Effect of new equipment, antifoulant on equipment performance

The *Fundamental models* are then used as "Surrogate" models, which are used as physically interpretable proxy models, to explain the results from Machine Learning models, which are often complex and opaque with respect to the why of an observation. The fundamental models used:

- Fundamental first principle Yield *models (reaction kinetics, mass & heat transfer, thermodynamics)*
    - COILSIM1D

- Steady state process simulation & optimization
    - Aspen suite/Hysys, FRNC/HTRI
- Computational Fluid Dynamics (CFD)
    - Fluent, OpenFOAM

Coking models are based on a combination of iterative runs between data driven models and Coilsim. Run lengths are predicted based on both Machine learning models as well as coking based models. Figure 4 shows a typical match between actual and predicted indicators of coking (COP and CPR). Furnace Coil Pressure Ratio predictions/forecasting supports furnace scheduling and corrective measures to improve/optimize run length

Figure 4

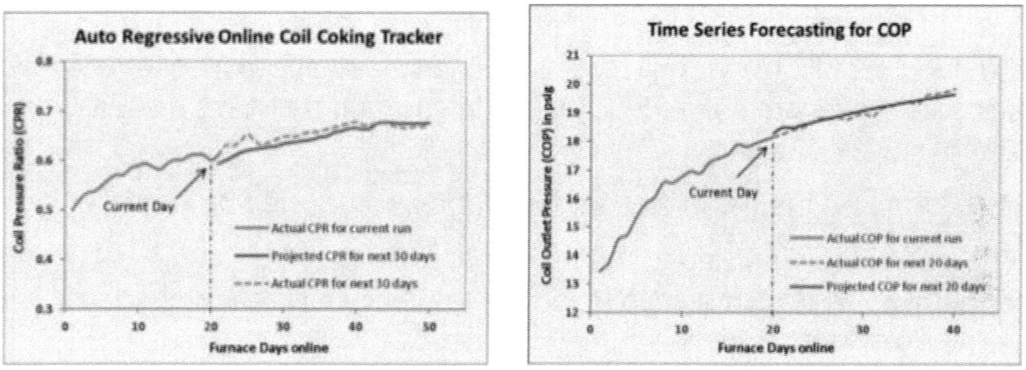

Predictive Coking Analysis

Utilization of models

Both the fundamental models and Machine Learning models are developed and then tuned to match the data collected. These combination of models are then used for:

1. **Descriptive analytics:** To provide a description of what happened. These are regular runs that would provide a daily material and energy balance, a reconciled daily production report, a reconciled yield and energy calculation report, processing costs down to the equipment level, faulty meters identification, equipment operating efficiencies, bottlenecks for the operation
2. **Predictive analytics:** To provide a prediction of what could happen, given the possible conditions that may occur. These are what-if type scenarios run using the models, giving the impact of changes, cause and effects.
3. **Prescriptive analytics:** What can be done to solve an issue being seen or expected to occur. Again typically a combination of what-if scenarios and optimization run using the models. Provides optimal operating conditions, opportunity quantification, capex justification, non-intuitive solutions, efficient capex options, value proposition assessment and justification of any operational change idea.
4. **Diagnostic analytics:** An analysis of why a certain something happened. Again based on what-ifs run using the models to do retro analysis to identify root causes, validation of

hypothesis, identification of key indicators for performance or a problem.

Examples of how the various types of models are used in conjunction with one another for various purposes for an ethylene plant application:

- Reaction kinetics model to compute yield and coking rate as a function of key parameters like Feed type and composition, Steam to Hydrocarbon ratio, Coil Outlet Temperature, Coil Inlet Pressure, Coil Outlet Pressure
    - ✓ Parallel data driven machine learning models allow early warning indications and quick primary adjustments, which can be cross verified by kinetic models
- Machine Learning models to predict and adjust run lengths to plan decoking cycle
- CFD to address flue gas channeling, flame impingement and hot spots on tubes and flow pattern in Transfer Line Exchanger inlet cone
- Combination of radiant and convection section models, aids to monitor and adjust burners to maximize run length and production while ensuring that tube metal temperatures are within allowable limits
    - ✓ Determines the parameter settings for maximum safe production
    - ✓ Control rate of fouling at the individual coil level
    - ✓ Further convection section models helps in overall thermal efficiency improvement (illustrative example in figure 5)
        - Prediction/evaluation of convection section individual pass performances
        - Monitoring of fouling in individual passes for both flue gas and process side
        - Identification of possibility of channeling of flue gas

Figure 5

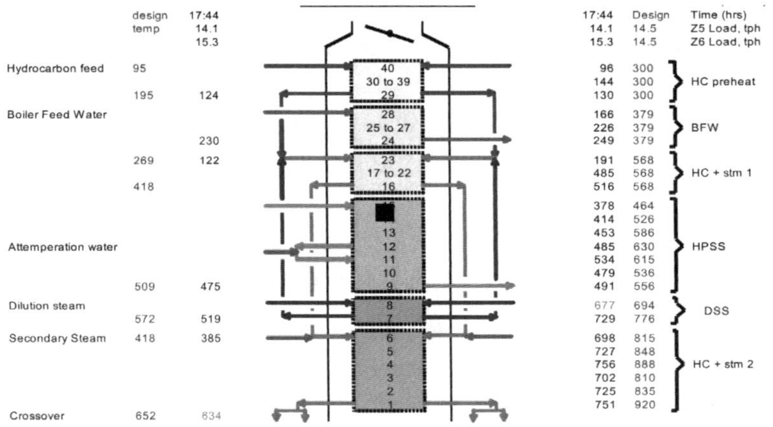

Capacity limitation: Dilution Steam Superheater (DSS) design skin temperature

Convection Section Model Utilization Example

(allows maximum throughput without exceeding design tube skin temperature

Presentation of results: Intelligent dashboards and interfaces

Data analytics is only as good as your ability to understand and share it, which behooves the need for effective presentation static, interactive and collaborative dashboards and interfaces. The Furnace Reliability Dashboard, a snapshot of which is shown in Figure 6, is an example of how the abstract calculations from the models are translated into anomaly detection intelligence.

Figure 6

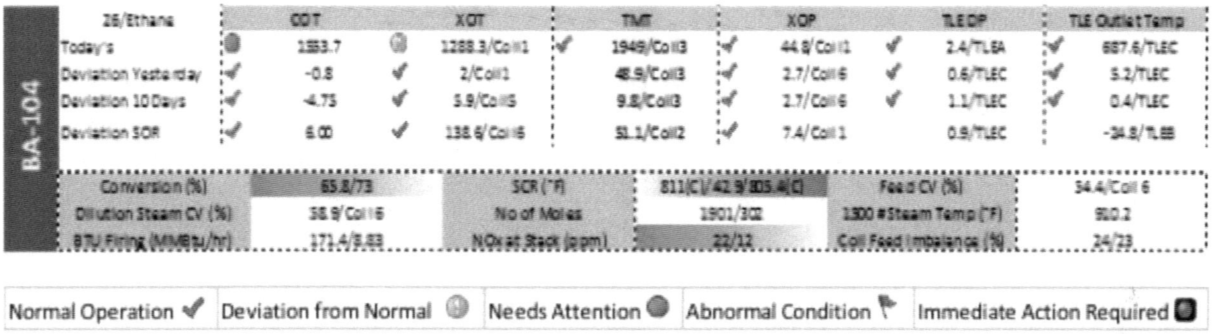

Furnace Reliability Matrix

On this dashboard, key furnace operating parameters like Coil Outlet Temperature, Tube metal temperatures, TLE pressure drops, etc. are monitored versus calculated parameters from the predictive machine learning models. The deviation limits itself are variable as per furnace cracking feed slate and number of days into the furnace run. Parameters are highlighted based on level of significance and criticality, as shown in the Figure, classified as normal operation, needing attention, flagging of an abnormal condition and where immediate attention required.

Case examples of model based data analytics at an ethylene plant

Case Example 1: Furnace Operation/Coil Balance

Key ethylene furnace parameters were optimized based on coil-wise results from the yield model and radiant section flux model to balance the overall furnace operation resulting in substantial improvement. Prevailing scenario:

Imbalanced Furnace Operation					
	CIP	Feed	Severity	Yield	COT
Coil AB	33.3	6.4	Max: 67.8% Min: 65.3% Avg: 67.1%	▲ 52.0	1560
Coil BA	34.5	6.3		▲ 51.6	1560
Coil BC	31.8	6.2		▲ 52.1	1560
Coil CB	40.3	6.6		▼ 50.3	1560
Coil CD	34.9	6.5		▭ 51.4	1560
Coil DC	35.1	6.2		▲ 51.5	1560

- Imbalanced furnace operation
    - ✓ Resulted in differential coking in coils (Coil Inlet Pressure & TLE ΔP deviations)
    - ✓ Was not controlled and lead to early termination of furnace run length
    - ✓ Affected furnace yields, operational efficiency and coil life
- Comprehensive Coilsim results and flux model were used to initiate actions to balance furnace operation
    - ✓ Selection of analyzer location
    - ✓ Setting uniform heat flux profile through burner adjustments
    - ✓ Adjustment in fuel gas bias, Coil Outlet Temperature bias, instrumentation corrections etc.

Corrective actions based on the analysis helped in improving overall furnace balance, resulting in sizeable performance improvement

Balanced Furnace Operation					
	CIP	Feed	Severity	Yield	COT
Coil AB	29.2	5.9	Max: 68% Min: 67.5% Avg: 67.8%	▲ 53.0	1556
Coil BA	30.9	5.9		▼ 52.5	1556
Coil BC	29.2	5.9		▬ 52.8	1556
Coil CB	29.2	5.9		▲ 52.8	1556
Coil CD	29.4	5.9		▲ 52.8	1556
Coil DC	29.3	5.9		▲ 52.9	1556

- ✓ Feed deviation reduced from 6% to 1%
- ✓ Difference in min-max conversions reduced from 2.5 to 0.5
- ✓ Coil inlet pressure rise observed to be uniform and rate of rise controlled well

- Benefits: Balanced operation resulted in uniform coking in coils, which helped in improving furnace performance
    - ✓ Scope for extending furnace run length and/or severity
    - ✓ Reduced coking in coils and fouling in Transfer Line Exchangers (TLEs)
        - ✓ Lower Coil inlet pressures and Coil Outlet Pressures
        - ✓ Lower coil Tube Metal Temperatures
        - ✓ Shorter decoke period
    - ✓ Improvement in ethylene yield by 1.3 wt%

Case Example 2: Early Detection Of Measurement Drift

Abnormal deviation in furnace operating parameters was accurately detected and subsequent actions helped in reducing differential coking and thereby extending furnace run length using Principal Component Analysis (PCA) analysis as part of the Machine Learning analysis. This approach saves the effort of looking at multiple trends for multiple variables and judging faults or abnormal operation based on "mental" models

Model predictions ~ 10th day: Normal Performance

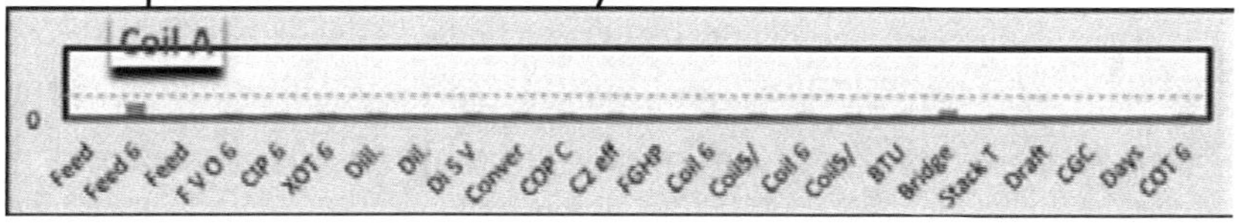

Model predictions ~ 15th day: Some Variation in Performance

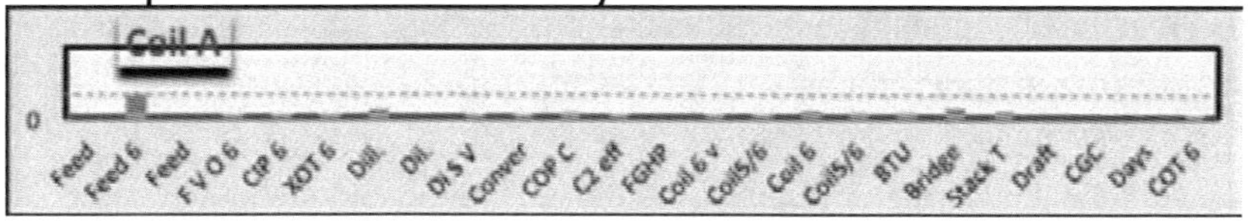

Model predictions ~ 25th day: Differential coking observed

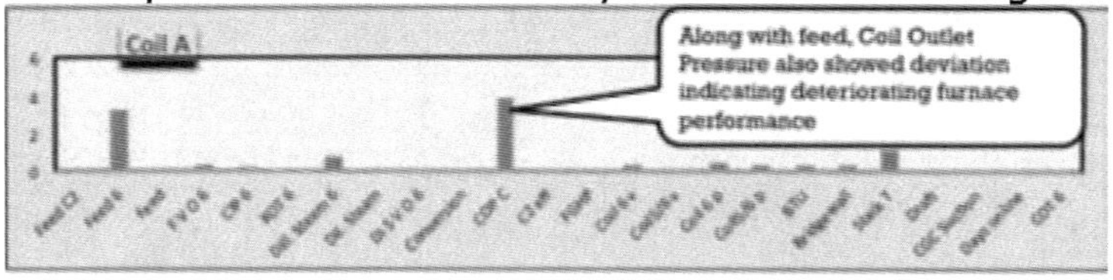

- ✓ Model predictions on ~25th day of furnace operation indicated:
  - ➢ Higher rate TLE fouling, which may affect furnace run length
  - ➢ Deviation in coil feed flow possibly affecting coil performance/coking
- ✓ Based on further analysis, it was identified that drift in indicated coil outlet temperature (COT) which controls the coil feed flow was causing higher coking
- ✓ Instrumentation error in COT indication was corrected, additional actions like heat flux balancing was carried out
- ✓ Facilitated in extending the furnace run length considerably
  - ➢ Differential coking could have restricted run length to 35 – 40 days
  - ➢ Early identification and corrective actions extended furnace run length to 55 days

Key associated elements of an analytics process for success and sustainability

Ease of Use

The algorithms and models used and the results of the analysis itself are often complex and there would be very few who will be able to use it as is. While the development and configuration requires significant expertise levels, combining Data Science, Process and Thermodynamic modeling and process operations, the following are aspects that are necessary for sustainability and to operationalize the analytics:

- ✓ Intelligent dashboards to present the results providing a bird's eye view with context for identification of problems and enabling of action, reducing the cycle time from problem identification to solution and action.
- ✓ Building an abstraction layer that hides the technical complexity of the algorithm development process from the analysts who would use it regularly – dumb it down so that the analysts can get their job done faster, to prevent them from getting bogged down in data engineering and preparation work and allow them to run complex queries

- ✓ Utilize the Internet of Things (IoT), to take data to the experts versus bringing the experts to the data. IoT is not about a particular product, but about how people can benefit from use of connected devices on the Internet, enabling better utilization of technology (5,6).
    - o The modeling and analytics can be more effectively and efficiently done a convenient location, remote from the manufacturing plant, where data scientists and analysts are readily available and can be focused, while data inputs and outputs can reside onsite at the production facility.
    - o Capability of the Internet and security measures available today, makes this a feasible proposition.

The follow through process

The follow through process and aspects of the analytics as shown in Figure 7 are equally important for sustainability:

- ✓ Tracking of results from analysis
- ✓ Capturing lessons learnt into a Knowledge management (KM) database to accumulate prior intelligence, thus reducing the time on subsequent analysis of similar situations
- ✓ Utilization of rules based on the KM system to feed the Intelligent dashboards
- ✓ Correcting or improvising previous understanding as learnt over time or as stated in an SOP
- ✓ Implementation log and tracking system
    - o If implemented, what is the result and how does it compare with prediction?
    - o If not implemented, why not and follow through?

Figure 7

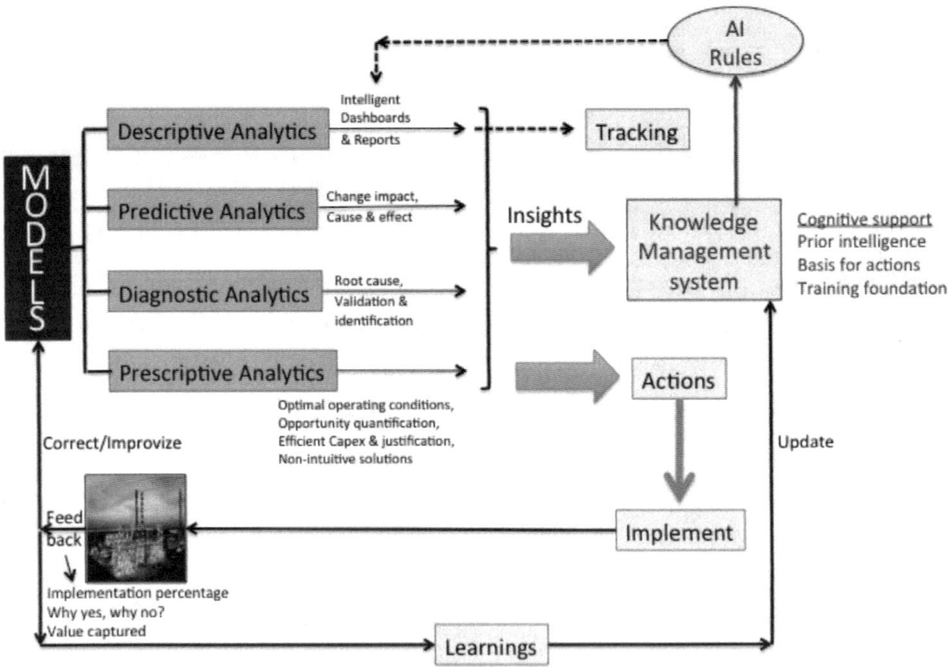

The follow through and execution of an analytics process

Case examples of results from application of an integrated big data analysis process at multiple ethylene plants

This Integrated analytics process was implemented at three ethylene sites of a major ethylene producer and has shown consistent value addition year after year. Table 2 shows a summary of some of the value derived in 2014.

Table 2

Results for three plants for a Major Ethylene Producer in 2014

Parameter	Facility A	Facility B	Facility C
Furnace Runlength	**2 x Design**	**+100%**	**+50%**
Yield	Increased	Increased	+2.3%
CGC Performance	+60 tpd	RCA solved problem	Improved
Capacity Increase	Debottleneck study	Best ever (@107%)	Max capability
Other	C2 Splitter optimized / Exchanger improve	Same + RCA	Decoke and C3 splitter improve
Significant What-if Results	25 proactive	8 proactive + 31 queries response	Numerous

Improvement in furnace run-lengths has been a consistent value gleaned at these sites as well as others where this process has been implemented, as summarized in Table 3.

Table 3

Consistent improvement in furnace run-lengths

#	Technology Supplier	Feed Type	Furnace runlength before IPOG (In days)	Furnace runlength with IPOG (In days)
1	Technip	Ethane	50	> 90
2	Technip	Ethane	30	60
3	SWEC	Naphtha	35	55
4	SWEC	C5 and C6	45	60
5	KBR	NGL	10	20
6	SELLAS	Ethane	25	45
7	LUMMUS	Ethane/Propane	55	Sustained with 55 -60 days with 8% Feed Increase

Results of performing such analytics at an ethylene facility in the Middle East are summarized in Figure 8.

Figure 8

Case Example of benefits from an integrated analytics program at a Middle East Ethylene facility

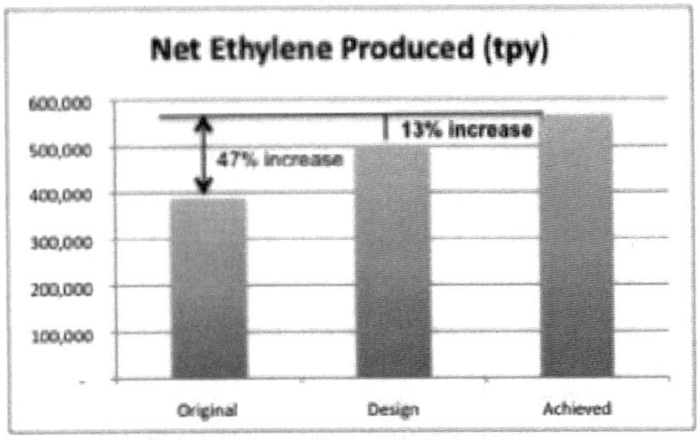

Increased capacity without capex

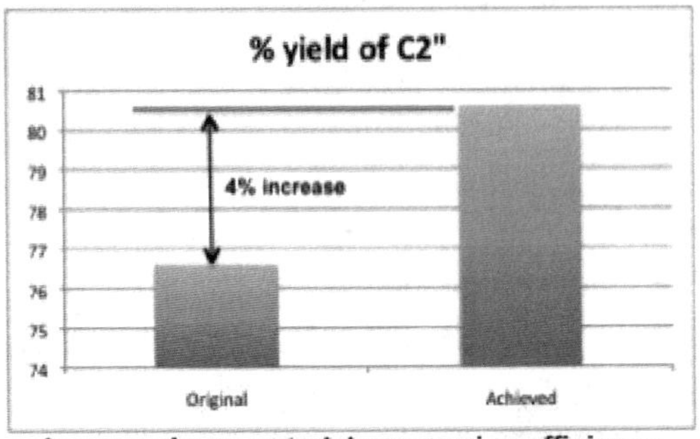

Increased raw material conversion efficiency

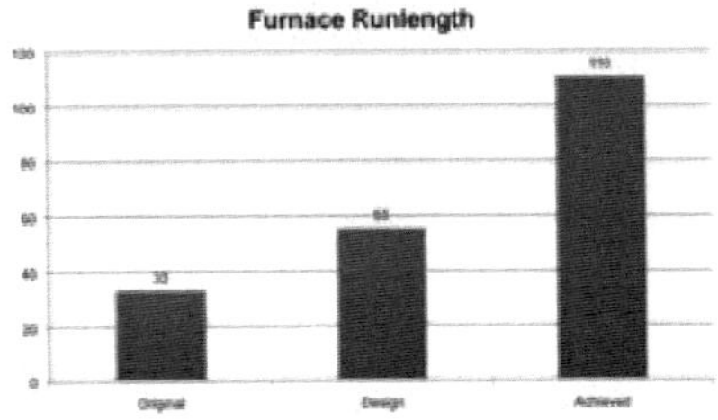

Furnace run lengths doubled

- Cumulative measured value addition of $200 million over 5 years
- Returns ~70 times the cost to put solution in place and sustain
- Helped run the plant an extra year before a turnaround, optimally
- Early detection of furnace tube leak to avert tube failure
- Failure predictions and residual life analysis prevented multiple safety/maintenance events
- Debottlenecking study at 50% cost and time for 25% capacity increase
- Multiple cases of early detection of anomalous conditions prevents a plant upset

References:

1) Laney, D., (Feb 6, 2001) "3D Data Management: Controlling Data Volume, Velocity and Variety", META Group, Inc.
2) L.F.Colegrove, M.B. Seasholtz, C. Khare., (March 2016) "Big Data – Getting Started on the Journey", CEP, pp. 41-45
3) P. Hall, W. Phan, K. Whitson., (May 2016) "The Evolution of Analytics – Opportunities and Challenges for Machine Learning in Business", Published by O'Reilly Media, Inc.
4) "Driving the IoT Journey: 10 Trends to watch", ABI Research publication, June 2016
5) P. Nair, V. Save, K. Patil, T. Sakhalkar., (Oct 2002) "Continuous operations performance improvement using an integrated model guidance system", Aspen World presentation
6) P. Nair, (March 2007) "Large amounts of data requires an outside management system", Houston Business Journal.

MODEL-BASED REAL-TIME MONITORING OF ETHYLENE CRACKING FURNACES

Stepan Spatenka[1], Abduljelil Iliyas[2], Constantinos Pantelides[1], Ying-Sheng Cheng[1], Tareg Mohammed Alsoudani[3], Koji Tomida[4], Filipe Calado[1], Sreekumar Maroor[1], Mark Matzopoulos[1] and Steve Hall[1]

(1) Process Systems Enterprise Ltd., London, United Kingdom,

(2) SABIC Technology Centre, Saudi Basic Industries Corporation, Riyadh, Saudi Arabia,

(3) SABIC Technology Centre, Al-Jubail, Saudi Arabia,

(4) Eastern Petrochemical Company, A SABIC Affiliate, Al-Jubail, Saudi Arabia.

Contact: Stepan Spatenka, Tel: +44 20 8563 0888, Email: s.spatenka@psenterprise.com

1. Introduction

Olefins plants are at the heart of the petrochemicals value chain. They convert gas feedstocks such as ethane and liquid feedstocks such as naphtha to ethylene, propylene and butylene. These products are then further processed into a wide range of valuable performance chemicals such as polymers, resins, esters, glycols and solvents. The conversion is performed via a high-temperature energy-intensive process taking place in thermal cracking furnaces.

In 2015, annual olefin production was around 250m tonnes, involving 277 plants in 60 countries. Production rate is now increasing significantly as a result of the shale gas revolution in the USA[1]. It is the single most energy-consuming process in the chemical industry and modern plants have capacities up to 1.5 million tonnes/year ethylene. At this scale, small improvements in efficiency can yield significant annual savings. Olefin plant operation is thus a prime candidate for optimisation aimed at maximising economic performance.

Two fundamental related challenges in the operation of a cracking furnace are (a) obtaining a reliable measurement of conversion and product yield - a key measure of plant performance and an important variable for severity control and (b) the ability to monitor the degree of coking in the furnace, which has a direct effect on residence time and thus product yields and tube metal temperatures. Both of these are critical to the economic operation of a plant.

This paper describes a new approach for accurate soft-sensing of key cracking furnace variables including conversion, product yields and degree of coking, based on detailed predictive models coupled with dynamic state estimation techniques. The approach, in which the coking calculation is effectively self-tuning, has been shown to provide significantly better estimates than previous approaches based on using a pre-determined coking kinetic model.

2. Challenges to Operational Excellence – yield analysis and coking

Measuring the conversion and product yields from furnaces is a constant challenge, as physical instruments which use extractive sampling techniques often fail or are subject to fouling, requiring frequent maintenance and recalibration. Conversion and yield change with

---

[1] Where will all the new US olefins go ?, by Michael McCafferty, S&P Global Platts website blog, June 29, 2016

feedstock and furnace operating conditions and also, on a day-to-day basis, with the degree of coking. An accurate measure of conversion is critical for effective furnace severity control.

Coking starts from day 1 of coil operation and continues until the effects are serious enough to warrant taking the unit out of service for de-coking. The rate of coking can be exacerbated by poor operation. The effects of coke build-up increase over time and include reduced heat transfer leading to higher metal skin temperatures; increasing pressure drop; and in general lower efficiency of operation with more energy required per unit of production. This leads to loss of yield and shortened run length. More serious is during upset operation, for example power failure, where there is significantly higher potential to damage the cracking tubes with resulting lost production and replacement costs that can run to tens of millions of dollars. Such damage can also impact safety.

The net effect of both these challenges is lower efficiency with loss of yield resulting in more recycled material, consequent reduction in throughput, greater energy requirements and shortened run length.

A central part of any Operational Excellence strategy should thus be to not only improve the accuracy of yield prediction but also to monitor the level of coking in the system at any point.

2.1 Current practice – limitations and drawbacks

The use of physical instruments such as process gas chromatographs for composition measurement is challenging, in part because of the extractive sampling approach required and due to tars and contaminants created in the furnaces which can contaminate the sample lines and analyser. These affect sampling rates and sample conditioning and as a consequence effluent analysers are frequently out of order or poorly calibrated. In most cases it is not easy to maintain analysers during normal operation. In addition, each furnace will likely have its own analyser and there may, in the case of ethane/propane mixed feed furnaces, be feed analysers too. All these analysers provide many potential sources of error.

There are several existing simulation-based approaches that attempt to *predict* rather than *measure* yield in real-time. Typically, yield models might use feed flow and composition, coil inlet pressures and temperatures. Often there is a limited attempt to take into account the degree of coking as it occurs throughout the production run and because coking affects conversion, the results regularly have to be 'interpreted' by experienced operators. In addition, uncertainty in any of the measured variables from the plant is not taken into account, mainly because there is no way using conventional approaches to incorporate such uncertainty into any predictive calculations. Hence there is a risk of inaccurate values being calculated for critical parameters such as yield, pressure drop and maximum tube metal temperature,

Sometimes yield models are used in conjunction with coking models, with coking over the run length simulated by a semi-empirical coking model using presumptive coking kinetic parameters fixed beforehand. The available models do not generally predict coking accurately, especially given that coke build up varies radially, axially and over time. Coke predictions tend to depart from reality over the normal course of a run, particularly if there is deviation from standard operation. Consequently such models lack the ability to predict yield and tube metal temperature to a sufficient degree of accuracy.

In some cases, soft sensors based on statistical models[2] are used. Again, these reflect 'normal' operation but have no predictive capability, so can deviate from reality if there is some type of upset in the operation.

The lack of accurate and timely yield information can have a significant impact on the ability to control furnace operation, leading to non-optimal operation. Given the immense cash flows in olefins plant operations, any non-optimal operation can represent a substantial missed revenue or cost-saving opportunity.

The degree of coking affects virtually all aspects of operation and an accurate, up-to-date prediction of coking is a critical pre-requisite for optimisation and yield accounting.

3. Advanced process modelling approach for accurate yield prediction and coking monitoring

Advanced process modelling (APM) combines high-fidelity predictive process models with mathematical optimisation and other advanced numerical solution techniques.

The technology described in this paper is based on a novel APM approach[3] that couples a detailed predictive furnace / plant model with regularly-updated plant data. The predictive furnace model includes first-principles representation of coil geometry, material and heat balances, cracking kinetics and coking kinetics. The plant data used by the model are measured variables such as feed rate, feed composition, coil inlet temperature and coil outlet temperature, together with any additional available measurements that can be used for real-time calibration of the model, such as pressure drop over coils or tube metal temperatures. Uncertainty is assigned to these additional measured variables and these are considered in the overall reconciliation of the plant data.

The resulting system is then solved using dynamic state estimation techniques based on an Extended Kalman Filter approach to provide up-to-date estimates of process state variables, selected process parameters and other key performance indicators (KPIs). The use of Kalman filtering as feedback in process control is not new. However, the novelty here comes from combining the use of Kalman filters with rigorous process simulation models.

The approach exploits data redundancy features within the data system and the applied confidence limits to measured variables to reduce overall model uncertainty, i.e. the approach makes maximum use of all the data available to it. This powerful technique allows the model to be used, if necessary, beyond its initial operating window. Kinetic parameters of the coking model are adjusted periodically to match observed reality as new evidence arrives in the form of actual plant measurements such as pressure drop over the cracking coils. Effectively this results in a *self-calibrating coking model* that can accurately predict the current state of coking along the length of the coil and take this into account in the yield prediction calculation.

---

[2] Soft sensors for processing plants, by G. D. Gonzalez, published in the Proceedings of the Second International Conference on Intelligent Processing and Manufacturing of Materials, July 10-15, 1999, Honolulu, Hawaii

[3] First-principles modelling for optimization of process operations, Costas Pantelides, Foundations of Computer Aided Process Operations / Chemical Process Control, Tucson, Arizona, January 8-12, 2017

3.1 State estimation

The operation of a state estimator in conjunction with a process model is illustrated in Figure 1. At the beginning of the first interval, the plant model has what it believes are correct values for the plant state (e.g. pressures, state of coking) and various measurements. It then calculates new values for the next time interval (orange dots) based on this understanding. There is an element of error in the new prediction, because no model is ever perfect.

In the meantime, new measurements (grey rectangles) arrive from the plant sensors. These will typically differ from the values calculated by the model; they also carry a degree of uncertainty, reflecting, for example, instrument error.

The state estimator then uses mathematical and statistical techniques to find the best 'reconciled' values for all the variables that it is allowed to adjust (blue dots). This effectively simultaneously takes into account uncertainty in plant measurements as well as possible errors in the model representation (for example, resulting from an inaccurate coking model), to provide a reconciled plant model. This process continues for each subsequent time-interval.

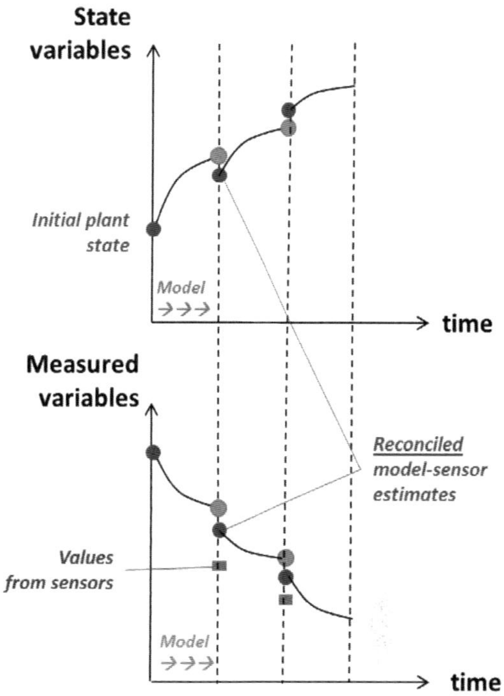

Figure 1 – State estimator operation

At the end of each interval, the model contains the best up-to-date reconciled information that can be achieved for the process, subject to the accuracy of the model representation and the uncertainty in the measurements.

3.2 Furnace model

3.2.1 Requirements

In order to be deployed effectively by the state estimator, the underlying furnace model requires a certain minimum level of fidelity. In particular it is important that it has a high degree of *predictive* capability. The model needs to include, at a minimum:

- **Furnace geometry.** The coil geometry needs to be represented accurately, along with its material properties such as thermal conductivity.
- **Heat input to cracking coils.** It was found in offline studies that using a representative relative heat release profile according to the furnace design together with a severity specification provides sufficient fidelity on modelling the firebox side of the furnace when the focus is on prediction of cracking yields and coking rates along the tubes. Imposing a relative heat flux profile can therefore be used as a much simpler and faster approach that is very effective for representing the actual heat input, as opposed to using more rigorous methods such as zonal methods or computational fluid dynamic-like approaches.
- **Detailed cracking kinetics**. Accurate kinetics for the cracking reactions are critical to reliable modelling of the furnace and indeed the whole ethylene plant performance. There are many ways of supplying kinetics: kinetic calculations supplied with the modelling package; commercially-available kinetics packages; or via users' own

kinetics, perhaps fitted to plant or laboratory data. In the modelling tool used, supplied kinetics based on literature sources can be adjusted to match observed plant or pilot plant data.

- **Coking model**. Again, the coking model is important as coking starts from day one of a production run and affects virtually all aspects of operation and economics. Given the many variables in furnace operation, it is virtually impossible to predict kinetics accurately from a static model; this is a key reason for the approach described here. Well-established kinetic models from literature were used, with parameters such as rate constants periodically updated based on observed plant performance, using dynamic state estimation techniques.

3.2.2 Implementation

A furnace model fulfilling these requirements was implemented in PSE's gPROMS Olefins Cracking Monitor tool using its Olefins model library. Figure 2 shows some of the furnace configuration screens related to coil selection and definition of the coking kinetic model.

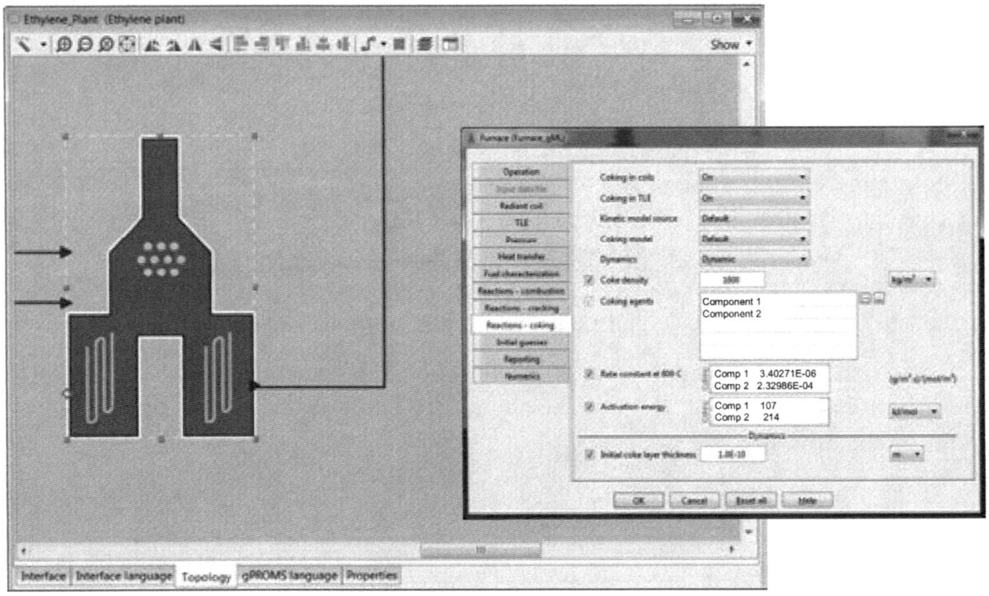

Figure 2 – Ethylene furnace setup showing coking kinetics definition dialogs

4. Application

The technology has been applied to a large ethylene plant in Saudi Arabia processing mainly ethane and propane feedstocks.

4.1 Approach

In implementing the gPROMS Olefins Cracking Monitor, the first step was to extensively test the performance of the model using offline data taken from the site's data historian, in order to establish:

- the overall functionality of the data transfer and control mechanisms
- the ability of the state estimator to detect and handle missing data or data with significant (gross) errors
- the accuracy of the yield predictions from the state estimation model

- the applicability of the coking model for different feedstocks.

Once these various elements of the Cracking Monitor had been tested to satisfaction, a team of engineers then implemented the tool in the operations centre on-site, integrating it within the DCS.

4.2 Results

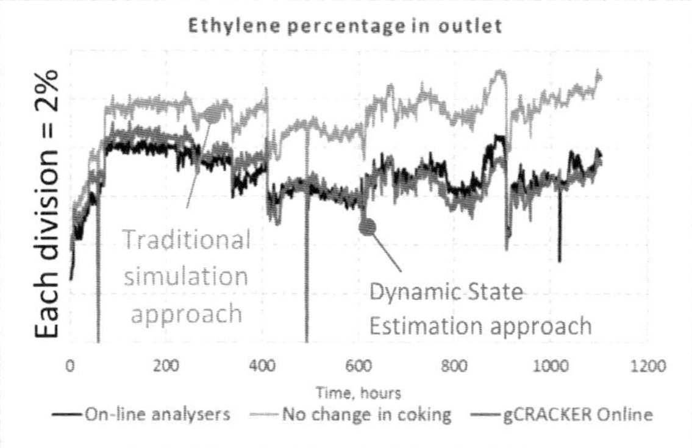

Figure 3 – Ethylene prediction (red and grey line) compared with plant data (black line)

The plot in Figure 3 shows results from both the Cracking Monitor and a simulation case compared to plant data. The data were taken from a period of operation where reasonably reliable ethylene yield measurements were available and could therefore be used to provide verification of model predictions.

During this performance run, great care was taken to ensure the relevant analysers were working well, in a calibrated state with a working sampling system. Two sets of results are compared against the on-line analyser data (black line) over a 50-day run.

The grey line shows the ethylene yield when the coking is simulated assuming fixed coking parameters and no change to coking rate. The conditions in the furnace are simulated as the run progresses, taking into account actual feed flowrate, controls, etc., to determine coking and ethylene yield. This approach, which is used by current commercial software tools, works well in the early stage of the run. However, the inability of the coking model to accurately reflect all events affecting coke deposition over the course of the run eventually causes the calculated coking state and hence ethylene yield prediction, to deviate significantly from reality.

By contrast, the red line shows the prediction using the dynamic state estimation techniques. This is more accurate than simply simulating the operation, as the plant data are used to adjust the prediction of coking progressively as the run unfolds – i.e. the coking model is effectively *self-tuning*, meaning that the degree of coking reported is always accurate and up-to-date.

Note that the ethylene yield measurement recorded by the on-line process analysers was *not* included in the state estimation approach so the ethylene concentration is a purely *predicted* value. It can be seen that the predictions reflect the actual measured values very closely.

Figure 4 provides details on the actual coke build-up in one of the coils. Figure 4(a) shows the cumulative amount of coke deposited over the 50-day run. Figure 4(b) goes further, to show the time-based development of the coke thickness at every point along the coil length over the course of the run. Note that that the estimate of coke deposited at any point may be adjusted as new plant measurements arrive and are taken into account by the state estimator. Such information, calculated automatically during the reconciliation step, provides valuable new insights that can be used to identify specific points for tube skin temperature measurements, inform control approaches and ensure asset integrity.

5. Conclusions

With the furnace model and state estimator connected to the plant's data acquisition system, plant data is processed in real time. The approach provides:

- highly accurate predictions of conversion and product yields, which can be used for furnace severity control
- accurate estimates of the current degree of coke deposition in each section of the furnace coil, for use in optimising de-coking intervals
- accurate soft-sensed measurements of furnace KPIs or any other variables of interest in the plant that can be used for reporting or for improved control
- an up-to date model of each furnace that, when included with a high-fidelity model of the surrounding process incorporating updated plant parameters, can be used for whole-plant optimisation (online or offline), look-ahead simulation and many other activities.

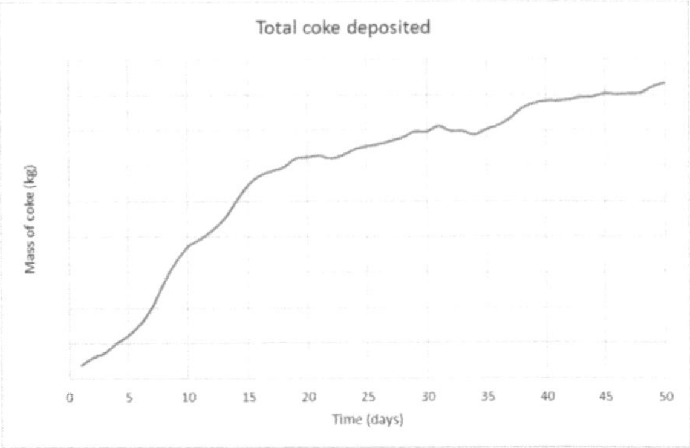

Figure 4(a) – cumulative mass of coke deposition

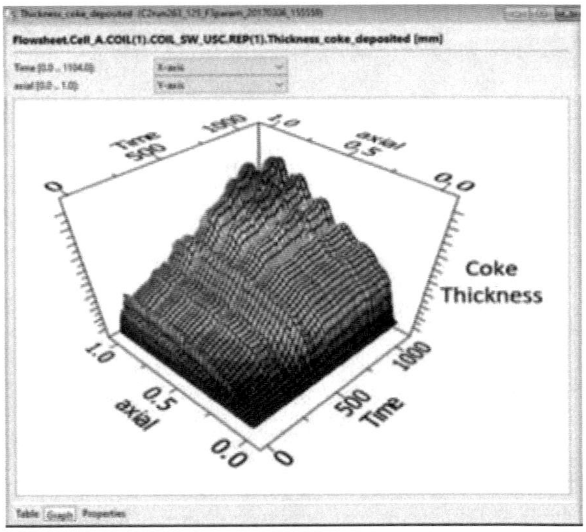

Figure 4(b) – reported coke thickness over the length of the coil over time

In terms of benefits, this results in:

- significantly improved confidence in the reliability of furnace effluent composition values
- the ability to accurately control the furnace based on conversion as opposed to, for example, Coil Outlet Temperature
- improved furnace control, leading in turn to higher throughput and profitability
- better control and management of coking, leading to higher efficiency and longer run lengths
- early warning of excessive coking, helping to avoid split tubes and other unplanned outages which can be very costly
- better, up-to-date process models on which to base optimisation, leading to better, more profitable operation
- accurate soft-sensed values for all variables within the scope of the process model, leading to greater understanding and monitoring of operation
- improved consistency between off-line and on-line models which aid in accurate predictions for production planning and control.

A key further application of the approach outlined here is in providing an up-to-date plant model – including accurate coking prediction and adjusted reaction kinetics – for a real-time plant optimiser. This is a crucial requirement for extracting the maximum value from online optimisation calculations.

AIChE Paper Number: 2D

RIGOROUS NON-LINEAR OPTIMIZATION IN DOW'S HYDROCARBONS BUSINESS

Edo Kingma
Senior Technology Manager
Dow Benelux B.V.

Prepared for presentation at the 2017 Spring National Meeting
San Antonio, TX, March 26-30, 2017

RIGOROUS NON-LINEAR OPTIMIZATION IN DOW'S HYDROCARBONS BUSINESS

Edo Kingma
Senior Technology Manager
Dow Benelux B.V.

ABSTRACT

This paper presents an overview of Real Time Optimization in Dow Hydrocarbons as applied on the bigger part of Dow's cracker fleet: the context of RTO, the implemented support model and performance metrics. Also the use of plant models and the modeling approach is outlined. The usage of rigorous non-linear models is justified with data from several models.

Real Time Optimization in Dow Hydrocarbons

Introduction

Dow's cracker fleet consists of 14 Ethylene plants (JV's excluded). The geographic spread is N- and S-America and Europe. Feedstocks are ranging from ethane through condensates, while most plants are capable of handling multiple feedstocks.

Closed Loop Real Time Optimization has been installed on 12 plants. The current lifetime of these optimizers is more than a decade. Optimization loops include furnace feed rates, severities, dilution steam ratios, suction pressure of CGC- and refrigeration compressors, key column compositions, in total 50 – 120 depending on number of furnaces. Typical implementations of RTO always come with Advanced Process Control. The rationale is that the role of RTO is to optimize the plant's operating conditions, and APC's role is to provide stable control at these conditions. The coherence of the different layers of control is depicted in the so-called Control Pyramid. The picture is an illustration

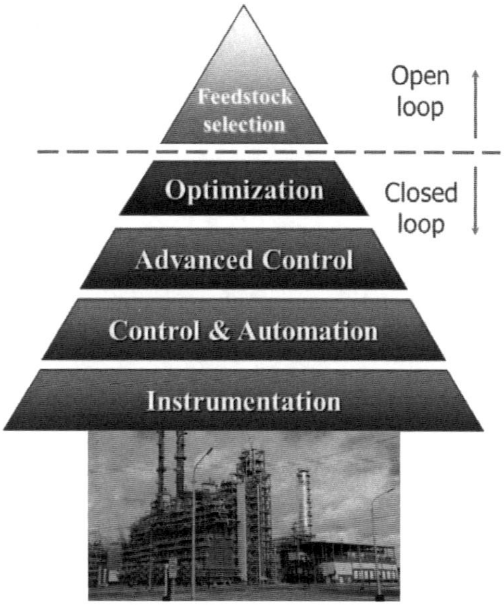

Figure 1 - Control pyramid

how higher levels of control depend on lower levels of control. A higher level of control receives information from a lower level, and is sending information (e.g. setpoints) to a lower level. Also, if there is a problem in a lower layer, all control levels above will suffer from this.

AC&O support

Experience in Dow is that adequate support of Advanced Control and Optimization is of vital importance for sustained success. WIth the implementation of the AC&O projects in the early 2000's it was well understood that an adequate support structure was necessary to achieve this. Support tasks include regular checking if the system is working properly (and troubleshoot if needed), make sure that the model continues to represent the plant accurately (and make changes if needed) and answer questions from operations about optimizer moves. At the local plant level there are typically two people designated to support the applications, one for RTO and one for APC. At the business level also two people to provide support for the local plant people, also one per discipline. Then, at the company level there are two people supporting AC&O in all businesses. All people in the support structure have RTO or APC as their prime responsibility.

RTO performance metrics

To monitor performance of the RTO-applications, there have been three KPI's defined: Availability, Capability and Utilization. Each of the KPI's has its' own metric. The metric for Availability is the Service Factor, which is the fraction of time that the RTO system is up and running. For Capability, there is a metric Health Factor, which is an indicator of how well the system is capable to provide high quality optimization modeling. And finally for Utilization the Fraction of Loops On-Optimization is tracked. Performance reporting of the three metrics is done on a regular basis.

RTO technology

Dow's Hydrocarbons business is using model-based optimization. A steady state model is developed for each of the plants. The models are first principles models; mass- and energy balances are obeyed, thermophysical properties are used, as well as kinetics for the pyrolysis reactions. The scope of the model is from furnace feed system through product rundown. Everything that is needed to model economics and constraints – all major equipment, all furnaces individually with multiple zones if relevant – is included. The total model size is over 100,000 equations which are solved simultaneously.

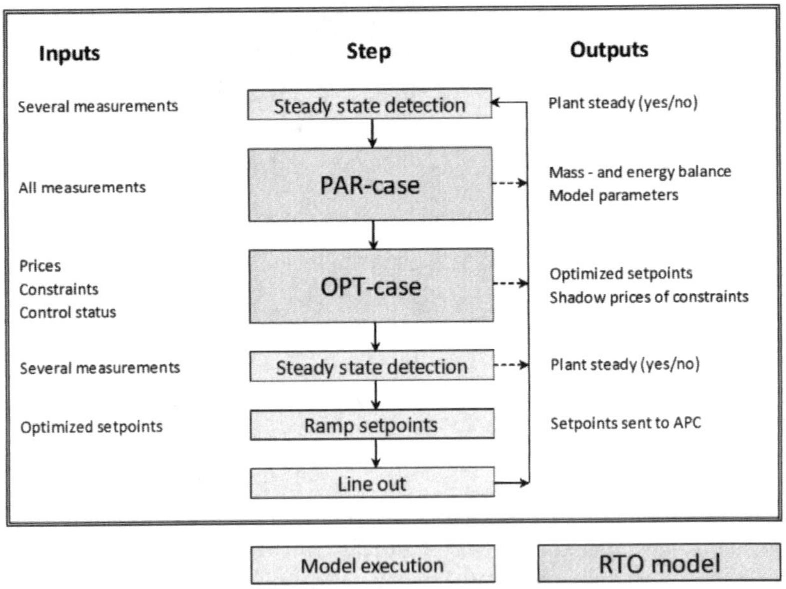

Figure 2 - CLRTO cycle

Model execution is done in Closed Loops Real Time Optimization cycle as shown in the picture. The model is solved in two different modes: the PAR- and the OPT-case. In the first one, current operating conditions are entered and the model is parametrized to prepare for the OPT-case calculations, in which optimal setpoints are calculated. Depending on model size, the model calculations may take several minutes to about half an hour to complete. The cycle frequency is once every few hours, mainly determined by the line out period.

Constrained Optimization

The solution algorithm for constrained optimization is described in text books, like e.g. *Encyclopedia of Mathematics* [1]. At the solution of the optimization calculation, the following equation must be satisfied for each degree of freedom:

$$\frac{\partial P}{\partial x_i} = \lambda_i + \sum_j \lambda_j \frac{\partial c_j}{\partial x_i}$$

In which:

P	Profit function – to be maximized
x	Degree of Freedom (DOF) - an independent variable
λ	Lagrange multiplier, also called Shadow Price
c	Active inequality constraints $c_{1...j}$ – a subset of all dependent variables

The partial derivatives of the objective (profit) function and of the constraints exhibit non-linearities, depending on the nature of the degree of freedom in question. In the two (similar) tables below an overview of the partial derivatives of the profit function with respect to the degrees of freedom before (Pre sensitivity) and after (Opt sensitivity)

Degree of Freedom	Opt Move	Pre sensitivity	Opt sensitivity	Increase
F04_SIDE_A_SHC_RATIO	0.00013	43.416	20.208	-53%
F04_SIDE_B_SHC_RATIO	0.000108	42.918	19.728	-54%
F04_SIDE_A_SEV	0.5	3.3984	2.9796	-12%
F04_SIDE_B_SEV	0.5	3.5484	3.1343	-12%
F04_SIDE_A_HC_FEED_F	10.44358	0.092417	0.088355	-4%
F04_SIDE_B_HC_FEED_F	5.89701	0.092168	0.088124	-4%
F07_SIDE_A_SHC_RATIO	0.006164	16.755	-17.153	-202%
F07_SIDE_B_SHC_RATIO	0.007003	25.946	-5.9506	-123%
F07_SIDE_A_SEV	0.146067	16.877	17.226	2%
F07_SIDE_B_SEV	0.070654	14.723	15.195	3%
F07_SIDE_A_HC_FEED_F	191.6058	0.17422	0.16752	-4%
F07_SIDE_B_HC_FEED_F	223.4744	0.17469	0.16777	-4%
K201_1ST_SUCT_P	1E-06	578.53	906.91	57%

Degree of Freedom	Opt Move	Pre sensitivity	Opt sensitivity	Increase
B09_ZONE_A_SEV	0.010808	1411	-734.23	-152%
B09_ZONE_B_SEV	-0.00979	1212.2	-174.28	-114%
B09_ZONE_A_SHC_RATIO	1E-06	390.23	307.13	-21%
B09_ZONE_B_SHC_RATIO	1E-06	377.98	393.56	4%
B09_TOTAL_FEED_F	250	0.3059	0.21405	-30%
B10_ZONE_A_SEV	0.002393	-2555.8	-2266	-11%
B10_ZONE_B_SEV	-0.00385	-2603	-2207.5	-15%
B10_ZONE_A_SHC_RATIO	-0.00019	25.248	-6.3807	-125%
B10_ZONE_B_SHC_RATIO	-0.00019	8.3229	-19.223	-331%
B10_TOTAL_FEED_F	2.734375	0.19007	0.31574	66%
K500_1ST_SUCT_P	1.1E-06	21183	-1866.2	-109%

Figure 3 – Partial derivatives of the objective function.

the optimization calculation is shown. These derivatives are not fixed, but are rather calculated from the model in each iteration towards the solution. The data shown here are from two different models, and in both only a small selection of all degrees of freedom is made. The last column shows the percentage change in the derivative before and after optimization. Close to 0% means fairly linear, less than -100% means a sign change. As might be expected, furnace feed rates behave relatively linear, but other degrees of freedom appear to be very non-linear.

A similar study can be done for the constraint derivatives. The following tables show an overview of partial derivatives of a constraint with respect to the degrees of freedom. For this purpose just one constraint in the finishing section of the plant was selected. Also here, the data are from two different models and cover only a small subset of all degrees of freedom, and the last column is the percentage change of the constraint derivative before optimization (Pre sensitivity) and after optimization (Opt sensitivity). The furnace feed rates behave not necessarily linear, but quite stable. However, other degrees of freedom show significant variation.

	Dependent	C152_TOP_DP		
Degree of Freedom	Opt Move	Pre sensitivity	Opt sensitivity	Increase
F07_SIDE_A_SHC_RATIO	0.000816	0.0035568	0.0058897	66%
F07_SIDE_B_SHC_RATIO	0.000887	0.003859	0.0068531	78%
F07_SIDE_A_SEV	-0.38683	0.00073889	0.0011994	62%
F07_SIDE_B_SEV	-0.5	0.00082817	0.0013564	64%
F07_SIDE_A_HC_FEED_F	145.8654	4.9969E-06	7.9978E-06	60%
F07_SIDE_B_HC_FEED_F	319.7269	0.000005026	8.0354E-06	60%
F08_SIDE_A_SHC_RATIO	0.00087	0.007409	0.012847	73%
F08_SIDE_B_SHC_RATIO	0.001114	0.0079778	0.0131	64%
F08_SIDE_A_SEV	0.251259	-0.00060746	-0.0010497	73%
F08_SIDE_B_SEV	0.161774	-0.00061235	-0.0010332	69%
F08_SIDE_A_HC_FEED_F	200.3136	9.6947E-06	0.00001563	61%
F08_SIDE_B_HC_FEED_F	68.12159	9.6719E-06	0.000015592	61%
K201_1ST_SUCT_P	-1E-06	-0.18767	0.39346	-310%

	Dependent	T920_INT_RFLX_F		
Degree of Freedom	Opt Move	Pre sensitivity	Opt sensitivity	Increase
B01_ZONE_A_SEV	8.92E-05	-0.61286	-0.14619	-76%
B01_ZONE_B_SEV	0.006275	-0.49054	-0.026246	-95%
B01_ZONE_A_SHC_RATIO	0.023716	0.0049755	0.07652	1438%
B01_ZONE_B_SHC_RATIO	-0.00913	0.016613	0.085871	417%
B01_TOTAL_FEED_F	4.302734	0.00019858	0.00020533	3%
B03_ZONE_A_SEV	0.012733	-0.60665	-0.2398	-60%
B03_ZONE_B_SEV	0.002966	-0.93528	-0.51419	-45%
B03_ZONE_A_SHC_RATIO	0.015816	-0.0083446	0.055199	-761%
B03_ZONE_B_SHC_RATIO	0.015896	-0.059409	0.010824	-118%
B03_TOTAL_FEED_F	93.42773	0.00019724	0.00020515	4%
B05_ZONE_A_SEV	-0.00372	-0.13808	0.29458	-313%
B05_ZONE_B_SEV	0.014619	-0.15063	0.26924	-279%
B05_ZONE_A_SHC_RATIO	0.03	0.11857	0.16435	39%
B05_ZONE_B_SHC_RATIO	0.03	0.12321	0.16314	32%
B05_TOTAL_FEED_F	14.8185	0.00021297	0.00021817	2%

Figure 4 – Partial derivatives of selected constraints.

The data demonstrate the relevance of having non-linear models to optimize plant operation.

Summary

Real Time Optimization has been implemented on the majority of Dow's cracker fleet. Critical success factor is adequate support for the optimization systems, as well as for the associated Advanced Process Control. RTO plays its' distinct role in the total plant control system.

The nature of the optimization model is described, as well as the way how RTO operates.

In constrained optimization the economics balance with the constraints. Both economics and constraints can behave very non-linear. Therefore it is important to acknowledge the non-linear nature of the process. For process control purposes, furnace feed rate responses may be approximated with linear gains.

References

1. *"Lagrange multipliers"*, in Hazewinkel, Michiel, Encyclopedia of Mathematics, Springer, ISBN 978-1-55608-010-4.

Ethylene Producers' Committee

Cooling Water Injection and Monitoring System to Efficiently Prevent Fouling in Centrifugal Compressors

Sergio Ghezzi

Centrifugal Compressors Product Manager

GE Oil&Gas

Francesca Monti

Petrochemicals Platform Manager

GE Oil&Gas

Daniele Panara

Centrifugal Compressor and Expander Technical Leader

GE Oil&Gas

Massimiliano Insinna

Researcher

Università di Firenze

Simone Salvadori

Researcher

Università di Firenze

Prepared for Presentation at the 2017 Spring National Meeting

San Antonio, Texas, March 27th 2017

Ethylene Producers' Committee

Cooling Water Injection and Monitoring System to Efficiently Prevent Fouling in Centrifugal Compressors

Sergio Ghezzi

Centrifugal Compressors Product Manager

GE Oil&Gas

Francesca Monti

Petrochemicals Platform Manager

GE Oil&Gas

Daniele Panara

Centrifugal Compressor and Expander Technical Leader

GE Oil&Gas

Massimiliano Insinna

Researcher

Università di Firenze

Simone Salvadori

Researcher

Università di Firenze

Abstract:

The impact of fouling on ethylene plant equipment can be substantial. The high possibility of gas polymerization occurring in the cracked gas compressor can lead to efficiency loss and unplanned outages with associated lack of production. Injecting cooling water into the process gas compressor can help producers reduce fouling by lowering the temperature.

Cooling water effectiveness is highly dependent on water atomization and gas stream interaction. A series of numerical simulations has been performed to validate the modeling approach by comparing the simulation results with available experimental data and a test case.

The outcome depends on the selection of the appropriate models for heat transfer, particle coupling and injection definition. The goal is to develop an enhanced cooling system analysis, design and monitoring system. The results of this investigation allow for a more accurate prediction of the actual temperature values inside the compressor, throughout the flow path. Consequently, the dimensioning of the water injection system, and more specifically the liquid flow rate, pressure drop and droplet sizes will be optimized.

An advanced stage-by-stage water injection system coupled with fully removable nozzles during compressor operation and an efficiency monitoring and control system, will reduce operating cost and risk of production loss by providing operators with continuously updated equipment performance.

Fouling Formation, Reduction and Prevention

Fouling is one of the major causes of lost production in ethylene plants. The possibility of polymerization occurring within the cracked gas compressor is quite high and will lead to detrimental impact on plant operation. Charge gas compressor fouling can lead to increased steam consumption by the steam turbine driver, increased compressor suction pressure resulting in lower ethylene yield, increased use of antifouling chemicals, as well as shorter compressor life. Fouling can increase the probability of unplanned outages due to unpredictable rotor imbalance that can lead to excessive vibrations and even trip the machine.

The traditional choice of ethylene producers to minimize fouling impact in the charge gas compressor has been to inject water and wash oil. Water has the positive effect of controlling gas discharge temperature and oil can solubilize polymeric deposits.

Ethylene producers are constantly on the lookout for cost savings. For a charge gas compressor, cost savings can be realized from longer maintenance intervals, reliable and smooth operation and reduced usage of antifoulants.

In early 2016 GE conducted many interviews with ethylene plant personnel as well as surveys of the installed fleet to better understand the details of all issues related to compressor performance and fouling. One objective was to find an optimized solution for water washing and to increase its effectiveness. As expected, cooling effectiveness is strongly influenced by the interaction of the two fluids in contact, i.e. water and gas.

Fouling formation is heavily dependent on temperature and occurs at a not linear rate. As shown in the plot below, the rate of polymerization increases rapidly with temperature and is in the critical range for temperatures above 90°C /194°F

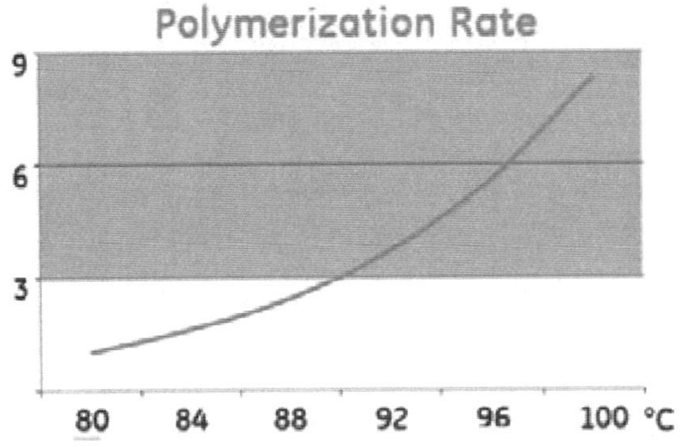

Figure 1 - Typical Charge Gas Polymerization Rate

Polymerization cannot be eliminated, however, an OEM can pursue the following two mitigation approaches:

1. Passive Approach - with an emphasis on increasing compressor tolerance to fouling

This is achieved by:

- Rotoric abradable seals on impellers and balance drums to avoid fouling stagnation inside labyrinths and consequent loss of sealing effectiveness. Figure 2 below show a comparison between these two types of seals.

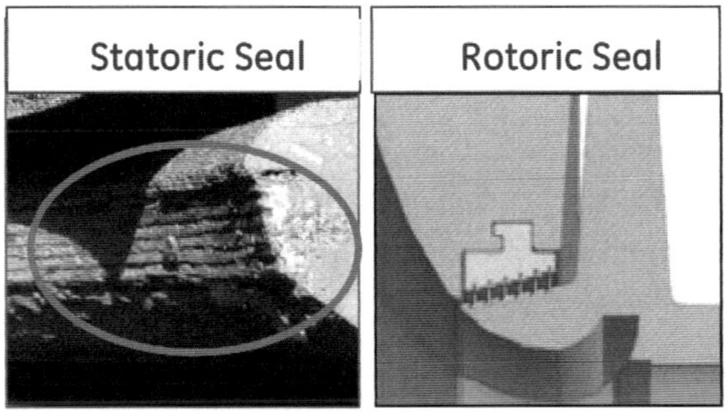

Figure 2 - Statoric vs. Rotoric seal

- Dedicated anti-fouling coatings, to improve friction coefficient and hydrophobicity characteristics of internal components in contact with gas. This will reduce the fouling buildup as well as increase coating resistance to liquid erosion.

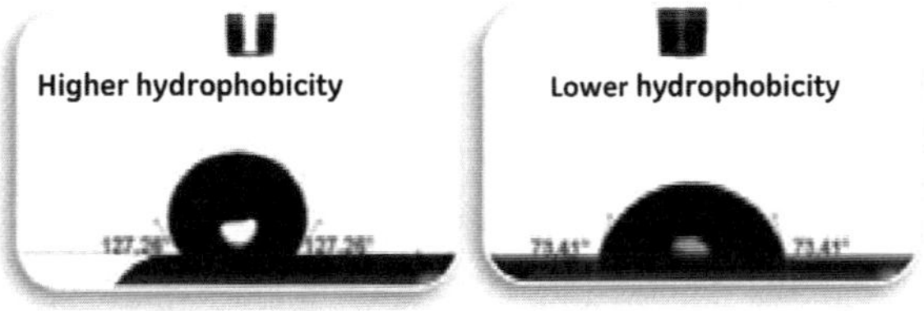

2. Active Approach - with an emphasis on preventing/reducing fouling formation

This is mainly achieved by controlling the temperature below a specified value to achieve an acceptable polymerization rate. The injection of water directly inside the flow path reduces the temperature mostly by absorbing the latent heat of vaporization from the hot gas.

The objective of the present work is to analyze this second approach through better characterization of the heat exchange phenomena using a fundamental model based analysis and lab test validation. The goal is to attain improved compressor operability and reliability by utilizing dedicated fluid injection hardware along with remote efficiency monitoring to control and reduce fouling.

Description of the Test Case

The test rig selected for the present activity is the Wet Gas Test Facility (WGTF) jointly developed by the Department of Industrial Engineering of the University of Florence and GE Oil & Gas – Nuovo Pignone. The schematic layout of the WGTF is shown in Figure 3, where the original and the new configurations are shown.

In the WGTF, air enters the preliminary section of the circuit through two screw compressors, and the flow rate is modulated by means of the admission valve (point 1 in Figure 3). The air then flows through a heater, which can heat the air up to 150°C / 302°F. Once heated (point 2), air enters the actual wet gas test facility through a pipe and is gradually decelerated across a diffuser to a settling chamber (point 3). After that, air flows through a short converging nozzle and enters the test section, where the mixing of air and water takes place (between point 4 and point 5).

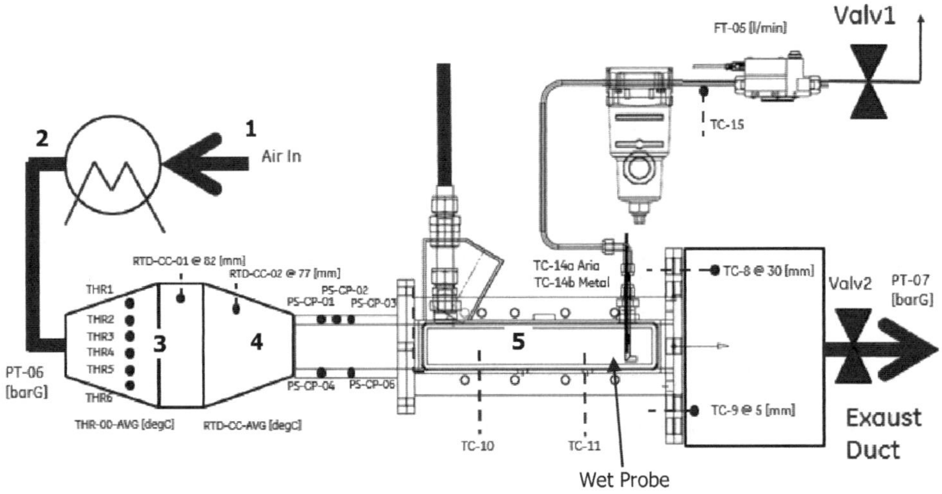

Figure 3 - WGTF schematic view

For a water injector positioned next to the inlet section (lateral optical access), this test rig arrangement allows the necessary space for the addition of two thermocouples (TC-10 and TC-11), positioned close to the lower wall and one "wet probe" positioned in the center of the channel.

In the present configuration, air enters at a total pressure of 2 bar / 29 psi and a total temperature of approximately 100°C / 212°F. Considering a main-flow velocity of around 100 m/s or 330 ft/s (constant for all the investigated cases) it is estimated that the Mach number in the test section is never higher than 0.4. For the present work, a Schlick injector (model 100, size 3) was selected. Considering a borehole of 5 mm / 3/16", it will guarantee a flow rate of 13.80 l/min or 3.6 gpm at 6 bar / 87 psi with a nominal spray angle of 78° (in the hollow cone configuration). From the large amount of available experimental data obtained by varying air and water flow rates, five different Liquid Mass Fraction (LMF) values were considered for the numerical simulations. The liquid mass fraction is defined as the ratio between the water mass flow rate and the overall flow rate:

$$LMF = \frac{\dot{m}_w}{\dot{m}_w + \dot{m}_{air}}$$

Increasing liquid mass fractions equal to 0.052, 0.236, 0.297, 0.347 (nominal point) and 0.382 characterize the operating points tested numerically. The operating points with LMF=0.297 and LMF=0.382 are characterized by an air mass flow that is higher than the other cases by a factor 1.33, but maintain the same Mach number of about 0.3.

Once exiting from the test section, the mixture enters a diffuser and after a corner bend it reaches the back-pressure valve and the water/air separator, which has a water removal efficiency of 95 %. The largest part of the water flow is recovered and collected in the lower water recycling tank. The residual water flow (less than 1%) is exhausted with the humid air stream through a vent.

The aim of the present work is to validate the numerical methodology used for the analysis of the WGTF. As will be explained later, some of the injection parameters, namely the injection velocity and the cone angle, will be tuned to match the experimental data for the nominal point (LMF = 0.347) and then scaled to analyze the other working points. Since the nozzle characterization is not provided by the manufacturer for a cross-flow configuration and in the present case the incoming cross-flow air greatly affects the performance of the nozzle this approach is considered acceptable.

Description of the Numerical Approach

The performance of the selected test case is numerically analyzed using the commercial code ANSYS® CFX® V17.1 1. Steady Reynolds-averaged Navier-Stokes (RANS) equations are solved with a finite volume approach. The test section, reported in Figure 4, is modelled as a rectangular box with no probes. A structured computational mesh is generated using the commercial ANSYS® ICEM CFD® tool, with variable sizing of the element size in the direction of the main-flow.

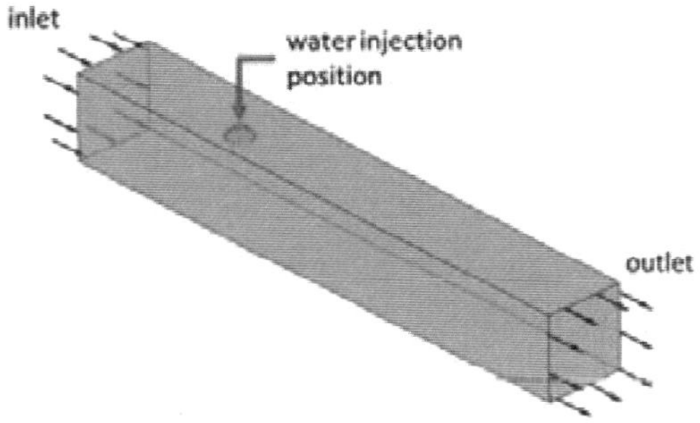

Figure 4 - Computational Domain

Total pressure and total temperature are used at the air inlet section while mass-flow and temperature are used for the injected water. Turbulence inlet parameters are assumed because no information is available. Outlet pressure is iteratively modified through an automatic procedure to match the sum of the air and the vapor mass-flow, thus considering the impact of evaporation.

Particle injection is modelled through a Lagrangian approach, which studies the droplet deformation and evaporation along its trajectory. A homogeneous mixture of water and vapor is injected in a variable composition ideal mixture of air and vapor (the latter is always treated as ideal gas). With this approach the simulation will consider the mass transfer from a dispersed liquid phase to a gas phase containing a mixture of air and vapor.

Air and water droplets are fully coupled, with a surface tension coefficient evaluated at the water injection temperature. Drag force for the momentum transfer is considered through the Ishii-Zuber approach[4] while heat transfer is modelled through the Ranz-Marshall correlation for a sphere[3]. The Cascade Atomization and Breakup model (CAB)[6], which is based on the Taylor analogy, is considered for the particle breakup. Based on the local Weber number, the model identifies the breakup regime and evaluates the dimension of the child droplet using an exponential law.

Water is injected through an area which is representative of the actual experimental configuration and is positioned at the same location as in the experiments. For the present case, a hollow cone has been considered with a cone angle set at zero, in keeping with the experimental evidence. Particle size is defined through a Rosin-Rammler distribution where mean size and the spread exponent are representative of the data obtained by the supplier of the nozzle.

Results are compared with the available experimental data in terms of air temperature drop, as measured by the three thermocouples. Although three possible data are available for comparison, this paper is focused on the data obtained from the aspirated probe positioned in the mid-section of the channel downstream of the injection point.

Description of Test Conditions

Because the purpose of the investigation was to evaluate cooling effects in ethylene charge gas compressor, the test environment has been designed to simulate flow characteristics in similitude with real charge gas thermodynamic conditions. To account for different process conditions associated with different feedstocks ranging from ethane to naphtha the graph below was developed based on the following non-dimensional parameters:

PHI = Flow Coefficient
Mu = Impeller Tip Mach Number

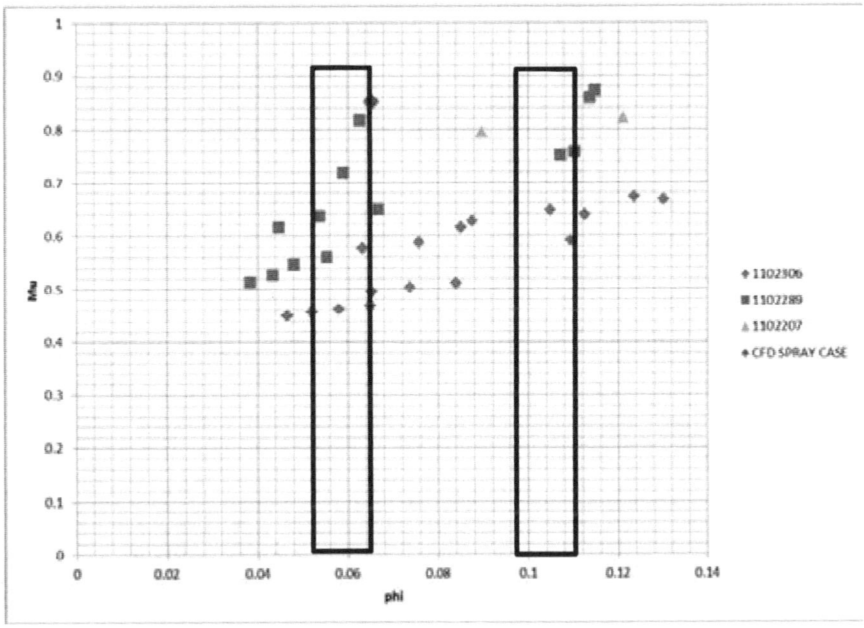

Figure 5 - Charge Gas Parameters Setting the Test Condition

This similitude is important because the cooling effectiveness is greatly influenced by the droplet break up and the evaporation phenomena is dependent on the available surface for heat exchange. The droplet break up is a function of the ratio between the linear momentum of the cooling water and air and matching the Mach values is critical to obtain results comparable to the real charge gas environment.

The temperature measurement inside the flow channel where the gas is mixed with liquid is important. A special probe has been designed to measure only the gas temperature without interference from the liquid.

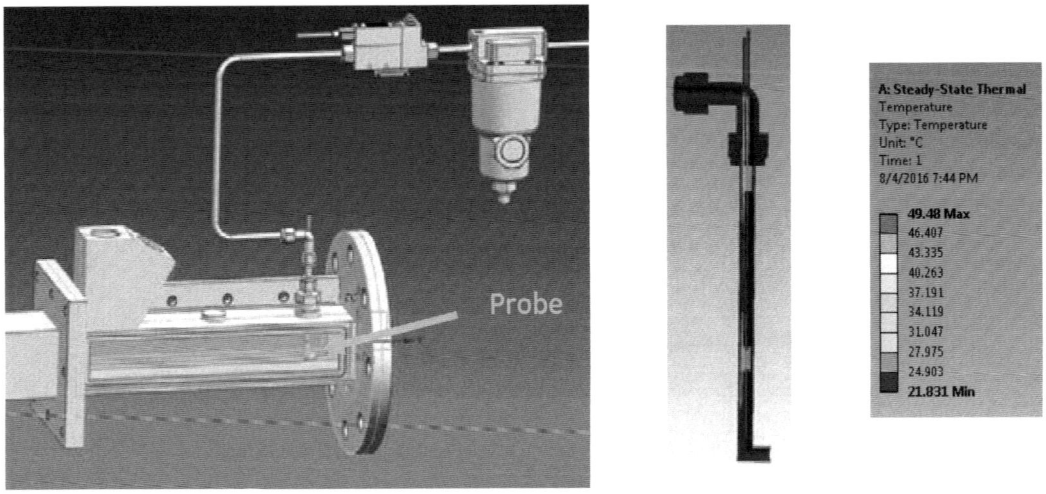

Figure 6 - Probe Layout and Simulation

The probe is continuously purged by bleeding a small air flow from the channel, enough to avoid gas stagnation in the probe but not enough to entrain liquid droplets. Two thermocouples are employed, the first one on the probe wall measures the temperature of the water that runs over it and the second one is coaxial with the probe vertical vented pipe. The second thermocouple is not touching the walls and, therefore, measures only the gas temperature. This temperature is not affected by probe wall because of the continuous air purge.

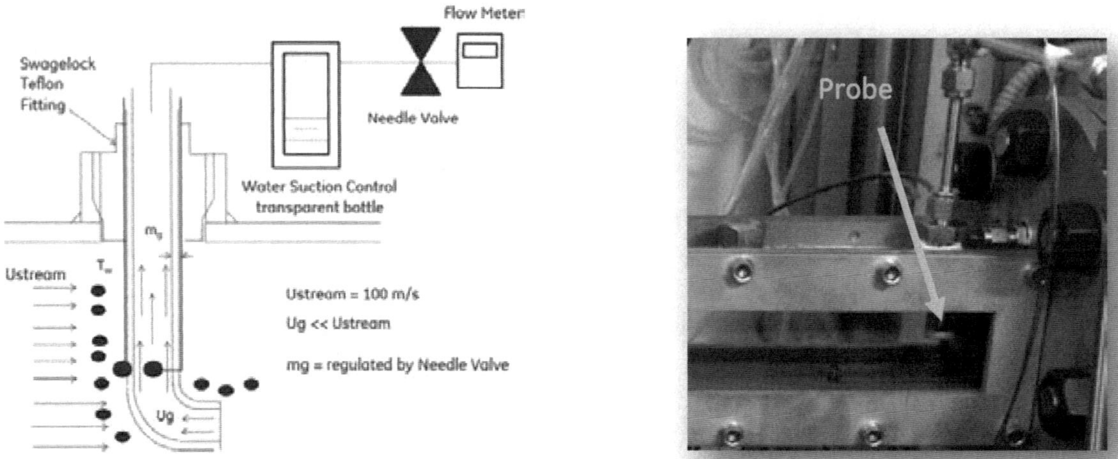

Figure 7 - Wet Gas Temperature Probe

Finally, several spray nozzles have been tested to identify the most effective ones under different flow conditions. Droplet size is very important and significantly influences the actual cooling effect. The tests highlight how different flows and nozzle angles significantly affect water atomizing and substantially deviate from the specification provided in the spray nozzle datasheet and consequently from the cooling prediction.

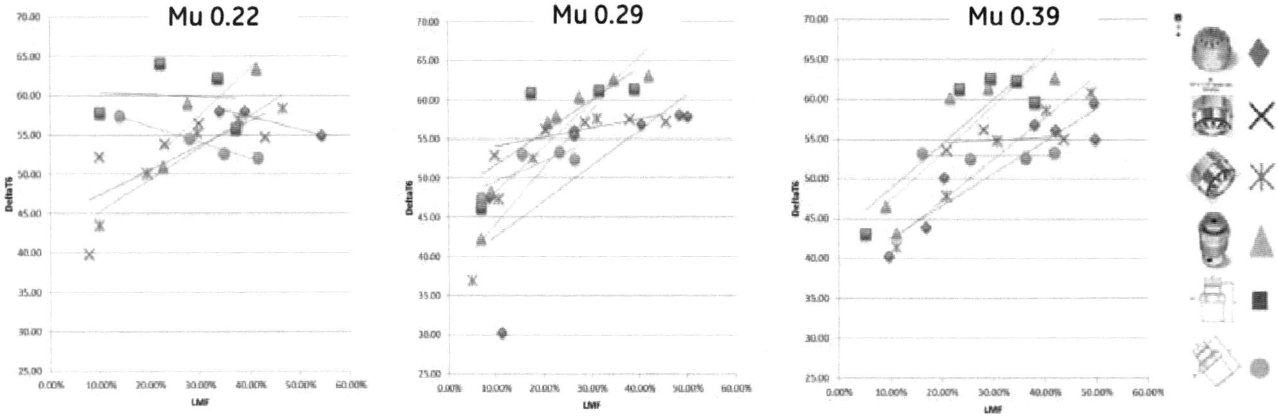

Figure 8 - Cooling Data Regression

$$\Delta T_6^0 = T_s - TC\_14\_A$$

The parameter above shows how much temperature drops below the static flow temperature.

The higher the differential the better. As can be seen in Figure 8 above, it is sensitive to variation in Mach number.

Discussion of Results

Figure 9 reports the non-dimensional temperature fields evaluated on a longitudinal section passing from the injection point for various liquid mass fractions (LMF). The non-dimensional temperature T_{nd} is defined as:

$$T_{nd} = \frac{T_0 - T_{w,in}}{T_{0,in} - T_{w,in}}$$

This parameter ranges from zero to one. It is equal to one where no droplets are present and decreases to zero as T0 approachs the water temperature. It allows a clear definition of the zones of the domain where an effective cooling of the main flow is obtained.

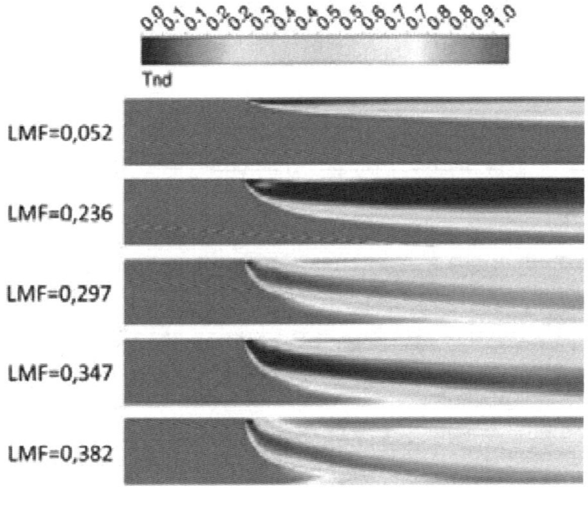

Figure 9 - Temperature Distribution Passing Through the Injection Point

As observed in Figure 9, the increase of LMF leads to an increase of the jet penetration inside the main flow. At the lowest LMF, the low temperature region does not reach the center of the channel and remains confined near to the upper wall. The increase of LMF from 0.052 to 0.236 generates a remarkable extension of the low temperature zone in the vertical direction downstream of the injection position. With LMF=0.236, the momentum of the water jet does not allow the cooling of the main flow along the entire span of the channel before the exit section of the domain. Conversely, the exit sections of the domain for cases LMF=0.297 and LMF=0.347 are completely reduced to a total temperature lower than that of the main stream. The heat sink effect produced by the water is effectively redistributed along the entire span. A further increase of the water flow rate to LMF=0.382 leads to a pronounced penetration, with the coolest region impinging on the lower wall.

Figure 10 below shows the normalized static temperature profiles on a line at the axial position of the aspirated probe on the meridian plane of the channel. The aspirated probe is positioned about 2.6 times the channel height downstream of the injection position. Only one measurement point is available but it allows a proper validation of the CFD results. The vertical position of the point of the profile characterized by the minimum temperature can be considered representative for evaluating the jet penetration. The increase of jet penetration consequent to increases of LMF described before can be observed also in Figure 10.

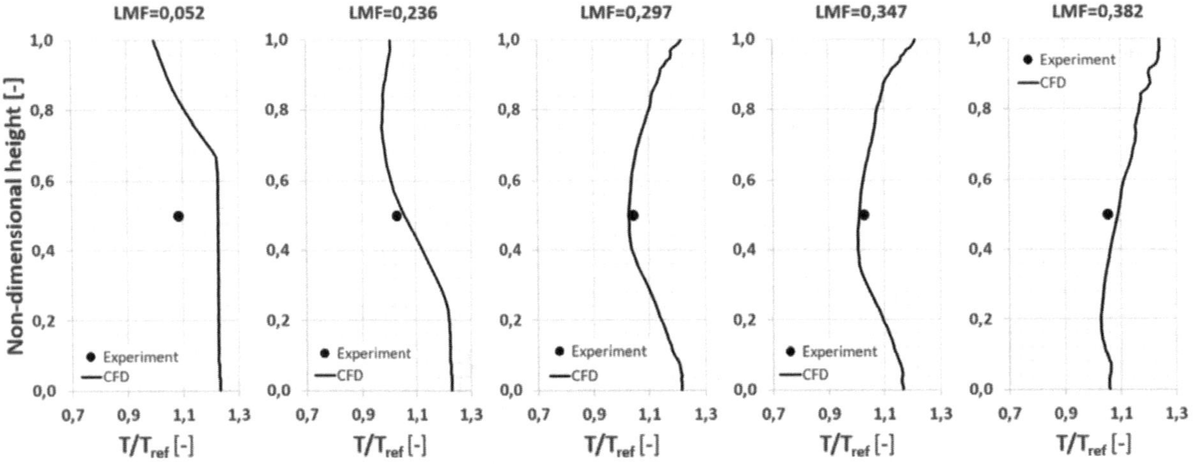

Figure 10 - Non-Dimensional Temperature Profiles at the Axial Position of the Aspirated Probe

Generally good agreement between the numerical and experimental results is observed in Figure 10 except for the case with LMF=0.052. For this case, a remarkable overestimation of the gas temperature is noted. The reason for such overestimation is due to an underestimated penetration of the at this low water flow rate. The limitation of the RANS approach in modelling cross flow jets is well known in literature. In fact, the unsteady effects are related to turbulence enhanced mixing and diffusion mechanisms. The time-averaged behavior of such effects is reproduced only with limited accuracy in a steady RANS simulation. Therefore, it is reasonable to ascribe the inaccuracy to this unsteady phenomena.

Figure 11 reports the non-dimensional temperature difference between the static temperature of main flow measured at the inlet section and the local temperature measured through the aspirated probe. The trend of the numerical curve is close to the experiments especially in the region between LMF=0.297 and LMF=0.347 where the plateau is well reproduced. The considerations noted previously regarding the unsteady effects explain the behavior of the numerical curve away from the nominal point.

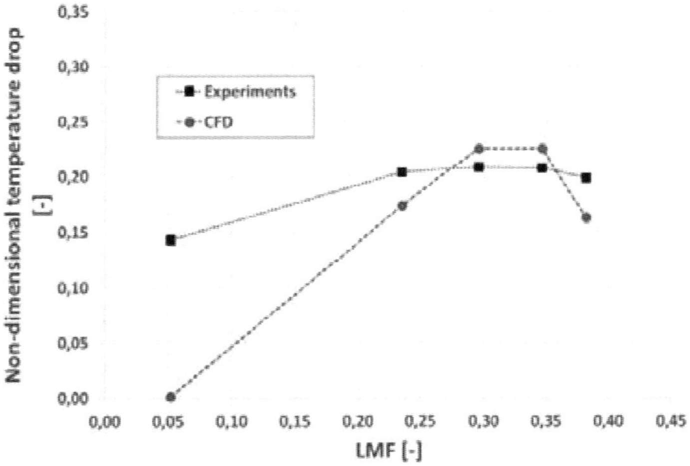

Figure 11 - Non-Dimensional Temperature Reduction Evaluated at the Point of the Aspirated Probe

Thermodynamic simulation results can support operational decisions about water spraying inside charge gas.

An additional tool to improve water mix operation is a retractable nozzle that can be assembled in the charge gas casing. A description of this tool follows.

Retractable Nozzles

Figure 12 - Retractable Nozzle 3D drawing and assembled for test

This apparatus is mainly composed of a hand wheel (1), a movable shaft inside an outer pipe (2), a body (3) where water or oil is driven into the compressor through a sprinkler, an interception industrial ball valve (4) to ensure the sealing, and a venting valve (5) to empty the device before removal.

The system utilizes a retractable spray apparatus for centrifugal compressors or any pressurized vessels which can be dismantled for maintenance without the need of vessel depressurization.

The interception industrial ball valve is mounted on the compressor casing. On top of the interception valve, the retractable injector body is mounted and connected to the water line and the venting line through the venting valve and the water inlet threaded connection.

The shaft can be operated by the hand wheel from the fully extended condition (operating condition) to the most retracted condition where the ball valve can be closed allowing safe removal of the injection nozzle.

During pressurized assembly and disassembly, the outer pipe is disconnected and the mechanism body (3) remains fixed without the need to disassemble the venting and the water supply lines, greatly simplifying maintenance operations.

Efficiency Monitoring

As mentioned in the paragraphs above, there are different methods to prevent fouling inside charge gas compressors: injecting water to control discharge temperature, injecting wash oil to solubilize polymeric deposits as well as other specialty proprietary anti-fouling formulations.

Multiple methods have been analyzed to monitor and evaluate equipment performance and have led to mixed results. Due to the confounding of different factors, it is not easy to correctly measure the compressor efficiency.

In general, data used to monitor compressor efficiency, as well as fouling degree, can be collected by an onsite monitoring system. These data can be process data (flow rate, gas composition, pressure, temperature, etc.), static data (performance curves) and dynamic data (train speed, valve status, etc.).

Compressor polytropic efficiency primarily depends upon gas density, flow rate, suction and discharge temperature, suction and discharge pressure, as well as fouling degree. All the sensors necessary to measure such parameters can be located inside the compressor to record and track all variations. The calculation can be applied to one stage only or to all compressor stages. Monitoring of all stages is the preferred solution to understand and prevent performance degradation.

Regardless the number of stages that are monitored and the kind of monitoring system, the calculation of polytropic efficiency is based on the following equation:

$$\eta = (k-1)/k * ln(P_2/P_1) / ln (T_2/T_1)$$

where:

K=Cp/Cv (ratio of constant pressure specific heat over the constant volume specific heat)

T1, T2 = Suction (T1) and discharge (T2) temperatures

P1, P2 = Suction (P1) and discharge (P2) pressures

Suction and discharge conditions can be related to one phase only or to the whole compressor.

The k value can be derived from gas chromatograph measurements of the gas composition, while pressure and temperature can be directly measured by pressure gauges and thermocouples. All the measurements can be collected at a certain frequency or on continuous basis.

To remove the influence of other variables that are not related to fouling, like feed gas composition, that can change due to variation in the feedstock and cracking process, it is very important to normalize the polytropic efficiency. This can be done through statistical models or through the design value. The degree of fouling in the charge gas compressor can be measured as the difference between the actual polytropic efficiency and the expected one, derived during the design stage from mathematical tools and models.

Operators can receive the information via DCS or any other display or tool available on site. The degree of fouling so measured will be useful to take decision about additional water injection, wash oil or any other chemical injection as well as measuring the operation risk at that specific moment.

Conclusions

To effectively counteract fouling in a charge gas compressor, gas cooling is becoming an almost essential feature. The complexity of heat, mass and momentum transfer, coupled with the compressor geometry and operating conditions necessitate more detailed modeling, testing, validation and evaluation to accurately predict the cooling effect.

Accurate modelling and prediction allows the optimization of the cooling system. This results in a substantial increase in compressor reliability due to minimization of fouling as well as the detrimental effects of erosion. In fact, water cooling can often be designed to target a fully saturated gas to obtain minimum temperature after cooling. In this scenario, if the required liquid amount is determined with simplified heat-exchange models, without considering nozzle type, spray angles, droplet size and water temperature the result can be significantly different from actual conditions. If a significant amount of the injected water consistently remains in the liquid phase at the inlet of the impeller it will lead to erosion.

This erosive environment can be aggressive, even for the metallic substrate of the impellers, but will mainly destroy the standard coatings (PTFE based surface treatments) used in charge gas units.

Looking at the evolution of ethylene processes over the last 10 years, it is evident that the temperatures inside the cracked gas compressor have increased from an average of 90 - 95°C (194 - 203°F) to 100 - 105°C (212 - 221°F). This has led to requiring continuous water injection rather than on an intermittent basis. Under these more demanding operating conditions erosion becomes an increasingly important effect which must be carefully evaluated in the design phase and will influence the choice of the coating.

The concurrent adoption of a model that is more accurate in calculating the temperature inside the compressor, the proper nozzle selection, utilizing hot-replaceable nozzles and remote monitoring of actual efficiency allows for for appropriate adjustments and obtain the desired results listed below:

- Reduced fouling and its consequences
- Maximum compressor availability
- Optimized cooling injection system
- Optimized washing oil injection system
- Minimize erosion and enhance durability of anti-fouling coatings

References

1. ANSYS Inc., ANSYS CFX-Solver Theory Guide, April 2016.
2. Poling, B.E., Prausnitz, J.M., and O'Connell, J.P., "The Properties of Gases and Liquids", McGraw-Hill, New York, 2001.
3. Ranz, W.E. and Marshall, W.R., Chem. Eng. Prog. 48(3), p. 141, 1952.
4. Ishii, M. and Zuber, N., "Drag Coefficient and Relative Velocity in Bubbly, Droplet or Particulate Flows", AIChE J., 25, 843-855, 1979.
5. Menter, F.R., "Two-equation eddy-viscosity turbulence models for engineering applications", AIAA-Journal., 32(8), pp. 1598 - 1605, 1994.
6. F.X. Tanner, Cascade atomization and drop breakup model for the simulation of high-pressure liquid jets. SAE Paper 2003-01-1044, 2003.
7. Greco, G. – Systems and methods for determining a level of fouling of compressors - US 2014/0229121 A1, 2014.

Nomenclature

$\dot{m}$	mass flow rate [kg/s]
T	temperature [K]
y^+	non-dimensional wall distance [-]

Subscripts

0	stagnation quantity
Air	referred to air
In	referred to inlet
Nd	non-dimensional
ref	reference quantity
w	referred to water

Acronyms

CAB	Cascade Atomization and Breakup
CFD	Computational Fluid Dynamics
LMF	Liquid Mass Fraction
WGTF	Wet Gas Test Facility

Turnaround Planning for Turbomachinery in Petrochemical Service
A Successful Approach

by

Tomás Sanchez
Global Service
Field Service Regional Manager – Gulf – Americas
Elliott Group
Houston, TX

Abstract

Shutdowns, Turnarounds, and Outages (STOs) are typically large projects executed under firm time constraint with severe financial impact for delays. When olefin plants are shut down, millions in lost revenue are at stake, and operators incur extraordinary costs for labor, tools, equipment, and parts. Careful planning enables operators to reduce STO downtime and mitigate the financial impact.

STOs and Turbomachinery

Plant outages may be planned or unplanned, and in both cases, part or all of the plant is taken out of service. There are various types of outages. The focus of this paper is planning for an olefin turnaround. An olefin turnaround is a planned, periodic shutdown of one or more olefins processing units, and eventually the entire olefin equipment fleet, to perform maintenance, inspection, and repair of equipment.

Typically, turnarounds for key olefins processing units are planned for every three to five years. A turnaround may involve one to two years of advance planning, sometimes even more when major changes are needed, using dedicated teams from the company, as well as outside contracting and engineering firms. While the objective of advance planning is to minimize the time a unit is offline, the turnaround may result in several weeks to months of plant shutdown. During this time there is an overnight transition from a familiar routine to frenetic activity conducted by hundreds, possibly thousands of contractors and subcontractors unfamiliar with the olefin facility. The work schedules are demanding and unyielding. The chances of injury and error are high. Only extreme good fortune or careful planning with clear communication can mitigate the risks of a major turnaround resulting in on-time completion of the project. Smart operators choose careful planning.

Among the more critical and challenging elements of a major turnaround in an olefin plant are the overhaul and upgrade of the turbomachinery equipment. These machines are integral to processes throughout the plant. They comprise a broad array of complex equipment and auxiliaries from many different manufacturers, including centrifugal compressors, steam turbines, electric motors and generators, pumps, couplings, lubrication systems, seal systems and control systems.

Turnaround Planning and Scope

The success of a major turnaround relies primarily on planning, execution, and a defined scope of work. Planning and preparation for unexpected events, including changes in the scope, is crucial. This is fundamental for turbomachinery since it is not easy to assess its current health while the equipment is operating, and therefore, to properly define the scope of work. The scope of work for turbomachinery can range from doing nothing at all, to a full overhaul, or even replacing an entire unit. Equipment that shows little or no deterioration in performance might require only a minor inspection and the replacement of wearing parts such as bearings and seals. However, while doing this work, the rotor may be found to be severely worn, in which case the scope of work would increase to a major overhaul.

In order to minimize scope changes, the scope of work needs to be agreed upon well in advance of the STO event, with a 'freeze' date of no later than six months before the execution date. Planning for a turnaround project may start eighteen months in advance of the event. [2]

In fact, it is never too soon to begin planning for the next turnaround. As soon as a turnaround is completed and the plant resumes full production, the turnaround team should conduct a post-turnaround review. The review must begin with an evaluation of the turnaround objectives and the degree of success in meeting them.[5]

Per Paquin [4] the top reason for poor STO performance can be traced back to poor requirements, in other words, poor planning. Figure 1 identifies the main reasons why STOs are poorly executed in a survey of over 195 asset-intensive industries.

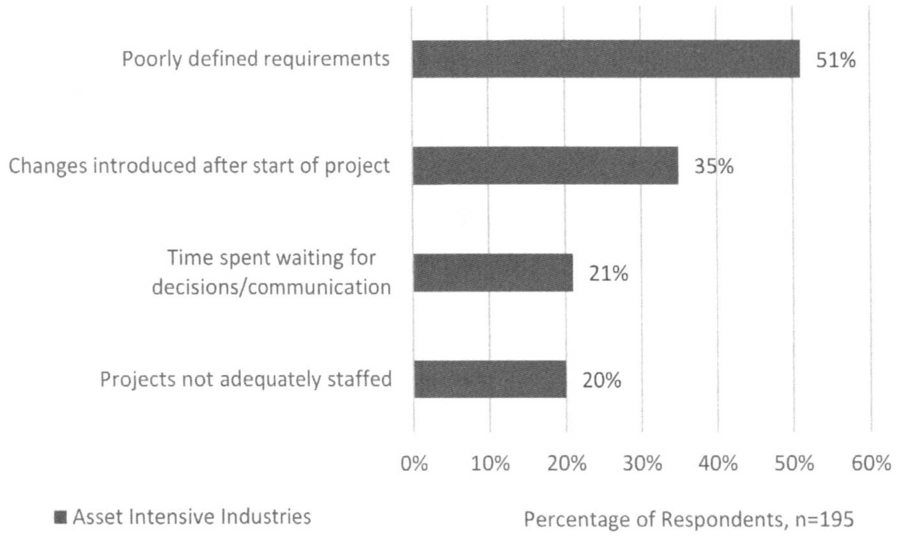

Figure 1: Top reasons STOs are poorly executed [4]

The second most common reason for poor STO performance is the introduction of changes to the scope of work after the start of the project. While it is true that in most cases, the full view of requirements cannot be determined, it is crucial to plan for the worst and arrange for appropriate resources to execute the most unexpected scenario. The third reason for poor STO performance is a deficit of troubleshooting and decision making capabilities in asset-intensive industries. This is often the case when the fourth reason comes into play In this study, 60% of surveyed companies feel they are understaffed in their high-skill positions; thus the success of a well-executed STO relies on the tribal knowledge and experience of their senior employees.

Per Sanders, [3] the number of full-time staff whose primary role is outage management, planning, or scheduling, continues to decline as economic pressures and industry demographic trends converge. In general, the availability of skilled resources may be the biggest problem affecting outage execution as industry-wide constraints affect the availability of both skilled and supervisory personnel.

The availability of skilled labor is even more important for the critical rotating equipment that is the heart of olefin processing. To tackle the issue of skilled labor availability, operators often outsource it to professional companies who manage turbomachinery turnarounds on a frequent basis. Original equipment manufacturers (OEMs) are a good choice, not only because of their experience, but also because of their engineering knowledge and support. Their expertise helps to mitigate the risks when rotating equipment deterioration is more severe than anticipated during STO planning and scope definition. Changes in the scope of work and lack of timely decision making are the third reason cited in Figure 1 for STOs not being successful. For this specific reason, OEM participation can be vital to an STO's success.

When the plan and scope of work are well established, and a third party is chosen to participate in the STO, the next logical step is to define roles and responsibilities. However, per Spothelfer and Johnson [5], roles and responsibilities should be clearly identified during plan and scope definition to ensure a consistent understanding of the project's objective and to allow sufficient time to consider and address all the work that must be done to achieve the objectives. Changes and additions to the plan become increasingly difficult to coordinate as the turnaround date draws near. [5]

Turbomachinery overhauls proceed more smoothly and quickly when the work is well planned and unforeseen work is minimized. For this to happen, the following steps are crucial:

- Planning
- Defining the scope of work
- Assigning roles and responsibilities among the various involved parties

A Successful Approach

With decades of experience managing an average of 160 outages per year that include turbomachinery from Elliott and other suppliers, Elliott has established a systematic approach which has proven successful in managing STOs.

This approach was recently used during a turnaround of a mega-ethylene plant. The scope of work comprised 17 compressor strings with steam turbine drivers and included rerates of the cracked gas compressor train driver, the ethylene heat pump compressor driver, the high- and low-pressure propylene refrigerant compressors with steam turbine driver, and the high-pressure ethylene refrigerant compressor with steam turbine driver. Initial discussions for the rerates began five years in advance of the turnaround, and detailed engineering and manufacturing began two years before the turnaround. The turnaround was completed on schedule, and each rerated equipment string achieved its new performance objective: cracked gas train(+9%), heat pump compressor (+12%), propylene refrigerant train (+14%), ethylene refrigerant train (+15%).

The approach follows a typical timeline such as the one in Figure 2; however discussions for re-rates often start earlier. This depends on the complexity and quantity of foreseen work.

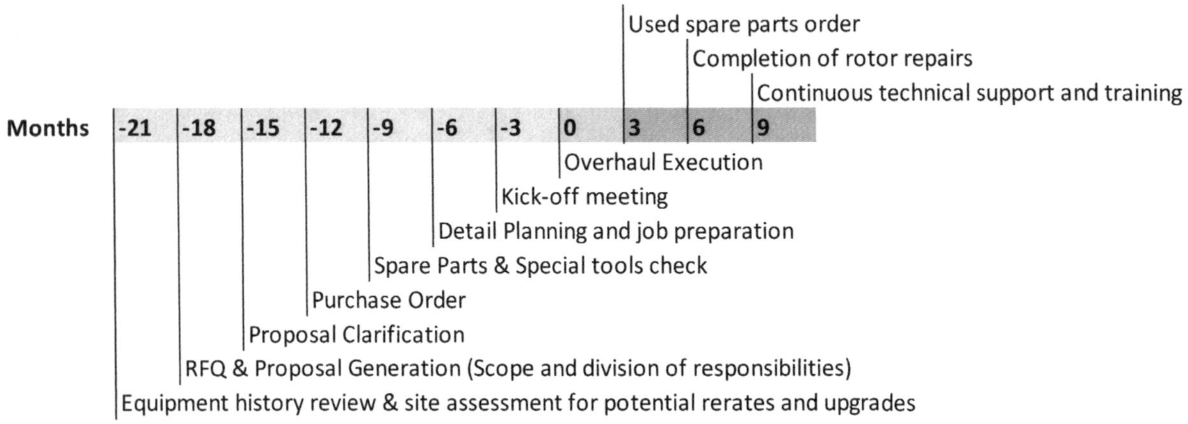

Figure 2: Suggested time line for a successful turbomachinery overhaul

As discussed earlier and suggested by other authors, eighteen months ahead of the STO appears to be the ideal time to start the planning process. The operator submits a Request for Quotation (RFQ) to contractors, and the contractor quotes the scope of work and defines the roles and responsibilities. Clarification meetings help to distinguish and divide responsibilities and to trim the scope of work. The result is a comprehensive quotation which defines the number of services that are to be provided by the contractor during the turnaround.

Next, the operator awards the project and issues a Purchase Order (PO) to the contractor who will execute the work. Ensuring the appropriate contracts are in place at the right moment is as fundamental as early planning of the turnaround. Issuing the PO a minimum of twelve months ahead of the STO is recommended to ensure skilled labor availability and to guarantee sufficient resources for the work awarded.

Once the contract is awarded, the spare parts and special tools check should follow. While this typically starts after the award of the contract, for larger turnarounds it is recommended that this inspection begin as early as possible, at least nine months prior to the turnaround start (see Figure 2).

For the turnaround at the mega-ethylene plant previously mentioned, Elliott was responsible for providing more than 5000 discreet parts for 22 Elliott and non-Elliott units located in seven areas of the plant where more than 5000 people would be working during the turnaround. The risk of losing or misplacing individual parts was high, and the field service crew could not afford to squander time searching through myriad crates for the correct parts. Elliott devised a system to ensure quick traceability and localization of the parts. Each order was color coded and labeled with the machine serial number and the customer's tag number to aid identification onsite. In addition, schematics of the parts were included in each box to make it easy for the crew to verify they had the parts they needed. Packing cases were similarly labeled and marked with "DO NOT REMOVE PARTS" warnings in several languages. As a result of this careful planning, the replacement parts arrived correctly organized for each unit.[5]

As the turnaround date approaches, the contractor creates a detailed plan for executing the work and discusses it with the operator. It is recommended that the detailed plan be finalized no later than six months ahead of the turnaround, and discussed with the customer during the kick-off meeting at least three months ahead of the turnaround.

After the turnaround, both parties should meet to share lessons learned and to discuss both positive and negative aspects of the turnaround to continuously improve. In addition, it is recommended that the plant operator order a spare set of the parts used in the turnaround and to proceed with the repair of spare rotors. This will ensure availability of critical parts in the case of an unexpected emergency plant shutdown.

Establishing a single point of contact between the two parties aids efficiency, eases communications, and helps to develop positive relationships. Elliott Group uses this model with a great degree of success.. Figure 3 illustrates a single point of contact model in which a project manager is assigned for the life cycle of the turnaround, including the preparation and execution phases.

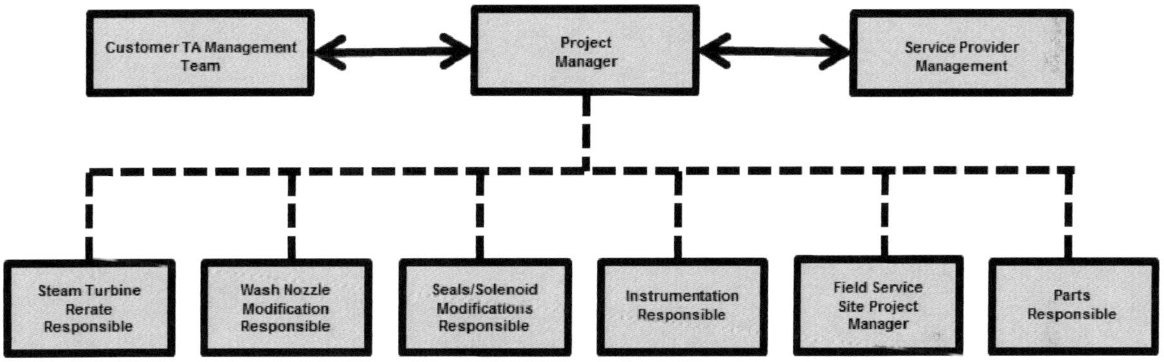

Figure 3: Single point of contact

Elliott has applied this approach on various turnarounds which have met and exceeded customer expectations where time and budgets were tight, and no injuries or safety incidents have occurred.

Summary

Completing STOs on time and on budget is crucial for olefin plant operators who are dealing with stringent environmental and safety standards that pose unique challenges to these large and complex projects. Early and meticulous planning to assess, quantify, and prioritize the risks of executing the scope of work is key to a successful turnaround. This approach has been successfully and systematically used by Elliott's field service group to win the work and earn the trust of olefin plant operators around the globe.

References

[1] US DEPARTMENT OF ENERGY: *Refinery Outages: Description and Potential Impact on Petroleum Product Prices*, Office of Oil and Gas, Washington, DC, March 2007

[2] SHAHI, A: *Best Practices for the Planning and Execution of Turnaround Projects in the Oil and Gas Industry*, Coreworx Project Information and Cost Control

[3] C. SANDERS, M: *Revisiting Planned Outage Execution Practices*, Altran North America

[4] PAQUIN, R: *Optimizing STO management to achieve operational excellence*, Aberdeen Group, September 2015

[5] SPOTHELFER, F & JOHNSON, R: *Turbomachinery Require Precise Planning*, Elliott Group, June 2016

Piping Forces and Moments on Compressors and Steam Turbines

Satoru Yoshida
Application Manager
Mitsubishi Heavy Industries Compressor International Corporation

Atsushi Higashio
Application Manager
Mitsubishi Heavy Industries Compressor International Corporation

Satoshi Hata
Vice President
Mitsubishi Heavy Industries Compressor International Corporation

Prepared for Presentation at the 2017 Spring National Meeting
San Antonio, TX, February 10, 2017

Piping Forces and Moments on Compressors and Steam Turbines

Satoru Yoshida
Application Manager
Mitsubishi Heavy Industries Compressor International Corporation

Atsushi Higashio
Application Manager
Mitsubishi Heavy Industries Compressor International Corporation

Satoshi Hata
Vice President
Mitsubishi Heavy Industries Compressor International Corporation

ABSTRACT

At recent chemical plants and LNG plants, the increase of plant capacity has caused the size of turbomachinery to increase and larger nozzle sizes are being utilized. . Especially for offshore facilities, a simpler piping arrangement is required despite the larger capacity trend because the space around the turbomachinery is constrained and therefore the piping design allows for less flexibility in its arrangement. Additionally, the heavier total weight of the turbomachinery and the larger size of piping nozzles make it more difficult for plant contractors to handle external piping forces and moments. Although NEMA SM 23 (1994) or API Standard 617 (2002) have been conventionally referred to for allowable loads in turbomachinery piping arrangement designs, the increase in plant capacity and piping size with larger machine installations for offshore facilities has heightened the need to relax this specification.

In this study, the possibility of relaxing the maximum allowable load on centrifugal compressors and steam turbines is investigated based on the general design philosophy of turbomachinery, our experience and analysis of quantitative influence using FEM (Finite Element Method) of typical machines with a piping load of 1.85 times the allowable limit specified by NEMA SM 23. Various key components including nozzle deformation, stress level, clearance and misalignment of shaft ends are examined and, as a result, a bottleneck (i.e., the limiting factor) which should be taken into account for centrifugal compressors and steam turbines is clarified in the case of excess piping load. Finally, applicable solutions from the machine manufacture's side are suggested in order to relax this specification for piping loads on turbomachinery.

INTRODUCTION

The allowable piping load on turbomachinery is an important issue to be agreed upon between plant constructors and machinery manufactures. Since API Standard 617 stipulates that 1.85 times the NEMA SM23 value should apply for the allowable piping load on compressor casing nozzles, this limit has been conventionally proposed as the allowable piping forces and moments on our turbomachinery. As is often the case with this topic, relaxation of the criteria is requested at later stages of projects by plant constructors, which causes exhausting discussions. It is not an exaggeration to say that the tendency of the Oil and Gas industry to value reliability based on past experience has made machine manufacturers quite conservative and cling to international criteria. In addition, it is not preferable to use different limits for various cases from the viewpoint of modular design because it may increase the number of design criteria and cause superfluous complexity.

New applications such as floating LNG (FLNG) or mega plants, however, seem to need higher allowable limits due to limited piping space. It can be a great opportunity as well as an ordeal for our business to meet such challenging requests from our customers. This is the motivation of this study and suggestions to solve this question based on bottleneck analyses are provided in this paper.

Purpose of this study

Our goal is to elicit solutions for the relaxation of the allowable limits of piping forces and moments to meet severer requirements in new fields such as FLNG from a technical point of view instead of disputing contractual agreements. For this purpose, it is important to clarify what should be considered as key issues for this problem, what the bottleneck is and how the manufacturers should modify their turbomachinery designs. It is hoped that this will encourage discussion of practical and viable solutions between manufacturers and plant constructors.

Standard requirement on piping load

NEMA SM23 is the standard code that covers single stage and multistage mechanical drive steam turbines. API standard 617, the standard code for centrifugal compressors for petroleum, chemical and gas service industries, quotes NEMA SM23 for the stipulation of piping loads and recommends 1.85 times the allowable limit of NEMA SM23. According to API standard 617, the forces and moments acting on compressor casing nozzles should normally be limited as below.

- Total resultant forces and moments should not exceed the limitation of any nozzle:

$$F_R+1.09M_R \leq 54.1D_e \quad (1)$$

- The combined resultants of the forces and moments of nozzles resolved at the centerlines of the largest connection should not exceed the limitations:

  For resultants:

$$F_C+1.64M_C \leq 40.4D_C \quad (2)$$

  For components:

$$F_x \leq 16.1D_C,\ F_y \leq 40.5D_C,\ F_z \leq 32.4D_C \quad (3)$$

$$M_x \leq 24.6D_C,\ M_y \leq 12.3D_C,\ M_z \leq 12.3D_C \quad (4)$$

Where, F_R and M_R mean resultant force and moment, F_C and M_C are the combined resultant of inlet, sidestream, and discharge forces and moments, D_e is equivalent pipe diameter of the connection, D_C is diameter of one circular opening equal to the total areas of openings. F_x, F_y, F_z, M_x, M_y and M_z show the individual components of resultants. The constants are "1.85 times NEMA".

Since there is no specified stipulation on relaxation for steam turbines like compressors, the same, 1.85 times NEMA critera, have been proposed for mechanical drive steam turbines as compressors due to the uniformity of train design.

APPROACH

What determines the allowable limit of piping forces and moments on turbomachinery? It seems quite difficult to answer this question because the bottleneck may depend on various factors. Although the products belong to the categorized design models, every machine is different because there are many different combinations of machine type, size, design pressure and temperature, nozzle location, service and so on. In addition, all the directions and distributions of piping forces and moments during the detailed design of the turbomachinery are assumptions as the directions and distributions are normally specified by the plant contractor at the later stages of projects.

In this study, all the possible risks are listed and examined to know how much influence each risk has on turbomachinery design, which risk can be a bottleneck and how the bottleneck can be removed. The following five possible risks are considered in this study:

1. Lift condition
2. Contact condition
3. Excess misalignment
4. Excess stress of casing
5. Gas leakage

First of all, no lift condition of turbomachinery shall be one of the criteria because no vertical movement upward is presumed in our design. If lift condition should happen, a lot of unexpected phenomena would arise and reliability would be no longer assured. Secondly, since piping load causes deformation of the casing, contact between moving parts and static parts should also be avoided. Contact condition is most likely to occur at labyrinth seals. Thirdly, misalignment should be within the mechanical limit of coupling. Fourthly, the stress of casing materials caused by piping load must be considered because the limit depends on the material used. Finally, the possibility of gas leakage at flanges should be investigated for horizontal split type compressors.

For lift condition evaluation, a statistical approach is adopted with a simplified model and thus the result can be used as empirical knowledge. For the other four risks, however, it is difficult to check all the possible variations. Therefore, FEM analyses for typical machines were performed as the first step in order to capture the influence of those risks and elicit the key issues to be examined carefully during further investigation.

FEM ANALYSIS

Two things should be underlined in the FEM analysis; one is the fact that this analysis is just a case study. It is important to derive implication applicable for general machines from the few results for specific machines. The other is decoupling the influence of piping loads from that of other thermal and mechanical factors. For instance, the margin for contact condition shall be shared by not only piping load but also the other factors including thermal growth and internal pressure.

Our goal is not to form a database by collecting detailed behavior of specific machines, but to provide information about what may become a bottleneck and how it can be removed.

Compressor

The selected type of compressor is double-flow type, which has a fabricated casing, two 60 inch suction nozzles and one 54 inch discharge nozzle. Casing feet as well as the compressor body is considered in the calculation. The underside of the foundation is assumed to be fixed in all directions. The behavior of the compressor body and feet is calculated when piping forces and moments are applied. The direction and distribution of piping forces and moments are shown in Table 1. Individual components of the resultants are fixed at 1.85 times NEMA. The forces and moments at each nozzle are allocated in proportion to each nozzle area. The FEM analysis of compressor was performed with the following approach. Firstly, deformation characteristics were examined with single force or moment on a single nozzle. Secondly, several cases for combined forces and moments were simulated to check if linearity of the behavior is applicable. It is helpful to capture the whole picture of deformation in order to estimate the quantitative influence of piping load. Finally, the actual operating conditions were simulated considering the machine's own weight, thermal growth, bolting, internal pressure and piping forces and moments.

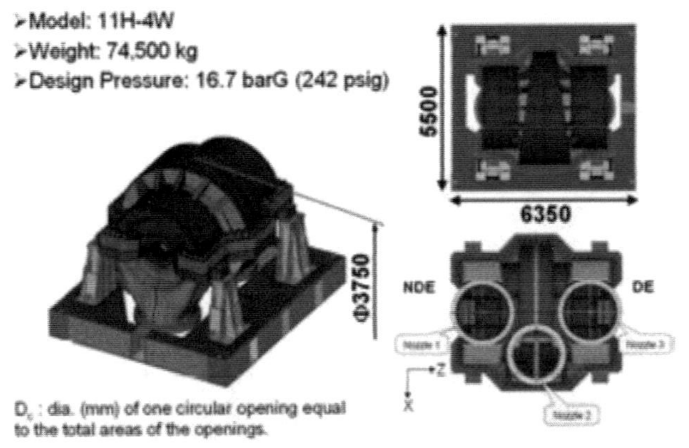

Figure 1. FEM model for compressor

Table 1. Piping load on compressor nozzles

	Nozzle 1	Nozzle 2	Nozzle 3
F_x	5,522 N (1,241 lbf)	4,466 N (1,004 lbf)	5,522 N (1,241 lbf)
F_y	13,891 N (3,123 lbf)	11,234 N (2,526 lbf)	13,891 N (3,123 lbf)
F_z	11,113 N (2,498 lbf)	8,988 N (2,021 lbf)	11,113 N (2,498 lbf)
M_x	8,437 Nm (6,222 lbf-ft)	6,824 Nm (5,032 lbf-ft)	8,437 Nm (6,222 lbf-ft)
M_y	4,219 Nm (3,111 lbf-ft)	3,412 Nm (2,516 lbf-ft)	4,219 Nm (3,111 lbf-ft)
M_z	4,219 Nm (3,111 lbf-ft)	3,412 Nm (2,516 lbf-ft)	4,219 Nm (3,111 lbf-ft)

Two types of steam turbines were considered with an FEM analysis, those being back pressure type and condensing type. The condensing type, which is typical of larger turbines, is subject to attracting force due to vacuum pressure on the exhaust side. This vacuum condition is different from a back pressure type, which is applied mainly to small size and light weight turbines.

The FEM analysis of steam turbine was performed considering turbine weight, heat and pressure load at operating conditions and piping loads simultaneously mainly in order to evaluate contact condition because it is empirically known that contact condition often takes place with an excess piping load. As to piping forces and moments on the nozzles, both distributed and concentrated conditions were simulated. The magnitude of each component was determined as 1.85 times NEMA. Figure 2 and 3 show FEM models for steam turbine.

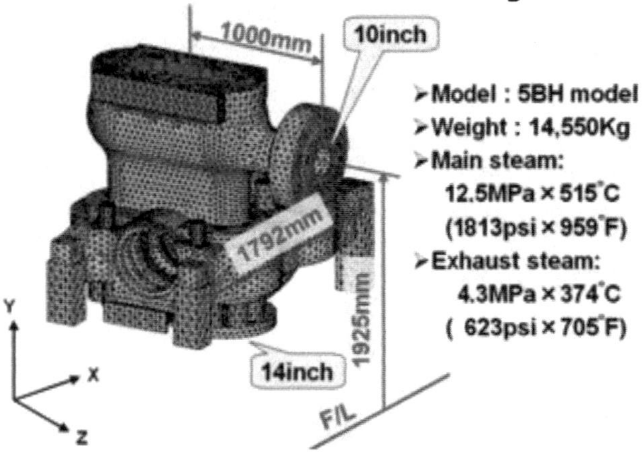

Figure 2. FEM model for steam turbine (back pressure type)

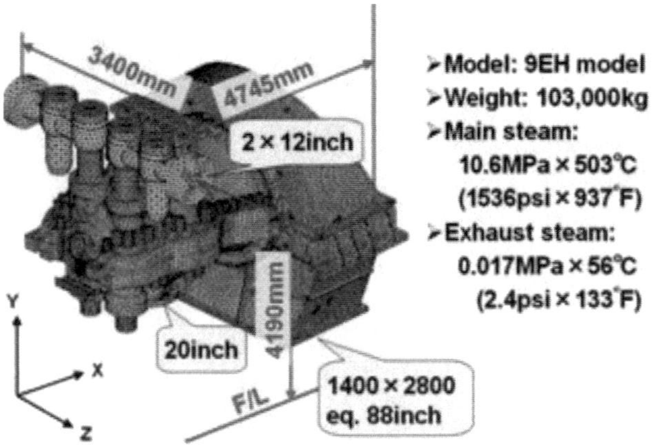

Figure 3. FEM model for steam turbine (condensing type)

Table 2. Piping load on steam turbine (back pressure type)

	Inlet	Exhaust
F_x	2,260 N (508 lbf)	2,608 N (586 lbf)
F_y	5,590 N (1,257 lbf)	6,450 N (1,450 lbf)
F_z	4,400 N (989 lbf)	5,078 N (1,142 lbf)
M_x	12,740 Nm (9,395 lbf-ft)	3,843 Nm (2,834 lbf-ft)
M_y	5,958 Nm (4,394 lbf-ft)	1,922 Nm (1,417 lbf-ft)
M_z	1,044 Nm (770 lbf-ft)	1,922 Nm (1,417 lbf-ft)

Table 3. Piping load on steam turbine (condensing type)

	Inlet Mouth 1	Inlet Mouth 2	Extraction	Exhaust
F_x	2,432 N (547 lbf)	2,432 N (547 lbf)	3,128 N (703 lbf)	9,032 N (2,031 lbf)
F_y	6,016 N (1,353 lbf)	6,016 N (1,353 lbf)	7,737 N (1,739 lbf)	22,342 N (5,023 lbf)
F_z	4,736 N (1,065 lbf)	4,736 N (1,065 lbf)	6,091 N (1,369 lbf)	17,589 N (3,954 lbf)
M_x	-9,623 Nm (7,096 lbf-ft)	13,238 Nm (9,762 lbf-ft)	4,609 Nm (3,399 lbf-ft)	13,310 Nm (9,816 lbf-ft)
M_y	-2,828 Nm (2,086 lbf-ft)	6,413 Nm (4,729 lbf-ft)	2,305 Nm (1,700 lbf-ft)	6,655 Nm (4,908 lbf-ft)
M_z	880 Nm (649 lbf-ft)	880 Nm (649 lbf-ft)	2,305 Nm (1,700 lbf-ft)	6,655 Nm (4,908 lbf-ft)

BOTTLENECK ANALYSIS

Bottleneck analyses for piping forces and moments on compressors and steam turbines were performed as per the above-described approach.

Lift condition

By comparing machine weight and vertical component of F_C as shown in Figure 4 and 5, a critical value for lift condition can be roughly estimated. For simplification, the distance between the acting vertical force and supporting point is assumed to be twice that between the center of gravity and the same supporting point of the machines. The $21.9D_C$ ($=40.5D_C/1.85$) indicates 1.0 times NEMA of the vertical component of F_C. Multiplication constant number at critical lift condition is calculated with all the past machines. The constant X means that X times NEMA is the allowable limit for lift condition.

Figure 6 and 7 show the allowable limit of lift condition with machine size. The first finding is that larger size machines have a larger allowable limit for lift condition. Needless to say, the business principle should be recognized that machinery be designed with minimized cost to attain the required performance. Another finding is that dispersion of the critical values even for the same size arises due to various design factors including design pressure and machine type. For instance, turbomachinery for higher pressure application should have larger weight and hence a higher limit for lift condition.

When the universal value is applied as standardized criteria for all varieties of machine, the criteria shall be determined with small and light machines and therefore the criteria will become quite conservative. The critical values will be around 2.5 times NEMA for both compressor and steam turbine. Setting aside using the standardized criteria, however, there is room for relaxing allowable limits on lift condition especially with large and heavy machines as shown in Figure 6 and 7.

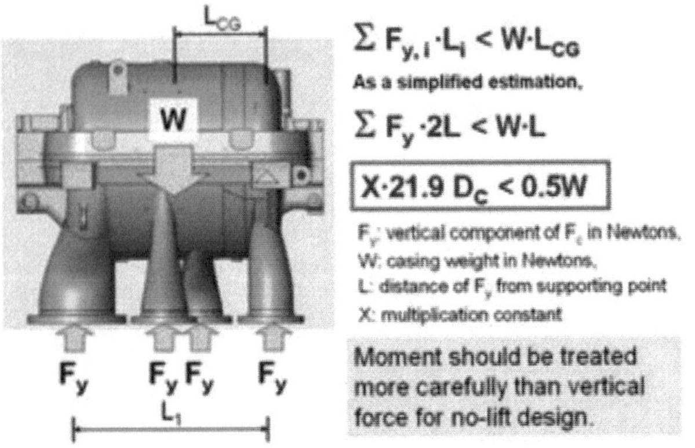

Figure 4. Model of lift condition for compressor

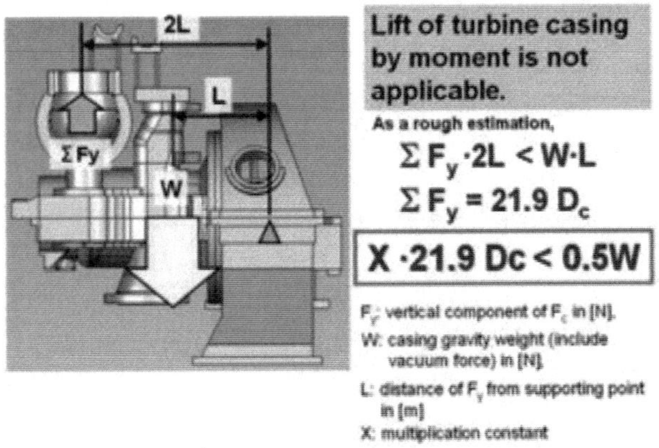

Figure 5. Model of lift condition for steam turbine

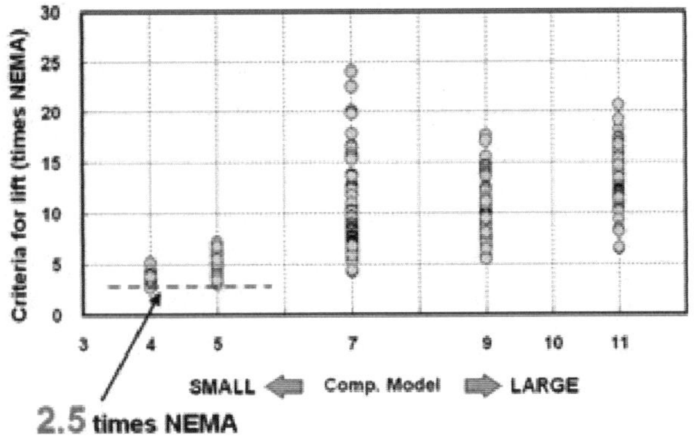

Figure 6. Critical lift condition for compressor

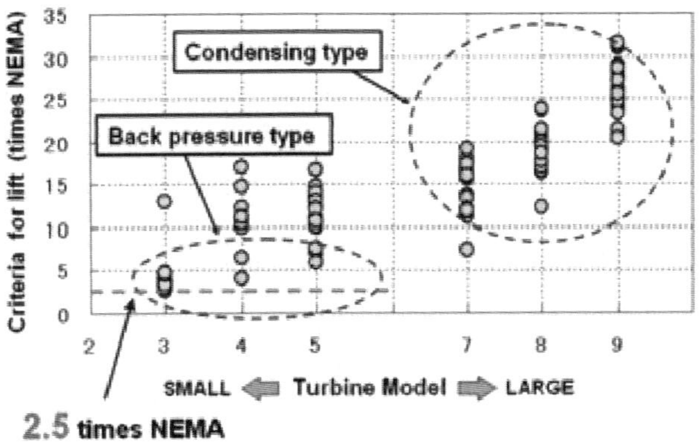

Figure 7. Critical lift condition for steam turbine

Contact condition

Both compressors and steam turbines are carefully designed considering the possible deformation of casing and rotor. Contact during operation would cause unexpected vibration or result in devastating damage of the machines. Contact is most likely to occur at a small clearance of the labyrinth seal structure as shown in Figure 8. The labyrinth seal structure is normally used for gas separation seals, shaft seals and interstage seals.

Figure 9 shows the typical deformation characteristics of compressors caused by single piping force. The maximum deformation, which is not equal to change of clearance, seems to demonstrate that deformation depends on the direction of the applied piping load. Change of clearance can be calculated based on an estimation of temperature distribution of the casing and internal parts (diaphragm, impeller and shaft), initial bending due to rotor weight and tolerance of the assembly. The contribution of each factor on the change of clearance is shown in Figure 10. It seems that the influence of piping forces and moments on clearance change is much smaller than those of the other factors and therefore can be regarded as negligible for the centrifugal compressor. In addition, it is possible to estimate the actual stiffness values of each piping nozzle under a postulation of linearity as shown in Figure 9. It would be helpful for plant engineers to use the actual stiffness values instead of a typical stiffness value of 1.8E12 N/mm (1E12 lb/inch) which is generally used in many programs. Since such a high stiffness value means that the turbomachinery nozzle is almost rigid, many problems can be eliminated by using more realistic stiffness values.

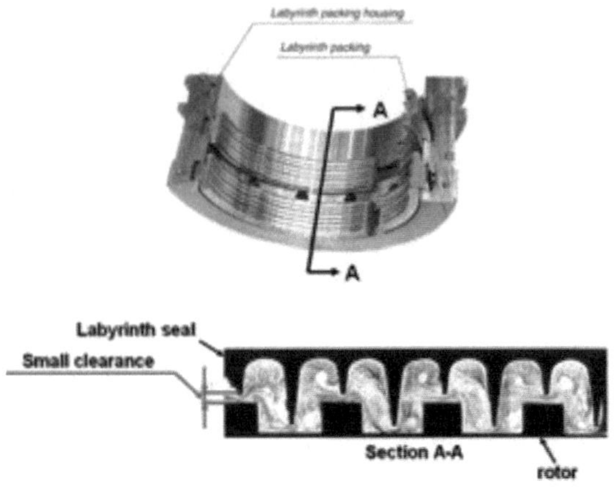

Figure 8. Typical labyrinth structure

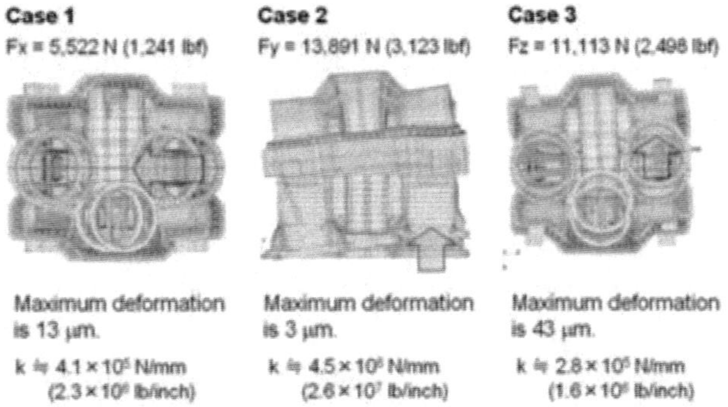

Figure 9. Typical deformation of compressor by single piping force (1.85 times NEMA)

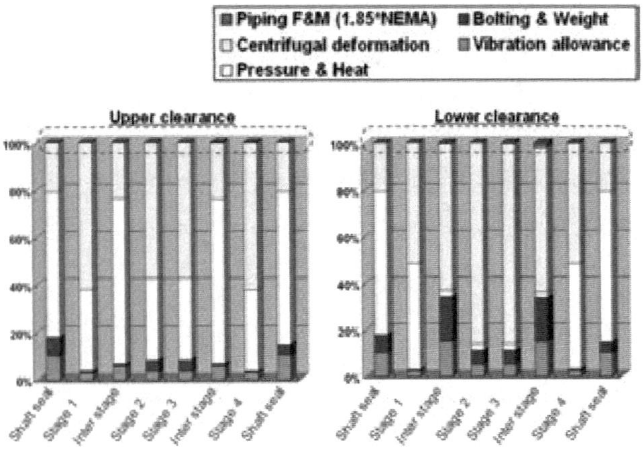

Figure 10. Contribution to the change of clearance (compressor case; influence of piping forces and moments is highlighted with a dotted line.)

Evaluation of the risk of contact condition was performed for two types of steam turbine. According to the simulation results, it is found that vertical deformation is significant and thus should be observed on back pressure type steam turbines; on the other hand, horizontal deformation is more significant for the condensing type because vertical deformation is constrained by the attracting force brought on by vacuum pressure.

Figure 11 and 12 are the evaluation results for contact condition of a back pressure type steam turbine. Figure 11 demonstrates that clearance change at front gland is largest. The critical contact condition can be estimated by linear extrapolation of the result of 1.85 times NEMA. The result is presented in Figure 12. Relaxation of the allowable limit of piping load on a back pressure type steam turbine will be preceeded by no lift condition first.

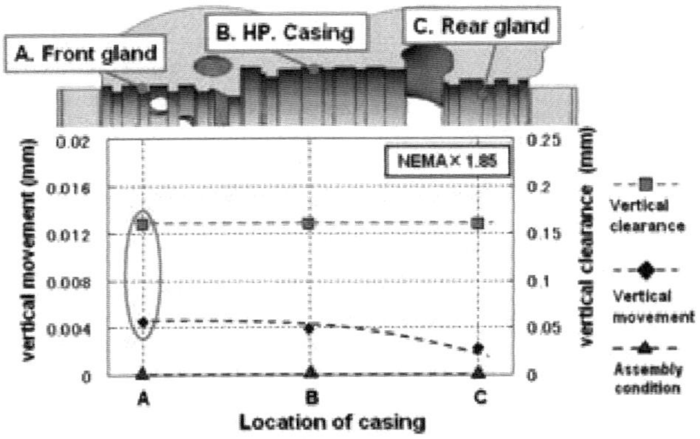

Figure 11. Clearance and its change due to piping load (back pressure type steam turbine case)

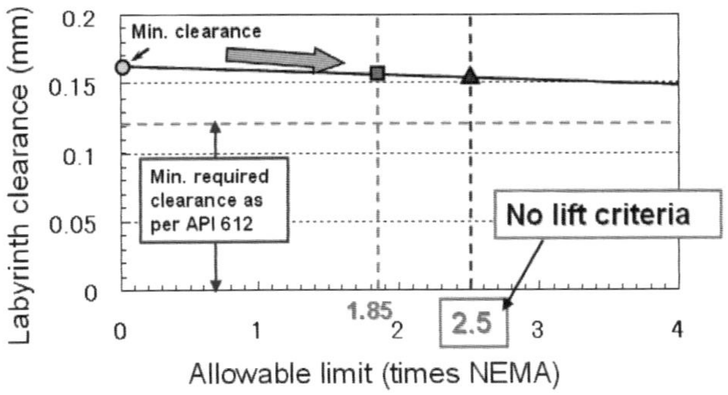

Figure 12. Critical contact condition (back pressure type steam turbine)

Figure 13 and 14 show the results of evaluation for a large size condensing type steam turbine. The change of clearance at MP (mid-pressure) stage is the largest. In the severest case (when a concentrated load is applied), the relaxation limit appears to be up to 6 times NEMA according to linear extrapolation when the minimum required clearance is considered as per API 612 Standard.

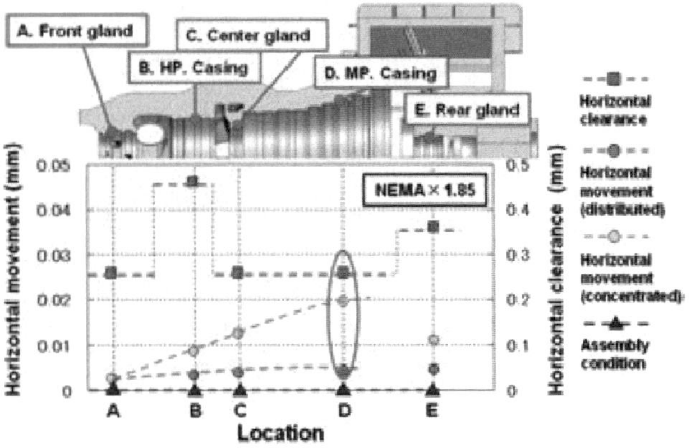

Figure 13. Clearance and its change due to piping load (large size condensing type steam turbine)

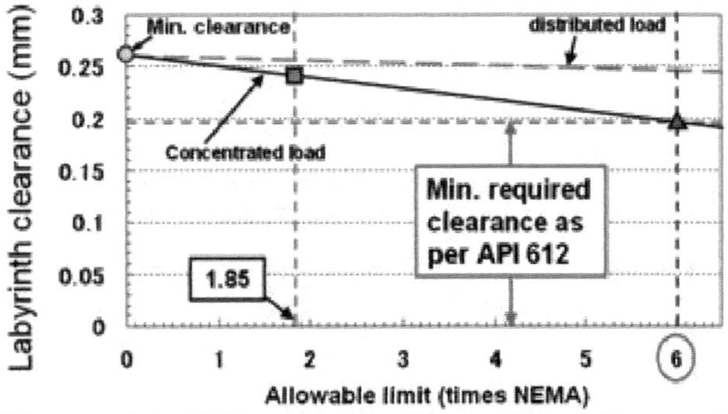

Figure 14. Critical contact condition (large size condensing type steam turbine)

The case study analysis for contact condition implies that contact condition can be a potential risk with large steam turbines when a relaxation of 1.85 times NEMA is attempted. Since clearance is a controllable design parameter, mitigation of the risk is possible by changing clearances in consideration of the trade-off between clearance and performance.

Casing stress, misalignment and gas leakage

These items are evaluated with FEM analysis for compressors in the following manner. Two cases are calculated; with and without a piping load of 1.85 times NEMA. Comparing the results of the two cases, the sole influence of piping load on turbomachinery can be highlighted. Typical results are shown in figure 15, 16 and 17. Figure 15 shows additional stress on the compressor casing caused by piping load. Maximum stress is observed at the place where compressor body and foot are in contact. The stress level of 1 MPa is much smaller than the casing strength. Figure 16 shows additional alignment change caused by piping load. The change is not so large as to jeopardize the mechanical limit of the coupling design. Figure 17 demonstrates that surface pressure between a compressor's horizontal mating flanges is not significantly affected by piping load.

As a consequence, the influence of piping load pertaining to these items is much smaller than that of the other factors and thus seems to be negligible for large compressors when piping forces and moments of 1.85 times NEMA are applied. Since an elastic body is presumed in the FEM analysis, the influence of a larger piping load can be inferred with linear extrapolation of the result. It is implied that stress of casing, misalignment and gas leakage do not become critical for the relaxation of allowable limits on not only compressors but also steam turbines in advance of lift and contact conditions.

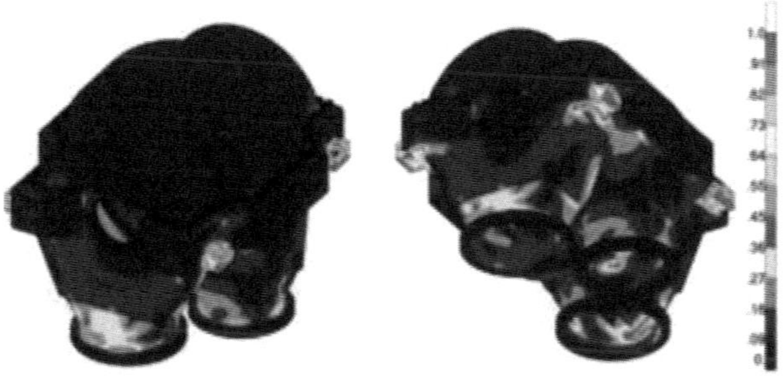

Figure 15. Casing stress distribution of compressor

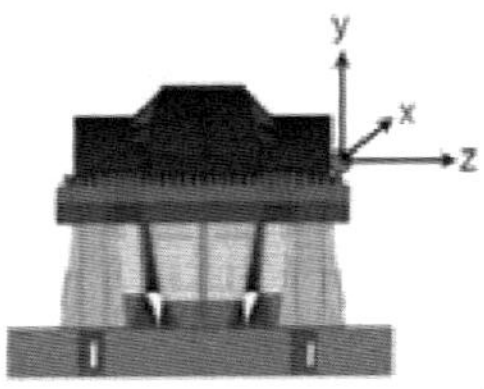

	Misalignment (mm) δ_y	Misalignment (mm) δ_x
• Bolting + Weight	-0.141	-0.010
• Bolting + Weight +Pressure	-0.096	0.014
• Bolting + Weight +Pressure + Heat	-0.673	0.066
• Piping forces & moments	0.001	-0.071
Total	0.672	-0.005
Parallel offset	3.570	3.570

Figure 16. Misalignment by piping forces and moments on compressor

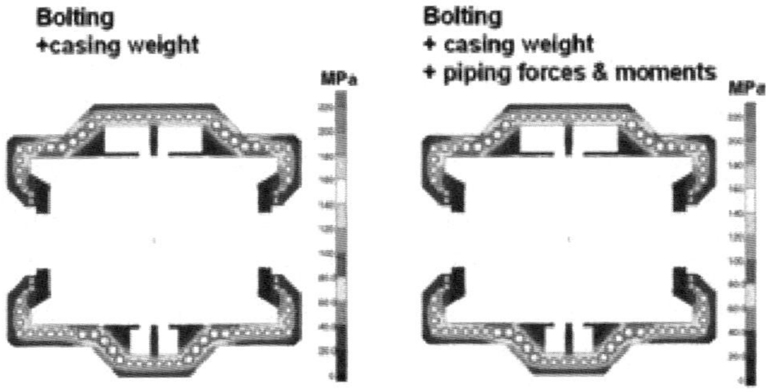

Figure 17. Surface pressure between horizontal flanges of compressor

SUMMARY AND SUGGESTIONS

As potential risks of excess piping forces and moments on centrifugal compressors and steam turbines, five relevant items are investigated in this study and some important information was derived as follows:

Firstly, it appears that the key issues are lift condition for small-size compressors and steam turbines (back pressure type) and contact condition for large size steam turbines (condensing type). The case study indicates that 2.5 times NEMA is the critical lift condition for small-size machines and 6 times NEMA for large size condensing type steam turbines. Since the FEM analysis is just a case study, it would be wrong to conclude that these results are universally valid. The results of case studies should be used to prioritize the concerned items rather than specifying critical values or preferable directions.

Secondly, it is inferred that a directional selection of piping load be effective because deformability is sensitive with direction of the piping load.

Therefore, for the relaxation of the allowable piping load limit from 1.85 times NEMA, lift and contact conditions shall be carefully considered in turbomachinery design and appropriate modifications depending on the target limit shall be applied for actual practice. Such modifications are exemplified as follows:

No lift design

- Additional weight should be put at appropriate positions on the machine casing.
- Casing thickness should be increased to surpass the piping load upward at the worst-case lift condition.

No contact design

- Clearance where contact is likely to occur should be increased. Trade-off relationships between increase in clearance and decrease in performance should be considered.
- Deformability of the machine casing and rotor is controlled mechanically to avoid worst-case contact condition.

The above-mentioned modifications are not easily accomplished on the manufacturer's side only because of the uncertainty of direction and distribution of piping load stipulated in NEMA SM23. The same problem exists on the plant contractor's side; the directions and distributions of piping load can be fully specified only after finishing entire plant engineering. If this timing mismatching problem is solved with mutual concessions and breakthrough ideas, relaxation of the allowable piping load limit would be much more attainable. With a specified piping load, it would be viable to perform a verification test in the manufacturer's test bed.

It should be added that stiffness values of each piping nozzle can be estimated by a deformation analysis under a postulation of linearity. It would be helpful for plant engineers to use these stiffness values instead of a typical stiffness value of 1.8E12 N/mm (1E12 lb/inch) in their piping design programs because such a high stiffness value is likely to cause excessive reactive forces in the piping system.

CONCLUSIONS

The 1.85 times NEMA has been conventionally proposed as the allowable limit of piping forces and moments on centrifugal compressors and steam turbines as recommended in API 617 Standard. In reaction to the increasing pressure for drastic alleviation of the allowable limit of piping load on turbomachinery, possible risks were presumed and the possibility of such alleviation was examined with a statistical approach based on the general design philosophy and experience and a case study approach using FEM analyses. As a result, important implications were elicited in this study as follows:

- Lift and contact condition may be bottlenecks and should be examined as priority issues for the drastic relaxation of the allowable piping load limit.
- Directional selection of piping forces and moments would be effective for tailor made design of the turbomachinery.
- If all the directions and distributions of piping forces and moments are specified in advance of the detail design stage, the criteria would be better assured with easier tailor made design.
- Use of the actual stiffness values of piping nozzles for piping design programs would eliminate many problems caused from a postulation of de facto rigidity of piping nozzles.

In order to confirm the critical conditions, FEM analyses should be performed with the actual compressors and steam turbines. The information elicited in this study, however, can be utilized to reduce the number of items to be examined. It is hoped that this study would be of help to provide answers to the frequently asked and challenging questions about piping loads on turbomachinery.

NOMENCLATURE

F_R	Resultant force
M_R	Resultant moment
F_C	Combined resultant of inlet, sidestream and discharge force
M_C	Combined resultant of inlet, sidestream and discharge moment
D_e	Equivalent pipe diameter of the connection
D_C	Diameter of one circular opening equal to the total areas of openings
F_x	Individual component of resultant force in x direction
F_y	Individual component of resultant force in y direction
F_z	Individual component of resultant force in z direction
M_x	Individual component of resultant moment in x direction
M_y	Individual component of resultant moment in y direction
M_z	Individual component of resultant moment in z direction

L	Distance from supporting point
L_{CG}	Distance of center of gravity from supporting point
W	Machine weight
X	Multiplication factor
k	Stiffness value
▪$_x$	Horizontal movement at shaft end
▪$_y$	Vertical movement at shaft end

REFERENCES

Bloch, H. P., 1998, *Improving Machinery Reliability*, Gulf Professional Publishing, pp. 329-338.

ANSI/API Standard 612, 2005, "Petroleum, Petrochemical and Natural Gas Industries—Steam Turbines—Special purpose Applications," American Petroleum Institute, Washington, D.C.

API Standard 617, 2002, "Axial and Centrifugal Compressors and Expander-compressors for Petroleum, Chemical and Gas Industry Services," American Petroleum Institute, Washington, D.C.

NEMA SM 23, 1994, "Steam Turbine for Mechanical Drive Service," National Electrical Manufacturers Association, Washington, D.C.

AIChE Paper Number: 36a

ENHANCING STRATEGIES FOR FEEDSTOCK QUALITY MANAGEMENT

Cristina Orta
Research Engineer-in-Training
NOVA Chemicals Corporation, Centre for Applied Research, Calgary, AB, Canada

Marie-Claire Campbell
Research Engineer-in-Training
NOVA Chemicals Corporation, Centre for Applied Research, Calgary, AB, Canada

Vasily Simanzhenkov
Research Engineer
NOVA Chemicals Corporation, Centre for Applied Research, Calgary, AB, Canada

Prepared for Presentation at the 2017 Spring National Meeting
San Antonio, TX, March 26 – 30, 2017

ENHANCING STRATEGIES FOR FEEDSTOCK QUALITY MANAGEMENT

Cristina Orta
Research Engineer-in-Training
NOVA Chemicals Corporation, Centre for Applied Research, Calgary, AB, Canada
Marie-Claire Campbell
Research Engineer-in-Training
NOVA Chemicals Corporation, Centre for Applied Research, Calgary, AB, Canada
Vasily Simanzhenkov
Research Engineer
NOVA Chemicals Corporation, Centre for Applied Research, Calgary, AB, Canada

Abstract:

With the continued introduction of new and diverse feed sources to ethylene manufacturing facilities and a heightened awareness of the impact that impurities may have on process operations, it has become more important to screen new feeds for quality assurance purposes. Stante et al. (2014) have previously outlined strategies to address sampling and analysis for feedstock contaminants. These practices have been applied at NOVA Chemicals Corporation when faced with the advent of a new feed. A comprehensive protocol to assess feed quality was also recently established to provide guidelines on sample collection, an analytical program, and how to communicate results. This paper reviews recommendations for the development of such a protocol. It also details approaches for sampling and analyzing elusive contaminants as well as processing samples limited in size. Finally, the importance of tracking contaminants to monitor feedstock quality is highlighted.

Introduction

A limited number of contaminants are monitored online at petrochemical facilities to confirm the quality of feed and process streams, and subsequently ensure smooth downstream operation. Focus is typically placed on compounds that are listed on the specification sheets or on contractual agreements which outline the maximum allowable limits that can be tolerated by the facility. The most common reason for a contaminant to be excluded from online monitoring is that it simply is not expected to be present. A contaminant may then be discovered as a result of process issues arising, such as fouling or corrosion.

Grab or spot samples are typically used for identifying contaminants that may have contributed to a process issue, or to proactively identify unexpected components. The custody chain experienced by the sample during spot sampling is outlined in Figure 1 below.

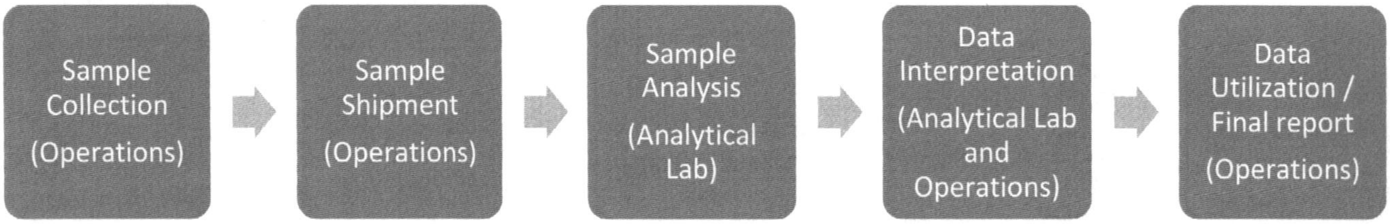

Figure 1. Flow diagram illustrating the custody chain for grab samples

Stante et al. have previously outlined the importance each step has on the quality of results and ultimately on the final "Data Utilization" step. It can be seen in Figure 1 that the "Operations" group is responsible for a number of tasks including sample collection, sample transport management from the field to the analytical lab, and data interpretation for arrival to a final decision. In turn, the "Analytical Lab" is responsible to prepare the sample for analysis, to calibrate, maintain, and operate the analytical equipment, and to interpret the results before conveying them. In addition, if the sample is derived from a third-party facility, then involvement from a "Commercial" party may be required. It becomes evident that multiple groups share the responsibility of obtaining quality results that can be used to make an informed decision. Given the cross-functional nature of this work, it is critical to establish clear expectations and communication pathways for all involved stakeholders. A feed quality assessment protocol can aid in the consistency of sampling efforts, thus leading to increased confidence in the results. Recommendations for the development of such a protocol are presented herein.

This paper also describes examples of enhanced sampling strategies which may prove helpful in tackling unusual grab sample collection and analyses situations. Note that any analyses program based on the collection of grab samples can only be reliable if the collected sample is representative of the bulk stream conditions. The sampling location, sampling point design, and sampling technique should be kept constant throughout the whole program to ensure the accuracy of results. Even then, it is best that offline analysis not be used for quality assurance of feed or process streams. Rather, it should be viewed as a valuable and cost effective tool in investigating an issue or in proactively preparing for the unexpected.

Development of a Feed Quality Assessment Protocol

Greater consistency in conducting feed quality assessments can be achieved by having a formal strategy in place for sampling, analyses, and communication of results. This will ensure critical elements are not missed, provide internal parties with clear guidelines on their roles and expectations, and allow for better comparison of feed quality data. An internal analytical protocol for assessing the quality of new feedstocks can thus be developed to address this need. While such a document aims to apply a degree of standardization to each new feed quality assessment, it is important to note that it is a dynamic document – analytical practices, risk tolerance levels, and other factors may change over time, thus affecting the strategies outlined in the protocol.

The approach to drafting the protocol is largely based on the vision and needs of stakeholders directly involved in feed acquisition, feed analysis, and operation of the facilities receiving the feeds. The document should thus first and foremost present a purpose statement developed following the input of the various stakeholders. Each section thereafter should aim to address this overall objective.

The first few sections in the protocol outline best practices for both taking and analyzing samples. This is important to capture succinctly so that these practices can be easily adopted and consistently applied, allowing for a higher degree of comparability and reliability in the analytical results. These best practices can be derived from industry standards [2] – [5], published papers, or the experience of subject matter experts.

After the objective and best practices are defined, the bulk of the document establishes the sampling, analytical, resource, and communication plans associated with different phases of feedstock quality analysis work. Note the use of the term "phases" of work – instead of focusing solely on analysis of a new feed prior to its arrival at a manufacturing facility or blending with other feed supplies, the protocol should also include strategies for monitoring baseline quality levels of select primary and ancillary streams before and after new feeds have been introduced. This provides baseline data for comparison when assessing the potential impact, if any, of a new feed on plant operations.

Regardless of how many phases of work apply, it is important to establish the core team members involved in each phase of work and define their roles and responsibilities. This will ensure that everyone is clear on their expectations and work can be conducted in an effective manner. It may be worthwhile applying a formal project management tool, such as the RACI matrix, to establish who should be Responsible, Accountable, Consulted with, or simply Informed of the work.

With team members' roles and responsibilities defined, strategies for analyses, sampling, resourcing, and communication can be developed for each phase of the work. The following should be addressed by each plan, noting there may be some overlapping factors or case-by-case considerations to be made:

- Analytical Plan:
  - Determine which species to analyze for and the suitable analytical technique(s).
  - If atypical species may be present in the stream, identify what sampling provisions may be necessary, and confirm if these species can be detected with the selected techniques.
  - For each species, determine the anticipated concentration range and confirm the analytical technique detection limits are adequate.
  - Consider if external lab resources can assist with the analysis.

- Sampling Plan:
  - Determine if spot or composite samples should be collected [1], or if there are reliable online analyzers that could be used.
  - Determine the number of samples to be collected, including how many rounds or sets are to be completed and whether samples should be collected in duplicate or triplicate.
  - Establish the sampling frequency, considering the mode of operation of the manufacturing facility (e.g. if a plant is in recycle mode, does this change the sampling strategy?).
  - Identify sampling locations and evaluate the set-up to ensure quality samples can be collected.
  - Incorporate any special sampling or delivery arrangements into the sampling plan (e.g. time-sensitive, additional preparation work required, sampling can only be done under certain conditions, etc.).

- Resource Plan:
  - Identify who is responsible for providing sampling supplies, taking samples, sending them, and then finally analyzing them.
  - Determine the type and quantity of sampling supplies required.
  - Establish how much advance notice is required to ensure resources are available.

- Communication Plan:
  - Establish what internal meetings should take place, their timing, and who should be involved.
  - Determine how, when, and to whom analytical results must be communicated.
  - Identify the individual responsible for managing third-party communications.

Another important factor to contemplate – one that is not addressed above but can be integrated into the aforementioned plans or is perhaps deserving of its own separate section – is what to do when a quality concern arises. For example, if a contaminant is discovered outside of feed quality specifications or internal tolerance levels, what steps should be taken? Is it first necessary to validate the finding? This might be particularly important if grab samples were taken, as the higher potential for error associated with this sampling method may rule out its dependability in matters of quality assurance [1]. In other words, grab samples may be better suited to flagging potential quality concerns instead of proving a concern is present. If a validation program is required, it may be preferable to contract a third-party to assist in the analyses.

Once all these details have been discussed and plans clearly outlined in the protocol, reviews can be conducted. It is recommended that all members of the core team are offered an opportunity to review the final protocol document and provide feedback. Only once consensus from all parties involved in the work is obtained will there be likelihood that the protocol is valued and consistently referenced when conducting feedstock quality assessments.

Finally, consideration should be given to using elements from the protocol to assist in generating clear work guidelines for any third-party members who may be involved. For example, if a feedstock supplier is providing samples, an external sampling protocol can be created using sampling best practices and planning information from the internal protocol. With this external protocol in hand, feed suppliers will have written guidelines on how best to sample, how many samples to take, when to take samples, etc. Additional information, such as safety features of any sampling supplies provided and contact information for shipping samples, can also be included. This will extend the benefits of the internal protocol to all parties, thereby improving consistency and accuracy of the feed quality data.

Sampling Strategies

Stante et al. have outlined a number of strategies pertaining to sampling and analyses for feed quality assessment purposes [1]. This section offers examples of how these strategies can be built upon in order to address unconventional situations.

Safe Sampling Design for Streams Containing a Corrosive Contaminant

As discussed by Stante et al., it is important to consider the nature of the vessel being used for sample collection and whether all elements are fully compatible with not only the bulk components in the stream, but potential contaminants as well [1]. An inappropriate selection of materials of construction can not only compromise the accuracy of the analytical results but, more importantly, the integrity of the vessel itself.

Consider the case of sampling a gas or liquefied gas containing a corrosive impurity into a sample cylinder. Though inert coatings can be applied to cylinders in order to avoid undesirable contaminant-vessel wall interactions, such coatings often cannot be directly applied to all internal surfaces of a safety relief device, leaving them susceptible to corrosion and hence failure. In such a situation, the relief element then proves a safety risk rather than a protective feature as the cylinder contents could be unexpectedly released to the surroundings despite maintaining sufficiently low pressures within the vessel. Fortunately, it is often possible to develop safe options for sampling without a safety relief device.

The first resource that should be consulted when considering the removal of a critical safety element are the current regulations for the transportation of dangerous goods. These regulations vary depending on jurisdiction, but generally, exemptions from using relief devices on pressurized vessels are available so long as certain constraints can be met. These constraints often pertain to the chemical nature of the hazardous material, the amount of material being transported, the physical state/conditions of the sample taken, and the specifications of the cylinder itself.

Most exemptions rely in part on restricting the maximum pressure a container can be exposed to during filling and subsequent transportation. When this is the case, a pressure gauge or regulator can be used to ensure that pressure limitations are not exceeded during filling. The impact of temperature fluctuations on pressure can be simulated beforehand to ensure that the fill pressure is adjusted according to sampling temperature and that, should the ambient temperature during transportation reach a high extreme, the pressure within the vessel will remain below the maximum allowable level. A safety factor should also be incorporated in the calculations to accommodate for the potential inability of the simulation to fully model all real-world aspects of the problem. Note also that it is important to confirm that the lower fill pressures associated with sampling without a relief device still allow for collection of sufficient sample to run the subsequent analyses required.

Once all aforementioned checks have been conducted, then changes to the final sampling design can be implemented. As discussed, this may involve modifying the sampling station to introduce a permanent pressure gauge or regulator for the purposes of controlling the fill pressure when taking samples without a pressure relief device. Recommendations on sampling port design listed by Stante et al. [1] imply that it would be more appropriate to place additional elements downstream of the cylinder assembly to reduce the potential for additional contaminants to enter the sampling vessel. As such, a pressure gauge or backpressure regulator could be installed prior to a vent line leading from the exit of the cylinder filling location. Notably, having a vent line allows for purge-through of the cylinder, thereby further ensuring the sample collected is representative of the bulk material [1].

The sampling procedure must also be modified to reflect the changes made to the sampling station. It is recommended to introduce a reference table of simulation results showing the fill pressure required for given temperatures. Operators should then be trained on the new procedure and set-up, with emphasis placed on the risks associated with transporting samples in a vessel without overpressure protection and the necessity of complying with the procedural demands. By following all these steps, it is possible to successfully collect samples in a manner that complies with safety regulations.

Small-Scale Simulation for Sample Collection

Collecting samples – whether in liquid, gaseous or solid state – is rarely a trivial task and many factors can affect the success of this activity. Representative samples can be difficult to obtain if care is not given to prevent cross-contamination, over- or under-representation of certain species, or inadvertent phase changes during sample collection. Stante et al. have already outlined the importance of sampling port and vessel design in collecting a representative sample [1]. At times, however, conventional techniques prove unsuitable and innovation is required.

This is often the case when targeting the collection of elusive contaminants; that is, contaminants that may only manifest themselves under specific conditions. For example, fouling deposits may accumulate when a stream experiences a phase change. It can be challenging to collect a sample of the solid contaminant as it could be dissolved in the liquid phase or precipitate out of the vapour phase as it is injected into an analyzer.

In order to tackle such a situation, it becomes necessary to devise a sampling system that effectively collects the fouling material. Given that it is difficult to sample the solid material in either the liquid phase or the vapour phase, a small-scale sample collection device can be created to replicate the phase change in the process stream across a filter. A slip stream can be diverted through the filter and the pressure drop monitored to ensure a phase change is triggered as planned. The solid fouling will then be gathered on the filter surface and submitted for analysis to confirm the presence and amount of the contaminant of concern.

It is important to highlight that many times samples can be very small in size and often difficult to retrieve from the collection device, making it challenging to analyze them. Consider, however, that some analytical methods require very small amounts of sample for analysis, at times within the microgram or milligram ranges. Additionally, analytical techniques can use effective solvents to extract the sample into a liquid medium that can then be processed by various analytical methods such as Gas Chromatography, Ion Chromatography, and many other techniques mentioned in literature [1]. If a sample cannot be retrieved from a collection device, it is possible to submerge the entire material in a solvent aliquot that could then be analyzed to detect the contaminant of concern. It is thus important to keep in mind that sample size need not always be a limiting factor in analysis.

Collection of Field Blanks

Another important aspect to keep in mind when sampling is the potential for cross-contamination from the sampling tools and vessels themselves, through human interaction, or even from contact of the sample with the environment. The purpose of sampling is to characterize the material as it exists in the system of interest. Every attempt must be made to keep the sample as close to the bulk material as possible. However, it is not uncommon for samples to become contaminated prior to sampling. Many times, process equipment must be flushed with chemical solutions before opening them in order to render them safe. This cleaning could alter the chemical composition of the samples or simply skew analytical results. It is important for the analyzing party to be made aware of such possible contamination in order to ensure correct interpretation of results. If possible, a sample of any potential cross-contaminant should also be provided. This material can be analyzed along with the sample to subtract its effect from the results.

Similarly, sampling efforts can affect the sample itself through operator actions such as contact with dirty tools, vials, gloves, or even environmental conditions [1]. Preventing sample contact with external sources of contamination should always be attempted; however, at times this cannot be achieved. In such cases, field blanks can also provide a helpful solution. These blank samples can take up many forms, from the glove that was used while a sample was collected, to an unused filter sample if solid material cannot be detached from the used filter. Once again, the analysis of these field blanks can provide helpful insight on which species present in the sample could be a result of cross-contamination and should thus be disregarded.

A rarely considered form of cross-contamination can take place from contact of the sample with air and the contaminants present within. Typically, contact with air is inevitable and may have a negligible impact on the sample composition. However, when measuring very low-level contamination (low ppb, or sub-ppb levels) it is recommended to give some thought to the potential for air contaminants to skew the results of the analysis. A good example is the measurement of low ppb levels of sodium in process water. As always, care must be taken to prevent cross-contamination, but this becomes particularly difficult when it comes to minimizing air exposure. A way to quantify this impact, is to collect an additional sample, using distilled water, and subject it to the same air exposure as the process water. Two additional samples can then be analyzed, distilled water from the lab that was not exposed to air at the sampling location, and the distilled water that was. This way it is possible to quantify the impact that the sampling itself could have on the contaminant levels and subtract any cross-contamination levels from the final result.

Contaminant Monitoring Database

The collection of data concerning contaminants entering production facilities through various sources, accumulating in specific areas, and potentially causing operational issues can aid in knowledge building efforts which ultimately lead to cost avoidance and improved plant reliability. Contaminants can have significant effects in plant operations by promoting fouling, foaming, emulsion formation, decreasing product quality, temporarily or permanently poisoning catalysts, etc. However, data collection by itself does little to help understand and mitigate these issues. In order to flag trends resulting from the data, it is suggested that measured values be mapped at each process area and data added as it becomes available over time. That way, any anomalies worth investigating, including potential areas of contaminant accumulation, can be detected easily and tackled appropriately.

Another consideration is whether any impurities entering the system will inadvertently interact with materials present within, such as feeds, products, co-products, chemical additives, metallurgy, etc. At times, a seemingly innocuous substance may adversely interact with a contaminant to create unexpected materials that could lead to process upsets.

Creating a diagram or database where this data is stored and readily available can provide opportunities for synergy with other data collection efforts. A good example, previously described by Kim et al., is the creation of a database or diagram to map the use of chemical additives throughout the manufacturing facility and understand their fate [6]. Such a diagram can be further improved by overlapping data collected from grab samples and highlighting the levels of contaminants of concern present in process areas. This can serve in understanding any potential interaction between these chemicals, accumulation points, and potentially aid in root cause investigations for process issues [6].

Conclusion

As new challenges arise with the change in diet and the introduction of unconventional feeds to ethylene manufacturing facilities, it becomes a key activity to monitor the quality of these streams, the presence of contaminants, and the impact impurities could have on the operation of these assets. The creation of a feed quality assessment protocol can greatly aid in the standardization of procedures and in establishing effective communication pathways. This can lead to a streamlined process where communication is clear and each of the parties involved understands the importance of collecting representative samples and the impact they have in the success of these efforts. This protocol should be considered a dynamic document which will evolve as it is used and new learnings arise from sampling efforts.

Through experience in carrying out sampling efforts it becomes evident that traditional sampling strategies may not always be suitable and innovative techniques must be devised to collect samples of elusive contaminants and to analyze materials despite limited sample size. Moreover, awareness of opportunities for cross-contamination should be identified prior to or during sampling, and efforts must be made to minimize its impact or, if that cannot be achieved, quantify it through the use of field blanks.

Finally, the purpose of the sample collection and analysis efforts must always be kept in mind. This data can only be meaningful and valuable if it is used to make timely decisions that help ensure the reliability of the process. The development of a system, such as a diagram or a database, to compile contaminant measurements can aid in identifying trends and potential areas of concern before issues arise.

References

1. M. Stante, T. Giese, S. Morrow, Y. Kieboom, K. Nadeau, S. Christner and V. Simanzhenkov, "Feedstock Impurities - How To Find Them, If It Is Not Known What They Are?," in *AIChE 2014 Spring National Meeting*, New Orleans, 2014.
2. Standard Practice for Sampling Liquefied Petroleum (LP) Gases, Manual Method, *ASTM Standard D1265*, 2011.
3. Obtaining Natural Gas Samples for Analysis by Gas Chromatography, *GPA Standard 2166*, 2005.
4. Collecting and Handling of Natural Gas Samples for Custody Transfer, *API Standard 14.1*, 2016.
5. Natural Gas - Sampling Guidelines, *ISO Standard 10715*, 1997.
6. G. Kim and M. Stante, "Inventory and Fate of Chemical Additives to Improve Plant Reliability," in *AIChE 2014 Spring National Meeting*, New Orleans, 2014.

AIChE Paper Number 36b

COMPRESSOR FOULING MONITORING METHODS IDENTIFY CHANGES IN FOULING MECHANISM TO DIRECT MITIGATION PERFORMANCE IMPROVEMENT

Roger Metzler
Technical Services Manager
Baker Hughes Incorporated

Hua Mo
Senior Chemist
Baker Hughes Incorporated

Prepared for Presentation at the 2017 Spring National Meeting, San Antonio, TX, March 26th-30th, 2017

COMPRESSOR FOULING MONITORING METHODS IDENTIFY CHANGES IN FOULING MECHANISM TO DIRECT MITIGATION PERFORMANCE IMPROVEMENT

Roger Metzler
Technical Service Manager
Baker Hughes Incorporated
Hua Mo
Senior Chemist
Baker Hughes Incorporated

Abstract:

Fouling within the olefins plant cracked gas compressor is a relatively common occurrence. Many gas crackers report some concern with compressor fouling. Significant percentages of liquid and mixed feed crackers also report fouling concerns. In some plants, the fouling occurs to a degree that imposes minimal operational limitations. The primary concern in these plants is an increase in energy consumption, and the fear of limitations if or when something causes the rate of fouling to increase. However, other plants consistently struggle with fouling that limits operational reliability and throughput capacity. A variety of fouling mitigation methods are available: internal coatings, specialty wash-oils and antifoulant additives. In most cases, the fouling is caused by a reactive polymerization mechanism. Quite often, this is a free radical mechanism, but Diels Alder mechanisms also occur. Changes in feedstock and/or changes in cracking severity can change or shift the predominance of the foulant mechanism. As the fouling mechanism changes, the mitigation method effectiveness may also change. In this paper, the fundamental monitoring methods to aid the evaluation of compressor performance and possible changes in fouling mechanism to direct changes in fouling mitigation techniques. Laboratory tests on process stream samples and operational data analysis procedures will be demonstrated. Two case histories will be presented to demonstrate how this methodology was used to facilitate performance improvement with an operational optimization and an additive antifoulant mitigation method.

Introduction

Fouling within the olefins plant cracked gas compressor is a relatively common occurrence. In the cracking process, a number of byproducts are generated that can react in the compressor and create fouling deposits. Many gas crackers report some concern with compressor fouling. Significant percentages of liquid and mixed feed crackers also report fouling concerns. In some plants, fouling occurs to a degree that imposes minimal operational limitations. The primary concerns in these plants are an increase in energy consumption and the fear of processing limitations, if or when something causes the rate of fouling to increase. However, other plants consistently struggle with fouling that limits operational reliability and throughput capacity. A variety of fouling mitigation methods are available: internal coatings, specialty wash-oils and antifoulant additives. In most cases, the fouling is caused by a reactive polymerization mechanism. Quite often, polymer is formed by a free radical mechanism, but Diels Alder mechanisms also occur. Changes in feedstock and/or changes in cracking severity can change or shift the predominance of the fouling mechanism. Inorganic foulant deposition is less common. Iorganic foulants may be introduced via import or recovery/recycle streams, corrosion, or salts due to acid-base neutralization reactions. Inorganics can accelerate both initiation and propagation of some polymerization reactions as well. As the fouling mechanism changes, the mitigation method effectiveness may also change. This paper explores the use of fundamental monitoring methods to aid the evaluation of compressor performance and changes in fouling mechanism in order to direct changes in fouling mitigation techniques. Laboratory tests on process stream samples and operational data analysis procedures will be discussed. Two case histories will be presented to demonstrate how this methodology was used to facilitate performance improvement with an operational optimization and an antifoulant additive mitigation program.

Fouling Mechanisms

Cracked gas is compressed in each stage of the compressor system and sent to the downstream processes (see Figure 1). During the compression process, heat as well as condensed liquid from cracked gas is generated. These condensed liquid droplets generate fouling reactions primarily initiated via the heat of compression.

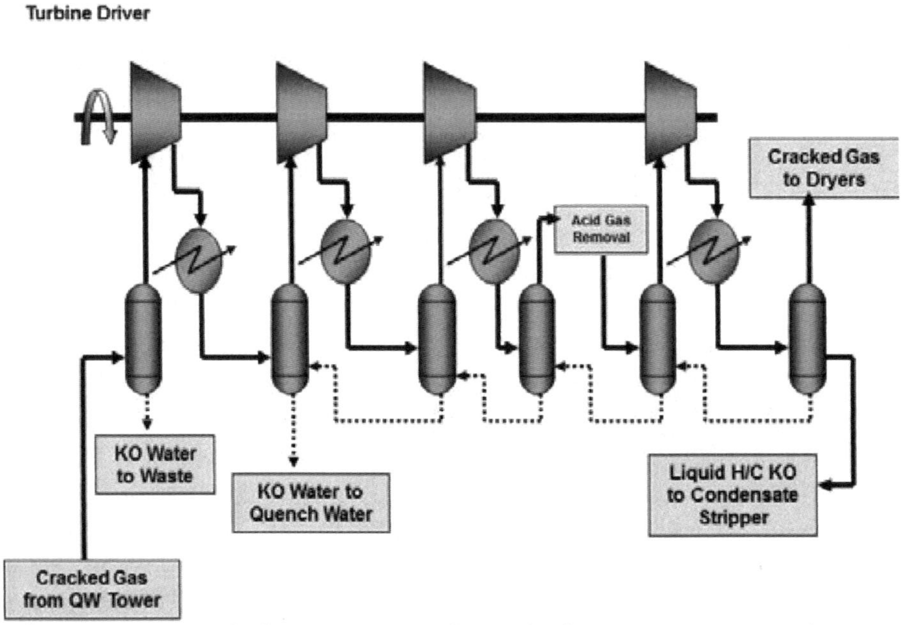

Figure 1. Typical Flow Diagram of Cracked Gas Compressor System

Different impurities in the cracked gas contribute to different reaction mechanisms. There are two major organic fouling mechanisms that are observed: free radical polymerization and Diels Alder reactions. The presence of conjugated vinyl monomers, such as styrene and butadiene, etc., can contribute to free radical polymerization. The presence of cyclopentadiene derivatives may contribute to Diels Alder reactions. There can be additional deposition reactions due to corrosion of the system, which is caused by the condensation of organic acids. Organic acids in the condensed water phase will cause corrosion reactions. When a foulant forms, it may agglomerate and deposit in the compressor.

Mitigation of fouling can be approached in a variety of means. Often, a multi-prong approach is taken and/or needed. Internal coatings need to be designed, engineered, and applied offline. Their objective to is reduce surface/foulant interactions and provide a relatively "non-stick" surface for the foulant to adhere to. Washoil mitigation can be applied continuously or intermittently. Washoils can provide solvency to dissolve low molecular weight non-cross-linked gums/polymers. Washoils can also provide surface wetting to reduce adherence of the foulant deposit. Water injection can be used to reduce compressor temperatures to retard polymer initiation and rate of reaction of polymer formation. When these methods fail to provide sufficient mitigation inhibitor and antifoulant additives can additionally be employed.

Free Radical Polymerization

There are three steps of free radical polymerization reaction, as shown in Figure 2.

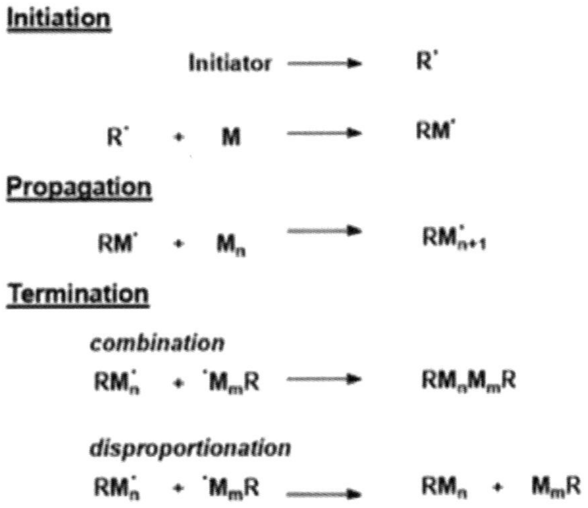

Figure 2. Free Radical Polymerization Mechanism

Initiation is required to generate reactive free radicals. Propagation allows growth of oligomer/polymer chain length. Termination finishes free radical polymerization. The presence of free radical initiators, such as peroxides, can initiate the reactions at lower temperature conditions than would be required without the presence of such initiators. The presence of conjugated olefins products provides the monomers for chain propagation. Polymer chain length and rate of polymerization depends on temperature, stability of intermediates, monomer concentration, and residence time in the process. Free radical polymerization can be mitigated by different types of inhibitor additives.

Diels Alder Reaction

The general mechanism of Diels Alder reactions is shown in Figure 3. Unlike free radical polymerization, the Diels Alder reaction is self–initiated, which can occur at relatively low temperatures. Diels Alder precursors are conjugated dienes such as 1,3-cyclopentadiene derivatives. The chain length of the polymer depends on the temperature and monomer concentration. Polymerization inhibitors usually do not help to mitigate the fouling issues associated with Diels Alder reactions. Mitigation is typically accomplished with dispersant antifoulant additives.

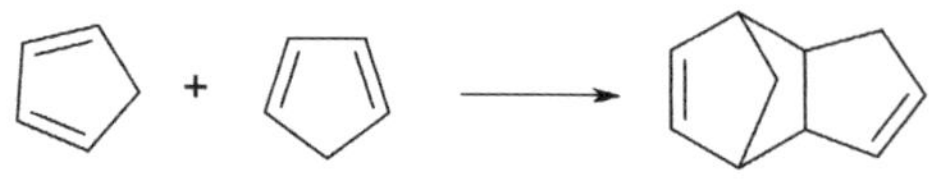

Figure 3. Diels Alder reaction

Corrosion Reaction

Before gases pass through the caustic tower to remove acidic gas, some organic acids can be condensed in the compressor. In the presence of water, the acid yields protons and causes corrosion issues. The general corrosion mechanism is shown in Figure 4.

$$HA \longrightarrow A^- + H^+$$

$$Fe \xrightarrow{H^+} Fe^{2+}$$

Figure 4. Acid Corrosion

To mitigate corrosion, neutralizers are added to form amine salts. Organic acid-amine salts at compressor temperatures will not typically form a solid. However, they can form "molton" or liquid salts that can still be quite corrosive if adequate aqueous phase is not present to keep them in dilute solution. Corrosion products can serve as polymerization initiators.

Different compressor systems experience different fouling mechanisms, and many mechanisms may co-exist in the same compressor. However, there may not be a single mechanism responsible for the bulk of the fouling in a given compressor system. To ensure successful and efficient treatment, each fouling mechanism and/or combination of fouling mechanisms must be identified.

Identification of Fouling Mechanism

Baker Hughes has developed a series of test protocols to identify fouling mechanisms. The tests are comprised of the following:

- Knock out pot sample analysis
  - Hydrocarbon sample analysis
  - Water sample analysis
- Deposit analysis (collected at equipment openings, filters/strainers, or filtered from liquid

samples)

- Wash oil analysis

Knock Out Pot Sample Analysis

The knock out pot is used to collect the condensed liquid after compressor inter-stage coolers among the different stages of compression. The condensed liquid could either be hydrocarbon, aqueous, or both. The fouling precursors, low molecular weight organic foulant and corrosion by-products could be measurable components in the condensed liquid. These fouling precursors can demonstrate the potential for foulant formation. Therefore, by monitoring and analyzing the hydrocarbon and aqueous samples from knock out pots, the presence of various fouling mechanisms can be identified. Changes in concentration of the foulant precursor caused by changes in operating conditions, feedstock changes, and changes in additive dosage or chemistry can also be evaluated. This information can help guide selection and evaluation of additive programs to improve the performance of compressor fouling mitigation.

The knock out pot analysis includes hydrocarbon analysis and water analysis. The hydrocarbon analysis is primarily used to identify the organic fouling mechanisms, while the water analysis is primarily used to identify the possible corrosion issues in the compressors.

1. Hydrocarbon Sample Analysis

Organic fouling precursors, such as styrene and cyclopentadiene, are concentrated in the hydrocarbon phase. Existent gums/polymers can also be identified. These fouling precursors, without any treatment, will contribute to organic deposits in the compressor. The fouling tendency is directly related to the type and concentration of fouling precursors.

The hydrocarbon sample analysis is used to identify the type and concentration of organic fouling precursors, and the fouling potential of the tested streams. There are two types of hydrocarbon analyses: component analysis and gum analysis. The component analysis analyzes the type and concentration of fouling precursors. Gum analysis is used to determine the fouling potential in the streams by existent gum and heat induced gums. Examples of these analyses are shown below:

- ***Component Analysis***

An example of hydrocarbon component analysis of samples from compressor knock-out pots is shown in the Table 1. The purpose of the component analysis is to identify the primary fouling precursors.

Table 1. Component Analysis

Component	Free Radical Polymerization Precursors	
	2nd stage (weight %)	3rd stage (weight %)
Styrene	10.58	5.13
Methyl Styrene	5.6	1.47
Indene	2.41	0.79
Others (butadiene, etc)	0.38	1.76
Component	Diels Alder Polymerization Precursors	
	2nd stage (weight %)	3rd stage (weight %)
1,3-cyclopentadiene	0.51	0
methylcyclopentadiene	0.26	0.42
Dimethylcyclopentadiene	0	1.39

The above analysis indicated that the hydrocarbon contained significant amount of free radical polymerization precursors, such as styrene derivatives, butadiene, etc., and Diels Alder reaction precursors, such as cyclopentadiene derivatives. Therefore, both free radical polymerization and Diels Alder reactions could potentially contribute to fouling in this compressor.

- ***Gum Analysis***

Reactivity of the same hydrocarbon samples was evaluated by way of existent and heat induced gums testing. The existent gum indicates the baseline of foulant (polymer) in the streams. Heat induced gums show the reactivity of the streams. The fouling potential can be assessed by the difference between heat induced gum and the existent gum. These test methods can be used to determine the performance of polymer inhibitor chemistries. A polymerization inhibitor was evaluated. If the free radical polymerization contributed to the fouling in the compressor, the inhibitor additive would significantly reduce the formation of heat induced gum. Test results are shown in the Figure 5 for a compressor knock-out sample.

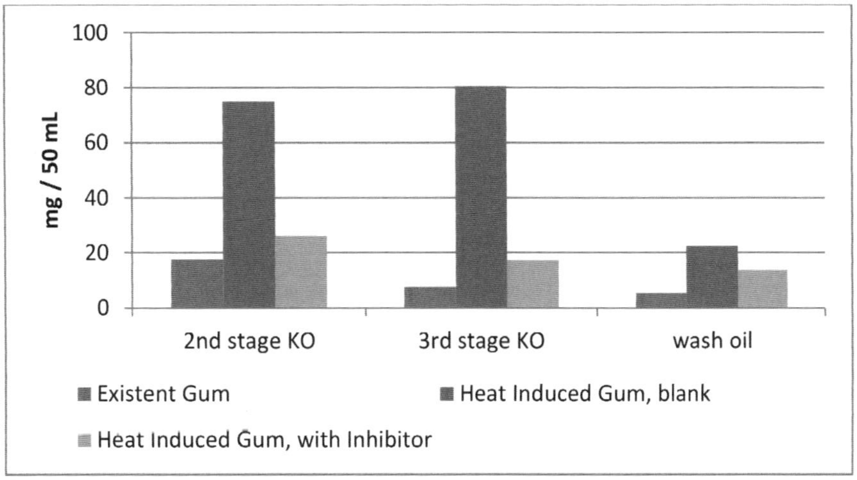

Figure 5. Gum Analysis of Hydrocarbon Sample from a Compressor Knock-Out Pot

Results indicated that the samples had significant potential to form polymeric deposits. The heat induced gum was significantly higher than the existent gum. Since the inhibitor additive significantly reduced gum/polymer formation, the free radical polymerization was likely to be the major contributor to the fouling. However, the inhibitor additive did not completely eliminate polymer formation as shown by the difference between the existent gums and the treated test result. These results indicated that other mechanisms, such as Diels Alder reactions, could be present.

The component analysis supported the results of the gum analyses. Both analyses indicated the presence of free radical polymerization and Diels Alder reaction mechanisms. The gum analyses indicated that free radical polymerization was the major contributor. When the plant feedstock changed, the reaction mechanism changed (further discussed in Case Study #2).

2. Water Analysis

Corrosion precursors, such as organic acids, are condensed and concentrated in the aqueous phase. Corrosion by-products, such as soluble iron species, can be found in the aqueous phase. Therefore, water sample analyses are used to identify possible corrosion precursors, such as organic acids, amines, and corrosion by-product (iron). An example of a water analysis is shown in Table 2.

Table 2. Species Identified in a Water Sample Analysis

Organic Species	
Acid Component	Concentration (ppm)
Acetic	12
Propionic	2
Formic	2
Butyric	<0.4
Thiosulfate	0.5
Base Component	Concentration (ppm)
NH3	< 0.2
Other amine species	2
Inorganic Species	
Inorganic Species	Concentration (mg/L)
Sodium (Na)	13
Potassium (K), mg/L	0.05
Magnesium (Mg), mg/L	0.05
Calcium (Ca), mg/L	0.3
Iron (Fe), mg/L	0.09

The pH of this water sample was 7. Results indicated that the water sample contained acidic species, amines, and inorganic species, such as sodium, calcium and iron. The cations may have come from the wash water in this particular application. The presence of iron indicated a possible corrosion issue, which was confirmed by the equipment inspection during turnaround. The presence of amines and acidic species in the water sample indicated that salt corrosion could be a concern. Injection of boiler condensate water to this compressor confirmed presence of salts (further demonstrated in Case Study #1).

Deposit Analysis

Deposits collected from the compressor, knock-out drum, and interstage cooler provide information on the nature of foulant. Organic-type fouling mechanism, the deposit contains functional groups of reactive organics–, such as styrene, butadiene, cyclopentadiene, etc.. Inorganic fouling mechanisms will be demonstrated by the presence of mineral scales, irons, and acid functional groups. Several instrumentation techniques such as Thermogravimetric Analysis (TGA) weight loss, elemental analysis, spectroscopy, and chromatography are used to identify the detailed nature of deposits. An example of deposit analysis from a compressor is shown in Table 3.

Table 3. Example of Compressor Deposits Analysis

Organic Elemental Analysis	**Weight %**
Carbon (C)	71
Hydrogen (H)	7
Nitrogen (N)	6
H/C mole ratio	1.18
Inorganic Elemental Analysis by XRF	
Oxygen (O)	11
Sulfur (S)	1
Iron (Fe)	2
Weight Loss at 800 °C	98
Chloroform Soluble	11

This deposit is primarily organic because the sample lost 98% of its weight at 800°C. Organics evaporate above 500°C.

An empirical method—C, H and N elemental analysis—is used to show possible functional groups of the deposits. If all the C, H, and N elements are from polymers, the calculated H/C mole ratio of deposit is 1.18, which is similar to the known ratio for cyclopentadiene type polymer (H/C ratio = 1.2), and styrene type polymer (H/C ratio = 1.1). The deposit would likely be a mixture of cyclopentadiene and styrene derivatives. The detail calculation of H/C mole ratio is shown in the Table 4.

Table 4. Calculation of CHN Elemental Analysis

Elements	Weight %	Element Molecule Weight	Element Mole @ 100 g deposit
C	71	12	5.92
H	7	1	7
N	6	14	0.03

The presence of iron and sulfur in the deposit indicated potential corrosion in the system. In addition to the elemental analysis, pyrolysis GC/MS is used to investigate the organic polymeric materials in the deposit matrix. Under the pyrolysis conditions (> 550°C), the polymer backbone is broken down to monomer fragments and identified by GC/MS. Pyrolysis GC/MS of the above sample revealed that the backbone of the deposit contained 1,3-cyclopentadiene and styrene. This analysis confirmed the CHN elemental analysis.

The presence of cyclopentadiene indicated the possibility of Diels Alder reactions. The presence of styrene indicated the possibility of free radical polymerization reactions. Combined, the deposit analysis results indicated that both free radical polymerization and Diels Alder reactions contributed to the compressor fouling.

Wash Oil Analysis

Wash oil is injected to mitigate fouling in many compressors. The wash oil will wet compressor internals and dissolve foulant deposits. The wash oil should be relatively non-volatile at compressor conditions. Although there is no general standard for compressor wash oil, the commonly considered "good" wash oils will have low filterable solids, low fouling precursors, and more aromatics. Baker Hughes has developed methods to analyze different types of wash oil. The wash oil analyses include determination of aromatic content and volatility. Additional tests, such as existent gums, heat induced gums, and solubility of foulant in the wash oil are also used to investigate the suitability of wash oils.

Three types of wash oils were analyzed. Two wash oils (Wash Oil #1 and Wash Oil #2) are commercial products. Wash Oil #3 is an in-house wash oil, which is a hydrogenated pygas. The application of wash oil #3 will be discussed in case study #2. Analyses of the aromatics content of several wash oils are shown in Table 5; analyses of the volatility of the wash oils by comparing distillation recovery and temperature are shown in Table 6.

Table 5. Aromatic Content Analyses

Component	Weight %		
	Wash Oil #1	Wash Oil #2	Wash Oil #3
Paraffin	0.62	1.11	<0.01
Olefin	0	0	67.27
Aromatic	95.48	68.22	0.36
Naphthalene	0.08	0.14	25.58
Unclassified	3.82	30.53	6.79

Table 6. Distillation Recovery (Weight %) vs. Temperature (F)

Components Distilled	Tested Boiling Point (F)		
	Wash Oil #1	Wash Oil #2	Wash Oil #3
Initial Boiling Point	349.2	352.0	321.44
5% Rec	360.1	385.7	329.54
10% Rec	361.4	389.8	332.78
20% Rec	363.5	393.8	339.44
30% Rec	365.4	399.0	343.94
40% Rec	367.6	404.8	345.74
50% Rec	370.1	410.0	348.26
60% Rec	373.1	417.9	350.42
70% Rec	376.8	430.5	353.48
80% Rec	383.3	453.4	356.72
90% Rec	394.2	521.6	360.86
95% Rec	406.2	583.2	364.82
Final Boiling Point	428.0	606.9	382.64

Case Studies of Improved Compressor Performance by Analytical Methods

The complex analyses discussed can not only help to identify the fouling mechanisms, but also help to improve compressor performance. Two examples are provided to demonstrate the application of the methods to improve the compressor performance.

Mitigate Under Deposit Salt Corrosion

At this plant, wash water was being injected to the compressor to lower the operation temperature. A previous water analysis indicated the presence of materials that could contribute to under deposit salt corrosion (See the *water analysis* in knock out pot sample Analysis). The hypothesis was confirmed by observations from the turnaround. Small corroded areas were found on the compressor wheels as shown in Figure 6.

Figure 6. Corrosion of the compressor wheel during the inspection

Although the current treatment program was more than adequate to meet the customer's turnaround goals, some operations were adjusted to provide a more effective treatment program. The plant wished to better mitigate the corrosion in the compressor to reduce maintenance costs. One of the operational adjustments was performing online cleaning by injecting boiler condensate to the compressor. A monitoring and testing plan to follow this effort was developed. Knock-out pot water sample analysis was used to monitor the efficiency of the condensate injection. Several samples were obtained for analysis. The samples were from 2nd and 3rd stage compressor knock-out pot before and after condensate water wash. A sample of condensate water was also sent as a reference. If salt deposits were washed by the condensate, the ion content in the knock-out samples would increase. The analytical results are shown in Table 7.

Table 7. Comparison of Organic Species before / after Condensate Wash

Acid Species (ppm)					
	2nd stage before Cond.	**2nd stage after Cond.**	**3rd stage before Cond.**	**3rd stage after Cond.**	**Condense Water**
Acetic	78	59	170	210	ND*
Propionic	21	13	24	27	ND*
Formic	4	3	8	12	ND*
Base Species (ppm)					
NH_3	0.4	<1	<2	<1	ND*
EA	19	7	32	54	ND*
DiEA	96	65	340	300	0.7

* ND: Not Detected in the sample

The water analysis revealed that the boiler condensate water used in the online cleaning was free of organic species. Therefore when the condensate water was injected, a dilution effect of -the knock-out water sample would happen. The salt and acid concentration in the 2nd stage decreased due to this dilution effect. Since the condense water removed salt deposits in the 3rd stage, the salt concentrations increased after the condensate injection in this sample. The water analysis confirmed the possibility of salt deposition in the compressor. Results also indicated that injection of condensate water will mitigate corrosion by removing salt deposits.

Fouling Caused by Switching Feedstock

After an ethylene cracking plant switched to a light feed stock from heavier liquid feed, the observed compressor efficiency significantly decreased. Even with increased inhibitor dosage, the efficiency did not recover. To bring back the efficiency of the compressor, a complex analysis was carried out. The purpose of the analysis was to monitor the change of compressor operations, determine the fouling mechanism, and recommend a treatment program to improve the efficiency of the compressor.

A hydrocarbon analysis was used for samples from the knock-out pots. The component analysis did not show a significant change in the paraffin, olefin, aromatic and naphthalene concentration. However, the heat induced gum analysis of hydrocarbon indicated a probable change of fouling mechanism due to the change of plant feed. Unlike the sample before the feedstock changes (see Figure 5), the heat induced gum did not significantly decrease when treated with free radical inhibitors, as shown in Figure 7. The results indicated that significant non-free radical polymerization contributed to the fouling and lost compressor efficiency. Therefore, adjusting the dosage of free radical polymerization inhibitors would not help to mitigate the fouling in the compressor.

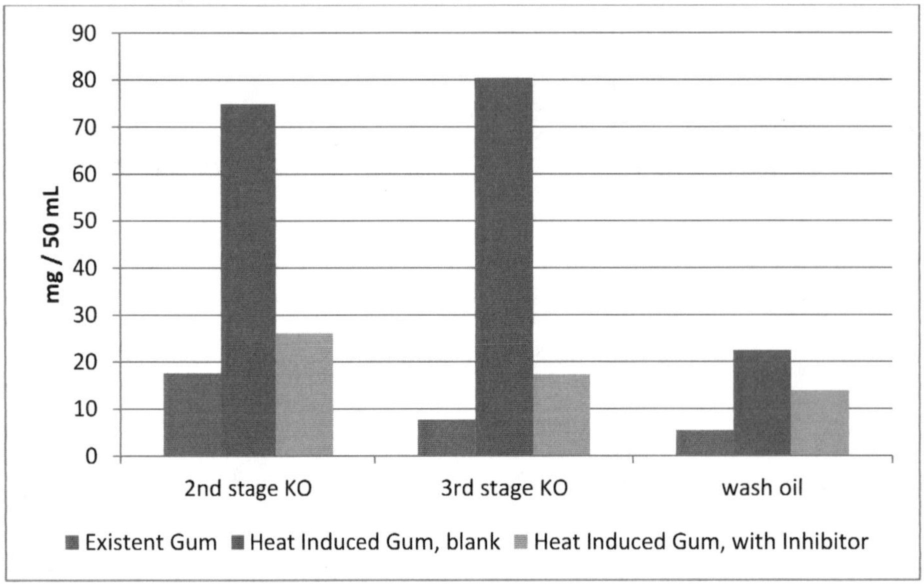

Figure 5. (reprint) Gum Analysis of Hydrocarbon Samples before Switching Feedstock

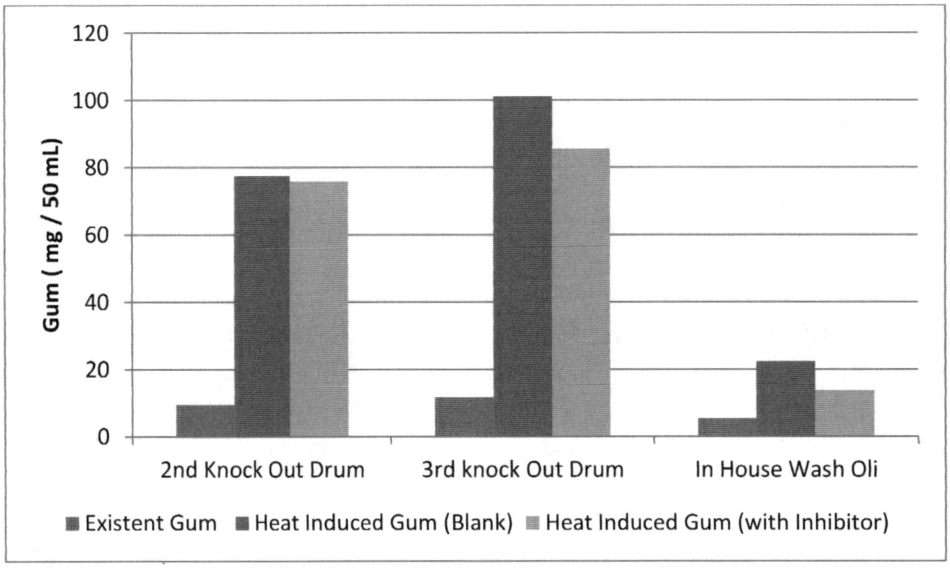

Figure 7. Gum Analysis of Hydrocarbon Samples after Switching Feedstock

Further process analyses indicated that compressor efficiency loss related to lower concentrations of heavier aromatic components. When the plant feedstock becomes lighter, lower amounts of heavy aromatics, such as gas oil, are generated as shown in Figure 8. Therefore the ability of the wash oil to wet / dissolve or remove the foulant by hydrocarbon liquid decreased. Elevated concentrations of light gas components also increase the work load of compressors.

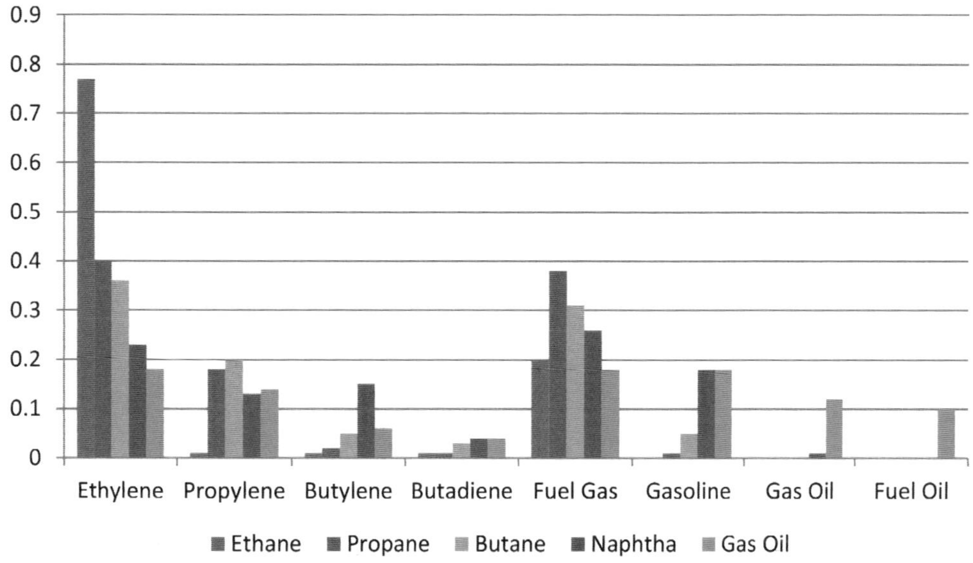

Figure 8. Comparison of Individual Components in Different Plant Feedstocks

Since non-free radical polymerization, such as Diels Alder reaction, significantly contributed to the fouling, a dispersant antifoulant additive was recommended to mitigate the fouling. Recognizing the deficiency of heavy aromatics in the cracked gas, Baker Hughes also recommended the addition of a dispersant additive to the make-up in house wash oil. The idea of this treatment program was to control accumulation / agglomeration / deposition of the fouling materials in the compressors. This particular dispersant additive also had wetting agent functionality to improve the wetting ability of the washoil.

An in-house wash oil was analyzed for the aromatic contents and volatility (See wash oil #3 in Table 5). The component analysis indicated that this wash oil sample did not contain a significant amount of aromatics. The volatility analysis indicated that the wash oil was more volatile than others in use at different plants. The reactivity of the in-house wash oil was also analyzed. The heat induced gum test indicated that wash oil was not reactive (see Figure 7). Further lab tests indicated that the wash oil could dissolve polymer formed in the heat induced gum test. The analyses indicated that although the wash oil could be used, higher injection rate would be needed due to the low aromatic contents and high volatility.

After the treatment program (wash oil + dispersant) was used in the compressor, the efficiencies of compressor were quickly recovered as shown in Figure 9. The recovery of compressor efficiency indicated that the monitoring program could help to improve the performance of compressor operation by identifying the fouling mechanism.

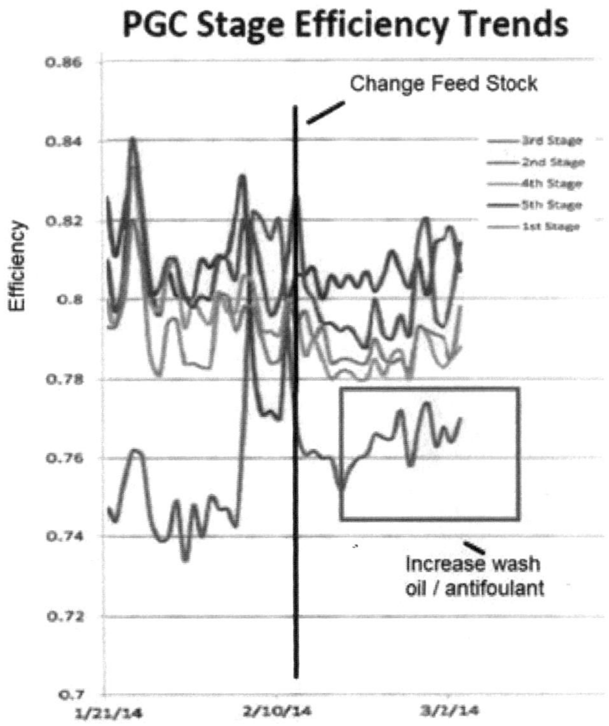

Figure 9. PGC stage efficiency Trend after Treatment Change

Conclusion

In the paper, a complex, process is used to analyze the compressor fouling mechanisms. The process consisted of knock-out pot sample analyses, deposit analyses and wash oil analyses. Two examples were presented to illustrate how the method helped improve the performance of a compressor system.

Acknowledgement

The authors wish to extend their gratitude to the analytical group at Baker Hughes for their help in providing the analytical work. We also wish to express our appreciation of our customer unit personnel for their collaboration, communication and discussions of compressor.

AIChE Paper Number: 36c

OXYGEN-CONTAINING CONTAMINANTS AND STEAM CRACKING: UNDERSTANDING THEIR IMPACT USING COILSIM1D

Ismaël Amghizar

Laboratory for Chemical Technology

Ghent University

Andrés E. Muñoz G.

R&D Manager

AVGI

Marko R. Djokic

Laboratory for Chemical Technology

Ghent University

Kevin M. Van Geem

Laboratory for Chemical Technology

Ghent University

Guy B. Marin

Laboratory for Chemical Technology

Ghent University

Prepared for Presentation at the 2017 Spring National Meeting

San Antonio, Texas, March 26-30, 2017

OXYGEN-CONTAINING CONTAMINANTS AND STEAM CRACKING: UNDERSTANDING THEIR IMPACT USING COILSIM1D

Ismaël Amghizar
Laboratory for Chemical Technology
Ghent University
Andrés E. Muñoz G.
R&D Manager
AVGI
Marko R. Djokic
Laboratory for Chemical Technology
Ghent University
Kevin M. Van Geem
Laboratory for Chemical Technology
Ghent University
Guy B. Marin
Laboratory for Chemical Technology
Ghent University

Abstract:

Depleting oil reserves and increasing environmental concerns have enhanced the interest in alternative and renewable feedstocks for the production of base chemicals. Consequently, significant efforts are being carried out aiming at the fundamental understanding and the development of micro kinetic models of the thermal degradation of oxygen containing renewable feeds and oxygen-containing contaminants, which are of interest as feedstocks for olefin production via steam cracking. Undoubtedly these oxygen-containing species can affect the product distribution and, therefore, accurately describing the thermal decomposition of these components is important. Besides the yields of high valuable chemicals (small olefins), it is necessary to predict the yields of the small oxygen containing components, e.g. methanol and aldehydes, which affect the operation of the process: They can lead to problems in the downstream separation train, such as off-spec products, and are known to lead to gum formation.

This paper focusses on the (co)-cracking of renewable feeds or feeds with a considerable amount of oxygen-containing contaminants. COILSIM1D is a well-known tool for accurate yield prediction of industrial steam crackers. Recently, its microkinetic model has been extended to account for various common oxygen-containing compounds such as methanol and dimethyl ether. The fundamental kinetic model explicitly accounts for the elementary steps and the corresponding kinetic and thermodynamic parameters are derived from high level ab-initio calculations. Therefore, the model has a broad validity range in terms of feeds and operating conditions. To illustrate this, a case study of steam cracking of oxygen containing species in a hydrocarbon mixture matrix will be presented. Particular attention will be paid to the small pyrolysis products which are important contaminants of the downstream separation section. The results of the COILSIM1D simulations are compared with experimental data for validation purposes.

Introduction

Throughout the last decade, the petrochemical industry has been faced with several challenges, largely determined by economics, politics, environmental concerns and technical developments. Throughout all these scenarios, the interest on feedstock flexibility has been a major priority, as it can significantly help producers to better adapt to the ever-changing environments in which they have to operate.

Steam crackers are often considered to be the heart of a petrochemical plant. Therefore it is very important to ensure that the available feeds can be processed adequately. Besides the usual feeds (i.e. ethane, propane, naphthas, ...), crackers are now presented with the possibility of (co-)cracking bio-derived feeds such as hydrogenated waste oils or tall oils[1, 2], plastic waste pyrolysis oil or shale gas/oils. This can be of great benefit depending on economics and availability, and can represent a new step towards a more sustainable process for light olefin production. Nevertheless, this also poses significant technical challenges for producers, because of the relatively high heteroatoms content of these feeds. Various studies have shown that these atypical feeds contain significant amounts of impurities, when compared to typical fossil-derived feeds[3-6]. These impurities act as contaminants and can have a negative impact in the process.

Oxygen-containing compounds are among the most problematic contaminants, as they have been shown to cause gum formation and to significantly affect the downstream separation train, which can lead to off-spec products[7]. Additionally, the presence of oxygenates can enhance the production of CO, CO_2, methanol and small acids. Consequently, the formation of these components needs to be carefully monitored in order to prevent corrosion, as it can significantly reduce the lifetime of different process units. Furthermore, CO is known to act as a poison for the catalyst used downstream for acetylene hydrogenation[7, 8].

Currently, a large amount of the operations in a petrochemical plant are carried out with the guidance of simulation packages. They aid on the prediction and optimization of the different units of the process. When it comes to steam cracking reactors, simulation packages can also be extremely useful but their development presents very important challenges. Besides a representative reactor model, it is crucial to count on a kinetic model that is broad enough as to account for all the species and reactions taking place in the reactor, whilst keeping the simulation time to a minimum. If willing to account for the effect of oxygen-containing components, the kinetic model must also include these species, the products of their thermal decomposition and all other intermediate (molecular and radical) species.

Creating and validating such a kinetic model is no easy task. Nevertheless, making use of high level ab-initio calculations for the calculation of reaction rate coefficients and state-of-the-art experimental equipment for validation, Ghent University and AVGI have been able to incorporate such a network into the steam cracking simulation package COILSIM1D, allowing users to explicitly account for the presence of oxygen-containing compounds.

This paper presents a case study, in which the effect of feeding a steam cracker with oxygen-containing species within a hydrocarbon matrix is shown. The simulation results are validated with experimental data, obtained with the pilot plant steam cracker of the Laboratory for Chemical Technology (LCT) of Ghent University.

Kinetic model for oxygenate/hydrocarbon mixtures

In this work, COILSIM1D is used to simulate the cracking reactor. This software package combines a 1D reactor model with a single event microkinetic model, called CRACKSIM (see Figure 1). These models can be used for any type of tubular reactor and are therefore applicable to practically any configuration, ranging from laboratory scale to full-size crackers. Making use of a dedicated algorithm, simulations can be solved within less than one second, giving users the possibility of exploring a broad window of conditions and feeds in a limited time. A very detailed description of COILSIM1D and its features can be found elsewhere in literature[9, 10].

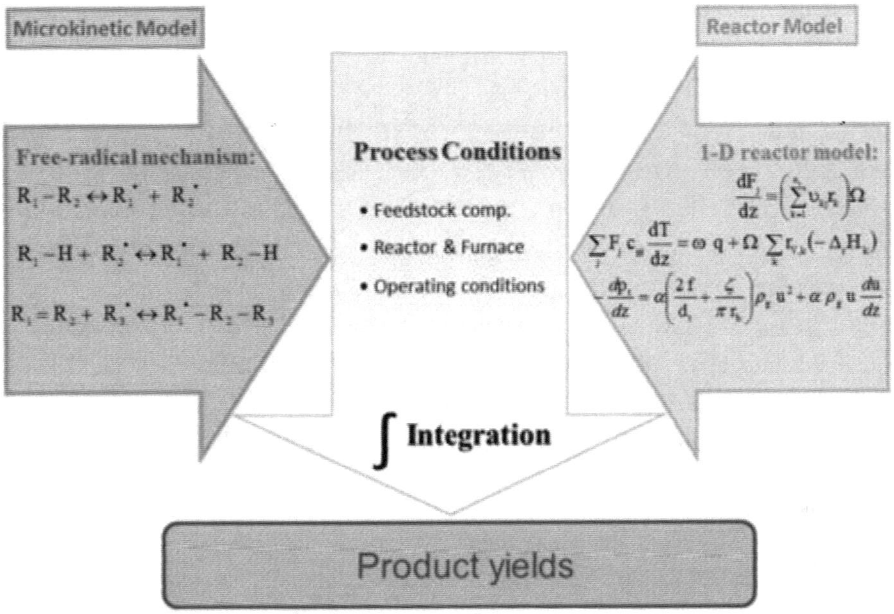

Figure 1. Interaction of microkinetic and reactor models in COILSIM1D[10]

Steam cracking proceeds through a free radical mechanism[11-13] and three important reaction families can be distinguished: carbon-carbon and carbon-hydrogen bond scissions of molecules and the reverse radical- radical recombination, hydrogen abstraction reactions, radical addition to olefins and the reverse β-scission of radicals[9, 14]. CRACKSIM follows a strongly fundamental approach, describing each of the reactions that occur in the reactor as a single step. The corresponding kinetic parameters are calculated by means of high level ab-initio calculations[15-18]. Currently, more than 720 species and 40 radicals are included. Additionally, the model is validated with an extensive set of data originating from industrial furnaces, as well as the pilot plant and bench-scale steam crackers of LCT, for several feeds ranging from light gases to very heavy fractions, such as gasoils, heavy waxes and even crude oil[10, 19].

As can be expected, accounting for this large number of components results in an very big network. To operate with such large networks, different hypothesizes such as the μ-hypothesis and quasi steady state approximation (QSSA) can be applied, in order to maintain an acceptable computational time without losing accuracy. The µ-hypothesis[20-22] divides the network into two sub models: the β-network and the monomolecular µ-network[9]. In the latter, monomolecular reactions dominate for radical species. Applying the QSSA on the µ-radicals, their concentrations can be solved analytically and thus eliminated from the final model equations. This approach leads to a significant reduction of the continuity equations that have to be solved. On the other hand, the β-network comprises the reactions involving β-species for which the µ-hypothesis does not hold. The complete network accounts for more than 1400 reversible β-reactions and 300.000 elementary single step reactions.

In the context of this paper, it is particularly relevant to mention that some of the most commonly found oxygenate feed contaminants such as methanol and dimethyl ether (DME) are also considered in the microkinetic model, next to various other oxygen containing species and associated radicals. This has been carried out using the network generation tool PRIM-(O)[1, 23, 24]. Specific decomposition schemes and methodologies for the calculation of kinetic parameters have been incorporated too, which determine the behavior of these feeds. In general, the type of reactions that oxygenates undergo during steam cracking are not essentially different than those that hydrocarbons undergo. Nevertheless, the presence of oxygen within the structure will affect the possible interactions between functional groups and hence the reactive moiety. This effect has been accounted for by considering these additional functional groups when applying the group additivity methodology implemented in CRACKSIM.

To better illustrate how the reactions proceed, a simplified example of the decomposition of hexane by hydrogen abstraction by an oxygen centered radical is presented in Figure 2. In a first step, a radical abstracts a hydrogen from the hexane molecule. This can occur at different positions in the carbon chain yielding a µ-radical. This µ-radical will undergo monomolecular reactions, as result of the µ-hypothesis, such as different hydrogen and carbon β-scissions, yielding a variety of products.

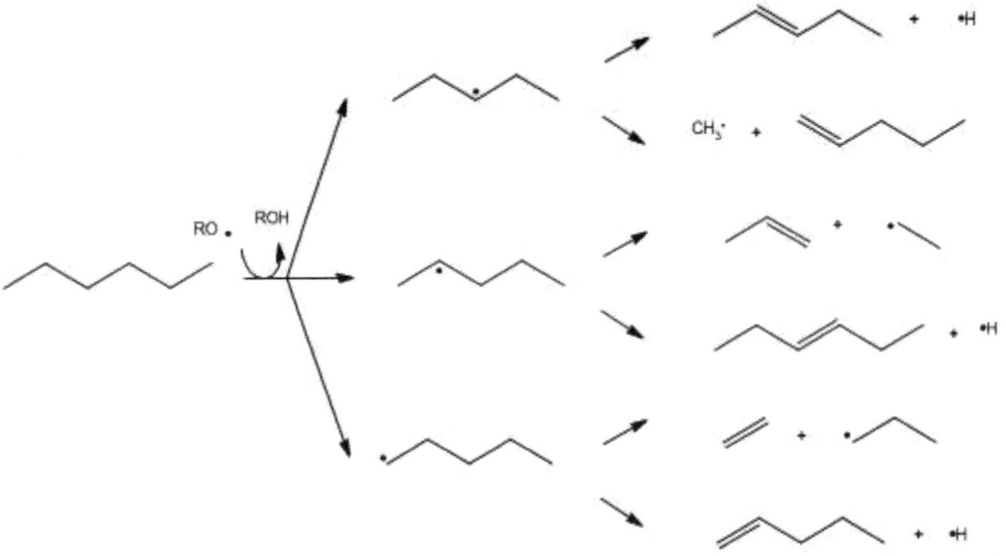

Figure 2. Simplified example of a decomposition scheme of hexane by H-abstraction by an oxygen centered radical

Pilot plant steam cracker

In order to validate the simulation results, experimental data obtained from the pilot plant steam cracker of LCT is used. This unit has been described in numerous publications[18, 25, 26]. Briefly summarized, it consists of a 12.8 meter long (9 mm ID) tubular reactor, vertically suspended inside a fire-heated furnace, divided into 7 independently heated cells. This allows to regulate each cell independently so that any type of temperature profile can be set. Ninety premixed gas burners are mounted on the side walls of the furnace, to provide uniform heating to the reactor. Twenty-two thermocouples and five pressure transducers are used, in order to measure the temperature and pressure of the process gas inside the reactor.

State of the art analytical equipment is used to determine and quantify the effluent composition at the outlet of the reactor. Hydrocarbons ranging from methane up to heavy Poly-Aromatic Hydrocarbons (PAHs) can be detected and quantified, using a combination of four gas chromatographs: A permanent gas analyzer, a refinery gas analyzer, a detailed hydrocarbon analyzer and a GCxGC-FID/TOF-MS. Additionally, a Light Oxygenates Analyzer and an infrared CO/CO_2 detector are used. Online sampling is carried out during the cracking runs, to ensure an accurate evaluation of the reactor effluent practically in real time. This combination of analytical equipment enables to cover the whole range of products that are formed during steam cracking of hydrocarbon feeds, even when oxygenates are included.

Simulation cases and results

In order to evaluate the effect of oxygen-containing molecules in the product distribution, this paper presents two cases that have been carried out using COILSIM1D. Methanol and DME have been selected as typical contaminants[7]. The two studied cases are: i) Naphtha with fractions of methanol and ii) Naphtha with fractions of DME. The concentration of oxygenates on the feed stream varied from 0 to 2 wt% for each case, as can be seen in Table 1.

Table 1. Concentration of oxygenates for the simulated cases

Case	Hydrocarbon	Oxygenate	Concentrations [wt%]
i	Naphtha	Methanol	[0, 0.025, 0.05, 0.075, 0.1, 0.5, 1.0, 2.0]
ii	Naphtha	DME	[0, 0.025, 0.05, 0.075, 0.1, 0.5, 1.0, 2.0]

The reactor simulations were carried out imposing typical operating conditions[27], i.e. a Coil Outlet Temperature (COT) of 840 °C, a Coil Outlet Pressure (COP) of 1.8 bar, and a dilution of 0.35 kg_{H2O} / $kg_{dry\ feed}$.

Table 2. Commercial indices of the naphtha used for simulations.

Commercial indices of the naphtha feed	
Specific density [kg/m³]	708.8
ASTM D86 [°C]	
IBP	39
50%	99
FBP	165
PIONA [wt%]	
Paraffins	36.5
Iso-paraffins	32.8
Olefins	0.0
Naphthenes	21.4
Aromatics	9.20

The hydrocarbon mixture that was fed to the process in the simulations was a typical full-range naphtha[19]. The main commercial indices of this feed are presented in Table 2. Its molecular composition was reconstructed using SIMCO[9, 10], a feed reconstruction tool included in the COILSIM1D package. The reconstructed naphtha had a carbon number ranging from C_4 to C_{12}.

Figure 3 and Figure 4 show the simulated yields of the most important cracking products at the reactor effluent for cases i) and ii), respectively. Under the simulated conditions, methanol exhibited a conversion of approximately 65%, whereas the conversion of DME reached values of 85%. These values are in line with those obtained during experiments under similar process conditions.

Under the steam cracking conditions here considered, the simulations indicated that methanol and DME decomposed to form mainly CO, CO_2 and formaldehyde. CO was the most abundantly formed product, with concentrations reaching values of 0.7 wt% and 0.3 wt% when the largest amounts of methanol and DME, respectively, were added to the feed. Such concentrations of CO are significantly higher than those typically observed when cracking a pure hydrocarbon feed (approx. 0.04 wt%)[27] and give a clear indication of the risks and operational complications that even small amounts of oxygen contaminants can pose to steam crackers.

Besides enhancing the formation of CO and CO_2, it is also important to note that the presence of these species has a negative impact on the yields of ethylene, propylene and butadiene. As can be seen from Figure 3 and Figure 4, the yields of these products are significantly lower when compared to a pure hydrocarbon feed, even when very small concentrations of oxygenates are considered. The ethylene yield is the most affected, with a decrease of more than 1 wt%, when the DME concentration is highest. The yields of propylene and butadiene also drop (approx. 1 wt% and 0.2 wt%, respectively) as the oxygenate concentration increases, although the latter is affected to a lesser extent. On the other hand, the yields of methane and hydrogen are higher than for the pure hydrocarbon feed. These simulation results follow the same trends that have been observed during pilot plant experiments carried out with spiked hydrocarbon feeds.

These differences in yields and selectivities can be mainly attributed to the introduction of additional oxygen-centered radicals in the radical pool. For example, hydroxyl and methoxy radicals can be formed by the C-O scission reaction of the DME and methanol. These radicals interact with the hydrocarbons and consequently affect the selectivity of the products. Therefore, properly accounting for the presence of these species, even in very small concentrations, is very important for an accurate evaluation of the economic value of a given feed. In this respect, COILSIM1D offers the possibility to explicitly account for contaminants such as methanol and DME, which clearly have a non-negligible effect on the final yields, and whose thermal decomposition products like CO can lead to important operational issues downstream.

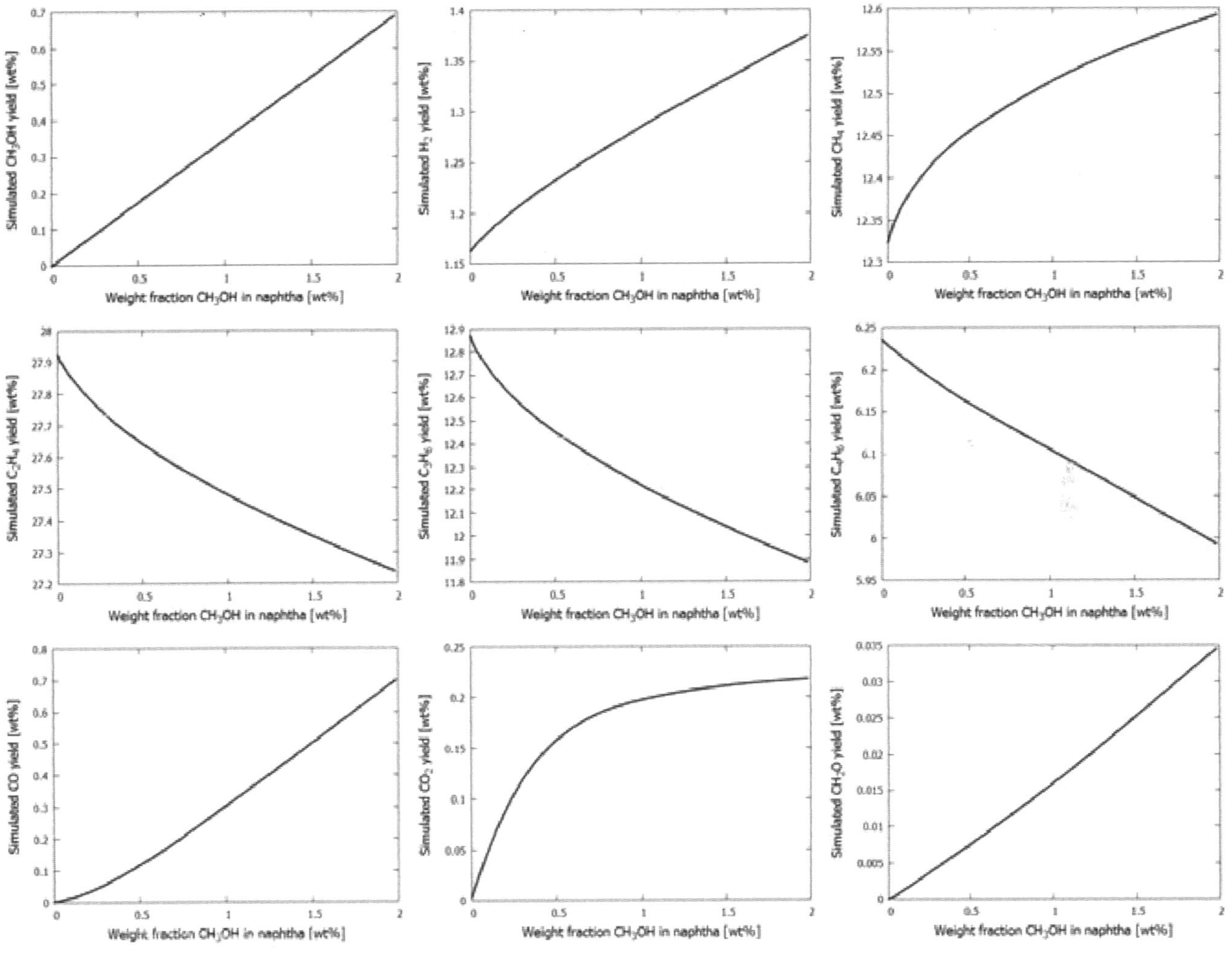

Figure 3. Main simulated product yields for of the steam cracking of a methanol-spiked naphtha as a function of the methanol weight fraction in the naphtha.

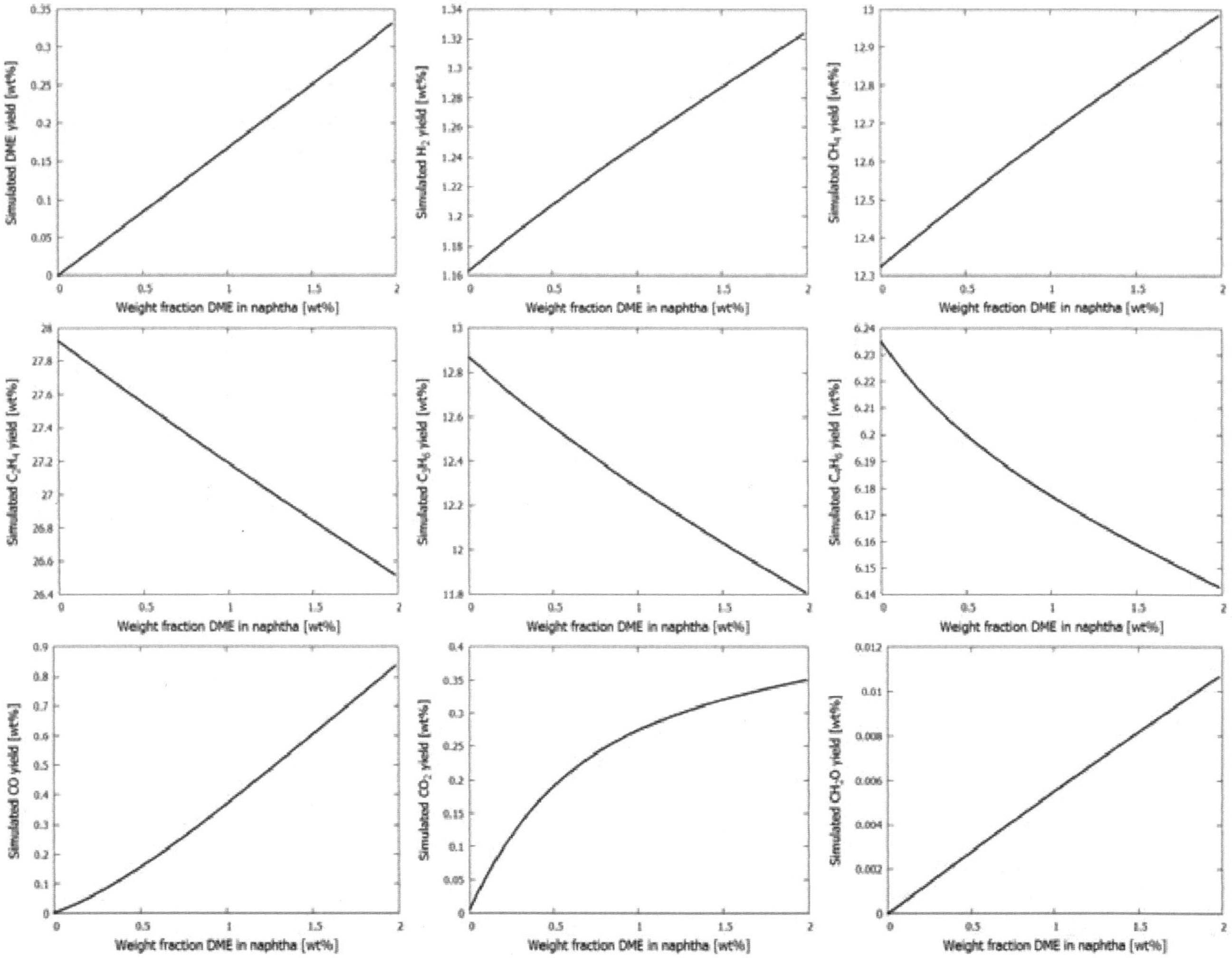

Figure 4. Main simulated product yields of the steam cracking of a DME-spiked naphtha as a function of the DME weight fraction in the feed.

In order to validate the simulations that were conducted for cases i) and ii), Figure 5 shows a parity plot comparing the simulated yields with those that have been experimentally measured at the pilot plant steam cracker of LCT for the conditions used in this study and, additionally, for other cases in which the COT ranged between 800 °C and 860 °C. A very good agreement is observed between the experimental and simulated values, indicating that the model is adequate and flexible, and is thus able to accurately predict the effect of oxygen-containing molecules present in the feed.

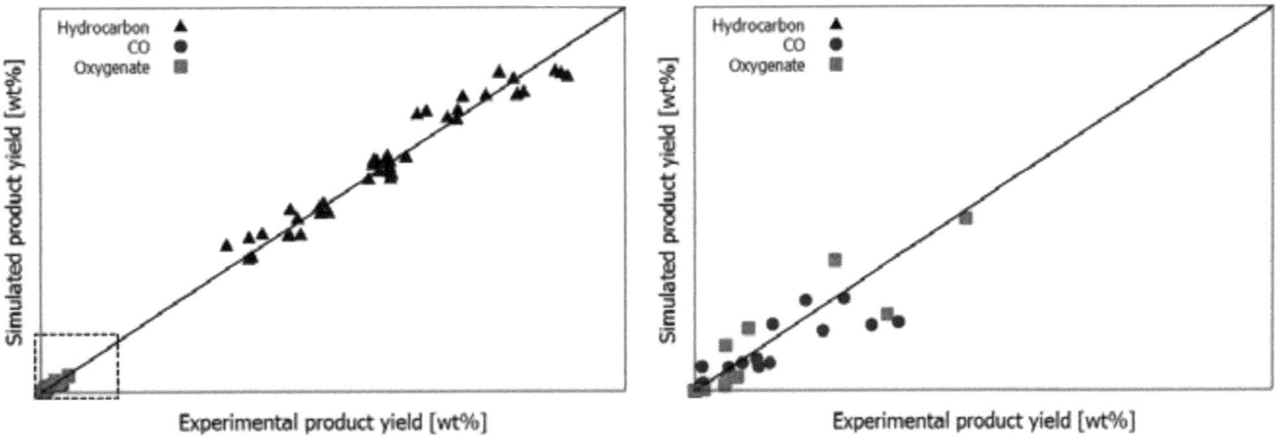

Figure 5. Parity plots comparing experimental and simulated product yields for steam cracking of methanol- and DME-spiked liquid hydrocarbon feeds. The dashed area in the left figure is augmented in the right figure.

Conclusions

CRACKSIM, the microkinetic model of Coilsim1D has been extended to take into account the effect of various oxygen containing contaminants accounting for the single step reactions, leading to an accurate simulation tool for a broad range of conditions and feeds.

This work shows the application of this model for steam cracking of a naphtha spiked with methanol and DME. In both cases, the simulations indicated that the yields of CO, CO_2 and formaldehyde significantly increased as a function of the oxygenate concentration in the feed, reaching values of up to 0.7 wt% for CO.

The yields of desired products such as ethylene, propylene and butadiene are considerably reduced by the presence of oxygenates in the feed. In the simulated cases, the yields of ethylene and propylene reached values approximately 1wt% lower than when cracking a pure naphtha. This phenomenon is attributed to the interaction of oxygen-centered radicals (formed by the decomposition of oxygenate species) with the hydrocarbons, affecting the reaction paths that lead to the formation of desired products.

The simulated results are well aligned with observations coming from pilot plant experiments. A parity plot of simulated versus experimentally measured yields (for a broad range of concentrations and temperatures) is presented in this paper. It indicates a very good agreement of both sets, showing that the microkinetic model is able to accurately predict the effect of oxygenate species in the reactor, and flexible enough as to account for various operating conditions.

Considering the detrimental effect for the overall operation of a cracker and downstream units of high CO concentrations in the reactor effluent, as well as the economic impact of lowered olefin yields, COILSIM1D proves to be a very useful tool to accurately predict the effect of oxygenate species in feeds and to consequently optimize operation in order to achieve larger economic and technical benefits for producers.

Acknowledgement

The authors would like to acknowledge the work of Dr. Thomas Dijkmans, who helped carrying out the experimental work at LCT – UGent, used for validation of this work

References

1. De Bruycker, R., et al., (2014), "Assessing the Potential of Crude Tall Oil for the Production of Green-Base Chemicals: An Experimental and Kinetic Modeling Study". Industrial & Engineering Chemistry Research. **53**(48): p. 18430-18442.
2. Pyl, S.P., et al., (2011), "Biomass to olefins: Cracking of renewable naphtha". Chemical Engineering Journal. **176**: p. 178-187.
3. Gaweł, B., M. Eftekhardadkhah, and G. Øye, (2014), "Elemental composition and Fourier transform infrared spectroscopy analysis of crude oils and their fractions". Energy & Fuels. **28**(2): p. 997-1003.
4. Dijkmans, T., et al., (2015), "Comprehensive compositional analysis of sulfur and nitrogen containing compounds in shale oil using GC × GC – FID/SCD/NCD/TOF-MS". Fuel. **140**: p. 398-406.
5. Djokic, M.R., et al., (2012), "Quantitative analysis of crude and stabilized bio-oils by comprehensive two-dimensional gas-chromatography". Journal of Chromatography A. **1257**: p. 131-140.
6. Toraman, H.E., et al., (2014), "Detailed compositional characterization of plastic waste pyrolysis oil by comprehensive two-dimensional gas-chromatography coupled to multiple detectors". Journal of Chromatography A. **1359**: p. 237-246.
7. Reid, J. and G. Nowowiejski, (2003), *Overview of oxygenates in olefin units in relation to corrosion, fouling, product specifications, and safety*. in *Proceedings of the 15th Ethylene Producers Conference*. American Institute of Chemical Engineers New York.
8. Reyniers, M.-F.S. and G.F. Froment, (1994), "The influence of metal surface and sulfur addition on coke deposition in the thermal cracking of hydrocarbons", Editor^Editors, American Inst. of Chemical Engineers, New York, NY (United States).
9. Van Geem, K.M., M.F. Reyniers, and G.B. Marin, (2008), "Challenges of modeling steam cracking of heavy feedstocks". Oil & Gas Science and Technology-Revue de l'IFP. **63**(1): p. 79-94.
10. Van Geem, K.M., M.-F. Reyniers, and G. Marin, (2008), *Taking optimal advantage of feedstock flexibility with COILSIM1D*. in *AIChE Spring Meeting: Ethylene producers conference*.

11. Rice, F., (1931), "The thermal decomposition of organic compounds from the standpoint of free radicals. I. Saturated hydrocarbons". Journal of the American Chemical Society. **53**(5): p. 1959-1972.
12. Dente, M., et al., (2007), "Kinetic modelling of pyrolysis processes in gas and condensed phase". Advances in chemical engineering. **32**: p. 51-166.
13. Froment, G., (1981), "Thermal cracking for olefins production. Fundamentals and their application to industrial problems". Chemical Engineering Science. **36**(8): p. 1271-1282.
14. Back, M.H. and R.A. Back,(1983), "Thermal decomposition and reactions of methane". Academic Press, New York.
15. Paraskevas, P.D., et al., (2014), "Kinetic Modeling of α-Hydrogen Abstractions from Unsaturated and Saturated Oxygenate Compounds by Carbon-Centered Radicals". ChemPhysChem. **15**(9): p. 1849-1866.
16. Sabbe, M.K., et al., (2009), "Carbon-Centered Radical Addition and β-Scission Reactions: Modeling of Activation Energies and Pre-exponential Factors". ChemPhysChem. **10**(12): p. 1951-1951.
17. Sabbe, M.K., et al., (2010), "Hydrogen Radical Additions to Unsaturated Hydrocarbons and the Reverse β-Scission Reactions: Modeling of Activation Energies and Pre-Exponential Factors". ChemPhysChem. **11**(1): p. 195-210.
18. Sabbe, M.K., et al., (2011), "First principle-based simulation of ethane steam cracking". AIChE Journal. **57**(2): p. 482-496.
19. Alexander Vervust, I.A., Andrés E. Muñoz G., Kevin M. Van Geem, Guy B. Marin, (2016), *Full Furnace Simulation and Optimization with COILSIM1D*. in *AIChE Spring Meeting: Ethylene producers conference*. Houston, Texas.
20. Warth, V., et al., (1998), "Computer-Aided Derivation of Gas-Phase Oxidation Mechanisms: Application to the Modeling of the Oxidation of n-Butane". Combustion and Flame. **114**(1–2): p. 81-102.
21. Clymans, P. and G. Froment, (1984), "Computer-generation of reaction paths and rate equations in the thermal cracking of normal and branched paraffins". Computers & chemical engineering. **8**(2): p. 137-142.
22. Ranzi, E., et al., (1983), "Initial product distributions from pyrolysis of normal and branched paraffins". Ind. Eng. Chem. Fundam.;(United States). **22**(1).
23. Pyl, S.P., et al., (2012), "A comprehensive study of methyl decanoate pyrolysis". Energy. **43**(1): p. 146-160.
24. De Bruycker, R., et al., (2015), "Microkinetic model for the pyrolysis of methyl esters: From model compound to industrial biodiesel". AIChE Journal. **61**(12): p. 4309-4322.
25. Van Geem, K.M., et al., (2010), "On-line analysis of complex hydrocarbon mixtures using comprehensive two-dimensional gas chromatography". Journal of Chromatography A. **1217**(43): p. 6623-6633.
26. Olahova, N., et al., (2016), "Thermal Decomposition of Sulfur Compounds and their Role in Coke Formation during Steam Cracking of Heptane". Chemical Engineering & Technology. **39**(11): p. 2096-2106.
27. Zimmermann, H. and R. Walzl, (2009), "Ethylene". Ullmann's Encyclopedia of Industrial Chemistry.

AIChE paper number 36d

Feedstock Contaminants in Ethylene Plants - 2017 Update

Mark Brayden
R&D Fellow
The Dow Chemical Company

Art Baumgartner
Senior Technologist
Shell Global Solutions

Steve Coleman
Senior Advisor
LyondellBasell

David Smith
Industry Technical Consultant
Nalco Champion

Lea Ann Nairn
Account Manager
Honeywell UOP

Joe Lally
Technical Development Manager
Evonik Corporation

Andre Bernard
Process Design Engineer
NOVA Chemicals (Canada) Ltd.

Prepared for Presentation at the 2017 Spring National Meeting
San Antonio, TX
March 26-30, 2017

Feedstock Contaminants in Ethylene Plants - 2017 Update

Mark Brayden
R&D Fellow
The Dow Chemical Company

Art Baumgartner
Senior Technologist
Shell Global Solutions

Steve Coleman
Senior Advisor
LyondellBasell

David Smith
Industry Technical Consultant
Nalco Champion

Lea Ann Nairn
Account Manager
Honeywell UOP

Joe Lally
Technical Development Manager
Evonik Corporation

Andre Bernard
Process Design Engineer
NOVA Chemicals (Canada) Ltd.

Abstract: Ethylene plant operators were polled for their responses on a wide range of feedstock contaminant issues. This paper presents the responses to this survey and an analysis of the gathered information. A special emphasis is placed on the impact of shale gas on the US Gulf Coast plants. Comparisons with the 2004 Feedstock Contaminant Survey are also included.

Introduction

This is the third time this survey has been conducted over the past 25 years. The first was in 1992 (1) and the second in 2003 (2). Since the 2003 survey was reported at the 2004 EPC, fracking has become a major source of cracker feedstock, primarily, in the form of natural gas, as well as some associated liquids. The purpose of conducting this survey at this time is to see if there are changes in the feedstock contaminants and the problems these cause. This data will be used to direct the activities of this Subcommittee during the coming years. This paper will present an analysis of the survey information and highlight any changes since the 2003 survey. This will not be exhaustive analysis, and so we also provide the raw data for further analysis by our subscribers.

The survey is structured based upon the 2003 survey. While there are some changes in emphasis, the structure is similar. The survey is divided into four sections, or questions. One major difference is a single company may have multiple inputs instead of only one, in recognition that many companies operate multiple plants in multiple parts of the world of varying vintage and size, as well as configuration. Each location will have its own unique issues with feedstock contaminants, and so these can be more easily identified by each cracker's size, location, configuration, and vintage.

The Introduction to the survey deals with identifying the size, vintage, and geographic area (continent), and the primary and secondary feed sources. We recognize there are many other niche feedstocks but concentrated on the major feeds to the US Gulf Coast plants where fracking is a major source of natural gas feedstocks. By identifying plants by other regions, this provides a basis for comparison, as well as highlighting issues that are both regional and global.

Question 1 pre-identified 31 common contaminants and asked the companies completing the survey to identify the impurities it encounters in the 4 commonly used feeds (naphtha, *ethane, EP mix, and propane), as well as offering the opportunity to comment on other feeds* like mixed butanes. In addition the participants listed the concern levels of these identified contaminants.

Question 2 addressed contaminants with known issues and asked the survey participants to identify those that this company location has experienced an issue.

Question 3 went into more detail on specific feedstock contaminants that have been issues in the past like mercury, asking about removal and for details on how this is done. Besides mercury, the other contaminants in this survey are salt, silicone, and ammonia/nitriles.

Question 4 concerned those other sources of feed and product that may be considered recycles to recover valuable monomers or as an alternative method to prevent waste.

The Feedstock Impurities Subcommittee appreciates the time and effort that its many participants put into this effort. The survey was not as user-friendly as we had hoped and appreciate the patience with using the tool.

2016 Survey – General

The survey was sent out using an electronic service, SurveyMonkey. The subcommittee received 59 entries with varying degrees of completeness. Some responded to all questions, while others were incomplete, and still others only answered with minimal input. For instance, on Question 1, of the 59 inputs 29 gave significant information in the form of impurity levels. This was consistent for all questions, as about 29-36 gave complete information on these questions in each section.

For the General section, the data used for analysis are only those 29 that gave complete information.

Plant Locations

One of the goals of this survey was to compare previous results with current for US Gulf Coast plants receiving the higher amounts of feedstock from fracking, as well as the responses from other parts of the world. As expected the highest area of representation is North America. With almost equal representation from Europe, South America, and Asia, each of these regions had about half as many responses as North America. Even so, this is much higher representation for areas outside of North America than the previous surveys. For instance, the 2003 survey showed almost 70% of the respondents from North America versus only 41% in this current survey.

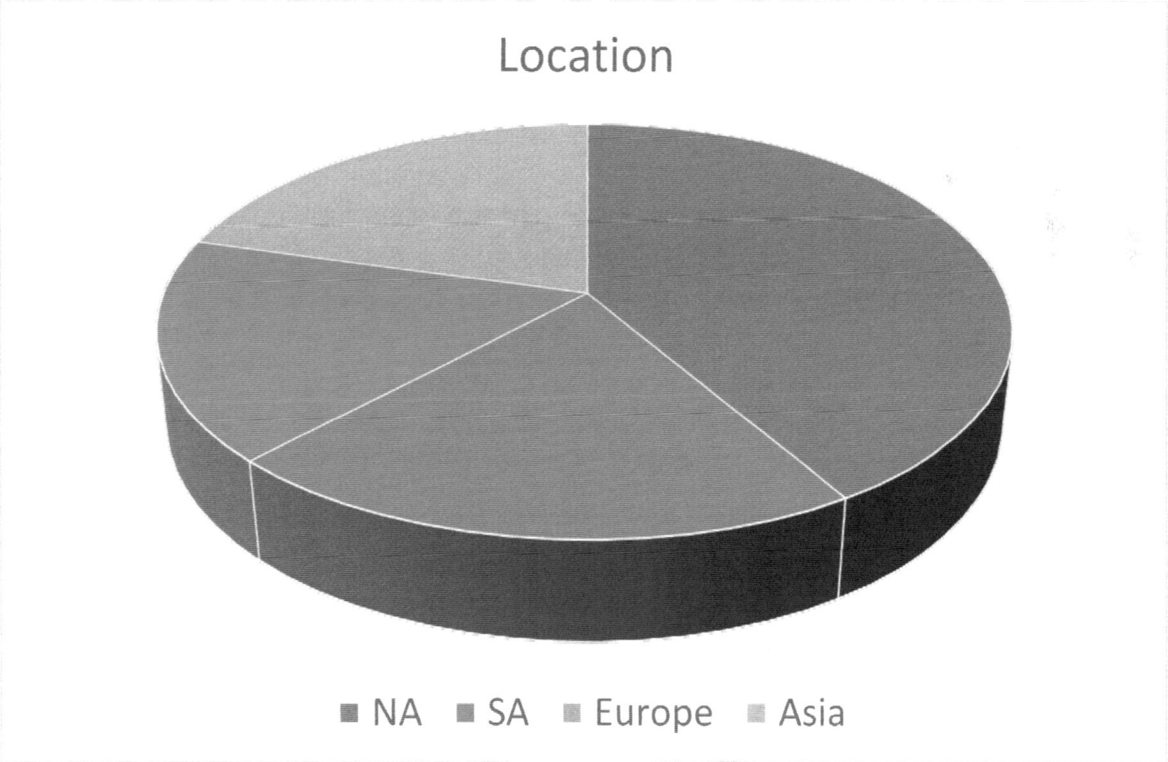

Figure 1. Plant Locations

Plant Age

For this survey the majority (62%) of the plants were built after 1980, but this still is a significant number of plants that were built before this.

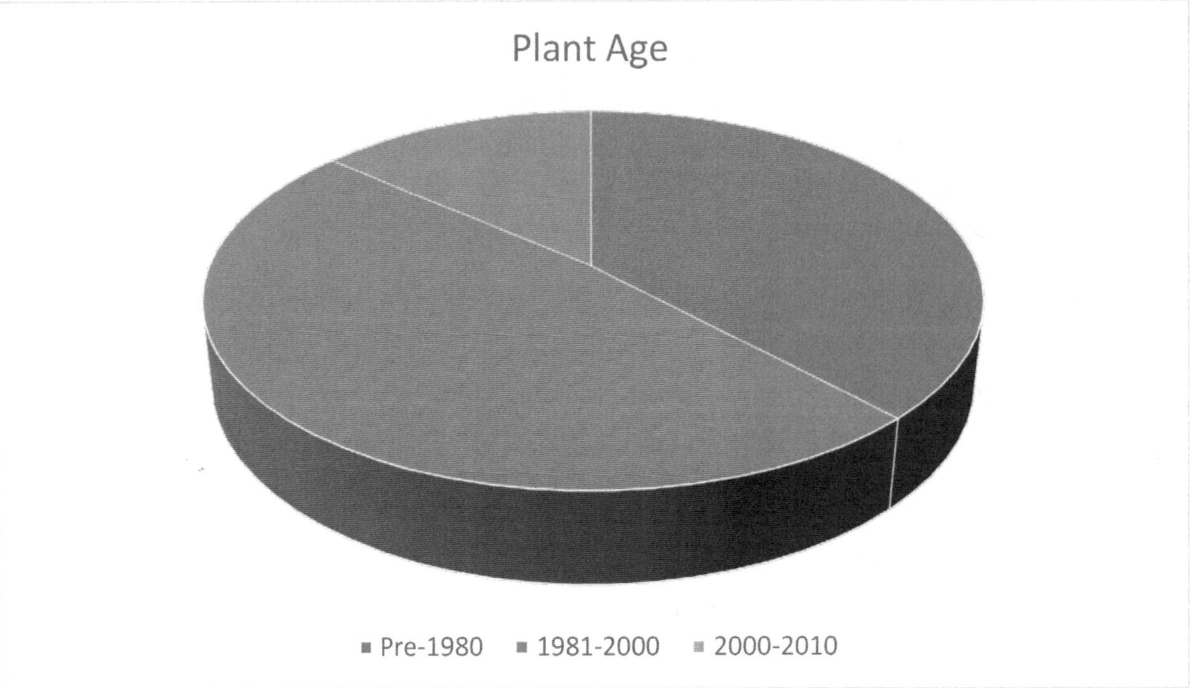

Figure 2. Plant Age

Cracker Feedstocks

The questionnaire asked this differently this year as opposed to the last survey. Most liquids crackers, we assumed, could crack liquids from both refinery and natural gas sources. (As pointed in the feedback we received, these should have been asked separately.) One interesting trend is the increase in the number of plants cracking multiple feedstocks. In the last survey, only 29% of the pool had this capability. In the 2017 survey, this increased to 41% based on how the 2003 survey categorized multiple feed plants. (For instance, a plant that cracked both ethane and propane was designated as an EP cracker.) A difference in the survey is that the 2003 survey differentiated between refinery (RCO, or Refined Crude Oils) and natural gas sources (GFC, or Gas Field Condensates) of liquid feeds. A comparison between the two surveys is shown in Table 1.

Table 1

Feed Type	2003	2017
EP	40 %	38 %
GFC + RCO	25	21
Multiple	29	41
No response	6	0

If we further breakdown the data in Table 1 for 2017, we get a result that might be expected, namely, of the 11 respondents that crack EP only, 9 reside in North America, while for liquids only cracking none of the 6 respondents is in North America.

A breakdown of the 2017 survey results also shows some interesting feed combinations, as shown in Table 2. While plants that crack ethane as a primary feed and liquids or butane as a secondary feed are represented by one respondent each, the highest number of respondents are more traditional in their feedstocks. Liquids cracking alone or with supplemental ethane, propane or butane represent 16 out of 29 of the respondents. For gas crackers these represent 11 out of the 29 respondents.

Table 2.

Liquids		6
Liquids	Butane	5
Ethane or E/P		4
Ethane	Propane	4
Liquids	Propane	4
Propane	Ethane	2
Propane	Liquids	1
Ethane	Butane	1
Ethane	Liquids	1
Liquids	Ethane	1

Configuration

This was not asked in the 2003 survey, but was included in the 2017 survey. This could be useful information when we investigate the impact of impurities. Front end deethanizer plants are not well represented in the survey, only 3 identified respondents who then gave adequate information for analysis of the impact of impurities is in this survey. While this under represents this configuration, this may have been a deficiency in the past surveys but went unidentified.

To illustrate how this data might be used, here is a simple case study using the data in this survey. Mercury is a known catalyst poison (3) and contributes to coldbox corrosion (4). Based on past data reported in EPC, mercury tends to follow C3 and heavier streams even though a small percentage (0-5 %) was reported to be in the C2 cut (5). Likewise, ethane from gas processing plants with no mercury removal, contains only about 5 % of the mercury, the least of all the products (6). Front end deethanizer (FEDE) plant acetylene hydrogenation catalyst, then, are minimally impacted by mercury in the feedstock (7). While there were only 3 respondents that had FEDE configuration, none reported an issue with catalyst poisoning or corrosion. Further, since Hg does follow the C3's to a greater extent, the impact on FEDP plants is significant. Of the 5 plants that reported mercury was a catalyst poison in Question 2, 3 were FEDP plants and 2 were FEDM. Likewise, of the 6 plants that reported "Corrosion" as a concern, 3 were FEDP. While there may be a concern in a FEDP plant about the coldbox, the acetylene hydrogenation catalyst acts as a good mercury removal unit. Also, 2 of the FEDP plants remove Hg with molecular sieve, and the other removes with activated carbon.

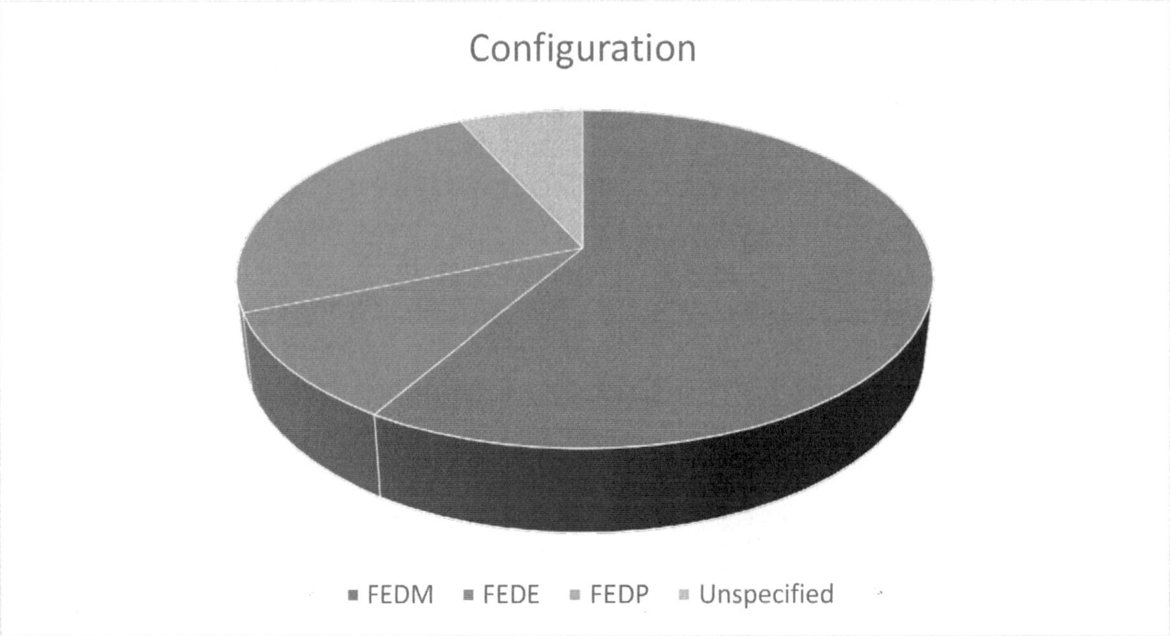

Figure 3. Configuration

Plant Size Data

The categories for this changed for the last survey to reflect the increasing size of ethylene plants over the years. While a direct comparison to the 2003 data is not possible, it is possible to say that there has been an overall increase in the plant size. In the 2003 survey, 40% of the respondents had plants of 600K mta or smaller. In the 2016 survey, the number of plants < 500K mta accounted for only 14 % of the respondents.

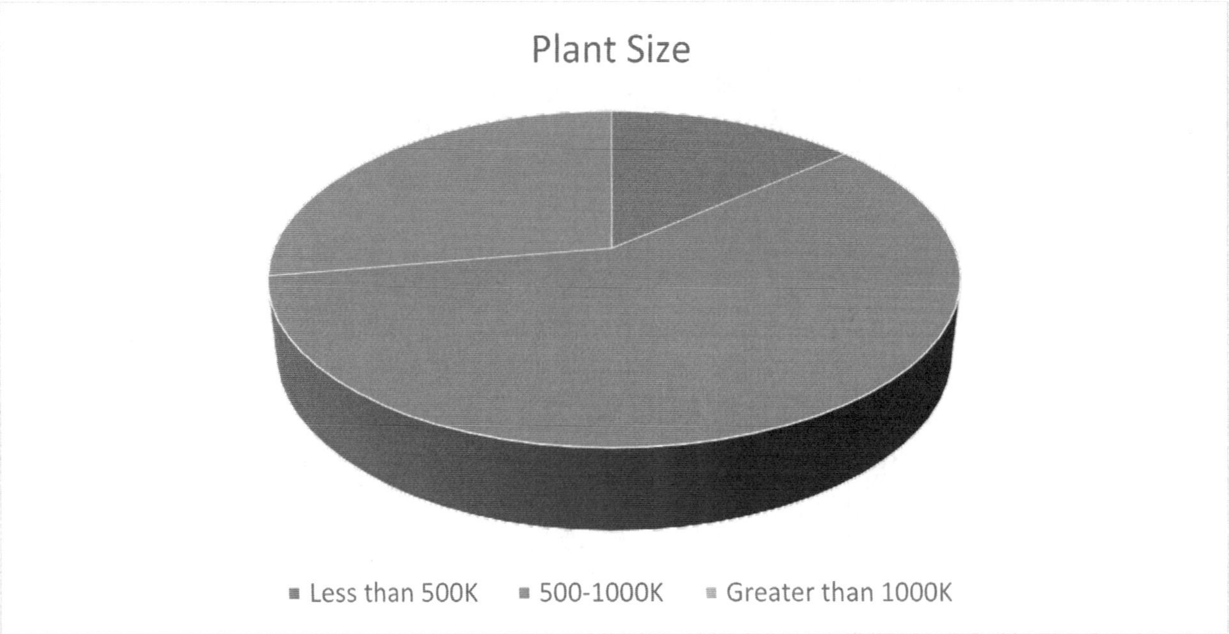

Figure 4. Plant Size

While we have presented some analysis of the data and a comparison to the 2003 survey data, there is much more analysis that can be done. For that reason our subcommittee has posted the raw data in the tables at the end of this report. We are also working on allowing access to the raw data for the subscribers to the EPC Proceedings via an Excel spreadsheet.

2016 Survey - Question One

The survey started with characterizing the feedstock and then led into the first set of questions exploring contaminants. This separated the feedstocks into 5 types:

1) Ethane,
2) E/P Mix (Ethane/Propane),
3) Propane and other NGL (Natural Gas Liquids),
4) Naphtha and Condensates, and
5) Other.

With the feedstock types selected the questions narrow to the contaminants experienced in particular plants based on the feedstock type.

Each page of survey Question 1 lists 31 contaminants and asked:

1. What is the level of concern?
2. Is it tested?
3. Is it detected?
4. What is the base line or normal value?
5. What is the maximum value?

The responses to question one are organized into 3 tables:

1. Level of concern for contaminants,
2. Contaminants tested for and detected, and
3. Normal levels and maximum permissible levels for the particular contaminant.

These tables are generated for each of the 5 feedstock types.

Table 3 is a summary of these results. When the contaminants are ordered according to highest percentages of the plants where there was at least a medium level of concern for the feedstock types, the following ranking of Important contaminants results, (note: ranking of 1 is the highest percentage value). For simplicity the Ethane and E/P mixes are combined in this summary.

The bulk of the data for Question 1 can be found in Appendix A. Tables 17, 18 and 19 show the level of concern or the amount of interest shown for each contaminant that is listed, reported by feedstock type. The responses are tabulated in terms of the percent of plants responding with data, where there was either a medium or high level of concern; responses with low concern are not counted. The tables show the medium and high levels individually and as a sum to identify the contaminants of concern.

Table 17 shows responses for Ethane and E/P Mix Types. Table 18 shows Propane and other NGL (Natural Gas Liquids), and Naphtha and Condensates. Finally, Table 19 shows "Other" where the "Other" feedstocks were written in and included Mixed Butanes and C Raffinate, SRGO (Straight Run Gas Oil), LPG (Liquefied petroleum gas), Hydowax (sic) and Hydrocracker residue, and recycle streams ranging from C4-C8.

This summary shows similar concerns expressed in previous contaminant surveys. Light acid gases, such as CO_2 that are difficult to remove, are of special importance in the light feeds. Heavier feeds have more interest in sulfur, although its rank has moved up for Ethane and E/P Mixes. Mercury did move up in rank for all feeds since the 2003 Survey

Tables 20, 21, and 22, also in Appendix A, contaminants tested for and detected are listed in terms of percentage of plants responding. As in Tables 4-6 these are broken up by feedstock types. Table 20 shows responses for Ethane and E/P Mix Types. Table 21 shows Propane and other NGL (Natural Gas Liquids), Naphtha and Condensates. Finally Table 22 shows "Other". For the Naphtha and Condensates there more testing for a variety of contaminants compared to the lighter feedstock types.

In final portion of this question the baseline and maximum permissible levels are tabulated to the plants that indicated a contaminant is a high or medium level of interest. The survey allowed responses in concentration by volume or weight which are shown with the units included. The results vary as maximums and ranges so a list of the results is shown. For each of the feed stocks the top 2 contaminants are listed first. Tables 23 thru 27 in Appendix A look at each feedstock type individually.

Table 3. Ranking of Impurities Identified by Feedstock

Ethane Or E/P Mix		Propane or Other NGL's		Naphtha/Condensate		Other	
Rank	Contaminant	Rank	Contaminant	Rank	Contaminant	Rank	Contaminant
1	CO2	1	Inorganic Hg	1	Total Sulfur	1	Organic Hg
2	Total Sulfur	1	Sodium	2	Inorganic Hg	2	Total Sulfur
3	Sodium	2	Total Sulfur	3	CS2	2	Inorganic Hg
4	Methanol	2	Methanol	4	Oxygenates	3	Sodium
5	H2O	2	Organic Hg	5	Arsenic	3	Oxygenates
5	Organic Hg	2	Oxygenates	5	Inorganic Cl	3	Arsenic
5	Oxygenates	3	Inorganic Cl	6	Sodium	3	NOX
5	H2S	3	Iron	7	Methanol	4	Methanol
6	Inorganic Hg	4	Arsenic	7	Organic Hg	4	Inorganic Cl
6	Inorganic Cl	4	AsH3/PH3	7	Organic Cl	4	Organic Cl
6	Organic Cl	4	Organic Cl	7	Si Derivatives	4	AsH3/PH3
6	Iron	5	CO2	8	Lead	4	O2
7	Arsenic	5	Fluorides	9	Olefins	5	CO2
7	AsH3/PH3	6	COS	10	Iron	5	H2O
8	COS	6	CS2	10	Vanadium	5	Iron
8	NH3 & Amines	6	H2S	11	AsH3/PH3	5	Fluorides
9	CS2	6	Lead	12	H2O	5	Lead
9	NOX	6	Olefins	12	H2S	5	Olefins
9	O2	7	H2O	12	NH3 & Amines	6	COS
10	Fluorides	7	NH3 & Amines	12	NOX	6	H2S

10	Lead	7	NOX	12	O2	6	CS2
10	CO	7	O2	13	COS	6	NH3 & Amines
10	Oils (pumps & compressor)	8	CO	14	Fluorides	6	CO
11	Phosphorus	8	Oils (pumps & compressor)	14	Phosphorus	6	Oils (pumps & compressor)
11	Radon	8	Phosphorus	14	CO	6	Si Derivatives
11	Si Derivatives	8	Radon	14	Oils (pumps & compressor)		
12	Vanadium	8	Si Derivatives	14	Calcium		
12	Calcium	9	Calcium	15	CO2		
		9	Vanadium	15	Radon		

2016 EPC Feedstocks Contaminant Survey – Question 2

Question 2 of the Survey focuses on various contaminants and their consequences for plant operations. There were 59 total responses, but 27 gave no response to any of the contaminants. Of the 32 that did list an issue: 12 were gas crackers, 18 were liquid or flexi-crackers, and 2 did not did not say; 12 of the plants were built pre-1980, 14 crackers began operation between 1981 and 2000, and 4 plants started up post-2001. 35 plants participated in the survey in 2003 and reported at the 2004 AIChE Spring National Meeting (2).

The raw data are presented in Appendix B, Tables 28, 29, and 30, located at the end of the paper. The following consequences were listed:

- Product Quality
- Corrosion
- Catalyst Poison
- Safety/Hygiene
- Environmental (air/water)
- Capacity Limit
- Operability
- Furnace Problems

Figure 5 shows the number of concerns recorded for each contaminant, broken down by type of feedstock and overall. The top contaminants of concern overall were Sulfur, Methanol/Alcohols, Chlorides, Water, Mercury, and Carbon Dioxide.

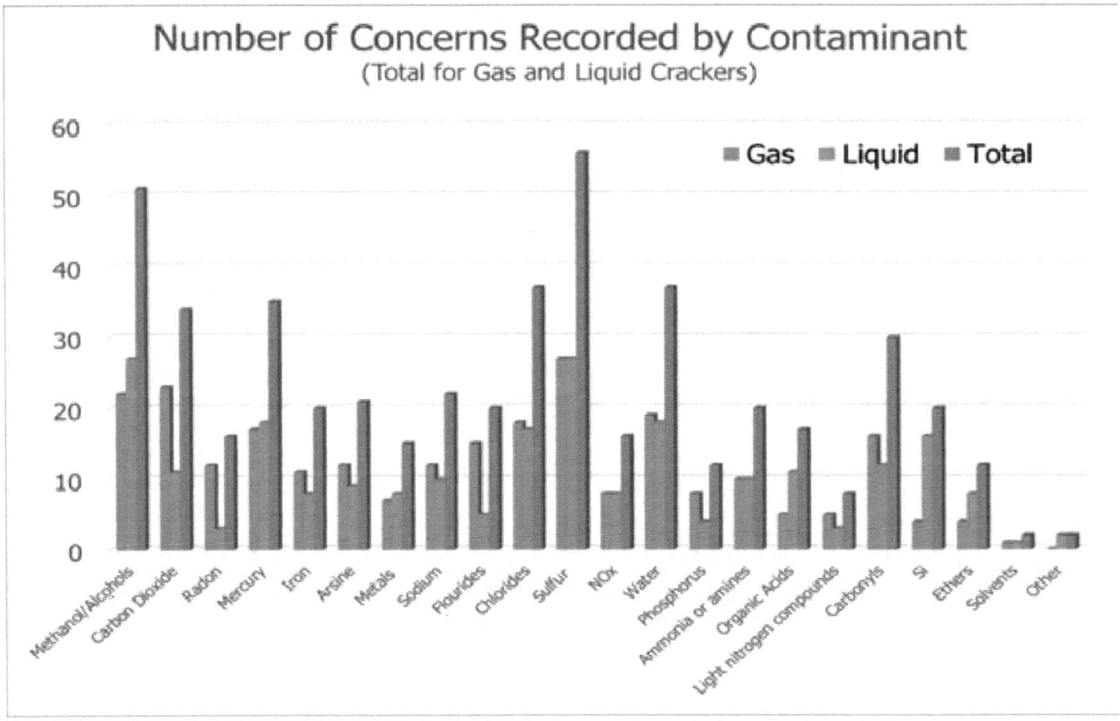

Figure 5

The data show a slight difference between the top concerns between gas crackers and liquid crackers: gas crackers had sulfur as the most frequently-cited issue, while liquid crackers more frequently cited methanol than did gas crackers.

Across vintages, all three categories cited sulfur concerns most prominently, with older plants more likely to cite methanol/alcohol concerns than plants built post-2001. Figure 6 shows the same data as **Figure Q2-1**, but broken down by plant vintage instead of feedstock.

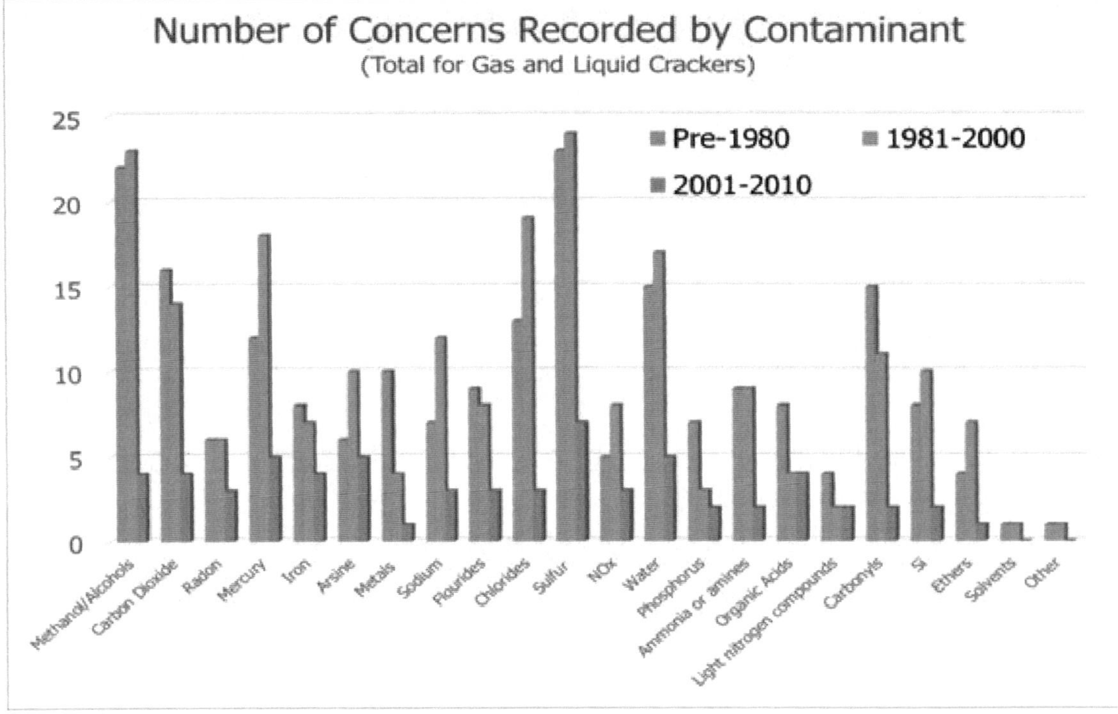

Figure 6

In terms of the consequences of feedstock contaminants, the main concerns from contaminants were Product Quality, Catalyst Poison, Furnace Problems, and Operability.

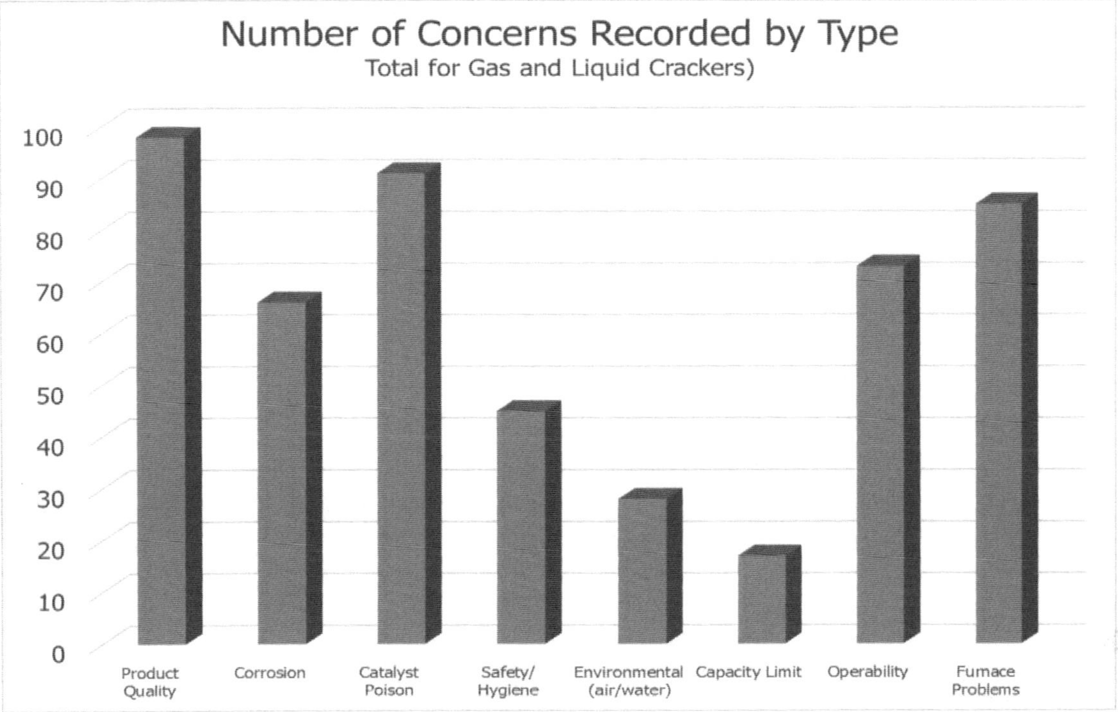

Figure 7

Some differences are noted comparing data from the 2016 Survey to that performed in 2003. While Sulfur, Methanol/Alcohols, and Chlorides remained key concerns, Sodium moved down the list. This is possibly due to the increased prevalence of replacing caustic with amine as the bulk neutralization chemistry for the dilution steam system. Mercury showed up in the 2016 survey, while CO_2 dropped off the list in favor of water.

Table 4. All Crackers - Top 5 Problem Contaminants

2003 Survey	**2016 Survey**
Sulfur	Methanol/Alcohols
Methanol/Alcohols	Sulfur
Sodium	Mercury
Chlorides	Water
CO_2	Chlorides

Table 5. Gas Crackers - Top 5 Problem Contaminants

2003 Survey	**2016 Survey**
Methanol/Alcohols	Sulfur
CO_2	Carbon Dioxide
Chlorides	Methanol/Alcohols

Sulfur	Water
Fluorides	Chlorides

Table 6. Liquid Crackers - Top 5 Problem Contaminants

2003 Survey	**2016 Survey**
Mercury	Methanol/Alcohols
Sulfur	Sulfur
Sodium	Mercury
Methanol/Alcohols	Water
Arsine	Chlorides

Finally, a comparison of the Concerns by Type show that while the top 3 concerns remained the same, Safety/Hygiene and Environmental concerns entered the list, replacing Furnace Problems and Operability, respectively. This could perhaps be the result of increased emphasis on Safety, Health, and Environmental issues industry-wide from both a cultural and a regulatory standpoint.

Table 7. Concerns by Type

2003 Survey	2016 Survey
Product Quality	Product Quality
Corrosion	Corrosion
Catalyst Poison	Catalyst Poison
Furnace Problems	Safety/ Hygiene
Operability	Environmental (air/water)

Question 3. Specific Feedstock Contaminants

Mercury - Question 3a

Question 3a requested information on mercury contamination in common ethylene plant feedstocks and how that contamination is dealt with in a plant.

Of the 59 plants surveyed, 31 responded to the question "Is Mercury a contaminant you encounter in your plant?" 20 of those plants, (65%) said yes and 11, (35%) said no.

In the 2003 contaminant survey, mercury was reported to be a common problem in plants cracking liquid feedstocks. The same was found in the 2016 survey, of the 20 plants that reported encountering mercury in feedstocks, 70% of them cracked liquid feedstocks.

The second question asked: how have you dealt with mercury? One half, (50%) of the plants that replied to this question stated they used feedstock specifications to deal with mercury. Other plants used various mercury adsorbents to remove mercury, either Activated Carbon, (3%), Sulfur Impregnated Activated Carbon, (12%), Molecular Sieve, (26%), or Alumina-Based Adsorbent, (9%). These results are shown in the table below.

Table 8. How ethylene plants deal with mercury

Method used to control mercury	Number of Plants	% of Plants Responding
Feed Specification	17	50
Adsorption onto Activated Carbon	1	3
Adsorption onto Sulfur Impregnated Activated Carbon	4	12
Adsorption onto Molecular Sieve	9	26
Adsorption onto Alumina-Based Adsorbent	3	9

The third question asked: Where do you remove mercury? Many of the plants, 43% of those responding to this question replied that mercury was removed from their plant in regeneration gas from dryers and/or catalyst beds. Others responded that they removed mercury from Feedstock, (17%); from Cracked Gas Upstream of Process Gas Dryers (4%); from Chilling Train Feed, (4%); and from Hydrogenation Reactor Feed, (13%). Less common places, (amounting to 17% of the total) included: removal from cracked gas downstream of dryers; from the condensate stripper; and from refinery off gas feed by adsorption downstream of dedicated refinery off gas dryers. The answers to the third question are listed in the table below.

Table 9. Where mercury is removed in an ethylene plant

Location	Number of Plants	% of Plants Responding
Feedstock	4	17
Cracked Gas Upstream of Process Gas Dryers	1	4
Chilling Train Feed	1	4
Hydrogenation Reactor Feed	3	13
Dryer / Reactor Regeneration Gas	10	43

Other less common locations	4	17

The fourth question asked: Is your method of dealing with mercury meeting your expectations and if not, why? Most of the plants responding to this question, (22 of 23) replied yes, their method of dealing with mercury is meeting expectations.

There were some interesting comments also included with the replies to this question on expectations. One plant also responded: "Typically we control it (mercury) by testing of the feedstock prior to introduction of material to the plant. Feedstocks have changed for the better and no mercury has been detected for years." Another plant added "We remove mercury in our turnaround (each 6 years)". Also, one plant provided advice on use of metal sulphide, they mentioned that while expectations are met, use of metal sulphide adsorbent downstream of their Gas Dryer resulted in moisture generation by reaction with H_2S (< 0.2 ppmv) with adsorbent.

The fifth question and final question on mercury asked: Are you aware of the EPA requirements for mercury emissions to air? Most plants, 23 of 27, (85%) that responded to this question answered yes.

Salt (Sodium Chloride) - Question 3b

Question 3b requested information on sodium chloride contamination in common ethylene plant feedstocks and how that contamination is dealt within a plant.

Of the 59 plants surveyed, 32 responded to the question: Is Salt (Sodium Chloride) a contaminant you encounter in your plant? Their responses were evenly split, 16 of those plants, (50%) said yes they encountered sodium chloride and 16 said no.

As in the 2003 survey, sodium chloride contamination was reported to be a common problem in plants cracking liquid feedstocks. The same was found in the 2016 survey, of the 16 plants that reported encountering sodium chloride in feedstocks, liquid feedstocks were the primary feedstock for 69% of them.

The second question asked: in what area of your plant has salt caused problems. Of the 25 plants that responded to this question, 11 plants reported Cracking Heater Operation (coking) problems; 12 plants reported Cracking Heater Reliability (equipment failure, salt deposition) problems; 1 plant reported Water Quench System / Process Water Quality problems; and 1 plant reported problems with Feed System pluggage of coalescer and DP issues across dryers.

These responses are reported in the table below.

Table 10. Areas in ethylene plants where salt caused problems

Description	Number of Plants	% of Plants Responding
Cracking Heater Operation (coking)	11	44
Cracking Heater Reliability (equipment failure, salt deposition)	12	48

Water Quench System / Process Water Quality	1	4
Feed System pluggage of coalescer and DP issues across dryers	1	4

The third question asked: How have you dealt with this feedstock contaminant? 29 plants responded to this question. Of those plants, 12 plants responded that they had dealt with salt contamination via Feed Specifications; 4 plants responded Water Washing and/or Coalescing; 13 plants responded Tank Draining Practices / "Good Housekeeping". Their responses are listed in the table below.

Table 11. How ethylene plants deal with salt

Method used	Number of Plants	% of Plants Responding
Feed Specification	12	41
Water Washing and/or Coalescing	4	14
Tank Draining Practices / "Good Housekeeping"	13	45

The fourth question asked: Do you monitor salt in Cracking Furnace Feed? Of the 21 plants that replied, 14 replied no, they don't monitor salt and 7 responded yes. Of those 7 plants that responded yes they monitor salt in feedstocks, 6 of them were liquid feedstock cracking plants.

The fifth and final question asked: Is your method of dealing with the contaminant meeting your expectations? If not, why?

Of the 19 plants that responded to this question, 15 plants replied yes and 4 plants replied no, expectations were not met. For those 4 that replied no, most explained further that salt contamination is an infrequent issue however it is important to note that one reported infrequent failures of furnace components are still experienced.

Silicon - Question 3c

Question 3c requested information on silicone, specifically whether it is encountered in contamination of common ethylene plant feedstocks and how that contamination is dealt with in a plant.

Silicon was not listed as a contaminant in the original 1993 Feedstock Contaminant Survey. In the 2004 update (2), silicon was included as a listed contaminant and was found to be a contaminant of medium or greater concern for plants cracking liquid feedstocks. For plants cracking naphthas and condensates, it ranked 10th, tied with MeOH and Arsine, while in plants cracking C4 and C5 raffinate streams, it ranked 6th, tied with Fluorides, MeOH, mercury and olefins. For plants cracking gas oils, it ranked 4th, tied with nitrogen and behind only sulfur, chlorides and mercury. For this reason, silicon was selected for more detailed examination in the 2016 survey.

Silicon may enter the plant in the form of a silicon-based antifoam, polydimethylsiloxane, or as the artifacts of thermal cracking of material containing this antifoam (8). These artifacts are primarily ring structures of alternating Si and O, with 6 or 8 member rings being predominant. They track through the ethylene plant with pyrolysis gasoline, and are a poison for pyrolysis gasoline hydrogenation catalysts (9). Feedstocks which might contain antifoam include raffinate or recycle streams from downstream C4 and/or C5 recovery units utilizing extractive distillation for separation of isomers and product purification (8) or from aromatics recovery units using solvent extraction (10). Gas field condensates may also contain antifoam added during fractionation (11). Finally, in delayed cokers upgrading refinery resid, silicon-based antifoam is frequently added to the coke drums where some thermally cracks, forming the above-mentioned silicon – oxygen rings which track with the coker gasoline, or naphtha (8). If unhydrotreated coker naphtha is part of a plant's feed slate, it represents a potential source of silicon.

Of the 59 plants surveyed, 32 responded to the question "Is silicone (for example, antifoams) a contaminant you encounter in your plant?" Of those plants responding, 17 (55%) replied that they did. Those responding positively represented all regions of the world: 4 from North America, 5 from South America, 4 from Asia and 3 from Europe (one did not provide a location.)

Responding to the second question, "How is silicone introduced into your plant?" (some indicated multiple sources), 12 indicated the source to be naphtha feedstock, 6 indicated pyrolysis gasoline as a source, 3 mentioned antifoam and 1 reported shale condensate.

The third and fourth questions address options for dealing with this contaminant: "How have you dealt with this contaminant?" and "Is your method of dealing with the contaminant meeting your expectations?" Responses are summarized in the following table:

Table 12. How does your plant deal with silicon?

	Number of Plants	Meeting Expectations?
Feed Specification	10	8 yes / 2 no
Removed as Product Contaminant	4	2 yes / 2 no
Deposition on Hydrogenation Catalyst	9	4 yes / 5 no

Not surprisingly, it appears as though the preferred means of dealing with silicone entering your plant is to keep it out. And if it does enter the plant as a feedstock contaminant, it either leaves as a product contaminant (typically gasoline) or is deposited on hydrogenation catalyst, neither of which is particularly desirable.

Question 5 addresses silicon deposition on hydrogenation catalyst: " Do you analyze your spent pygas catalyst?" There were 13 positive responses, which included 11 of the 14 that reported that silicon is a contaminant encountered in their plant, and 8 of the 9 that reported deposition on catalyst as its disposition within the plant (reduced catalyst life could be attributed to silicon without the supporting spent catalyst analysis.)

The final question expanded on the previous question: "What is the quantity of silicon found in pyrolysis gasoline as well as the quantity found on the various hydrogenation reactor catalysts and guard beds in your ethylene plant?" One respondent reported finding 0.02 wt% silicon on an acetylene hydrogenation catalyst, and no one reported finding silicon on C3 and C4 hydrogenation catalysts or associated guard beds. Only one reported on the concentration of silicon in pyrolysis gasoline: 0.6 ppm average, with values ranging from 0 to 4.5 ppm. There were 7 respondents providing information on the amount of silicon found on their pyrolysis gasoline hydrogenation catalyst. Results are summarized in the following table:

Table 13. Quantity of silicon found on pygas hydrogenation catalyst:

Respondent 1	10%
Respondent 2	0.4 wt%
Respondent 3	10-15 wt%
Respondent 4	10 wt%
Respondent 5	"silicon present in top layer"
Respondent 6	"0.4% after 4 years in operation"
Respondent 7	1%

Ammonia / Amines / Nitriles - Question 3d

Question 3d requested information on ammonia, amines and nitriles in ethylene plants, more specifically, do they cause problems in your plant, how are they introduced to the plant and how are they dealt with. This group of contaminants was also addressed in the 2003 contaminant survey update (2), allowing both direct comparisons and trend data.

Ammonia

Ammonia can enter an ethylene plant with some refinery gas feeds, such as FCCU off-gas, but the more common source of ammonia in ethylene plants is the breakdown in the cracking furnace of amines introduced to the plant. These amines may be process additives used for pH control (monoethanolamine, diethanolamine and others) or as treatment chemicals to reduce fouling (diethyl hydroxylamine, for example). Carryover from a gas scrubber using an amine solvent to absorb acid gas from the feed to a gas cracker is another source. Other sources of ammonia include the recycle gas from a pyrolysis gasoline hydrotreater (nitrogen compounds converted to ammonia) and the decomposition of nitriles such as acetonitrile, an extractive distillation solvent used in C4 and C5 separations, if recycled back to the ethylene plant (11)

Of the 59 plants surveyed, 30 responded to the question "Does ammonia cause problems or concerns in your plant?" Of the 30 who responded, 9 responded positively (30%). This compares to 16 of 35 respondents reporting issues with ammonia (46%) in 2003.

Of the 9 plants reporting ammonia concerns, 5 were North American, 3 were Asian and 1 was South American. The primary feedstock for 5 of the 9 was ethane or EP, while 4 were liquids crackers. There were an additional 5 respondents who provided data on ammonia but had no issues or concerns. These 5 were all liquids crackers, with 4 being in Europe and 1 in South America.

In response to the second question, "How is ammonia introduced into your plant?" These responses are reported in the table below (some indicated multiple sources). Also shown are the responses from the 2003 survey.

Table 14. How is Ammonia Introduced into your Plant?

	Number of Plants	Comparison: 2003 Results
Feedstock	7	8
Treatment Chemicals	11	14
Primary Fractionator Overhead	1	
Other (vents or other internal / recycle streams		3

The primary sources of ammonia entering the plant have not changed significantly over the past 13 years. Mentioned in 2003 by 1-2 respondents but not found in the 2016 responses as sources of ammonia were vent streams, NESHAPS system and solvent-contaminated raffinate from downstream C4 recovery unit.

Responses to the third question, "In what areas of your plant has ammonia been an issue?", the responses are summarized in the following table, with some indicating multiple areas:

Table 15. In what areas of your plant has ammonia been an issue?

	Number of Plants	Comparison: 2003 Results
Ethylene Product Quality	6	11
Process Water	3	5
Chilling Train	1	1
Fuel Gas Quality	1	

Again, responses in 2016 were comparable to those in 2003, with ethylene product quality being the primary variable negatively affected by ammonia. There were in 2003 a few additional areas mentioned by single respondents: Furnace convection section corrosion (dilution steam), catalyst and other product quality.

The fourth and fifth questions address options for dealing with this contaminant: "How have you dealt with this contaminant?" and "Is your method of dealing with ammonia meeting your expectations?" Results are summarized in the following table (some plants used more than one method for ammonia control). Note the variety of methods employed for dealing with ammonia, and the rather high success rate.

Table 16. How does your plant deal with ammonia?

	Number of Plants	Meeting Expectations?
Feed Specification	3	3 yes
Removal via Molecular Sieve or Other Adsorbent	5	4 yes / 1 no
Removal from Quench Water / Process Water	5	5 yes
Removal via Low pH Water Wash Absorber	2	2 yes
Removal via Dilution Steam Generator Blowdown	2	1 yes / 1 no

Of the 14 respondents providing information on how they deal with ammonia (9 of which reported issues with ammonia), all but two reported that their method of dealing with ammonia is meeting their expectations. Of the two cases where removal was not meeting expectations, the issue with removal via adsorbent was the impact that during regeneration of the adsorbent bed with fuel gas, the ammonia entered the fuel gas system and was converted to NOx in the firebox, resulting in high NOx in stack gas emissions.

In the 2003 survey, where 16 respondents reported ammonia issues (some used more than one):

10 removed ammonia from quench water / process water,
6 relied on feed specifications,
6 used an amine treater,
4 used molecular sieve adsorbent, and
1 used a low pH absorber.

Of the 18 respondents providing information on how they deal with ammonia (16 of which reported issues with ammonia), 4 reported that their method did not meet expectations, similar to the satisfaction level reported in 2016. However, the 2016 results indicate a shift to more operators using adsorbents. And while 6 respondents relied on an amine treater for ammonia removal in 2003, there was no mention of an amine treater in the 2016 responses.

Summarizing, ammonia continues to be a contaminant which is an issue in roughly 30% of the plants responding to the questions on ammonia, down from roughly 45% of the responders on ammonia questions in 2003. Ammonia was primarily an issue with ethylene quality, with process water being the second area affected, unchanged from 2003. The primary means of dealing with ammonia continues to be removal from quench water or process water, there is an increase in the use of adsorbents / molecular sieves for ammonia removal since 2003.

Amines

There were 23 responses to the question "Have amines caused problems in your plant?" Of these, 4 replied that they did. One reported that the result was product being off spec on ammonia. Another reported that overfeeding amines resulted in the formation of acetonitrile. A third was concerned with the potential for NO formation, particularly from amines containing an oxygen (diethyl hydroxyl amine, for example). The fourth did not respond as to the nature of the concern.

This is significantly lower than the level of concern expressed in 2003, where 9 of 18 respondents reported issues with amines, with the primary area affected being ethylene quality, and dilution steam / quench water being second.

To summarize, issues with amines as feedstock contaminants appear to have declined since 2003 and do not represent a concern to most ethylene producers. However, there is a greater level of concern relating to ammonia and means of entering the plant: feedstock and process chemicals. The nitrogen contaminant in the feed or nitrogen component of the process chemical may well be an amine, so their impact has already been captured in the section on ammonia.

Nitriles

There were 22 responses to the question "Have nitriles (acetonitrile / acrylonitrile) caused problems in your plant?" Of these, only 2 reported issues, one resulting from overfeeding amines (forming acetonitrile) and the other experiencing pH issues.

This compares to 4 of 20 respondents reporting issues with nitriles in the 2003 survey. One had issues with C4 byproduct contamination; no further information on the issues from the other three.

Summarizing, nitriles represent a minor contaminant to a very limited group of ethylene producers.

2016 Survey- Question Four

The fourth question of the survey addresses feedstocks that are introduced to the ethylene unit downstream of the furnaces. Six common streams were identified: Flare drum knockout liquids, Flare gas recovery, Vents from other plants, Refinery FCC or Coker unit dry gas, Wastewater/ NESHAPS or spent caustic stripper overhead, and Raw pyrolysis naphthas. An "Other" category was provided to capture any other streams. For each stream the survey had the following questions:

- Do you process?
- Specify origin, if applicable.
- Where does it enter the process?
- Give a brief description of any problems caused by the stream.

Flare Drum Liquids

Of the 59 respondents to the survey, 16 reported processing flare knockout drum liquids in their plants. Twelve of the sixteen (75%) were liquid crackers, well above the 56% liquid/ flexible feed crackers in the survey population, indicating the liquid crackers can produce more of the liquids collected in the plant's flare drums. Geographic location as well as plant vintage were in line with the survey population.

The flare drum liquids are typically reprocessed to the primary fractionator or water quench tower and PGC. The majority of respondents reported no issue with the flare drum liquids. The two most reported issues were separation issues and pH. The detailed responses are shown in Table 17. Question 1-1 Part 1 of 3

Feedstock Class	Ethane			E/P Mix		
Number of Plants	20 (31)			11 (18)		
Contaminant	% Medium	% High	% Medium + High	% Medium	% High	% Medium + High
CO2	35%	25%	60%	55%	27%	82%
Total Sulfur Compounds	30%	15%	45%	36%	18%	55%
Sodium	15%	20%	35%	18%	36%	55%
Methanol	20%	10%	30%	27%	9%	36%
H2O	20%	5%	25%	18%	18%	36%
Inorganic Hg	5%	20%	25%	0%	27%	27%
Organic Hg	0%	25%	25%	0%	36%	36%
Inorganic Chlorides	15%	10%	25%	0%	27%	27%
Total Oxygenates	20%	5%	25%	36%	0%	36%
COS	15%	5%	20%	0%	18%	18%
H2S	15%	5%	20%	45%	0%	45%
Organic Chlorides	10%	10%	20%	0%	36%	36%
Arsenic Compounds	5%	15%	20%	0%	27%	27%
Iron	10%	10%	20%	18%	18%	36%
Fluorides	10%	5%	15%	0%	9%	9%
Arsine/Phosphine	0%	15%	15%	0%	36%	36%
CS2	15%	0%	15%	9%	9%	18%
NH3 & Amines	10%	5%	15%	18%	9%	27%
NOX	0%	10%	10%	0%	27%	27%
Lead	0%	10%	10%	0%	18%	18%
Phosphorus	5%	5%	10%	0%	9%	9%
O2	5%	5%	10%	18%	9%	27%
CO	10%	0%	10%	18%	0%	18%
Oils (pumps & compressor)	5%	5%	10%	9%	9%	18%
Radon	5%	0%	5%	18%	0%	18%
Vanadium	0%	5%	5%	0%	9%	9%
Silicon Derivatives	5%	0%	5%	9%	9%	18%
Calcium	0%	5%	5%	0%	9%	9%
Olefins	0%	0%	0%	0%	0%	0%

Table 18. Question 1-1 Part 2 of 3

Feedstock Class	Propane or Other NGL's			Naphtha/Condensate		
Number of Plants	20 (35)			22 (36)		
Contaminant	% Medium	% High	% Medium + High	% Medium	% High	% Medium + High
CO_2	15%	10%	25%	5%	0%	5%
Total Sulfur Compounds	35%	10%	45%	68%	23%	91%
Sodium	25%	25%	50%	32%	18%	50%
Methanol	25%	20%	45%	27%	18%	45%
H_2O	10%	5%	15%	18%	5%	23%
Inorganic Hg	10%	40%	50%	18%	50%	68%
Organic Hg	0%	45%	45%	5%	41%	45%
Inorganic Chlorides	20%	15%	35%	36%	18%	55%
Total Oxygenates	20%	25%	45%	41%	18%	59%
COS	5%	15%	20%	14%	5%	18%
H_2S	15%	5%	20%	18%	5%	23%
Organic Chlorides	15%	15%	30%	23%	23%	45%
Arsenic Compounds	5%	25%	30%	23%	32%	55%
Iron	25%	10%	35%	23%	9%	32%
Fluorides	20%	5%	25%	0%	9%	9%
Arsine/Phosphine	5%	25%	30%	5%	23%	27%
CS_2	15%	5%	20%	27%	36%	64%
NH_3 & Amines	10%	5%	15%	14%	9%	23%
NOX	0%	15%	15%	0%	23%	23%
Lead	0%	20%	20%	14%	27%	41%
Phosphorus	5%	5%	10%	0%	9%	9%
O_2	10%	5%	15%	18%	5%	23%
CO	10%	0%	10%	9%	0%	9%
Oils (pumps & compressor)	5%	5%	10%	5%	5%	9%
Radon	10%	0%	10%	5%	0%	5%
Vanadium	0%	5%	5%	27%	5%	32%
Silicon Derivatives	5%	5%	10%	23%	23%	45%
Calcium	0%	5%	5%	5%	5%	9%
Olefins	10%	10%	20%	36%	0%	36%

Table 19. Question 1-1 Part 3 of 3

Feedstock Class	Other		
Number of Plants	9 (24)		
Contaminant	% Medium	% High	% Medium + High
CO2	0%	22%	22%
Total Sulfur Compounds	33%	22%	56%
Sodium	0%	44%	44%
Methanol	33%	0%	33%
H2O	11%	11%	22%
Inorganic Hg	22%	33%	56%
Organic Hg	22%	44%	67%
Inorganic Chlorides	11%	22%	33%
Total Oxygenates	44%	0%	44%
COS	11%	0%	11%
H2S	11%	0%	11%
Organic Chlorides	11%	22%	33%
Arsenic Compounds	11%	33%	44%
Iron	11%	11%	22%
Fluorides	22%	0%	22%
Arsine/Phosphine	11%	22%	33%
CS2	0%	11%	11%
NH3 & Amines	11%	0%	11%
NOX	22%	22%	44%
Lead	0%	22%	22%
Phosphorus	0%	0%	0%
O2	22%	11%	33%
CO	11%	0%	11%
Oils (pumps & compressor)	0%	11%	11%
Radon	0%	0%	0%
Vanadium	0%	0%	0%
Silicon Derivatives	0%	11%	11%
Calcium	0%	0%	0%
Olefins	22%	0%	22%

Table 20. Question 1-2 Part 1 of 3

Feedstock Class	Ethane		E/P Mix	
Number of Plants	20 (31)		11 (18)	
	# Testing	# Detecting	# Testing	# Detecting
CO2	12	13	7	9
Total Sulfur Compounds	8	9	5	6
Sodium	2	2	2	2
Methanol	4	4	2	3
H2O	3	4	2	3
Inorganic Hg	1	1	1	1
Organic Hg	1	1	2	2
Inorganic Chlorides	1	2	1	1
Total Oxygenates	2	1	0	0
COS	5	3	3	4
H2S	3	3	4	5
Organic Chlorides	1	1	2	2
Arsenic Compounds	1	2	1	1
Iron	1	0	0	0
Arsine/Phosphine	1	1	2	2
CS2	2	1	0	1
Lead	1	2	1	1
CO	0	1	1	2
Radon	2	2	1	1
Vanadium	1	0	0	0
Silicon Derivatives	1	1	1	1
Calcium	0	0	0	0
Olefins	7	7	4	5
Aluminum	0	1	0	0

Table 21. Question 1-2 Part 2 of 3

Feedstock Class	Propane or Other NGL's		Naphtha/Condensate	
Number of Plants	20 (35)		20 (36)	
	# Testing	# Detecting	# Testing	# Detecting
CO2	6	7	1	0
Total Sulfur Compounds	12	14	18	21
Sodium	4	4	5	4
Methanol	6	7	12	14
H2O	2	6	10	11
Inorganic Hg	2	4	12	14
Organic Hg	2	3	7	8
Inorganic Chlorides	2	4	8	11
Total Oxygenates	6	6	14	15
COS	6	7	2	2
H2S	6	7	9	11
Organic Chlorides	3	4	14	17
Arsenic Compounds	3	5	7	10
Iron	2	1	7	7
Fluorides	1	3	0	1
Arsine/Phosphine	3	3	1	2
CS2	2	0	12	13
NH3 & Amines	1	2	3	3
NOX	0	0	0	1
Lead	3	3	12	17
Phosphorus	0	0	1	2
CO	2	1	2	0
Radon	3	1	0	0
Vanadium	2	2	3	5
Silicon Derivatives	1	1	6	5
Calcium	0	1	3	3
Olefins	9	10	12	13

Table 22. Question 1-2 Part 3 of 3

Feedstock Class	Other	
Number of Plants	9 (24)	
	# Testing	# Detecting
CO2	3	5
Total Sulfur Compounds	5	7
Sodium	1	2
Methanol	1	1
Inorganic Hg	2	3
Organic Hg	3	4
Inorganic Chlorides	1	2
Total Oxygenates	0	1
COS	2	2
H2S	1	1
Organic Chlorides	1	2
Arsenic Compounds	1	2
Fluorides	1	0
Arsine/Phosphine	1	2
NOX	0	1
Lead	1	1
CO	1	1
Radon	1	1
Silicon Derivatives	1	1
Olefins	2	3

Table 23. Question 1-3 Part 1 of 5

Feedstock Class	Ethane		
Contaminant	Responses	Normal Concentration	Maximum Permissible
CO_2	11	<5 ppm Mole, 11-20 ppm Mole <50 ppm Mole <10 ppm Mole (2) 501-1,000 ppm Mole <1,000 ppm Mole 101-500 ppm Wt (2) <100 ppm Weight	20 ppm Mole 50 ppm Mole (2) 3,000 ppm Mole(2) 1,000 ppm Weight (6)
Total Sulfur Compounds	8	<50 ppb Mole <10 ppm Weight 1-10 ppm Weight (2)	2 ppm Weight 20 ppm Weight (2) 50 ppm Weight (3) 100 ppm Weight (2)
COS	3	<0.1 ppb Mole <50 ppb Mole <500 ppb Mole	10-2000 ppb Weight
H_2S	3	<0.1 ppb Mole <50 ppb Mole	<0.5 ppm Mole, 20 ppm Weight, 10 ppm Weight
H_2O	3	<50 ppm Mole <50 ppm Weight	500 ppm Weight 100 ppm Mole 500 ppm Weight
Inorganic Hg	3	<0.1 ppb Weight	1 ppb Weight (2) 0.5 ppb Mole
Organic Hg	4	<0.1 ppb Weight	1 ppb Weight (2) 0.5 ppb Weight 10 ppb Mole
Organic Chlorides	4	1-5 ppb Weight	<1 ppm Weight (2) 1 ppm Weight (2)
Inorganic Chlorides	3	1-5 ppb Weight	<0.5 ppm Weight 1 ppm Weight (2)
Fluorides	2		<5 ppm Weight <0.5 ppm Weight
Methanol	4	<10 ppm Weight 1-10 ppm Mole <100 ppm Mole <5 ppm Mole	50 ppm Mole 500 ppm Weight 20 ppm Mole
Sodium	3	<0.1 ppm Weight	<0.5 ppm Weight 2 ppm Mole 10 ppb Weight

Feedstock Class	Ethane		
Contaminant	Responses	Normal Concentration	Maximum Permissible
NOX	1		1 ppb Weight
Lead	2		<0.5 ppm Weight 50 ppb Weight
Phosphorus	1		10 ppb Weight
O2	2		10 ppm Weight <5 ppm Weight
CO	2	<0.1 ppb Mole	500 ppm Weight 50 ppm Mole
Vanadium	1		<0.5 ppm Weight
Olefins	1	<1,000 ppm Mole	
Silicon Derivatives	0		
Arsenic Compounds	3	1-5 ppb Weight	1 ppb Weight1 ppb Weight50 ppb Mole
Arsine/Phosphine	3	1-5 ppb Weight	1 ppb Weight (3)
Calcium	1		<0.5 ppm Weight
CS2	1	<2 ppm Weight	3 ppm Weight
Iron	1		<0.5 ppm Weight
NH3 & Amines	2		1 ppb Weight 10 ppm Mole
Total Oxygenates	2	1-10 ppm Mole	50 ppm Mole 20 ppm Weight

Table 24. Question 1-3 Part 2 of 5

Feedstock Class	E/P Mix		
Contaminant	Responses	Normal Concentration	Maximum Permissible
CO_2	9	<1 ppm Mole 51-100 ppm Mole <2 ppm Weight <100 ppm Weight 51-100 ppm Weight 101-500 ppm Weight 1,001-5,000 ppm Wt	20 ppm Mole 1,000 ppm Weight (7) 6,000 ppm Weight
Total Sulfur Compounds	6	<2 ppm Weight (2) 1-10 ppm Weight (2)	20 ppm Weight 50 ppm Weight (2) 100 ppm Weight (3)
COS	2	<1 ppm Weight <1 ppm Mole	10 ppb Weight 2 ppm Mole
H_2S	4	<1 ppm Weight <2 ppm Weight <10 ppm Weight	20 ppm Weight 1 ppm Weight <10 ppm Weight 100 ppm Weight
H_2O	4	<50 ppm Mole Saturated	100 ppm Mole 500 ppm Weight (2) 10 ppm Mole
Inorganic Hg	3	<0.1 ppb Weight	1 ppb Weight 50 ppm Weight 0.5 ppb Weight
Organic Hg	4	<0.1 ppb Weight 1-5 ppb Weight	1 ppb Weight 50 ppm Weight 0.5 ppb Weight 10 ppb Weight
Organic Chlorides	4	1-5 ppb Weight <1 ppm Weight	<0.5 ppm Weight 10 ppm Weight 1 ppm Weight <1 ppm Weight
Inorganic Chlorides	3	1-5 ppb Weight	<5 ppm Weight 1 ppm Weight 1 ppm Weight
Fluorides	1		<0.5 ppm Weight
Methanol	2	<100 ppm Mole <50 ppb Weight	500 ppm Weight 50 ppm Weight

Feedstock Class	E/P Mix		
Contaminant	Responses	Normal Concentration	Maximum Permissible
Sodium	4	<0.1 ppm Weight 1-5 ppb Weight	<0.5 ppm Weight 1 ppm Weight 2 ppm Weight 50 ppb Weight
NOX	2		1 ppb Weight 10 ppb Weight
Lead	2		<0.5 ppm Mole 50 ppb Weight
Phosphorus	1		10 ppb Weight
O2	2		<5 ppm Weight 10 ppm Weight
CO	2	<0.1 ppb Mole	500 ppm Weight 50 ppm Mole
Vanadium	1		<0.5 ppm Weight
Olefins	1	<1,000 ppm Mole	
Silicon Derivatives	1		1 ppm Weight
Arsenic Compounds	3	1-5 ppb Weight	0.1 ppb Weight 0.01 ppm Weight 1 ppb Weight
Arsine/Phosphine	4	1-5 ppb Weight (2)	1 ppb Weight (2) 50 ppb Weight (2)
Calcium	1		<0.5 ppm Weight
CS2	1		2 ppm Weight
Iron	1		<0.5 ppm Weight
NH3 & Amines	1		1 ppb Weight
Total Oxygenates	2		20 ppm Weight 100 ppm Weight

Table 25. Question 1-3 Part 3 of 5

Feedstock Class	Propane or Other NGL's		
Contaminant	Responses	Normal Concentration	Maximum Permissible
Inorganic Hg	7	<0.1 ppb Weight <1 ppb Weight <50 ppb Weight <0.05 ppm Weight	0.5 ppb Weight 1 ppb Weight 1 ppb Weight (2) 50 ppb Weight (3)
Sodium	3	<1 ppm Weight	<0.5 ppm Weight 1 ppm Weight 2 ppm Weight
Total Sulfur Compounds	9	<1 ppm Mole 21-50 ppm Mole (2) 1-10 ppm Weight (4) <50 ppm Weight	100 ppm Mole (2) 2 ppm Weight 20 ppm Weight 50 ppm Weight 100 ppm Weight (3) 500 ppm Weight
Methanol	4	<50 ppm Weight <100 ppm Mole 11-20 ppm Mole	100 ppm Weight 50 ppm Mole 500 ppm Weight 50 ppm Weight
COS	4	<5 ppm Weight <0.5 ppm Mole <1 ppm Weight <0.01 ppm Weight	<5 ppm Mole 0.01 ppm Weight 2 ppm Weight 5 ppm Weight
H_2S	2	<0.1 ppm Mole <10 ppb Mole	20 ppm Weight <10 ppm Mole
CO_2	4	<10 ppb Mole <100 ppm Weight	500 ppm Mole 1,000 ppm Weight (3)
H_2O	2	<50 ppm Mole	100 ppm Weight 100 ppm Mole
Organic Hg	6	<0.05 ppm Weight <0.1 ppb Weight 1-5 ppb Weight <50 ppb Weight	0.01 ppm Weight 0.5 ppb Weight 1 ppb Weight (2) 50 ppb Weight (2)
Organic Chlorides	5	<0.5 ppm Weight 1-5 ppb Weight	1 ppm Weight (2) 2 ppm Weight <5 ppm Weight 10 ppm Weight
Inorganic Chlorides	6	<0.5 ppm Weight 1-5 ppb Weight	1 ppm Weight (3)

Feedstock Class	Propane or Other NGL's		
Contaminant	Responses	Normal Concentration	Maximum Permissible
			2 ppm Weight <5 ppm Weight 100 ppm Weight
Fluorides	4	<2 ppm Weight	2 ppm Weight <5 ppm Weight (2) 5 ppm Weight
NOX	2		1 ppb Weight 10 ppb Weight
Lead	4	1-5 ppb Weight	10 ppb Weight 50 ppb Weight 1 ppm Weight <0.5 ppm Weight
Phosphorus	1		10 ppb Weight
O2	3		<5 ppm Weight 10 ppm Weight (2)
CO	2	<0.1 ppb Mole	50 ppm Mole 500 ppm Weight
Vanadium	1		<5 ppm Weight
Olefins	2	1,001-5,000 ppm Weight (2)	6,000 ppm Weight (2)
Silicon Derivatives	1		0.1 ppm Weight
Arsenic Compounds	5	1-5 ppb Weight (2) <50 ppb Weight	1 ppb Weight (2) 10 ppb Weight (2) 50 ppb Weight
Arsine/Phosphine	5	<50 ppb Weight 1-5 ppb Weight	1 ppb Weight (3) 50 ppb Weight (2)
Calcium	1		<0.5 ppm Weight
CS2	2	<2 ppm Weight (2)	2 ppm Weight
Iron	1		<0.5 ppm Weight
NH3 & Amines	2	<1 ppm Weight	1 ppb Weight 2 ppm Weight
Total Oxygenates	5	1,001-5,000 ppm Mole <50 ppm Weight 21-50 ppm Weight	50 ppm Mole 20 ppm Weight 100 ppm Weight (2) 500 ppm Weight

Table 26. Question 1-3 Part 4 of 5

Feedstock Class	Naphtha/Condensate		
Contaminant	Responses	Normal Concentration	Maximum Permissible
Total Sulfur Compounds	19	<1 ppm Weight (2) <50 ppm Weight 21-50 ppm Weight <100 ppm Weight (2) 51-100 ppm Weight (4) <500 ppm Weight (4) 101-500 ppm Weight (3)	100 ppm Weight (5) 500 ppm Weight (9) 1,000 ppm Weight (5)
Inorganic Hg	9	<1 ppb Weight 1-5 ppb Weight (4) 6-10 ppb Weight <50 ppb Weight <2 ppm Weight (2)	10 ppb Weight (4) 50 ppb Weight (3) 0.1 ppm Weight
CS2	11	<1 ppm Weight <2 ppm Weight <5 ppm Weight (6) <10 ppm Weight 1-10 ppm Weight (2)	10 ppm Weight (2) <10 ppm Weight 3 ppm Weight (4) 2 ppm Weight 5 ppm Weight <5 ppm Weight (2)
Total Oxygenates	11	<1 ppm Weight <10 ppm Weight (2) <50 ppm Weight (2) <100 ppm Weight 101-500 ppm Weight (2)	10 ppm Weight 20 ppm Weight 50 ppm Weight (4) 100 ppm Weight (5)
COS	1		<5 ppm Mole
H2S	4	<1 ppm Weight (2) <10 ppm Weight	<10 ppm Mole 1 ppm Weight 5 ppm Weight 10 ppm Weight
CO2	1		500 ppm Weight
H2O	2	<500 ppm Weight	500 ppm Weight 100 ppm Weight
Organic Hg	6	<1 ppb Weight 1-5 ppb Weight (2) <50 ppb Weight	1 ppb Weight 10 ppb Weight (2) 50 ppb Weight (2) 0.1 ppm Weight

Feedstock Class	Naphtha/Condensate		
Contaminant	Responses	Normal Concentration	Maximum Permissible
Organic Chlorides	8	1-10 ppm Weight <1 ppm Weight (6) <10 ppm Weight	10 ppb Weight 1 ppm Weight 5 ppm Weight (3) <10 ppm Weight 10 ppm Weight (2)
Inorganic Chlorides	8	<1 ppm Weight (2) 1-10 ppm Weight 101-500 ppm Weight (2)	10 ppb Weight 1 ppm Weight <5 ppm Weight (2) 5 ppm Weight 10 ppm Weight 100 ppm Weight (2)
Fluorides	5		<1 ppm Weight 4 ppm Weight 5 ppm Weight (2) 10 ppb Weight
Methanol	9	<2 ppm Weight (2) <10 ppm Weight (3) 1-10 ppm Weight (2)	<10 ppm Weight (2) 10 ppm Weight (4) 50 ppm Weight (3)
Sodium	5	<0.5 ppm Weight	<1 ppm Weight (2) 1 ppm Weight (2) <0.5 ppm Weight
NOX	2		1 ppb Weight 10 ppb Weight
Lead	6	<1 ppb Weight <1 ppm Weight 1-5 ppb Weight (2) <10 ppb Weight	10 ppb Weight (2) 50 ppb Weight (2) 0.1 ppm Weight <1 ppm Weight
Phosphorus	2	<0.1 ppm Weight	10 ppb Weight 0.1 ppm Weight
O2	2		<5 ppm Weight 10 ppm Weight
CO	1		500 ppm Weight
Vanadium	4	<10 ppb Weight (2) <0.5 ppm Weight	10 ppb Weight (2) 50 ppb Weight <0.5 ppm Weight

Feedstock Class	Naphtha/Condensate		
Contaminant	Responses	Normal Concentration	Maximum Permissible
Olefins	6	<100 ppm Weight (2) 1,001-5,000 ppm Weight (2)	100 ppm Weight (2) 6,000 ppm Weight (4)
Silicon Derivatives	3	<10 ppb Weight (2) <0.5 ppm Weight	0.1 ppm Weight <0.5 ppm Weight 1 ppm Weight
Arsenic Compounds	6	<10 ppb Weight (3) 1-5 ppb Weight	1 ppb Weight 10 ppb Weight 50 ppb Weight (3) 0.1 ppm Weight
Arsine/Phosphine	3	<1 ppb Weight	10 ppb Weight (2) 50 ppb Weight
Calcium	1	<1 ppm Weight	<0.5 ppm Weight
Iron	3	<50 ppb Weight <1 ppm Weight	50 ppb Weight (2) <0.5 ppm Weight
NH3 & Amines	5	<2 ppm Weight	1 ppb Weight 1 ppm Weight <10 ppm Weight 10 ppm Weight (2)

Table 27. Question 1-3 Part 5 of 5

Feedstock Class	Other		
Contaminant	Responses	Normal Concentration	Maximum Permissible
Organic Hg	3	<0.1 ppb Weight <1 ppb Mole <0.05 ppm Weight	0.5 ppb Weight 10 ppb Mole 50 ppb Weight
Inorganic Hg	2	<0.1 ppb Weight <0.05 ppm Weight	0.5 ppb Weight 50 ppb Weight
Total Sulfur Compounds	6	<2 ppm Mole 1-10 ppm Weight (2) 11-20 ppm Weight <50 ppm Weight	100 ppm Mole 50 ppm Weight 100 ppm Weight (2) 500 ppm Weight 3,000 ppm Weight
COS	1	<2 ppm Mole	<5 ppm Mole
H2S	1		10 ppm Weight
CO2	2	101-500 ppm Mole <100 ppm Weight	20 ppm Mole 1,000 ppm Weight
H2O	2	<50 ppm Mole	100 ppm Mole 100 ppm Weight
Organic Chlorides	2	1-5 ppb Weight	1 ppm Weight 10 ppm Weight
Inorganic Chlorides	2	1-5 ppb Weight	1 ppm Weight (2)
Fluorides	1		5 ppm Weight
Methanol	1		50 ppm Weight
Sodium	3	<0.1 ppm Weight	10 ppb Mole 1 ppm Weight 2 ppm Weight
NOX	1		50 ppb Weight
Lead	2		50 ppb Mole 50 ppb Weight
O2	1		10 ppm Weight
CO	1	<0.1 ppb Mole	50 ppm Mole
Olefins	2	1,001-5,000 ppm Mole Saturated	6,000 ppm Mole 6,000 ppm Weight
Silicon Derivatives	1		0.1 ppm Weight

Arsenic Compounds	3	1-5 ppb Weight	50 ppb Mole 1 ppb Weight 50 ppb Weight
Arsine/Phosphine	2	1-5 ppb Weight	1 ppb Weight 10 ppb Weight
CS2	1		2 ppm Weight
Iron	1		50 ppb Weight
Total Oxygenates	1		100 ppm Weight

Appendix B – Raw Data for Question 2

Table 28. Q2-1: Total Responses (All Plants)

Contam-nant	Qual-ity	Corro-sion	Catalyst Poison	Safety/ Hygiene	Environ-mental (air/ water)	Capacity Limit	Oper-ability	Furnace Proble ms
Methanol/ Alcohols	22	1	11	1	5	3	5	3
Carbon Dioxide	17	6	0	0	0	2	9	0
Radon	0	0	0	11	5	0	0	0
Mercury	0	7	6	11	6	1	4	0
Iron	0	5	3	0	0	0	1	11
Arsine	0	0	15	6	0	0	0	0
Metals	0	1	4	0	0	0	0	10
Sodium	0	2	1	0	0	0	0	19
Fluorides	6	5	1	0	2	0	0	6
Chlorides	5	13	2	0	1	0	3	13
Sulfur	12	5	16	5	6	2	6	4
NOx	0	1	0	9	1	1	3	1
Water	3	6	1	0	0	4	12	11
Phosphorus	0	0	7	0	0	0	0	5
Ammonia or amines	9	1	2	2	1	1	4	0
Organic Acids	0	9	1	0	0	1	6	0
Light nitrogen compounds	7	0	0	0	0	0	1	0
Carbonyls	11	2	3	0	0	1	12	1
Si	2	0	12	0	0	0	5	1
Ethers	2	1	6	0	0	1	2	0

Solvents	0	1	0	0	1	0	0	0
Other	2	0	0	0	0	0	0	0

Table 29. Q2-2: Responses from Gas Cracking Plants

Contaminant	Qual-ity	Corro-sion	Catalyst Poison	Safety/ Hygiene	Environ-mental (air/ water)	Capacity Limit	Oper-ability	Furnace Prob-lems
Methanol/ Alcohols	10	0	6	0	1	1	2	2
Carbon Dioxide	12	5	0	0	0	1	5	0
Radon	0	0	0	8	4	0	0	0
Mercury	0	2	3	6	3	0	3	0
Iron	0	4	1	0	0	0	0	6
Arsine	0	0	8	4	0	0	0	0
-Metals	0	0	1	0	0	0	0	6
Sodium	0	1	1	0	0	0	0	10
Fluorides	4	4	1	0	1	0	0	5
Chlorides	4	5	2	0	1	0	0	6
Sulfur	8	1	7	4	4	1	1	1
NOx	0	1	0	5	1	0	1	0
Water	0	4	0	0	0	2	7	6
Phosphorus	0	0	5	0	0	0	0	3
Ammonia or amines	6	0	1	1	0	0	2	0
Organic Acids	0	3	0	0	0	0	2	0
Light nitrogen compounds	5	0	0	0	0	0	0	0
Carbonyls	6	1	2	0	0	0	6	1
Si	0	0	3	0	0	0	1	0
Ethers	0	0	4	0	0	0	0	0
Solvents	0	1	0	0	0	0	0	0
Other	0	0	0	0	0	0	0	0

Table 30. Q2-3: Responses from Liquids Cracking Plants

Contaminant	Qual-ity	Corro-sion	Catalyst Poison	Safety/ Hygiene	Environ-mental (air/ water)	Capacity Limit	Opera-bility	Furnace Prob-lems

Methanol/ Alcohols	11	1	4	1	4	2	3	1
Carbon Dioxide	5	1	0	0	0	1	4	0
Radon	0	0	0	2	1	0	0	0
Mercury	0	5	3	5	3	1	1	0
Iron	0	1	1	0	0	0	1	5
Arsine	0	0	7	2	0	0	0	0
Metals	0	1	3	0	0	0	0	4
Sodium	0	1	0	0	0	0	0	9
Fluorides	2	1	0	0	1	0	0	1
Chlorides	1	7	0	0	0	0	3	6
Sulfur	3	4	8	1	2	1	5	3
NOx	0	0	0	4	0	1	2	1
Water	3	2	1	0	0	2	5	5
Phosphorus	0	0	2	0	0	0	0	2
Ammonia or amines	3	1	1	1	1	1	2	0
Organic Acids	0	5	1	0	0	1	4	0
Light nitrogen compounds	2	0	0	0	0	0	1	0
Carbonyls	4	1	1	0	0	1	5	0
Si	2	0	9	0	0	0	4	1
Ethers	2	1	2	0	0	1	2	0
Solvents	0	0	0	0	1	0	0	0
Other	2	0	0	0	0	0	0	0

Appendix C. Raw Data for Question 4

Table 31. Q4-1 Flare Drum Knockout Liquids.

Flare Gas Recovery

Five plants reported processing the flare gas recovery. The issues observed as a result of processing this stream include benzene and butadiene in fuel gas, trace O2 in cracked gas, and changes in caloric value. All three plant sizes were represented with the five responses as well as geographic location (with the exception of Asia). The detailed responses are shown in

Table 32. Q4-2 Flare Gas Recovery.

Vents from Other Plants

Seventeen respondents reported processing vent streams in the recovery section of their ethylene plants. The largest source of vent streams is polymer units (10). In most cases the vents are returned to the cracked gas compression train (8) with a few recovering the vent streams in the primary quench area (3).

The largest reported issue with reprocessing the vent stream was fouling either in the caustic tower or the cracked gas compression area. The second most reported issue was the negative effect on rates. Additional issues reported were nitrogen and oxygen contamination. The detailed responses are shown in Table 33. Q4-3 Vents from Other Plants.

Refinery FCC or Coker Unit Dry Gas

FCC or Coker Unit dry gas was processed by four of the respondents (the fifth reported a previous condition but no longer processes this stream). All reported issues, mostly with oxygen and NO contamination. One reported mercury in cold section. Results are included in Table 34. Q4-4 Refinery FCC or Coker Unit Dry Gas.

Wastewater/ NESHAPS or Spent Caustic Stripper Overhead

Seven of our fifty nine respondents process wastewater/ NESHAPS or spent caustic stripper overhead in their plants. The streams were reprocessed through the primary quench system or the deoiling vessel. The only issue reported with this stream was sulfur deposits in the economizer section. Additional details can be found in Table 35. Q4-5 Wastewater/ NESHAPS or Spent Caustic Stripper Overhead.

Raw Pyrolysis Naphtha

Six producers reported processing raw pyrolysis naphtha—five of those were located in North America. Four of the plants were built before 1980. There was a variety of sources reported for this stream. The few issues reported were fouling if introduced to first tray, water contamination, low solvency, and high sulfur leading to off spec benzene product. The responses can be found in Table 36. Q4-6 Raw Pyrolysis Naphtha.

Other

Additional streams fed to the recovery section of ethylene plants not covered in the above categories were contributed by five producers. There was (1) refinery grade propylene and (1) FCC C4 with no reported issues, (1) fresh ethane with fouling issues, (1) polyolefins source with resultant emulsion in quench, and (1) C4 raffinate reporting MeOH contamination. The detailed responses are shown in Table 37. Q4-7 Other.

2016 Survey: Conclusion

Comments on SurveyMonkey

The 2017 survey was conducted via SurveyMonkey. This was not an easy task and required diligence by many to complete. This was the first time our Subcommittee has used this tool, and there is certainly room for improvement. Most of the feedback dealt with the inability to supply precise limits for impurities; this is a deficiency of using the pre-populated pull-down menus in SurveyMonkey. One comment dealt with the fact that some impurities may be expressed in terms of the species measured instead of the specific compound, for instance, expressing CS2 as ppm S instead of ppm as CS2.

One person was very precise in her suggestions, and these are included verbatim, as this encompasses most of the issues with the survey and SurveyMonkey:

• It would be nice to have the ability to save and revisit the survey for further updating. Maybe that is available now, but it wasn't clear to me that if I exited the survey in the middle of entering the data that my previous choices would still remain. (Maybe that isn't feasible as there would also have to be a way to know which plant the data was being entered for to be able to go back and continue the entries for that facility.)

• For fluorides and chlorides, consider including a total option rather than just the breakdown between inorganic and organic.

• The units of measure are a little confusing for maximum permissible limit, i.e. what's the difference between entering <10ppm and 10ppm. To me, those are the same when talking about a max limit.

• In the section where the data is entered on each contaminant....it would be helpful if there were a way to put another column to the left of each row that said what feedstock that entry row was for. Each section has a header, but I found it was easy to get off a section when trying to scroll up and down. I would think I was entering data for propane and then realize somehow I had scrolled too far and was actually in the section for naphtha/condensate.

• Maybe consider separating out naphtha and condensate as far as the contaminant data entry and limits as we don't have the same limits for both.

• Maybe add option for higher weight percent limit/UOM (e.g. 3 wt% olefins in Mixed Butanes) or at least have an option for greater than the 6000ppm max on maximum permissible limit.

•Is there any way to order the limits and units of measure from low to high when displayed in the drop down boxes? When scrolling through the list, you might come across <1 ppm first and then a little further down see 0.01ppm.

• For people having to collect the data, I think it's helpful to have the data in a spreadsheet first before entering the data into the survey. Maybe you could consider blanking out that workbook I put together and sending that out as a template for users to use in collecting the data.

Summary

The 2016 Survey was conducted by SurveyMonkey, and of the 59 people who logged in, 29-36 gave sufficient data, depending on the question, to be included in this survey. One issue with SurveyMonkey discussed above is that if the survey is incomplete and then one logged out, then to input the data that person would have to start over. For an anonymous survey this might not be possible to correct.

The 2016 Survey was more representative of the world compared to the past surveys in 1992 and 2003.

While we have performed some analysis of the survey data, this study has not been exhaustive. To allow access to the data in a form that EPC Proceedings subscribers can easily do their own data analysis, the Subcommittee is working with the EPC to provide this access to an Excel spreadsheet with the data that can be downloaded from the website in the near future.

The authors appreciate all of the work that has gone into this survey, including all of the participants who collected the data and inputted it into SurveyMonkey.

The authors would like to give a big thanks to Tom Pickett, our Subcommittee representative on the Main Committee, and Deanne Smith of Technip Stone and Webster, who transferred the survey questions from a Word document into SurveyMonkey. We appreciate this tremendous effort, without which we would not have been successful.

References

1. Graham, Mark, "Selected Ethylene Feedstock Impurities: Survey Data", 4th Ethylene Producers' Conference, March 21-April 1, 1992.
2. Baumgartner, A. J., et al., "Feedstock Contaminants in Ethylene Plants – an Update", 16th Ethylene Producers' Conference, April 25-29, 2004.
3. Tiedtke, D., et al., "Chemicals Influencing the Activity of Palladium-Based Catalysts for the Selective Hydrogenation of Acetylene to Ethylene in Acetylene Converters", 13th Ethylene Producers' Conference, April 22-26, 2001.
4. Bell, Ronald N., "Understanding and Preventing Failure of Aluminum Equipment in the Presence of Liquid Mercury", 17th Ethylene Producers' Conference, April 10-14, 2005.
5. Sarrazin, P., et al., "Arsenic and Mercury Removal from Hydrocarbons Streams, 4th Ethylene Producers' Conference, March 31-April 1, 1992.
6. Gonzales, Thomas J, Zongxuan Hong, "Chevron Philips Chemical Company LP's Unit 24 Operating Experience with Front End E-Series Acetylene Converter Catalyst", 18th Ethylene Producers Conference, April 6-10, 2008.
7. Wilhelm, S. M, N. Bloom, "Mercury in Petroleum", Fuel Processing Technology, 1999 (in Gerard Subirachs Sanchez, MEng. Thesis, "Mercury in Extraction and Refining Process of Crude Oil and Natural Gas", University of Aberdeen, May 20, 2013).
8. Hood, D. and Coleman, S., "Minor Components, Major Issues: An Overview Contaminants in Ethylene Plants", 19th Ethylene Producers' Conference, April 24, 2007.
9. Britto, J. M., et al., "Troubleshooting Silicon Contamination on Hydrotreatment Catalysts – Sources, Impacts and Mitigation", 22nd Ethylene Producers' Conference, March 23, 2010.
10. Zinke, K. W., "Antifoam Impacts in an Olefins Plant: A Case History", 13th Ethylene Producers' Conference, April 22-26, 2001.

11. Lordo, S. A. and Smith, M. D., "Oil Field Additives and Ethylene Feedstock Contaminants", 25th Ethylene Producers' Conference, April 13, 2016.

Appendix A – Raw Data for Question 1

Table 17. Question 1-1 Part 1 of 3

Feedstock Class	Ethane			E/P Mix		
Number of Plants	20 (31)			11 (18)		
Contaminant	% Medium	% High	% Medium + High	% Medium	% High	% Medium + High
CO2	35%	25%	60%	55%	27%	82%
Total Sulfur Compounds	30%	15%	45%	36%	18%	55%
Sodium	15%	20%	35%	18%	36%	55%
Methanol	20%	10%	30%	27%	9%	36%
H2O	20%	5%	25%	18%	18%	36%
Inorganic Hg	5%	20%	25%	0%	27%	27%
Organic Hg	0%	25%	25%	0%	36%	36%
Inorganic Chlorides	15%	10%	25%	0%	27%	27%
Total Oxygenates	20%	5%	25%	36%	0%	36%
COS	15%	5%	20%	0%	18%	18%
H2S	15%	5%	20%	45%	0%	45%
Organic Chlorides	10%	10%	20%	0%	36%	36%
Arsenic Compounds	5%	15%	20%	0%	27%	27%
Iron	10%	10%	20%	18%	18%	36%
Fluorides	10%	5%	15%	0%	9%	9%
Arsine/Phosphine	0%	15%	15%	0%	36%	36%
CS2	15%	0%	15%	9%	9%	18%
NH3 & Amines	10%	5%	15%	18%	9%	27%
NOX	0%	10%	10%	0%	27%	27%
Lead	0%	10%	10%	0%	18%	18%
Phosphorus	5%	5%	10%	0%	9%	9%
O2	5%	5%	10%	18%	9%	27%
CO	10%	0%	10%	18%	0%	18%
Oils (pumps & compressor)	5%	5%	10%	9%	9%	18%
Radon	5%	0%	5%	18%	0%	18%
Vanadium	0%	5%	5%	0%	9%	9%
Silicon Derivatives	5%	0%	5%	9%	9%	18%
Calcium	0%	5%	5%	0%	9%	9%
Olefins	0%	0%	0%	0%	0%	0%

Table 18. Question 1-1 Part 2 of 3

Feedstock Class	Propane or Other NGL's			Naphtha/Condensate		
Number of Plants	20 (35)			22 (36)		
Contaminant	% Medium	% High	% Medium + High	% Medium	% High	% Medium + High
CO2	15%	10%	25%	5%	0%	5%
Total Sulfur Compounds	35%	10%	45%	68%	23%	91%
Sodium	25%	25%	50%	32%	18%	50%
Methanol	25%	20%	45%	27%	18%	45%
H2O	10%	5%	15%	18%	5%	23%
Inorganic Hg	10%	40%	50%	18%	50%	68%
Organic Hg	0%	45%	45%	5%	41%	45%
Inorganic Chlorides	20%	15%	35%	36%	18%	55%
Total Oxygenates	20%	25%	45%	41%	18%	59%
COS	5%	15%	20%	14%	5%	18%
H2S	15%	5%	20%	18%	5%	23%
Organic Chlorides	15%	15%	30%	23%	23%	45%
Arsenic Compounds	5%	25%	30%	23%	32%	55%
Iron	25%	10%	35%	23%	9%	32%
Fluorides	20%	5%	25%	0%	9%	9%
Arsine/Phosphine	5%	25%	30%	5%	23%	27%
CS2	15%	5%	20%	27%	36%	64%
NH3 & Amines	10%	5%	15%	14%	9%	23%
NOX	0%	15%	15%	0%	23%	23%
Lead	0%	20%	20%	14%	27%	41%
Phosphorus	5%	5%	10%	0%	9%	9%
O2	10%	5%	15%	18%	5%	23%
CO	10%	0%	10%	9%	0%	9%
Oils (pumps & compressor)	5%	5%	10%	5%	5%	9%
Radon	10%	0%	10%	5%	0%	5%
Vanadium	0%	5%	5%	27%	5%	32%
Silicon Derivatives	5%	5%	10%	23%	23%	45%
Calcium	0%	5%	5%	5%	5%	9%
Olefins	10%	10%	20%	36%	0%	36%

Table 19. Question 1-1 Part 3 of 3

Feedstock Class	Other		
Number of Plants	9 (24)		
Contaminant	% Medium	% High	% Medium + High
CO2	0%	22%	22%
Total Sulfur Compounds	33%	22%	56%
Sodium	0%	44%	44%
Methanol	33%	0%	33%
H2O	11%	11%	22%
Inorganic Hg	22%	33%	56%
Organic Hg	22%	44%	67%
Inorganic Chlorides	11%	22%	33%
Total Oxygenates	44%	0%	44%
COS	11%	0%	11%
H2S	11%	0%	11%
Organic Chlorides	11%	22%	33%
Arsenic Compounds	11%	33%	44%
Iron	11%	11%	22%
Fluorides	22%	0%	22%
Arsine/Phosphine	11%	22%	33%
CS2	0%	11%	11%
NH3 & Amines	11%	0%	11%
NOX	22%	22%	44%
Lead	0%	22%	22%
Phosphorus	0%	0%	0%
O2	22%	11%	33%
CO	11%	0%	11%
Oils (pumps & compressor)	0%	11%	11%
Radon	0%	0%	0%
Vanadium	0%	0%	0%
Silicon Derivatives	0%	11%	11%
Calcium	0%	0%	0%
Olefins	22%	0%	22%

Table 20. Question 1-2 Part 1 of 3

Feedstock Class	Ethane		E/P Mix	
Number of Plants	20 (31)		11 (18)	
	# Testing	# Detecting	# Testing	# Detecting
CO2	12	13	7	9
Total Sulfur Compounds	8	9	5	6
Sodium	2	2	2	2
Methanol	4	4	2	3
H2O	3	4	2	3
Inorganic Hg	1	1	1	1
Organic Hg	1	1	2	2
Inorganic Chlorides	1	2	1	1
Total Oxygenates	2	1	0	0
COS	5	3	3	4
H2S	3	3	4	5
Organic Chlorides	1	1	2	2
Arsenic Compounds	1	2	1	1
Iron	1	0	0	0
Arsine/Phosphine	1	1	2	2
CS2	2	1	0	1
Lead	1	2	1	1
CO	0	1	1	2
Radon	2	2	1	1
Vanadium	1	0	0	0
Silicon Derivatives	1	1	1	1
Calcium	0	0	0	0
Olefins	7	7	4	5
Aluminum	0	1	0	0

Table 21. Question 1-2 Part 2 of 3

Feedstock Class	Propane or Other NGL's		Naphtha/Condensate	
Number of Plants	20 (35)		20 (36)	
	# Testing	# Detecting	# Testing	# Detecting
CO2	6	7	1	0
Total Sulfur Compounds	12	14	18	21
Sodium	4	4	5	4
Methanol	6	7	12	14
H2O	2	6	10	11
Inorganic Hg	2	4	12	14
Organic Hg	2	3	7	8
Inorganic Chlorides	2	4	8	11
Total Oxygenates	6	6	14	15
COS	6	7	2	2
H2S	6	7	9	11
Organic Chlorides	3	4	14	17
Arsenic Compounds	3	5	7	10
Iron	2	1	7	7
Fluorides	1	3	0	1
Arsine/Phosphine	3	3	1	2
CS2	2	0	12	13
NH3 & Amines	1	2	3	3
NOX	0	0	0	1
Lead	3	3	12	17
Phosphorus	0	0	1	2
CO	2	1	2	0
Radon	3	1	0	0
Vanadium	2	2	3	5
Silicon Derivatives	1	1	6	5
Calcium	0	1	3	3
Olefins	9	10	12	13

Table 22. Question 1-2 Part 3 of 3

Feedstock Class	Other	
Number of Plants	9 (24)	
	# Testing	# Detecting
CO2	3	5
Total Sulfur Compounds	5	7
Sodium	1	2
Methanol	1	1
Inorganic Hg	2	3
Organic Hg	3	4
Inorganic Chlorides	1	2
Total Oxygenates	0	1
COS	2	2
H2S	1	1
Organic Chlorides	1	2
Arsenic Compounds	1	2
Fluorides	1	0
Arsine/Phosphine	1	2
NOX	0	1
Lead	1	1
CO	1	1
Radon	1	1
Silicon Derivatives	1	1
Olefins	2	3

Table 23. Question 1-3 Part 1 of 5

Feedstock Class	Ethane		
Contaminant	Responses	Normal Concentration	Maximum Permissible
CO2	11	<5 ppm Mole, 11-20 ppm Mole <50 ppm Mole <10 ppm Mole (2) 501-1,000 ppm Mole <1,000 ppm Mole 101-500 ppm Wt (2) <100 ppm Weight	20 ppm Mole 50 ppm Mole (2) 3,000 ppm Mole(2) 1,000 ppm Weight (6)
Total Sulfur Compounds	8	<50 ppb Mole <10 ppm Weight 1-10 ppm Weight (2)	2 ppm Weight 20 ppm Weight (2) 50 ppm Weight (3) 100 ppm Weight (2)
COS	3	<0.1 ppb Mole <50 ppb Mole <500 ppb Mole	10-2000 ppb Weight
H2S	3	<0.1 ppb Mole <50 ppb Mole	<0.5 ppm Mole, 20 ppm Weight, 10 ppm Weight
H2O	3	<50 ppm Mole <50 ppm Weight	500 ppm Weight 100 ppm Mole 500 ppm Weight
Inorganic Hg	3	<0.1 ppb Weight	1 ppb Weight (2) 0.5 ppb Mole
Organic Hg	4	<0.1 ppb Weight	1 ppb Weight (2) 0.5 ppb Weight 10 ppb Mole
Organic Chlorides	4	1-5 ppb Weight	<1 ppm Weight (2) 1 ppm Weight (2)
Inorganic Chlorides	3	1-5 ppb Weight	<0.5 ppm Weight 1 ppm Weight (2)
Fluorides	2		<5 ppm Weight <0.5 ppm Weight
Methanol	4	<10 ppm Weight 1-10 ppm Mole <100 ppm Mole <5 ppm Mole	50 ppm Mole 500 ppm Weight 20 ppm Mole
Sodium	3	<0.1 ppm Weight	<0.5 ppm Weight 2 ppm Mole 10 ppb Weight

Feedstock Class	Ethane		
Contaminant	Responses	Normal Concentration	Maximum Permissible
NOX	1		1 ppb Weight
Lead	2		<0.5 ppm Weight 50 ppb Weight
Phosphorus	1		10 ppb Weight
O2	2		10 ppm Weight <5 ppm Weight
CO	2	<0.1 ppb Mole	500 ppm Weight 50 ppm Mole
Vanadium	1		<0.5 ppm Weight
Olefins	1	<1,000 ppm Mole	
Silicon Derivatives	0		
Arsenic Compounds	3	1-5 ppb Weight	1 ppb Weight1 ppb Weight50 ppb Mole
Arsine/Phosphine	3	1-5 ppb Weight	1 ppb Weight (3)
Calcium	1		<0.5 ppm Weight
CS2	1	<2 ppm Weight	3 ppm Weight
Iron	1		<0.5 ppm Weight
NH3 & Amines	2		1 ppb Weight 10 ppm Mole
Total Oxygenates	2	1-10 ppm Mole	50 ppm Mole 20 ppm Weight

Table 24. Question 1-3 Part 2 of 5

Feedstock Class	E/P Mix		
Contaminant	Responses	Normal Concentration	Maximum Permissible
CO_2	9	<1 ppm Mole 51-100 ppm Mole <2 ppm Weight <100 ppm Weight 51-100 ppm Weight 101-500 ppm Weight 1,001-5,000 ppm Wt	20 ppm Mole 1,000 ppm Weight (7) 6,000 ppm Weight
Total Sulfur Compounds	6	<2 ppm Weight (2) 1-10 ppm Weight (2)	20 ppm Weight 50 ppm Weight (2) 100 ppm Weight (3)
COS	2	<1 ppm Weight <1 ppm Mole	10 ppb Weight 2 ppm Mole
H_2S	4	<1 ppm Weight <2 ppm Weight <10 ppm Weight	20 ppm Weight 1 ppm Weight <10 ppm Weight 100 ppm Weight
H_2O	4	<50 ppm Mole Saturated	100 ppm Mole 500 ppm Weight (2) 10 ppm Mole
Inorganic Hg	3	<0.1 ppb Weight	1 ppb Weight 50 ppm Weight 0.5 ppb Weight
Organic Hg	4	<0.1 ppb Weight 1-5 ppb Weight	1 ppb Weight 50 ppm Weight 0.5 ppb Weight 10 ppb Weight
Organic Chlorides	4	1-5 ppb Weight <1 ppm Weight	<0.5 ppm Weight 10 ppm Weight 1 ppm Weight <1 ppm Weight
Inorganic Chlorides	3	1-5 ppb Weight	<5 ppm Weight 1 ppm Weight 1 ppm Weight
Fluorides	1		<0.5 ppm Weight
Methanol	2	<100 ppm Mole <50 ppb Weight	500 ppm Weight 50 ppm Weight

Feedstock Class	E/P Mix		
Contaminant	Responses	Normal Concentration	Maximum Permissible
Sodium	4	<0.1 ppm Weight 1-5 ppb Weight	<0.5 ppm Weight 1 ppm Weight 2 ppm Weight 50 ppb Weight
NOX	2		1 ppb Weight 10 ppb Weight
Lead	2		<0.5 ppm Mole 50 ppb Weight
Phosphorus	1		10 ppb Weight
O2	2		<5 ppm Weight 10 ppm Weight
CO	2	<0.1 ppb Mole	500 ppm Weight 50 ppm Mole
Vanadium	1		<0.5 ppm Weight
Olefins	1	<1,000 ppm Mole	
Silicon Derivatives	1		1 ppm Weight
Arsenic Compounds	3	1-5 ppb Weight	0.1 ppb Weight 0.01 ppm Weight 1 ppb Weight
Arsine/Phosphine	4	1-5 ppb Weight (2)	1 ppb Weight (2) 50 ppb Weight (2)
Calcium	1		<0.5 ppm Weight
CS2	1		2 ppm Weight
Iron	1		<0.5 ppm Weight
NH3 & Amines	1		1 ppb Weight
Total Oxygenates	2		20 ppm Weight 100 ppm Weight

Table 25. Question 1-3 Part 3 of 5

Feedstock Class	Propane or Other NGL's		
Contaminant	Responses	Normal Concentration	Maximum Permissible
Inorganic Hg	7	<0.1 ppb Weight <1 ppb Weight <50 ppb Weight <0.05 ppm Weight	0.5 ppb Weight 1 ppb Weight 1 ppb Weight (2) 50 ppb Weight (3)
Sodium	3	<1 ppm Weight	<0.5 ppm Weight 1 ppm Weight 2 ppm Weight
Total Sulfur Compounds	9	<1 ppm Mole 21-50 ppm Mole (2) 1-10 ppm Weight (4) <50 ppm Weight	100 ppm Mole (2) 2 ppm Weight 20 ppm Weight 50 ppm Weight 100 ppm Weight (3) 500 ppm Weight
Methanol	4	<50 ppm Weight <100 ppm Mole 11-20 ppm Mole	100 ppm Weight 50 ppm Mole 500 ppm Weight 50 ppm Weight
COS	4	<5 ppm Weight <0.5 ppm Mole <1 ppm Weight <0.01 ppm Weight	<5 ppm Mole 0.01 ppm Weight 2 ppm Weight 5 ppm Weight
H2S	2	<0.1 ppm Mole <10 ppb Mole	20 ppm Weight <10 ppm Mole
CO2	4	<10 ppb Mole <100 ppm Weight	500 ppm Mole 1,000 ppm Weight (3)
H2O	2	<50 ppm Mole	100 ppm Weight 100 ppm Mole
Organic Hg	6	<0.05 ppm Weight <0.1 ppb Weight 1-5 ppb Weight <50 ppb Weight	0.01 ppm Weight 0.5 ppb Weight 1 ppb Weight (2) 50 ppb Weight (2)
Organic Chlorides	5	<0.5 ppm Weight 1-5 ppb Weight	1 ppm Weight (2) 2 ppm Weight <5 ppm Weight 10 ppm Weight
Inorganic Chlorides	6	<0.5 ppm Weight 1-5 ppb Weight	1 ppm Weight (3)

Feedstock Class	Propane or Other NGL's		
Contaminant	Responses	Normal Concentration	Maximum Permissible
			2 ppm Weight <5 ppm Weight 100 ppm Weight
Fluorides	4	<2 ppm Weight	2 ppm Weight <5 ppm Weight (2) 5 ppm Weight
NOX	2		1 ppb Weight 10 ppb Weight
Lead	4	1-5 ppb Weight	10 ppb Weight 50 ppb Weight 1 ppm Weight <0.5 ppm Weight
Phosphorus	1		10 ppb Weight
O2	3		<5 ppm Weight 10 ppm Weight (2)
CO	2	<0.1 ppb Mole	50 ppm Mole 500 ppm Weight
Vanadium	1		<5 ppm Weight
Olefins	2	1,001-5,000 ppm Weight (2)	6,000 ppm Weight (2)
Silicon Derivatives	1		0.1 ppm Weight
Arsenic Compounds	5	1-5 ppb Weight (2) <50 ppb Weight	1 ppb Weight (2) 10 ppb Weight (2) 50 ppb Weight
Arsine/Phosphine	5	<50 ppb Weight 1-5 ppb Weight	1 ppb Weight (3) 50 ppb Weight (2)
Calcium	1		<0.5 ppm Weight
CS2	2	<2 ppm Weight (2)	2 ppm Weight
Iron	1		<0.5 ppm Weight
NH3 & Amines	2	<1 ppm Weight	1 ppb Weight 2 ppm Weight
Total Oxygenates	5	1,001-5,000 ppm Mole <50 ppm Weight 21-50 ppm Weight	50 ppm Mole 20 ppm Weight 100 ppm Weight (2) 500 ppm Weight

Table 26. Question 1-3 Part 4 of 5

Feedstock Class	Naphtha/Condensate		
Contaminant	Responses	Normal Concentration	Maximum Permissible
Total Sulfur Compounds	19	<1 ppm Weight (2) <50 ppm Weight 21-50 ppm Weight <100 ppm Weight (2) 51-100 ppm Weight (4) <500 ppm Weight (4) 101-500 ppm Weight (3)	100 ppm Weight (5) 500 ppm Weight (9) 1,000 ppm Weight (5)
Inorganic Hg	9	<1 ppb Weight 1-5 ppb Weight (4) 6-10 ppb Weight <50 ppb Weight <2 ppm Weight (2)	10 ppb Weight (4) 50 ppb Weight (3) 0.1 ppm Weight
CS2	11	<1 ppm Weight <2 ppm Weight <5 ppm Weight (6) <10 ppm Weight 1-10 ppm Weight (2)	10 ppm Weight (2) <10 ppm Weight 3 ppm Weight (4) 2 ppm Weight 5 ppm Weight <5 ppm Weight (2)
Total Oxygenates	11	<1 ppm Weight <10 ppm Weight (2) <50 ppm Weight (2) <100 ppm Weight 101-500 ppm Weight (2)	10 ppm Weight 20 ppm Weight 50 ppm Weight (4) 100 ppm Weight (5)
COS	1		<5 ppm Mole
H2S	4	<1 ppm Weight (2) <10 ppm Weight	<10 ppm Mole 1 ppm Weight 5 ppm Weight 10 ppm Weight
CO2	1		500 ppm Weight
H2O	2	<500 ppm Weight	500 ppm Weight 100 ppm Weight
Organic Hg	6	<1 ppb Weight 1-5 ppb Weight (2) <50 ppb Weight	1 ppb Weight 10 ppb Weight (2) 50 ppb Weight (2) 0.1 ppm Weight

Feedstock Class	Naphtha/Condensate		
Contaminant	Responses	Normal Concentration	Maximum Permissible
Organic Chlorides	8	1-10 ppm Weight <1 ppm Weight (6) <10 ppm Weight	10 ppb Weight 1 ppm Weight 5 ppm Weight (3) <10 ppm Weight 10 ppm Weight (2)
Inorganic Chlorides	8	<1 ppm Weight (2) 1-10 ppm Weight 101-500 ppm Weight (2)	10 ppb Weight 1 ppm Weight <5 ppm Weight (2) 5 ppm Weight 10 ppm Weight 100 ppm Weight (2)
Fluorides	5		<1 ppm Weight 4 ppm Weight 5 ppm Weight (2) 10 ppb Weight
Methanol	9	<2 ppm Weight (2) <10 ppm Weight (3) 1-10 ppm Weight (2)	<10 ppm Weight (2) 10 ppm Weight (4) 50 ppm Weight (3)
Sodium	5	<0.5 ppm Weight	<1 ppm Weight (2) 1 ppm Weight (2) <0.5 ppm Weight
NOX	2		1 ppb Weight 10 ppb Weight
Lead	6	<1 ppb Weight <1 ppm Weight 1-5 ppb Weight (2) <10 ppb Weight	10 ppb Weight (2) 50 ppb Weight (2) 0.1 ppm Weight <1 ppm Weight
Phosphorus	2	<0.1 ppm Weight	10 ppb Weight 0.1 ppm Weight
O2	2		<5 ppm Weight 10 ppm Weight
CO	1		500 ppm Weight
Vanadium	4	<10 ppb Weight (2) <0.5 ppm Weight	10 ppb Weight (2) 50 ppb Weight <0.5 ppm Weight

Feedstock Class	Naphtha/Condensate		
Contaminant	Responses	Normal Concentration	Maximum Permissible
Olefins	6	<100 ppm Weight (2) 1,001-5,000 ppm Weight (2)	100 ppm Weight (2) 6,000 ppm Weight (4)
Silicon Derivatives	3	<10 ppb Weight (2) <0.5 ppm Weight	0.1 ppm Weight <0.5 ppm Weight 1 ppm Weight
Arsenic Compounds	6	<10 ppb Weight (3) 1-5 ppb Weight	1 ppb Weight 10 ppb Weight 50 ppb Weight (3) 0.1 ppm Weight
Arsine/Phosphine	3	<1 ppb Weight	10 ppb Weight (2) 50 ppb Weight
Calcium	1	<1 ppm Weight	<0.5 ppm Weight
Iron	3	<50 ppb Weight <1 ppm Weight	50 ppb Weight (2) <0.5 ppm Weight
NH3 & Amines	5	<2 ppm Weight	1 ppb Weight 1 ppm Weight <10 ppm Weight 10 ppm Weight (2)

Table 27. Question 1-3 Part 5 of 5

Feedstock Class	Other		
Contaminant	Responses	Normal Concentration	Maximum Permissible
Organic Hg	3	<0.1 ppb Weight <1 ppb Mole <0.05 ppm Weight	0.5 ppb Weight 10 ppb Mole 50 ppb Weight
Inorganic Hg	2	<0.1 ppb Weight <0.05 ppm Weight	0.5 ppb Weight 50 ppb Weight
Total Sulfur Compounds	6	<2 ppm Mole 1-10 ppm Weight (2) 11-20 ppm Weight <50 ppm Weight	100 ppm Mole 50 ppm Weight 100 ppm Weight (2) 500 ppm Weight 3,000 ppm Weight
COS	1	<2 ppm Mole	<5 ppm Mole
H2S	1		10 ppm Weight
CO2	2	101-500 ppm Mole <100 ppm Weight	20 ppm Mole 1,000 ppm Weight
H2O	2	<50 ppm Mole	100 ppm Mole 100 ppm Weight
Organic Chlorides	2	1-5 ppb Weight	1 ppm Weight 10 ppm Weight
Inorganic Chlorides	2	1-5 ppb Weight	1 ppm Weight (2)
Fluorides	1		5 ppm Weight
Methanol	1		50 ppm Weight
Sodium	3	<0.1 ppm Weight	10 ppb Mole 1 ppm Weight 2 ppm Weight
NOX	1		50 ppb Weight
Lead	2		50 ppb Mole 50 ppb Weight
O2	1		10 ppm Weight
CO	1	<0.1 ppb Mole	50 ppm Mole
Olefins	2	1,001-5,000 ppm Mole Saturated	6,000 ppm Mole 6,000 ppm Weight
Silicon Derivatives	1		0.1 ppm Weight

Arsenic Compounds	3	1-5 ppb Weight	50 ppb Mole 1 ppb Weight 50 ppb Weight
Arsine/Phosphine	2	1-5 ppb Weight	1 ppb Weight 10 ppb Weight
CS2	1		2 ppm Weight
Iron	1		50 ppb Weight
Total Oxygenates	1		100 ppm Weight

Appendix B – Raw Data for Question 2

Table 28. Q2-1: Total Responses (All Plants)

Contam-nant	Qual-ity	Corro-sion	Catalyst Poison	Safety/ Hygiene	Environ-mental (air/ water)	Capacity Limit	Oper-ability	Furnace Proble ms
Methanol/ Alcohols	22	1	11	1	5	3	5	3
Carbon Dioxide	17	6	0	0	0	2	9	0
Radon	0	0	0	11	5	0	0	0
Mercury	0	7	6	11	6	1	4	0
Iron	0	5	3	0	0	0	1	11
Arsine	0	0	15	6	0	0	0	0
Metals	0	1	4	0	0	0	0	10
Sodium	0	2	1	0	0	0	0	19
Fluorides	6	5	1	0	2	0	0	6
Chlorides	5	13	2	0	1	0	3	13
Sulfur	12	5	16	5	6	2	6	4
NOx	0	1	0	9	1	1	3	1
Water	3	6	1	0	0	4	12	11
Phosphorus	0	0	7	0	0	0	0	5
Ammonia or amines	9	1	2	2	1	1	4	0
Organic Acids	0	9	1	0	0	1	6	0
Light nitrogen compounds	7	0	0	0	0	0	1	0
Carbonyls	11	2	3	0	0	1	12	1
Si	2	0	12	0	0	0	5	1
Ethers	2	1	6	0	0	1	2	0

Solvents	0	1	0	0	1	0	0	0
Other	2	0	0	0	0	0	0	0

Table 29. Q2-2: Responses from Gas Cracking Plants

Contaminant	Qual-ity	Corro-sion	Catalyst Poison	Safety/ Hygiene	Environ-mental (air/ water)	Capacity Limit	Oper-ability	Furnace Prob-lems
Methanol/ Alcohols	10	0	6	0	1	1	2	2
Carbon Dioxide	12	5	0	0	0	1	5	0
Radon	0	0	0	8	4	0	0	0
Mercury	0	2	3	6	3	0	3	0
Iron	0	4	1	0	0	0	0	6
Arsine	0	0	8	4	0	0	0	0
-Metals	0	0	1	0	0	0	0	6
Sodium	0	1	1	0	0	0	0	10
Fluorides	4	4	1	0	1	0	0	5
Chlorides	4	5	2	0	1	0	0	6
Sulfur	8	1	7	4	4	1	1	1
NOx	0	1	0	5	1	0	1	0
Water	0	4	0	0	0	2	7	6
Phosphorus	0	0	5	0	0	0	0	3
Ammonia or amines	6	0	1	1	0	0	2	0
Organic Acids	0	3	0	0	0	0	2	0
Light nitrogen compounds	5	0	0	0	0	0	0	0
Carbonyls	6	1	2	0	0	0	6	1
Si	0	0	3	0	0	0	1	0
Ethers	0	0	4	0	0	0	0	0
Solvents	0	1	0	0	0	0	0	0
Other	0	0	0	0	0	0	0	0

Table 30. Q2-3: Responses from Liquids Cracking Plants

Contaminant	Qual-ity	Corro-sion	Catalyst Poison	Safety/ Hygiene	Environ-mental (air/ water)	Capacity Limit	Opera-bility	Furnace Prob-lems

Methanol/ Alcohols	11	1	4	1	4	2	3	1
Carbon Dioxide	5	1	0	0	0	1	4	0
Radon	0	0	0	2	1	0	0	0
Mercury	0	5	3	5	3	1	1	0
Iron	0	1	1	0	0	0	1	5
Arsine	0	0	7	2	0	0	0	0
Metals	0	1	3	0	0	0	0	4
Sodium	0	1	0	0	0	0	0	9
Fluorides	2	1	0	0	1	0	0	1
Chlorides	1	7	0	0	0	0	3	6
Sulfur	3	4	8	1	2	1	5	3
NOx	0	0	0	4	0	1	2	1
Water	3	2	1	0	0	2	5	5
Phosphorus	0	0	2	0	0	0	0	2
Ammonia or amines	3	1	1	1	1	1	2	0
Organic Acids	0	5	1	0	0	1	4	0
Light nitrogen compounds	2	0	0	0	0	0	1	0
Carbonyls	4	1	1	0	0	1	5	0
Si	2	0	9	0	0	0	4	1
Ethers	2	1	2	0	0	1	2	0
Solvents	0	0	0	0	1	0	0	0
Other	2	0	0	0	0	0	0	0

Appendix C. Raw Data for Question 4

Table 31. Q4-1 Flare Drum Knockout Liquids

Location	Age/ Vint-age of Plant	Size of Plant	Flare drum knock-out liquids			
			Do you process ?	Specify origin if applicable	Where does it enter the process?	Give a brief description of any problems caused by the stream
North America	1981-2000	>1000K mta	No			

Location	Age/ Vint-age of Plant	Size of Plant	Flare drum knock-out liquids			
			Do you process ?	Specify origin if applicable	Where does it enter the process?	Give a brief description of any problems caused by the stream
Europe	2000-2010	500 - 1000K mta	Yes	Mostly steam/condensate from flare steam. Can be some hydrocarbons, in case of upset or cleaning. All liquids come from own plant, drum/flare is not shared with other plants.	Water Quench Tower; depending on contents of flaredrum, some liquids can't be pumped cause they cause separation issues, these are removed by vacuum truck.	Separation issues, could contain C4's unable to use vacuum truck
North America	1981-2000	>1000K mta	Yes	Flare Knock Out	Quench Water Tower	
North America	Pre-1980	500 - 1000K mta	Yes	Flare Knock Out (collection from various areas)	Quench Water Tower	
North America	Pre-1980	500 - 1000K mta	Yes		Quench	
Europe	Pre-1980	500 - 1000K mta	No			
Asia	2000-2010	500 - 1000K mta	Y	Steam Cracker and co-located derivative unit	Gasoline Fractionator	Fouling
North America	1981-2000	>1000K mta	N			

Location	Age/ Vint-age of Plant	Size of Plant	Flare drum knock-out liquids			
			Do you process ?	Specify origin if applicable	Where does it enter the process?	Give a brief description of any problems caused by the stream
North America	Pre-1980	500 - 1000K mta				
Asia	1981-2000	500 - 1000K mta	Yes			
South America	2000-2010	500 - 1000K mta				
South America	2000-2010	500 - 1000K mta				
North America	Pre-1980	>1000K mta	Yes	ISBL	Quench	Ph
North America	Pre-1980	>1000K mta				
North America	Pre-1980	>1000K mta				
North America	Pre-1980	>1000K mta	Yes		Primary Fractionation Area	No major issues
North America	Pre-1980	>1000K mta				
North America	Pre-1980	>1000K mta				
North America	Pre-1980	>1000K mta				
North America	1981-2000	>1000K mta	yes		recycle back to quench/compe rssion	Unknown
North America	Pre-1980	>1000K mta				
North America	Pre-1980	>1000K mta				

Location	Age/ Vint-age of Plant	Size of Plant	Flare drum knock-out liquids			
			Do you process ?	Specify origin if applicable	Where does it enter the process?	Give a brief description of any problems caused by the stream
North America	Pre-1980	>1000K mta	Yes	Flare drum which supports the plant.	Quench Tower	On the rare occasion that methanol ends up in the flare drum, the amount returned to the Quench Tower has caused an emulsion in the quench water.
North America	Pre-1980	>1000K mta				
North America	1981-2000	>1000K mta				
North America	Pre-1980	500 - 1000K mta				
Europe	1981-2000	<500K mta	Yes	entrainments , condensed liquids from cracker only	to slops recycled to the quench oil column	None
South America	1981-2000	500 - 1000K mta	No			
Europe	1981-2000	<500K mta				
South America	1981-2000	500 - 1000K mta	Yes.		Quech water tower or spent caustic tank	water-gasoline interface disturbance
South America	Pre-1980	500 - 1000K mta				
South America	Pre-1980	500 - 1000K mta	No			

Location	Age/ Vint-age of Plant	Size of Plant	Flare drum knock-out liquids			
			Do you process ?	Specify origin if applicable	Where does it enter the process?	Give a brief description of any problems caused by the stream
South America	1981-2000	500 - 1000K mta	Yes.	internal knock-out drum	quech water tower	water-gasoline interface disturbance
Europe	Pre-1980	500 - 1000K mta	Yes		QWT	pH-Fluctuation; Problems in QWS separation process
Europe	Pre-1980	500 - 1000K mta				
Europe	Pre-1980	<500K mta	Yes		Quenchoil tower via Pygas slop	
Asia	1981-2000	500 - 1000K mta	No			
Asia	1981-2000	500 - 1000K mta				
North America	1981-2000	>1000K mta	No			
Asia	1981-2000	>1000K mta				
Asia	1981-2000	500 - 1000K mta				
Asia	1981-2000	500 - 1000K mta				
Asia	1981-2000	500 - 1000K mta	No			
Asia	1981-2000	500 - 1000K mta				

Location	Age/ Vint-age of Plant	Size of Plant	Flare drum knock-out liquids			
			Do you process ?	Specify origin if applicable	Where does it enter the process?	Give a brief description of any problems caused by the stream
North America	Pre-1980	500 - 1000K mta	No			
North America	1981-2000	500 - 1000K mta	Yes	Knock-out on hot flare header	Quench tower	
North America	Pre-1980	<500K mta	No			
Asia	2000-2010	<500K mta	No			
Asia	1981-2000	500 - 1000K mta				
Asia	1981-2000	500 - 1000K mta				
Asia	2000-2010	<500K mta				
North America	Pre-1980	500 - 1000K mta				
Asia	2000-2010	<500K mta				

Table 32. Q4-2 Flare Gas Recovery

Location	Age/ Vintage of Plant	Size of Plant	Flare Gas recovery			
			Do you process?	Specify origin if applicable	Where does it enter the process?	Give a brief description of any problems caused by the stream
North America	1981-2000	>1000K mta	No			
Europe	2000-2010	500 - 1000K mta	No			
North America	1981-2000	>1000K mta	No			
North America	Pre-1980	500 - 1000K mta	No			
North America	Pre-1980	500 - 1000K mta	No			
Europe	Pre-1980	500 - 1000K mta	Yes	Flare	burners firebox	change of caloric value
Asia	2000-2010	500 - 1000K mta	N			
North America	1981-2000	>1000K mta	N			
North America	Pre-1980	500 - 1000K mta				
Asia	1981-2000	500 - 1000K mta	No			
South America	2000-2010	500 - 1000K mta				
South America	2000-2010	500 - 1000K mta				
North America	Pre-1980	>1000K mta	No			
North America	Pre-1980	>1000K mta				
North America	Pre-1980	>1000K mta				
North America	Pre-1980	>1000K mta	None			
North America	Pre-1980	>1000K mta				

Location	Age/ Vintage of Plant	Size of Plant	Flare Gas recovery			
			Do you process?	Specify origin if applicable	Where does it enter the process?	Give a brief description of any problems caused by the stream
North America	Pre-1980	>1000K mta				
North America	Pre-1980	>1000K mta				
North America	1981-2000	>1000K mta	no			
North America	Pre-1980	>1000K mta				
North America	Pre-1980	>1000K mta				
North America	Pre-1980	>1000K mta	Yes	Flare system.	Charge gas compressor suction.	No problems of concern.
North America	Pre-1980	>1000K mta				
North America	1981-2000	>1000K mta				
North America	Pre-1980	500 - 1000K mta				
Europe	1981-2000	<500K mta	Yes	vents from cracker only	To the quench tower	introduction of O2 traces into cracked gas
South America	1981-2000	500 - 1000K mta	No			
Europe	1981-2000	<500K mta				
South America	1981-2000	500 - 1000K mta	No			
South America	Pre-1980	500 - 1000K mta				
South America	Pre-1980	500 - 1000K mta	No			

Location	Age/ Vintage of Plant	Size of Plant	Flare Gas recovery			
			Do you process?	Specify origin if applicable	Where does it enter the process?	Give a brief description of any problems caused by the stream
South America	1981-2000	500 - 1000K mta	Yes.	Flare gases from flare knock-out drum	Fuel gas system	Any problem reported.
Europe	Pre-1980	500 - 1000K mta	No			
Europe	Pre-1980	500 - 1000K mta				
Europe	Pre-1980	<500K mta	Yes	Flare system & gasoline tank vents	fuel gas system	benzene and butadiene content in fuelgas
Asia	1981-2000	500 - 1000K mta	No			
Asia	1981-2000	500 - 1000K mta				
North America	1981-2000	>1000K mta	No			
Asia	1981-2000	>1000K mta				
Asia	1981-2000	500 - 1000K mta				
Asia	1981-2000	500 - 1000K mta				
Asia	1981-2000	500 - 1000K mta	No			
Asia	1981-2000	500 - 1000K mta				
North America	Pre-1980	500 - 1000K mta	No			
North America	1981-2000	500 - 1000K mta	No			

Location	Age/ Vintage of Plant	Size of Plant	Flare Gas recovery			
			Do you process?	Specify origin if applicable	Where does it enter the process?	Give a brief description of any problems caused by the stream
North America	Pre-1980	<500K mta	No			
Asia	2000-2010	<500K mta	No			
Asia	1981-2000	500 - 1000K mta				
Asia	1981-2000	500 - 1000K mta				
Asia	2000-2010	<500K mta				
North America	Pre-1980	500 - 1000K mta				
Asia	2000-2010	<500K mta				

Table 33. Q4-3 Vents from Other Plants

Location	Age/ Vintage of Plant	Size of Plant	Vents from other plants			
			Do you process?	Specify origin if applicable	Where does it enter the process?	Give a brief description of any problems caused by the stream
North America	1981-2000	>1000K mta	No			
Europe	2000-2010	500 - 1000K mta	No			
North America	1981-2000	>1000K mta			Cracked Gas Compressor	

Location	Age/ Vintage of Plant	Size of Plant	Vents from other plants			
			Do you process?	Specify origin if applicable	Where does it enter the process?	Give a brief description of any problems caused by the stream
North America	Pre-1980	500 - 1000K mta	Yes	Poly Recycle	Cracked Gas Compressor	
North America	Pre-1980	500 - 1000K mta	Yes		Cracked Gas Compressor	
Europe	Pre-1980	500 - 1000K mta	No			
Asia	2000-2010	500 - 1000K mta	Y			
North America	1981-2000	>1000K mta	N			
North America	Pre-1980	500 - 1000K mta				
Asia	1981-2000	500 - 1000K mta	Polymer Plant		Quench Section	LDPE Off Gas
South America	2000-2010	500 - 1000K mta				
South America	2000-2010	500 - 1000K mta				
North America	Pre-1980	>1000K mta	Yes	Polymer Units	Compressors	Oxygen, Rates (cut feeds on high vent flow)
North America	Pre-1980	>1000K mta				
North America	Pre-1980	>1000K mta				
North America	Pre-1980	>1000K mta	None			

Location	Age/ Vintage of Plant	Size of Plant	Vents from other plants			
			Do you process?	Specify origin if applicable	Where does it enter the process?	Give a brief description of any problems caused by the stream
North America	Pre-1980	>1000K mta				
North America	Pre-1980	>1000K mta				
North America	Pre-1980	>1000K mta				
North America	1981-2000	>1000K mta	no			
North America	Pre-1980	>1000K mta				
North America	Pre-1980	>1000K mta				
North America	Pre-1980	>1000K mta	Crude C4's.	A variety of external sources.	Either the Depropanizer or Debutanizer, but typically the Debutanizer is used.	No problems of concern.
North America	Pre-1980	>1000K mta				
North America	1981-2000	>1000K mta				
North America	Pre-1980	500 - 1000K mta	Yes	Polymer plant recycle streams	Quench system and/or downstream distillationtowers	Vinyl acetate can cause issues with caustic tower fouling if not treated appropriately
Europe	1981-2000	<500K mta	No			
South America	1981-2000	500 - 1000K mta				

Location	Age/ Vintage of Plant	Size of Plant	Vents from other plants			
			Do you process?	Specify origin if applicable	Where does it enter the process?	Give a brief description of any problems caused by the stream
Europe	1981-2000	<500K mta				
South America	1981-2000	500 - 1000K mta	No			
South America	Pre-1980	500 - 1000K mta	Yes	polymers plant offgas recycle	chilling train upstream caustic tower	increase polymer formation in caustic tower
South America	Pre-1980	500 - 1000K mta	Yes	Polyethylene plant	Third stage of charge gas compressor	We believe these stream cause high fouling formation in caustic wash tower and spent caustic treatment unit.
South America	1981-2000	500 - 1000K mta	Yes.	Ethylene and propylene consumers.	Cracked gas compressor	Fouling in the cracked gas compressor.
Europe	Pre-1980	500 - 1000K mta	Yes	PE Bleeds	PGC	No Problems discovered
Europe	Pre-1980	500 - 1000K mta				
Europe	Pre-1980	<500K mta	HDPE offgas C2= rich/ HDPE offgas N2 rich	HDPE plant	C2= rich: Qenchwater tower liquid train/ N2 rich: firebox boiler	N2 rich: suddenly increasing heating value => CO

Location	Age/ Vintage of Plant	Size of Plant	Vents from other plants			
			Do you process?	Specify origin if applicable	Where does it enter the process?	Give a brief description of any problems caused by the stream
Asia	1981-2000	500 - 1000K mta	Yes			
Asia	1981-2000	500 - 1000K mta				
North America	1981-2000	>1000K mta	Yes	Waste gas from PSA hydrogen recovery	fuel to furnaces	None
Asia	1981-2000	>1000K mta				
Asia	1981-2000	500 - 1000K mta				
Asia	1981-2000	500 - 1000K mta				
Asia	1981-2000	500 - 1000K mta	Yes	PE Recycle	Cracked gas compressor	
Asia	1981-2000	500 - 1000K mta				
North America	Pre-1980	500 - 1000K mta	Yes	Polymer Purges	Demethanizer	Capacity Restrictions
North America	1981-2000	500 - 1000K mta	Yes	Polyethylene and polypropylene units at site	3rd or 4th charge gas suction drum	Suspicious of air/N2 contaminant but not proven
North America	Pre-1980	<500K mta	No			

Location	Age/ Vintage of Plant	Size of Plant	Vents from other plants			
			Do you process?	Specify origin if applicable	Where does it enter the process?	Give a brief description of any problems caused by the stream
Asia	2000-2010	<500K mta	No			
Asia	1981-2000	500 - 1000K mta				
Asia	1981-2000	500 - 1000K mta				
Asia	2000-2010	<500K mta				
North America	Pre-1980	500 - 1000K mta				
Asia	2000-2010	<500K mta				

Table 34. Q4-4 Refinery FCC or Coker Unit Dry Gas

Location	Age/ Vintage of Plant	Size of Plant	Refinery FCC or Coker Unit dry gas			
			Do you process?	Specify origin if applicable	Where does it enter the process?	Give a brief description of any problems caused by the stream
North America	1981-2000	>1000K mta	No			
Europe	2000-2010	500 - 1000K mta	No			

Location	Age/ Vintage of Plant	Size of Plant	Refinery FCC or Coker Unit dry gas			
			Do you process?	Specify origin if applicable	Where does it enter the process?	Give a brief description of any problems caused by the stream
North America	1981-2000	>1000K mta	No			
North America	Pre-1980	500 - 1000K mta	No			
North America	Pre-1980	500 - 1000K mta	No			
Europe	Pre-1980	500 - 1000K mta	No			
Asia	2000-2010	500 - 1000K mta				
North America	1981-2000	>1000K mta	Y	FCC, Coker	Compressor	Contaminants
North America	Pre-1980	500 - 1000K mta				
Asia	1981-2000	500 - 1000K mta				
South America	2000-2010	500 - 1000K mta				
South America	2000-2010	500 - 1000K mta				
North America	Pre-1980	>1000K mta	No			
North America	Pre-1980	>1000K mta				
North America	Pre-1980	>1000K mta				

Location	Age/ Vintage of Plant	Size of Plant	Refinery FCC or Coker Unit dry gas			
			Do you process?	Specify origin if applicable	Where does it enter the process?	Give a brief description of any problems caused by the stream
North America	Pre-1980	>1000K mta	None			
North America	Pre-1980	>1000K mta				
North America	Pre-1980	>1000K mta				
North America	Pre-1980	>1000K mta				
North America	1981-2000	>1000K mta	no			
North America	Pre-1980	>1000K mta				
North America	Pre-1980	>1000K mta				
North America	Pre-1980	>1000K mta	No			
North America	Pre-1980	>1000K mta				
North America	1981-2000	>1000K mta				
North America	Pre-1980	500 - 1000K mta				
Europe	1981-2000	<500K mta	Not anymore (was the case in the past)	FCC off-gas	charge gas compressor	Plugging of cold box with NOx
South America	1981-2000	500 - 1000K mta	Yes	Polyethylene Plant	Third stage of charge gas compressor	Spent caustic treatment unit
Europe	1981-2000	<500K mta				

Location	Age/ Vintage of Plant	Size of Plant	Refinery FCC or Coker Unit dry gas			
			Do you process?	Specify origin if applicable	Where does it enter the process?	Give a brief description of any problems caused by the stream
South America	1981-2000	500 - 1000K mta	No			
South America	Pre-1980	500 - 1000K mta	Yes	FCC and Coke Units from nearby refineries	A dedicated unit removes contaminants, NH3, NOx and separates a methane/h2 cut to fuel gas, a C2 cut to acetylene converter and a C3+ cut to depropanizer	Possible nitro guns accumulation in chilling trais and mercury carryoevr to chilling
South America	Pre-1980	500 - 1000K mta	No			
South America	1981-2000	500 - 1000K mta	No			
Europe	Pre-1980	500 - 1000K mta	No			
Europe	Pre-1980	500 - 1000K mta				
Europe	Pre-1980	<500K mta	FCC C3 / FCC off gas	BO Refinery Neustadt (FCC offgas)	FCC-C3 in C3 splitter/ FCC-offgas: chargegas compressor gas train	O2 entry, NO possible, but never measured

Location	Age/ Vintage of Plant	Size of Plant	Refinery FCC or Coker Unit dry gas			
			Do you process?	Specify origin if applicable	Where does it enter the process?	Give a brief description of any problems caused by the stream
Asia	1981-2000	500 - 1000K mta	No			
Asia	1981-2000	500 - 1000K mta				
North America	1981-2000	>1000K mta	No			
Asia	1981-2000	>1000K mta				
Asia	1981-2000	500 - 1000K mta				
Asia	1981-2000	500 - 1000K mta				
Asia	1981-2000	500 - 1000K mta	No			
Asia	1981-2000	500 - 1000K mta				
North America	Pre-1980	500 - 1000K mta	No			
North America	1981-2000	500 - 1000K mta	No			
North America	Pre-1980	<500K mta	No			
Asia	2000-2010	<500K mta	No			

Location	Age/ Vintage of Plant	Size of Plant	Refinery FCC or Coker Unit dry gas			
			Do you process?	Specify origin if applicable	Where does it enter the process?	Give a brief description of any problems caused by the stream
Asia	1981-2000	500 - 1000K mta				
Asia	1981-2000	500 - 1000K mta				
Asia	2000-2010	<500K mta				
North America	Pre-1980	500 - 1000K mta				
Asia	2000-2010	<500K mta				

Table 35. Q4-5 Wastewater/ NESHAPS or Spent Caustic Stripper Overhead

Location	Age/ Vintage of Plant	Size of Plant	Wastewater/ NESHAPS or spent caustic stripper overhead			
			Do you process?	Specify origin if applicable	Where does it enter the process?	Give a brief description of any problems caused by the stream
North America	1981-2000	>1000K mta	No			
Europe	2000-2010	500 - 1000K mta	No			
North America	1981-2000	>1000K mta	No			

Location	Age/ Vintage of Plant	Size of Plant	Wastewater/ NESHAPS or spent caustic stripper overhead			
			Do you process?	Specify origin if applicable	Where does it enter the process?	Give a brief description of any problems caused by the stream
North America	Pre-1980	500 - 1000K mta	No			
North America	Pre-1980	500 - 1000K mta	Yes	Waste Water Stripper	Quench	
Europe	Pre-1980	500 - 1000K mta	No			
Asia	2000-2010	500 - 1000K mta	N			
North America	1981-2000	>1000K mta	N			
North America	Pre-1980	500 - 1000K mta				
Asia	1981-2000	500 - 1000K mta				
South America	2000-2010	500 - 1000K mta				
South America	2000-2010	500 - 1000K mta				
North America	Pre-1980	>1000K mta	Yes	ISBL		
North America	Pre-1980	>1000K mta				
North America	Pre-1980	>1000K mta				

Location	Age/ Vintage of Plant	Size of Plant	Wastewater/ NESHAPS or spent caustic stripper overhead			
			Do you process?	Specify origin if applicable	Where does it enter the process?	Give a brief description of any problems caused by the stream
North America	Pre-1980	>1000K mta	Yes, condensed vapor		Returns to Hot Water Belt, Primary Fractionation	No major issues
North America	Pre-1980	>1000K mta				
North America	Pre-1980	>1000K mta				
North America	Pre-1980	>1000K mta				
North America	1981-2000	>1000K mta	no			
North America	Pre-1980	>1000K mta				
North America	Pre-1980	>1000K mta				
North America	Pre-1980	>1000K mta				
North America	Pre-1980	>1000K mta				
North America	1981-2000	>1000K mta				
North America	Pre-1980	500 - 1000K mta	Yes	spent caustic stripper overhead stream	Quench system	none
Europe	1981-2000	<500K mta	No			
South America	1981-2000	500 - 1000K mta				
Europe	1981-2000	<500K mta				

Location	Age/ Vintage of Plant	Size of Plant	Wastewater/ NESHAPS or spent caustic stripper overhead			
			Do you process?	Specify origin if applicable	Where does it enter the process?	Give a brief description of any problems caused by the stream
South America	1981-2000	500 - 1000K mta	Yes.	Spent caustic treatment unit.	Downstream deoiling vessel	No problems have been reported.
South America	Pre-1980	500 - 1000K mta				
South America	Pre-1980	500 - 1000K mta	No			
South America	1981-2000	500 - 1000K mta	Yes.	Spent caustic treatment unit	Deoiling vessel, downstream of the caustic tower.	Any problem reported.
Europe	Pre-1980	500 - 1000K mta	No			
Europe	Pre-1980	500 - 1000K mta				
Europe	Pre-1980	<500K mta	spent caustic stripper overhead/ wate water stripper overhead	spent caustic stripper/ waste water stripper overhead	scstr. :fire box superheater/ wwsttr. Quenchwater tower	sulphur deposits in economizer section (Claus reaction)
Asia	1981-2000	500 - 1000K mta	No			
Asia	1981-2000	500 - 1000K mta				

Location	Age/ Vintage of Plant	Size of Plant	Wastewater/ NESHAPS or spent caustic stripper overhead			
			Do you process?	Specify origin if applicable	Where does it enter the process?	Give a brief description of any problems caused by the stream
North America	1981-2000	>1000K mta	No			
Asia	1981-2000	>1000K mta				
Asia	1981-2000	500 - 1000K mta				
Asia	1981-2000	500 - 1000K mta				
Asia	1981-2000	500 - 1000K mta	No			
Asia	1981-2000	500 - 1000K mta				
North America	Pre-1980	500 - 1000K mta	No			
North America	1981-2000	500 - 1000K mta	No			
North America	Pre-1980	<500K mta	No			
Asia	2000-2010	<500K mta	No			
Asia	1981-2000	500 - 1000K mta				
Asia	1981-2000	500 - 1000K mta				

Location	Age/ Vintage of Plant	Size of Plant	Wastewater/ NESHAPS or spent caustic stripper overhead			
			Do you process?	Specify origin if applicable	Where does it enter the process?	Give a brief description of any problems caused by the stream
Asia	2000-2010	<500K mta				
North America	Pre-1980	500 - 1000K mta				
Asia	2000-2010	<500K mta				

Table 36. Q4-6 Raw Pyrolysis Naphtha

Location	Age/ Vintage of Plant	Size of Plant	Raw pyro naphtha (dripolenes, raw pygas, debutanizer bottoms)			
			Do you process?	Specify origin if applicable	Where does it enter the process?	Give a brief description of any problems caused by the stream
North America	1981-2000	>1000K mta	No			
Europe	2000-2010	500 - 1000K mta	No			
North America	1981-2000	>1000K mta	No			
North America	Pre-1980	500 - 1000K mta	No			
North America	Pre-1980	500 - 1000K mta	No			
Europe	Pre-1980	500 - 1000K mta	No			

Location	Age/ Vintage of Plant	Size of Plant	Raw pyro naphtha (dripolenes, raw pygas, debutanizer bottoms)			
			Do you process?	Specify origin if applicable	Where does it enter the process?	Give a brief description of any problems caused by the stream
Asia	2000-2010	500 - 1000K mta	N			
North America	1981-2000	>1000K mta	N			
North America	Pre-1980	500 - 1000K mta				
Asia	1981-2000	500 - 1000K mta				
South America	2000-2010	500 - 1000K mta				
South America	2000-2010	500 - 1000K mta				
North America	Pre-1980	>1000K mta	Yes	ISBL	Quench, Compressor	Low Solvency
North America	Pre-1980	>1000K mta				
North America	Pre-1980	>1000K mta				
North America	Pre-1980	>1000K mta	Yes when applicable		Aromatics Feedstock Tank	No major issues
North America	Pre-1980	>1000K mta				
North America	Pre-1980	>1000K mta				
North America	Pre-1980	>1000K mta				
North America	1981-2000	>1000K mta	No			we do recycle a slipsteam of Debut btms to quench for gravity

Location	Age/ Vintage of Plant	Size of Plant	Raw pyro naphtha (dripolenes, raw pygas, debutanizer bottoms)			
			Do you process?	Specify origin if applicable	Where does it enter the process?	Give a brief description of any problems caused by the stream
						purposes - internal stream
North America	Pre-1980	>1000K mta				
North America	Pre-1980	>1000K mta				
North America	Pre-1980	>1000K mta	Yes	A variety of external sources.	Preferred: Gasoline Fractionator w/ Quench Oil (mid-tower reflux); It is also possible to put it into Gasoline Fractionator reflux (1st tray).	Free radicals can cause fouling problems if fed to the 1st tray, thus making mid-tower reflux preferred. On rare occasions, water coming in with dripolene stream can cause problems (water typically left over in barges or tanks).
North America	Pre-1980	>1000K mta				
North America	1981-2000	>1000K mta				
North America	Pre-1980	500 - 1000K mta				
Europe	1981-2000	<500K mta	No			
South America	1981-2000	500 - 1000K mta				
Europe	1981-2000	<500K mta				

Location	Age/ Vintage of Plant	Size of Plant	Raw pyro naphtha (dripolenes, raw pygas, debutanizer bottoms)			
			Do you process?	Specify origin if applicable	Where does it enter the process?	Give a brief description of any problems caused by the stream
South America	1981-2000	500 - 1000K mta	No			
South America	Pre-1980	500 - 1000K mta				
South America	Pre-1980	500 - 1000K mta	No			
South America	1981-2000	500 - 1000K mta				
Europe	Pre-1980	500 - 1000K mta				
Europe	Pre-1980	500 - 1000K mta				
Europe	Pre-1980	<500K mta	raw pygas	cracker	pygas hydrogenation => benzene cut column, light pygas blended to butane feed	high sulfur content lead to offspec of benzene cut
Asia	1981-2000	500 - 1000K mta	No			
Asia	1981-2000	500 - 1000K mta				
North America	1981-2000	>1000K mta	Yes	RPG from another one of our facilities	Fractionation units	No

Location	Age/ Vintage of Plant	Size of Plant	Raw pyro naphtha (dripolenes, raw pygas, debutanizer bottoms)			
			Do you process?	Specify origin if applicable	Where does it enter the process?	Give a brief description of any problems caused by the stream
Asia	1981-2000	>1000K mta				
Asia	1981-2000	500 - 1000K mta				
Asia	1981-2000	500 - 1000K mta				
Asia	1981-2000	500 - 1000K mta	No			
Asia	1981-2000	500 - 1000K mta				
North America	Pre-1980	500 - 1000K mta	No			
North America	1981-2000	500 - 1000K mta	Yes	Debutanizer bottoms to produce light pygas cut	Gasoline stripper at 2nd stage charge gas	None
North America	Pre-1980	<500K mta	No			
Asia	2000-2010	<500K mta				
Asia	1981-2000	500 - 1000K mta				
Asia	1981-2000	500 - 1000K mta				

Location	Age/ Vintage of Plant	Size of Plant	Raw pyro naphtha (dripolenes, raw pygas, debutanizer bottoms)			
			Do you process?	Specify origin if applicable	Where does it enter the process?	Give a brief description of any problems caused by the stream
Asia	2000-2010	<500K mta				
North America	Pre-1980	500 - 1000K mta				
Asia	2000-2010	<500K mta				

Table 37. Q4-7 Other

Location	Age/ Vintage of Plant	Size of Plant	Other:			
			Do you process?	Specify origin if applicable	Where does it enter the process?	Give a brief description of any problems caused by the stream
North America	1981-2000	>1000K mta				
Europe	2000-2010	500 - 1000K mta				
North America	1981-2000	>1000K mta				
North America	Pre-1980	500 - 1000K mta				
North America	Pre-1980	500 - 1000K mta	No			
Europe	Prc-1980	500 - 1000K mta				

Location	Age/ Vintage of Plant	Size of Plant	Other:			
			Do you process?	Specify origin if applicable	Where does it enter the process?	Give a brief description of any problems caused by the stream
Asia	2000-2010	500 - 1000K mta	N			
North America	1981-2000	>1000K mta	N			
North America	Pre-1980	500 - 1000K mta				
Asia	1981-2000	500 - 1000K mta				
South America	2000-2010	500 - 1000K mta				
South America	2000-2010	500 - 1000K mta				
North America	Pre-1980	>1000K mta				
North America	Pre-1980	>1000K mta				
North America	Pre-1980	>1000K mta				
North America	Pre-1980	>1000K mta	Refinery Grade Propylene	Midstream Company	Process Gas Compression Area	No majors quality issues
North America	Pre-1980	>1000K mta				
North America	Pre-1980	>1000K mta				
North America	Pre-1980	>1000K mta				
North America	1981-2000	>1000K mta				

Location	Age/ Vintage of Plant	Size of Plant	Other:			
			Do you process?	Specify origin if applicable	Where does it enter the process?	Give a brief description of any problems caused by the stream
North America	Pre-1980	>1000K mta				
North America	Pre-1980	>1000K mta				
North America	Pre-1980	>1000K mta				
North America	Pre-1980	>1000K mta				
North America	1981-2000	>1000K mta				
North America	Pre-1980	500 - 1000K mta				
Europe	1981-2000	<500K mta	Bleed from polyolefin plants	PP and PE plants	Quench Tower	Emulsions in quench tower (oil entrainments)
South America	1981-2000	500 - 1000K mta				
Europe	1981-2000	<500K mta				
South America	1981-2000	500 - 1000K mta				
South America	Pre-1980	500 - 1000K mta				
South America	Pre-1980	500 - 1000K mta	Fresh Ethane	Natural Gas	Pyrolisis Furnaces	Little bite high TLE fouling rate due the heavy tail present in the ethane.

Location	Age/ Vintage of Plant	Size of Plant	Other:			
			Do you process?	Specify origin if applicable	Where does it enter the process?	Give a brief description of any problems caused by the stream
South America	1981-2000	500 - 1000K mta				
Europe	Pre-1980	500 - 1000K mta	No			
Europe	Pre-1980	500 - 1000K mta				
Europe	Pre-1980	<500K mta	FCC-C4		C4 hydrogenation unit => Butane feed to furnace	
Asia	1981-2000	500 - 1000K mta	Yes	C4 Raffinate from OCT, C5~C8 Raffinate from Aromatic (BTX) plant	Mixed with Naphtha in Furnace Feed	In case of high MTBE in C4 Raffinate produce MeOH lead to Quench Water Blow down. Plant 1 is equipped with Methanol Adsorber at upstream of Propylene Tower.
Asia	1981-2000	500 - 1000K mta				
North America	1981-2000	>1000K mta				
Asia	1981-2000	>1000K mta				

Location	Age/ Vintage of Plant	Size of Plant	Other:			
			Do you process?	Specify origin if applicable	Where does it enter the process?	Give a brief description of any problems caused by the stream
Asia	1981-2000	500 - 1000K mta				
Asia	1981-2000	500 - 1000K mta				
Asia	1981-2000	500 - 1000K mta				
Asia	1981-2000	500 - 1000K mta				
North America	Pre-1980	500 - 1000K mta	No			
North America	1981-2000	500 - 1000K mta				
North America	Pre-1980	<500K mta				
Asia	2000-2010	<500K mta				
Asia	1981-2000	500 - 1000K mta				
Asia	1981-2000	500 - 1000K mta				
Asia	2000-2010	<500K mta				

Location	Age/ Vintage of Plant	Size of Plant	Other:			
			Do you process?	Specify origin if applicable	Where does it enter the process?	Give a brief description of any problems caused by the stream
North America	Pre-1980	500 - 1000K mta				
Asia	2000-2010	<500K mta				

AIChE Paper Number 36e

1st Stage Pygas Nickel Based Catalyst Poisons Study

Anke Derking

Senior Technologist Aromatics

Shell Technology Centre Amsterdam

Dirk Lose

Technical Service Manager, Europe

CRI Catalyst Leuna GmbH

Ernesto J Angueira

Technical Service Manager, Americas

CRI Catalyst Company LP

Prepared for Presentation at the 2017 Spring National Meeting
San Antonio, TX, March 26 – 30, 2017

1st Stage Pygas Nickel Based Catalyst Poisons Study

Anke Derking

Senior Technologist Aromatics

Shell Technology Centre Amsterdam

Dirk Lose

Technical Service Manager, Europe

CRI Catalyst Leuna GmbH

Ernesto J Angueira

Technical Service Manager, Americas

CRI Catalyst Company LP

Abstract:

Chemical companies are continuously exploring opportunities to run their units more efficiently and limit the technical risks. A key risk for raw pyrolysis gasoline processing units are impurities present in the raw feedstock, which may adversely influence performance characteristics of its hydrogenation catalysts. A key aspect to Pygas unit operation is therefore a better understanding of the catalytic robustness and tolerance to the impurities typically present in crude Pygas feedstock.

This paper will focus on some common poisons found in 1st Stage pygas processing units and their impact on nickel based catalyst. The contaminants include carbon disulfide (CS_2), and common forms of organic nitrogen, silicon and lead. The paper studies the short term effect of these poisons on the catalyst performance.

A key risk for raw pyrolysis gasoline processing units are impurities present in the raw feedstock, which may adversely influence performance characteristics of its hydrogenation catalysts. A key aspect to Pygas unit operation is therefore a better understanding of the catalytic robustness and tolerance to the impurities typically present in crude Pygas feedstock.

There are multiple arrangements for a Pygas unit, but pygas hydrotreating typically involves two stages: A 1st stage trickle phase operating reactor – diolefin/alkenyl-aromatic removal followed by a 2nd stage vapor phase reactor for olefin and sulfur removal. Both stages prepare the feed for aromatic extraction and/or motor gasoline pool addition. The presented study is focused on 1st stage pygas hydrogenation.

An important aspect in pygas hydrotreating is loss of catalytic hydrogenation activity due to the presence of permanent (poisons) or temporary (inhibitor) contaminants, which accumulate on the catalyst surface blocking the active sites. The presented experimental program with CRI Nickel catalyst pygas hydrogenating catalyst was launched to better understand the impact of common pygas contaminants on the catalyst performance.

The series of tests aimed at providing insights on impact to catalyst lifetime and subsequent regeneration needed to restore the catalyst performance. The following poisons were evaluated: nitrogen (N), lead (Pb), sulfur (S) and Silicon (Si).

A bench scale testing unit was employed to determine the impact of various contaminants on catalyst activity. The unit was fully automated and utilizing a process controlled system. Liquid samples were taken with an automatic sampler. The testing unit consists of two independent reactor systems allowing for two tests to be performed in parallel. Two separate feed tanks allowed for easy switching between straight pygas and pygas spiked with the desired catalyst poison.

The adjusted testing conditions mimic the typical conditions of a first stage pygas unit, but no recycle flow of hydrogenated product was applied. The adjusted weighted average bed temperature (WABT) to achieve a defined start of run activity of fresh catalyst was kept unchanged after the switch to "poisoned" feed. At the end of each run initial conditions were adjusted to monitor the loss of activity (Point of return –POR). A hot hydrogen strip was performed to evaluate the regenerability of the catalyst after the exposure with contaminants.

The catalyst used for these experiments was a CRI first stage Ni catalyst. The shape of the catalyst was a 2.5mm trilobe extrudate. The catalyst was loaded in the presulfurized form and an in-situ reduction was performed at 750°F (400°C). To ensure a proper radial distribution and minimize wall effects, the catalyst was diluted with an inert material. A fresh catalyst sample was used for every test. The results of the poisoning experiments are benchmarked with a reference test of fresh catalyst without poisons.

The main criteria to evaluate the catalyst performance was the changes of hydrogenation activity of styrene. Liquid samples were analyzed to determine content of slipped contaminates. The spent catalyst samples were tested for the accumulation of poisons and coke formation.

The safety and health risks of the planned poisons were assessed prior to the beginning of the experiments. A risk assessment was conducted for every poison.

Impact of sulfur as Carbon disulfide

The first tests were performed with sulfur as CS_2 and the main results are depicted in Figure 1. After a startup period to adjust stable and reliable conditions the feed was switched to a feed with 20ppmw S as CS_2 corresponding to 24ppmw CS_2. The impact of the catalyst performance is pronounced. The conversion drops up to 75% of initial activity. The effect of CS_2 depends also on the concentration in the feed. After equilibrium is reached, the performance of the catalyst becomes more stable but steady deactivation can still be observed. A temperature increase (not shown in the figure) was required to match the activity in the reference experiment. The effect of CS_2 was found to be reversible. Switching to the standard feed resulted in almost complete recovery of the activity over time. After a hot hydrogen strip procedure was performed, the activity is completely restored.

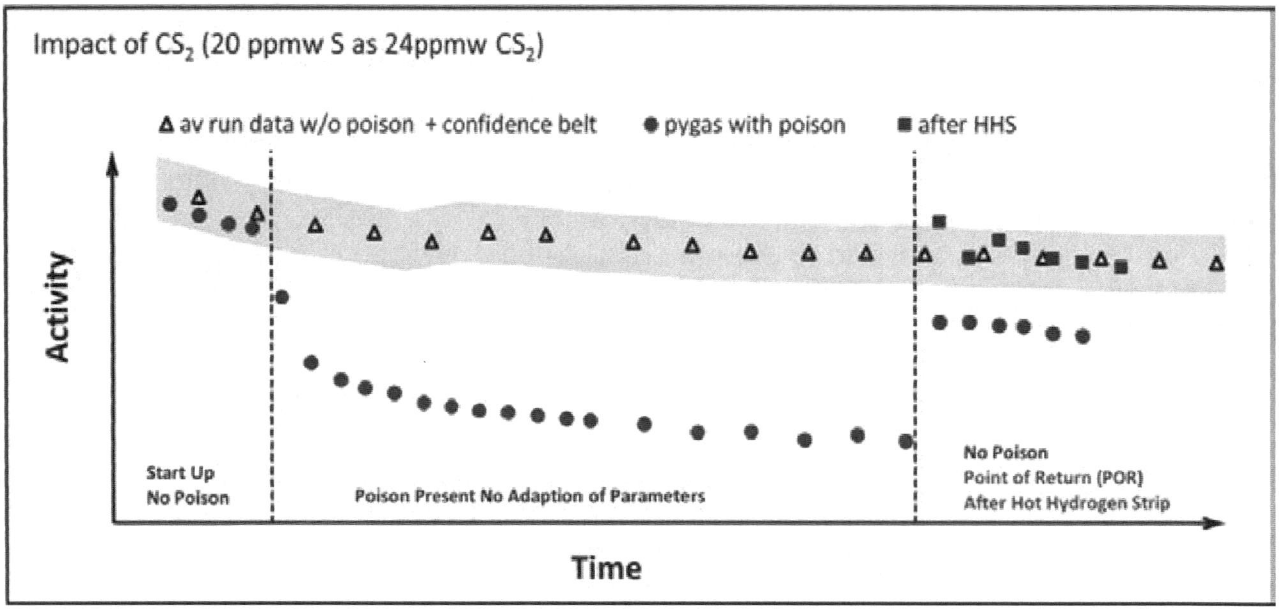

Figure 1. Impact of CS_2 (20 ppmw) on styrene conversion

It is demonstrated that the adsorption equilibrium of sulfur on the catalyst is established in the first hours of the run. During this period the sulfur partially adsorbs on the catalyst and partially passes the catalyst bed slipping with the product. After the adsorption equilibrium is established the complete amount of sulfur fed to the reactor was detected in the liquid product. Post-mortem analysis of the spent catalyst revealed that no additional sulfur accumulated on the catalyst and the residual sulfur found was in agreement with total sulfur amount slipping with the liquid product.

Key message: If a CS_2 spike can be expected, the drop of activity can be anticipated by a preventive increase of operating temperature.

Impact of (basic) nitrogen as primary amine

Basic amines are considered to be the most reactive components in the group of nitrogen containing compounds. Hexylamine as primary amine was tested and the effect differs from CS_2. The response of hexylamine is less strong and it requires more time to stabilize. Once the adsorption equilibrium is established, the performance of the catalyst stabilized. Monitoring of nitrogen content in the reactor product showed that the adsorption equilibrium is fully established very slowly after a long time on stream (TOS). During this period nitrogen is adsorbed on the fresh catalyst resulting in low concentrations of N in the liquid samples. Once adsorption is completed, all dosed nitrogen passes through the reactor and can be found in the liquid product. When dosing of the amine is stopped, hysteresis is observed. It takes about 30% longer to desorb the nitrogen. The initial performance is recovered after hot hydrogen strip as well as after resetting the reference conditions.

Key message: If a feed contains a high load of basic nitrogen, it is possible to counteract the deactivation with an increase of inlet temperature.

Impact of organic Silicon as siloxane

Hexamethyldisiloxane was tested because it is a good representative of the group of anti-foaming agents which are a typical source of silicon contamination of pygas feeds. Initial results show silicon to have a weaker poisoning effect on the presulfided nickel catalyst than expected. Due to the weak effect, a higher amount of Si was adopted to obtain a visible effect. The concentration could be increased up to 35 ppmw without a negative effect on the catalyst activity. Most of the Si slips through the first stage pygas reactor. Figure 2 shows the impact of Si on styrene conversion and after the start up with the un-doped feed the Si concentration was increased in steps from 1 to 35ppmw. Eventually the ability to perform a hot hydrogen strip after the treatment with Si was checked with no unexpected results about catalyst activity.

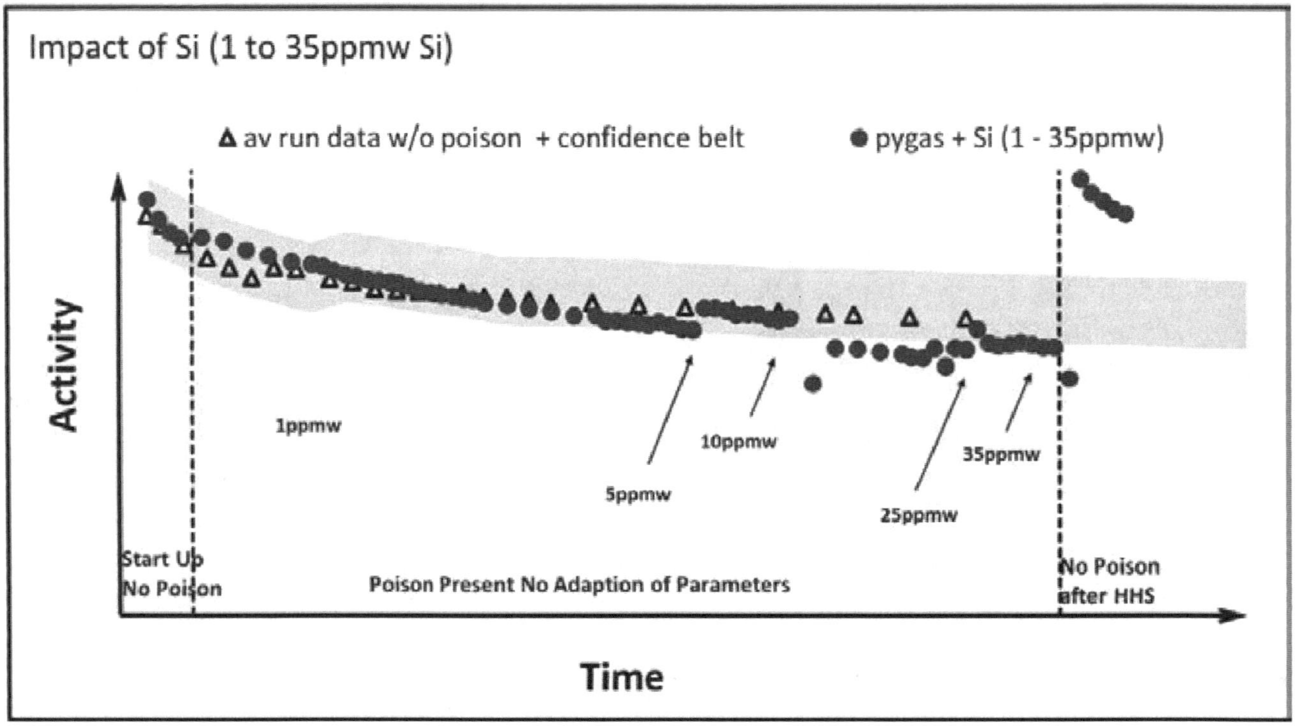

Figure 2. Impact of Si as siloxane (1-35ppmw Si) on styrene conversion

Key message: The effect of silicon as siloxane is low at first stage pygas conditions.

Impact of organic Lead

Lead is generally found in liquid feedstock's and natural gas liquid. It's rarely found as elementary lead, but most likely as organolead component such as Tetra Ethyl Lead (TEL) which ends up in the pygas fraction. TEL is considered to be a representative component for the experiments with other metallic contaminants and may possibly shed some light on the impact of other metals such as mercury which could not be tested due to potential hazards and manageability in the laboratory.

Based on first results, the amount of Pb was adopted to visualize the effect of lead more clearly and the results are shown in figure 3. After the start up with the fresh feed the Pb concentration was increased in steps from 0.1 to 1ppmw. The loss of activity builds up linearly with the amount of accumulated Pb. When the Pb amount is increased, the conversion drops below the reference data and continues to fall. The drop of activity is less severe when the Pb is constant at the level of 0.1 ppmw. It is possible to perform a hot hydrogen strip to restore the activity even when some accumulated lead is fixed on the active sites and was not be removed by regeneration. Spent catalyst analysis confirmed some Pb remains on the catalyst and after the switch back to pygas, the lead amount in the liquid samples totally disappears, which means the accumulated Pb is fixed on the catalyst. The liquid product contains about 20% of the dosed lead. This was an intriguing outcome about how the Pb is linked to the catalyst before or after the HHS. A possible theory is migration of the Pb to the Ni core during the HHS, but further investigation is needed.

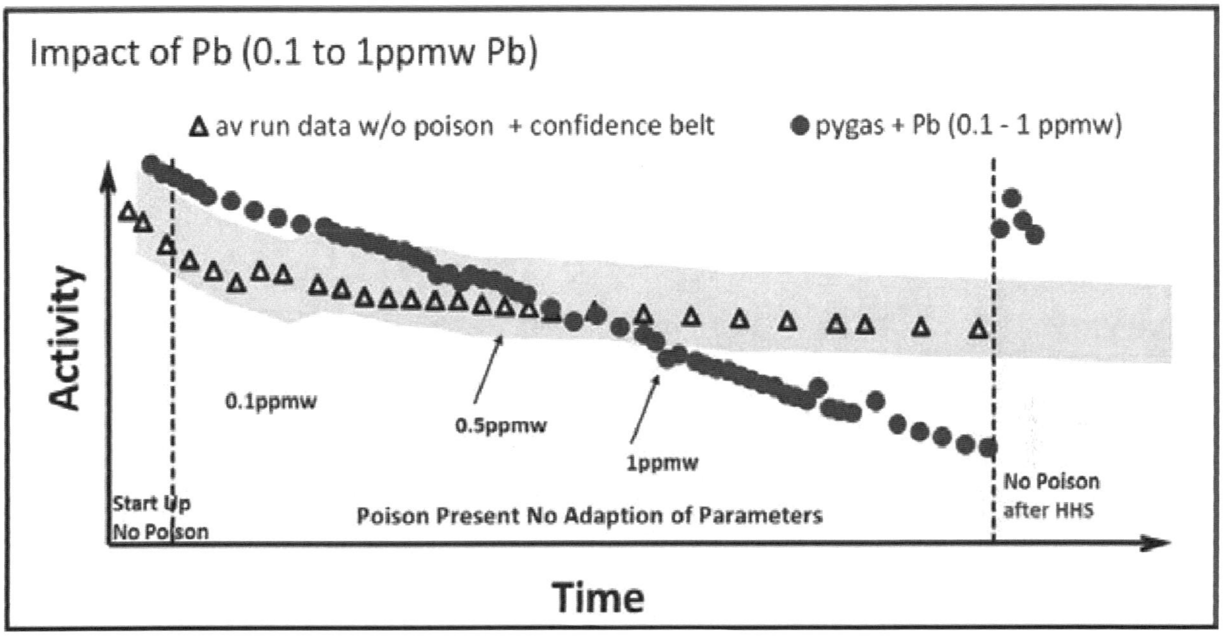

Figure 3. Impact of Pd as tetraethyl lead (0.1-1ppmw Pd) on styrene conversion

Key message: The tests revealed a progressive poisoning of the catalyst. The activity can be restored but poison remains on the catalyst.

Conclusion

The impact of several poisons on a typical sulfided nickel catalyst used in the 1st stage pygas hydrogenation system was evaluated. The selection of poisons for the study was driven by relevance or known incidents in the industry, availability of a representative component and related risk assessments.

This experimental program aimed at solidifying up the maximum load values for individual contaminants by dosing representative compounds to the pygas feed and monitoring catalyst performance as function of contaminant concentrations in the feed over time. A group of representative compounds of sulfur, nitrogen (basic), Pb, Si, and others were tested. The main areas studied in the testing include the effect on the catalyst activity, the effectiveness of hot hydrogen stripping to restore the activity as well as the accumulation of the contaminants and carbon on the catalyst surface.

In all cases, most if not all of the original activity was recovered by performing hot hydrogen stripping. The impact of the temporary poisons CS_2 and basic nitrogen can be mitigated by a proactive increase of temperature. Si as siloxane did not deactivate the catalyst and lead showed a linear accumulation on the catalyst leading to a shortening of cycle length over time.

AIChE Paper Number: 36f

Carbon Disulfide Removal to Help Meet New Specifications for Petrochemical Grade Naphtha

Jay Gorawara
Fellow R&D
Honeywell UOP

Stephen Caskey
Lead R&D Scientist
Honeywell UOP

John Griffiths
Senior Manager ES&E Innovation
Honeywell UOP

Prepared for Presentation at the 2017 Spring National Meeting
San Antonio, TX, March 26 – 30, 2017

Carbon Disulfide Removal to Help Meet New Specifications for Petrochemical Grade Naphtha

Jay Gorawara
Fellow R&D
Honeywell UOP

Stephen Caskey
Lead R&D Scientist
Honeywell UOP

John Griffiths
Senior Manager ES&E Innovation
Honeywell UOP

Abstract:
Ethylene producers have linked carbon disulfide (CS_2) in petrochemical grade naphtha (PCN) to corrosion in locations downstream of the cracker and to deterioration of pygas hydrogenation catalyst activity. This has led some producers to implement very low specifications (ppm levels) of CS_2 in petrochemical grade naphtha. Other customers have also indicated CS_2 as a primary contaminant particularly in some shale derived feedstocks. Thus CS_2 contamination can effectively limit feedstock flexibility and associated costs. Adsorptive removal of CS_2 represents an economically attractive solution.

To help meet this growing need, Honeywell UOP have developed a new adsorbent, MOLSIV™ CS2-100, for CS_2 removal from petrochemical grade naphtha/NGL. The new adsorbent can remove CS_2 efficiently thereby upgrading PCN/NGL feeds. Extensive pilot plant testing with a variety of commercial feeds has been conducted to demonstrate technical and economic viability of an adsorptive solution. Discussion will focus on technical options to reduce CS_2 in petrochemical grade naphtha, including Honeywell UOP's technology offering using the CS2-100 molecular sieve adsorbent that can selectively remove CS_2 from straight run naphtha/NGL streams to help meet the needs of PCN suppliers and ethylene producers.

Introduction

While ethylene producers have recognized carbon disulfide (CS_2) as a contaminant of concern for a number of years,[1] awareness of the impact of CS_2 has risen in recent years.[2,3] Restriction on CS_2 dates back to 2014 when the naphtha industry agreed to report CS_2 level – particularly influenced by Japanese isoprene producers.[4] In addition, other polymer producers, outside of the isoprene market, have recognized CS_2 as a catalyst poison and precursor of polymer chain defects.

CS_2 passes through the cracking furnaces largely unaffected. Since the standard caustic wash after the cracking furnace does not remove carbon disulfide, it concentrates in the C5+ pyrolysis gasoline (pygas) and particularly in the C5 and C6 hydrocarbon fractions due to its low boiling point.

After cracking, there are a few further processing routes in an ethylene plant:

- Extract C5's from pygas (isoprene extraction unit)
- Hydrogenate the C5s for recycle cracking
- Direct C5+ to pygas hydrotreater

The isoprene extraction and pygas hydrogenation units have experienced problems related to CS_2 in terms of corrosion and catalyst deactivation, respectively.[2] In order to maintain acceptable catalyst performance, the expensive catalyst needs to be replaced at higher rates when CS_2 is present.

With unprecedented global ethylene capacity expansion currently underway and the presence of high levels of CS_2 reported in some US shale derived natural gas liquids, the topic of CS_2 removal is especially relevant to ethylene producers. UOP's discussions with ethylene producers indicate that one of the major pain points is pygas hydrogenation catalyst deactivation leading to reduced activity and more frequent change outs/regenerations. The result is a high operational cost, which the petrochemical industry is no longer willing to overlook. Although some new CS_2 tolerant pygas hydrogenation catalyst solutions have been developed, these only soften the pain somewhat, not eliminating it.[2] Another relates to severe CS_2 related corrosion problems in the isoprene fraction for some ethylene producers. In shale derived natural gas liquids, CS_2 has become a growing concern. Some natural gas condensate users are evaluating low cost alternatives to hydrotreating to meet gasoline sulfur specs where CS_2 contamination can present a challenge.

Carbon disulfide is both found in hydrocarbon reserves and may be produced during some refinery reactions, Figure 1. In the majority of these, hydrogen sulfide is integral to the reaction, which suggests that in both natural sources and reducing atmospheres, those feedstocks rich in H_2S will likely be rich in CS_2. Given the stability of CS_2 at high temperatures, it is relatively difficult to convert to a more easily extracted molecule once formed.

$2H_2S + CH_4 \rightarrow CS_2 + 4H_2$	$\Delta H_R = 231.9 kJ/mol$
$COS + H_2S \rightarrow CS_2 + H_2O$	$\Delta H_R = 34.2 kJ/mol$
$CH_3SH + H_2S \rightarrow CS_2 + 3H_2$	$\Delta H_R = 159.8 kJ/mol$
$CH_4 + 4S \rightarrow CS_2 + 2H_2S$	$\Delta H_R = 1.9 kJ/mol$
$2SO_2 + 3C \rightarrow CS_2 + 2CO_2$	$\Delta H_R = 77.9 kJ/mol$

Figure 1: Reaction Pathways for CS_2 production[5]

Carbon disulfide levels vary greatly globally and levels are highly dependent on the hydrocarbon source. The carbon disulfide content of selected literature and UOP analyzed naphtha samples is reported in Table 1 below. The quantity of CS_2 varies widely, ranging from <1 ppm to 1800 ppm. Some shale derived sources of light naphtha have high CS_2 content. One possible explanation for this difference is that CS_2 may be a constituent of shale fracking solvent for enhanced oil recovery[6], this may enhance the level of CS_2 in the resultant liquids.

Table 1: Globally Reported CS_2 Content Ranges in Selected Naphtha Sample

	Glen Davis Shale (Australia)[7]	Mid-East Condensate	Americas Condensate	Eagle Ford Shale Liquid
CS_2 content (wppm)	300 – 1800	<1 – 350	<1 – 350	50 – 250

While traditionally classed with other sulfur species, the growing concern about CS_2 has led some industry consortiums to undertake detailed analysis of the naphtha, Figure 2. It is estimated from this data that the average level of carbon disulfide in petrochemical grade naphtha in the Far East region, those which exceed 1 ppmw, is about 3.6 ppmw. Therefore, in order to meet a 1 ppmw specification, many suppliers of petrochemical grade naphtha would be required to remove between 50% and 80% of the CS_2 prior to shipping. However, for certain sources the levels of CS_2 removal exceed 95%.

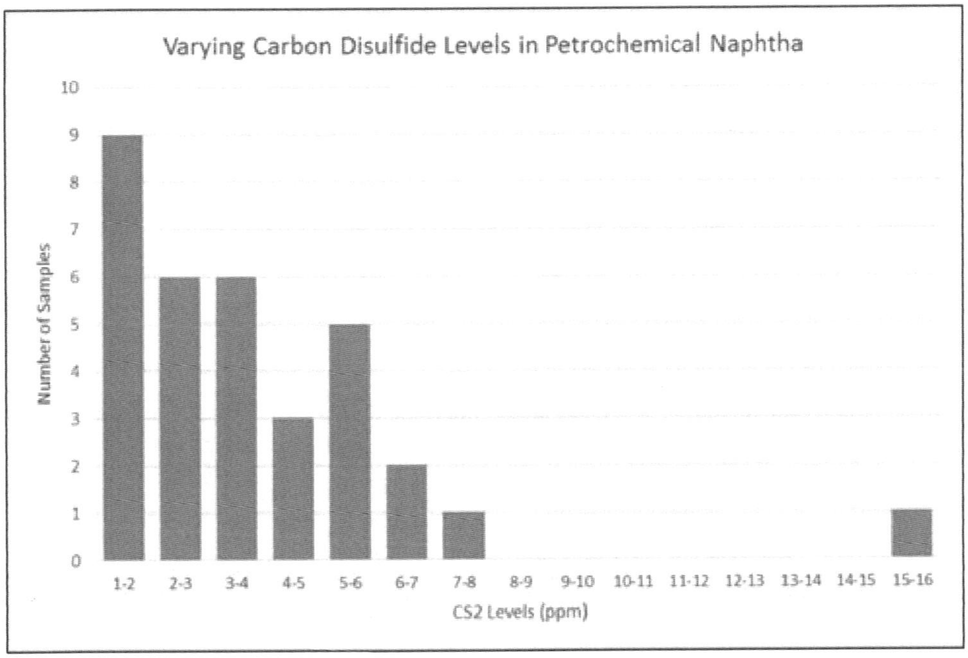

Figure 2: Carbon Disulfide Levels in Petrochemical Naphtha in the Far East[3]

CS_2 Removal from Petrochemical Naphtha Technology Options

Currently available and demonstrated options for removal of CS_2 from PCN include hydrotreating and distillation. UOP can provide the technologies for either of these options.

- Although a typical naphtha hydrotreater (NHT) unit can achieve ppm CS_2 levels, hydrotreating is capital and energy intensive. In the process of removing CS_2, standard hydrotreating will also remove most of the other contaminants such as oxygenates and organic sulfur species from PCN. In addition to increased hydrogen consumption, which is often in short supply in a refinery, removal of some of the organic sulfur species may not be necessary, or desired, for steam cracking operations.
- A fractionation separation is another possible option to remove CS_2 into a narrow C5 fraction which can be further processed in a refinery complex, while the other hydrocarbon components in the petrochemical naphtha (which are low in CS_2) are used for petrochemical naphtha (Figure 3). Carbon disulfide has a boiling point between isopentane and hexane. A reduction of CS_2 in the product naphtha stream also entails production of another hydrocarbon stream that will need to have an outlet elsewhere in the refinery complex. A divided wall distillation column and high efficiency trays may help reduce the energy requirements compared to a two columns or standard fractionation with a side draw. However the fractionation approach is particularly energy intensive, and unless there is an economically

viable outlet for the CS_2 contaminated C5 hydrocarbon fraction stream there will be a loss of the valuable C5 fraction from PCN which co-boils with CS_2.

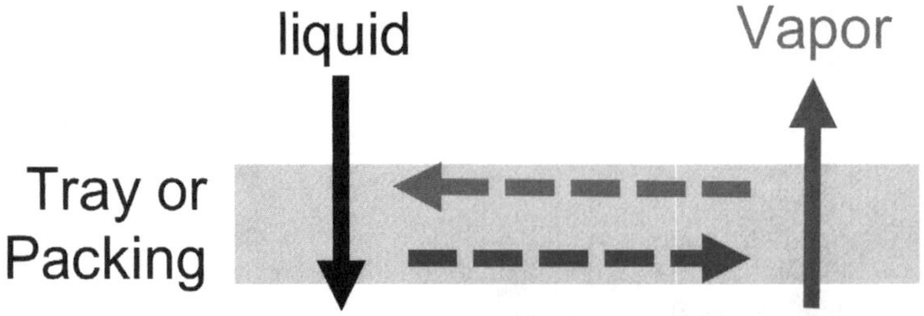

Chemical	Formula	Boiling Point*
Isopentane	C_5H_{12}	27.9°C
Pentene	C_5H_{10}	30.0 °C
Ethanethiol	C_2H_6S	35.0 °C
Pentane	C_5H_{12}	36.1 °C
Carbon Disulfide	CS_2	46.3 °C
Cyclopentane	C_5H_{10}	49.3 °C
2,2-Dimethylbutane	C_6H_{14}	49.7 °C
2-Propanethiol	C_3H_8S	52.6 °C
4-Methyl-1-pentene	C_6H_{12}	53.9 °C
2,3-Dimethylbutane	C_6H_{14}	58.0 °C
Methanol	CH_5O	65.0 °C
1-Propanethiol	C_3H_8S	67.0 °C
Hexane	C_6H_{14}	69.0 °C
1-Hexene	C_6H_{12}	63.4 °C

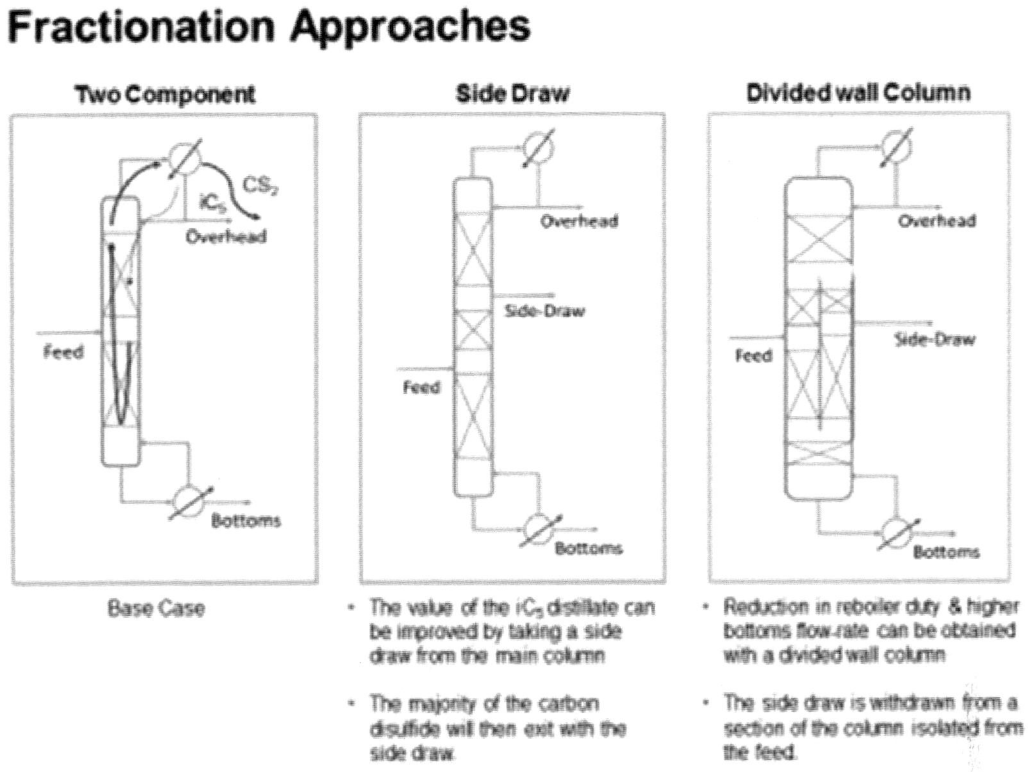

Figure 3: Fractionation options for CS_2 removal from petrochemical naphtha

Other potential options for CS_2 removal include hydrolysis, absorption with solvents and adsorption, including chemisorbents and physisorbents.

- Although carbon disulfide can be converted via hydrolysis to carbonyl sulfide and to hydrogen sulfide catalytically at the right process conditions, the thermodynamics are unfavorable to achieve low CS_2 levels at process conditions required for reasonable kinetic rates of conversion.
- There is some literature[9] for CS_2 removal via absorption. Solvents to remove CS_2 via several complex steps and regeneration of the solvents have been reported. However, these have not been demonstrated to be commercially viable.
- Traditional adsorption based solutions (chemisorbents[8] or physical adsorbents) for CS_2 removal from petrochemical naphtha are not inexpensive, and have been unable to economically handle highly contaminated feedstocks.
- In order to help refiners and ethylene producers address their desire for lower CS_2 content of their petrochemical naphtha streams, UOP has developed a unique regenerable adsorbent, MOLSIV™ CS2-100, which can remove CS_2 from petrochemical grade naphtha and natural gas liquids. The new adsorbent can remove CS_2 selectively thereby efficiently enabling PCN suppliers to meet ethylene producer's needs. Extensive dynamic pilot plant regenerative testing with a variety of commercial feeds, including Mid-East and Americas derived samples, has been

conducted to demonstrate technical and economic viability of an adsorptive solution for a variety of different feedstocks.

The amount of contaminant that an adsorbent can hold is dependent upon many factors, the key being:

- **Temperature**: In a temperature swing adsorption unit (TSA), at low temperatures the adsorbent has more contaminant capacity and at high temperatures a much lower capacity.
- **Concentration (or partial pressure) of the contaminant:** In general the higher the concentration the higher the capacity

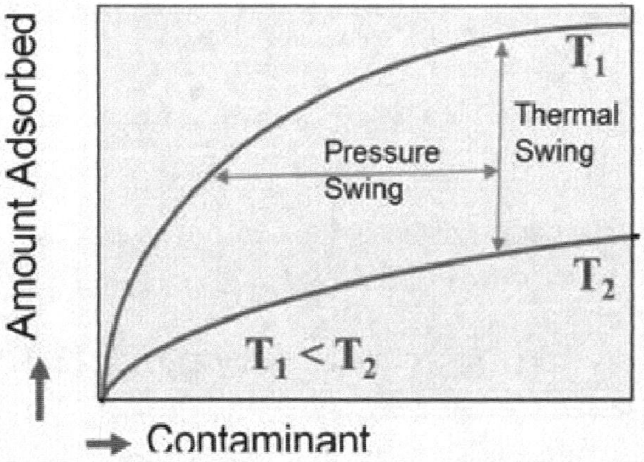

- This testing has allowed generation of robust design tools to enable a thermal swing adsorption (TSA) unit design. Spent adsorbent disposal and handling recommendations will also be available to our customers. Liquid TSA offers a number of advantages over a naphtha hydrotreating (NHT) unit. In the case of a liquid phase TSA if hydrogen is used only for regeneration purposes it could be "borrowed" from the hydrotreater or hydrocracker and used to convey heat to the adsorbent bed and to carry away the desorbed contaminants from the bed. The hydrogen is not consumed in a TSA and the spent regeneration hydrogen can be returned to the hydrotreater or hydrocracker.

Thermal Swing Adsorption Technology Development

A wide variety of adsorbents were screened for CS_2 removal from simulated naphtha. None of the traditional alumina based or molecular sieve adsorbents used in the purification or hydrocarbon separations were found to be suitable for selective CS_2 removal, due to a result of relatively weak CS_2 adsorption and hydrocarbon co-adsorption, or interference of other contaminants or sulfur species typically present in a naphtha fraction. UOP has developed a unique type of adsorbent that not only has excellent CS_2 adsorption capacity, but is also pore size selective so as to effectively exclude heavier hydrocarbons and contaminants thereby retaining an excellent selectivity for CS_2 in typical petrochemical naphtha streams.

UOP's CS2-100 adsorbent showed excellent capacity for carbon disulfide at low concentrations in simulated naphtha testing. Pilot plant testing of the new CS_2 adsorbent in simulated feed demonstrated its effectiveness in maintaining its regenerative capacity upon thermal regeneration. This screening test warranted further evaluation for CS_2 removal from commercial naphthas.

Adsorbent Pilot Plant Testing

UOP conducted extensive pilot plant testing with a variety of commercial naphtha feeds, including Mid-East and Americas derived samples and some customer specific feeds, to demonstrate the technical viability of an adsorptive solution for a variety of different naphtha feedstocks.

Commercial naphthas vary widely in composition depending upon the crude source and fractionation choices. Analysis of two commercial naphtha feed samples conducted by UOP is shown in Figures 4 and 5. A brief summary of the analysis conducted by UOP of a typical commercial naphtha feed sample:

- 1 – 300 ppmw CS_2
- 200 – 400 ppmw sulfur total
- Water saturated
- Boiling range of hydrocarbons: 0 – 185 °C
- Paraffin content: >70 wt%
- Olefin content: <0.5 wt%
- Naphthene content: 15 – 25 wt%
- Aromatic content: 5 – 10 wt%

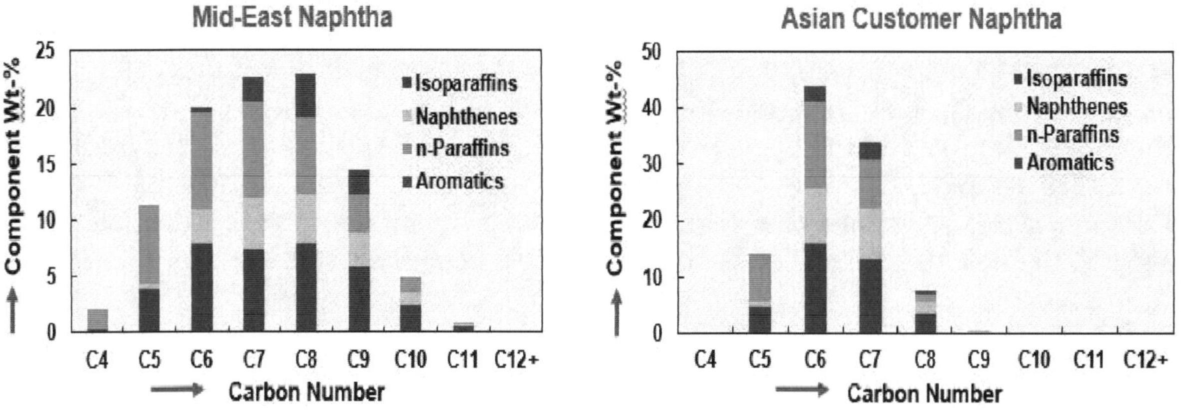

Figure 4: Carbon number and type distribution of Mid-East and Asian customer samples

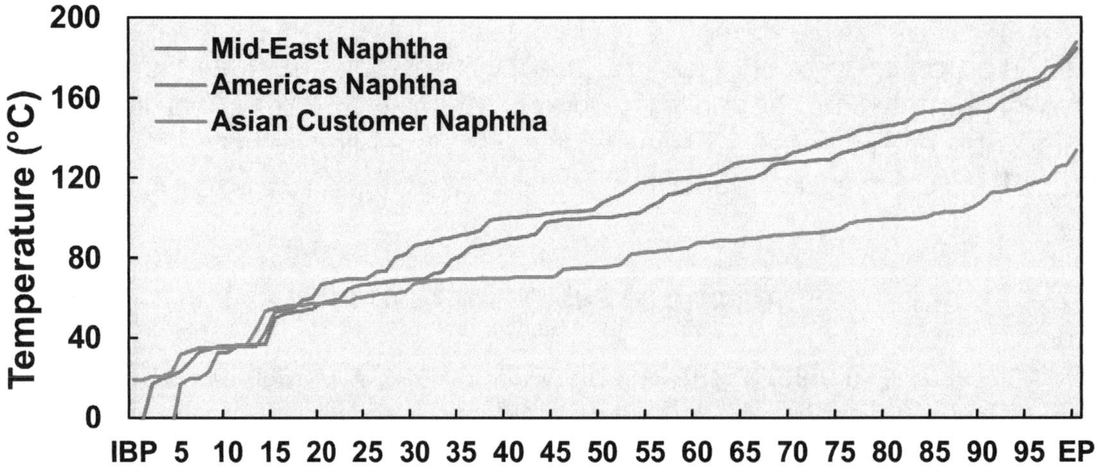

Figure 5: Boiling range comparison of Mid-East, Americas, and Asian customer samples

Accelerated pilot plant testing of the Asian customer PCN feed has further verified UOP's robust design tools by confirming the high capacity and regeneration concept of the adsorbent with commercial feedstocks. As received, the Asian PCN feedstock contained ~12 ppmw CS_2. This feed was spiked with additional CS_2 up to 80 ppmw. As shown in Figure 6, cyclic testing was conducted to evaluate the capacity and mass transfer properties of the adsorbent. In addition to testing a variety of commercial feeds, process variable studies were conducted to develop the necessary design tools to simulate performance of the adsorbent over a variety of CS_2 feed concentrations ranging from 1 ppm to 300 ppm CS_2. Dynamic performance of the adsorbent to achieve effluent CS_2 concentrations of <0.1 to 1 ppm have been evaluated. Stable cyclic adsorbent performance has been demonstrated by multiple pilot plant tests. Since the design tools developed are based on a wide variety of feedstocks, we are able to predict performance for the typical feedstock variation a refiner may expect to see over the course of operations. No adsorbent deactivation was observed in the pilot plant testing with the Asian customer or other commercial feedstocks tested. Deactivation would be identified by the loss of CS_2 capacity and/or lengthening of the mass transfer zone. The graphs (Figure 6 and Figure 7) show the excellent regenerability of CS2-100 in one commercial PCN feedstock. Spent adsorbents after multiple regeneration cycles in commercial feeds have shown no measurable accumulation of coke (carbon) on the adsorbent, additionally demonstrating adsorbent stability.

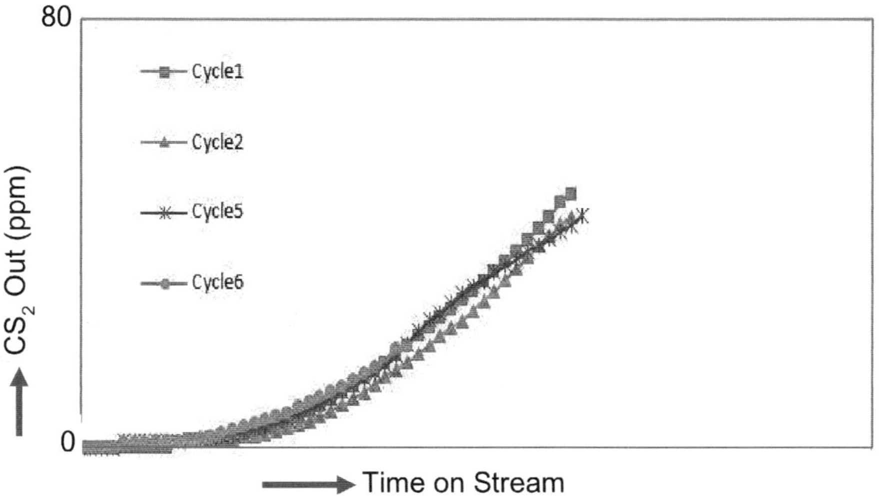

Figure 6: Selected pilot plant CS_2 breakthrough data on UOP MOLSIV™ CS2-100 at accelerated testing conditions

Furthermore, thermal swing adsorption can achieve very low effluent levels of <0.1 ppm CS_2 on a sulfur basis which is the limit of detection by ASTM D5623. Cyclic pilot plant data at this level is shown in the graph below. Even under accelerated testing conditions, <0.1 ppm CS_2 effluent concentration can be maintained for extended periods depending on typical run conditions such as LHSV, temperature, CS_2 concentration, and feedstock properties (boiling range, viscosity, density).

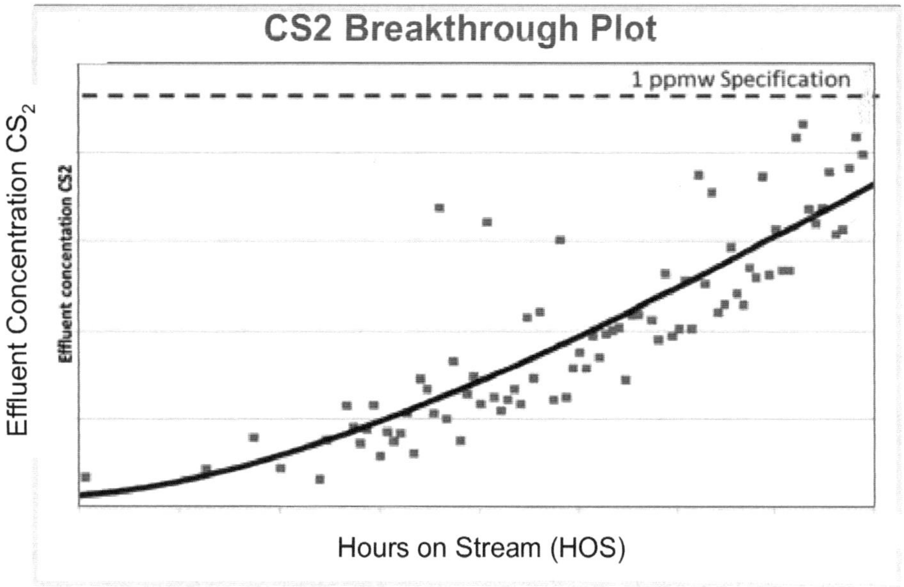

Figure 7: Pilot plant CS_2 breakthrough testing data under accelerated conditions showing extended periods of outlet concentration of < 0.1 ppm CS_2 as sulfur

The extensive pilot plant testing has allowed us to develop predictive design tools which account for differences in key feed physical properties and the concentration of other contaminants that may be present in the naphtha stream including organic sulfur species, oxygenates, and other species.

Regeneration

Adsorptive removal of CS_2 can be viewed as a two-step process (Figure 8). CS_2 must first be transferred from the naphtha to a regeneration stream. Then the CS_2 containing regenerant is sent to a reactor(s) which may convert the CS_2 so that the reaction effluent is 'clean', or the CS_2 containing regenerant is blended with a suitable process stream and sold as a marketable product, or used within the complex. The regenerant choice is driven by the specific customer configuration and varying market economics.

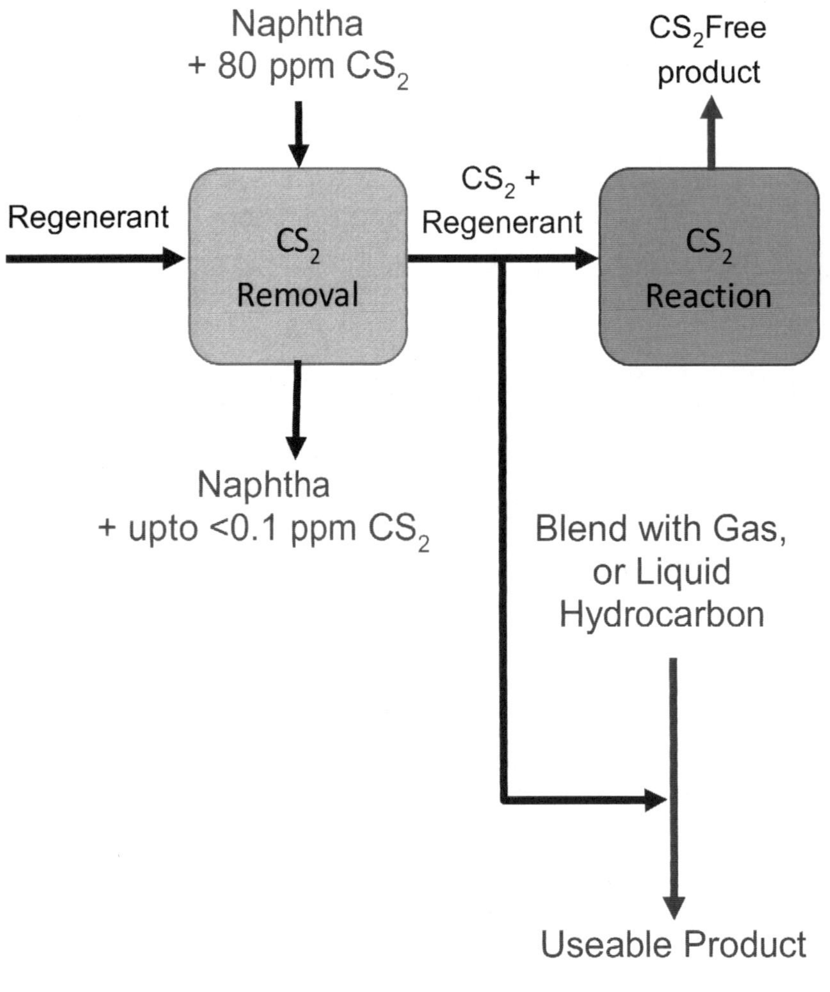

Figure 8: Adsorptive Removal Process for CS_2 from Petrochemical Naphtha

Many regeneration options are available. The choice of regenerant stream depends on availability and customer specific unit integration possibilities. The choices range from light gases such as hydrogen to hydrocarbons that can be vaporized at typical thermal regeneration temperatures. Hence refinery hydrogen streams, light naphtha, or inert gases such as nitrogen or clean contaminant free natural/ fuel gas may be acceptable regeneration media. UOP will work with customers to identify a suitable stream that can be integrated into this existing processes for their specific case. The heat duty of the regeneration cycle is calculated based on heating flow rate, type of regenerant, and heating duration.

As shown in Figure 10, a thermal swing adsorptive solution to remove CS_2 from petrochemical naphtha is a flexible technology that can handle varying feedstock composition as well as handle feed streams that may have varying CS_2 levels. In addition, the adsorption route to purification can achieve very low levels of CS_2 (<0.1 ppmw) in the product naphtha with only minor changes in operations, which may not be possible via other routes of CS_2 mitigation.

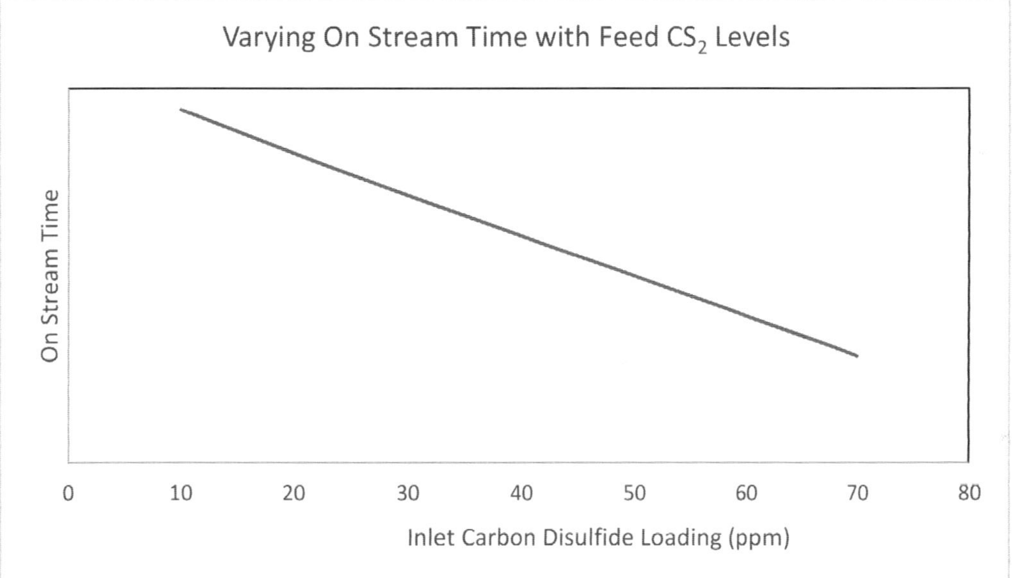

Figure 10: Changing adsorption Cycle Times with Varying Inlet CS_2 Loadings

Modular Thermal Swing Adsorption Process Solution

UOP's commercial experience in developing and manufacturing adsorbents, and commercializing hydrocarbon processing solutions has enabled us to develop a Liquid Thermal Swing Adsorption technology solution that can be provided as a modular unit to our customers to enable a refiner or ethylene producer to very quickly and cost effectively implement a system for carbon disulfide removal.

A pre-fabricated modular unit is easy to install, start-up, operate and maintain to help minimize capital and operating expenses. The unit is flexible and can accommodate the variation in feed contaminant levels. The system can be designed to meet CS_2 specifications of <0.1 ppmw product CS_2. The key decisions to be made are the type of regeneration stream (liquid or gas) and how to integrate that into the process flow scheme of a customer. Since, each case is somewhat unique, we have developed and can recommend a variety of ways to integrate the regeneration system into a customer's complex. A typical TSA unit block flow diagram is shown in Figure 9. There are several possible variations of such a flow scheme to maximize the adsorbent utilization and /or to minimize the amount of regenerant. We have also identified a variety of potential locations where the spent regenerant can be integrated in a typical refinery complex.

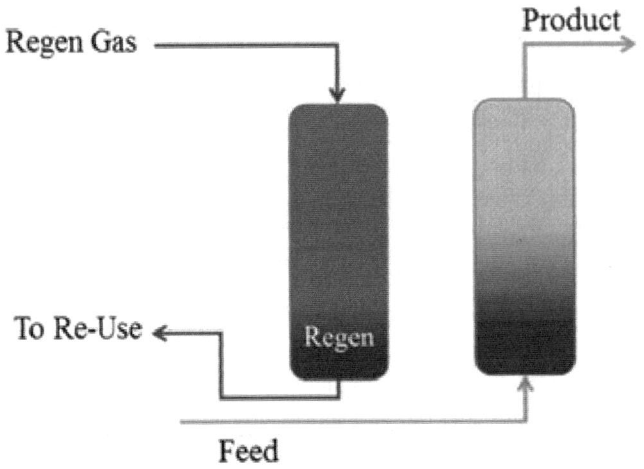

Figure 9: TSA system with Regeneration

Conclusions

1. In our assessment, hydrotreating is the highest capital cost option when compared to Distillation and TSA, and is more energy intensive process than the TSA option. Traditional hydrotreating can produce a naphtha product devoid of other sulfur species as well which may not be necessary (or even not desired in some cases) for the purpose of steam cracking. However, the hydrotreating option may be preferred by a customer for their particular unit and specific situations.

    - If an existing hydrotreater has sufficient available capacity, and H_2 is available, this is an excellent solution for CS_2 removal in most cases. UOP can assist with hydrotreating revamps, and grass-root design solutions for this option.

2. Distillation is another technically feasible solution for CS2 reduction in petrochemical naphtha. It is also the lowest capital cost option when compared to a new hydrotreater or TSA. However, distillation can be a highly energy intensive process, this when coupled with the need for an outlet for a CS_2 enriched C5 fraction which represents a key component of PCN may make it economically unattractive.

- The high energy requirements of distillation for CS_2 removal can be somewhat mitigated by use of high efficiency trays, in addition to use of a divided wall column approach. For certain customer specific flow schemes and objectives, distillation may be a suitable option for CS_2 removal from petrochemical grade naphtha.

3. The capital cost of a regenerable adsorptive solution (TSA) for CS2 removal is between that of distillation and hydrotreating.

    - TSA has lower operating costs than the other options. In addition, it is a flexible technology that is suited to handle variation in feedstock CS2 levels.

Table 2 summarizes the heater duty requirements for treating an 60,000 BPSD narrow boiling range naphtha having 80 ppm CS2 in order to achieve < 2ppm product CS2 product.

Table 2: Comparison of Heater Duty requirements of Hydrotreating, Distillation and TSA

Process	Relative Heater Duty	Waste Naphtha (% of Product)	Benefits
Conventional Naphtha Hydrotreater	**32**	**0%**	Leverage existing asset
Two Product Fractionation	**65**	**48%**	Easy installation
Side-Draw Fractionation	**154**	**43%**	May generate high value streams
Divided Wall Column	**109**	**38%**	May generate high value streams
Liquid TSA – Naphtha Regen	**5**	**10%**	Flexible to inlet CS_2 concentration
Liquid TSA – Hydrogen Regen	**1**	**0.2%**	Flexible to inlet CS_2 concentration

In summary, a new liquid phase Thermal Swing Adsorption process has been developed for removal of CS_2 from petrochemical naphtha. We believe that this solution provides a reliable, cost effective solution to the damaging downstream effects of carbon disulfide on ethylene producers' isoprene extraction and pyrolysis gas hydrogenation units. Based upon extensive testing of commercial naphtha feedstocks, UOP is positioned to offer a modular thermal swing adsorptive solution for carbon disulfide (CS_2) removal in petrochemical grade naphtha and natural gas liquid streams to adequately meet, and exceed the CS_2 specifications in petrochemical naphtha.

Note that we do not recommend using this adsorbent in the more reactive (and smaller volume of) cracked liquid or pygas streams in an ethylene cracker. Those streams are extremely reactive and at this stage CS_2 separation from pygas streams is not practical. Although there are other known options to reduce CS_2 in a narrower isoprene stream those are very specific and not practical for either petrochemical naphtha or pygas streams.

Alternative technologies for mitigating the CS2 to achieve < 2ppm in petrochemical naphtha include traditional hydrotreating and various distillation options, including a divided wall column approach that can minimize the high utility requirements of traditional fractionation approach. UOP can assist with technical and economic evaluation of various options for CS2 removal from PCN based on the unique situations of individual petrochemical producers or refiners.

References

1. Reid, J. A. "Introduction to Sulfur Compounds and Sulfur Removal as Applied to Olefin Unit Design and Operation," Proceedings of the 10th Ethylene Producers Conference, April 1998 combined AIChE and EPC meeting, pp. 714 – 754.
2. Coenen, Pascal and Ernesto Angueira. "Carbon Disulfide (CS_2): A New Challenge for Steam Cracker Complexes?," Proceedings of the 28th Ethylene Producers Conference, April 2016 combined AIChE and EPC meeting.
3. Open Spec Naphtha 2016 Meeting Documentation, http://osn.sg/.
4. Chiam, Clarice "Asia naphtha industry agrees to state carbon disulfide levels in Sikka, Western cargoes," Platts Singapore, April 2014.
5. National Institute of Standards and Technology, US Department of Commerce (NIST)
6. Mesters, Carolus and Ronald Jan Schoonebeek, "Process for the Manufacture of Carbon Disulphide and Use of a Liquid Stream Comprising Carbon Disulphide for Enhance Oil Recovery", US 8,097,230.
7. Mapstone, G. E., J. S. Durham, and H. D. Taskis. "Studies in Shale Oil Part III. Some Constituents of Shale Naphtha", Journal of the Institute of Petroleum, 38, 1952, pp. 172 – 178.
8. Hisataka, K., Koichi, S,, and Masao, F. "Purification of C5 Hydrocarbons" Japanese Patent Application 57-045814
9. Weissheimer, F. and Link, J. "Use of Polyamines as a CS2 scavenger in Isoprene production" WO2015026649A1

AIChE Paper Number:

LARGE STEAM TURBINE OVERSPEED TRIP MITIGATION IN LARGE OLEFINS PLANTS

Samuel Snively
Bladed Machinery Engineering Manager
Elliott Group

Michael Martys
Bladed Machinery Product Engineer
Elliott Group

Steven Sykes
Business Development Manager
Tri-Sen

Sam Snively is the Bladed Machinery Engineering Manager for Elliott Group in Jeannette, Pennsylvania. Responsibilities within the bladed group include core application engineering, demand aerodynamics, product engineering and unit design for new equipment. Mr. Snively received his B.S. degree (Mechanical Engineering) from the University of Pittsburgh in Pittsburgh, Pennsylvania and M.B.A. from Seton Hill University in Greensburg, Pennsylvania.

Michael Martys is a product engineer in the Bladed Machinery group for Elliott Group in Jeannette, Pennsylvania. Responsibilities within the bladed group include steam turbines, axial compressors, and hot gas expanders. Mr. Martys received his B.S. degree (Mechanical Engineering) from the University of Pittsburgh in Pittsburgh, Pennsylvania.

Steve Sykes is the Business Development Manager at Tri-Sen Systems in Webster, TX and has over 25 years of experience in TurboMachinery control and protection systems. Steve received his B.S. degree (Electrical Engineering) from Tennessee Technological University.

LARGE STEAM TURBINE OVERSPEED TRIP MITIGATION IN LARGE OLEFINS PLANTS

Samuel Snively
Steam Turbine Engineering Manager
Elliott Group
Michael Martys
Steam Turbine Product Engineer
Elliott Group
Steven Sykes
Business Development Manager
Tri-Sen

Abstract

Steam turbine overspeed events in large olefins plants may happen from a sudden loss of load of the driven equipment or an instantaneous loss of coupled inertia as a result of complete coupling or shaft end failure. This overspeed event occurs as load is rapidly depleted and steam is still being supplied to the rotor system. The largest contributor to steam turbine overspeed is entrapped steam and trip set point. This paper will offer suggestions for OEM and customer collaboration of steam turbine design with special consideration for ancillary components and supporting system parameters. Means of initiating a trip event such as rate of acceleration will be offered as alternatives to traditional trip set points as a function of rated speed.

This paper presents findings regarding the most important considerations in the design, engineering, and implementation of electronic overspeed trip systems, demonstrated through models and first-principles simulation to maximize system responsiveness, reliability, and availability.

Introduction

Mechanical drive steam turbines typically operate under steady power and speed requirements. These steady-state conditions can change rapidly and without warning as a result of a load dump or mechanical failure. A properly functioning overspeed trip system and well thought out steam piping can help to ensure the safety of personnel in the immediate area and minimize the damage to critical equipment.

Calculating overspeed enables OEMs to predict the maximum speed a rotor system can attain. This value is compared to the various mechanical speed limits in a given rotor. These limits include, but are not limited to, locking pin shear, base of vane tensile loading, and disc tear out forces. Mechanical failure at these locations, among others, can cause catastrophic failure of a stage's buckets or an entire rotor system. Therefore accurately predicting overspeed is crucial in rotor system design.

Power Test Code (PTC) 20.2 is the industry-accepted calculation for estimating maximum speed attained and the dominant contributing factor is the entrapped steam volume downstream of the trip valve and/or non-return valves for induction or extraction lines. Effectively minimizing the peak rotor speed in an overspeed event is achieved by designing effective overspeed protection systems and choosing valve locations that serve to minimize the amount of entrapped steam available to the rotor.

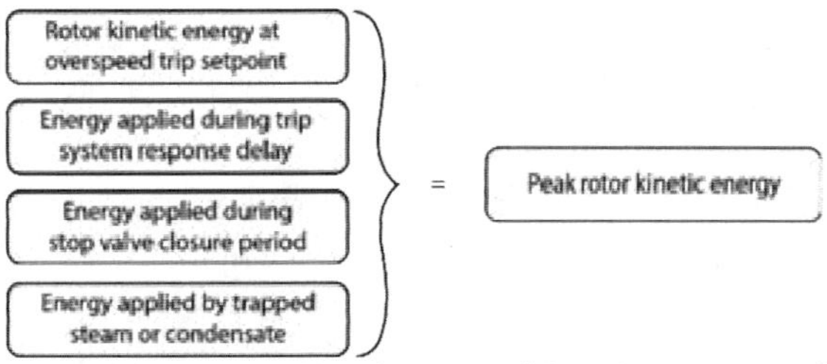

Figure 1: Peak rotor kinetic energy blocks (American Petroleum Institute, 2014)

Explanation of PTC-20.2

Contribution of Rotor Inertia (Kinetic Energy)

Rotor inertia is one of the four contributing factors to predicting turbine overspeed according to PTC 20.2 for an instantaneous loss of coupled inertia. Equation 1 shows the factors involved in this calculation.

$$E_t = 2.31 \times 10^{-7}(WR^2)(N_t)^2 \tag{1}$$

where:
E_t is the kinetic energy at trip initiation in kilowatt-seconds
WR^2 is the rotational inertia of the rotor assembly in pounds-feet squared
N_t is the trip speed of the turbine in rpm

Contribution of Signal Delay Time

Another factor in estimating overspeed requires finding the energy added to the system during the signal time delay. Equation 2 shows the contributing factors.

$$\Delta E_1 = T_s P_{g,tot} \quad (2)$$

where:
ΔE_1 is the energy added during trip signal delay in kilowatt-seconds
T_s is the trip latency in seconds
$P_{g,tot}$ is the combined full load power from the front and back ends in kilowatts

Contribution of Trip Valve Closing Time

The added energy during the closing of the trip valve is shown in Equation 3.

$$\Delta E_2 = f T_v P_g \quad (3)$$

where:
ΔE_2 is the energy added during trip valve closure in kilowatt-seconds
f is the valve closing factor in seconds
T_v is the valve closing time in seconds
P_g is the turbine rated power of the respective end and valves in units of kilowatts

The valve closing factor is an approximation of the characteristic curve of a given trip and throttle valve. Each type of valve will vary, however a conservative number from the valve supplier is considered for ease of calculation.

Contribution of Entrapped Steam

The final component in estimating turbine overspeed is entrapped steam that exists between trip and non-return valves upstream of the exhaust. Additional terms will have to be added in the case of an extraction turbine. Equation 4 shows the additional terms added for an extraction turbine.

$$\begin{aligned}\Delta E_3 = 1.056[&\eta_i(W_{1i}u_{1i} - W_{2i}u_{2i} - (W_{1i} - W_{2i})h_{2i}) \\ &+ \eta_x(W_{1x}u_{1x} - W_{2x}u_{2x} - (W_{1x} - W_{2x})h_{2x})]\end{aligned} \quad (4)$$

where:
ΔE_3 is the energy added from entrapped steam in kilowatt-seconds
η_i is the front end full load efficiency

Front end:
W_{1i} is the weight of the steam in the steam chest in pounds-mass
u_{1i} is the energy of the steam in the steam chest in BTU per pound-mass
W_{2i} is the weight of the steam remaining after isentropic expansion in pounds-mass
u_{2i} is the energy of the steam after isentropic expansion in BTU per pound-mass
h_{2i} is the enthalpy of remaining steam in BTU per pound-mass

Back end:
η_x is the back end full load efficiency
W_{1x} is the weight of the steam that will enter the back end of the turbine in pounds-mass

u_{1x} is the energy of the steam that will enter the back end of the turbine in BTU per pound-mass
W_{2x} is the weight of the steam remaining after isentropic expansion to exhaust conditions in pounds-mass
u_{2x} is the energy of the steam remaining after isentropic expansion to exhaust conditions in BTU per pound-mass
h_{2x} is the enthalpy of the steam remaining after isentropic expansion to exhaust conditions in BTU per pound-mass

It is implicit that correct conditions will be used for all calculations. Equation 5 shows the method for solving energy of the steam, which is used in Equation 4.

$$u = h - Pvk \tag{5}$$

where:
h is the enthalpy of the steam found using the steam tables in BTU per pound-mass
P is the pressure of the steam in psia
v is the specific volume of the steam in cubic feet per pound-mass
k is the conversion factor of $\frac{1{,}728\ ft^3/in^3}{9{,}338\ lb_f in}$

Sensitivity Analysis Tool

A tool designed in Excel was used to conduct an analysis using PTC-20.2 calculations in order to determine how sensitive the maximum speed reached on an overspeed event (N_{max}) was to variations in the multiple contributors. Contributors were varied by +/- 50% in 10% increments to analyze their respective sensitivities.

Initial assumptions:

- Turbine parameters such as rotor inertia, horsepower, rated speed, isentropic efficiency, etc., are fixed
- Trip speed multiplier initial value (K_{trip}) = 15.5% of turbine rated speed
- Trip Latency (T_s) = 200 ms
- Valve flow characteristic (f) = .86
- Valve closure time (T_v) = 300 ms

Inlet Conditions:		
Inlet Temperature (T1) =	950.00	F
Inlet Pressure (P1) =	1,537.40	psia
Steam Chest Volume (V1) =	33.00	ft^3
Steam Mass (w1) =	65.08	lb
Find steam properties:		
Specific Volume (v1) =	0.51	ft^3/lb
Internal Energy (u1) =	1,313.76	btu/lb
Entropy (S1) =	1.575	btu/lb.F
Enthalpy (h1) =	1458.019	btu/lb
Machine:		
Trip Speed (N) =	4,484.87	rpm
First Section hp =	44,076.71	hp
First Section Efficiency =	0.783	
Second Section hp =	55,229.99	hp
Second Section Efficiency =	0.871	
Rotor Inertia (WR^2) =	24,370.00	bl.ft^2
Inlet Valve:		
Signal Delay Time (Td) =	0.2	sec
Power reduction factor (f) =	0.86	
Valve Stroking Time (Tv) =	0.3	sec

Extraction Conditions		
Pressure (P2) =	616.400	psia
Isentropic Entropy (S2is) =	1.575	btu/lb.F
Total Extraction Volume =	175.00	ft^3
Steam Mass (w2) =	162.40	lb
Find Steam Isentropic Properties:		
Isentropic Enthalpy (h2is) =	1339.80	btu/lb
Isentropic Temp (T2is) =	684.12	F
Specific Volume (v2is) =	1.03	ft^3/lb
Internal Energy (u2is) =	1222.77	btu/lb
Entropy (s2is) =	1.575	btu/lb.F
Find Steam Actual Properties:		
Actual Enthalpy (h2ac) =	1365.45039	btu/lb
Actual Temp (T2ac) =	727.72	F
Specific Volume (v2ac) =	1.08	ft^3/lb
Internal Energy (u2ac) =	1242.53	btu/lb
Entropy (s2ac) =	1.60	btu/lb.F

Exhaust Conditions		
Pressure (P3) =	2.18	psia
Isentropic Entropy (S3is) =	1.597	btu/lb.F
Exhaust Steam Mass (w3) =	227.48	lb
Find Steam Isentropic Properties:		
Isentropic Enthalpy (h3is) =	930.69	btu/lb.F
Temp (T3is) =	128.75	F
Specific Volume (v3is) =	134.08	ft^3/lb
Internal Energy (u3is) =	876.70	btu/lb
Entropy (s3is) =	1.597	btu/lb.F
Find Steam Actual Properties:		
Actual Enthalpy (h3ac) =	986.77	btu/lb
Extraction Actual Temp (T3ac) =	128.75	F
Specific Volume (v3ac) =	143.09	ft^3/lb
Internal Energy (u3ac) =	929.15	btu/lb
Entropy (s3ac) =	1.692	btu/lb.F

Figure 2: Sensitivity Analysis Base Conditions

Explanation of Contributors

Rotor Inertia

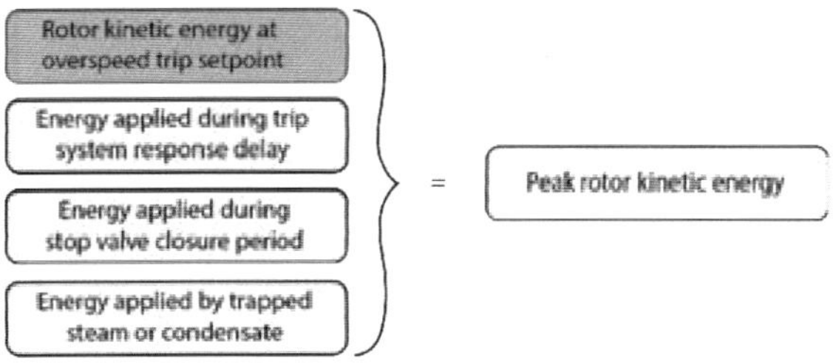

Simply adding mass to increase rotor inertia will help to lessen the speed achieved but that additional mass does not add any value to the turbine design for normal operation. Heavier rotors require larger bearings with additional lubrication and increase the potential for rotor stability issues.

Increasing the main shaft diameter will increase the inertia but also increase the annulus size of the inter-stage diaphragm shaft seals resulting in performance degradation.

Rotordynamics is greatly affected by rotor inertia. Increasing rotor weight is directly proportional to shaft stiffness which increases the speed of the first mode excitation. For a flexible shaft design, the first mode must be at a speed less than minimum governor speed. An increase in mass may bring this first critical speed within the running speed of the rotor.

Hydro-dynamic oil film bearings are designed to maintain a maximum operating load. Exceeding this limit can cause issues such as increased babbitt wear and increased bearing metal temperatures.

Bearing load design permits temperature rise up to 212°F. Increasing the weight of the rotor can cause the temperature to rise above this design limit, given the same bearing geometry. Therefore larger bearings may need to be implemented in order to keep the temperature from reaching this point. This would add cost and possibly change the characteristics of the rotor's critical speeds.

A reduction in rotor weight reduces inertia which allows steam to spin the rotor more easily. In an overspeed scenario, this translates to increased speeds which can have catastrophic effects.

In addition to increased overspeed, a light shaft is a more flexible shaft. Therefore the second critical speed of the rotor is now closer to the maximum continuous speed of the unit. Approaching second critical speed can cause the rotor to oscillate in such a way that damages any close-fitting inter-stage seals at the nodal points. Damaged inter-stage seals allow more stage leakage which can reduce the overall efficiency of the unit.

Overspeed Protection System

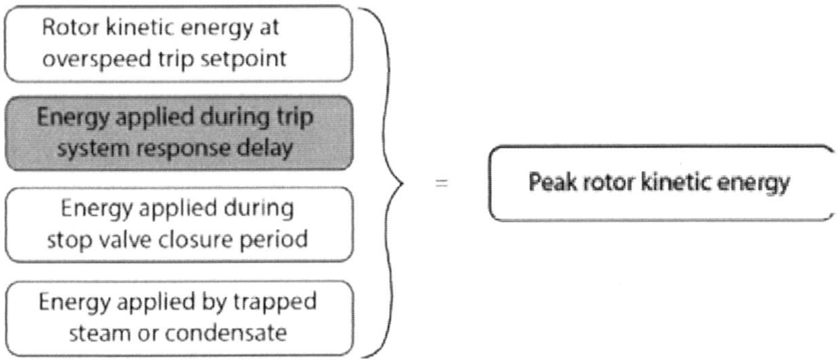

API 670 specifies an overspeed protection system must consist of an overspeed detection system (OSD) as well as other components specified in API 612.

The OSD system can be easily categorized as belonging to either the mechanical variety (MOSD) or the electronic variety (EOSD). Most turbine OEMs typically provide an EOSD as original equipment for large turbine driven compressor sets. The focus will be on the EOSD type on which API 670 is heavily related.

The EOSD typically interfaces with speed sensors, multi-tooth speed sensing surfaces on the input side, and solenoid valves, either directly or through interposing relays, on the output-side. (Machinery Protection Systems, API Standard 670, 2014, p. Annex O)

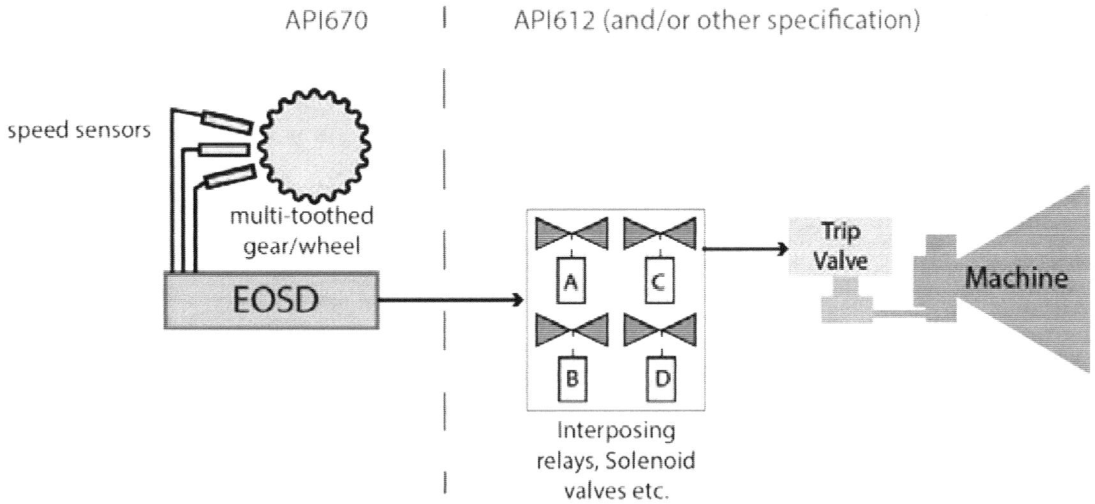

Figure 3: Overspeed protection interface (adapted from API 670, 2014, p. Annex O)

Speed sensors in proximity to the multi-tooth speed sensing surface transmit a low-voltage pulse signal to the EOSD each time a tooth/gap on the speed sensing surface passes the end of the sensor.

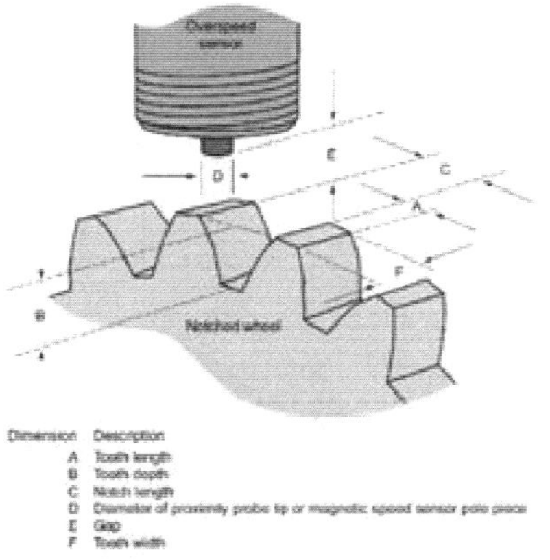

Figure 4: Typical speed measurement interface

The EOSD interprets the speed signal and compares it to a pre-determined trip set point. If the measured speed exceeds the trip set point, the EOSD typically de-energizes its outputs (there are some applications where users have opted for energizing to trip; however, this practice is discouraged). The output side of the EOSD is typically connected to an array of solenoid operated valves (SOVs) that dump the hydraulic trip oil used to keep the trip valve open. During an overspeed event, a properly functioning EOSD de-energizes the power to the SOVs, opening the SOVs which dumps the hydraulic trip oil and closes the turbine trip valve.

Component redundancy is necessary to maximize system reliability and availability pertaining to the individual parts that make up the EOSD and the hydraulic dump interface.

API has taken a prescriptive approach on redundancy requirements for both the EOSD and the hydraulic dump interface. API 612 (7th edition, 2014) 12.3.2.1 states for the EOSP, "An overspeed detection system based on three independent measuring circuits and two-out-of-three voting logic in accordance with API 670 shall be supplied." API has several additional requirments; the EOSP must incorporate functionality such that an end user can test and repair the EOSP while the turbine is on-line. Additionally, API 670 stipulates that "combining the overspeed system with any other control, protection, or monitoring systems shall not be allowed." An exception to this rule is allowed when the EOSD is combined with an emergency shutdown system, provided the failure of the shutdown system does not affect the EOSD system.

Overspeed Protection System (OPS) Responsiveness

OPS response time is the trip latency time (T_S) in PTC-20.2 described earlier (equation 2). API 670 defines this time as "the response time of the electronic overspeed shutdown device as well as the response time of all the components between the electronic overspeed shutdown device and the steam shutdown valve such as the hydraulic shutdown block, hydraulic piping runs, and solenoid valves."

Trip latency time is comprised of three separate and equally weighted delays:

- EOSP response time
- Hydraulic trip interface (SOV) response time
- Trip header translation time (time it takes for pressure to drop enough to activate trip valve

EOSP Response Time

API 670 defines the EOSP delay as being made up of sensor delays, logic solver I/O scan rate delays, and logic solver program scan rate delays × 2.[1]

API 670 (2014) 8.4.4.1 provides the following guidelines on EOSP response time. "The system shall sense an overspeed event and change the state of its output relays within 40 ms or as determined in Annex O, whichever is more restrictive."

Maintaining compliance with API 670 requires an EOSD with a translation time that is equal to or faster than 40ms input to output (screw to screw), and dependent upon our model, results may need to be faster still. This requirement significantly limits any EOSP-dependent variation to the trip latency time (T_S) of the overspeed calculation described earlier.

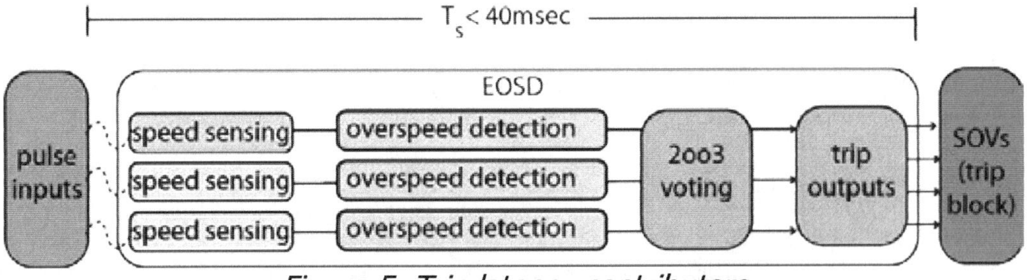

Figure 5: Trip latency contributors

[1] The logic solver program scan rate is multiplied by two in case the overspeed event happens in the middle of a program scan (conservative).

Hydraulic Trip Interface (SOV) Response Time

The second delay in the overall OPS responsiveness is the hydraulic trip interface (SOV) response time. Unlike API 670 compliant EOSDs, there can be variation in SOV response times. For a given solenoid, "its response time might be significantly affected by maintenance; replacement of coils, and the addition of components such as fly back or noise suppression, diodes or interposing relays"[2] (Taylor & Smith, 2009). For this reason, hydraulic trip interface delays can range from 15 ms to as long as 200 ms.

Trip Header Translation Time

Trip header translation time is a function of both the hydraulic trip interface (C_V) and the actual volume of the trip header. The trip header translation time is measured from the time the hydraulic trip interface (SOVs) transition to an open position, dumping the hydraulic trip oil, until the trip valve begins to close and shut off steam to the turbine.

Trip Valve Closing Time

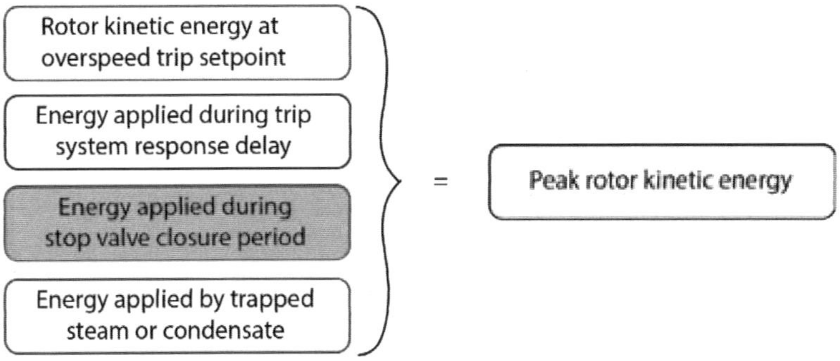

Trip valve closing time is a period where high-energy steam is being supplied to a rotor that potentially has very little loading other than inertia and bearing losses. A closure time of 250 ms is typical for today's industrial trip valves. However, this analysis assumes a conservative time of 300 ms. While it may be possible to move a valve stem faster via stiffer springs, the potential for internal damage to the valve and valve seat upon impact increase with the additional force. A slower reaction time simply means more energy for a longer period of time and should be avoided.

---

[2] Suppression diodes, MOVs, etc. are added to the SOV circuit to suppress and protect the EOSD contact outputs from the inductive surge associated with SOV energization.

Entrapped Steam

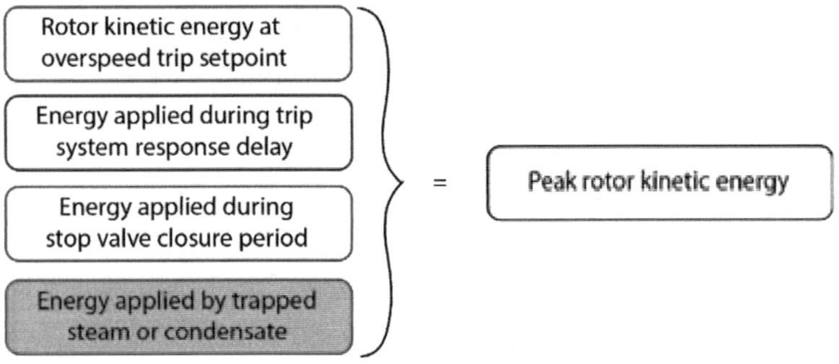

Entrapped steam is perhaps the largest contributor to turbine overspeed. The total entrapped volume includes the pipe downstream of the inlet trip valve, steam chest, pipe between an extraction casing flange and non-return valve, and extraction casing volume.

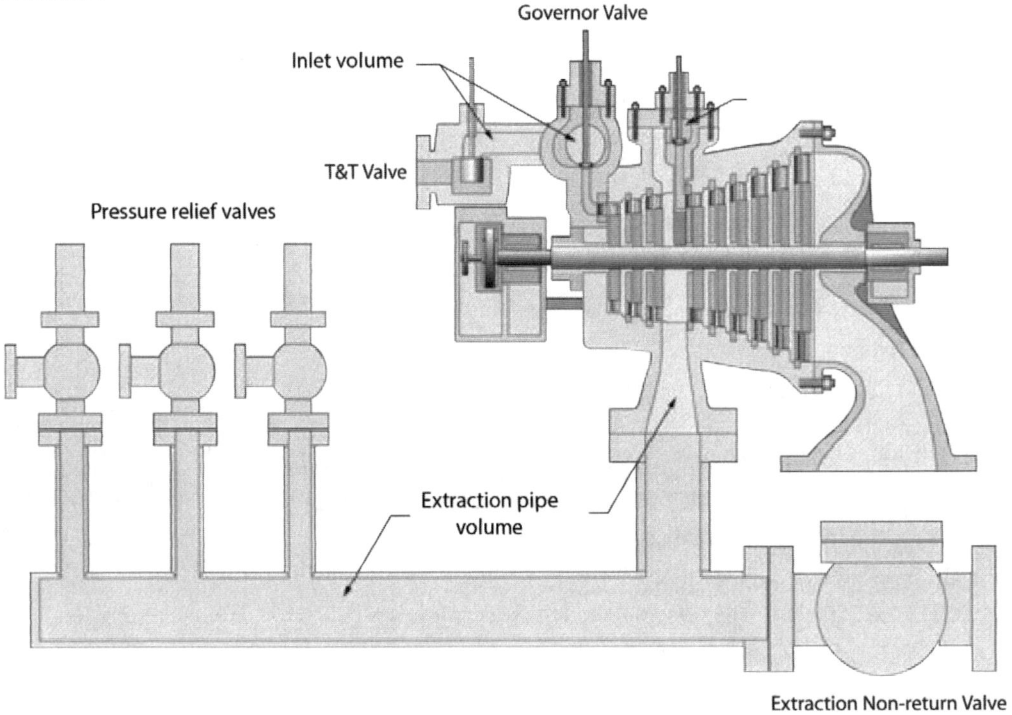

Figure 6: Entrapped steam diagram

It is important to note the exhaust volume is not included since no steam energy is added to the rotor system once steam exits the final stage buckets.

Sensitivity of Contributors to Final N_{max}

Control Time Delay (T_s)

We started our analysis with an initial T_s of 200 ms. Each of the three contributors that make up T_s (EOSD delay, SOV delay, and trip header translation delay) were weighted equally; that is, any reduction in any of the three contributors resulted in a corresponding decrease in T_s (e.g. a 5 ms reduction in the trip header translation time is equal to a 5 ms reduction in T_s). Our analysis showed that a 10% decrease in the value of T_s (e.g. reducing the EOSD delay time from 40 to 20 ms) resulted in a 21 RPM reduction in the maximum rotor speed. This reduction was nearly linear, and a 20% reduction (40 ms) in T_s led to a 43 RPM reduction in maximum rotor speed.

Table 1: Sensitivity of control time delay to overspeed

	Ts	
	Sec	**Nmax**
50%	0.10	6033
60%	0.12	6055
70%	0.14	6076
80%	0.16	6098
90%	0.18	6119
100%	0.20	6141
110%	0.22	6162
120%	0.24	6184
130%	0.26	6205
140%	0.28	6226
150%	0.30	6247

Given the relatively strict requirements from API regarding the performance of the EOSD (invariability between API 670-compliant EOSD solutions), most influential improvements in trip latency will have to be achieved through improvements to the SOV response and trip header translation times. For example, a 50% delay time improvement in an API 670-compliant EOSD will result in a decrease in the maximum rotor speed by 21 RPM for the turbine described in our model. A 50% reduction in the SOV delay and trip header translation time (160 ms to 80 ms) will result in a decrease in the maximum rotor speed of 86 RPM.

Valve Stroke Time (T_v)

Valve stroke time is the time the trip valve takes to travel from fully open to fully closed. Our model shows that a 10% decrease in T_v, from .3 to .27, results in a lower maximum speed attained from 6141 to 6113 RPM (28 RPM).

Table 2: Sensitivity of valve stroke time to overspeed

	Tv	
	Sec	**Nmax**
50%	0.15	6001
60%	0.18	6029
70%	0.21	6057
80%	0.24	6085
90%	0.27	6113
100%	0.30	6141
110%	0.33	6169
120%	0.36	6196
130%	0.39	6223
140%	0.42	6251
150%	0.45	6278

Valve Flow Characteristic (f)

The contribution of (*f*), the fraction of maximum steam flow that passes through the stop valve during the valve closure period, to the maximum speed attained was evaluated in the same way as T_V. Our model shows that a 10% decrease in f, from .86 to .77, results in a lower maximum speed attained from 6141 to 6113 RPM (28 RPM).

Table 3: Sensitivity of valve closing factor to overspeed

	f	
	Fraction	**Nmax**
50%	0.43	6001
60%	0.52	6029
70%	0.60	6057
80%	0.69	6085
90%	0.77	6113
100%	0.86	6141
110%	0.95	6169
120%	1.03	6196
130%	1.12	6223
140%	1.20	6251
150%	1.29	6278

Overspeed Shutdown Trip Set Point (K_{trip})

The contribution of the overspeed trip set point setting (K_{trip}), was also evaluated in the same way as the other contributers. Our model shows that a 10% decrease in K_{trip}, from 15.5% to 14.0%, results in a lower maximum speed attained from 6141 to 6097 RPM (44 RPM).

Table 4: Sensitivity of trip speed to turbine overspeed

	Ktrip Factor	**Nmax**
50%	0.078	5925
60%	0.093	5967
70%	0.109	6010
80%	0.124	6054
90%	0.140	6097
100%	0.155	6141
110%	0.171	6185
120%	0.186	6229
130%	0.202	6274
140%	0.217	6319
150%	0.233	6364

Inlet Steam Volume

The contribution of inlet steam volume, from the trip and throttle valve to the nozzle banks, to maximum speed attained was evaluated similarly to the other contributors. Our model shows that a 10% decrease in volume from 33 ft^3 to 30 ft^3 results in a lower maximum speed attained from 6141 to 6116 RPM (25 RPM).

Table 5: Sensitivity of inlet volume to turbine overspeed

	Inlet Volume ft^3	**Nmax**
50%	17	6005
60%	20	6031
70%	23	6056
80%	26	6082
90%	30	6116
100%	33	6141
110%	36	6166
120%	40	6200
130%	43	6225
140%	46	6249
150%	50	6282

Extraction Steam Volume

The contribution of extraction steam volume, the fraction of maximum steam flow that returns from the stop valve after the valve closes completely, to the maximum speed attained was evaluated similarly to the other contributors. Our model shows that a 10% decrease in volume from 175 ft^3 to 158 ft^3 results in a lower maximum speed attained from 6141 to 6076 RPM (65 RPM).

Table 6: Sensitivity of extraction volume to turbine overspeed

	Extraction Volume ft^3	**Nmax**
50%	88	5802
60%	105	5869
70%	123	5940
80%	140	6007
90%	158	6076
100%	175	6141
110%	193	6209
120%	210	6272
130%	228	6339
140%	245	6401
150%	263	6466

Methods for Reducing Overspeed

Reducing Entrapped Steam

Main inlet trip valves should be bolted directly to the steam turbine inlet chest. Relocating these valves away from the turbine casing for convenience or because of space constraints can greatly increase the volume of entrapped steam.

The non-return valves (NRVs) should also be located close to the turbine casing. While it is not always possible to locate these valves immediately downstream of the turbine flange, efforts should be made to minimize this distance. Depending on the scope of supply, location of the NRVs may not be in the control of the turbine manufacturer. This requires additional collaboration between the OEM and the customer to ensure that the valve location does not adversely impact the predicted overspeed.

Traditionally the pressure relief valves (PRVs) are located upstream of the NRV and sized to pass full inlet flow in the event of a trip and NRV failure in the closed position. PRV valve locations depend upon layout of the equipment and auxiliary equipment. Placing these valves away from the turbine to minimize clutter negatively affects overspeed by adding additional steam volume.

Placing the PRVs downstream of the NRV will eliminate an excess volume of steam because the NRV will prevent that steam from entering the turbine. However, the ability of the safety valves to function will now rely on the successful operation of the NRV. NRVs are notorious for sticking in the open position due to deposit buildup on the stem and are less likely to stick in the closed position.

API requires that the "full-capacity safety relief valve shall be installed in the exhaust piping between each exhaust connection and exhaust block valve to prevent overpressure and possible rupture of the turbine casing". It is assumed that the valves must be placed upstream of the NRV because the block valves are typically out of scope for the turbine manufacturer and they are not aware of the entire steam system design. Open discussion and collaboration between the OEM and the customer can ensure optimal location of the valves.

Alternative Methods of Overspeed Monitoring

The importance of the overspeed trip set point setting has not been lost to most EOSD providers. Moreover, some EOSD providers have built in quasi-overspeed anticipation logic.

The first type of implementation of this acceleration-based overspeed trip logic depends on an average acceleration generated by taking the change in angular velocity over time and comparing it to a maximum allowable set point. If the calculated average acceleration is above the maximum acceleration set point, a trip command is generated. The weakness of this approach is the inability to separate acceleration monitoring from signal noise without using excessive filtering.

The second type of implementation of this acceleration based overspeed trip logic also depends on an average acceleration generated by taking the change in angular velocity over time. The difference in the second approach is that the logic adjusts the overspeed trip setpoint proportionally to the acceleration rate. This method is preferred due to its prevention of false trip initiations.

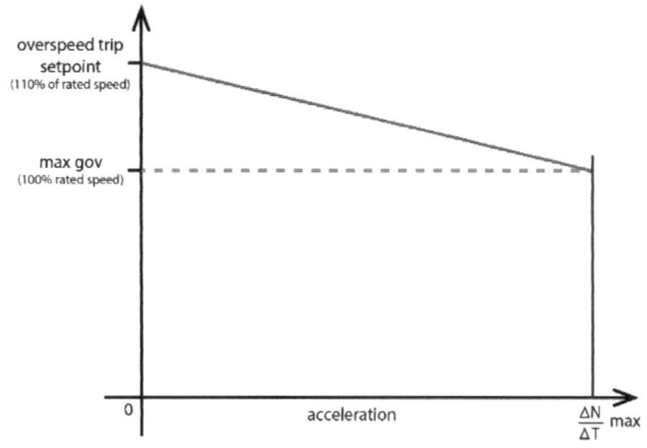

Figure 7: Acceleration compensated trip set point

Summary

The sensitivity analysis results showed that the least significant contributor to the maximum speed excursion was the latency time contribution from the EOSD, SOV, trip header, and trip valve pilot. The trip valve flow characteristic, valve closure time, and inlet volume all had nearly the same impact on the maximum speed attained. The overspeed trip set point and extraction volume both made the largest contributions.

The following figure summarizes our analysis. From the base case where an overspeed event resulted in a maximum speed excursion (N_{max}) of 158%, a 50% reduction in extraction volume lowered N_{max} to 149.4%. A 50% decrease in K_{trip}, lowered the N_{max} to 152.6%. A 50% decrease in either T_v, *f*, or Inlet Volume lowered N_{max} to about 154%. A 50% decrease in T_s, lowered the N_{max} to 155.4%.

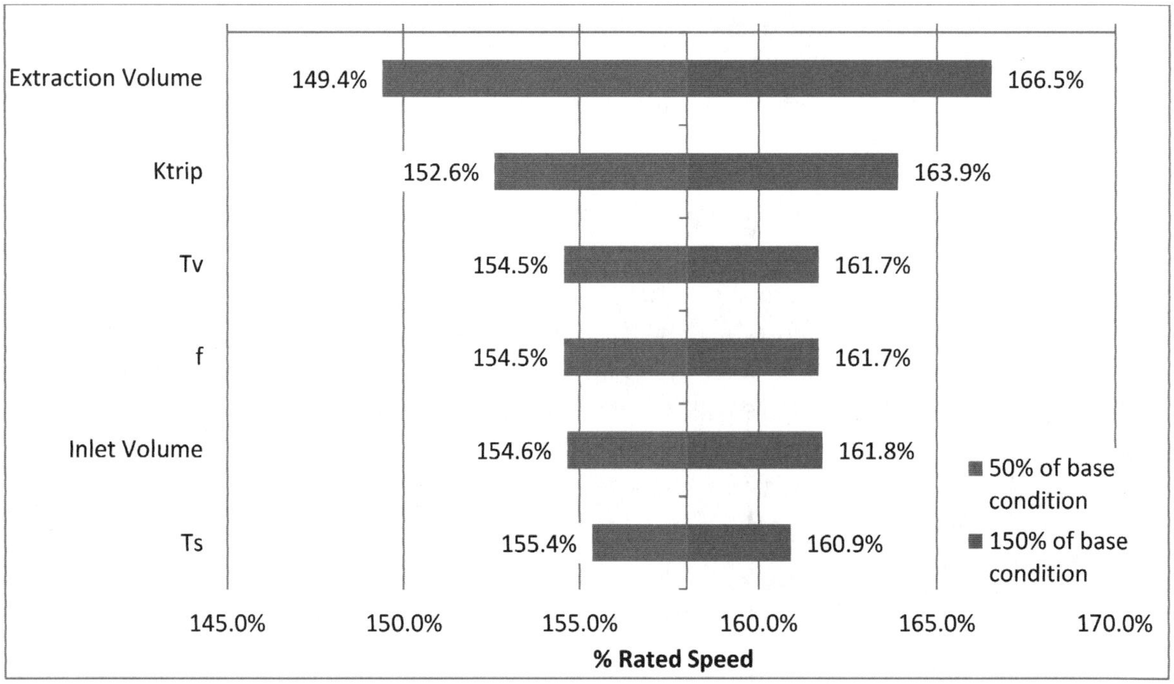

Figure 8: Trip Latency contributors and % rated speed achieved upon overspeed event

While it is evident that improving (reducing) trip delay is beneficial to reducing the maximum speed excursion during an overpeed event, our analysis implies that performance differences in API 670-compliant EOSDs are not very significant to reducing maximum speed excursions.

Careful consideration should be made to the performance, sizing, and distance of SOVs from the trip valve due to its variability. Additionally, due to SOV failure rates and failure modes, any perfomance evaluation should also take into consideration the on-line testing and on-line repairability of the hydaulic trip interface.

As mentioned previously, our model also suggests that the trip valve characteristics (flow and speed), and the inlet volume, contribute to the overspeed excursion roughly equally and slightly more significantly than the EOSP delay.

As an example, a 10% reduction in the overspeed trip percentage (from 10% to 9%) had the same effect as a 20% reduction in T_S. This suggests that when evaluating the overspeed trip system, careful consideration of the trip valve performance, as well as the overspeed trip set point, should be made.

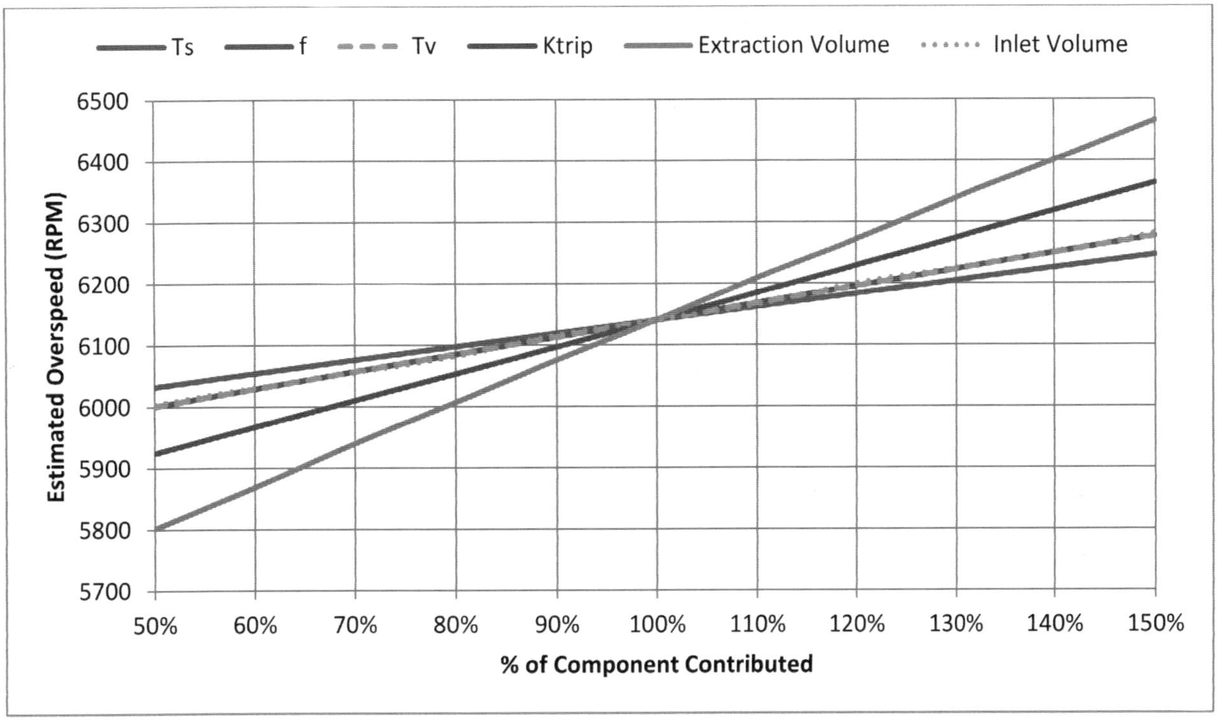

Figure 9: Estimated overspeed achieved upon varying contributors

The example used for this analysis is based on a typical turbine capable of making 75 MW of shaft power. Given the nature of larger turbines, achieving the API 127% maximum overspeed becomes very difficult, so it is important to design the rotor to be capable of safe momentary speeds above this point. For this example, the overspeed is predicted to be 158% of turbine rated speed.

Incorporating a lower trip set point, as well as minimizing entrapped steam volumes, will greatly reduce the maximum speed attained. In this example, reducing the trip set point as well as the inlet and extraction volumes by 50% reduces the predicted overspeed to 140% of rated speed. This translates to an overall reduction in overspeed by about 719 RPM.

Conclusion

The two largest contributors to turbine overspeed during an instant loss of load scenario are the entrapped steam volumes and trip set point. A reduction in both of these contributors will greatly reduce the maximum speed attained. Accurately predicting this speed is incorporated in rotor design which will ensure the safety of the immediate surroundings as well as the lifespan of the turbine. In most cases, the unit will be able to start up without inspection in the event of a coupling failure. This allows customers to restart their process with minimal down time.

These up-front considerations will allow OEMs to design rotor systems that can utilize a higher speed operating range. Operating steam turbines at a higher speed means the driven centrifugal compressor can be smaller and still achieve the same gas compression.

Understanding the design constraints and speed limitations early in a project will also allow the OEM to confirm the rotor design early, and avoid costly design modifications later in the manufacturing process.

The future of mechanical drive steam turbines in large olefins plants will continue to be driven by higher energy steam supply (pressure and temperature) as well as higher horsepower and speed. Collaboration between the customer and OEM is critical to ensure the equipment is properly designed for safe and efficient operation.

References

1. ASME PTC-20.2, 1965, Reaffirmed 1986, "Overspeed Trip Systems for Steam Turbine-Generator Units", American Society of Mechanical Engineers, New York, New York.
2. API Standard 670, 2014, "Machinery Protection Systems", Fifth Edition, American Petroleum Institute, Washington, D.C.
3. API Standard 612, 2014, "Petroleum, Petrochemical, and Natural Gas Industries – Steam Turbines – Special-purpose Applications," Seventh Edition, American Petroleum Institute, Washington, D.C.
4. Taylor, S. L. and Smith, S. S., 2009, "Turbine Overspeed Systems and Required Response Times", *Proceedings of the Thirty-Eight Turbomachinery Symposium*, pp. 157 – 167

TECHNICAL CHALLENGES OF COMPRESSORS AND STEAM TURBINES FOR EFFICIENT AND SUSTAINABLE OPERATIONS IN MEGA ETHYLENE PLANTS

Satoshi Hata, Ph.D.
VP, Engineering
Mitsubishi Heavy Industries Compressor International

Matt Walton, P.E.
Engineering Manger
Mitsubishi Heavy Industries Compressor International

Kyoichi Ikeno
Mechanical Engineering, Turbine Design Section
Mitsubishi Heavy Industries Compressor Corporation

Akinori Tasaki
Mechanical Engineering, Compressor Design Section
Mitsubishi Heavy Industries Compressor Corporation

Prepared for Presentation at the Year 2017 Spring National Meeting San Antonio, TX, March 26 – 30, 2017

ABSTRACT

Turbomachinery equipment manufacturers have been moving forward to apply state of the art technologies for critical components in order to provide more efficient and reliable machines corresponding to the high power requirements for ethylene services. The authors introduce the technology development history map in the past and future for typical designs, verification tests, and application results in terms of transient fluid dynamics, thermodynamics, rotor dynamics, and blade vibration strength evaluation. In addition, after commissioning turbomachinery performance and reliability tends to deteriorate during long-term operation. The typical damage and deterioration map for machinery is introduced, and the authors explain practical technologies that are applied, such as flow path surface treatment, effective on-line washing, combination of anti-corrosion, erosion prevention and stage performance enhancement by partial component replacement, NDE techniques for both compressors and steam turbines, including a unique casing replacement technique on the same footprint for increasing capacity.

INTRODUCTION

During this decade, the capacity of ethylene plants has tended to increase continuously. Consequently, flow rates in the high- and low-pressure stages of the steam turbine driver have also increased. To meet this capacity increase, it has been necessary to apply larger sub-components including large flow governing valves, longer blades, and longer bearing span rotors. For compression equipment, larger impellers with longer bearing spans are required. Compressor and turbine trains with capacities in excess of 1.5 MTPY have already been designed, manufactured, and are successfully operating. Therefore, in order to minimize power consumption and downtime losses, it is necessary to conduct machine monitoring for prevention and diagnosis of performance deterioration.

Turbomachinery equipment manufacturers have been moving forward to apply state of the art technologies for critical components in order to provide more efficient and reliable machines corresponding to the high power requirements for ethylene services. The authors introduce the technology development history map in the past and future for typical designs, verification tests, and application results in terms of transient fluid dynamics, thermodynamics, rotor dynamics, and blade vibration strength evaluation. In addition, after commissioning turbomachinery performance and reliability tends to deteriorate during long-term operation. The typical damage and deterioration map for machinery is introduced, and the authors explain practical technologies that are applied, such as flow path surface treatment, effective on-line washing, combination of anti-corrosion, erosion prevention and stage performance enhancement by partial component replacement, NDE techniques for both compressors and steam turbines, including a unique casing replacement technique on the same footprint for increasing capacity.

Mega Ethylene Plant Trend

In the past decade the capacity of ethylene plants has tended to increase continuously as shown in Figure-1. Consequently, flow rates in the high- and low-pressure stages of the steam turbine driver have also increased. To meet this capacity increase, it is necessary to apply larger sub-components including large flow governing valves, longer blades, and longer bearing span rotors. For compression equipment, larger impellers with longer bearing spans are required. Compressor and turbine trains with capacities in excess of 1.5 MTPY have already been designed, manufactured, and are successfully operating.

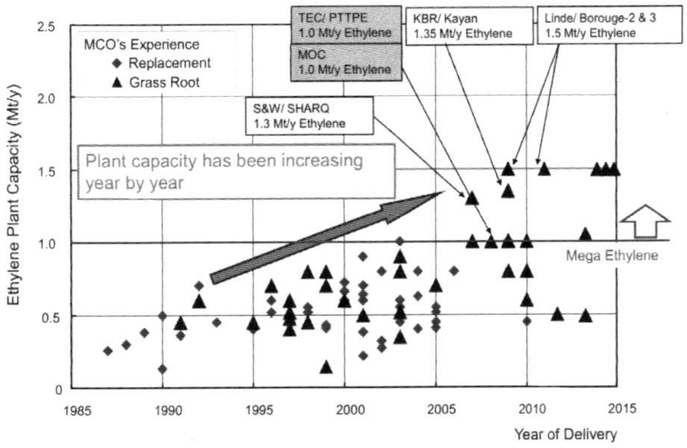

Figure.1 Ethylene Plant Capacity Increase Supported by Upgraded Equipment and Technology

Figure 2 shows the relationship between MCR (max. continuous rotating speed) and steam turbine maximum power. This relation is basically exponential. The maximum required power of a steam turbine for a Mega ethylene plant charge gas compressor (CGC) MP/LP/HP driver is 95MW, and can potentially be as high as 120MW. An optimized MCR for steam turbine and compressor performance and rotor dynamics is around 4,000 rpm.

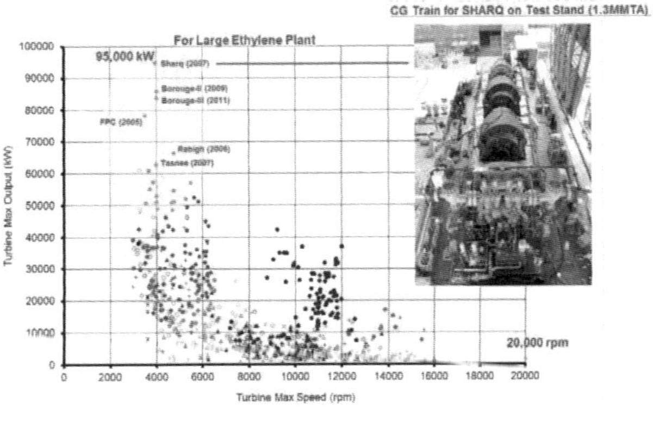

Figure.2 Recent Steam Turbine Range for Mega Crackers - Simple Single Train Solution

In Figure 3, below, the application range of each service and its corresponding Mitsubishi Advanced Compressor (MAC) frame size is shown on a compressor suction flow and discharge pressure map. Large frames (9H and larger) are applied for mega ethylene plants. These same sizes are used in other applications, such as LNG main refrigerant compressors.

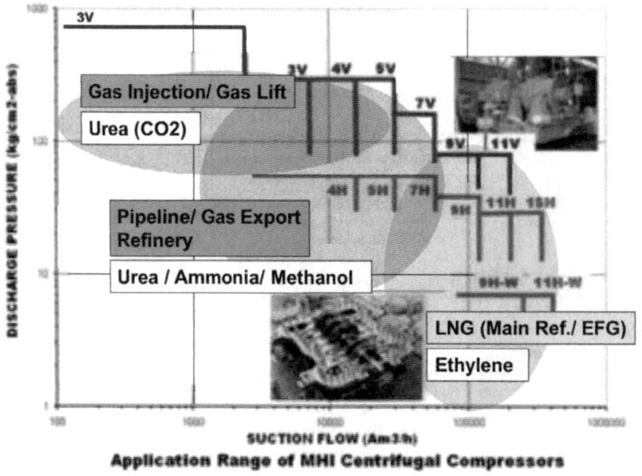

Figure.3 Application Range for Mitsubishi Advanced Compressor frame sizes

Figure.4 Recent Steam Turbine Range for Mega Crackers - Simple Single Train Solution

Every compressor's performance, including rotor dynamics, is measured and verified as a part of the manufacturing process. In addition, the steam turbine and compressors can be coupled and the complete train is tested to confirm combined mechanical performance, such as vibration and bearing condition, on a large test stand at prior to shipment. Figure 4 shows a 1.5MMTA class mega ethylene

CGC train in the same cooperation condition as the site. Through these processes the guarantee of high quality machines is provided to the clients.

Challenging Technologies for Critical Components

As plant capacities become larger and process flows increase the corresponding compressor impeller and casing size must become larger. The required power also increases. Controlled manufacturing processes for 1 piece or 2 piece impellers are also important to meet to the design and quality requirement for these large components. Figure.5 shows key technologies such as 3D impellers and FFD (Flexible Frame Design) which are utilized in compressor design. To increase the rotor dynamic stability a swirl canceling labyrinth seal and a special damper are applied depending on rotor rigidity.

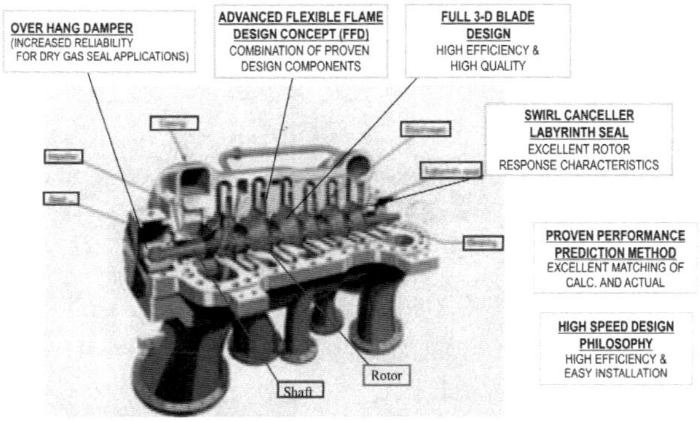

Figure.5 Compressor Key Technologies

To produce the increase power the required inlet steam flow (HP section) and exhaust steam flow (LP section) become larger and a larger component size for the flow path is necessary. Figure.6 shows key technologies to meet the requirement for large plant and machines which are used in steam turbine design. A reaction optimized impulse blade is applied for the inlet and middle stage flow paths in order to increase efficiency and minimize the thrust load due to differential pressure across each blade inlet and exit.

The last 3 stages in the LP section utilize a z-lock integral shroud and high rigidity blade profiles are applied to reduce resonance vibration stresses in the blade and to increase stage efficiencies.

A critical design concept in modern rotor design is to increase rigidity as much as possible by optimizing blade stage number, bearing span, and HP/LP section efficiency at the required operating speed. A high rigidity rotor can reduce vibration response due to unbalance and increase vibration stability.

Recently, operating intervals of over 8 to 10 years have been required to meet customer demand. For this requirement, several new technologies have been developed and applied to existing machines.

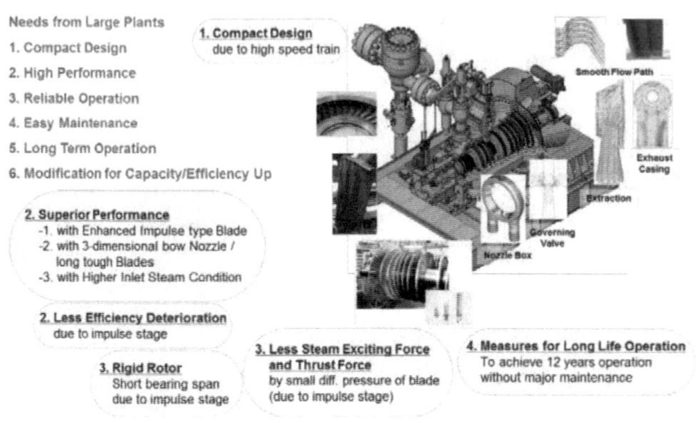

Figure.6 Steam Turbine Key Technologies

Figure 7 shows the technical matrix for present and future steam turbine designs. For long term operation and increased reliability, several surface treatments are applied to protect the flow path against environmental damage from solid particles or drain attack on wet stages.

During operation, the flow path tends to develop deposits caused by degraded steam quality and insufficient control. These deposits cause performance deterioration and potential reliability issues. On-line wash technology can clean and wash the flow path in full power operation in less than one hour.

To simplify the governing valve control system, the oil free governing valve was developed.

Application of Key Technologies (Today) for steam turbines results in higher performance, strength, and reliability as shown below. MCO applies new technologies for the continuous improvement of their designs.

	Performance	Strength	Reliability
Turbine	High speed application	-	On-line washing
Casing	Application for super high press / temp steam		No leakage
TTV	Quick shut down		No stick
Rotor	Higher rigidity	Solid disk	No crack
Blades	Hybrid improved impulse blading	ISB	Ceramic coating PTA welding Hybrid coating of fluorine resin & ceramics
Diaphragms / Nozzles		No bending	Boronize coating Hybrid coating of fluorine resin & ceramics
Governing system	Higher controllability	-	Oil Free governing valve control system

Figure.7 Technical Matrix for Present and Future Steam Turbine Designs

Technological Development History Map

Figure 8 shows the latest technologies for steam turbine development history from 1980 to the present. Flow path and profile design concepts and procedures have been developing from 2D to 3D by improvement of computational technologies and computer calculation speed/data capacity. Due to larger machine requirements, a longer blade is necessary for the last stage. Integrally shrouded blades (ISB) for the high pressure section were developed. This ISB grouping concept for impulse type blading is different from conventional twisted designs and creates one robust blade grouping. TTV and governing valves were developed for larger sizes and to meet higher inlet steam temperatures. Direct lubricant tilting pads for journal and thrust bearings instead of flood types reduce mechanical losses and reduce pad metal temperatures in high velocity situations. Rotor repair technologies have been developed for planned and emergency outages to minimize the risk of no spare rotor. Several types of seals have been developed to reduce leakage and increase efficiency. For long term operation, surface treatment technologies are applied. NDE techniques were also developed to inspect internal damage on rotor disk grooves. One of the latest technologies is PAUT, phased array ultra-sonic testing. Based on measured crack length, location, and orientation the allowable residual operational life can be estimated for turnaround and asset planning.

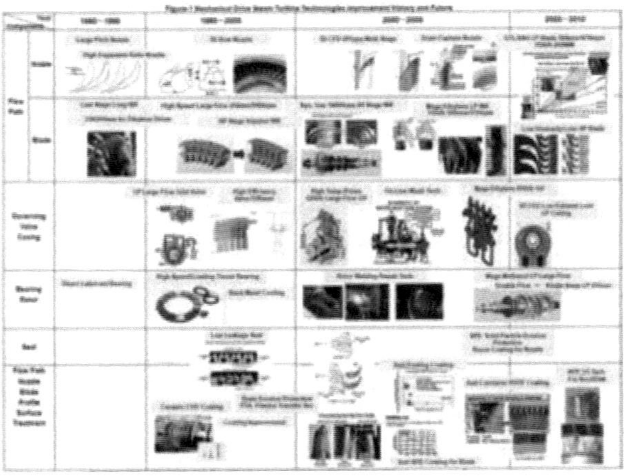

Figure.8 Latest Technologies for Steam Turbine and History

Figure 9 shows a centrifugal compressor research and development map. Similar to the steam turbine, flow path and profile design concepts and procedures have been developing from 2D to 3D through the improvement of computational technology. Compressor rotor stability is critical in high pressure applications in large bearing span machines. To increase rotor rigidity and maintain efficiency, the high boss ratio impeller was developed. To increase the damping of the rotor/bearing/seal vibration system and rotor stability a mechanism for gas excitation force reduction and a high damping seal were developed. In terms of manufacturing processes, new methods were developed such as 1 piece impellers by 5-axis machining for large high flow coefficient impellers and EDM/ECM for small throat, low flow coefficient designs.

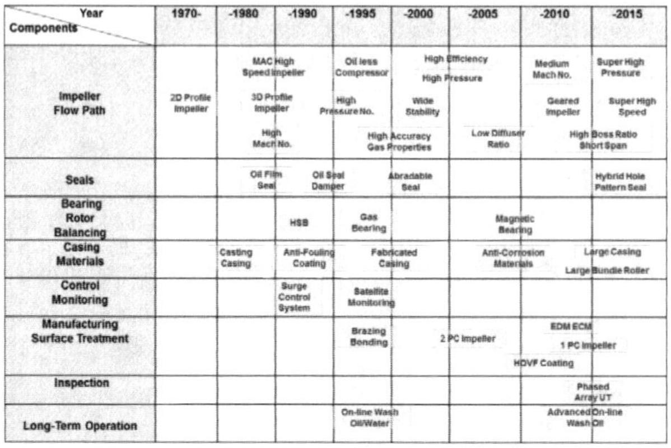

Figure.9 Centrifugal Compressor Research and Development Map

Design, Verification Tests, and Application in terms of Transient Fluid Dynamics, Thermodynamics, Rotor Dynamics and Blade Vibration Strength Evaluation

Machine components are designed to meet the required specification which is decided by process data. In some cases, the compressor impellers and steam turbine blades need to be newly developed. Here, several design and performance verification case studies are explained to show the typical development process.

Compressor Development

Figure 10 shows the Compressor Flexible Frame Design (FFD) Concept. The fixed frame consists of flow path suction parts, discharge parts and bearings. These are high reliability and well-proven components. Flow path aerodynamics design parts are flexible to achieve optimum performance with high efficiency.

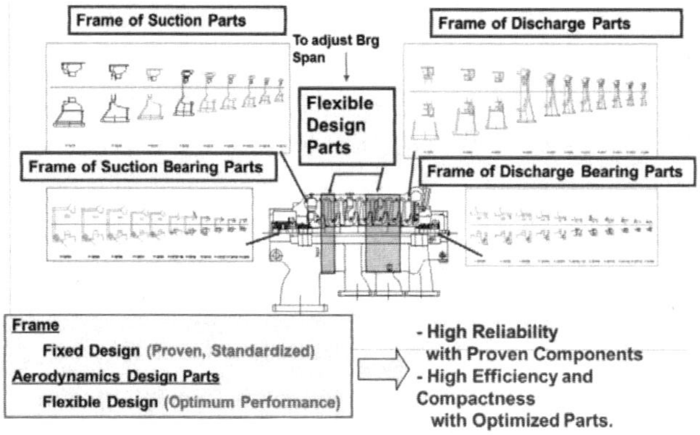

Figure.10 Compressor Flexible Frame Design (FFD) Concept

Figure 11 shows the verification process for a new impeller for a compressor. As the first step, the impeller profile, disk and cover structure dimensions are decided according to the results of performance predictions by CFD and static/dynamic strength evaluation by FEA. The actual impeller size is scaled to the required diameter using the similarity rule of turbomachinery flow paths (velocity triangles) and strength (vibration, stress, power). Next, a standard size test impeller is manufactured according to this similarity rule. In the third step, a single stage impeller performance test is conducted to characterize the performance and compare it to calculated results as shown in Figure 12. This data is added to the impeller performance database and is used for actual application design in FFD process.

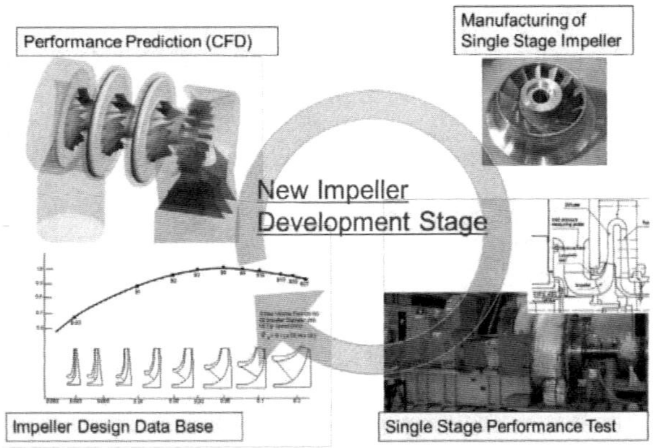

Figure.11 Verification Process for New Impeller

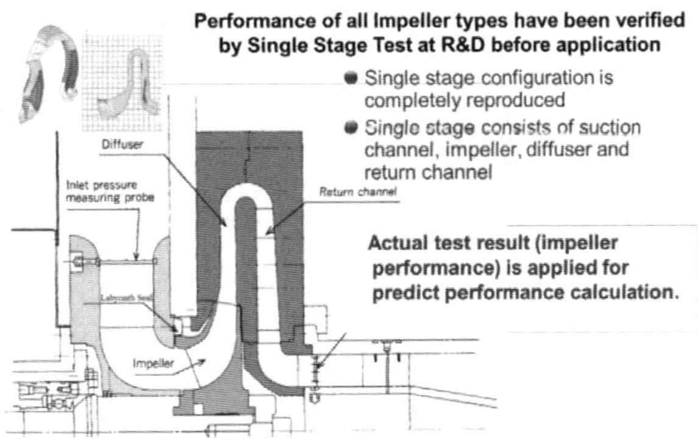

Figure.12 Development / Improvement of Impeller

Case Study: The High Boss Ratio Impeller

The latest impeller design concept combines a high boss ratio and short span as shown in Figure 13. In order to increase the rotating speed for compact designs the shaft rigidity needs to be larger for low vibration response and high stability. In addition, the bearing span needs to be shorter while the same number of stages remains the same. In order to realize both requirements, a new impeller standard series was developed and is in service in operational machines.

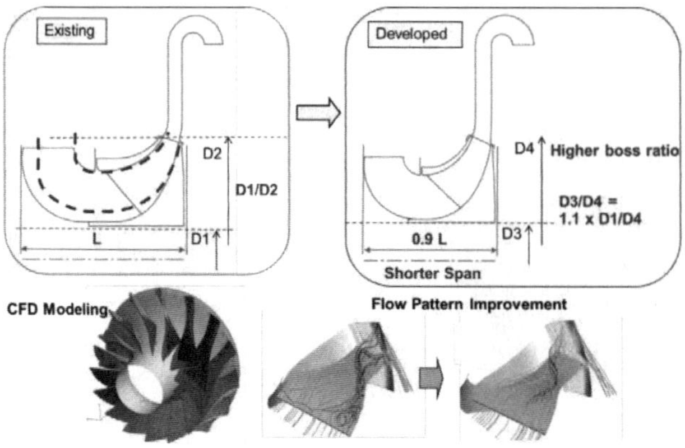

Figure.13 High Boss Ratio/Short Span Impeller Design

Figure 14 shows the profile modification for high boss ratio and short span impeller design and corresponding performance test results. Two concepts are applied: a reduced blade count for reduction of friction losses and a modified blade shape for optimum flow line. This design has a confirmed polytropic efficiency gain over a conventional impeller.

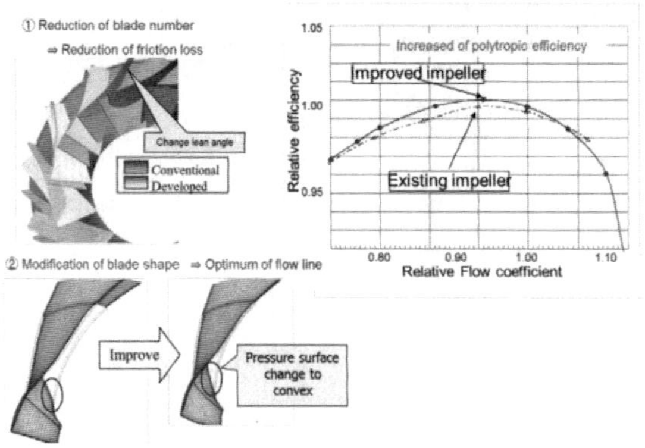

Figure.14 Profile Modification for High Boss Ratio and Short Span Impeller Design and performance test results

The advantage of the high boss ratio/short span impeller for an actual machine application is shown in Figure 15. A Mega ethylene plant PRC train is selected as a typical case. The number of compressor casings can be reduced from two to one; in addition, the weight and train length can be reduced by roughly 20% and 50% respectively. This represents significant cost savings through reduced raw material requirements.

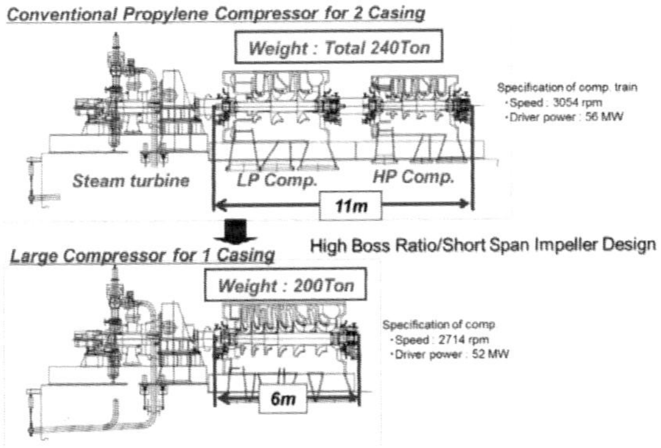

Figure.15 Advantage of High Boss Ratio Impeller

Steam Turbine Development

For steam turbines, Figure16 shows the high performance of the hybrid impulse blade design used to improve the efficiency of the HP section of the turbine. It is possible to minimize the friction losses of the steam path by conducting blade and nozzle CFD analyses and the actual performance is verified by a test machine and actual machine at the shop.

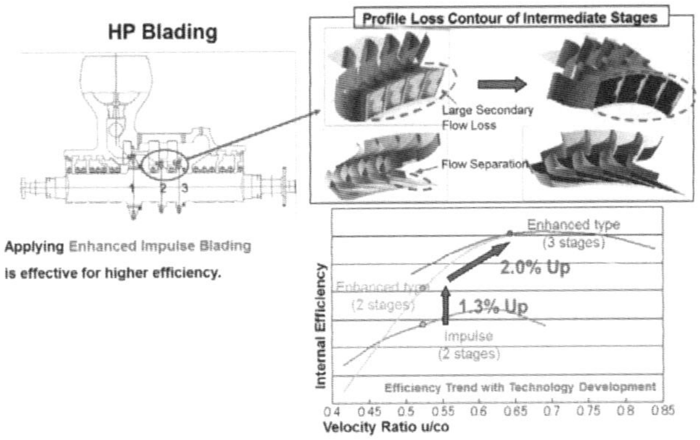

Figure.16 High Performance of Hybrid Impulse Blade Design

Figure 17 shows the high performance of the LP section long blades. This example details improvements to efficiency by optimizing the steam flow using CFD to prevent secondary flows between blades and the high Mach flow area. It was possible to improve the LP section efficiency by 8% in this case study.

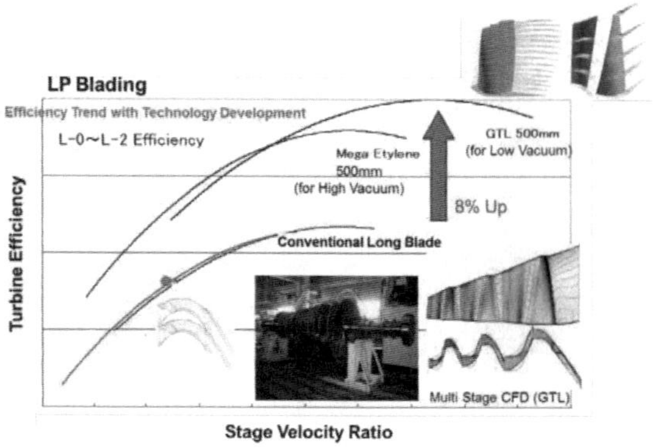

Figure.17 High Performance of LP section Long Blade

Figure18 shows the application for higher inlet steam conditions and this is the approach to achieve higher steam flow utilization.

Increasingly end users request improved steam consumption. Higher pressure and temperature steam reduces the steam consumption of the turbine. High pressure/temperature steam has a higher enthalpy and a higher theoretical maximum efficiency. Figure 18 is a relevant case study from a recent project. One challenge is to apply advanced materials for the high pressure/temperature steam and design structure while maintaining the same dimensions as the original. The stress analysis is done to evaluate thermal stress under static and transient conditions comparing design criteria including creep strength.

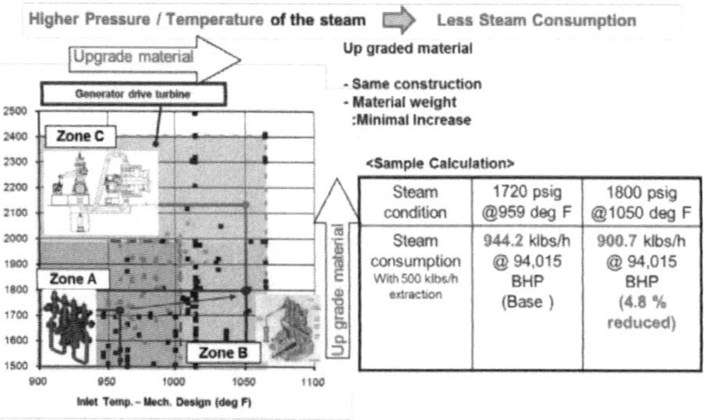

Figure.18 Application for Higher Inlet Steam Condition

Finally, Figure 19 shows the reliable operation for rotor dynamics. API standards are widely used in the petrochemical industry. Typically steam turbine designs must comply with this API standard, including do a maximum shaft vibration of 25.4 micron peak-to-peak during shop mechanical running tests. Figure 19 shows historical vibration records during mechanical running tests in the shop demonstrating the API criterion.

Mechanical drive steam turbines are designed based on impulse type blades. Typically an impulse turbine has a shorter bearing span than a reaction type blade design of similar power. Shorter bearing spans can increase the rotor rigidity and this higher rigidity contributes to low vibration levels during stable operation.

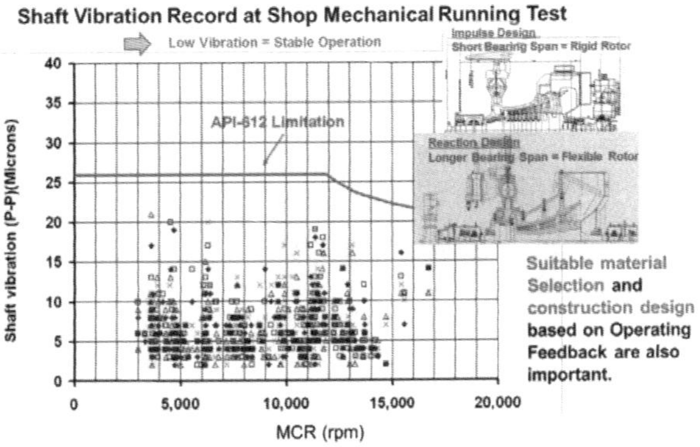

Figure.19 Reliable Operation for Rotor Dynamics

Case Study: LP Blading

In the following case study, the authors conducted an excitation test on two LP blade rows, using an actual scale model, and an actual load test with a scale model to verify the LP blades' reliability. With these tests, the authors could complete the verification program for the large steam turbine. Figure 20 shows the verification test for integrity of actual scale blades and confirmation of the principle of similarity. The rotating blades of steam turbines are essential elements in the conversion of fluid force to mechanical force, and high reliability in operation is a critical requirement. During the design/development stage, vibratory characteristics and strength of the actual blades were checked and verified with a rotating excitation laboratory test. However, an actual load test in a test turbine is the best method of verifying blade reliability. Due to the limited quantity of steam flow in the test facility, the actual loading test is usually carried out using a scale model turbine instead of a full scale model. Furthermore, mechanical and aerodynamic laws of similarity can be applied to the scale model turbine. This figure shows a summary of the principles of similarity.

Item	1/n Model	Note	
Rotating Speed (Ω)	n	Adjusted by tip speed	
Acceleration	n	$r\Omega^2$	$(1/n \times n^2 = n)$
Mass (m)	$1/n^3$	Volume	
Centrifugal Force (F)	$1/n^2$	$mr\Omega^2$	$(1/n^3 \times n = 1/n^2)$
Centrifugal Stress (σ)	1/1	$mr\Omega^2/A$	$(1/n^2 \times n^2 = 1/1)$
Twist Angle	1/1	$\rho\Omega^2 \ell^2/G$	$(n^2 \times 1/n^2 = 1/1)$
Elongation (Δℓ)	1/n	$\ell mr\Omega^2/AE$	$(1/n \times 1/n^2 \times n^2 = 1/n)$
Displacement	1/n		
Length (ℓ)	1/n		
Area (A)	$1/n^2$		
Velocity (V)	1/1	Adjusted by mach number	
Weight Flow (G_s)	$1/n^2$	AV	$(1/n^2 \times 1/1 = 1/n^2)$
Heat Drop (H)	1/1	Adjusted by mach number	
Turbine Output (L)	$1/n^2$	$G_s H$	$(1/n^2 \times 1/1 = 1/n^2)$
Turbine Torque (T)	$1/n^3$	L/Ω	$(1/n^2 \times 1/n = 1/n^3)$
Resonant Stress (σ_r)	1/1		
Blade Frequency (ω)	n	$\frac{1}{\ell^2}\sqrt{\frac{I}{A}}$	$(n^2 \times \sqrt{1/n^2} = n)$
Reduced Frequency (ω_r)	1/1	$\omega \cdot C/V$	$(n \times 1/n = 1)$
Reynolds Number (Re)	1/n	$\ell \times V$	$(1/n \times 1/1 = 1/n)$
Ω_R/Ω_N	1/1	Resonant RPM Ω_R /Blade Natural Frequency RPM Ω_N	
Tolerance	1/n		
Damping (δ)	1/1	Examined in this paper	
Stimulus (s)	1/1		

r : Radius , G : Shear Modulus , ρ : Mass Density
E : Elastic Modulus, C : Blade Cord Width, I : Moment of Inertia

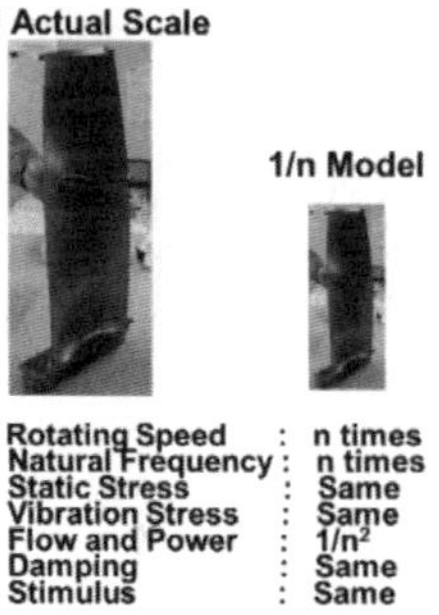

Figure.20 Verification Test for integrity of actual scale blades

Figure 21 shows the verification test for integrity of actual scale blades. Before a mechanical rotating test in a vacuum chamber typically used for high speed balance and the actual load test, a tapping test of the LP blade rows in the free-free condition was carried out. The authors conducted a 3-D model study of the LP blades (actual/ scaled model) to confirm the natural frequency mode shape of the scaled low pressure stage blades between analyses and actual measurements. As a result, a measured value for each mode shows a ±3% variation, compared with the calculated value.

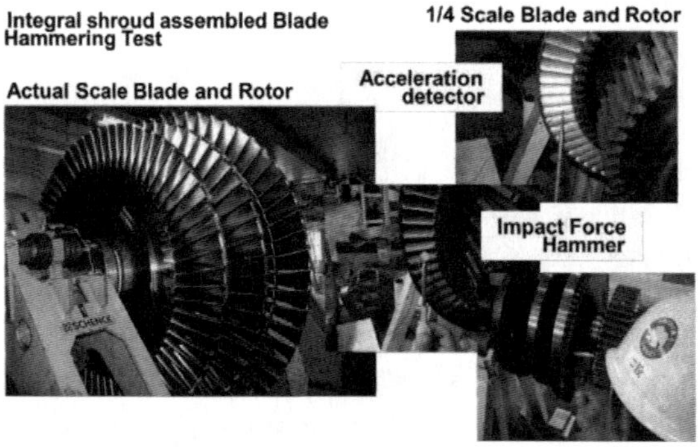

Figure.21 Verification Test for integrity of actual scale blades

Test Turbine

Figure 22 shows the test facility set-up schematic used for the scaled LP blade row verification. The major specifications of the test turbine and details of the manufactured test turbine and compressor are shown in this figure.

The test turbine was planned as a 5-stage straight condensing type, with 1 control stage and 4 LP stages. Its size was selected considering a steam flow for blade loading at the most severe case. For the inlet steam condition adjustment, a pressure control valve and temperature decrease devices were installed on the inlet steam line and controlled accordingly. Furthermore, the exhaust pressure was adjusted, and a vent valve was applied on the condenser. A 2-stage centrifugal type impeller was selected for the test compressor and air was used as the operating fluid in a closed loop.

Special Measurements

During the test, the authors took some special measurements for verification of the LP blade rows, including excited blades frequencies, vibration stress and flow path performance, in addition to normal measurements (speed, steam flow, pressure, temperature on piping, shaft vibration, etc.). In particular, to verify the LP blade rows vibration response, the authors computed blade vibration stress using strain gauges and thermocouples installed on the blades with a telemetry system. Figure 8 shows the special wiring for the verification test. Strain gauges were installed on the blade profile, tip or mean edge. Six strain gauges were installed on each stage (L-0 to L-3 stage). The authors adopted the latest available technology, an induced electromotive type telemetry system (no battery/self-generation type). Figure 23 shows the test rotor with the installed telemetry system. For reference, the cross section of the telemetry system is shown. The vibration response of the blades was available during testing through this system. Figure 24 shows a photo taken during measurement of the vibration response.

For verification of LP blade performance, the authors measured the Pressure and Temperature of each stage in the turbine casing and the turbine power through a coupling torque meter.

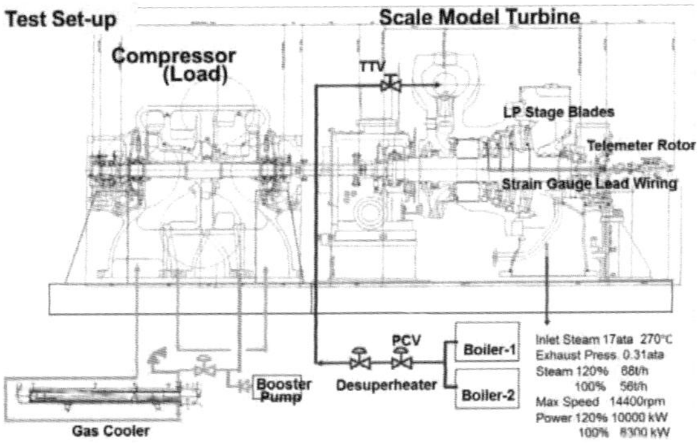

Figure.22 Verification Test for integrity of actual scale blades
Full Load Blade Excitation Test

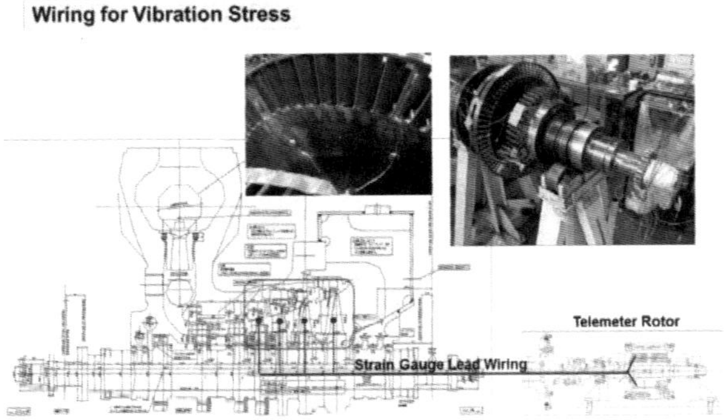

Figure.23 Verification Test for integrity of actual scale blades
Full Load Blade Excitation Test

Test Conditions

The applicable operation range of the type 14 LP blade rows was evaluated for the case of a 1/4 scale test. The measured points were covered at almost all the applicable operating ranges. Turbine power and exhaust pressure at MCR (14,400 rpm) were adjusted and a stable operating condition was confirmed. The speed was then decreased to Minimum Governor Speed (MGS, 11,520 rpm), and then increased to MCR again and stable operating conditions were confirmed. During the stable condition at MCR, performance measurement was conducted, and as the speed changed, blade vibration response was measured (twice for each measurement)

These measurements were performed for a total of 50 hours of operation over seven days. Additionally, 1.5 hours of 110% load test at harmonic resonant speed was conducted to verify blade stress cycles in excess of 1×10^7.

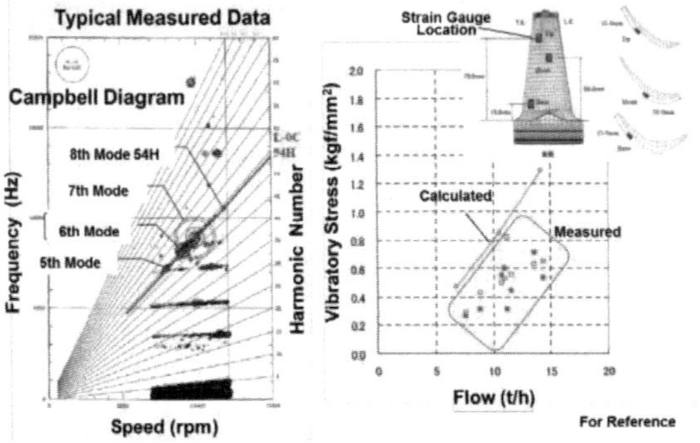

Figure.24 Verification Test for integrity of actual scale blades
Full Load Blade Excitation Test

Test Results

Figure 24 shows the measured Campbell Diagrams of the L-0 last stage blades. In these diagrams, the size of the circle represents the vibration stress levels. By changing conditions during the verification test the authors evaluated a variety of vibration response stress levels including harmonic resonance, low vibration modes (below 20 harmonic), nozzle wake resonance, high vibration modes, and random vibration levels from a low exhaust vacuum condition (L-0 stage).

The authors then checked and obtained vibration response from harmonic resonance to low vibration mode (below 20 harmonic) test results. The diagram shows a comparison between calculated and measured results. This confirmed the natural frequency of the low vibration mode of the scaled blades, which agreed with the study results. (Other LP stage results also agreed.)

The largest resonant stress measured during the verification test for each vibration mode was tabled, and a safety margin of L-0 stage for each vibration mode was calculated based on an allowable stress and weakest point stress. This was considered as an actual stress measured point (tip or mean edge) in each vibration mode. This enables the evaluation of the safety margin for all blades. In the case of the L-0 stage, the safety margin of the 1st mode was the lowest, but met design requirements.

Typical Damage and Deterioration Maps

Figure 25 shows the typical damage modes of mechanical drive steam turbines. Steam flow path parts such as nozzles, diaphragms and blades are eroded by solid particles in the HP section and water droplets in the LP section. Corrosion fatigue failure can occur in the blade of the LP section at the transition zone of wet and dry conditions in a corrosive environment. Excessive load due to drain carry over and poor quality of lubricant oil can induce melting damage of bearing pads.

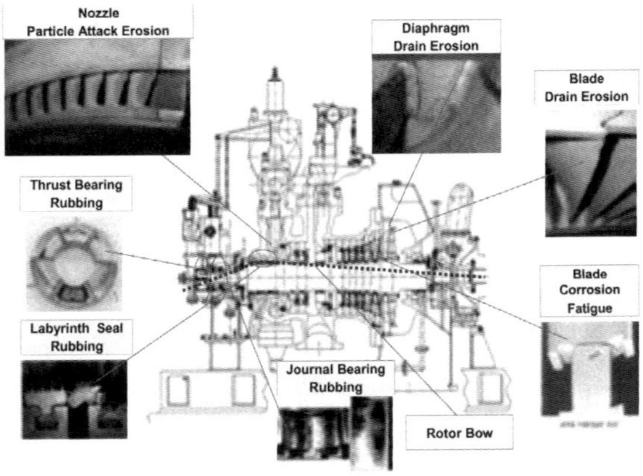

Figure.25 Steam Turbine Deterioration Map

Figure 26 shows the compressor deterioration map. Depending on the actual operating conditions, process gases can become wet and corrosive. As a result, the impeller tends to have flow path erosion, cracking at stress concentration critical areas and heavy chemical fouling which causes performance deterioration. At the balancing line, the differential pressure changes from design condition, excessive thrust force acts on the thrust pads and thrust bearing pads will see high metal temperatures. In a worst case scenario pads can be damaged and compressor train may need to be tripped. Dry gas seals can have damage due to oil mist invasion or lower than dew point temperature of supply.

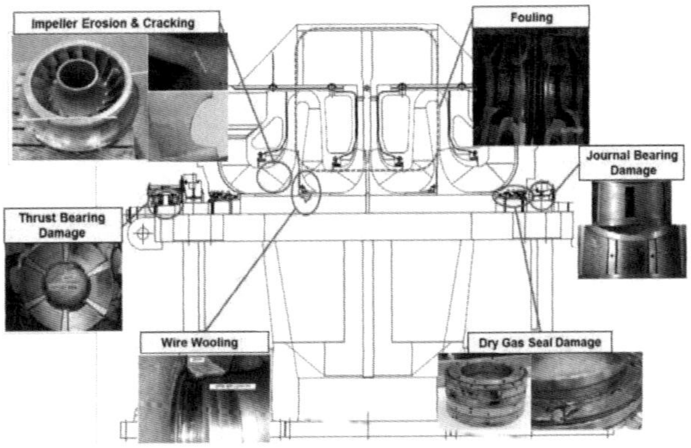

Figure.26 Compressor Deterioration Map

Practical Countermeasure Technologies Applied

Flow Path Surface Treatment

Figure 27 shows the applicable technologies for long term operation. Key technologies for steam turbine reliability improvement such as special blade coatings have been developed. In addition, applications of stainless steel materials to protect against drain erosion on the diaphragms, and the development of hybrid coatings of fluorine resin and ceramic have been utilized. One of the essential factors for long-term operation is coating technology. Regarding the deterioration phenomenon of internal parts, the two major causes are solid particle erosion and drain erosion. In the case of solid particle erosion, boronizing heat treatments are applied for high pressure stage nozzles, particularly the 1st stage nozzle. On the other hand, in case of drain erosion, Cr-TiN coatings and PTA welding are applied.

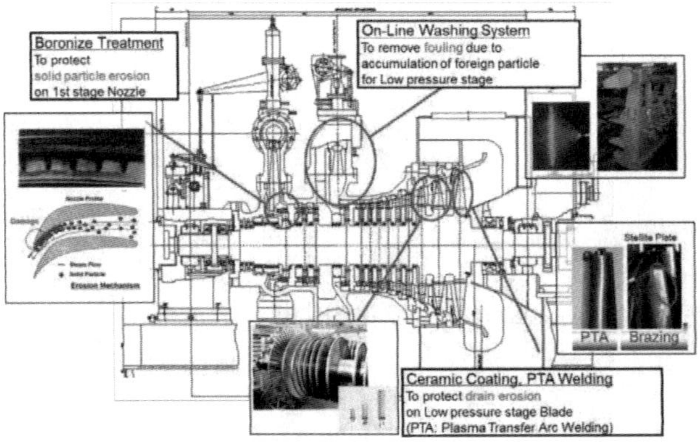

Figure.27 Applicable Technologies for Long Term Operation

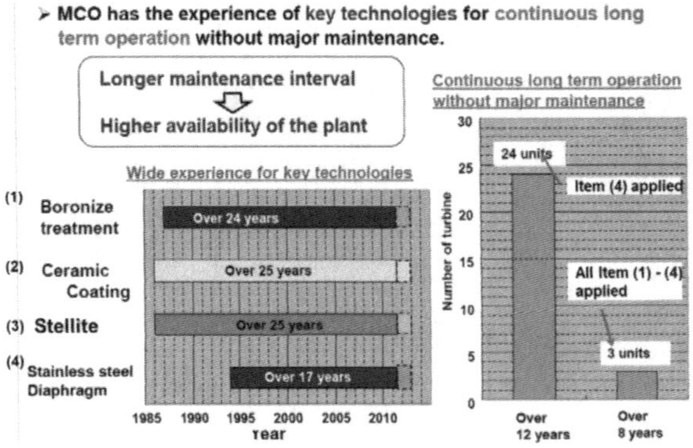

Figure.28 Experience for Long Term Operation

Figure 28 shows experience using applied technologies for long term operation. Key technologies, such as boronizing heat treatment, ion plating, Stellite brazing and stainless steel diaphragms, have decades of operational experience. Machines utilizing these countermeasures have achieved long continuous operation. Without major maintenance, 24 units, including drivers used in LNG plants, are running over continuous periods of 12 years, and three units run over eight years. Therefore, the turbines can be operated with longer maintenance intervals which will lead to achievement of higher availability of the plant.

Effective On-Line Washing

To compete in today's economic climate, petrochemical plants are strategizing on continuous long-term operation to reduce maintenance costs and increase productivity. This strategy has led some plants to go from eight years between turnarounds to 10 years. For rotating machinery such as mechanical drive steam turbines, one factor that affects this strategy is heavy deposition on steam turbine internals, caused by impurities in the steam. These impurities result in fouling on the blade and nozzle path surfaces due to contaminated materials such as silica and sodium as shown in Figure 29. As a result, turbine performance tends to deteriorate gradually.

As a countermeasure to performance degradation an innovative on-line washing technique can be utilized to minimize the impact caused by fouling of the steam path for large multi-stage condensing steam turbines. This technique, although applied here to extracting-condensing turbines, is also applicable to large condensing turbines. The new technology has water injection nozzles located in the steam chest of the extraction valve rack. The injection nozzles are plumbed to a water supply source, which controls a set point temperature by controlling the water injection rate. The objective is to directly wash off deposits adhering to the blades and nozzles on the low-pressure side with minimal power turndown, and without impacting the turbine's long term performance. Erosion damage and thermal stress of internal parts such as chest valves and blades due to the injected water had to be taken into consideration. To properly achieve this objective, and considering the potential for damage during the online wash, a new extraction valve box had to be designed.

The new design had to consider the effects of optimizing the mixing zone of the steam and water injection to generate a specific particle size, moisture propagation through the condensing section, mechanical deflection of stationary components, and the overall thermodynamic analysis of each stage during the on-line wash. A prototype model was built and several experiments were carried out based on the practical operating conditions of actual steam turbines. Analysis (FEA) was used to evaluate the strength of the internal parts during actual online washing. The final design was a compact extraction box that could replace existing models without any machining of the casing.

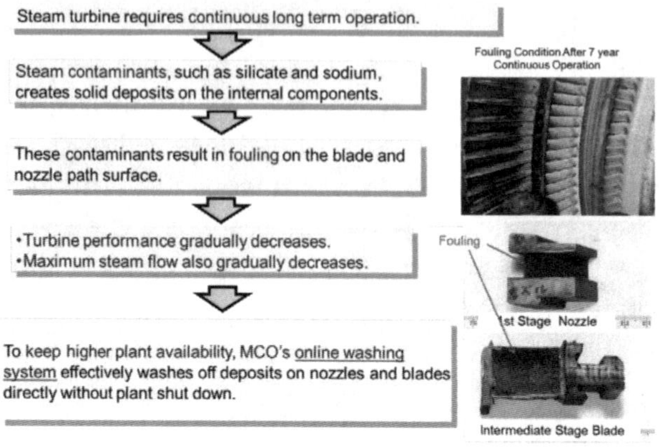

Figure.29 Background for development of On Line Washing

To keep plant availability, the online washing system is effective to wash off the deposits without a shut down. The washing system can be furnished on steam turbine itself, not on the inlet steam piping; it does not damage any other function, but improves the expected steam turbine performance. Figure 30 shows a drawing a map of typical online washing procedure. The steam turbine online washing system has already been used and applied over 30 existing turbine applications. First, hot steam comes from the high pressure section and the steam is mixed with injected water through spray nozzles in the turbine valve chest. The mixed steam is controlled in order to avoid the saturated condition, and fouling material on the nozzle/blade rows is washed by the steam in the wet condition.

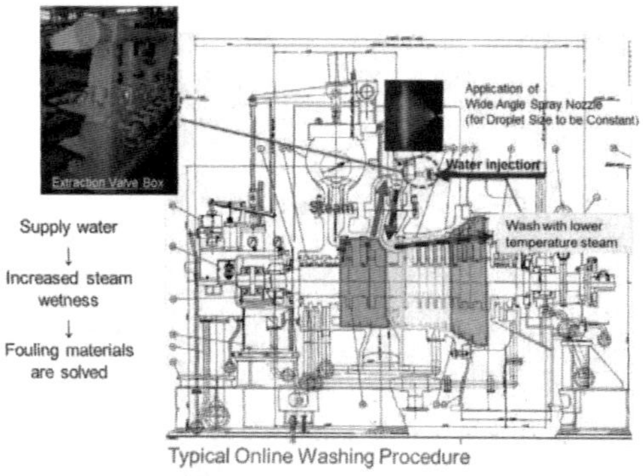

Figure.30 Online Washing for Steam Turbines

Figure 31shows typical operation data during online washing. The lower axis is steam flow, and the left axis is after 1st stage pressure for LP section. Before online washing, after 1st stage pressure is over the designed line, due to fouling condition. Though pressure gradually decreases during online washing, the pressure becomes stable in below the designed line after on-line washing. It means that performance was improved by on-line washing.

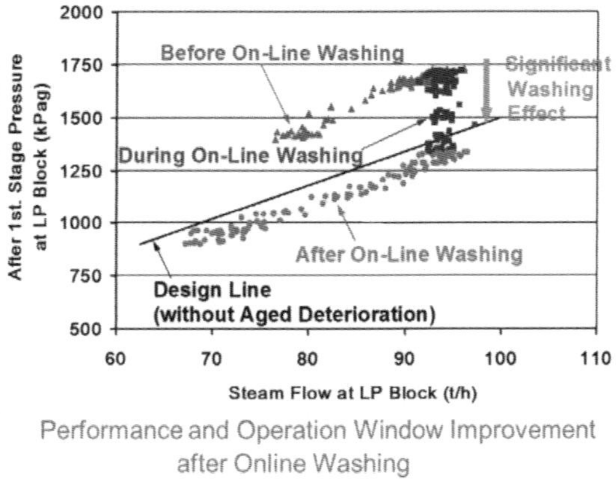

Figure.31 Performance Recovery by Online washing

Figure 32 shows the typical effect of steam turbine online washing at field execution. The result shows high power improvement after online washing, roughly a 6% recovery when compared to before washing.

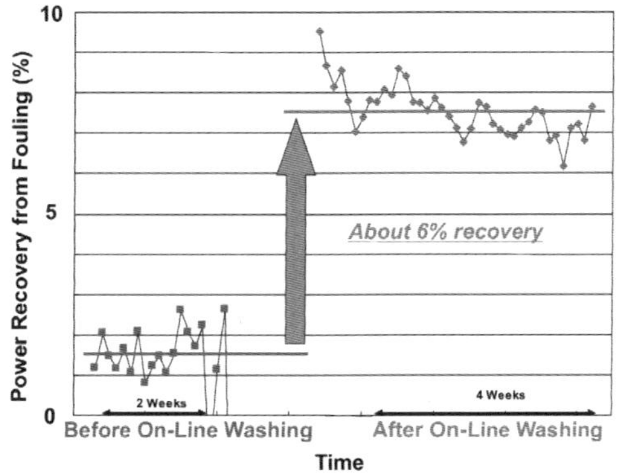

Figure.32 Typical Effect of Steam Turbine On-line

For the compressor, Figure.33 is a typical fouling condition for a charge gas compressor in an ethylene application. Heavy hydrocarbon deposits will accumulate at the flow passage if the gas temperature rises above 90 °C. This fouling causes large flow path losses and a reduction in efficiency, and requires more power from steam turbine driver to maintain output.

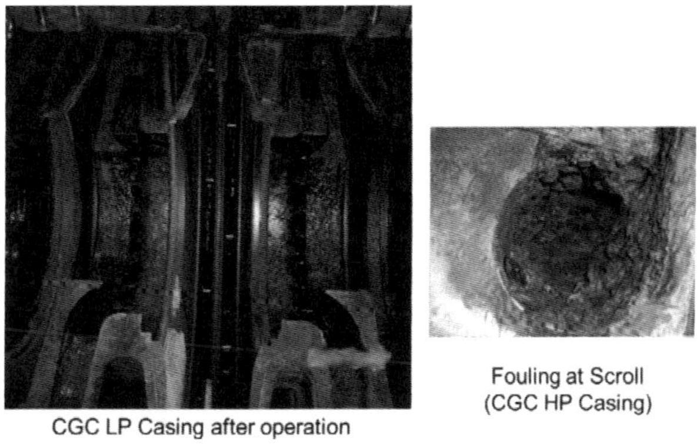

Figure.33 Typical Fouling in Charge Gas Compressor

CFD analysis was conducted in order to characterize the effectiveness of wash oil injection from the return bend, shown in Figure 34. As an initial result, the injected oil does not distribute remarkably. The effective oil wash area is only 3.6% of the return channel even if the quantity is increased. Another more effective injection method is necessary.

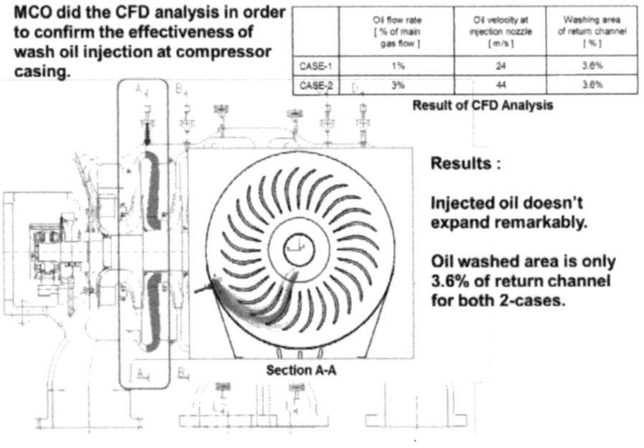

Figure.34 Verification of Oil injection at Return Bend

Figure 35 is the expected scenario of oil injection. First, oil is injected from return bend, but oil coverage is not expanded remarkably. However, when the oil reaches the impeller it washes the fouling deposit on the rotating impeller. After that the oil particles are scattered by the impeller. Therefore, the oil coverage is expanded widely to the next stage.

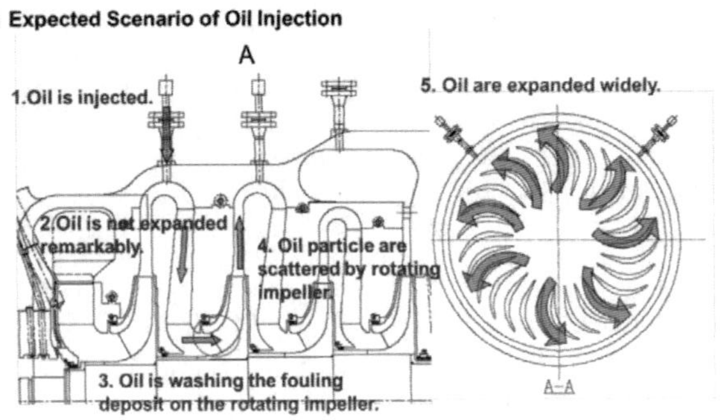

Figure.35 Expected Scenario of Oil injection at Return Bend

Figure 36 shows results of fouling survey at site successful oil/water injection. This is the result of a fouling survey at site. Due to effective oil injection, fouling of internal passages was minor. Internal hardware was very clean as shown on these pictures.

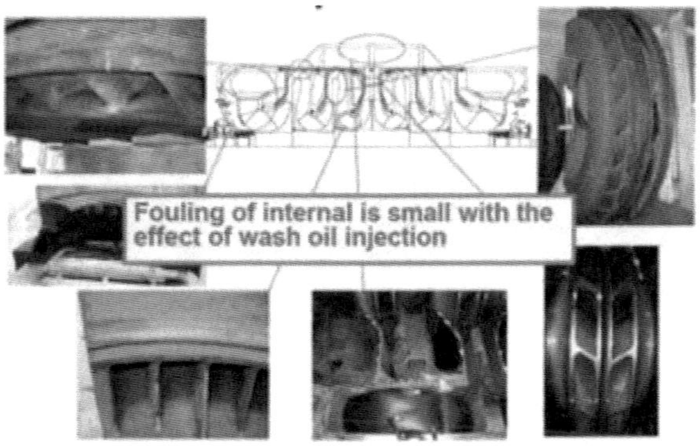

Figure.36 Results of Fouling Survey at Site Successful Oil Wash Injection

Combination of Anti-Corrosion and Erosion Prevention

A typical example of applications for robust technologies is shown in Figure 37. Application specific coatings are used to provide additional improvement against steam turbine efficiency deterioration. ISB designs were applied for all stages (impulse and reaction types) to increase overall efficiency and prevention of corrosion fatigue in the dry/wet transient zone. For the inlet 1st stage nozzles a boronizing heat treatment was applied to prevent solid particle erosion and L-1 and last stage blade surfaces were treated with TiN ion plating and PTA (plasma transfer arc welding) respectively to prevent drain erosion. In addition, wide pitch nozzles were applied for all stages to increase stage efficiency and to minimize after stage pressure due to fouling. In Figure 40, the difference in after stage pressure increase due to deposits on the nozzle profile is explained by comparing a fine pitch nozzle and wide pitch nozzle. If deposit thickness increase rate (mm/year) is the same, the nozzle area reduction rate of wide pitch nozzles can be decreased by half because of the smaller number of nozzles or smaller profile surface area than the fine pitch nozzle. The after-stage pressure increase is proportional to steam flow and the turbine can have a wider margin for operating time before reaching the pressure limitation under fouling conditions, while maintaining the required power.

Furthermore, if online washing is executed periodically, the operating time can be further extended by performance recovery. In this case study, the power recovery ratio after online washing is assumed to be 80%. The power decrease rate due to after-stage pressure increase caused by fouling is half of that of a fine pitch nozzle when using a wide pitch nozzle application. If newly developed coating and other improvements, including online washing and wide pitch nozzles as explained above, are applied systematically the turbine integrity and performance can be maintained to the same as the original during operation in excess of 8 years.

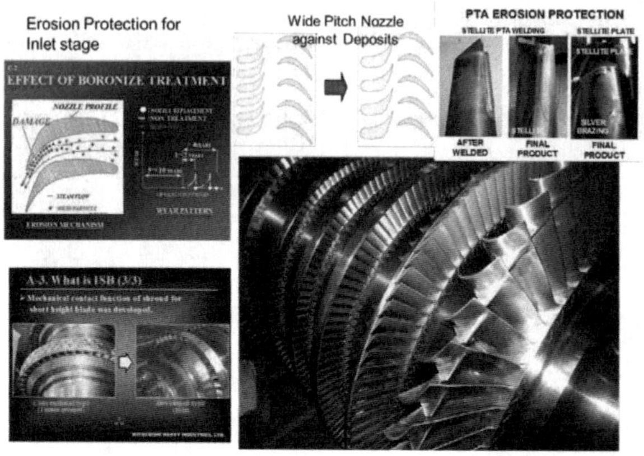

Figure.37 Surface Treatment Application for Steam Turbines

Figure38 is the typical fouling condition for charge gas compressors in ethylene service. To avoid fouling, water injection is done to keep the gas temperature low by water vaporization. A hydrophobic, low surface friction layer is applied as an anti-fouling coating for compressor flow paths to avoid the accumulation of polymer on internal surfaces for minimum efficiency losses. Oil (wash media) injection is done to clean the compressor internals and retain high performance without overhauls.

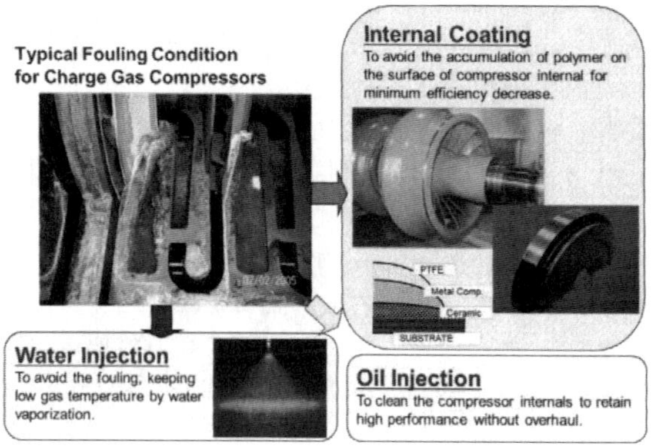

Figure.38 Anti-Fouling Coating for Compressor

Figure 39 is a typical example of an anti-erosion coating applied by high velocity oxygen fuel (HVOF) spraying. After four years of operation the 4th and 6th stage impellers had corrosion damage. Before changing the materials to anti-corrosion materials, for the spare rotor impellers of these stages an anti-corrosion coating by HOVF was applied.

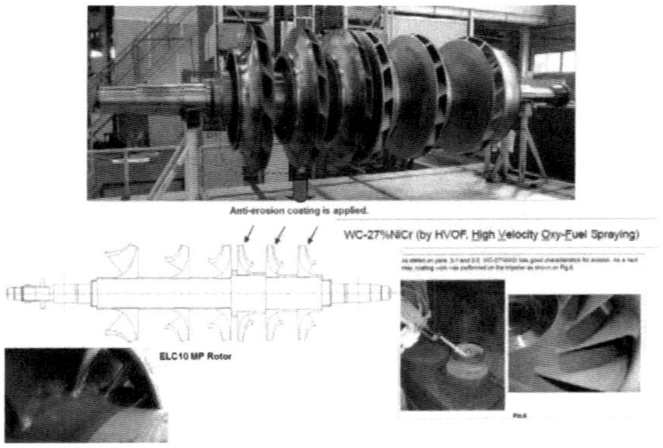

Figure.39 Anti-Erosion Coating HVOF Running Rotor

Figure.40 shows the evaluation results of anti-erosion factor for applied HVOF coating performance. The best composition of coating was decided from the point of view of anti-corrosion and erosion.

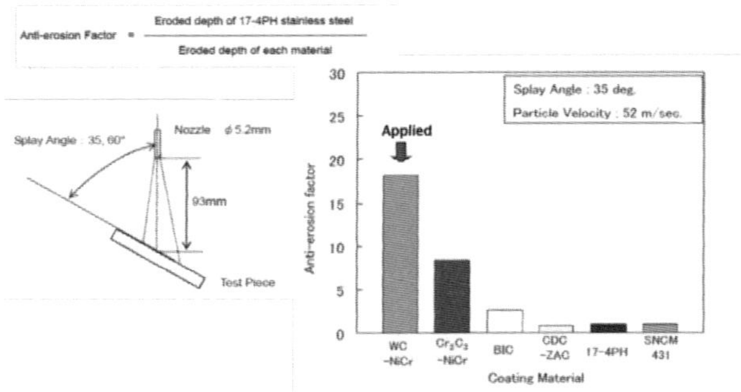

Figure.40 Evaluation of Anti-Erosion Factor for Applied Coating HVOF Performance

Stage Performance Enhancement by Partial Component Replacement

In case that existing machines need to be modified according to a desired process change or plant capacity increase, two options are proposed: one is a footprint replacement, casing with internals replacement; the other is the internal replacement of the rotor with impellers and stationary parts such as diaphragms. Figure 41 demonstrates the case of retrofit design of a flow field for a steam turbine LP section. Only the nozzle (stator) profiles of the diaphragms of the last 3 stages are modified to increase LP section efficiency. In this case study, total efficiency could be improved by 1.5%.

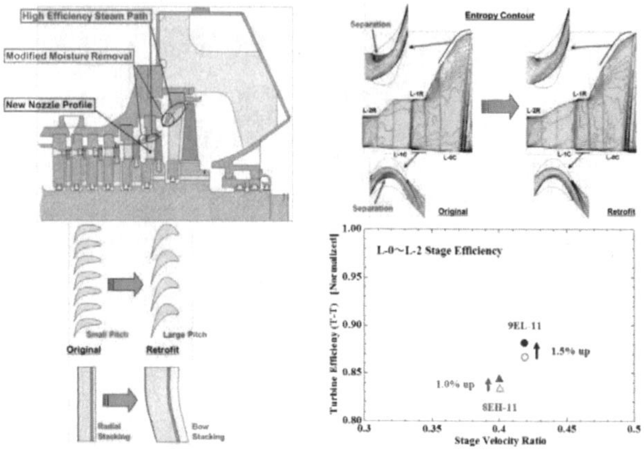

Figure.41 Retrofit Design of Flow Field for Steam Turbine LP Section

Figure 42 shows drain erosion prevention for a steam turbine LP section. The last stage moisture is about 10% to 14% and 20% to 30% of this moisture becomes water. Small water droplets of 50 to 300 micrometer diameter scatter to the blade with a high velocity and attack the blade profile surface. The amount of droplets has an adverse effect on the turbine last stage efficiency. If the amount of droplets is reduced, the drain damage is attenuated and efficiency can be increased. This nozzle profile has a slit to gather the condensation on the nozzle surface and an intake hole is provided in order to extract the drain outside of the flow path.

The structure geometry such as slit and hole locations are decided by analysis of the results of a 2-phase CFD (steam and water droplets).

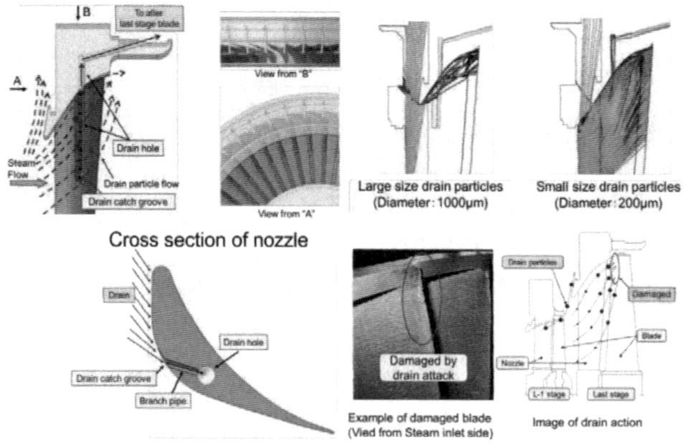

Figure.42 Drain Erosion Prevention for Steam Turbine LP Section

NDE Techniques

For long term operation, surface treatment technologies are applied. NDE techniques are also developed to inspect the internal damage of the rotor disk groove. One of the latest technologies is PAUT, phased array ultra-sonic testing. In contrast to traditional straight-beam UT, phased array sends and receives multiple UT signals simultaneously, allowing for greater detail and resolution. Based on measured crack length, location, and orientation the allowable residual operational life can be estimated for turnaround planning. Figure.43 shows the results of a PAUT inspection for steam turbine stress corrosion cracking (SCC) at the inside of the groove under the blade assembly.

Figure 44 shows a section of part of steam turbine disk and an actual stress corrosion crack fracture analysis result. According to these results, the PAUT measurement accuracy of crack length and location is verified.

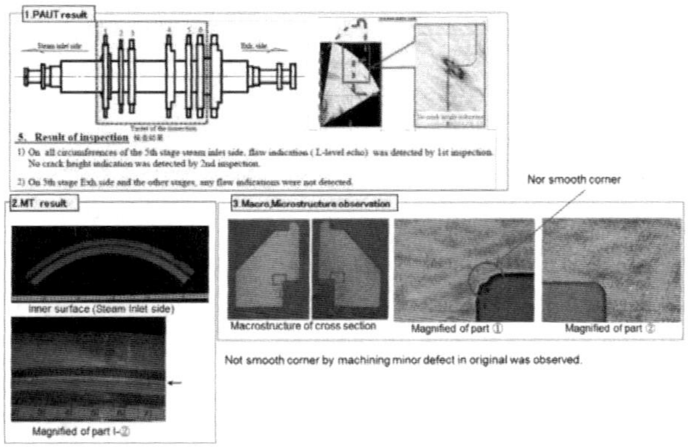

Figure.43 PAUT Phased Array Ultra Sonic Test for Steam Turbine SCC

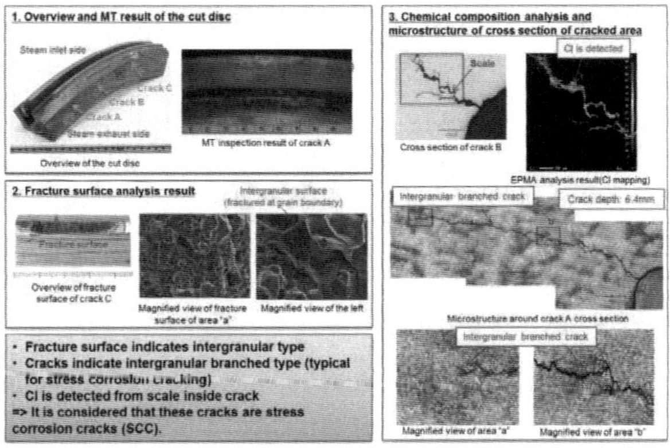

Figure.44 Typical SCC Crack for Steam Turbine Rotor Disk

Unique Casing Replacement Technique on Same Footprint

Figure 45 shows the basic steps for plant revamping considerations. As a first step, the end user should consider whether or not process improvements alone can satisfy their rerate targets. If so, this can be accomplished without equipment modifications. If process adjustments alone do not satisfy the requirements, the second step in this process is to consider re-rotoring the existing equipment. However, when mechanical revamping of the existing equipment will not achieve the plant rerate target, the end user will most likely need to consider equipment replacement and/or parallel train applications as shown in the slide.

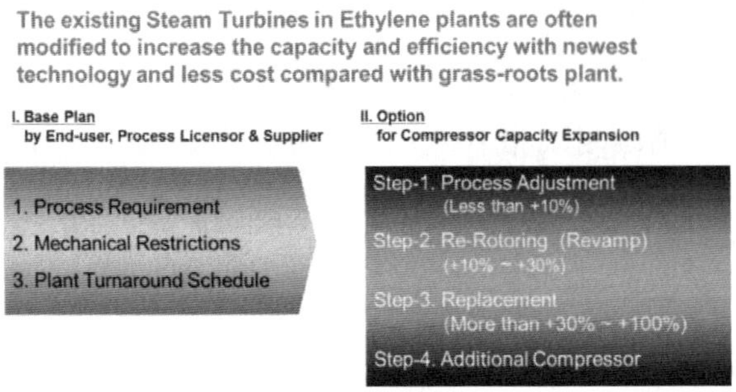

Figure.45 Basic Steps for Plant Revamping

In the replacement approach, an existing compressor is replaced with a new high efficiency design. In this case shown in Figure 46, it is important that the new large compressor is designed to allow the casing to be put on the existing foundation and the nozzles to set in the opening of the existing foundation. When a standard design compressor is applied to a replacement project, often nozzle positions will not fit in the existing main gas pipe positions and/or do not set in the narrow openings of the existing foundation. Modification of foundations significantly increases the difficulty, time, and cost of the installation.

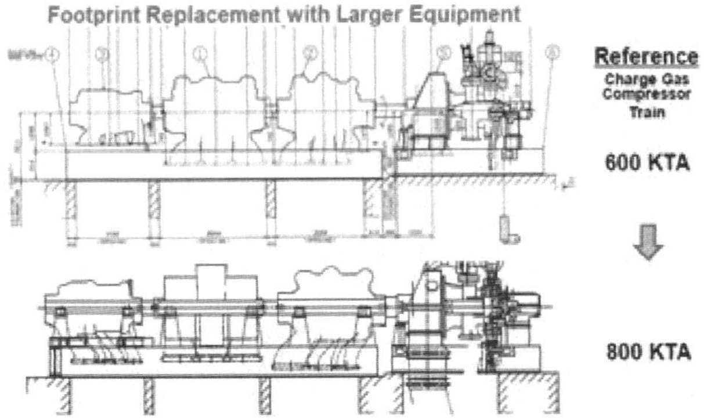

Figure.46 Replacement for Plant Capacity Increase

In terms of the cost side of revamping versus replacement, Figure 47 shows a comparison of initial investment cost based on 300 kiloton per annum ethylene plant. It is easily imaginable that the equipment cost of replacement is higher than that of revamping.

However, it is less well known that the total cost difference is not so large - only 18%. In the meantime, a 40% capacity increase could be achieved by replacement, while only a 20% capacity increase is achieved by revamping. So the initial investment cost difference may be able to absorbed by the profit difference from the production increase between the two cases. For the case at hand (300 KTA plant), production increase by replacement is 120 kiloton per annum.

Example of Initial Investment Cost (300 KTA Ethylene Plant)

Item (Increasing Ratio)	Revamping (+20%)	Replacement (+40%)
Planning / Engineering	18 %	18 %
Equipment	70 %	86 %
Compressor & Driver	(20 %)	(28 %)
Furnace, Others	(50 %)	(58 %)
Field Installation / Start-up Work	12 %	14 %
Total Initial Investment Cost	**100 %**	**118 %**

Production Increase by Replacement = 300 KTA x 0.4 = +120 KTA ⇨ Profit x 1,200 $/ton = +144 mil$/year

Figure.47 Comparison of Initial Investment Cost

Figure.48 Flexible Casing Nozzle Design for Compressor

Figure 48 shows a flexible casing nozzle design for a replacement compressor. The nozzle design is very important for compressors. Compressor cases are typically formed by casting. Therefore, the configuration of the nozzle is very flexible if the location requirements are considered early in the design process. This is especially useful for a footprint replacement project. If the replacement is properly done, the base plate can be reused and the process piping positions remain fixed. Therefore, the compressor nozzle has to be flexible in order to replace the existing compressor to larger one without significant and costly modifications. The flow of the nozzle should be analyzed by CFD analysis in order to ensure the design is successful.

Lessons learned in Mega Ethylene applications

Over the past years MHI Compressor has gained valuable practice experience in mega ethylene applications around the world. Mitsubishi technologies have been verified at sites through long-term operation. These technologies include steam turbine and compressor anti-fouling technologies, reduction in bearing metal temperatures, and supporting customer's demands for shorter outage intervals through innovative tooling solutions.

MHI performed systematic monitoring of installed trains in order to evaluate machinery performance deterioration in steam turbines and compressors. Actual torque measurements between steam turbine drivers and compressors were used – in some cases, between the driver and each compressor case individually. These torque measurements allowed for precise calculation of performance. In the case of a steam turbine, the performance is calculated according to HP and LP section operation, including measured pressures and steam consumption, compared with measured power output. The compressor performance is calculated by actual process data including molecular weight for each section.

Washing

Turbine deterioration due to fouling, foreign object damage, erosion, or corrosion can all lead to reductions in performance. However, MHI has observed that the primary problem for steam turbine performance is fouling. Deposits on nozzle and blade profiles can cause a reduction in flow area. Changes to the flow profile and increased surface roughness from deposition can change stage efficiencies. In some cases, this type of fouling can require a limit on the inlet steam flow in order to keep the first stage pressure below design limitations, causing a reduction in available power output. Similarly, conditions in compressors can cause adherence of fouling particles on flow path components. In some cases fouling can adhere even after the use of continuous water injection or intermittent oil washing. Internal efficiencies can suffer dramatically in a very short period.

Mitsubishi's new patented online washing system for steam turbines has been applied on over 30 turbines. This system injects water directly into the turbine valve chest, without requiring a reduction of power or speed. Washing has been shown to restore the turbine to design performance values. MHI has also designed and tested advanced washing methods for oil injection in compressor diaphragms, leading to increased wash oil coverage in stationary and rotating components.

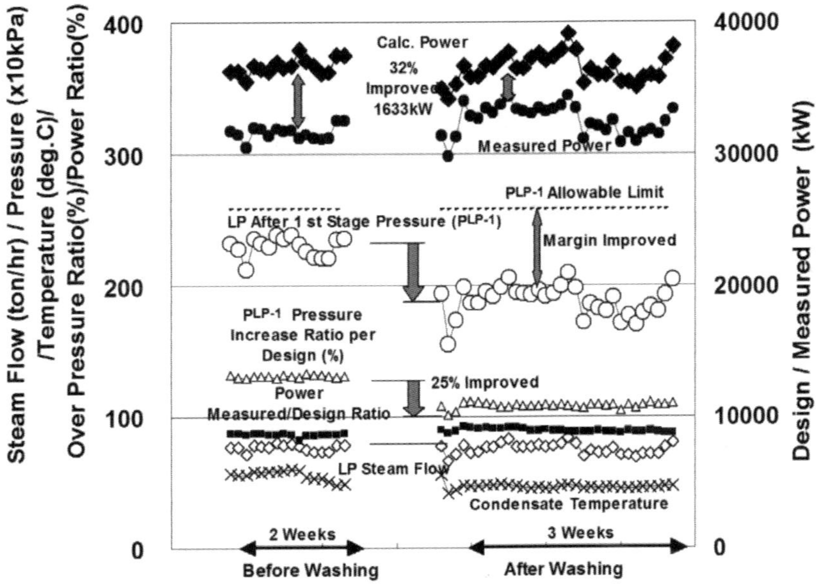

Compressor Anti-fouling technology

Two mega ethylene plants were shipped in recently with similar equipment and sizing (1.3 vs 1.5 Mega/year, operation since 2010). One machine utilized Sermalon coating on the charge gas compressor, while the other did not. Sermalon coatings use a PTFE sealing coat over a metal composite-ceramic base coat. At the first overhaul after four years of continuous operation, Company A's rotor and diaphragms were found to be clean for cracked gas service, with only moderate signs of polymer contamination.

Company B's rotor and diaphragms were not coated, and heavy fouling was observed. This polymer contamination resulted in reduced performance during operation and required multiple days of high pressure washing to remove during the overhaul. It was found that the injection nozzle was also plugged due to polymer accumulation. Other plants have reported similar findings when the application of Sermalon coatings is utilized

First O/H after 5 years operation ; Company B

Bearing metal temperature reduction

Some mega ethylene plants have reported high bearing metal temperatures when operating continuously at maximum output conditions. To address these issues, MCO has developed two modifications to reduce these temperatures.

Bearing journal pads using Cr-Cu backing (upgraded from steel) has better heat conductivity, allowing the pad to better transfer heat to the lubricating oil and away from the babbitt metal.

Additionally, the tilt pad bearing key was upgraded to a crown-machined style. In this modification, the alignment and contact surface is rounded, allowing the pad to be self-aligning both circumferentially and axially. The result is a more uniform oil film, reduced "hot spots" in the babbitt, and overall decreased journal temperatures.

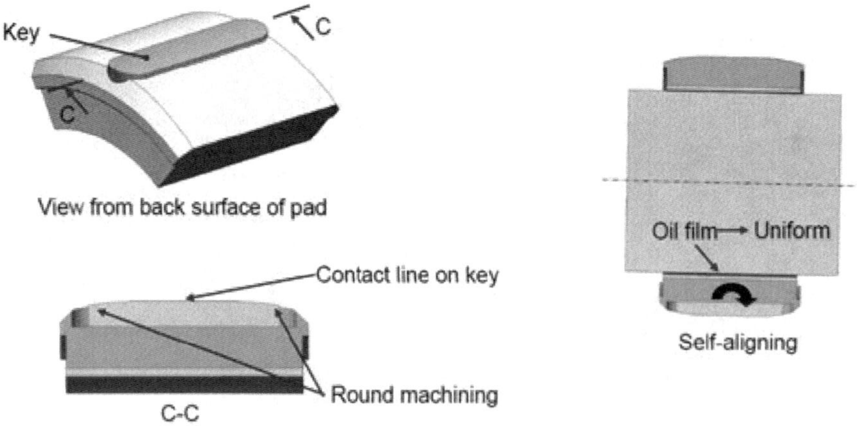

Outage reduction

MHI has developed a high frequency induction bolt heater to reduce turnaround durations when bolt heating is required. By placing the bolt within the flux generated by the induction coil, the entire bolt can be heated rapidly and simultaneously. The rapid nature of the heating reduces thermal expansion of the flanges, resulting in proper fastening clamping forces. The induction heater is also safer than other alternatives, as the coil remains at room temperature and the case casting increases in temperature only slightly. The induction heating coil has a broad range of applications, using a bendable coil that was further designed to avoid heating threaded portions of the stud bolt.

The data below is based on test bolt of 4 " O.D., 800mm length, 20mm hole diameter and 12% Cr material.

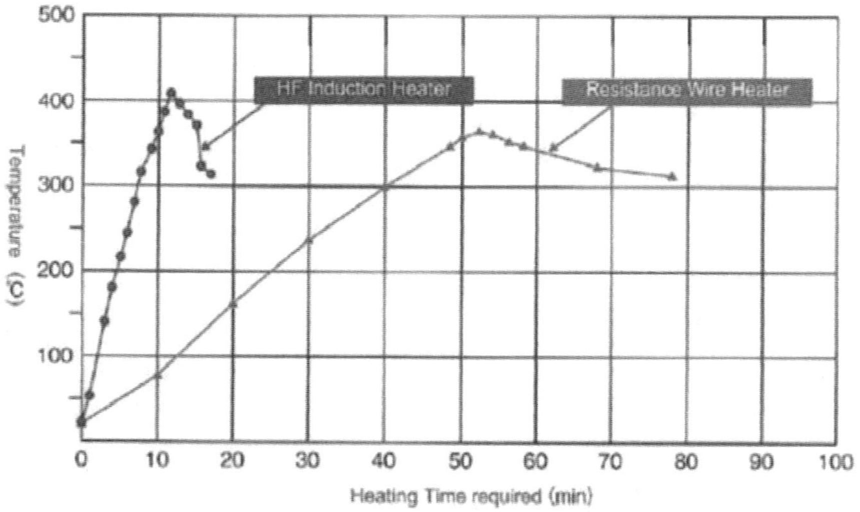

Remarkably reducing working time compared to resistance wire heater
:1/2 with 2 sets of type 4025
:1/3 with 2 sets of type 3050

When compared with a resistance wire heater, the HF induction heater was able to reduce bolt heating time by over half, in some cases up to 75%. This results in a reduction in time spent opening stubborn case split-line bolts.

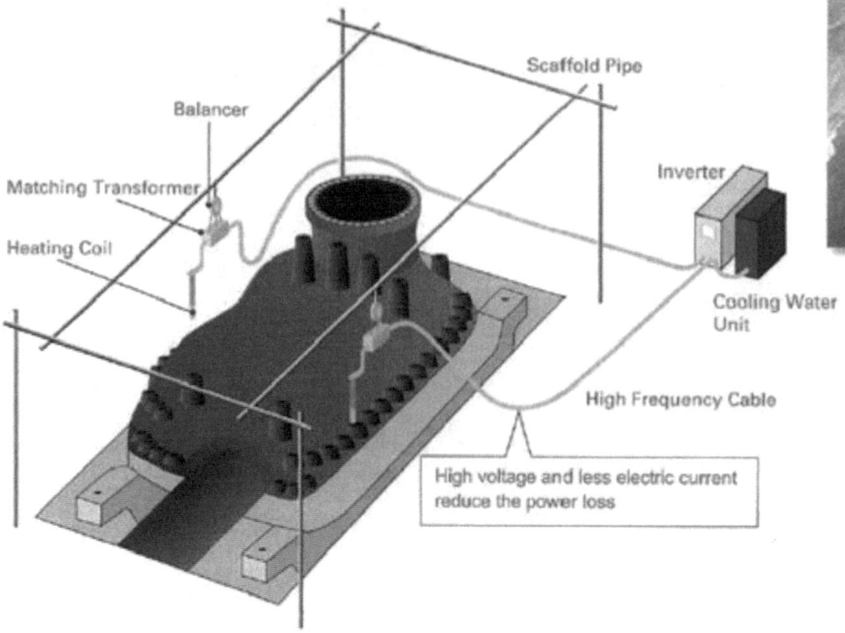

CONCLUSIONS

Turbomachinery manufacturers have been going forward to apply challenging technologies for critical components in order to provide highly efficient and more reliable machines corresponding to the requirement of higher power for larger ethylene services. The authors introduce the technology development history map in past and future for typical design, verification test and application results in terms of transient fluid dynamics, thermodynamics, rotor dynamics, and blade vibration strength evaluation. In addition, after commissioning, turbomachinery tends to have deterioration of performance and reliability for long-term operation. The typical damage and deterioration map for machine are introduced and the authors explain the practical technologies actually applied, such as flow path surface treatment, effective on line washing, combinations of anticorrosion, erosion prevention and stage performance enhancement by partial component replacement, NDE techniques for both compressors and steam turbines, and unique casing replacement techniques on an existing footprint for revamps to increase capacity.

REFERENCES

[1] Saga.M, Hata. S. et al., Repair Technologies of Mechanical Drive Steam Turbines for Catastrophic Damage, Proceedings of the 34th Turbomachinery Symposium (2005), pp. 15-23.
[2] The ASME Handbook on Water Technology for Thermal Power Systems p12
[3] J Jonas Otakar, Understanding Steam Turbine Corrosion 84 (1984) pp55 2 -55 28
[4] Ebara Ryuichiro, et al., Corrosion Fatigue Behavior of 13Cr Stainless Steel for Turbine Moving Blade, Mitsubishi Heavy Industry Technical Bulletin Vol.15 No.3 May 1978, pp1-10

[5] Naumann Hekmut G., Steam Turbine Blade Design Options: How to Specify or Upgrade, Proceedings of 11th Turbomachinery Symposium, pp29-49
[6] Kihara Hiroshi, Fractography
[7] Wagner L.F, Griffin J.H, Forced Harmonic Response of Grouped Blade Systems: Part I -Discrete Theory, Transaction of the ASME, Vol.118, January 1996, pp130-136
[8] Ebara Ryuichiro, et al., Corrosion Fatigue Strength of 13 Cr Stainless Steel for Steam Turbine Blades, Mitsubishi Heavy Industry Technical Bulletin Vol.19 No.5 September 1982, pp537-543
[9] Ebara Ryuichiro、Japan Steel Society , Nishiyama Memorial Technical Session, Environment Crack of Steel, October, 1982
[10] Ebara Ryuichiro, et al., Corrosion-Fatigue Behavior of 13Cr Stainless Steel in Sodium-Chloride Aqueous Solution and Steam Environment, American Society for Testing and Materials, Corrosion-Fatigue Technology, pp155-168
[11] Steltz W.G, et al. Verification of Concentrated Impurities in Low Pressure Steam Turbines, Transactions of the ASME, Vol.105, January 1983, pp192-198
[12] Bischoff James L, et al. Liquid-Vapor Relations for the System NaCl-H2O: Summary of the P-T-x Surface from 300℃ to 500℃, American Journal of Science, Vol.289, 1989, pp217-248
[13] Allmon.W.E. et al., Deposition of Salts from Steam, International Water Conference (1983), pp127-138
[14] Maday M.F. et al., Stress Corrosion Cracking Behavior of Two Ni-Cr-Mo-V Steels in Caustic Solutions and Pure Oxygenated Water, Corrosion Vol.45, No.4, pp273-281
[15] Ebara Ryuichiro .et al., Corrosion Fatigue Process of 12 Cr-Stainless Steel, ISIJ Inter-national, Vol.30 (1990), No.7, pp535-539
[16] Katayama Kazuso, et al., High-Speed and Large Capacity Compressor-Driving Turbines for Chemical Plants, Mitsubishi Heavy Industry Technical Bulletin Vol.14 No.5 September 1977, pp1-９
[17] Bhat. G.I., Hata. S. et al., New Technique for Online Washing of Large Mechanical–Drive Condensing Steam Turbines, Proceedings of the 33rd Turbomachinery Symposium (2004), pp.57-65.
[18] Hata. S. et al., Recent Technologies for the Reliability and Performance of Mechanical-Drive Steam Turbines in Ethylene Plants, Proceedings of the 34th Turbomachinery Symposium (2005), pp.15-23.
[19] Hata, S., Sasaki, T., Ikeno, K., New Technologies of Synthesis Gas Compressor Drive Steam Turbines for Increasing Efficiency and Reliability, Proceedings of the 31st Turbomachinery Symposium, Houston, (2002), pp.75-83.
[20] Hata, S., New Technique for Online Washing of Large Mechanical–Drive Condensing Steam Turbines, Proceedings of 2007 AIChE Spring National Meeting Ethylene Producer Conference, Houston, (2007).
[21] Hata, S., Case Study: Assessing a New Technique for Online Washing of Large Mechanical–Drive Condensing Steam Turbines, Proceedings of ROTATE 2007, UAE Abu Dhabi, (2007)
[22] S. Hata, T. Hirano, N., T. Wakai, H. Tsukamoto, New On Line Washing Technique for Prevention of Performance Deterioration due to Fouling on Steam Turbine Blades (1st Report: Fouling Phenomena, Conventional Washing Technique and Disadvantages), Transaction of JSME Div. B, Vol.72, No.723, November 2007, pp2589-2595.
[23] S. Hata, T. Hirano, N., T. Wakai, H. Tsukamoto, New On Line Washing Technique for Prevention of Performance Deterioration due to Fouling on Steam Turbine Blades, (2nd Report: Basic Experiment and Analysis for Actual Operation of Washing System), Transaction of JSME Div. B, Vol.72, No.724, December 2007, pp2970-2979.
[24] S. Hata, T. Miyawaki, N. Nagai, T. Yamashiata, H. Tsukamoto, Study on Corrosion Fatigue Phenomena of Low Pressure Blades and Integrity Improvement for Mechanical Drive Steam Turbines

(1st Report: Relation of Corrosive Chemicals Enrichment Zone and Corrosion Fatigue), Turbomachinery, Vol.35, No.2, February 2007, pp8-16.
[25] S. Hata, T. Yasui, K. Yamada, H. Tsukamoto, Study on Corrosion Fatigue Phenomena of Low Pressure Blades and Integrity Improvement for Mechanical Drive Steam Turbines (2nd Report: Operation Life Extension by Blade Surface Treatment and Coating), Turbomachinery, Vol.35, No.3, March 2007, pp50-59.
[26] S. Hata, Blades Improvement for Mechanical Drive Steam Turbines, Coating Technologies applied for Nozzle and Blade of Mechanical Drive Steam Turbine, Turbomachinery, Vol.29, No.5, May 2001, pp40-47.

ACKNOWLEDGEMENTS

The authors gratefully wish to acknowledge the following individuals for their contribution and technical assistance in analyzing and reviewing results, and for great suggestions and leading of practical applications and testing at site: The team members of the steam turbine design section and compressor design section of Mitsubishi Heavy Industries Compressor Corporation.

MOVE THE WORLD FORWARD MITSUBISHI HEAVY INDUSTRIES GROUP

Steam Turbine 101/201

Ethylene Producers Conference March 27, 2017
San Antonio, Texas

2017.2.7

MITSUBISHI HEAVY INDUSTRIES COMPRESSOR INTERNATIONAL

Presenter/Author Bios

MITSUBISHI COMPRESSOR

Bill Poppe is a Project Engineer for Mitsubishi Heavy Industries Compressor International Corporation in Houston, TX. He is the subject matter expert for reverse engineering and has experience in overhaul and maintenance of steam turbines and compressors. Mr. Poppe has a B.S. of Mechanical Engineering from the University of Nebraska – Lincoln.

Atsushi Nakajima is the Technical Advisory Manager for Mitsubishi Heavy Industries Compressor International Corporation in Houston, TX. He is the steam turbine and compressor design expert along with providing expertise in troublehshooting, maintenance, and repair. Mr. Nakajima has a M.S. of Mechanical Engineering from Okayama University in Japan.

Satoshi Hata is the vice president, engineer in Mitsubishi Heavy Industries Compressor International Corporation, doctor of engineering at Mitsubishi Heavy Industries Compressor Corporation, in Hiroshima, Japan. He is the specialist of detail design of rotating components in mechanical drive steam turbines and compressor has experience with R&D for them.

Introduction

1. Societal Role and History
   1.1 Steam Turbine Development History
   1.2 Major Steam Turbine Components

2. Steam Turbine Classification
   2.1 Flow Direction
   2.2 Application
   2.3 Steam Condition
   2.4 Blade Type
   2.5 Typical Steam Balance System of an Ethylene Plant

Introduction

3. Design Improvements for Mechanical Drive Steam Turbines
   3.1 Variable Speed Control and Power Output
   3.2 Blade Design for Variable Speed
   3.3 Rotor Design for Variable Speed
   3.4 International Standards

1. History

1.1 Steam Turbine Development History

Reaction

Heron(BC120) Parsons(1884) Brush Parsons(1915) Modern

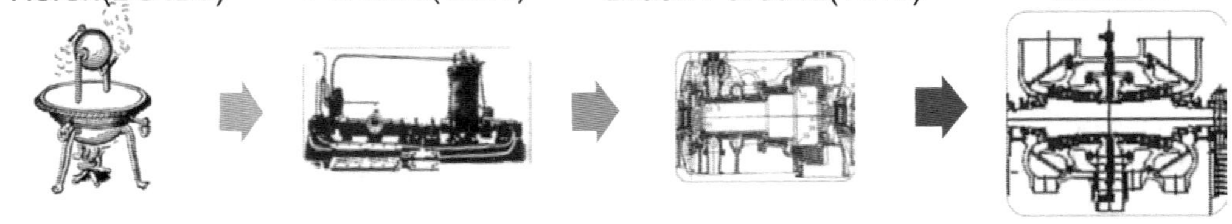

Impulse

Branca(1629) De Laval(1882) Rateau(1908) Modern

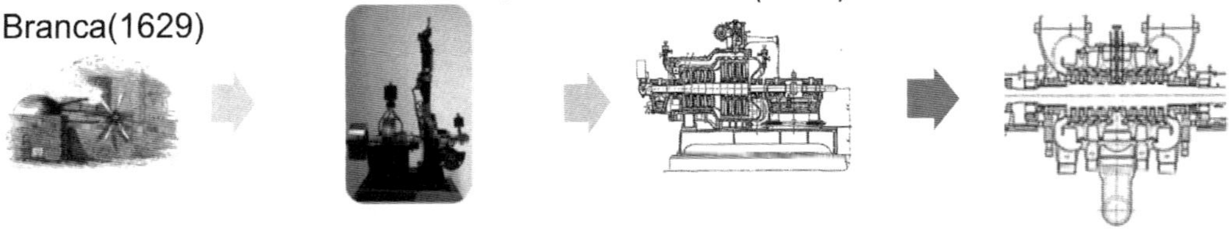

In modern steam turbines, the reaction and impulse types have become more similar due to the improved seal system of reaction type steam turbines and the increased reaction and stage number of the impulse type.

1.2 Major Steam Turbine Components

MITSUBISHI COMPRESSOR

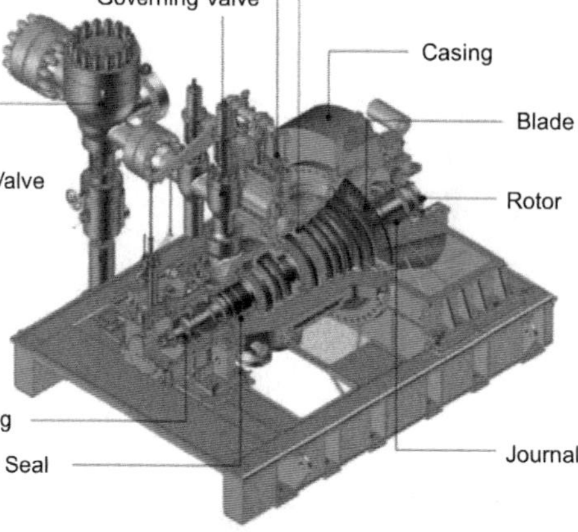

 7

MOVE THE WORLD FORWARD MITSUBISHI HEAVY INDUSTRIES GROUP

2. Steam Turbine Classification

2.1 Flow Direction

(1) Axial flow turbine

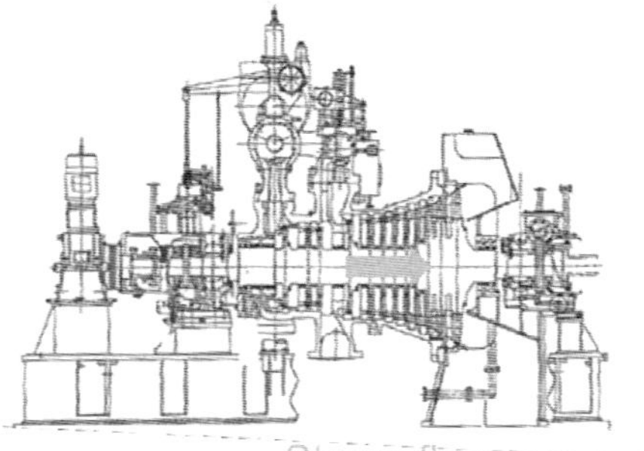

(2) Radial flow turbine

2.2 Application

(1) Engine for transportation systems
Main turbine for marine vessel, locomotive

(2) Engine for stationary machine
Drive turbine for generator,
blower, compressor, pump

2.3 Steam Condition

- Back pressure turbines
- Condensing turbines
- Extraction turbines
  - Extraction condensing turbines
  - Extraction back pressure turbines
- Mixing turbines
  - Mixing condensing turbines
  - Mixing back pressure turbines
- Extraction and mixing turbines
- Regenerating condensing turbines
- Reheating turbines
- Moisture separation turbines

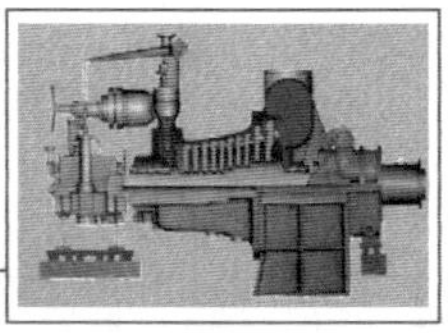

2.3 Steam Condition

Back pressure turbine

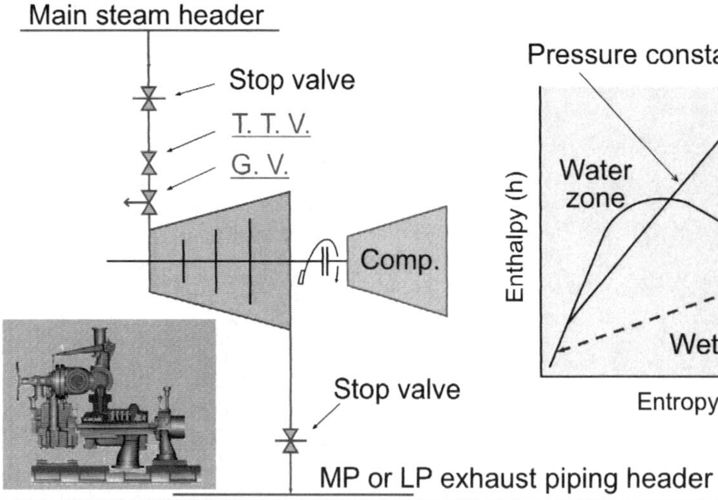

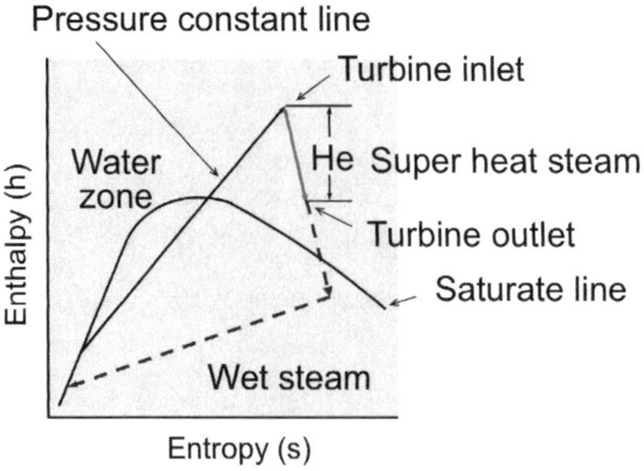

2.3 Steam Condition

Condensing turbine

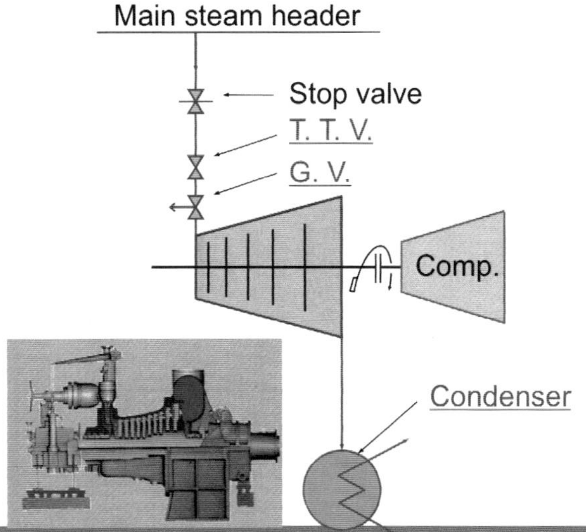

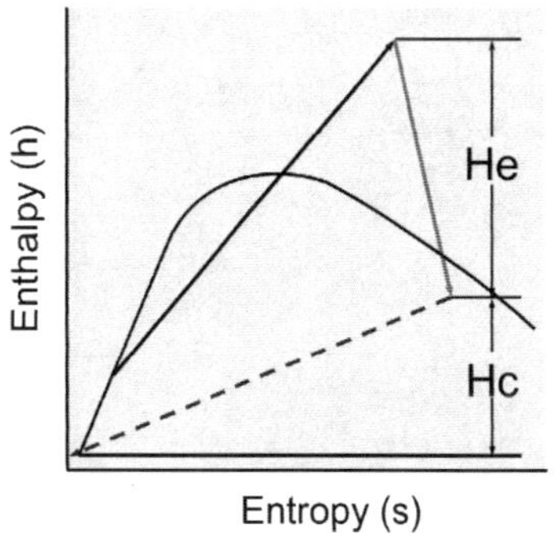

2.3 Steam Condition

Extraction condensing turbine

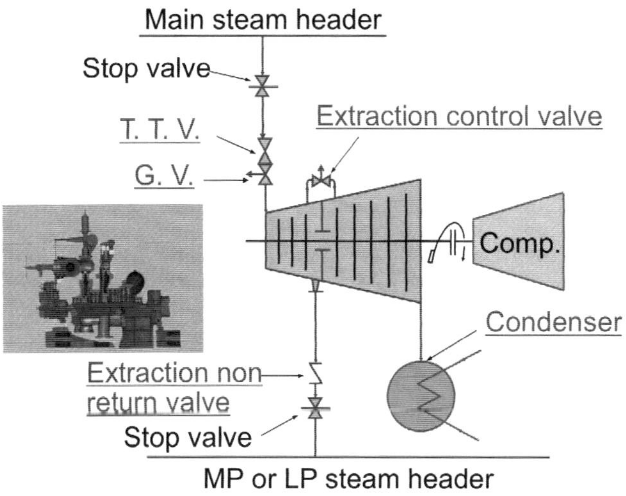

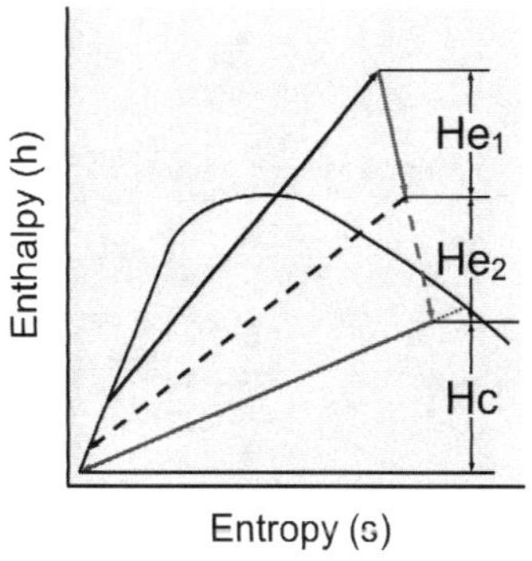

2.3 Steam Condition

Mix pressure turbine

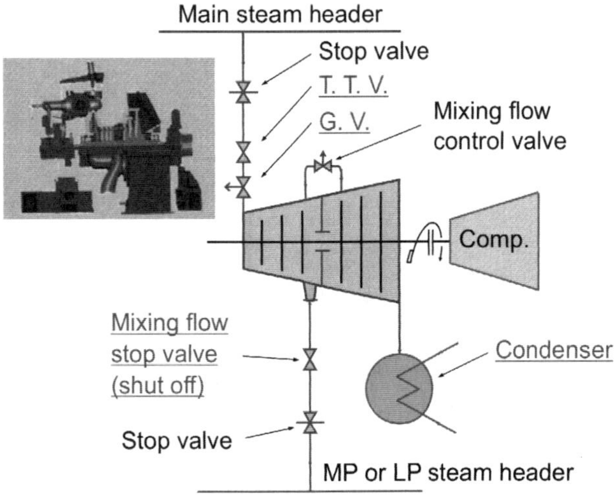

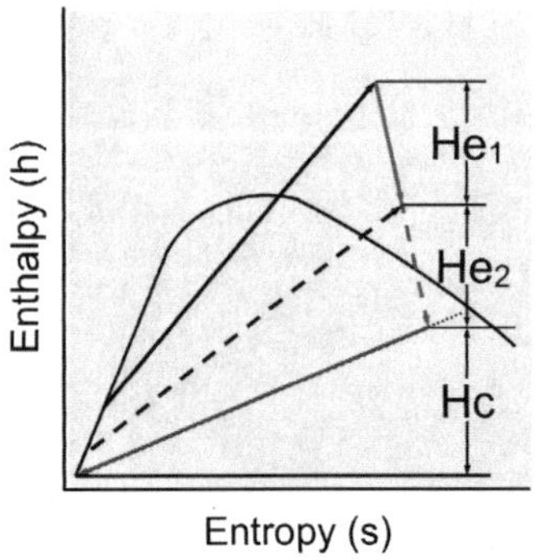

 15

2.4 Blade Type

Impulse and reaction turbines

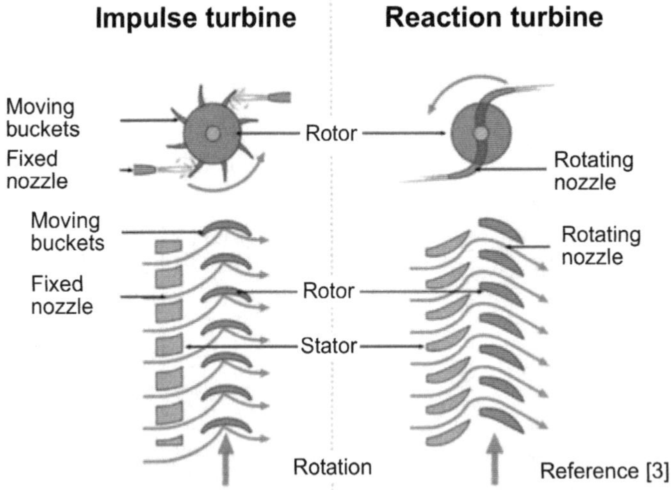

 16

2.4 Blade Type

Impulse and reaction turbines

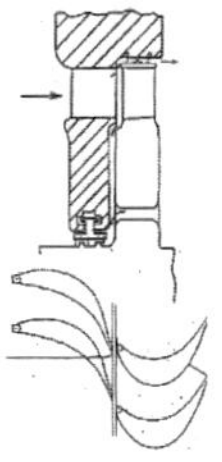

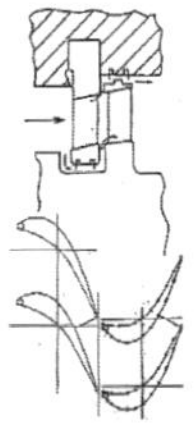

Steam turbine stage
- Impulse stage
  - Curtis (1896 USA)
  - Rateau (1897 France)
  - De Laval (1882 Sweden)
  - Zoelly (1903 Switzerland)
- Reaction stage
  - Parsons ρ= 0.5 (1884 Great Britain)
  - Reaction ρ＞0.5

2.4 Blade Type

Comparison of impulse type with reaction type

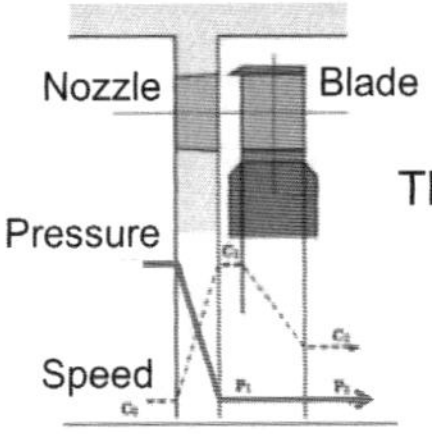

Impulse type

Thermal expansion
↓
Only nozzle

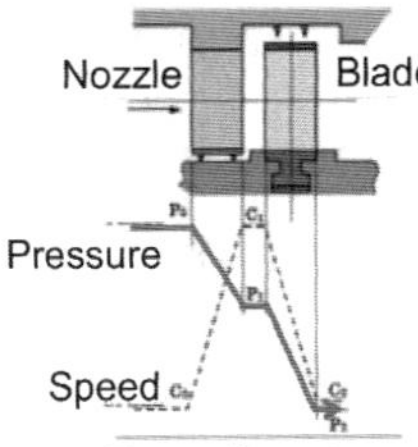

Reaction type

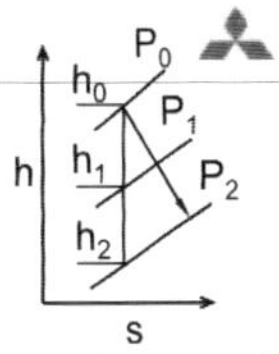

Thermal expansion
↓
Nozzle & Blade

$$\text{Reaction} = \frac{h_1 - h_2}{h_0 - h_2}$$

Characteristics of impulse, reaction blade

		Impulse	Reaction
Structure	No. of stage	Less	Many
	Rotor (type)	Disk	Drum
	Seal fin	Less	Many
Thrust force		Small	Big
Effect of efficiency	Reheat effect	-	Better
	Leakage loss	Better	-
	Secondary loss	-	Better

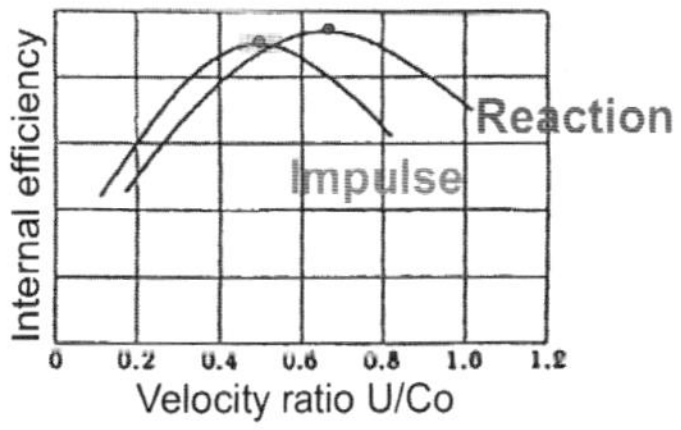

2.4 Blade Type

Impulse and reaction turbine

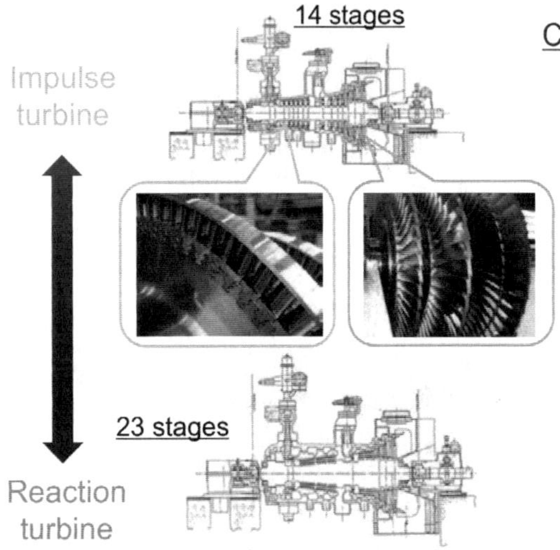

	Control stage	Inter stage	Last stage
Impulse turbine	Impulse	Impulse	Impulse
	Impulse	Impulse	Reaction
	Impulse	Reaction	Reaction
Reaction turbine	Reaction	Reaction	Reaction

2.5 Typical Steam Balance System of Ethylene Plant

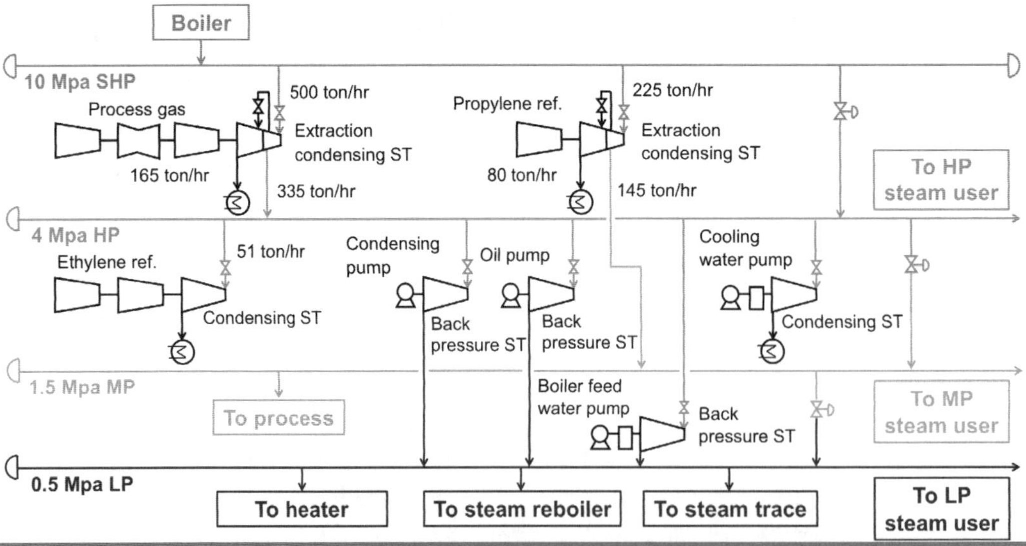

MOVE THE WORLD FORWARD MITSUBISHI HEAVY INDUSTRIES GROUP

3. Design Improvements for Mechanical Drive Steam Turbines

3.1 Variable Speed Control and Power Output

MITSUBISHI COMPRESSOR

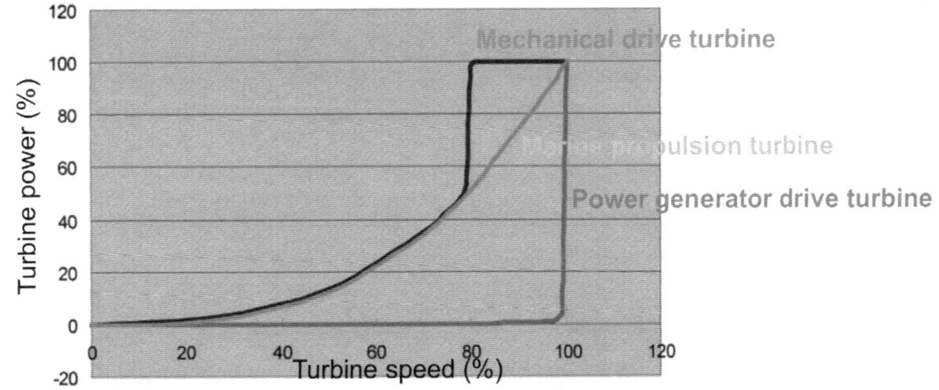

	Mechanical drive	Power generator drive
Speed at load operation	Variable (Wide speed range)	Constant
Blade	Un-tuned	Tuned
Rotor	Enough separation margin from the wide operation speed	Enough separation margin from a constant speed

3.2 Blade design for variable speed

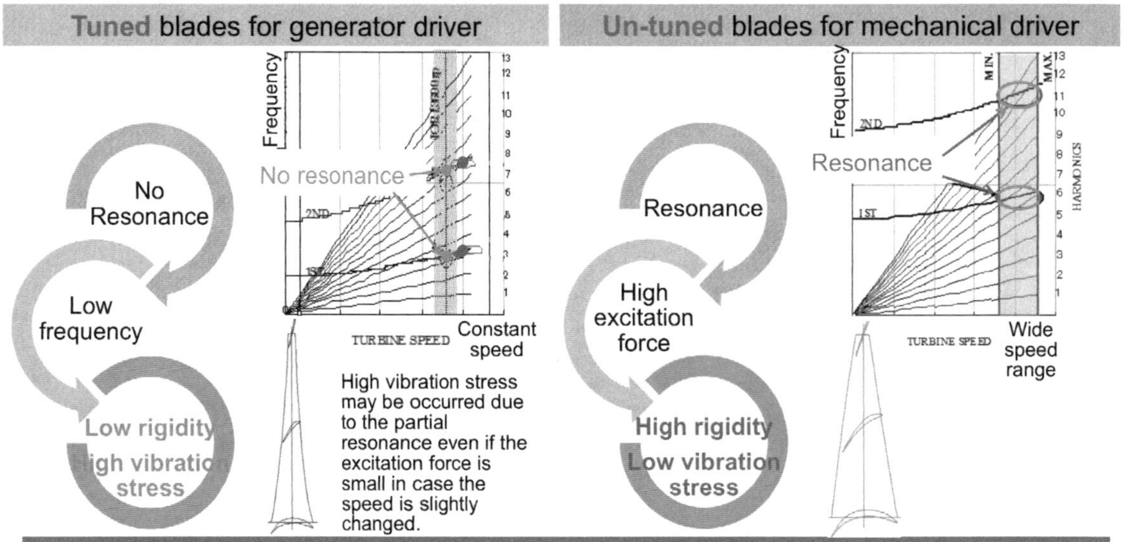

 23

3.3 Rotor design for variable speed

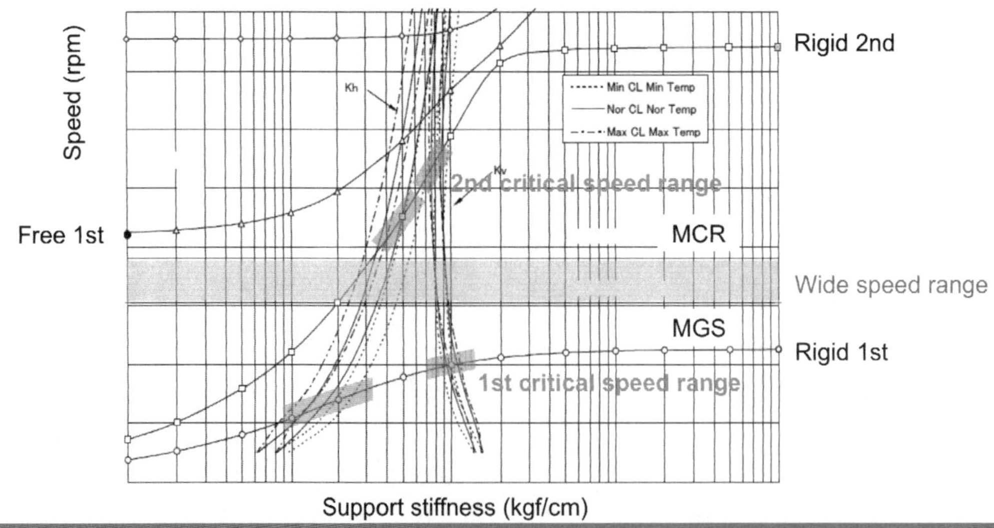

 24

3.4 International standards

MITSUBISHI COMPRESSOR

Design Code		API 611 / 612 (*1)	ASME	ATEX / IECEx
Application	Power Generation		✓	
	Chemical Plant	✓	✓	✓
	Oil & Gas	✓	✓	✓
Description		• Applied broadly to petrochemical plant. • Regulated turbine materials, operating speed range, rotor dynamics evaluation and mechanical running test, etc.	• Classed for piping and flanges. • Applied pressure vessel code (Section VIII) and welding process (Section IX), etc.	• Explosion protection standard for electrical apparatuses in hazardous areas.

(*1)

- API 611 : ISO 10436, General-Purpose Steam Turbines for Petroleum, Chemical, and Gas industry services
- API 612 : ISO 10437, Petroleum, Petrochemical and Natural Gas Industries- Steam Turbines- Special Purpose Applications

MOVE THE WORLD FORWARD

MITSUBISHI HEAVY INDUSTRIES GROUP

AICHE PAPER NUMBER:

OIL-FREE SCREW COMPRESSORS IN ETHYLENE STORAGE BOIL OFF GAS SERVICE

Sven Strache
Applications and Sales Engineer
MAN Diesel & Turbo North America

Ali Vakili
Principal Rotating Equipment Engineer
Bechtel Corporation

Prepared for Presentation at the Year 2017 Spring National Meeting San Antonio, TX, March 26 – 30, 2017

OIL-FREE SCREW COMPRESSORS IN ETHYLENE STORAGE BOIL OFF GAS SERVICE

Sven Strache
Applications and Sales Engineer
MAN Diesel & Turbo North America
Ali Vakili
Principal Rotating Equipment Engineer
Bechtel Corporation

Abstract

Major ethylene production facilities utilize large atmospheric tanks to store liquid ethylene at its cryogenic saturation temperature of -104 °C (-155 °F). Depending on the ambient conditions and operating scenarios, rate of ethylene vapor coming out of the ethylene tank varies. In order to recover the vaporized ethylene, the gas is routed through a liquefaction process. In this process, the boil off gas compression unit captures this ethylene gas and compresses it. The gas is liquefied and sub-cooled by passing through a condenser and economizer and expansion valve and is then recycled back to the tank. The compressor used in this service should maintain the pressure ratio in various suction volumetric flow rates. Thus, positive displacement compressors, mostly reciprocating machines, have traditionally been used for this purpose. Alternatively, oil-free screw compressors can be used in this service as a positive displacement machine with much higher reliability and proven record of significantly less required maintenance. This paper will describe which circumstances, i.e. flow rates and pressures, favor the use of oil-free screw compressors. Technical details will be given that describe the features of this compressor type.

Introduction

Ethylene is an important petrochemical product and the most consumed intermediate petrochemical product in the world [1], [3]. Its production rate may represent the scale of petrochemical industry in a country and even economy wellness in this industry. The reason is that this product is being used to make a wide variety of other petrochemical products and as a result is a feedstock to a diverse range of end-users.

In 2009 it was estimated that the world demand for ethylene was over 140 million tons per year, with an approximate yearly increase of 3.5% [2]. In recent years, due to increase of production shale gas since 2011 and significant increase in the availability of ethane, several new ethylene production plant constructions or expansions of existing plants were announced and completed [1]. Mass production of ethylene is achieved by steam cracking of a wide variety of hydrocarbons. In Asian and European countries, cracking of naphtha and hydrocarbon condensates is a commonly used method. This method will result in production of some side products as well. In the Middle East and North America, cracking of ethane and propane has been employed for commercial production ethylene.

The produced ethylene should be stored and transferred to the downstream petrochemical plants for production of other petrochemical products, such as PVC, HDPE, LDPE, etc. The most cost effective and technically justified method of storage and transportation of ethylene in very high quantities is to store liquid ethylene at atmospheric pressure. Since the liquid ethylene at atmospheric pressure is at cryogenic temperature of -104 °C` (-155 °F) [4], transfer of the heat from atmosphere to the ethylene is inevitable, which results in vaporization of some amount of ethylene. The nominal boil off rate is about 0.15 to 0.38% of the tank capacity per day [3, 5]. The vaporized ethylene, if not handled and removed from the tank, will increase the pressure and temperature of the ethylene.

Considering the size of the tank, the amount of ethylene boil off gas (BOG) will be quite considerable and from financial and environmental perspective, it is not feasible to vent or flare such rate of gas continuously. With flaring or venting not being an option, the only way for cargos and other ethylene storage facilities to handle the ethylene BOG is to re-liquefy it. In the ethylene tanks which are onshore and located in the petrochemical plants, there might be other options available, such as using ethylene in other units and processes. However, even in such applications, it is well justified to re-liquefy the ethylene and restore it into the tank.

This paper provides an insight into the liquefaction process of ethylene boil off gas with focus on the type of compressor being used in this service. Traditionally, reciprocating compressors have been utilized for this process. The use of screw compressors is being studied in such services and the benefits and disadvantages of the screw compressors versus reciprocating machines in this service are investigated.

Process description

For the re-liquefaction process of ethylene, cascade liquefaction processes have been used frequently. In this process, ethylene gas is first compressed to a certain pressure. The final pressure of ethylene at the discharge of the last stage of the boil off gas compressor unit will be the condensing pressure of the ethylene. Downstream the BOG compression unit an ethylene condenser will liquefy the ethylene by dissipating the ethylene latent heat to a refrigerant. The refrigerant used in the refrigeration loop determines the ethylene condensing temperature and subsequently the BOG compressor discharge pressure. In order to liquefy most of the ethylene boil off gas the condensed ethylene to be subcooled. For this purpose, the most reliable approach is to use an economizer to sub-cool the condensed ethylene. Downstream the economizer, the sub-cooled ethylene stream passes through an expansion valve to reach the atmospheric pressure and cryogenic temperature of -104 °C (-155 °F).

A cascade ethylene re-liquefaction process is illustrated in Figure 1. This system has been widely used in cargo carriers for sea transport of ethylene. Similar systems have been used in onshore ethylene storage tanks for re-liquefaction purpose.

R22 and R290 are frequently used for the refrigeration cycle. However, a broad range of refrigerants can be potentially used for this service. Ahmed Oadha and Boumedienne M. Beladjine performed an exergy based performance analysis based on using various types of refrigerant [5]. Depending on the boiling point of the refrigerant at atmospheric pressure and the desired approach temperature in the ethylene condenser cycle, the discharge pressure from the ethylene BOG compressor can be determined. Table 1 lists the boiling point of some of the commonly used refrigerants.

Table 1: Refrigerant boil off temperature

Refrigerant type	Mole Weight [g/mol]	Boiling Point at 0.12MPA, T_b [° C] / [° F]	Ethylene Liquefaction Temperature [° C] / [° F]
R22	86.5	-37 / -34.6	-30 / -22
R290 (Propane)	44.09	-38 / -36.4	-31 / -23.8
R1270 (Propylene)	42.1	-43.8 / -46.84	-38.8 / -37.84

Considering the above refrigerant boil off temperature and an approach temperature of 7 °C (13 °F), the ethylene condensation temperature can be determined. On this basis and also based on ethylene phase map diagrams, the operating pressure of ethylene in the condenser will be approximately between 15 to 19.3 bara (217.5 – 279.9 psia). Considering 1 bar pressure drop in the interconnecting piping and the heat exchangers, the ethylene BOG compressor discharge pressure is normally between 16 to 20.3 bara (232 – 294.4 psia).

In order to achieve such pressure ratios with relatively low flow rates, a positive displacement compressor is a suitable selection.

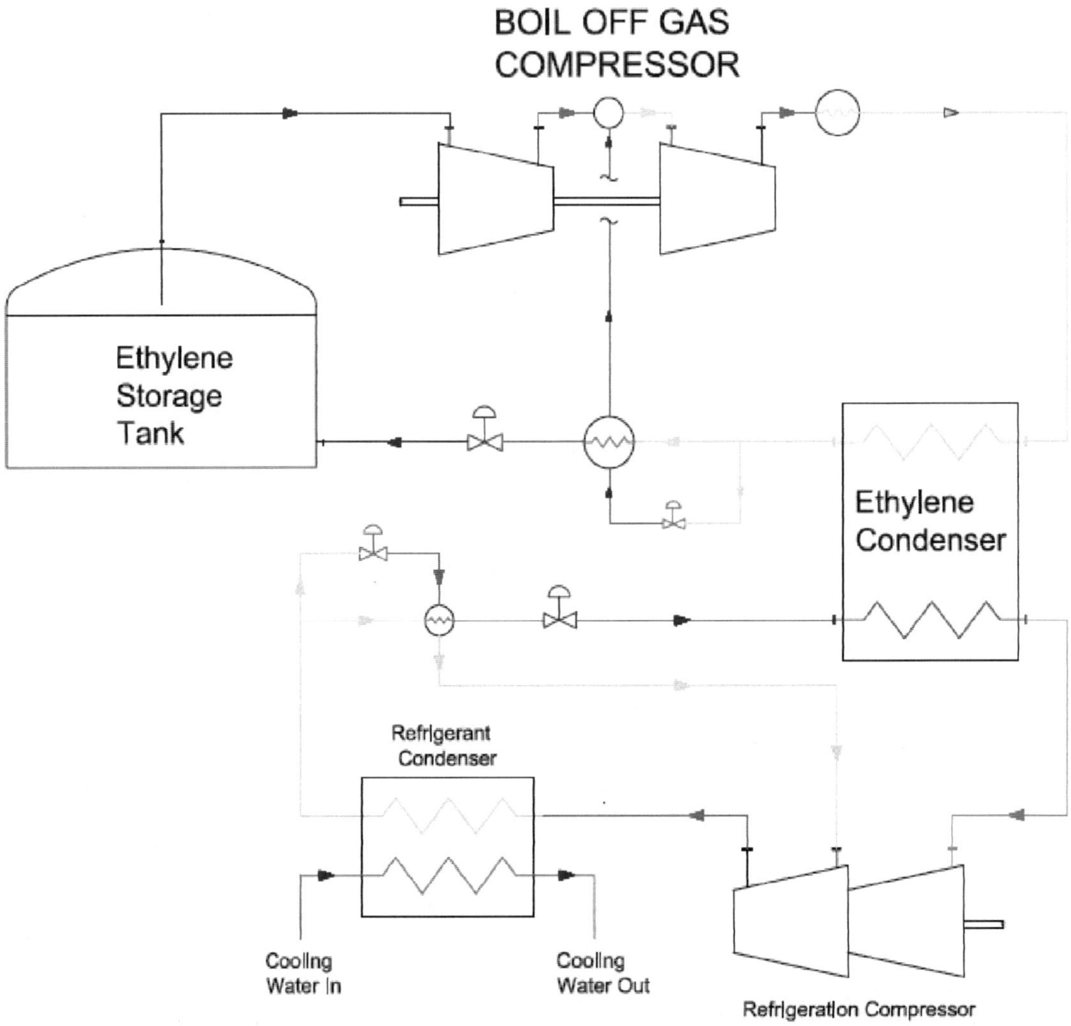

Figure 1: Ethylene boil off gas re-liquefaction process

Compressor technology

Requirements from the process to compressors

As explained in the previous section, the ethylene boil off gas is generated due to introduction of heat into the ethylene tank from atmosphere. The amount of vapor production varies based upon fluctuation in ambient temperature and exposure of the tank to the sun, however there is always going to be some boil off gas, which needs to be compressed. Consequently the boil off gas compression unit reliability is the most critical parameter in the overall reliability of the reliquefaction unit. As a result, the selection of the compressor type and redundancy of the compression units is a critical parameter in determination of the availability of the entire unit.

Considering the production of the boil off gas mainly depends on the ambient condition and in some cases various operating scenarios, it can widely vary from full load to a very low turn down throughout a year of operation. As a result, a boil off gas compressor should be able to handle 0% to 100% of the rated load. In petrochemical plants, there are operating cases in which the level of the liquid ethylene in the tank may rapidly rise or decrease in the storage tank. Along with prompt variation of the liquid ethylene level in the tank, the BOG compressor load will rise or decrease and the compressor must be able to respond to such rapid variation.

Another important issue in the design of the ethylene boil off gas compressor is consideration of a widely changing inlet temperature during start up versus normal operation. During normal operation, a cryogenic gas at -104 °C (-155 °F) and atmospheric pressure enters the first stage of the compressor. During the start-up phase the gas will be compressed at much higher temperatures. Therefore, careful considerations need to be taken for the design of the compressors so that all scenarios can be operated safely and reliably.

Introduction to compressor types

For a general classification, thermodynamic compression principles that are most commonly used in process gas applications can be categorized into axial, centrifugal, reciprocating and screw type applications. Each compression principle is employed in different variants of compressors. For instance, integrally geared and conventional inline centrifugal compressors both rely on the principle of using impellers in order to accelerate the gas by means of centrifugal forces and thus produce a pressure rise. Both oil-free and oil-injected screw compressors have intermeshing rotors that, thru rotation, trap the gas and compress it in between the rotor and casing clearances. The same can be said about the different types of reciprocating compressors (lubricated, non-lubricated, etc.). Despite these differences that result in various flow and pressure limits within those categories it is fair to make the following observations:

- volume flows beyond appr. 535,000 m^3/h (315,000 ACFM) can only be handled by axial type compressors
- very high pressures greater than appr. 700 bar (10,000 psi) can only be handled by reciprocating compressors
- very small flows lower than appr. 170 m^3/h (100 ACFM) fall typically within the range of reciprocating compressors
- an intermediate flow range 425 m^3/h to 50,000 m^3/h (250 to 29,500 ACFM) can usually be covered by all types of compressors

These numbers are further illustrated in Figure 2.

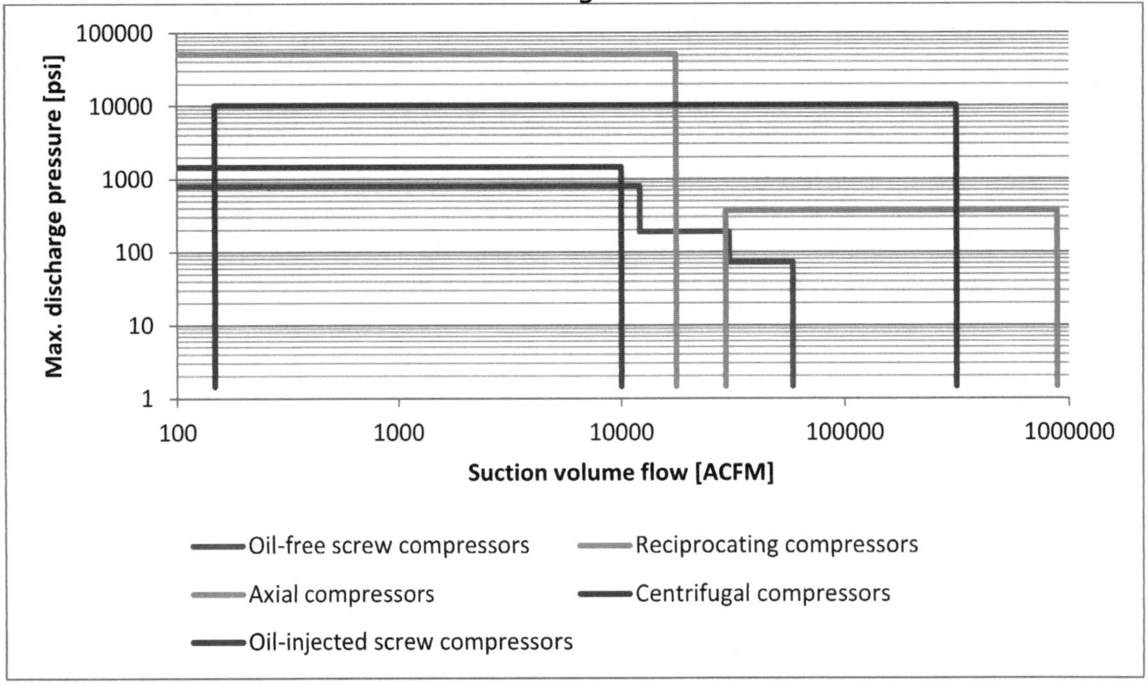

Figure 2: Performance map of different compressor types

Based on the typical flows and pressures as well as the minimum suction temperatures for ethylene boil off gas applications, oil-free screw and reciprocating compressors are typically considered. Both of these compressor types are classified as positive displacement machines (see Figure 3) and thus encompass very similar performance behaviors.

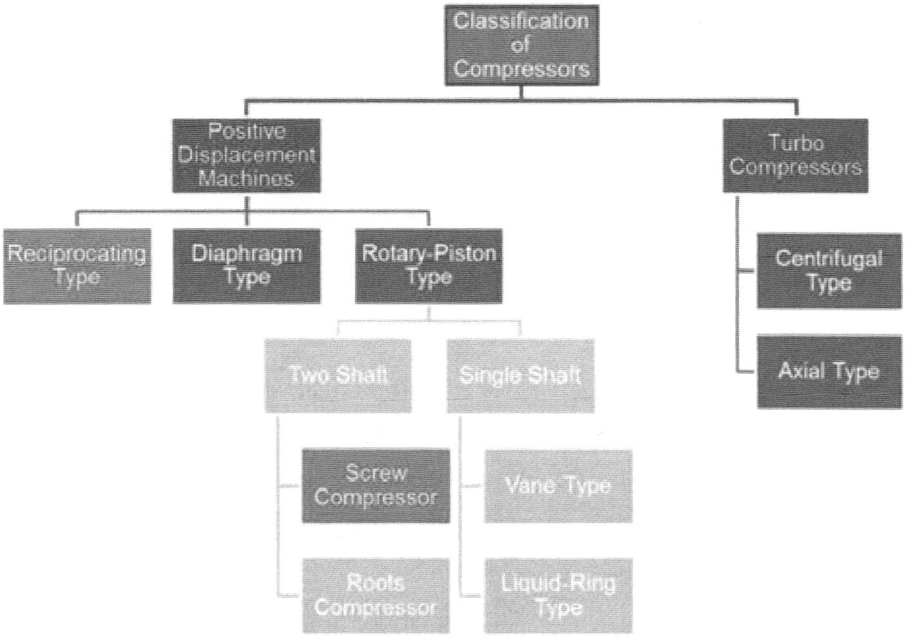

Figure 3: Classification of compressors

As described in previous chapters the rate of boil off gas can vary greatly depending on factors such as ambient temperatures or in some cases, the operating scenario of the plant. The boil off gas compressors should therefore be able to handle great variations in flow while being able to maintain the discharge pressure required to fully condense the compressed gas. As will be described in subsequent chapters, both compressor types are equally capable of performing as required, however oil-free screw compressors show much higher reliability numbers and lower operating costs than any type of reciprocating compressors.

Working principle and design of oil-free screw compressors

Dry or oil-free screw compressors, these terms are used interchangeably, can be dissected into few crucial components. Below figure 4 shows an overview of a vertically split machine:

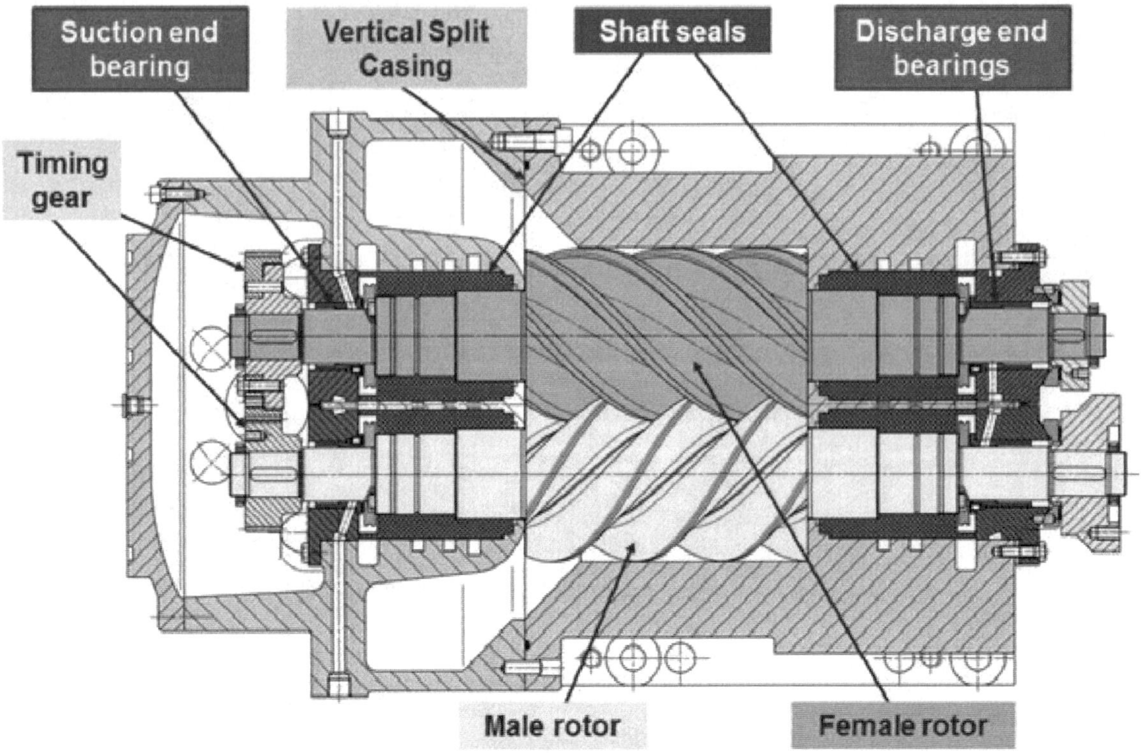

Figure 4: Sectional drawing of an oil-free screw compressor (courtesy of MAN Diesel & Turbo)

In an oil-free screw compressor the male rotor is usually driven by an electric motor (variable or fix speed) through a speed-increasing gear box and occasionally direct driven by a steam turbine. Through a timing gear the female rotors' speed is being synchronized with the male rotor which ensures that both rotors never touch and at the same time optimizes the compressor efficiency by minimizing the gap between them. Just like reciprocating compressors, screw compressors work in three cycles: suction, compression and discharge. This sequence happens in each lobe of the rotor profile so that for a typical rotor profile (four lobes on the male and six flutes on the female rotor), as shown in Figure 5, four compression cycles will occur with every 360° rotation of the rotors.

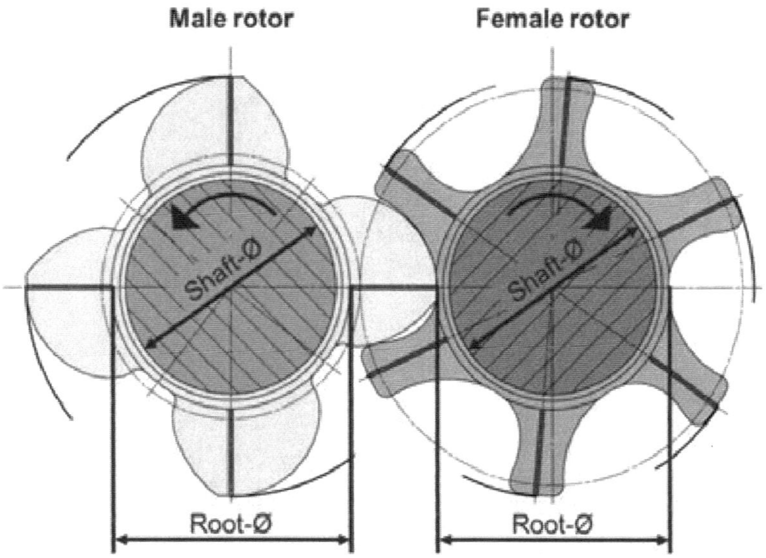

Figure 5: Typical 4/6 screw compressor rotor profile

Through rotation of the screw compressor rotors the gap between male and female as well as its surrounding casing opens up to a maximum (suction). Further rotation will shut off the gas supply to the specific gas pocket as it passes by the opening of the inlet nozzle and compressions starts (compression). Figure 6 shows then how this trapped volume is further reduced, the gas is being compressed until it finally reaches the outlet port of the compressor and all gas is being discharged to the downstream system (discharge). The size of this outlet port determines the pressure ratio across the compressor. The smaller this port is, the longer the compression occurs and the higher the discharge pressure will be. Reversely, the discharge pressure will be lower when the port is bigger so that the gas is discharged earlier. A clear definition of all operating scenarios in early project design phases is crucial in order to optimize and adapt this so-called built-in pressure ratio. A too big pressure ratio leads to the phenomenon of over-compression (pressure internal of compressor higher than pressure in downstream system) and a too low ratio to under-compression (downstream system pressure higher than pressure internal of compressor). Both will lead to pulsations which result in vibrations and in extreme cases damage of connected piping. Again, knowing all operating conditions upfront will help compressor OEMs to design the machines and pulsation dampeners accordingly and ultimately avoid undesired operating behavior.

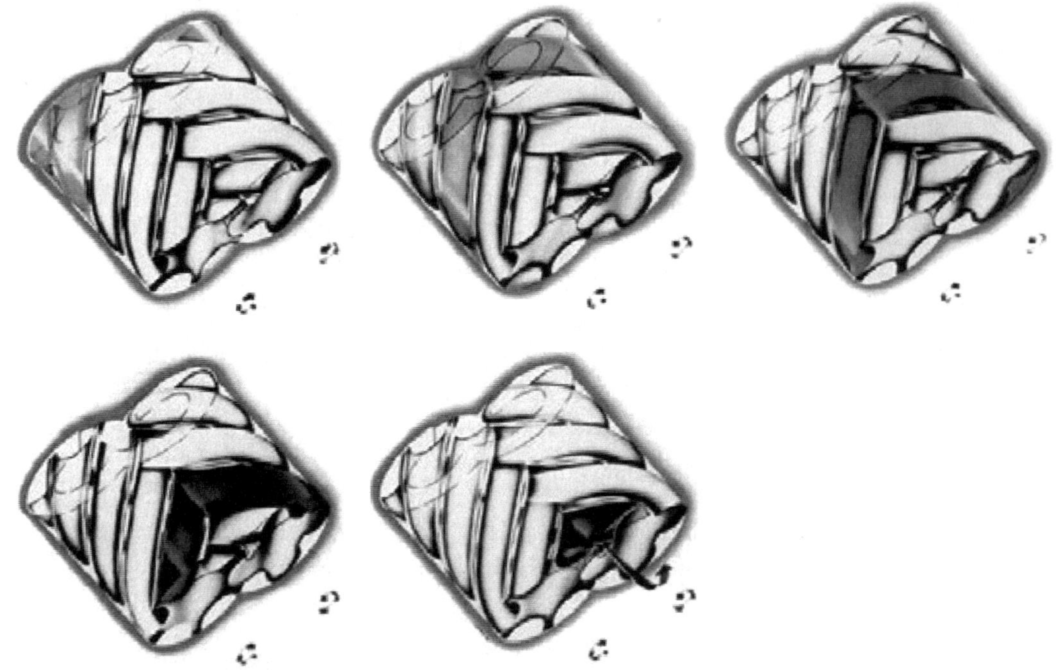

Figure 6: Illustrated working principle in screw compressors

The term "dry" or "oil-free" refers to the absence of any cooling or injection oil in the compression chamber. Unlike oil-injected or oil-injected screw compressors, oil-free screw compressors have shaft seals (see figure 4) installed between the compression chamber and bearings so that the bearing lube oil remains completely uncontaminated by process gas and vice versa. Therefore absolutely no oil is carried over into the downstream system and further conditioning of the process gas is not required.
In machines compliant with API 619, which is the governing standard for process gas screw compressors, the rotors are supported by hydrodynamic journal and thrust bearings for superior reliability and maintainability over the roller bearings, frequently used in non-API machines. Unlike for centrifugal compressors a variety of shaft seals are available and still used. As described in first edition of API619, screw compressor "are employed when no other compressor will work" [10]. Until today this produces rather exotic seal solutions. Depending on the severity of the application (dust content, polymerizing gases, use of permanent water injection, etc.) following various types of seals may be used: nitrogen buffered carbon rings, water or steam buffered floating ring seals, oil-cooled mechanical seal, and dry gas seals. For ethylene boil off applications the latter two are acceptable seal concepts and have been applied in the past.

Oil-free screw compressor unit design

As described previously, the pressure ratio across a screw compressor body is adapted to the system requirements by changing the size of its discharge port. Just like for any other compressor type, there are certain design limits to oil-free screw compressors. These are either of mechanical (deflection of rotors, stresses in bearings) or thermodynamic nature (high temperatures causing unallowable high thermal growth of rotors and casing). Actual limits change on a case by case basis so that pressure ratios per stage can be as high as 10 where the discharge temperature is limited by means of permanent water injection or as low as 3 when compressing gases with a high cp/cv value (for instance air, nitrogen or hydrogen) and where the compression heat cannot be dissipated by water injection because it is not desired or not compatible with the gas.

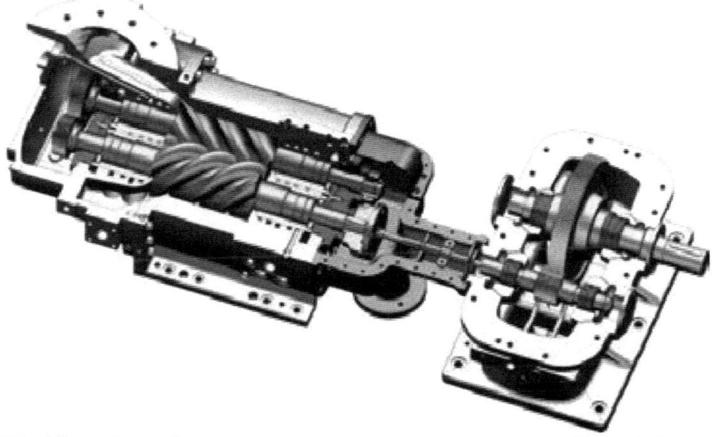

Figure 7: Single-stage screw compressor unit with gear box

The current discharge pressure limit for oil-free screw compressors is referenced by an installation in a refinery in the Netherlands and sets the mark at 51.5 bar (750 psi) [7]. Several OEMs report the discharge temperature limit to be at around 250 °C (480 °F). The highest referenced suction volume flow for a unit in a styrene vent gas application is 77,000 m^3/h (45,300 ACFM) [7]. The same source quotes the highest referenced power rating to be around 9 MW (12,000 HP).
As long as these absolute limits are observed, the overall compression ratio across an oil-free screw compressor unit can always be increased by adding another compressor stage and by cooling of the gas stream in between the compressors (Figure 8).

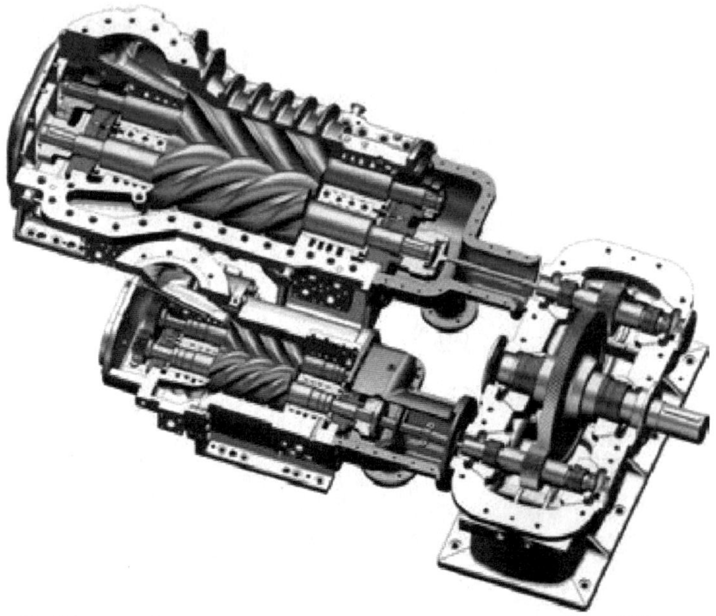

Figure 8: Two-stage screw compressor unit with gear box

In rare cases, even a third stage can be added that would be attached to and directly driven by one of the lower pressure stages (Figure 9). Such a configuration would be able to compress ethylene from atmospheric pressure to around 26 bar (380 psi).

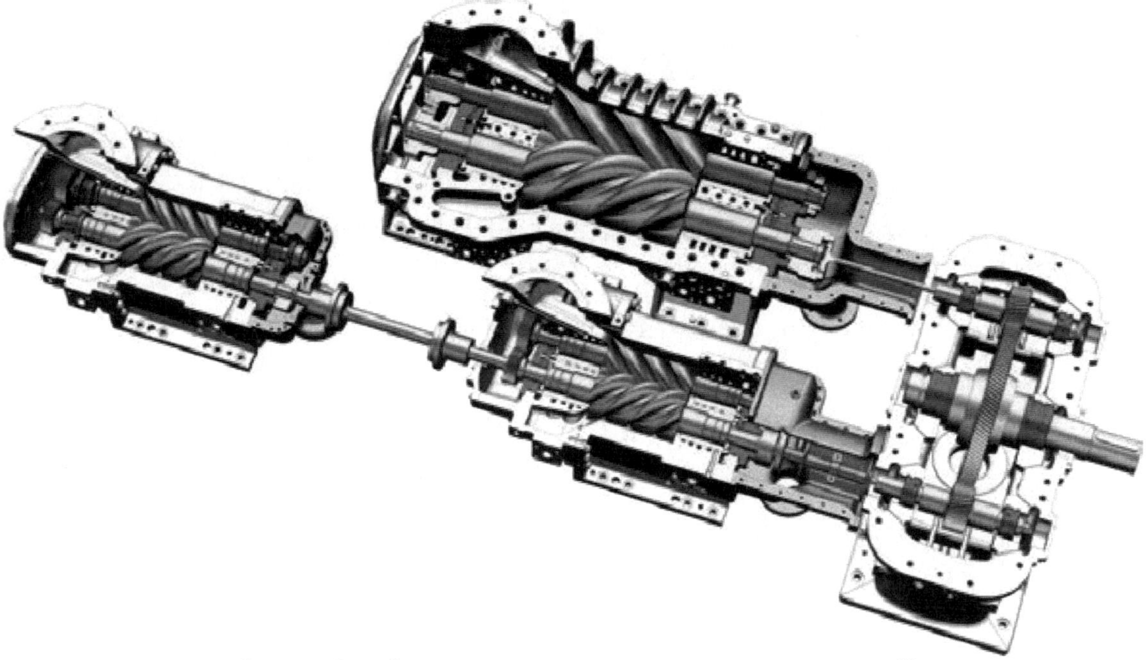

Figure 9: Three-stage screw compressor unit

Screw compressors vs. reciprocating compressors

Due to the relatively low flow and strict temperature requirements, reciprocating compressors were traditionally used for ethylene boil off gas. To the best of the authors knowledge, the first screw compressor used for compressing ethylene

at -104 °C (-155 °F) was as early as 1982. In the following the similarities and differences of both types of compressor types are discussed.
The screw compressor has only rotating elements so no oscillating masses have to be considered for the foundation design. Similar to a centrifugal, only the deadweight load of the compressor unit has to be considered. In the design of reciprocating compressors, the foundation is subject to much higher oscillating loads because of the nature of a reciprocating compressor motion
Given the same operating conditions, the overall footprint of an oil-free screw compressor package is less than the footprint of a reciprocating compressor package. Experience shows that a screw compressor skid can be around 50% smaller varying upon pressure ratio and flow rate.
Both oil-free screw and reciprocating compressors produce pulsations that have to be mitigated carefully in order to avoid damage to connected piping. In a reciprocating compressor, a pulsation analysis needs to be performed to determine the size of the pulsation bottles and evaluate the level of pulsations throughout the suction and discharge piping. Also a mechanical analysis is required to determine the response of the suction and discharge piping system to the excitations. The excitations coming from reciprocating compressors are low frequency and more likely to interfere with the natural frequency of the piping and supporting structures. As a result, it can result in required changes in the pipe routing and supports. In screw compressors, the pulsation and mechanical analysis is normally only focused on the discharge piping up to the compressor after cooler. Because of high frequency of excitations in oil-free screw compressors, the vibration waves which are at multiples of pocket passing frequency can propagate at oblique angle with respect to pipe walls. As a result, the excitations are more likely to interfere with the natural frequency of the pipe in the radial direction, known as cross wall natural frequency. To avoid such interference, in a screw compressor pulsation and MNF analysis the pipe diameter and pipe schedule may be affected rather than isometrics.
As mentioned previously, the boil off gas compressor turndown may vary from 100% to 0% capacity. When it comes to changes in the suction volume flow, both compressor types can be controlled via cooled spill back lines. However, this does not change the power consumption as the same amount of work is still performed. Other common capacity control options available to the reciprocating compressor are unloader valves and clearance pockets for long-term changes in flow. These mechanisms provide step control of the capacity and therefore should be combined with bypass loop and control valve to provide controllability of the flow in between those steps. Recently, stepless control for reciprocating compressors was introduced. This mechanism helps to save the operating power and installation costs. For screw compressor applications, varying the speed of the electric motor with a variable frequency drive is a common method of volume control for oil-free screw compressors. Depending on the specific application, the range of speed variation can be as wide as 35-105%. The effectiveness and efficiency of this method is demonstrated by Figure 10 and Figure 11. Simply put, these charts can be summa-

rized as follows: 50% speed equals to 50% volume flow and 50% power consumption.

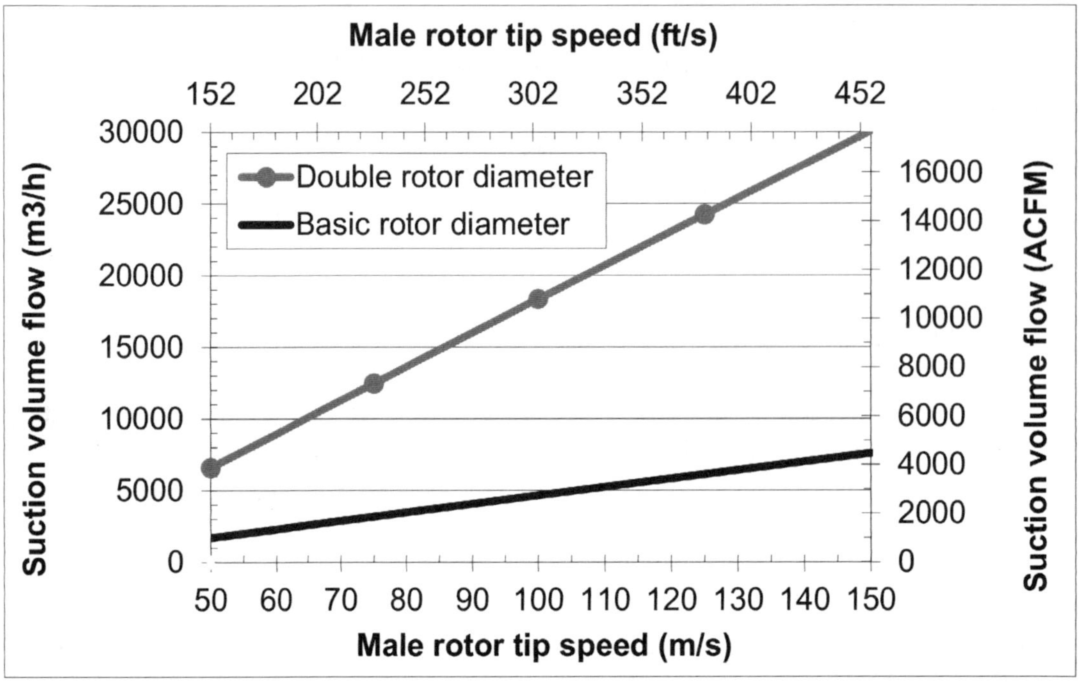

Figure 10: Effects of speed control on suction volume flow of an oil-free screw compressor

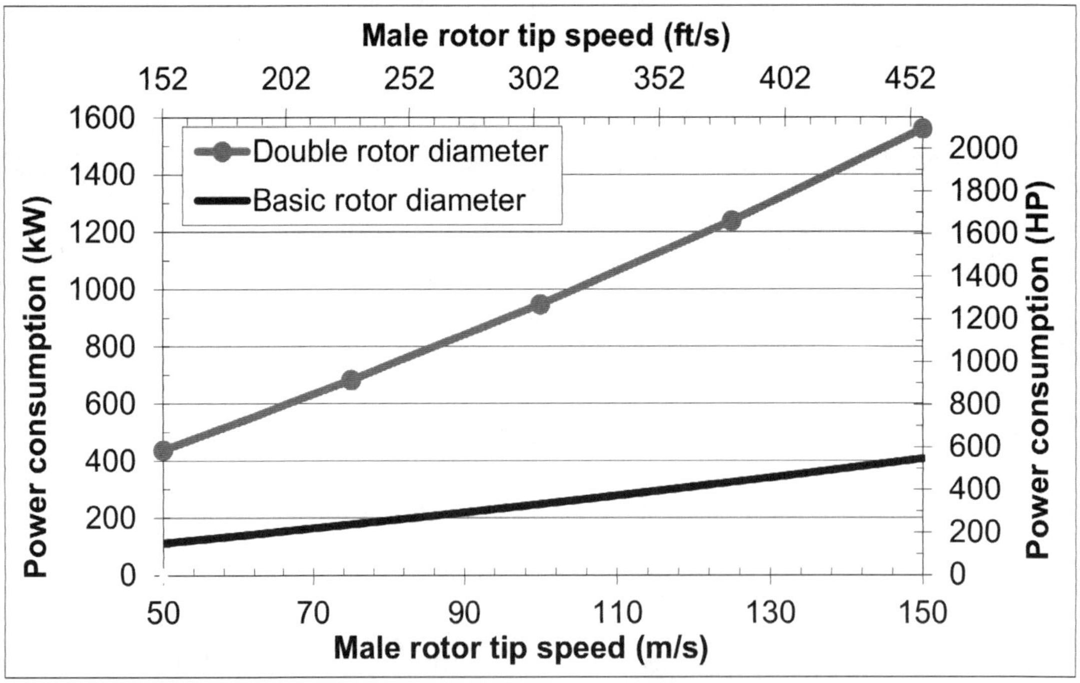

Figure 11: Effects of speed control on power consumption of an oil-free screw compressor

When primarily focused on CAPEX, reciprocating compressors show usually some benefits over oil-free screw compressors, however, especially for low temperature services this gap closes as non-lubricated, slow speed reciprocating are used which tend to be bigger in size and more complex. Nordquist et al. [6] summarized their maintenance and OPEX cost evaluation of different compressor types as shown in Table 2.

Discharge pressure range **[bar] / [psi]**	**Costs** **[cost index/ (hp and year)]**
Reciprocating	
20 – 24.5 / 300 - 500	833
200 – 245 / 3,000 - 5,000	1170
690 – 2070 / 10,000 - 30,000	1670
Centrifugal	
0.35 – 35 / 5 - 500	100
20 – 70 / 300 - 1,000	133
Screw	
0.6 – 24.5 / -5 - 300	200

Table 2: Maintenance costs for different compressor types [6]

Compared to an oil-free screw compressor, reciprocating compressors have a multitude of moving parts that are prone to wear such as valves, piston rings, crossheads, piston pins, etc. These parts are exposed to mixed friction and require regular and frequent maintenance. Where a 2x100% arrangement is typically required for reciprocating compressors due to reliability concerns, this philosophy is rarely applied for oil-free screw compressors which are to be designed for at least 3 years of uninterrupted operation [8]. In January 1999 the Hydrocarbon Processing magazine published an article [9] that summarized various reliability parameters for different types of mechanical equipment in the hydrocarbon industry. An overview of these findings is given in Table 3:

Compressor Type	Availability [%]	Reliability [%]	IMR&O [hr/year]	Forced DT [hr/year]	MTBF [year]
Centrifugal, clean service	99.7	99.8	24.8	18.8	8.0
Centrifugal, fouling service	99.0	99.5	90.6	40.6	3.7
Recip., lubricated	97.3	97.8	237.2	189.2	0.5
Recip., conv. non lube	91.3	92.3	766.1	670.1	0.3
Recip., labyrinth piston	97.6	98.3	207.2	147.2	2.0
Screw, oil-free	**99.0**	**99.7**	**90.0**	**30.0**	**5.0**
Screw, oil-injected	97.7	98.8	199.9	99.9	1.5

IMR&O: Inspection, Maintenance, Repair & Overhaul
MTBF: Mean Time between Failures
DT: Down Time

Table 3: Machinery reliability data for different compressor types

Figure 12 shows a range chart of typical volume flows and maximum discharge pressures of both reciprocating compressors and oil-free screw compressors (note the logarithmic scale). Furthermore it shows known references in ethylene boil off gas services for oil-free screw compressors. It is obvious that all of these references are comfortably within the range of both compressor types in terms of the suction volume flow and discharge pressure. A final decision on which compressor type to use can therefore exclusively be made based on experience, initial investment and long-term operating costs.

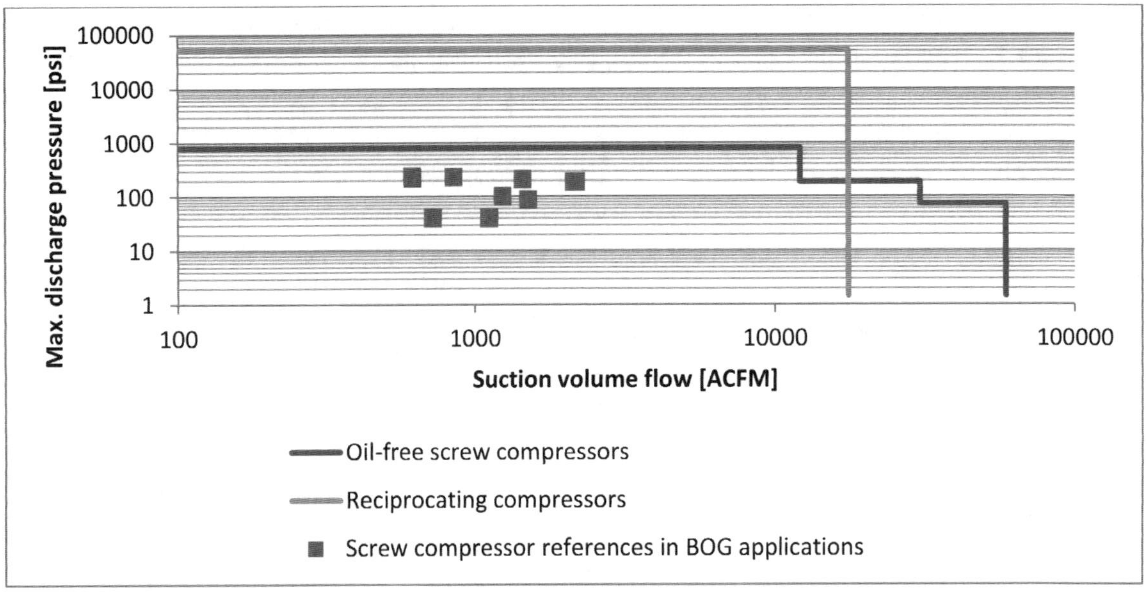

Figure 12: Range chart for different compressor types

Conclusion

Even though the re-liquefaction unit of ethylene plants is only auxiliary to the main processing equipment, its reliability is a crucial factor to the whole plant. With flaring or venting of the boil off gas not being an option, the availability of the compressor should be a priority to the plant design. The paper described how to determine the discharge pressure of the BOG unit and further showed that both reciprocating as well as oil-free screw compressors can be used for this application. Traditionally reciprocating compressors are used for this service. However the paper demonstrated that oil-free screw compressors not only show a similar, for this application desired, performance behavior but are also of a simpler design resulting in less maintenance costs and a much higher reliability. This compressor type is well referenced in the industry and has proven to be a reliable alternative choice for boil off gas applications.

[1] Warren Wilczewski (2015), Growing U.S. HGL production spurs petrochemical industry investment; U.S. Energy Information Administration

[2] Gregory Cameron, Linda Le, Julie Levine, Nathan Nagulapalli (2012), Process Design for Production of Ethylene from Ethanol, Pen Libraries, University of Pennsylvania, Department of Chemical & Biochemical Engineering,

[3] Yajun Li, Guang Jin, Zhiliang Zhong (2012), Thermodynamic Analysis-Based Improvement for the Boil-Off Gas Reliquefaction Process of Liquefied Ethylene Vessels; Chemical Engineering & Technology,

[4] Compressed Gas Association (1990), Handbook of Compressed Gas, Chapman & Hall

[5] Ahmed Ouadha, Boumedienne M. Beladjine (2015), Exergy Analysis of an Ethylene BOG Re-Liquefaction System, The 24th IIR International Congress of Refrigeration ICR2015

[6] Nordquist, Glen B. and Bielskus, Peter A. and Clayton, Russ (1992), Dry Screw Compressors in Process Gas Applications Including Maintenance Considerations, Proceedings of the twenty-first turbomachinery symposium

[7] Wennemar, Jürgen (2009), Dry Screw Compressor Performance and Application Range, Proceedings of the thirty-eighth turbomachinery symposium

[8] API619 5th ed. (2010), Rotary-type Positive Displacement Compressors for Petroleum, Petrochemical and Natural Gas Industries 5th edition, American Petroleum Institute

[9] Bloch, H. P. and Geitner, F. K. (1999), Use equipment failure statistics properly. Hydrocarbon Processing

[10] API619 1st ed. (2004), Rotary-type Positive Displacement Compressors for Petroleum, Petrochemical and Natural Gas Industries 5th edition, American Petroleum Institute

Operating Unstable Catalyst – What We've Learned

Jane Liang-Weissgerber
Jonathan Lunn
The Dow Chemical Company

Outline

- Incident Background
- Catalyst Instability
- Summary

DRIVE TO ZERO

Reactor Setup

- Front end de-ethanizer cracker
  - Lead bed with 98% conversion
  - Cleanup bed with 2% conversion
  - Interbed cooler to control conversion into guard bed
- Catalyst was replaced in 2012 after **16** years of service
- No issues with catalyst performance

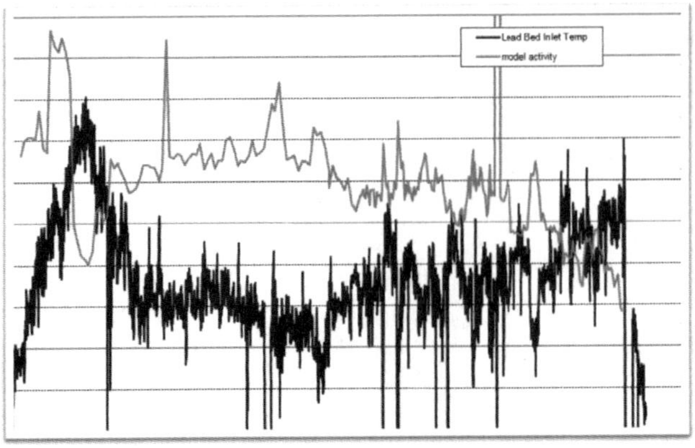

DOW

DRIVE TO ZERO

Incident Background – Reactor Exotherm

- **2014 Major Plant Incident** – *Plant restart **6** days later*
  - During reactor start-up flow could not be established, leaving pressurized cracked gas in a warm bed
  - **Excessive exotherm** occurred in clean-up bed
  - Signs of ethylene de-composition
- **Problems.....Catalyst Change Out**
  - 4 months post plant incident (return to full rates) – *dP RISING*
  - 1 month later after return to full rates – *EXOTHERM*
  - Started planning for catalyst change out only after **2** years of service

DOW

DRIVE TO ZERO

Incident Background – Reactor Exotherm

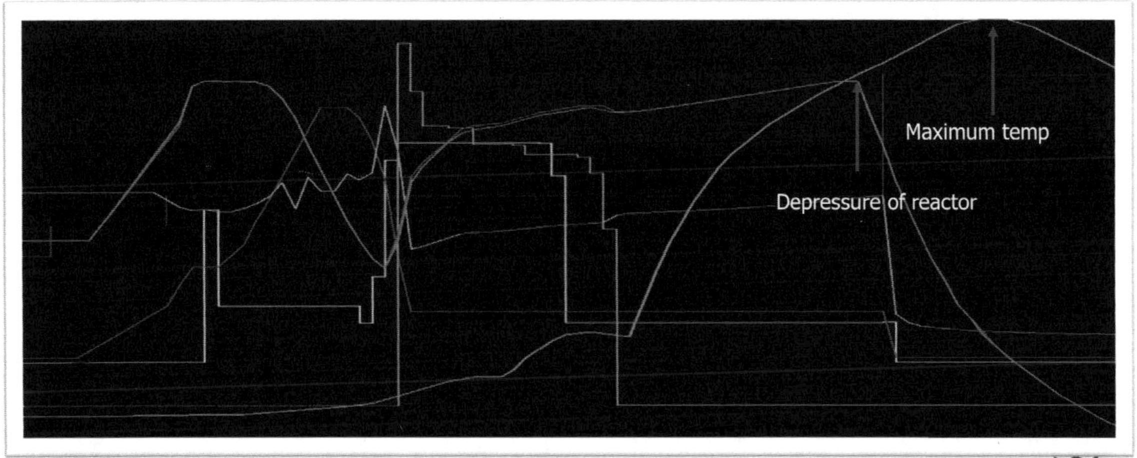

DOW

DRIVE TO ZERO

Incident Background – Root Cause

- ***Process Control***
  - High temperature trip logic did not include certain start-up steps
- ***Operating Discipline***
  - Reactor start-up procedure was not followed
- ***Contributing Causes***
  - The valves were not aborted in the critical start-up step
  - Reactor was started up too warm

DOW

DRIVE TO ZERO

Reactor Exotherm Mechanics

- Adsorption on catalyst governed by H_2 > Acetylene > CO >>> C_2H_4
  - $C_2H_2 + H_2 \rightarrow C_2H_4$
  - $C_2H_4 + H_2 \rightarrow C_2H_6$
  - Other reactions with MAPD and butadiene → not as relevant with front end de-ethanizer configuration
- Operating Window influenced by
  - CO
  - dT (i.e. adiabatic load)
  - Flow (GHSV)
  - Impurities, age of catalyst

DOW DRIVE TO ZERO

Unstable Catalyst – Multiple Exotherm Incidents

- Additional exotherms (4) initiated by small changes in
  - Feed rate
  - CO fluctuations
  - Radial dT drifts
- Further symptoms
  - dP on the guard bed kept rising
  - Same area as original exotherm would develop hot spot

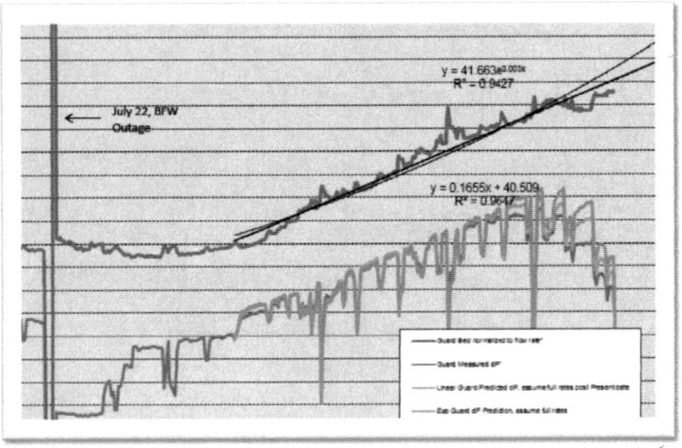

DOW DRIVE TO ZERO

Comparison of dP Changes

Post Exotherm Incident

16 Years of Operation

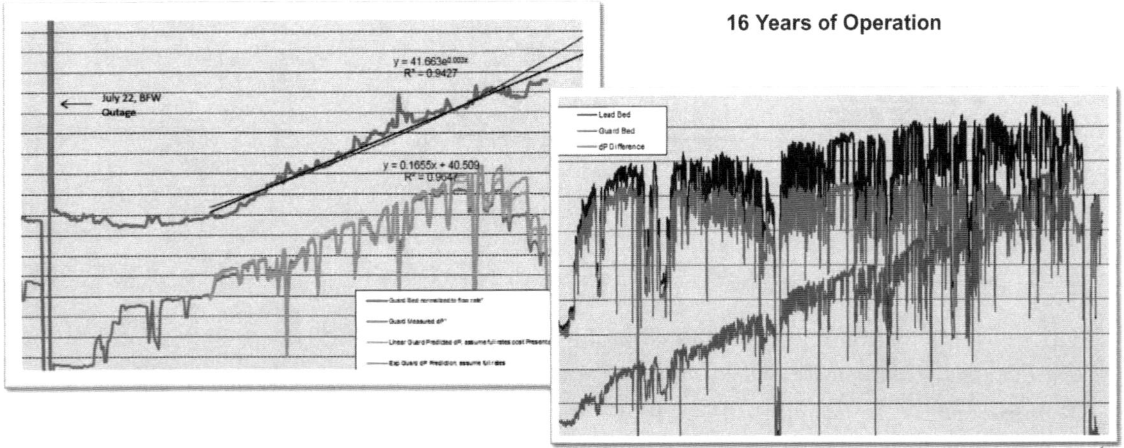

Dow

DRIVE TO ZERO

Why was the Catalyst Unstable?

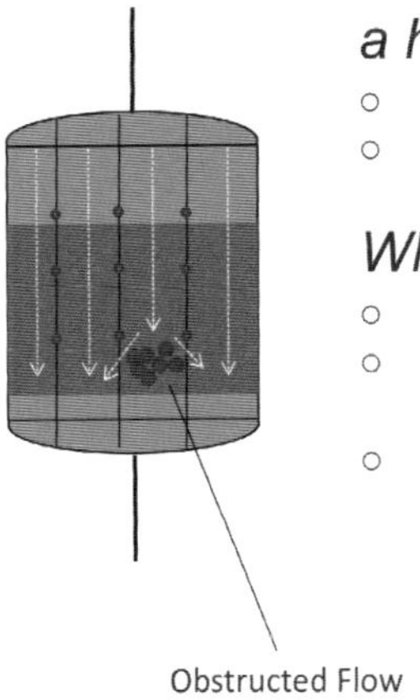

What evidence have we seen that would lead us to believe a hot spot is present?

- Numerous dT "blips" when running lower spec on guard bed outlet
- Runaways/exotherms in the same area as first incident

What could have caused it?

- Ethylene decomposition conditions created during major plant upset event
- Macroscopic coke/carbon likely obstructing flow in that section of the reactor (poor flow ? hot spot formation)
- Climbing dP indicative of loss of void space in reactor that is continuing to worsen

Dow

DRIVE TO ZERO

Potential Hot Spot?

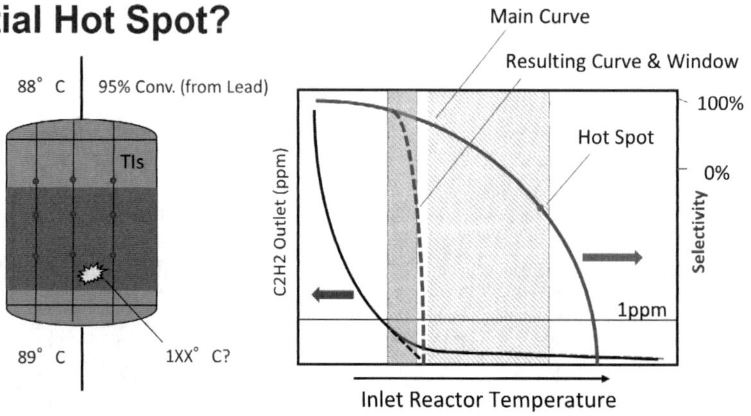

- A small hot spot may not be visible on outlet T (avg. T of all the gas passing through)
- If the hotspot T is near the edge of the operating window, small changes can push it over causing the hot spot to rapidly increase in T and spread across the bed

DOW DRIVE TO ZERO 11

Operating Unstable Catalyst - Overview

- Review Ethylene Specification
- Alarms
- Training and Response Guidelines

DOW DRIVE TO ZERO

Lead Bed vs Guard Bed

- Reactors are setup with inter-bed cooler
- Lead bed 98% conversion, Guard bed 2%
- Moved lead bed closer to runaway conditions to convert more of C2H2
- Fine balance of runaway conditions in both lead and guard bed

DOW

DRIVE TO ZERO

Operating Unstable Catalyst – C2H2 Spec

- Pipeline specification = 5 ppm
  - Normal operations operated below *1 ppm*
  - *New normal > 3 ppm → derivative plants not as efficient*
  - Mitigate runaway conditions on the reactor
- Developed a new protocol to deal with "off-spec" ethylene
  - Ethylene distribution is integrated across two sites
  - Mitigate big changes on the reactor
- Derivative plants managed to operate with higher/offspec C2H2
  - Plants continued to operate
  - Increased understanding of boundaries for the derivative plants

DOW

DRIVE TO ZERO

Operating Unstable Catalyst - Alarm

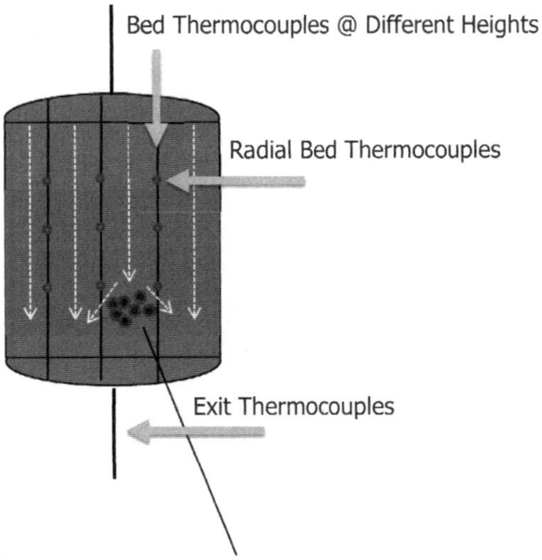

- Exit temperatures – would not detect local hot spot
- Radial temperature alarm to compare drift across thermocouples
- Implemented a monitoring system

DOW DRIVE TO ZERO

Training Operations – First Line of Defence

- Understand the factors that influence operating window
  - CO
  - dT
  - Furnace load (i.e. feed rate)
- Temperature guidelines on changes
- Reviewed with all operators one on one

DOW DRIVE TO ZERO

Training Operations – Responding to Temperature Excursions

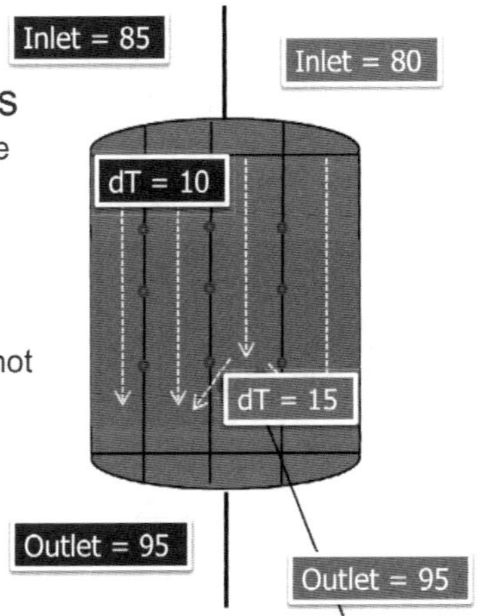

- ❖ Minimizing temperature moves on reactors
  - Allowed C_2H_2 on the outlet to rise to spec limit before needing to respond
- ❖ Monitoring actual dT across entire bed during temperature changes
  - dT would change rapidly because the exit temp had not changed compared to the inlet
- ❖ Temperature spread between beds
  - Maintain temperature between beds to allow for response to excursions

Training Operations – Responding to Temperature Excursions

- ❖ Normal response to offspec or change in uncontrollable conditions on reactor – cut rates!
- ❖ Not intuitive to **NOT** cut rates as an immediate response
  - Less flow = more residence time in reactor
- ❖ Cutting heat only on the guard bed
  - Cutting heat on the lead bed would increase C_2H_2 to guard bed
  - Hot spot already exist on guard bed, excess C_2H_2 would increase more temperature
  - Saved compressor trips

What we found when unloading the catalyst….

DOW

DRIVE TO ZERO

Summary

- **Everyone** understands the operating conditions and limits
- **Develop** operating discipline for response
- **Collaborate**

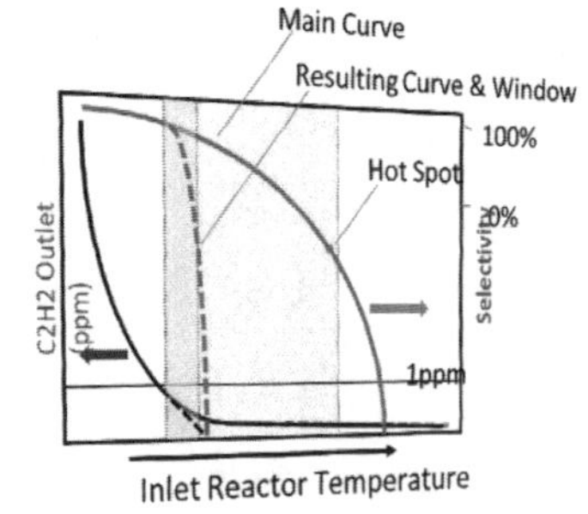

DOW

DRIVE TO ZERO

QUESTIONS?

Acknowledgements

- Megan Finstad – Lead LHC Improve Engineer
- Jonathan Lunn – Reactor SME
- Allison Wing - Reactor SME
- Lin Luo - Reactor SME
- Mariam Rasi – LHC Cold End Run Plant Engineer
- Jane Liang-Weissgerber – LHC Senior Run Plant
- Shawn Lefebvre – Cold End Tech Advisor
- Kate Heymans - LHC Improve Engineer

DOW

DRIVE TO ZERO

Thank
You

485223 Safe Operation in Trickle Phase Reactors

Abstract:

In the Chemical/Refining industry many units operate with high activity nickel (Ni) catalysts. Safe operation of these units is the number one priority. In this presentation we will discuss how to safely operate in trickle phase hydrogenations reactors, share temperature excursion experiences and the lessons learned.

The full and selective hydrogenation units, like C4 butadiene, Pygas 1st stage and aromatics saturation reactors, operate in a range of temperatures and a proper review of the maximum application temperature compared to the onset temperature of methane generation - and potentially run away cracking - is key in determination of a safe operating window for the unit. As with any high activity catalyst defining the desired operating conditions will ensure safe and successful performance.

During the presentation commercial data of multiple units will be shared which will show safe operations and the risks of temperature runaways.

The presentation will highlight the importance of a tailored design, will propose solutions and cover the key hardware requirements for proper flow distribution in the trickle phase flow regime to ensure safe operation.

IMPROVING OVERALL ETHYLENE PLANT PERFORMANCE BY OPTIMISING C2 HYDROGENATION

Authors: Hattachai Aeowjaroenlap[1], Kritsada Chotiwiriyakun[1], Nattawat Tiensai[1], Alejandro Cano[2], Stepan Spatenka[3], Sreekumar Maroor[3], Amit Goda[2]

1. *SCG Chemicals, Thailand; 2. Process Systems Enterprise Inc., New Jersey, U.S.A. 3. Process Systems Enterprise Ltd, England, U.K.*

Introduction

SCG Chemicals's affiliates produce olefins from naphtha and LPG feedstocks via steam cracking. The cracking process produces a certain amount of acetylene, which is reacted with hydrogen to produce ethylene, helping to maintain ethylene product quality as well as to increase ethylene yield. The reaction is typically carried out in a 2 or 3-stage fixed-bed catalytic reactor with inter-stage cooling.

Side reactions include the production of ethane, which needs to be recycled to the cracker, reducing overall plant throughput and increasing energy use. In tail-end C2 hydrogenation units, acetylene oligomerisation also produces so-called green oil, an oligomer of acetylene that deposits on the catalyst surface and causes catalyst de-activation and loss of selectivity. The challenge for C2 hydrogenation reactor operators is to maximise overall selectivity to ethylene while maintaining product quality.

At ethylene prices of around $1000 per tonne, every additional percent of ethylene production in a 1000 kilotonne per year plant can represent millions of dollars per year of improvement in plant economics. C2 hydrogenation plants are thus prime candidates for optimisation aimed at maximising economic performance.

This paper describes the application of an advanced process modelling and optimisation approach to a typical C2 hydrogenation tail-end process in order to determine optimal inter-bed conditions that maximise the selectivity of the desired reaction. By altering the inlet conditions and thus the conversion in each of the three beds it was possible to realise a 13% increase in ethylene gain from the process and a 10% improvement in process economics.

Approach

In a tail-end C2 Hydrogenation process, C_2H_2 present as an impurity in a C_2H_4 stream is reacted with H_2 to yield C_2H_4 Undesirable parallel reactions include the hydrogenation of C_2H_4 to C_2H_6, and the formation of green oil. The reaction is typically carried out in a 2 or 3-stage fixed-bed catalytic reactor consisting of multiple stages with inter-stage cooling and inter-stage hydrogen injection.

In such a setup, most of the conversion takes place in the first (top) reactor bed, with subsequent beds providing protection to ensure that the acetylene is consumed as far as possible. In an ideal situation the acetylene is all converted to ethylene, with minimum conversion of ethylene to ethane. However, under certain conditions selectivity to production of ethane or green oil is favoured.

The production of green oil in the course of normal operations leads to catalyst deactivation, which decreases the selective hydrogenation of acetylene while increasing the undesired hydrogenation of ethylene. Decrease in activity towards acetylene hydrogenation can be compensated by operating the reactor under more aggressive conditions (higher temperature, higher hydrogen feed rates), but this in turn can cause catalyst deactivation to accelerate and undesired ethane production to increase. Eventually, the degree of catalyst deactivation is such that continued operation is uneconomical and it becomes necessary to take the reactor out of service to regenerate the catalyst. Optimal determination of operating policy throughout a catalyst deactivation cycle can improve process economics by increasing selectivity towards the main reaction and lowering catalyst deactivation rates.

Model-based optimisation

The approach described here involves creating a high-fidelity model of the reactor catalyst beds capable of predicting how reaction rates and catalyst deactivation rates are affected by composition, temperatures and other important attributes to a high degree of accuracy.

Having validated the model against experimental or plant data to ensure its accuracy across the range of conditions to be investigated, the validated model is then used within a mathematical optimisation framework to determine optimal operating conditions. In this case, these are the inter-bed conditions that maximise selectivity of the desired reactions by altering the inlet conditions (temperature, hydrogen feed rates).

The objective of the optimisation is to *maximise economic gain* by *maximising ethylene production and minimizing undesired ethane production*. The decision variables used are the time-varying inlet temperature to each bed, the hydrogen flow to each bed and the run length, with constraints including the maximum bed temperatures and outlet C_2H_2 concentration.

Key aspects to be considered

In order to provide the right level of predictive accuracy, the C2 hydrogenation reactor model needs to include a full representation of bed heat and material balances that take into account the variation in conditions along the length of each bed, as well as the following reaction kinetic aspects:

- The acetylene and ethylene hydrogenation reactions, considering different catalyst active sites for acetylene and for ethylene hydrogenation and reaction mechanism known to depend on reaction conditions (high v. low acetylene concentrations)
- C_4s formation (acetylene dimerisation)
- C_6+ and green oil formation (acetylene hydro-oligomerisation)
- Catalyst deactivation during run, to quantify the loss of activity
- Selectivity change during run, to quantify the loss of selectivity
- Irreversible activity loss due to imperfect catalyst regeneration cycles.

One of the main challenges was that none of the published literature captures sufficiently catalyst deactivation and loss of selectivity, so a semi-empirical model had to be developed and validated against a large amount of historical data.

Determining the kinetic mechanisms

A key challenge in building an effective reactor model is to determine the reaction kinetics (rate laws and kinetic parameters). Two reaction kinetics schemes from literature[1] were used as the basis for the rate laws, and the kinetic parameters in the rate expressions were fitted to plant data. Figure 1 shows a typical reaction scheme formulation for the set of reactions. The main reactions are described here.

The desired reaction is the conversion of acetylene to ethylene:

$C_2H_2 + H_2 \rightarrow C_2H_4$ (Acetylene hydrogenation)

This results in 'ethylene gain' – the production of additional ethylene product.

Undesired reactions are the conversion to ethane, which then needs to be recycled

$C_2H_4 + H_2 \rightarrow C_2H_6$ (Ethane hydrogenation)

and the production of green oil, an oligomer of acetylene

$2C_2H_2 + xH_2 \rightarrow C_4H_{4+2x}$ (Green Oil formation)

The production of ethane effectively leads to 'ethylene loss' as the recycled ethylene limits the ability to process fresh feed. Green oil deposits on the catalyst surface and causes catalyst de-activation and loss of selectivity. The target of all operation is to maximise the ethylene gain, i.e. the selectivity to ethylene.

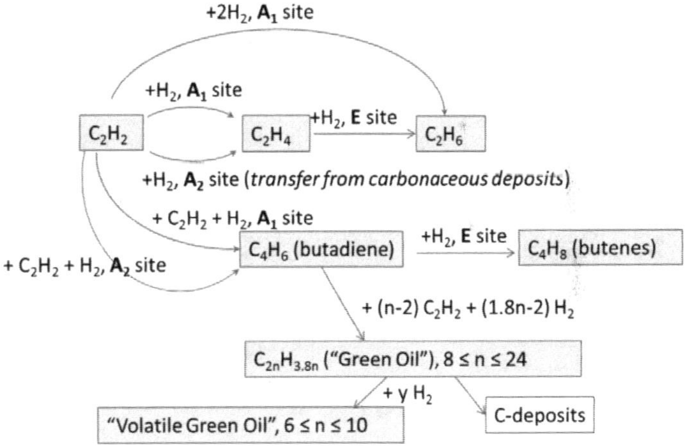

Fig 1. Example reaction schemes from literature

Estimation of kinetic parameters

Once a reaction scheme and rate equations had been selected, the next step was to determine the reaction kinetic parameters. Kinetic parameters cannot be derived from first principles, as they are different from one catalyst to another.

[1] A. Borodzinski and A. Cybulski, Applied Catalysis A: General 198 (2000), 51–66 and A. N. Bos et al., Chem. Eng. Proc. 32 (1993) , 53

They need to be estimated from observed data. This is typically done using a model of the experimental setup, including the proposed reaction set.

Parameter estimation in this context involves the process of fitting empirical parameters, such as the activation energies and pre-exponential constants in reaction rate equations, to observed data. Modern parameter estimation techniques are capable of estimating multiple parameters using data from multiple steady-state and dynamic experiments. The estimation process produces quantitative measures of the degree of confidence associated with each estimated parameter value, providing a quantitative measure of the accuracy of the reaction set model.

The best approach for this is to perform laboratory experiments with a small sample of catalyst pellets, in well-defined conditions, to determine scale-invariant parameters. Plant data typically do not have sufficient variability in operation conditions and there are too many extraneous effects – for example, non-uniform flow and temperature distributions – that mask the true kinetics. It was not possible to perform laboratory experiments in this case, so an additional challenge was how to derive complex reaction kinetics without small-scale experimentation. SCG decided to perform a detailed model-based analysis of historical data, taking account of most relevant guidance from literature.

Parameter estimation was performed on a limited set of kinetic parameters from historical plant data representing six cycles from two reactors using the parameter estimation capabilities in the gPROMS ProcessBuilder modelling software. All in all, over 20 kinetic parameters were estimated for reaction kinetics, where beds 1 and 2 were assumed to have different mechanism than bed 3.

The controlled variables in the plant data included time profiles of the feed rate, feed composition and pressure, as well as the inlet temperature and H2 flow to each bed. The measured variables included time profiles of C_2H_2 and H_2 concentrations at outlet from each bed, temperature measurements at outlet and inside each bed and the C_4s and C_6s product flowrates.

Because parameters were estimated from plant data rather than small-scale experiments, the validated model has limited predictive capability outside current range of operation conditions, but nevertheless is still fit-for-purpose for giving directional advice for determining the optimal operating strategy.

Estimation was performed using a subset of the available data, with other data reserved for comparison of predictions using the fitted parameters. Comparisons with plant data showed that the model reproduced historical cycles sufficiently well to be used in subsequent optimisation. Figure 2 shows the prediction of the validated model (solid line in each case) against plant data for C_2H_2 at the outlet of beds 1 and 2 respectively over a run.

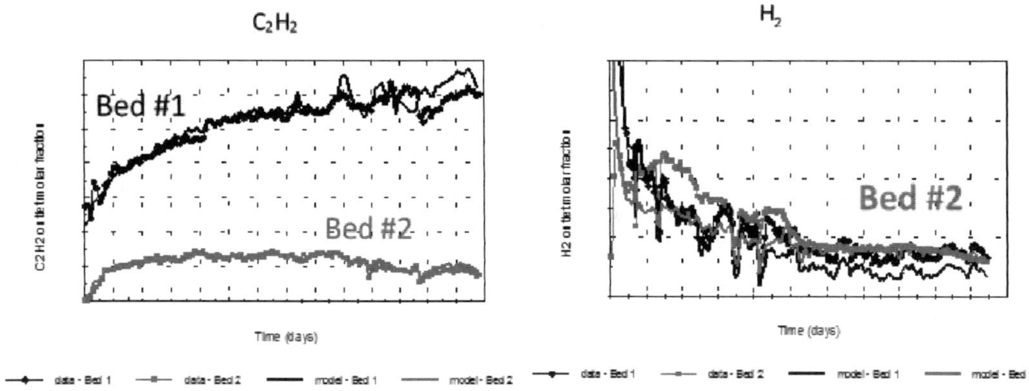

Fig 2. Typical fits to sample data

Constructing the full high-fidelity reactor model

Detailed multi-scale reactor models are increasingly being used to optimise reactor design and operation. Such models typically include high-fidelity sub-models for catalyst particles and the tubes, as well as in some cases shell side hydrodynamics for multitubular reactors. The mass & heat transport phenomena typically taken into account are shown in Figure 3(a); Figure 3(b) shows an example of catalyst geometry and properties definition. Tube and catalyst models are combined with the reaction set models, including the estimated parameters, to provide the full calculation.

Typical applications of such models are:

- Scale-up from laboratory to pilot to commercial plant scale, with a key objective being the maintenance of performance over scales

- Ensuring cost efficiency in investment and operation
- Analysing thermal stability via elimination of hot spots and adjustment of catalytic bed properties (length, activity, shape of particles, etc.)
- Design of cooling system; and catalyst lifetime extensions, through management of catalyst de-activation over the operational cycle.

The detailed SCG catalytic reactor model was developed using the Fixed Bed Catalytic Reactor library within the gPROMS ProcessBuilder advanced process modelling software. A lumped catalyst model was used rather than a full pellet distribution; this was considered to be fit-for purpose given that parameters were being fitted from plant data rather than laboratory experiments. The flowsheet model is shown in Figure 4.

The inputs to the reactor model were feed flow and composition, with manipulated parameters being temperature and H_2/C_2H_2 ratio. The key outputs calculated were product composition, acetylene conversion, ethylene gain, temperature profile in the beds, catalyst activity profiles in the beds, and economic gain.

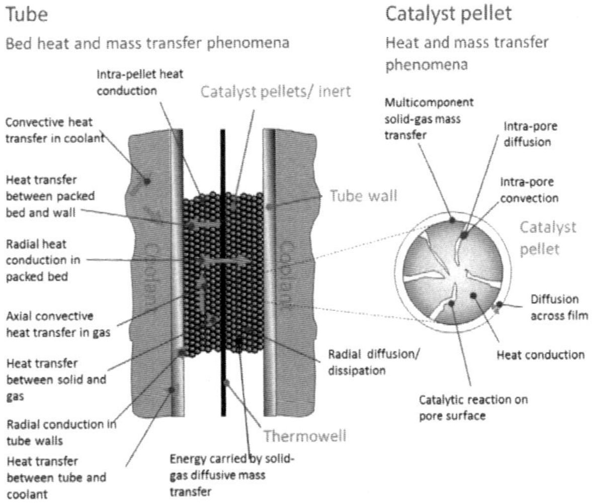

Fig 3(a). High-fidelity reactor models: heat and material flow phenomena taken into account

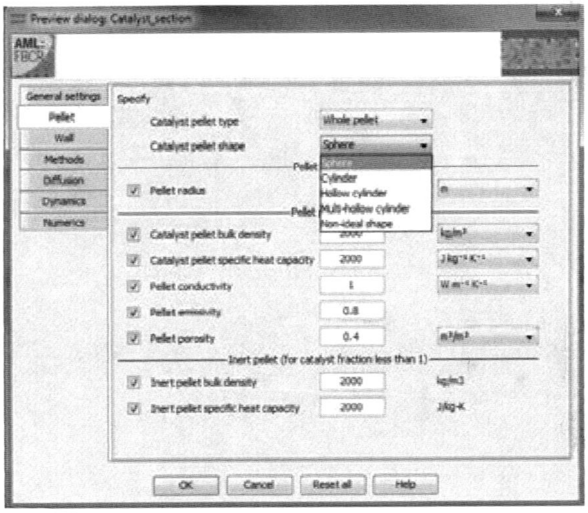

Fig 3(b). Catalyst definition dialog

Simulation analysis and results

Once the model was set up as shown in Figure 4, and the full reaction set incorporated, it was initially used in dynamic simulation mode to explore various operating scenarios, in particular to look at the effect of changing key manipulated variables.

For example, the inlet temperature profile and feed C_2H_2 to H_2 ratio were varied to gauge the effect on product stream compositions, the product outlet conditions, catalyst performance and in particular ethylene gain.

Figure 5 shows the effects of maintaining the same temperature profile, but decreasing the H_2 to C_2 ratio in the feed. This results in an ethylene gain of 5% in the initial stages of the run.

The results in Figure 5 show that lower H_2/C_2 ratio can initially result in a higher ethylene gain (higher selectivity), but it also enhances catalyst deactivation and therefore affects the selectivity later in the cycle. As such the dynamic optimisation problem needs to consider the whole operation cycle.

Optimisation

Having ascertained the key factors affecting ethylene gain through trial-and-error simulation analysis, a full optimisation run was set up. The objective of the optimisation was to maximise the economic performance over a run, by determining optimal settings for variables such as the inlet temperature, H2 feed rate at each stage (thus the C_2H_2:H_2 ratio entering each bed), the acetylene ppm amount at the bed 2 outlet, and the run length itself.

Once a model is available, it is relatively straightforward to perform an optimisation. This is done by defining an objective function to maximise or minimise, then selecting decision variables and, optionally, defining constraints. The major advantage of dynamic optimisation is that it allows rapid exploration of the full decision space, for many decision variables, in a way that cannot possible be achieved using trial-and-error dynamic simulation.

The objective function can comprise any variables in the model. It is typically an economic objective that sums the values and costs. Here the objective function is defined in terms of the economic gain from C2 reactor:

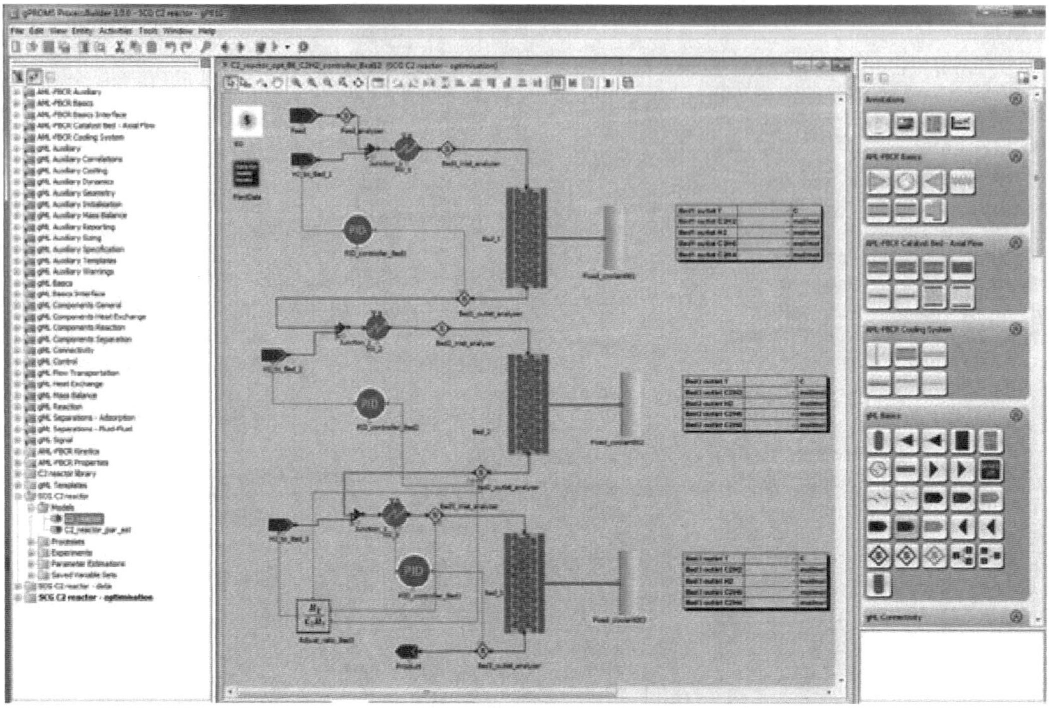

Fig 4. The C2 hydrogenation reactor flowsheet

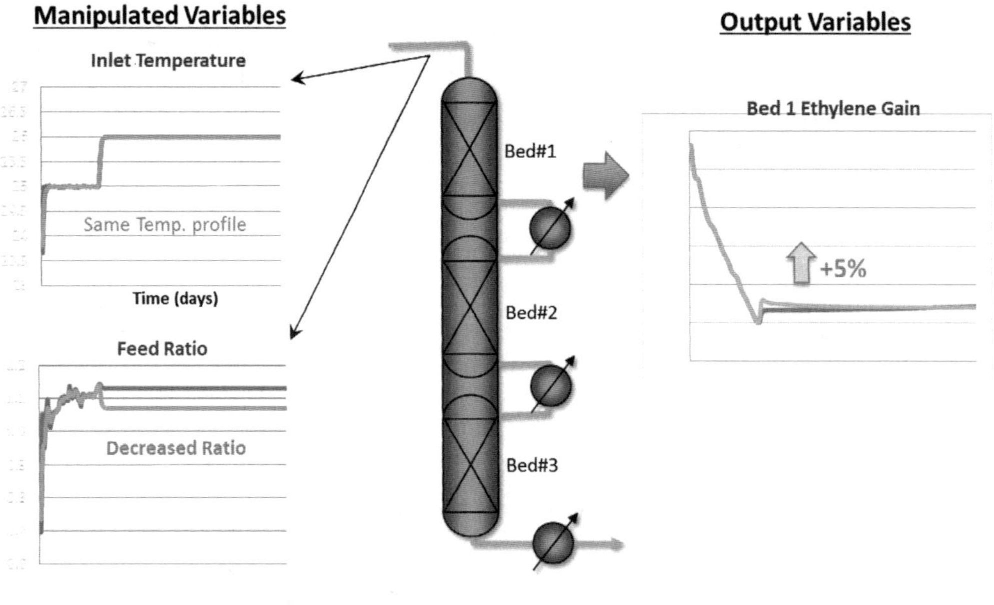

Fig 5. Simulation-based analysis

$$ECG\,[\$/day] = (P_{C2H4} \times Ethylene\ produced) + (Y \times P_{C2H4} \times Ethane\ produced) - (P_{H2} \times Hydrogen\ consumed) - C_{reg} - C_{cat} - C_{sw}$$

Where

P_{C2H4}	Ethylene price
P_{H2}	Hydrogen price
Y	Furnace yield from ethane
C_{reg}	Cost of regeneration per cycle
C_{cat}	Cost of fresh catalyst per cycle
C_{sw}	Switching cost (loss of ethylene) per cycle

Decision variables can be any specified variable in the model. They can be continuous variables, in which the value can vary in a continuous manner over the course of the optimisation (e.g. a tube length in a design optimisation), or integer variables, in which the decision variable may take only integer or discrete values (e.g. standard pipe diameters, or number of tubes in a reactor).

Decision variables used in this simulation are:

- Run length, discretised into time intervals
- Time-varying controls over run length
- Inlet temperature to each bed
- H2 flow to each bed

The following constraints were implemented:

- Maximum bed temperatures
- Outlet C2H2 concentration

Optimisation approach

Initially the approach was demonstrated on historical cycles. Inputs (flowrate, feed composition, pressure) were read from historical data, then the controls optimised for whole run length (Figure 6, top plot). This approach demonstrated what would be economic gain if selected historical cycles were operated optimally, including the optimal run length choice. Results are shown in Figure 6.

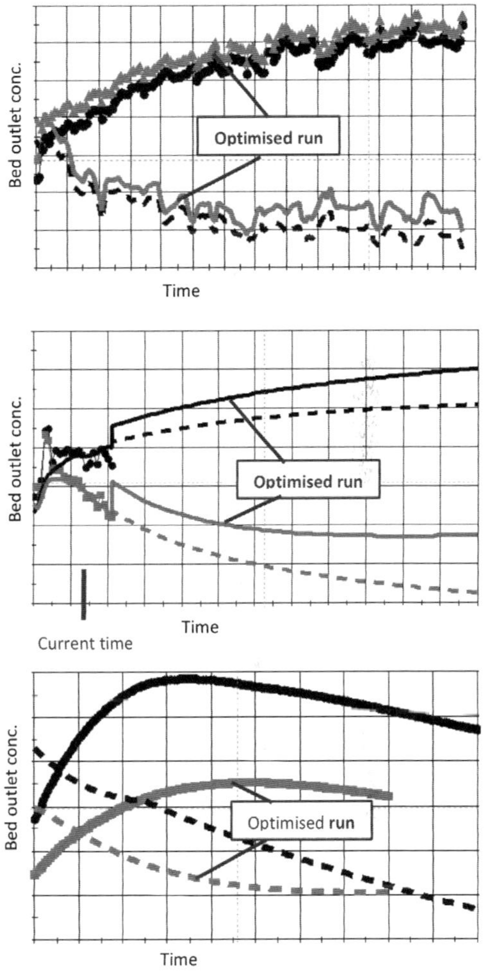

Fig 6. Demonstration on historical cycles

For an offline optimisation of the current (ongoing) cycle, inputs are read from historical data until the current date. From current time onwards, the model inputs are assumed fixed (feed rate, feed composition) and controls are optimised (Figure 6, centre plot). This way an advice can be provided to the operation of the remainder of the run, including optimal run length prediction.

For the optimisation of a new cycle, the time profile of key inputs, such as flowrate and feed composition, need to be assumed and the controls optimised during whole run (Figure 6, bottom plot).

Optimisation results

The optimisation run was executed in order to optimise the temperature and feed ratio profiles. The temperature decision variable was defined as a trajectory over a number of time intervals, and the optimiser allowed to choose the optimum temperature for each interval.

Figure 7 shows the optimised profile. Here it can be seen that the optimal profile is to operate at a lower temperature (brown line) than the usual operating policy (blue line) for a number of days before ramping to the final temperature as conversion and selectivity in the bed decrease. The lower temperature increases selectivity in the first bed, thus reducing production of unwanted ethane and green oil. It also reduces conversion, but this is taken up by spare capacity in the second bed. The lower temperature and corresponding lower rates of reaction in the bed has the effect of reducing catalyst deactivation, hence extending catalyst lifetime. The second plot shows the fixed feed ratio.

The results obtained in the optimisation allowed development of new operating policy that generally reduced loss of ethylene by reducing acetylene conversion in the first bed and increasing the conversion in the second bed.

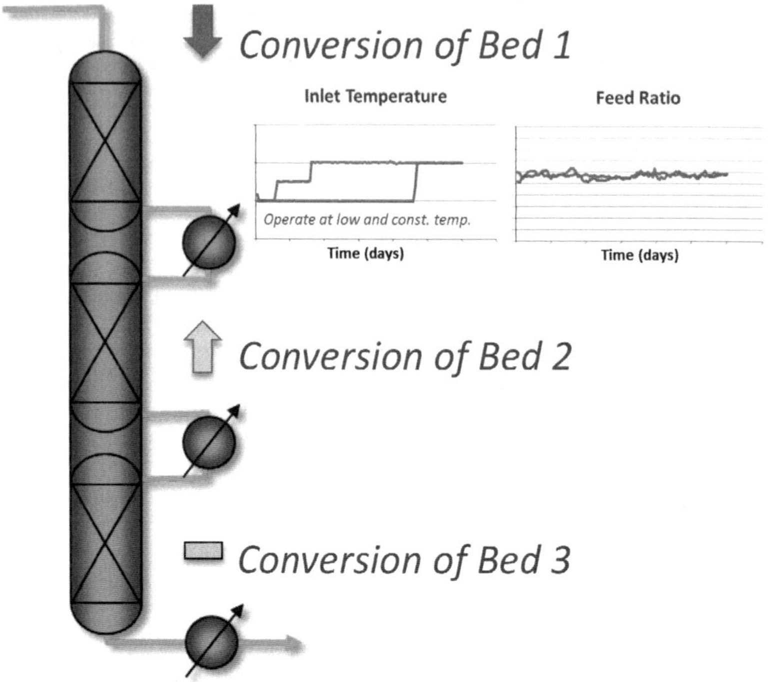

Fig 7. Optimisation-based analysis.

Figure 8 shows the results of application to a real cycle, in which the optimiser was allowed to adjust the feed temperature, feed ratio and run length. Here it can be seen that the ethylene gain increases over the run, to reach achieved a gain of 13% over the standard operation by the end of the run. The corresponding economic gain (determined from the objective function calculation) is 10% improvement.

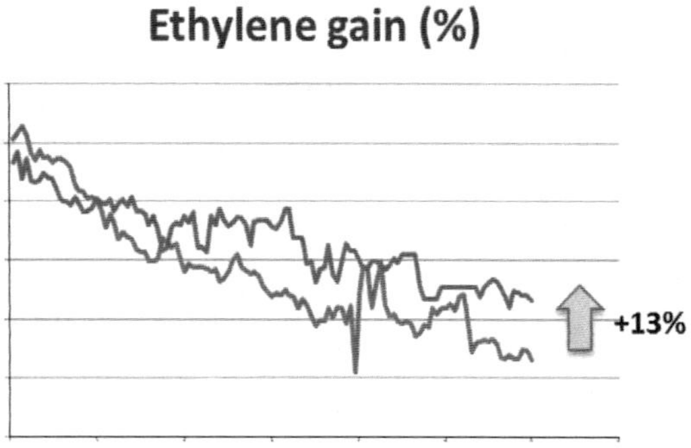

Fig 8(a). the Ethylene gain

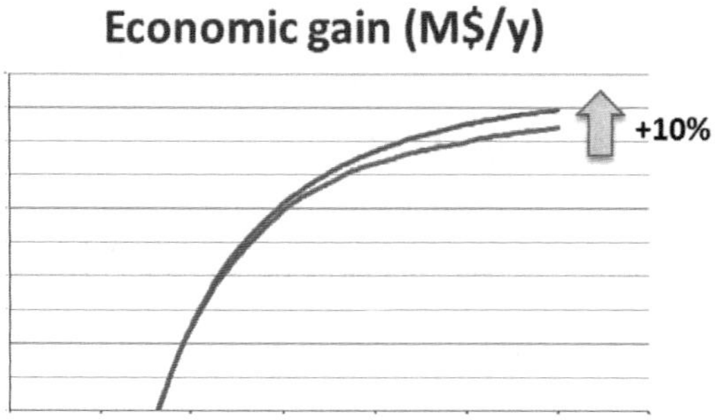

Fig 8(b). Economic gain

Conclusions

In conclusion, a rigorous model of the acetylene hydrogenation reactor was successfully developed in gPROMS ProcessBuilder. This is capable of explaining the reactor behaviour and helping to create new, optimal operating policy.

The major benefit of applying the model is that it identifies significant improvements in economic gain without the need for capital expenditure. The knowledge gained, as well as the model itself, can be applied to other fixed bed reactors within SCG Chemicals' operations, such as C3 hydrogenation.

There is also potential to connect the model to the process historian for real-time monitoring of catalyst activity and green oil accumulation. The optimisation model can be used for weekly production planning, process condition improvement and process risk assessment.

481101 FHR Air Dryout after 2015 Major Outage

Abstract:

Author: Joseph Bell, Flint Hills Resources Production and Optimization Engineer

Sponsor: Johnny Spell, Flint Hills Resources Technical Manager

The Flint Hills Resources Light Olefins Unit in Port Arthur, TX went through a major outage to execute significant cold train modifications and other TA related repairs / maintenance. Work was executed in the Demethanizer, Cold Train, C2 Recovery, Ethylene Refrigeration, and Propylene Refrigeration systems. After work completion, each of these systems was dried with air in order to achieve a safe, successful startup. This presentation will cover the following:

1. Dry air supply and header plan
2. Dewpoint measurement strategy
3. Process Safety concerns of O2 contaminating the flare system
4. Overall execution results and learnings

AIChE Paper Number: 81e

To Err is Human: Mitigating the Hazards of Dryer Switching Operations

Rob D. Fox
Senior Process Engineer
Equistar Chemicals, LP

Prepared for Presentation at the 2017 Spring Meeting
San Antonio, TX, March 27-30, 2017

To Err is Human: Mitigating the Hazards of Dryer Switching Operations

Author

Rob D. Fox, Equistar Chemicals, LP

Abstract:

Within the chemical process industries regenerable dryer systems have long been used to remove moisture and contaminants from process streams. The design and optimization of these mature technologies is well understood. In these systems, limited run lengths lead to frequent regenerations and it is during the period of transition from one operating mode to another that there is an increased likelihood of a production upset or process safety event to occur if appropriate mitigation steps are not taken. This effect is further amplified by the high level of interaction between operators and the process during switching operations, which may result in higher risks for these activities than previously identified. This paper will explore several of the higher consequence hazards associated with switching process dryers, touch on the frequency of operator error, and provide risk mitigation strategies for dryers with both automated and manual valves.

Overview

In the chemical process industries (CPI) the trend amongst operating companies is to strive for more automation, and less manual operations to improve efficiency. One of the most prevalent manual operations that remain in these processes is the switching of regenerable dryer systems between operation and regeneration modes. These activities typically require multiple steps to be completed in the correct order, and often require the input of both console and outside operators. This high degree of interaction with the process requires considerable communication, and when coupled with the high frequency of dryer regenerations presents an elevated risk for operator error to initiate a process safety event.

All systems are affected differently by human error, so the consistent use of industry validated rates of error within process hazard analyses (PHAs) is critical to understand the risks of dryer systems, and to properly mitigate these risks to acceptable levels. This paper will explore several typical risks of dryer systems in the olefins process, and propose effective mitigations for the hazards faced with both manual and automated valves.

Human Error

One of the most widely used forms of PHA is a layer of protection analysis (LOPA), where the unmitigated frequency and consequence of an event are evaluated to determine the unmitigated risk, and then the independent protection layers (IPLs) are credited to determine the final mitigated risk of a hazard. The Center for Chemical Process Safety (CCPS) has studied the topic of human error in depth; the quotes excerpted below are the basis for the frequency recommendations in their published guidelines [1]:

> "Human error is associated with a significant percentage of incidents in the chemical industry. As such, it is important to consider scenarios initiated by human error in LOPA"
>
> "The frequency at which an error can be expected is a function of both the predicted error rate per task and how many times the task is performed within a given time period."
>
> "The less often a task is performed, the fewer opportunities an individual has to make a mistake. Thus, the annual error rate would tend to decrease as the task frequency decreases. However, when a task is performed less often, the individual is typically less skilled in the task, tending to increase human error rate."

Based on these factors the CCPS recommends using a frequency of human error between 1:1 and 1:100 based on the frequency of the task [1]. In addition to the CCPS recommendations, individual company guidelines should be consulted when evaluating the risks associated with operator error during dryer switching operations.

Another important concept promoted by the CCPS is that of the "hierarchy of controls", which ranks the different types of IPLs based on their effectiveness. Figure 1 on the next page details the hierarchy, where inherently safer designs are considered the most effective process safety mitigations, while administrative controls (e.g., procedures, car seals) are the least. Two categories from the hierarchy that are used widely in the CPI are passive safeguards, which are protective layers that prevent an event from occurring (e.g., certain interlocks), and active safeguards, that react to an event to prevent the consequence (e.g., a pressure relief device (PRD) preventing overpressure). [2]

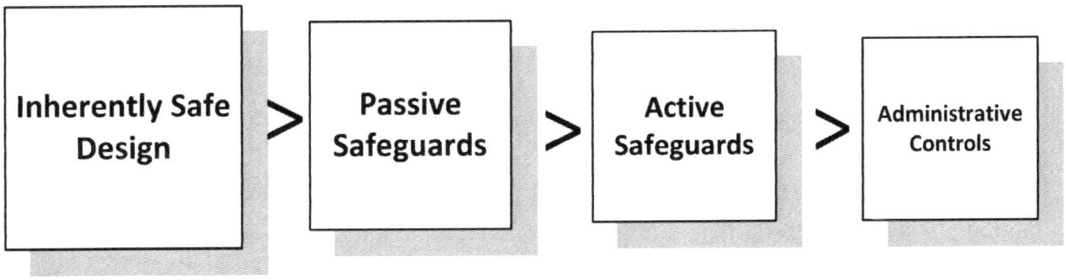

Figure 1: The "hierarchy of controls" as defined by the CCPS. [2]

Hazards of Automated Valve Misalignment

The most prevalent hazards of inadvertent valve operation in an olefins unit are seen with automated valves, where opening the wrong valve can be as simple as hitting an incorrect button. For a better understanding of this hazard, a case will be evaluated for the process gas dryers. A diagram for a typical three dryer system is provided below in Figure 2, where Dryer A is shown in regeneration while Dryers B & C are in service to dry the process gas that is routed forward to the cold box. A typical process downflow/regen upflow arrangement is being utilized for this example. This arrangement presents four distinct valve alignment hazards, which will all result in similar consequences:

- The process inlet valve (A) on the regenerating dryer is opened
- The process outlet valve (B) on the regenerating dryer is opened
- The regen inlet valve (C) on an operating dryer is opened
- The regen outlet valve (D) on an operating dryer is opened

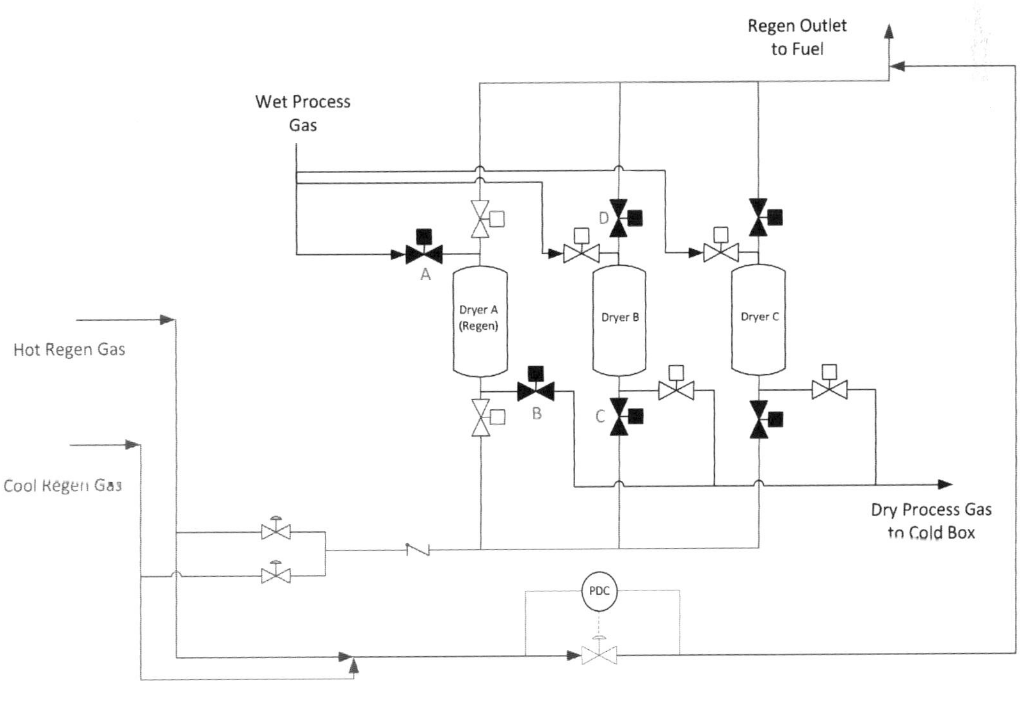

Figure 2: Typical diagram of the process and regeneration connections for process gas dryers.

For all of the misalignment cases above the end result would be routing a significant flow of process gas to the fuel system. Some of the cases above would result in backwards flow through the regen inlet check valve, but unless this valve is specifically designated as safety critical equipment (SCE) and monitored with an inspection and test plan (ITP) caution should be taken before claiming IPLs for this protection. In the unmitigated case of process gas flow to the regen supply system the process gas will still route to fuel, as the regen differential pressure controller will open in an attempt to maintain setpoint.

The unmitigated consequences expected from any of these events are high, with the potential for damage to the cracking furnaces and overpressure of the fuel gas system. The primary driver for the potential damage in the cracking furnaces is the dramatic change in heat input that will be experienced when the fuel gas is replaced with process gas. The higher density and heating value of the process gas will nearly double the heat input to the furnaces and place the furnace tubes at risk of high temperature failure. Typical furnace interlocks will mitigate this risk, but failure of furnace tubes upon interlock initiation is relatively common.

Mitigations for Automated Valves

With the initiating event frequency and potential consequences now understood, the LOPA next determines the required IPLs to reach an acceptable mitigated risk. Individual company risk management standards will vary, but typical industry guidelines would suggest three IPLs are needed for this scenario.

The most straightforward way to mitigate an overpressure hazard is through the use of pressure safety valves, although for the scenario of inadvertently opening the process gas to the fuel system this may not be a feasible solution. The problem lies with the significant flow that must be relieved at low pressure: at a minimum, the fuel system PRDs will need to relieve the full forward flow of process gas from the process gas compressor. If the regen lines are sized large enough, depressurization of the chilling train may be experienced, adding significantly to the required relief rate. For a 1000 kTa ethane cracker (front-end demethanizer arrangement) the minimum relief protection for this case would be approximately 90 in^2 of relief area (API basis) using pressure relief valves (PRVs), requiring four 8T10 PRVs to provide the necessary protection. To evaluate the potential depressurization of the chilling train a dynamic simulation would be needed, and the area required for the expected peak rates may be 2-3 times the steady state minimum.

To minimize the area requirement of these PRVs it may be possible to utilize restriction orifices (RO's) in the regen lines to limit flow, but this comes with consequences for normal regeneration of the dryers. Since either the regen inlet or outlet valve may be inadvertently opened, RO's are required on both lines and their combined effect may limit the flow of regen gas for the dryer.

While it may be possible to mitigate the overpressure hazard with pressure relief devices, the risks to the furnaces and the ensuing process upset will still be experienced during the event. A better solution is to provide a combination of safeguards, so a passive safeguard prevents the event from occurring, and if this fails then the impact is mitigated by an active safeguard.

For a dryer system with automated valves, the most effective passive safeguard is a pressure-based interlock (permissive) that will inhibit movement for both the regen and process valves at the wrong point in the sequence. For a process gas dryer there is a large difference between operating pressure and regeneration pressure, so the pressure of the dryer can be used to determine which valves are appropriate to operate at any point in the sequence; see Table 1 below:

Table 1: Process gas dryer valve permissive logic.

Mode	Regen Valve Movement Allowed?	Process Valve Movement Allowed?
Operating	N	Y
Regenerating	Y	N

The type of automated valve used for the dryer can have an impact on the available SIL rating of the permissive installed. For motor operated valves (MOVs) it may only be possible to claim one IPL for the permissive since a single point of failure exists when using relays to break the circuit providing power to the motor (e.g., relays in series can be fused closed by the same event). A higher integrity may be available for air or hydraulically powered valves due to the higher availability of the valves to vent the motive fluid.

If one IPL is available from the permissive, then two more IPLs must be provided to appropriately mitigate the hazard. The use of a SIL-2 trip to isolate the dryer upon detected high pressure in the regen system is an effective active safeguard, as it will contain the high pressure within the dryer and minimize the impact on the regen system. See Figure 3 for a typical installation of both the pressure permissive and the interlock trip.

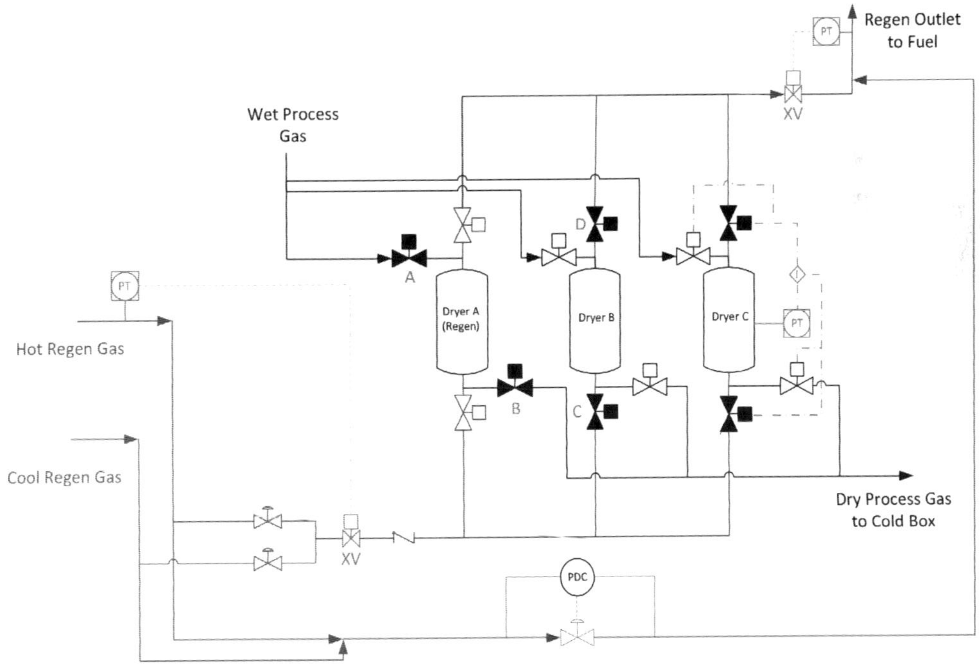

Figure 3: Typical process dryer system with valve permissive and interlock installed.

A positive feature of the system above is that by utilizing common trip valves for multiple dryers the cost of the interlock will be minimized. A simplification made for the sketch above is the required number of pressure transmitters; the final detailed design should incorporate company standards for both availability and reliability to prevent spurious trips. A specific note related to retrofitting this system into an existing unit: the piping specification break from a nominal 300# to a 150# system is typically taken at the outlet flange of the regen out MOV, so the regen outlet line up to the outlet XV will need to be replaced with 300# piping. The regen inlet line should be checked for a similar specification break concern when adding the inlet XV.

The interlock installation diagrammed above is capable of reaching a SIL-2 rating with only single trip valves if an appropriate testing frequency is utilized. The testing will likely be more frequent than once a year, but the system is available for testing any time the dryers are not actively in regeneration so no operational flexibility will be lost.

Hazards of Manual Valve Misalignment

When the valves of a dryer system are manually operated the potential hazards of misalignment are somewhat changed from the cases for automated valves, and the mitigations must change as well. While an operator may begin to open the wrong valve on a process gas dryer during regen, it is highly unlikely that the operator will fully open the valve and present the type of overpressure hazard to the fuel system that was previously discussed.

An interesting example case for the hazards of manual valve misalignment can be found during the regeneration of propylene dryers. To begin the regen sequence the liquid-phase propylene dryers are first drained to an appropriate system and then depressurized to fuel system pressure before starting a cool purge with regen gas (typically tailgas, a mixture of hydrogen and methane). Typically, there is sponge liquid left on the desiccant at the end of the draining/depressurizing process, and the liquid will begin to vaporize when the regen gas purge is started. The light gas atmosphere of the regen gas presents an autorefrigeration hazard to typical propylene dryers, which have historically been built with minimum design metal temperatures (MDMTs) in the range of 30°F to 60°F. At fuel system pressures the saturation temperature of propylene is approximately 20°F, but temperatures as low as -40°F can be experienced when vaporizing propylene in a methane/hydrogen environment.

A necessary tool for evaluating this type of hazard is a minimum allowable temperature (MAT) curve, determined using the criteria of API 579. This curve provides the minimum temperature the vessel can safely withstand at any expected pressure, instead of only at a single value like the minimum design metal temperature (MDMT). An example MAT curve for a typical propylene dryer is given on the next page in Figure 4.

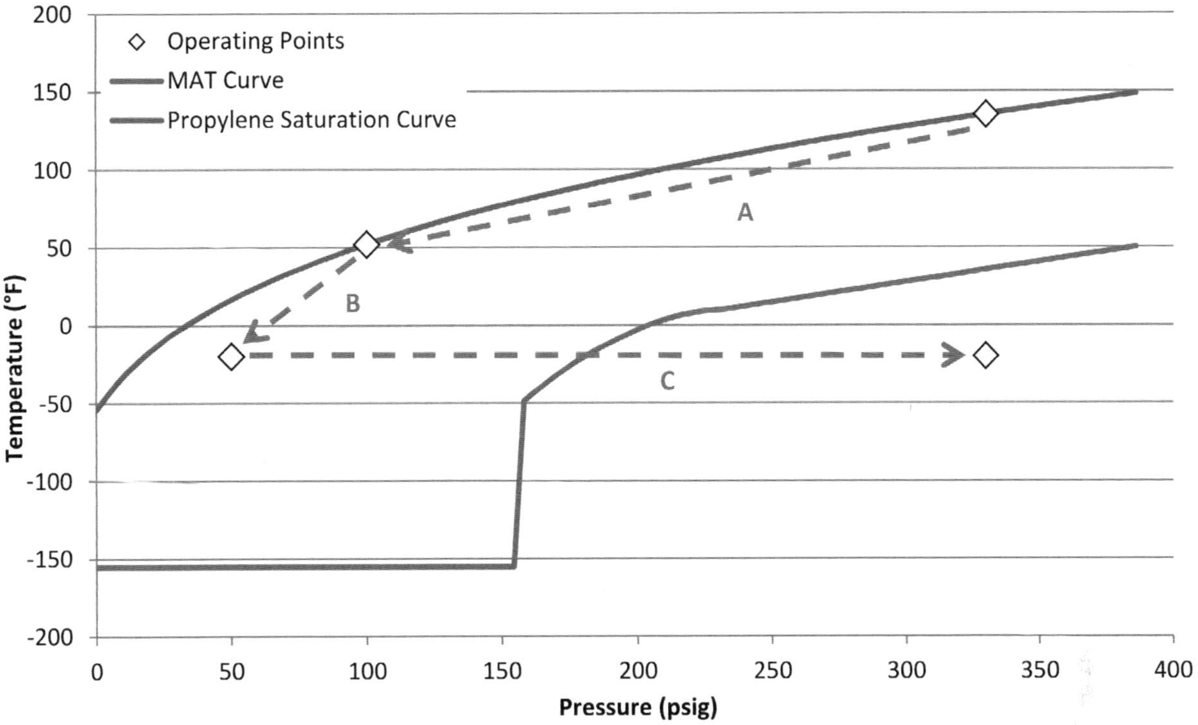

Figure 4: Thermodynamic path of propylene dryer depressurization and start of regen. The temperature follows the saturation curve while the dryer is drained and depressurized (A), becomes colder in the light gas atmosphere once regen gas purging begins (B), and would cross the MAT curve upon repressurization (C) if it were allowed.

The full thermodynamic path of the regen autorefrigeration hazard is shown above; the cold temperatures possible upon the admission of regen gas are acceptable at the low pressure during regeneration, but an operator error during the next step of the regen could open the process back to the low temperature dryer and increase the pressure, crossing the MAT curve. It should be noted that a conscious decision to repressurize the dryer at this point could cause the hazard as well.

Mitigations of Manual Valve Misalignment

For this example it is difficult to prevent the low temperature from occurring, so the only method to ensure a potential brittle fracture event is prevented is to prevent repressurization by inhibiting the movement of the manual process valves. This can be accomplished with a high degree of availability (typically SIL-2 equivalent) by utilizing a captive-key interlock.

With a captive-key interlock, devices are placed on the manual valves that prevent them from being operated unless they have been unlocked with a special key. Multiple keys are utilized in sequence so that the proper valves are operated at all points of the regen. A low temperature permissive can be incorporated into the sequence by utilizing a solenoid box that receives the input signal. A diagram of a typical system for protecting a propylene dryer is on the next page in Figure 5.

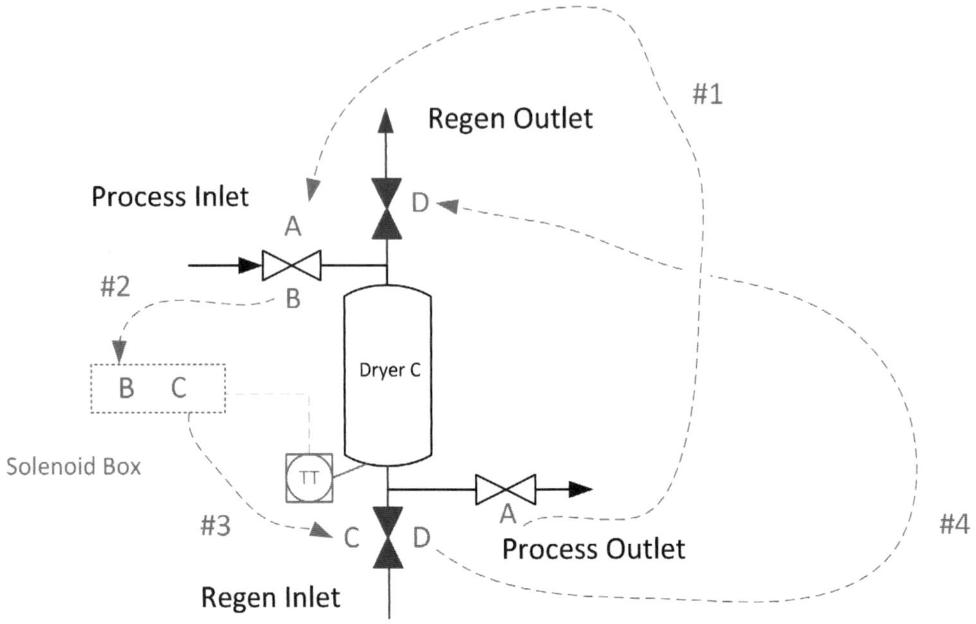

Figure 5: Diagram of a captive-key interlock for protecting a propylene dryer from autorefrigeration hazards during regeneration.

The key sequence proceeds in series through the regeneration, following the basic outline below:

- Close the process outlet valve, which allows Key A to be removed
- Insert Key A in the process inlet valve, close valve and remove Key B
  - Note: Small drain/depressure lines (not depicted) are used to ready dryer for regen gas purge
- Insert Key B in the solenoid box, if temperature is greater than the trip setting, Key C can be removed
- Insert Key C in the regen inlet valve, open valve and remove Key D
- Insert Key D in the regen outlet valve, open valve and proceed with regen

The safety function preventing repressurization is active when attempting to reverse this sequence: if the temperature is too low, Key B will not be released and the process valves cannot be operated, thus preventing the possibility of the repressurization crossing the MAT curve.

Another application where captive-key interlocks can be useful for mitigating manual valve misalignment hazards is for spare reboilers on distillation towers, where the valve sequencing can ensure that the process lines are open back to the tower before allowing the heat supply to be opened.

Summary

The reduction of process safety risk is a common goal within the chemical process industries, and one that requires a drive for continuous improvement. Thorough process hazard analyses play a significant role is driving down process safety risk, and a key factor in accurately assessing hazards is understanding the element of human error. Strong procedures and well-trained operators are vital to the safe and reliable operation of a chemical process, but human error will never be fully eliminated as an initiating event. The hazards presented by valve misalignment must be scrutinized and mitigated. By utilizing consistent frequencies of operator error as initiating events during a LOPA the hazards of operations like dryer switching can be understood. By using interlocks, both automated and captive-key, and permissives the hazards associated with both manual and automated valves can be properly mitigated to an acceptable risk.

References

1. Center for Chemical Process Safety/AIChE. (2014). Guidelines for Initiating Events and Independent Protection Layers in Layer of Protection Analysis. Center for Chemical Process Safety/AIChE.

2. Center for Chemical Process Safety. (2009). Inherently Safer Chemical Processes - A Life Cycle Approach (2nd Edition). Center for Chemical Process Safety/AIChE.

AIChE Paper Number 81f

DECOKING AND CRACKING IN A TWIN CELL FURNACE – DESIGN AND OPERATIONAL CONSIDERATIONS

Harry Wang[1]

Senior Process Engineering Supervisor

TechnipFMC, Claremont, CA

Joel Guillaume

Furnace Technology Manager

TechnipFMC, Houston, TX

Bruce Evans

Director of Pyrolysis Technology

TechnipFMC, Claremont, CA

Prepared for Presentation at the 2017 Spring National Meeting

San Antonio, Texas, March 26-30, 2017

[1] Corresponding author: harry.wang@technipFMC.com

DECOKING AND CRACKING IN A TWIN CELL FURNACE – DESIGN AND OPERATIONAL CONSIDERATIONS

Harry Wang

Senior Process Engineering Supervisor

TechnipFMC, Claremont, CA

Joel Guillaume

Furnace Technology Manager

TechnipFMC, Houston, TX

Bruce Evans

Director of Pyrolysis Technology

TechnipFMC, Claremont, CA

Abstract: Cracking furnaces are the heart of the ethylene plant. As crackers has increased in size, so has the cracking furnace capacity. Ethylene producers purchase various feed stocks ranging from "Gas" feeds to "Liquid" feeds and up to "Heavy Liquid" feeds. Economically, it is important to minimize the number of furnaces in the design of a modern ethylene plant, to have multiple feedstock flexibility and also to have the features to decoking in one cell and cracking in another cell in a twin cell furnace design.

Co-cracking or blended cracking is often used to handle two or more feed stocks in a single furnace. Segregated or "hybrid" cracking (two or more feeds cracked separately in a single cell or in a twin cell furnace) becomes important in modern furnace design for achieving feedstock flexibility while maintaining furnace availability, when it is desired to crack each feed at a specific conversion or severity. In addition, due to the large capacity of modern twin cell furnaces, it is important to consider decoking/cracking in the same twin cell furnace to increase furnace availability.

In 2011, TechnipFMC presented a paper in A.I.Ch.E. Spring Meeting (EPC Conference, Paper Number 96E) titled "Co-cracking vs Hybrid (Segregated) Cracking in Individual Furnace – Design and Operational Considerations". This paper is an extension (Part 2) of the paper presented in 2011 to the even more aggressive scenario of designing, simulating and operating cracking furnaces for the extreme situation of operating one cell in the steam cracking mode and the other cell simultaneously in the steam-air decoking mode.

Simulation of simultaneous decoking and cracking in a twin cell furnace is an important advancement to the simulation of hybrid cracking or co-cracking in an individual furnace. A significant challenge in the design of simultaneous decoking and cracking in a twin cell cracking furnace is how to properly simulate the complete furnace performance due to a large difference in the furnace fired duty and flue gas flow from each cell, as well as a large difference in bridge wall temperature from each cell. In addition to the radiant coils, simulation of furnace firing, flue gas mixing and performance of convection banks also present a significant challenge. Convergence of the simulation of the complete system presents a challenge due to the complexity of the system. How to control temperatures in the decoking cell during this operation is also an important consideration. Material selection of process convection banks which supply the decoking cell is to be carefully evaluated.

For decoking and cracking in the same twin cell furnace, a cracking furnace model has been developed by TechnipFMC called SPLIT-MIXING to simulate decoking and cracking using an integrated EFPS (firebox and radiant section) model and PROVISION (heat exchanger) software. An improved methodology is to utilize TechnipFMC's new generation of furnace simulation software: SPYRO® Suite 7, which is capable to rigorously simulate decoking and cracking in the same twin cell furnace in a single converged file. Additionally, CFD is utilized to verify flue gas flow patterns.

This paper will also discuss key operational aspects of individual cell decoking.

Introduction:

To maintain competitiveness in the current petrochemical market, the ethylene industry has continued to push the economy of scale. As crackers have increased in size, so has the cracking furnace capacity. Current modern furnace designs are routinely 200KTA+ in capacity and will soon be pushing to 300 KTA and beyond. Another large factor in cost competitiveness is the ability to crack a variety of feeds in the ethylene plant. On a recent project there were over 15 individual feeds ranging from ethane, propane, naphtha and up to hydrocracker residue. In addition, recycle streams include ethane and propane, and additionally C3, C4 and C5 streams.

With so many feeds and a limited number of installed furnaces, how can the end users capture both economies of scale and feed stock flexibility with 300KTA+ increments of Mega furnace capacity?

TechnipFMC Process Technology (TechnipFMC) has utilized many methods and techniques in their furnace design over the years to capture economy of scale by maintaining furnace flexibility and maximizing availability. These techniques include Cocracking, Pass/Cell Hybrid Cracking, Independent Cell Operation Decoking (ICD) or called Cracking and Decoking.

This presentation covers the concept of Independent Cell Cracking/Decoke (ICD) from process considerations to SPYRO® Suite 7 modeling, mechanical design considerations and operations.

History of TechnipFMC Process Technology Independent Cell Decoking (ICD) Technology:

TechnipFMC has a long history as a leader in the engineering, design, procurement and construction of twin cell ethylene cracking furnaces. In 1994, the first twin cell furnace with cracking and independent cell decoking (ICD) was in operation. This operation allows the end user to increase both the furnace availability and flexibility. Since then more than 50 twin cell furnaces have been designed and operated with this feature by TechnipFMC.

However, a major issue of this operation is how to accurately simulate this mode in addition to the challenges for the design and operation due to the very complicated system in nature.

In the past, the furnace convection section and radiant firebox had to be decoupled to have the overall furnace simulation with cracking in one cell and decoking in another cell converged. Once the operating conditions are changed, it also takes more time to have the simulation re-converged due to "trial and error" of this complicated system.

Instead of taking this "trial and error" approach to making this complex simulation, TechnipFMC has developed calculation methodology to simulate twin cell furnace with ICD using the latest SPYRO® Suite 7 program which enables simulation of the whole complicated operation in a single converged file. It also becomes much easier- and much faster - to re-converge the file with the change of operating conditions.

Figure 1 is a typical mega twin cell cracking furnace designed by TechnipFMC.

Figure 1. Large Capacity Twin Cell Cracking Furnaces

Types of Independent Cell Decoking (ICD):

The basic types of ICD operation in a twin cell furnace are summarized as follows:

- One cell in cracking mode and another cell in decoking mode
- One cell in cracking mode and another cell in HSSB mode
- One cell in cracking mode and another cell in transition between cracking and HSSB mode and vice versa
- Decoking or HSSB operation can be either Decoke/HSSB with effluent returned to firebox or Decoke/HSSB with effluent to a decoke cyclone and then vented to atmosphere.

Advantages of Independent Cell Decoking (ICD):

ICD mode allows the twin cell furnace design to crack feed at the full capacity and the designed conversion/severity in one cracking cell while the other cell is simultaneously in decoking or HSSB mode.

The advantage of ICD operation is that with such a flexibility of operation, the "very expensive spare" mega furnace may not necessarily be required for the mega plant design thereby minimizing total investment for the mega furnaces. This could be a huge benefit from the investment point of view.

The another advantage of ICD operation is to minimize total cracked gas flow fluctuation downstream of the plant and minimize the fluctuation of feed flow to other furnaces as compared with full furnace decoke/HSSB. This is particularly important for plant expansion such as adding one large capacity twin cell furnace next to many small sizes of existing furnaces. In that case, it is important to have ICD design feature for this large capacity twin cell furnace.

When furnace cracks two different feed stocks from each cell of a twin cell furnace such as cracking ethane in one cell and cracking propane in another cell, the run length from each cell may be quite different (for example, relatively longer run length in propane side and relative shorter run length in ethane side). The advantage of ICD operation is to extend the run length of the cracking cell processing propane feed to increase furnace availability as compared with decoking the full furnace (and bring it out of the production mode) even if the end of run criteria in the propane side has not been reached.

Simulations of Twin Cell Furnaces with ICD operation:

There are significant challenges in the design of ICD twin cell cracking furnace by the proper simulation of the complete furnace performance due to the following reasons:

- Process convection banks and radiant coil banks must process hydrocarbon feed(s), feed(s)+dilution steam in one cell and decoke air, decoke air+steam in another cell, separately, whereas utility banks (economizers, HP steam superheaters) are normally in common. Accurate prediction of heat transfer, process side and flue gas side temperature profiles in the convection and radiant sections is a challenge.

- Fired duty in each cell of a twin cell furnace is quite different during ICD operation. Challenges in the design are the prediction of flue gas mixing in the transition zone from the radiant section of each cell to the common convection section and associated impact on the firebox such as the process crossover temperatures. Prediction of the performance of the coils and passes of the convection section including tube wall and fin temperatures is essential. The same challenges are also applied to the parallel quench exchangers for cracked gas and decoke air/steam.

- Separate CFD modelling in the transition region between firebox and convection can be considered to refine the design.

For ICD operation, a new cracking furnace model has been developed by TechnipFMC called SPLIT-MIXING to simulate this mode using the latest version of TechnipFMC's furnace simulation/optimization software: SPYRO® Suite 7 model, which is capable of simulating ICD twin cell furnace directly. An example of the split-mixing ICD model is illustrated in Figure 2. Alternatively, this system can also be simulated using cracking furnace model such as Firebox/ SPYRO® program linked with other common simulation tools such as ASPEN or PROVISION.

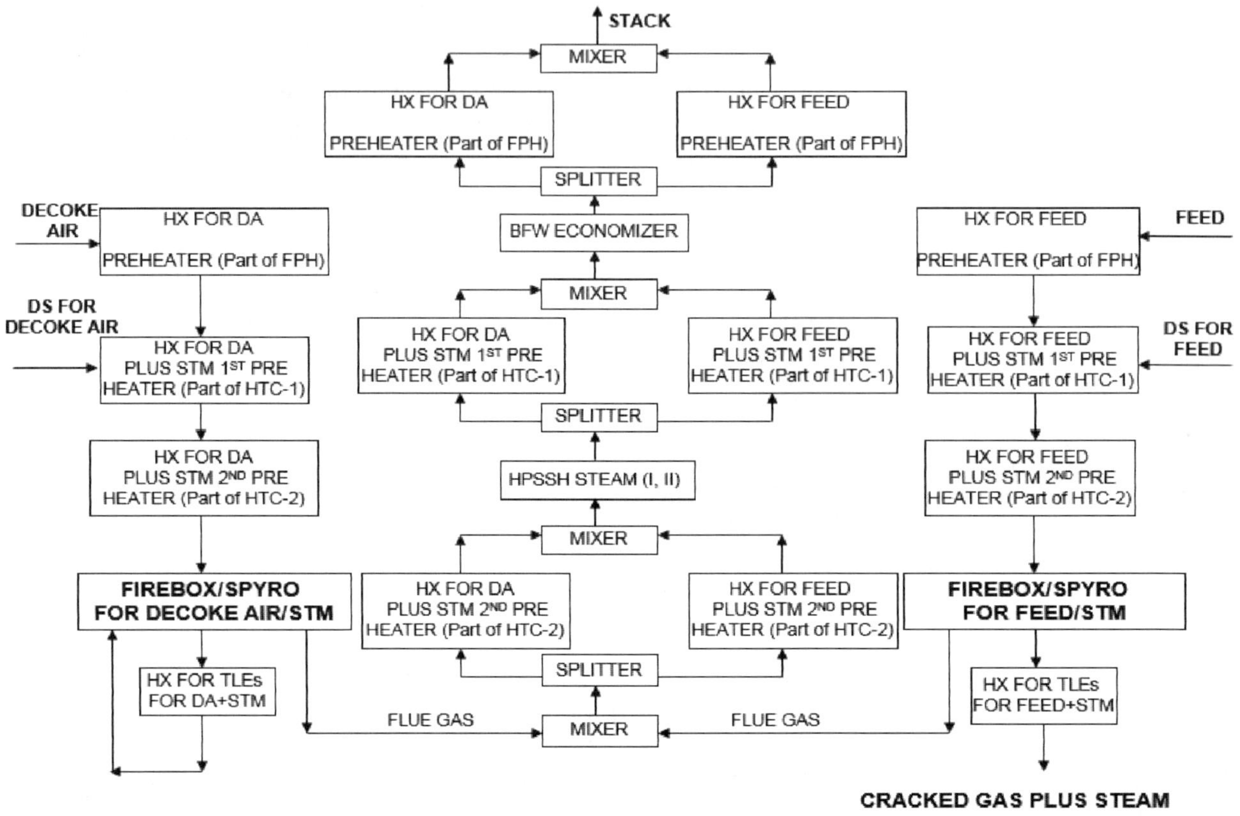

Figure 2. Split-Mixing Cracking/decoking (ICD) Model

SPYRO® Suite 7 is an equation-based simulation-optimization program to simulate almost any type of steam cracking furnace. The program enables flexible simulation of multiple steam cracking furnaces, convection sections, fireboxes and transfer-line exchangers through the flowsheet oriented structure of the program which includes common models like mixers, splitters, adiabatic (cracking) volumes etc.

The salient advantages with SPYRO® Suite 7 simulations for ICD cracking are summarized as follows:

- Rigorous convection section and TLE heat transfer calculations for cracking with ICD can be done automatically and there is no need to link with other external programs such as PROVISION or ASPEN.
- Simulation of cracking and decoking in all the convection section banks and TLE sections is accomplished with SPYRO® Suite 7.
- It can handle simulation of different rows or types of burners and different excess air amount in each cell of a twin cell firebox.
- Program can handle and converge large differences in fired duty between decoke cell and cracking cell.

It has the option to set cracking conversion for gas cracking or severity for liquid cracking in cracking cell and coil outlet temperature in decoke/HSSB cell separately.

- Once a model is built in the SPYRO® Suite 7 environment, calculation due to the changes of operating conditions is much faster. The convergence is relative stable.
- The program can also handle flue gas split and mixing in the convections section banks by the program itself.
- The program can converge all the above in one single file.

A SPYRO® Suite 7 graphic for the sample ICD scheme is given in Figure 3 for main overall flowsheet and Figure 4 for convection section flowsheet.

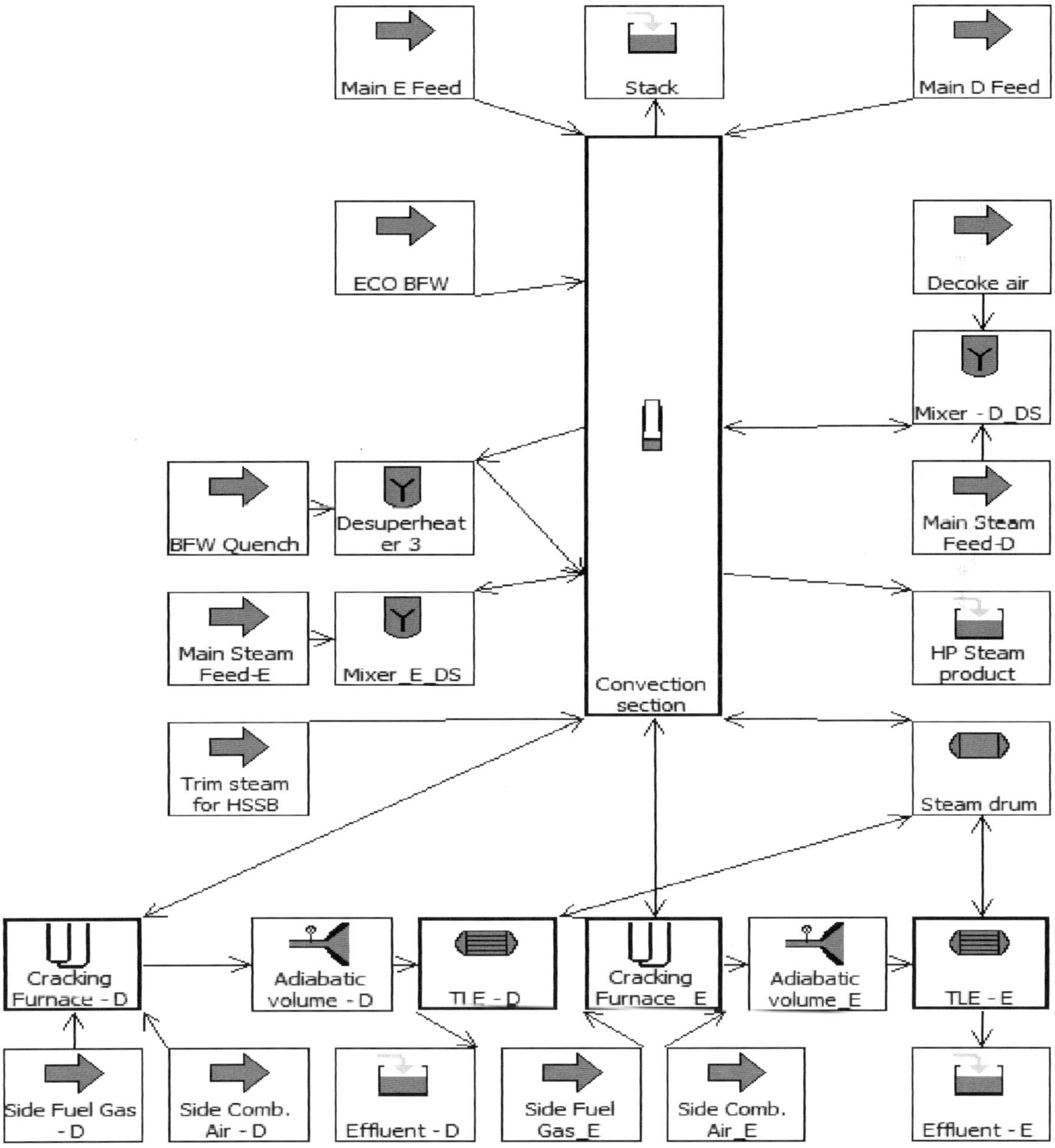

Figure 3. Main Flowsheet of ICD in SPYRO® Suite 7

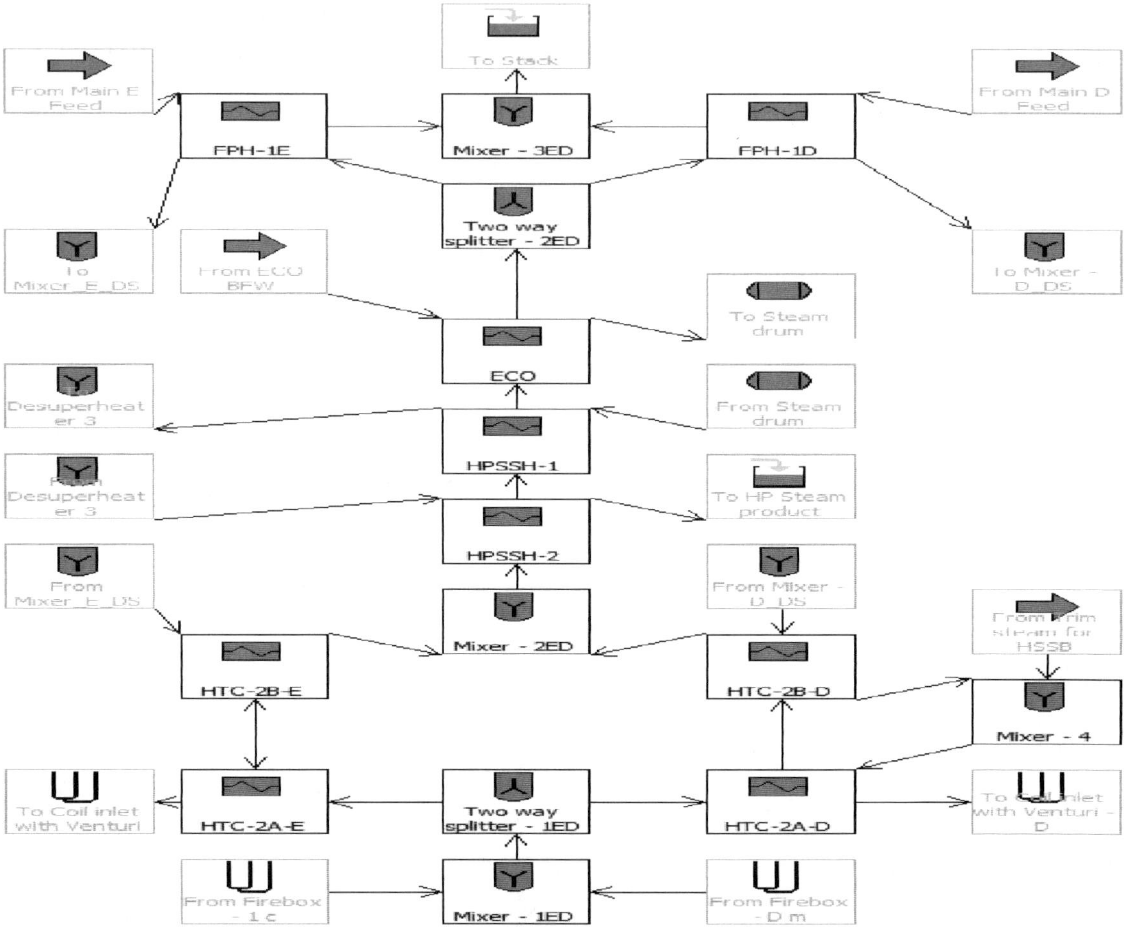

Figure 4. Convection Section Flowsheet of ICD in SPYRO® Suite 7

ICD Design Issues/Considerations:

In actual design of ICD twin cell furnaces, the challenges remain and are to be carefully considered during process design and engineering:

1. Process Design Aspects:

- Process separate convection section passes, radiant coil passes, parallel TLE passes. Ensure all the streams do not combine with each other during this mode of operation.
- How to properly split process banks in the convection section is a consideration. Flue gas interaction at the bridge wall is a consideration. Technip uses CFD and experience from the field data to refine the model.
- Tube metallurgy in the convection section is to be checked for ICD operation. Some material upgrades may be required.

- For both gas and liquid cracking with ICD, consider parallel cracked gas MOVs and decoke MOVs.
- For liquid cracking with ICD, feed vaporization in FPH coil needs to be considered in this operation mode as the flue gas temperature to feed preheater coil is lower for ICD operation than for normal cracking operation.
- For liquid cracking with ICD, it is necessary to consider parallel quench oil fittings.
- For gas cracking with ICD, if BFW preheat in the last transfer line quench exchangers is required, it is important to split BFW flow to the parallel quench exchangers for cracking and decoking/HSSB properly to avoid cracked gas condensation in last TLE. In addition, how to split BFW flow during transition from ICO to ICD and vice versa is also a consideration for smooth transition.

2. Process Control Aspects:

- Provide separate control stations for feed, decoke air, dilution steam (or clean steam) and DMDS to each cell of the furnaces.
- How to control the crossover temperature in the decoking cell from overheating is a consideration since decoke air + steam flow is much lower than the feed + steam flow. The preheating of decoke air + steam is greater than preheating feed+steam in the convection bank due to relatively higher UA.
- During ICD operation, it is also important to adjust the burner air dampers to control excess air to the furnace in the decoking cell with effluent returned to the firebox to prevent from overheating some parts of the convection section.
- It may also be necessary to inject some additional cold steam in the decoking passes of the convection section to cool down the crossover process temperature of the decoking cell.
- Watch firing in the cracking cell to prevent high firing during ICD. Some adjustments may be required for proper control.
- Draft is normally set at the cracking side.
- During ICD operation, separate control of firing to each cell is required as the firing rate from cracking cell and decoking cell is significantly different.

3. Process Safety Aspects:

- Provide positive isolation (EBV or ABV based on design standard) and interlock for feed, decoke air and DMDS during one cell in cracking mode and the other cell in decoking mode to ensure these streams do not combine with each other.
- Provide separate convection section passes, radiant coil passes, parallel TLE passes for decoke air/decoke air+steam and feed/feed+steam/cracked gas. Ensure all the passes do not combine with each other.
- Provide parallel MOV's for decoke air+steam and cracked gas. Provide separate MOV interlocks for each cell for MOV transfer between cracking and decoking within a cell.
- During ICD operation, depending on the design, the effluent can be routed to either a decoke cyclone and eventual venting to atmosphere or directed to the cell's firebox for

combustion. It is necessary to control O2 % at the arch to avoid overheating in the lower process bank feeding decoke air + steam.

- During decoke back to firebox, it is also important to evaluate flame stability while initially introducing steam to prevent from flame blow out, especially utilizing natural gas as fuel.

ICD Operational Issues/Consideration:

Figure 5 is a typical ICD operation in a gas cracking twin cell furnace.

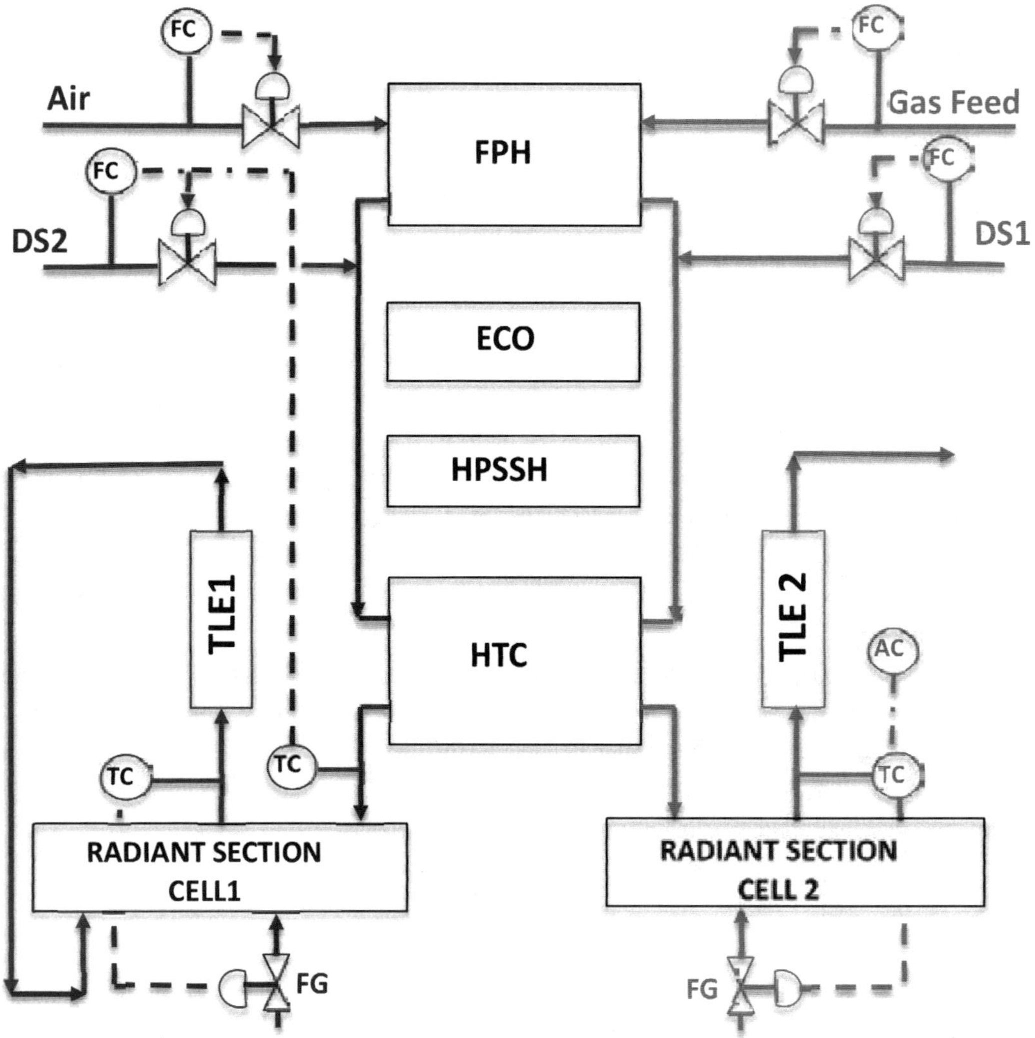

Figure 5. Typical ICD Operation in A Gas Cracking Twin Cell Furnace

Figure 6 is a typical ICD operation in a liquid cracking twin cell furnace.

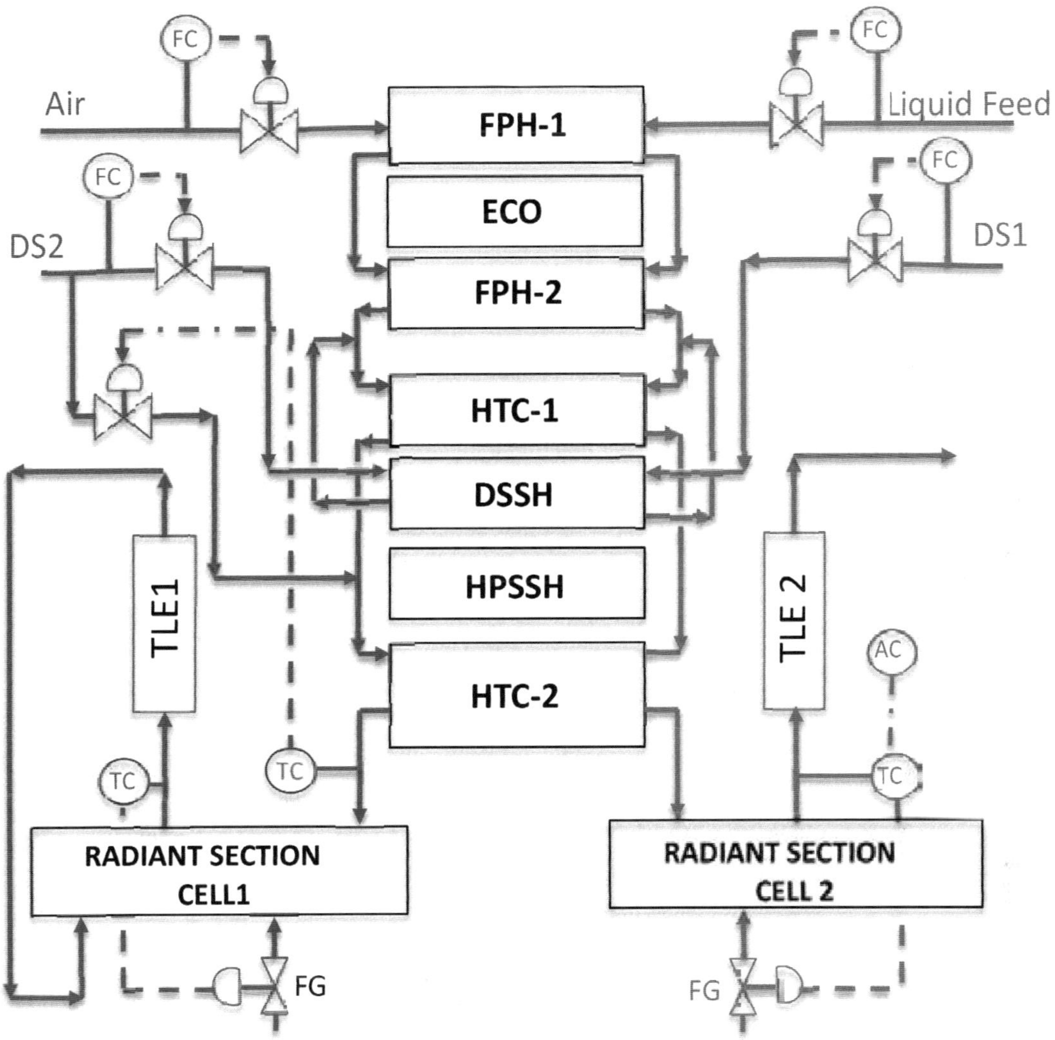

Figure 6. Typical ICD Operation in A Liquid Cracking Twin Cell Furnace

During ICD operation, the following points are recommended for consideration:

- In cracking cell, firing rate is controlled by the once through conversion for gas cracking or severity for liquid cracking resetting the coil outlet temperature which cascade to adjusting accordingly the fuel flow to the burners of that cell.
- In decoking cell, firing rate is controlled by coil outlet temperature.
- The process crossover temperature is limited by the decoke steam flow during decoke steam or trim decoke steam flow injection prior to HTC-2 bank (See Figure 5 and 6).

- Decoke steam/decoke air flow ratio in the decoking cell from start of decoke to end of decoke is a consideration in order to prevent local hot spots along the coils and also higher process crossover temperature in the decoking cell.
- Burner air registers are to be tuned for ICD operation, particularly for decoking back to firebox mode.
- It is important to monitor tube metal temperature in decoking cell. If any local hot spots occur, reduce the air flow in the decoking cell. Some adjustments of the burners nearby may also be required to avoid local hot spots.
- Increase the decoke air slowly. Monitor closely CO2 amount in the decoke effluent. If CO2 amount exceeds the required limit, reduce the decoke air to ensure CO2 is under control.
- Follow decoke procedure properly. The decoke procedure for ICD can be different from the procedure for a full furnace decoke.
- Watch flame stability for the mode of decoking to firebox and in particular with natural gas fuel. Some design improvement for the burners may be required to prevent from flame out when introducing steam to the firebox.
- Firing in the cracking cell will be higher than normal cracking due to lower crossover temperature during ICD operation. A reduction of feed flow or severity may be required.
- During ICD operation with 600 # steam system, BFW vaporization% at the outlet of economizer will be increased. Increased blowdown may be required to avoid too high vaporization within economizer bank.

An example of cracking with independent cell decoking (ICD) is illustrated in Figure 7.

In this case, ethane is cracking in one cell and steam-air decoke is in another cell with decoke effluent to a decoke cyclone and then vented to atmosphere. An ICD scheme was developed. There are parallel passes of the convection section with half of the passes supplied with ethane feed and half of the passes supplied with decoke air. There are also separate firing controls in each cell of the furnace. In this furnace design, ethane feed flow is fixed and ethane conversion is also fixed, which is the same as normal cracking mode. The result is that the firing in the cracking cell increases by about 10% due to lower bridge wall temperature than normal cracking in that cell. The ratio of decoke steam flow to decoke air in the decoking cell is set in order to control the crossover temperature at 1400 °F. Burner air registers remain unchanged in the cell for decoking. The relative firing rate between cracking and decoking cell is from 5 to 1 to 10 to 1. The BFW flow split between cracking and decoking is about 2 to 1.

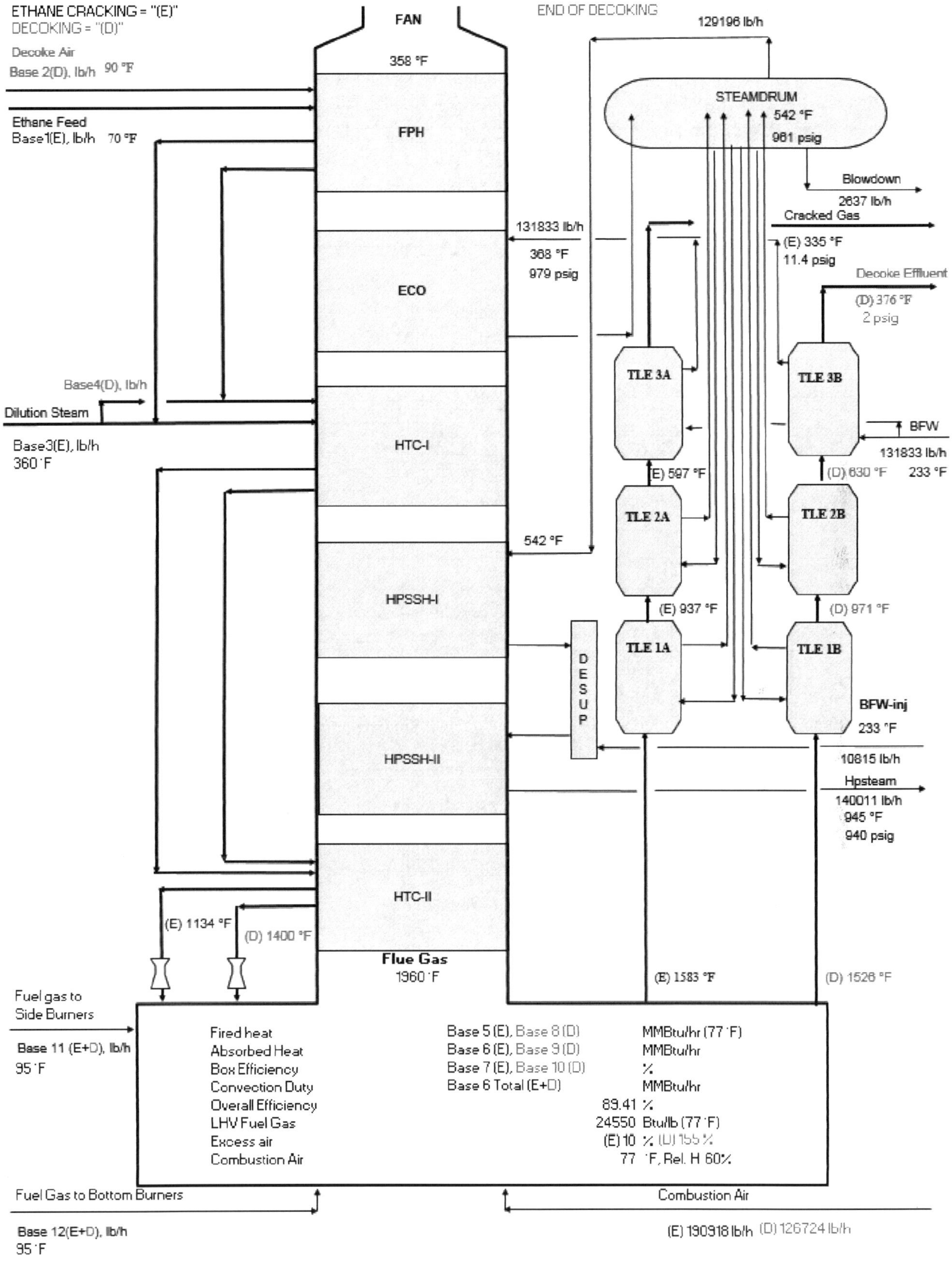

Figure 7. Furnace Performance for Ethane Cracking with ICD

Design Versus Operation:

So far we have discussed design and operating considerations, but how well does this translate to actual field operation? While many applications have been installed, a complete set of operating data that matches the process design parameters is difficult to obtain. Table 1 contains information from a twin cell furnace with ICD. The design data is for an end of decoke scenario, while the operating data is for a start of decoke scenario (HSSB).

Table 1. ICD Furnace Performance Comparison Between Design and Operation

	Design Values (end of Decoke)	Operating Data (start of decoke)
Cracking Cell Feed Flow (t/h)	26.07	26.22
Cracking Cell DS flow (t/h)	13.03	12.2
Decoking Cell DS flow (t/h)	18.75	11.07
Decoking Cell COT (°C)	900	750
Cracking Cell COT (°C)	828	800
Cracking Cell XOT (°C)	566	585
Decoking Cell XOT (°C)	742	740
Trim DS Flow to Decoke Cell (t/h)	0 (6 t/h max)	2.0
Bridgewall Temperature (°C)	1028	950
BFW to Attemperator (t/h)	4.0	2.0

It should be noted that the data is from an operating unit in day to day operation, it is not from a tuned test run with input from the licensor. So what can we glean from this data? First since it is not an exact match to the design process conditions, we have to look at trends but not direct comparison.

The cracking conditions are fairly close to design data, but the decoke conditions are much less severe. Essentially the furnace operation is in a HSSB to firebox mode. The flow of dilution steam to the decoking cell is only 60% of design and the COT is 150 °C less than design. Directionally this will lead to:

- Higher XOT for the decoking cell due to lower flow but same surface area.
- Lower bridgewall temperature due to lower firing in decoking cell
- Less BFW for attemperator due to less flue gas

As can be seen in the operating data the XOT is less than the design, but 2 t/h of cold DS is used to limit the temperature. As expected the bridgewall temperature and BFW low to attemperator are both less than design.

Summary and Conclusions:

For economic reasons, mega furnaces must include features to achieve feed flexibility. Feed flexibility not only includes capability to crack a range of feed stocks, co-cracking or hybrid cracking capability but also ICD features incorporated in a twin cell mega furnace design.

An advantage of applying ICD design feature is that an "expensive" spare furnace may not be required, and furnace availability may be improved. The decision to consider these modes of operation will be influenced by the desired final olefin ratios, individual fresh feedstock consumption, furnace availability and also by the capability of each individual furnace.

Simulation of these cracking modes is challenging due to the geometrical separation of the different fluid paths and the interactivity of these paths. In particular, there is a very large difference in the firing duty from each cell during ICD operation. TechnipFMC Process Technology has developed methodologies to accurately carry out these process designs and simulations to meet client needs.

Safe operation of furnaces with ICD features can be achieved with proven process design, control and operational aspects.

Reference:

[1] Wang, H, Evans, B, Van Goethem, M, "Co-cracking vs Segregated "Hybrid" Cracking in Individual Furnaces – Design and Operational Considerations". Proceedings of 23th Ethylene Producers Conference, March 13-17, 2011, Chicago, Illinois.

479666 Tray Operating Window - Description and Applications

Abstract:

Stable and effective operation of distillation trays are characterized by vapor and liquid loads falling within a number of limitations imposed by the tray geometry. Flooding, limits capacity of trays and can be induced by vapor and/ or liquid rates. At the low operating end, weeping, dumping and/ or flow regime transitions can severely affect the ability of a tray in achieving the desired separation. Significant amount of research has been done over the years for correlating those limitations to operating conditions, physical properties of fluids involved and tray geometry. Tray designers rely strongly on those correlations for optimizing the design of a tray based on a given set of process loads and physical constraints. Those correlations can be used to plot tray operating limits as a function of liquid and vapor loads for a given tray geometry and are often referred to as tray operating windows or stability diagrams. Those diagram are extremely useful to process engineer in setting up operating conditions or making process adjustments that ensure effective fractionation in tray towers. They are also a valuable tool for a trouble shooter when assessing poor performance of distillation towers. This paper describes the different limits plotted on typical tray operating windows and their applications in process design and operation through a few practical examples.

477798 Hidden Ties! - an Explanation for O'Connell's Success

Abstract:

Hidden Ties! – An explanation for O'Connell's success

Markus Duss, Sulzer Chemtech, Winterthur - Switzerland
Ross Taylor, Clarkson University, Potsdam, NY - USA

O'Connell's method has been widely used to estimate the performance of distillation columns equipped with trays and can be expressed as follows:

$$\eta_{Section} = 0.503 \cdot (\mu_L \cdot \alpha)^{-0.226}$$

A few attempts have been undertaken attempting to explain O'Connell's success. However, no prior explanation is entirely convincing. This presentation will show that O'Connell's simple, empirically derived correlation has ties to a more fundamental mass transfer based approach. A traditional approach determining the tray efficiency from the number of transfer units, point efficiency, tray efficiency and section efficiency is adapted and it is shown that there is less than 10% deviation from the O'Connell correlation when:

a) the number of transfer units are equal for the gas and liquid phase
b) the number of transfer units depends only on the liquid viscosity

A simple modification of the alpha-term is proposed:

$$\eta_{Section} = 0.503 \cdot \mu_L^{-0.226} \cdot \alpha^{-0.08}$$

The proposed modification leads to almost perfect agreement between a theoretical model and this modified O'Connell model when the above assumptions hold true and when $\alpha < 5$.

O'Connell's correlation should now be seen in a different light.

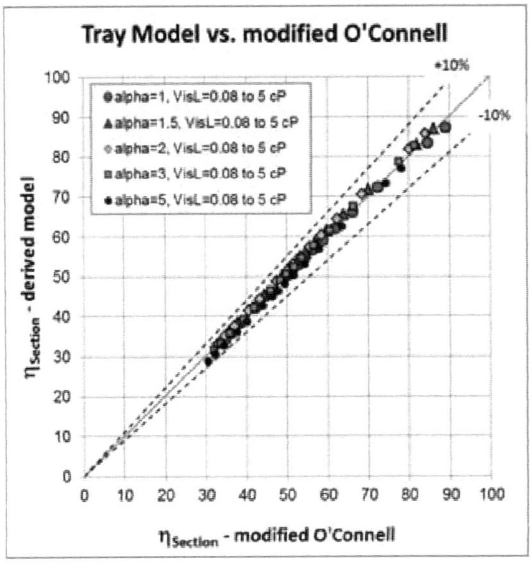

481687 Bubble Cap Tray Vapor Turndown

Abstract:

The UFM tray has been accepted globally as the next generation movable valve. It is only logical that this outstanding device be added to the Fixed Valve Arena for those applications that have a tendency to foul yet still need the high performance and high capacity that is provided by the UFM valve. This paper will show the capacity, performance and pressure drop of the new fixed UFM valve.

481688 Marinization of Mass Transfer Columns for FLNG Applications

Abstract:

Still by today, there is limited experience available about the impact of column motion on the fluid dynamic and mass transfer efficiency in industrial sized marinized columns. Raschig and UOP has started in 2010 a new approach to evaluate the performance of random and structured packings under column motion conditions. The experimental test results have been validated by CFD modeling. On the bases of CFD gas and liquid flow performance can be predicted at any elevated position mass transfer columns may have under combined heave, sway, surge, list, trim, roll, yaw and pitch conditions.

The paper offers new information about column operation for floating LNG applications.

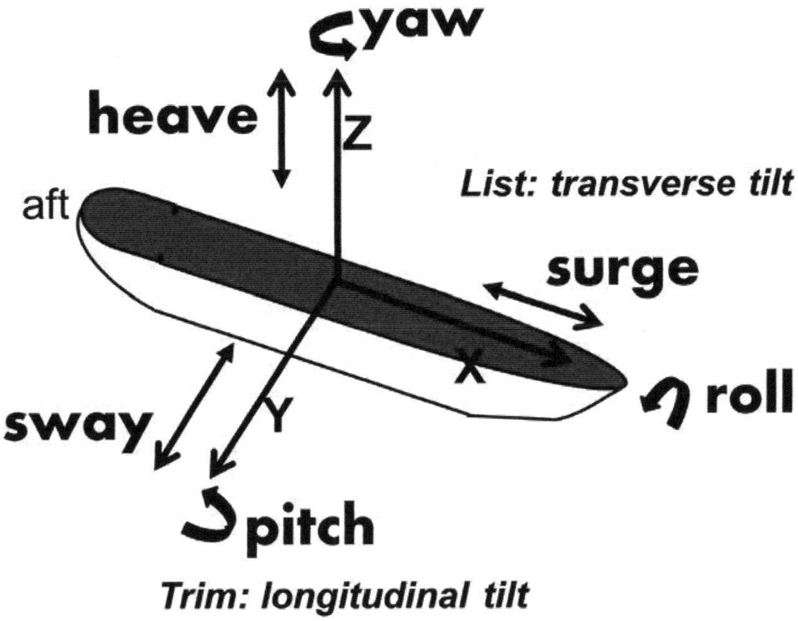

Fig. 1: Nomenclature for moving vessels

481689 Flow-Induced Tray Vibration and Its Prevention

Abstract:

In chemical processing industries (CPI), flow induced tray vibrations occurred with distillation and absorption columns using either sieve trays or valve trays. Numerous severe structural failures of industrial sieve or valve trays were attributed to flow induced vibrations. This paper discusses the basics of tray vibrations, and identifies tray parameters and tray hydraulics that affect the flow induced vibrations. Criteria and guidelines are proposed to prevent tray vibrations in new column designs or column revamps, and have been successfully applied to the industrial case of tray vibrations. The details of this industrial case and measures taken to prevent vibrations are presented and discussed.

479752 High Performance Trays for a Fractionator in an Aromatics Plant

Abstract:

Distillation is a very important separation technology in refineries and downstream chemical processing. Mass transfer internals in distillation columns can be equipped with trays, structured packing or random packing. Trays are generally the preferred option in columns with high liquid rates or in services with high fouling tendencies. After the acquisition of Nutter Engineering in 1998, Sulzer Chemtech expanded its tray product portfolio with inclusion of the acclaimed BDHTM rectangular float valves and the V-GridTM fixed valves. The technological alliance with Shell Global Solutions since 2000 has brought Shell high performance trays into Sulzer Chemtech's range of tray products. This case study highlights an integrated refinery complex in the East Asia, where various tray options were evaluated and Shell tray technology was eventually applied for a new xylene rerun column in an aromatic unit.

ADVANCE IN HIGH PERFORMANCE SPRAY TRAY DESIGN AND THEIR INDUSTRIAL APPLICATIONS

Zhuxiang Wang, Enxia Shang, Liming Zhang, Hong Ma, Tianjin ChuangJu Technology Co. Ltd, Tianjin, P. R. China

Yan Zhuang, Advanced Column Technology, Perth, Australia

Tianjin ChuangJu Technology Co., Ltd specializes in separation processes and equipment design, fabrication and installation, providing services to the chemical, oil and gas, pharmaceutical and environmental protection industries. ChuangJu developed its proprietary CJST series spray tray products and technologies with dozens of patents and know-hows, and successfully commercialised the spray tray technologies in nearly 1,800 industrial columns in 30 provinces of China and overseas, providing services to the chemical, petrochemical, oil and gas, coal chemical, methanol production, synthetic fibre production as well as pharmaceutical and other industries.

Key Words: spray tray, CJST, high performance, oil and gas, separation process

How does a Spray Tray Work

The spray tray utilizes a mechanical phase contactor, such as the vertical sieve cap as in New-VST spray tray, together with certain fluid dynamic conditions, to create a spray regime. In a vertical sieve cap contactor, the gas rises through the opening in the tray, while the liquid enters the cap through a gap between the cap and the tray floor. Inside the cap the liquid is drawn up, by the dynamic lifting of the rising vapor, as well as the liquid phase hydrostatic pressure on the tray. The liquid phase is broken into droplets, impinges to the top of the cap, which is the first stage of intensive phase contact; and then the vapor and liquid mixture is sprayed horizontally out through the vertical sieve holes, while the liquid droplets fall back to the tray and vapor rise up to the tray above, completing the second stage of phase contact. The liquid is the dispersed phase and the vapor is the continuous phase.

The spray type of contactor offers higher mass and energy transfer efficiency due to the following reasons:

- Increased total areas of liquid-gas interface by dispersing the liquid phase into a large amount of small droplets, which creates more interfacial contact areas;
- Longer phase contact time by two stages of phase contact ie inside the cap and outside of the cap;

- Faster interface renewal due to intensified phase contact, and
- Shorter diffusion path for liquid molecules within a small droplet compared with that in the bulk liquid.

Comparison of Conventional Tray vs Spray Tray

Froth Regime, Valve Tray Spray Regime, Sieve Tray Spray Tray, New-VST

Figure 1: Comparsion of Phase Contact of Different Tray Type and Operation Regimes

Column Tray Types

Conventional Trays	High Performance Spray Trays	
Bubble Cap	New-VST	
Sieve		CJST®
Valve		MPST I ®
Dual Flow	CJST Series Spray Trays	MPST II ®
		PFST®
		Low Back-mixing Spray Tray (LBST®)

New vertical sieve tray（New-VST）

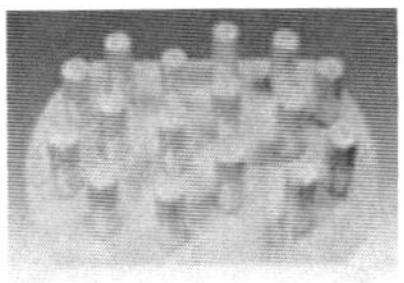

The New vertical sieve tray（New-VST） technology was developed by Japan's MItsui Engineering & Shipbuilding (MES) in 1963-1968. In 1971, the New-VST was tested by the Fractionation Research Inc. and the results indicated higher mass transfer efficiency, higher capacity and lower pressure drop, compared with the conventional bubble cap, sieve or valve trays[1][2].

Since 1980s in China, the New–VST spray tray technology has been successfully applied to benzol stripping column, benzol refining column, ammonia distillation column, tar oil distillation column in coking industry, high boiling point distillation column and low boiling point distillation column of vinyl chloride rectification, saturated hot water column, ammonia distillation column, hybrid-carbonization column of the fertilizer industry, pre-rectifying column, pressure distillation, atmospheric distillation column, recovery column of methanol distillation, catalytic distillation etc[3].

Case Study for New-VST：

1、Coke Gas to Methanol

Jiantao Stage 2 Methanol Project Revamp using the New-VST Tray

The Jiantao Stage 2 Methanol Project was designed and supplied by another company using a float valve design. The atmospheric distillation column suffered several column internals break downs and the product quality failed to meet the design specifications. The facility could not reach design capacity. After investigation, ChuangJu concluded that the controlling step is the atmospheric distillation column. By replacing the float valve tray with the New-VST tray and retaining the original down comer design, ChuangJu solved the problem with rather low revamp cost. The revamped column passed the 24 hour test successfully.

Atmospheric Methanol Distillation Column	**Before Revamp**	**After Revamp**
Tray Type	Valve	New-VST
Methanol Production	16t/h	17.5t/h
Reflux Rate	17.9m^3/h	13m^3/h
Steam Consumption	1.25t/t product	1.0-1.1t/t product
Ethanol content in Methanol Product	<30ppm	<10ppm (meeting USA AA grade)
CH_3OH content in bottom discharge	0.4%	0.05%
Side withdraw of alcohol mixture	66.6t/h	0

2、 **Methanol to Olifin Project**

Inner Mongolia Fenghui Chemical Co., Ltd. Methanol to Olefins Project

The 100,000 tonnes / year Methanol to Gasoline Facility, built by the Inner Mongolia Fenghui Chemical Co., Ltd., started commissioning at 7pm on the 5th of November 2014, and produced on-spec gasoline (with RON up to 93) in the early morning on 7th of November. The facility production achieved 100% design capacity in less than 72 hours after starting up. The facility, consisted of fractional distillation unit include de-ethanizer, absorption tower, de-butanizer, and gasoline tower, was designed and manufactured by ChuangJu using the New-VST spray trays.

	Actual Operation
Tray Type	New-VST
Capacity	100,000 t/a
RON number	93
Bottom product of de-ethanizer C2-	≤2kg/h
Bottom product of De-butanizer C1-4 content	≤0.02%
Top product of De-butanizer C5+	≤2.7%
1234MBZ content light alkane stabilization column top product	≤0.5%
Single Tray Pressure Drop	400Pa

Radial side-guiding spray tray（CJST®）

Radial side-guiding spray tray（CJST®） is an award-winning patented product and the patent number is ZL 2006 2 0025314.8. The product has since been successfully commercialized in coking, chlor-alkali, fertilizer, petrochemical and fine-chemical industries, etc. This tray design can handle high throughput. The column pressure is lower compared with a valve tray design, is fouling resistant, and especially suitable for large diameter columns. The liquid is spayed out toward the downcomer direction which reduces the liquid back-mixing tendency.

Case Study for CJST®: Atmospheric and Vacuum Oil Distillation Columns Xi'an Petrochemical Ltd. Sinopec Corp

The capacity of Xi'an Petrochemical's atmospheric distillation column was 700,000t/a. Due to the market requirement the capacity of the unit needed to increase to 1,100,000t/a.

The original column used a valve tray. When the throughput increased to 800,000t/a, the column separation efficiency sharply dropped. In 2011, the column was revamped using ChuangJu's CJST® tray. After the revamp, the throughput increased to 1,200,000 t/a, a 71% increase compared with that before the revamp. Moreover the column pressure drop was significantly reduced from a maximum of 55kPa to 25kPa.

2、 Natural Gas Industry

No.1, 2 and 3 Natural Gas Treatment Plants of Changqing Oil Field of PetroChina

This project was one part of China's West to East Natural Gas Transmission Project. The natural gas sweeting plants all were imported from overseas. The columns used floating valve trays by Kohn-Glitch. During start-up, the following problems occurred:

- Severe foaming caused flooding;
- The amine loss was high;
- The throughput is lower than design capacity;
- The gas output quality fluctuates;
- Severe corrosion problems in pipework, and
- Poor product quality in the sulfur recovery unit.

- From 2007, ChuangJu revamped all the columns in the three gas sweeting plants with its CJST® trays with successful results.

Gas Sweeting Column	Before Revamp	After Revamp
Column tray design	Kohn-Glitch floating valve	CJST® spray tray
Processing throughput	1.5 million-2.3million m^3/d	3million m^3/d
Single Tray Pressure Drop	1700Pa	750-900Pa
Steam Consumption	1.08t steam/10,000 m^3	0.86t steam/10,000m^3
Flooding occurrence/month	3 times	0
Anti-foaming Agent Dosage	1000-3000mL/d	0

Low resistance membrane spray tray（MPST Ⅰ®）

The low resistance membrane spray tray (MPST I) patent number is ZL200620152162.8. This tray was developed for using in the ammonia recovery column from the copper wash waste water in the ammonia fertilizer industry.

<6>Membrane spray without back-mixing tray（MPST Ⅱ®）

This technology is a new product developed by modifying the above mentioned (MP Ⅰ ®) to solve the problem of liquid back-mixing on the tray. It is suitable for process conditions with high liquid-vapor ratio or fouling susceptible systems. In this design the liquid that has been contacted in the membrane spray contactors is separated from the fresh liquid from the tray above, thus eliminated the problem of liquid back mixing on top of the tray.

Main characteristics：

（1） No liquid back-mixing and concentration gradient on the tray,

（2） High mass transfer efficiency,

（3） Low pressure drop. The tray resistance can be controlled to below 20mmH_2O.

（4） Strong anti-fouling capability

（5） Suitable for vacuum distillation or vacuum absorption, and absorption with heat exchange requirements.

Case Study： Qilu Petrochemical Technology Ltd 's Wastewater Stripping Column

The original waste water strip column was a DN2000, with 44 valve trays, designed to treat 80-120m3 waste water containing S components at 17,000 mg/L and NH_3-N at 12,000mg/L.

Before the revamping, the valves were blocked within one or two months of operation, and the treated waste water did not meet the design specification. ChuangJu revamped the column using its CJMP II and solved the problem for the Client. The results before and after revamp are summarized in the table below. .

	Before Revamp	**After Revamp**
Tray Type	Floating Valve	MP II
Processing Throughput	$110m^3/h$	$140m^3/h$
Ammonia Content	300 -400mg/L	80mg/L
Steam Consumption saving		10-15%
Maintenance shutdown	Once per month	Once per 12 month

2、Amine Regeneration Columns

Qilu Petrochemical Refinery Ltd.

Qilu Petrochemical Refinery's FCC Unit Amine Regeneration Column (DN2800) had been suffering flooding and liquid entrainment since October 2012, which limited the column's throughput to only $210m^3/h$. After replacing the float valve trays with ChuangJu's MP trays, the throughput has been increased to $240m^3/h$ with steady operation since May 2013.

	Before Revamp	After Revamp
Column Internals	Float Valve	MP II
Column Throughput	$210m^3/h$	$240m^3/h$
S content in Regenerated Solvent	N/A	≤1.6g/L
Steam Consumption	N/A	<16.5t/h
Single Tray Pressure Drop	2480Pa	＜440Pa

3、Water Treatment

Shaanxi Coal Chemical Group Shenmu Tianyuan Chemical Co., Ltd. 100t / h waste water treatment project

High phenol wastewater, called semi-coke wastewater, comes from coal distillation units and crude aromatic hydrogen production unit, which contains high concentrations of oil, organic components with phenol, ammonia nitrogen et.al. The semi-coke waste water contains much higher COD content compare with wastewater generated from other coking plant, especially the oil content can be up to 10,000ppm with the majority as emulsified oil. The traditional process is no longer applicable to the treatment of semi-coke wastewater.

ChuangJu developed a process specifically for the treatment of the semi-coke wastewater using its proprietary CJST® anti-fouling spray tray technologies. The process can effectively improve the efficiency of equipment, reduce the amount of emulsified oil, and recover ammonia in the wastewater to produce anhydrous ammonia with 99.9% purity, as well as oil, phenol and sulfur in the waste water.

	Design Value	Achieved Value
Column Internals	MP II	MP II
COD concentration	≤3500mg/L	2300mg/L
NH3-N	≤200mg/L	100mg/L
pH	5-7	7
Volatile phenol	≤400mg/L	220 mg/L
Oil content	≤180mg/L	150 mg/L
Sulfide content	≤80mg/L	50 mg/L

ChuangJu has successfully design or revamped over 70 columns using its own proprietary MP series spray tray technologies in the Coal Chemical Industry since 2009.

Partitioned Flow Spray Tray（PFST®）

The PFST® tray is designed with a sandwich plate to catch the liquid that has been sprayed out of the vertical sieve caps after mixing with the vapor. The liquid is then channeled down to the next tray by a multiple down comer design to handle high liquid load, thus eliminates the problem of liquid back mixing on the tray. With this design, the vapor and liquid phases come to contact counter-currently with the maximum mass transfer driving force, everywhere on the tray, without any liquid flow and concentration gradients. The spray action of the vertical sieve cap reduces the foaming tendency, thus reducing the tendency of vapor phase back mixing.

Case Study for PFST® ：

Raw C4 Downstream Processing

1、Luoyang Petrochemical Hongda Industry's Raw C4 Downstream Processing (100×10^3 t/a)

In September 2013, the Louyang Petrochemical Hongda Industry Ltd started its 100,000t/a Raw

C4 Downstream Processing Facility. The facility includes an extractive distillation, a de-gassing and a solvent regeneration column. ChuangJu designed and manufactured all the three columns. The extractive distillation column and the de-gassing column used ChuangJu's PFST® tray. The facility reached the design capacity and met or exceeded all products specification within 65 hours of start-up with steady operation.

Parameters		Design Value	Actual Operational
Temperature℃	Extractive Distillation	132	118.8
	De-gassing	172.9	165
Steam Consumption kg/h	Extractive Distillation	3,117	2,700
	De-gassing	5840	5000
Isobutane Product Concentration %		≥96%	98%
Olefins Concentration %		≥60%	66%

2、**Ningbo Haode Chemical Industry Ltd.**

Ningbo Haode Chemical Industry Ltd. owns and operates a facility that produce 47,000t/a of isobutene. The facility consists of a C4 separation, n-butane isomerization unit, a MTBE synthesizing unit and a MTBE cracking to isobutene unit. The capacity of the C4 separation unit is 120,000t/a, using a solvent system of MEK and NFM. This system uses a high liquid to vapor load and is susceptible to foaming.

The original design of the columns used a fixed valve tray. After the start up in March 2016, the unit could only run at 70% of the design capacity. ChuangJu was invited to investigate and solve the problem. After site visit with ChuangJu's experienced technical personnel, they found the issues included: the liquid flow cross section is too small; low the tray separation efficiency is low, and there is severe foaming which makes the column easily flooded.

ChuangJu chose their PFST® tray to revamp the two extractive distillation columns and the solvent degassing column. The revamped columns were started up successfully in June 2016.

The achieved throughput is increased by 70% compared with that before the revamp, and has been running at 30% above its design capacity since then.

Parameters		**Before Revamp**	**After Revamp**
Type of tray		Fixed Valves	PFST®
Feed rate, kg/h	**Distillation Columns A/B**	12,560	21,600
	De-gassing Column	124,660	226,930
Temperature℃	**Distillation Columns A/B**	142.8	140.3
	De-gassing Column	156.8	154.4
Steam Consumption kg/t feed	**Distillation Columns A/B**	520	295
	De-gassing Column	53.8	37.5
Solvent entrainment, ppm	**Distillation Columns A/B**	1,000	0
	De-gassing Column	1,000	0
Product Butane Content %		96%	96%
Total olefin content%		90%	90%
Scheduled maintenance times per year		4	1

Low Back-mixing Spray Tray (LBST®)

In 2015, in order to increase mass transfer efficiency and reduce liquid phase back mixing, ChuangJu developed the Low Back-mixing Spray Tray (LBST®) as an improvement to the New-VST technology. The LBST® tray modified the down comer design and added a liquid distribution trough to feed fresh liquid to each individual vertical sieve caps on the tray, so each cap will have the same mass transfer driving force to achieve a uniform efficiency. The patent No is 201521131476.5.

The main advantages of the LBST® tray:

- **Uniform and high mass transfer efficiency**: With its unique down comer and liquid distribution through design each of the vertical sieve cap is surrounded by an equal level layer of fresh liquid from the tray above, so each of the caps will have the same high mass transfer efficiency, thus a high tray efficiency.
- **Low liquid phase back-mixing**: Due to the unique down comer design, most of the liquid that sprayed out of the caps is separated from the fresh liquid from tray above, so the liquid back-mixing is very low.
- **Wider Operation window**: The liquid and vapor contact remains the same when the liquid flow rate varies due to the unique down comer design.

Case Study for LBST® Tray :

Coal to Methanol Technology

Revamping of the Pressure Distillation Column of ENN's 600,000 t/a Coal to Methanol Plant

At the request of ENN Xinneng Energy, ChuangJu was asked to modify the pressure distillation column which could not reach the design throughput. The pressure distillation column used valve trays, which could not handle the required liquid load and run into flooding and poor separation efficiency, especially at the rectification section. After investigation and modolling, ChuangJu decided only to replace the 10 trays in the rectification section with their LBST® spray trays, thus improve the whole column (has 69 trays in total) operation and solved the problem.

	Before Revamp	After Revamp
Tray Type	Valve	LBST®
Pressure distillation column feed rate	90m^3/h	130 m^3/h
Ethanol content in Methanol Product	600ppm	<100ppm

References:

[1] Li, Zhiqiang, Xu, Qingmao, etc., New-VST Technology, "The Journal of Chemical Industry Development", 2001, 20 (10): 58-60

[2] Du, Peiheng, Liu, Jincheng, The development and review of New-VST, "The Journal of Petrochemical Equipment", 1988, 17 (1): 27-31

[3] Du Jianting, Du Peiheng, The Fluid Dynamics and Mass Transfer Properties' Study on New-VST in China, "The Journal of Petrochemical Equipment", 2002, 31 (5): 35-40

AICHE PAPER NUMBER 123A

CASE STUDY EVALUATION OF END OF LIFE CONDITION OF CRACKING TUBES WITH ANTICOKING BARRIER TECHNOLOGY

Michael Györffy

R&D Metallurgist

Kubota Materials Canada Corp.

Matthew Webb

Technology and Engineering Senior Manager

Kubota Materials Canada Corp.

Eric Hendriks

R&D Metallurgist

Kubota Materials Canada Corp.

Harvey Fraser

R&D Metallurgical Technologist

Kubota Materials Canada Corp.

Prepared for Presentation at the 2017 Spring National Meeting

San Antonio, Texas, March 27-30, 2017

CASE STUDY EVALUATION OF END OF LIFE CONDITION OF CRACKING TUBES WITH ANTICOKING BARRIER TECHNOLOGY

Michael Györffy

R&D Metallurgist

Kubota Materials Canada Corp.

Matthew Webb

Senior Manager Technology and Planning

Kubota Materials Canada Corp.

Eric Hendriks

R&D Metallurgist

Kubota Materials Canada Corp.

Harvey Fraser

R&D Metallurgical Technologist

Kubota Materials Canada Corp.

Abstract: This paper examines the end of life condition of ethylene cracking tubes which use an anticoking barrier technology. As new technologies are applied to improve cracking tube performance, it is difficult to evaluate the maintenance requirements for coil replacement. Case studies show the effect of operation on a spinel anticoking surface tube technology. Operating temperature, location and carburization are compared from various coils in an ethane feed.

ANK 400

Ethane steam cracking is the most energy intensive process in the petro-chemical industry[1]. The generation of coke in the ethane cracking furnace tube results in higher energy consumption due to reduced heat transfer in the radiant section of the furnace. Coke generation is the cause of carburization the primary tube failure mode. Coke deposits constrict the flow of the product stream throughout the coil increasing the pressure drop across the coil and insulating coke decreases the heat transfer into the process. Maintenance is then required to assess the level of carburization, coil performance, and planning of a coil turn-around to maintain plant capacity.

ANK 400 is an alloy with a stable inert oxide on the inside surface of the tube. Having a surface free of nickel or iron catalytic particles, and a stable and robust oxide is beneficial to reducing coke formation[2]. The benefit from less catalytic coke formation is that the overall tube metal temperature (TMT) is lower, and therefore kinetically formed coking mechanisms are also less active. The combination of high temperature operation and severe cyclic oxidizing and reducing environments limits the available materials suitable for this process. The ANK 400 material is one such suitably designed alloy and is novel in that it has an engineered oxide surface upon installation.

Coke is a complex deposit and the result of many mechanisms. The four primary processes are considered to be catalytic coking, pyrolytic coking, condensation of high boiling point components and mechanical deposition from upstream sources[3]. The first is due to the presence of nickel and iron that catalytically increase the kinetics of coking[4]. The coke is characterized by the formation of carbon whiskers with a nickel bearing catalyst tip[5]. Pyrolytic coking is caused by over-cracking of hydrocarbon at high temperature and is due to the thermodynamic equilibrium achieved by heating olefins. All forms of coke deposit carbon on the inside surface of the tube, and contribute to the migration of carbon into the tube across that surface. Minimization of coke formation is therefore beneficial to improving tube life. Elimination of catalytic coking with the use of coatings has been shown to reduce the coking rate by up to 90%[6].

NOVA Chemicals Corporation and Kubota Corporation have developed a method to reduce both catalytic and pyrolitic coke with ANK 400. This specialized metallurgy separates the heat resistant high nickel tube metal from the hydrocarbon product stream with the inert ANK 400 spinel that is on the tube inner diameter (ID). This spinel layer is an inert manganese chromium oxide and is shown in the scanning electron micrograph in Figure 1.

Figure 1: Scanning electron micrograph of the ANK 400 spinel surface.

Conventional alloys form a Cr_2O_3 oxide which is well documented at 1080°C (1976°F) to decompose into volatile chromium carbides in carbonaceous conditions[7]. This results in a mechanism of chromium loss in the alloy, as well as a loss of protection to carburization. The ANK surface is a stable oxide up to 1100°C (2012°F). Reduced coking rates results in better heat transfer and lower tube metal temperatures. The ANK 400 metallurgy utilizes an industry standard heat resistant 45% Ni, 35% Cr alloy base which has known mechanical strength and weldability. The stability of the inner surface provides a reduced rate of surface degradation. The inert spinel has stability at higher operating temperatures than conventional material and acts as a barrier to carbon migration into the substrate metal.

To demonstrate the stability of the oxide; an oxidation test at 1100°C (2012°F) for 100hrs followed by furnace cooling visually demonstrates the stability of the oxide in oxidizing (decoke) conditions. Figure 2 shows the effect of the decoking simulation on the spinel surface of the ANK. Grain growth of the spinel is inhibited in the ANK surface, where conventional oxides grow to over 50µm size the ANK 400 oxide crystals remains less than 10µm in size. This improves their adhesion and prevents their spallation and loss. With this surface stability in runtimes is achievable.

Figure 2: Grain growth stability demonstrated by simulated decoking in ANK 400. The grain size remains stable and small, therefore maintaining an effect barrier to carburization.

It is recognized that oxides are a barrier to carburization[8]. Carbon cannot migrate across an oxide boundary at an appreciable rate. Many new alloys with novel oxides are being brought to the market. This paper examines the observations of the ANK 400 engineered surface and its relationship to the 1100°C (2012°F) temperature limit, as well as the impact on coil maintenance and coil life.

Application to NOVA Chemicals Furnaces

The performance of three different ANK 400 furnaces is shown in Figure 3. This figure shows the run times using coils with the spinel oxide on the inner surface. The typical furnace run-time prior to the use of this technology was approximately 33 days with conventional materials. Currently, ANK 400 is achieving repeatable run times of over 400 days. Many of these runs were terminated due to plant limitations or reasons not related to coke formation in the coil. The columns indicate successive runs, and the end of run criteria is indicated by the colour of the column.

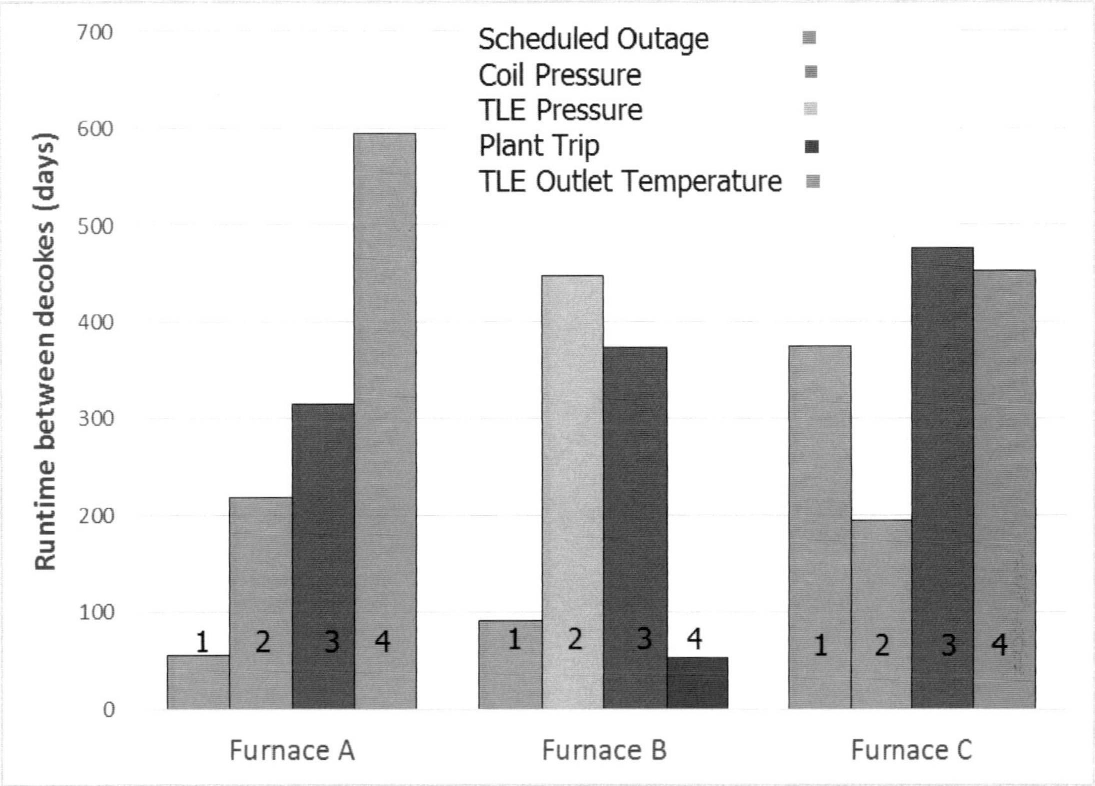

Figure 3: Extended run lengths due to the use of ANK 400.

Carburization

Carburization is a hot corrosion phenomenon. The alloy used for cracking tubes is based on an industry standard 35Cr-45Ni alloy. This material is a cast austenitic alloy structure with chromium carbides, and microalloying additions. It is optimized to provide a balance of carburization resistance to creep strength. When assessing the carburization resistance of bulk ANK 400 alloy to a comparable 35Cr-45Ni standard material, called KHR 45A, there is no significant difference in bulk carburization resistance. There is a negligible improvement based on these lab tests using a packed carburization test at 1100°C for 500 hours. The curve in Figure 4 shows the carbon profile measured in 0.5mm increments from the ID of the tube on the left side of the graph. It visualizes the migration of carbon into the tube, and is a relative test to compare the two alloys. The carbon content is incrementally measured on LECO Carbon combustion analysis equipment.

Another method of destructive evaluation is metallography. Sectioning, polishing and etching the ID of the tube can show subjectively the increase in the carbide phase in the tube. Comparison of the midwall to the ID can subjectively identify the tube metal temperature, based on the carbide structure, and the depth of carburization at the ID based on coarsening of the major phases. Reference microstructures of aged material can be used to estimate the

TMT and carburization kinetics. Of course, an isothermally aged sample is not exactly an aged sample from cyclic operation in the furnace; however the dissolution of secondary carbides and coarsening of primary carbides is known to occur at specific temperatures and is a good estimate of sample history.

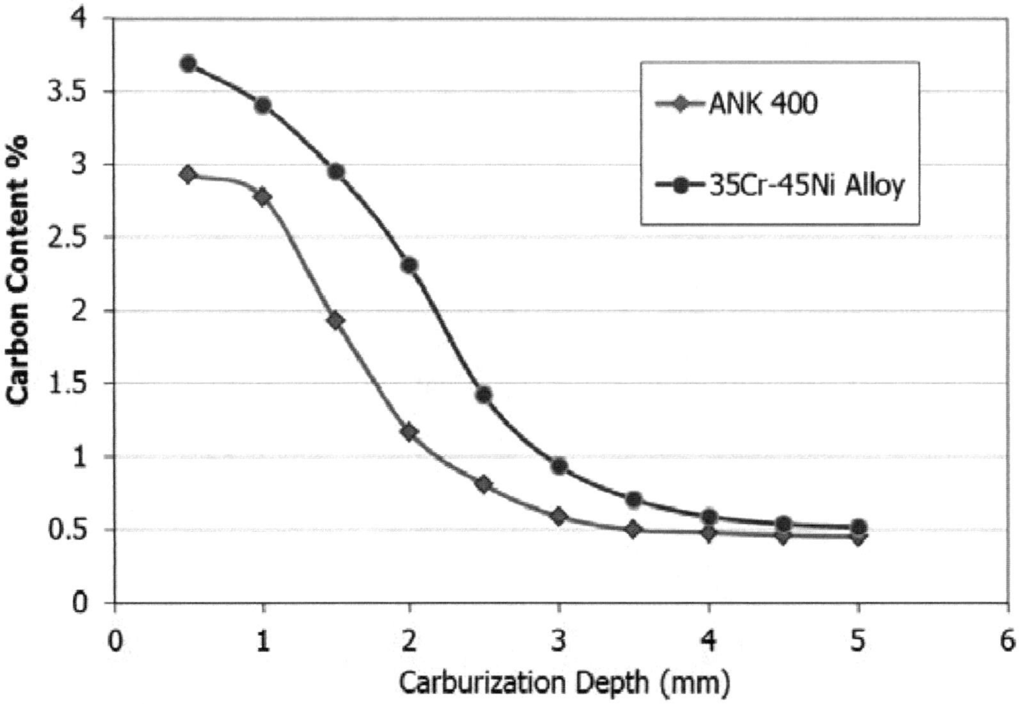

Figure 4: Pack carburization test at 1100°C for 500hrs comparing ANK to a 35Cr-45Ni alloy (KHR45A). The results show roughly the same bulk resistance to carburization.

Observing carburization can be performed by non-destructive means. New material is non-magnetic, but becomes magnetic as it carburizes. A magnetometer can be used as a calibrated tool to understand the location and depth of carburization in a coil. For this investigation a Kubota Carburization Detector, model KDC 100X was used for this purpose. A probe with an instrumented magnet was used to measure and scan for carburization on a used coil. The pull from the magnet can be correlated to the depth of carburization, and it acts as tool to investigate relative differences across the coil. The operation of the sensor is demonstrated in Figure 5, and for 35Cr-45Ni type material only the X sensor is used. The approximate relationship between the X sensor measurement and carburization depth as a function of tube wall thickness can be correlated however for this investigation the relative values are enough to determine consistency between different samples.

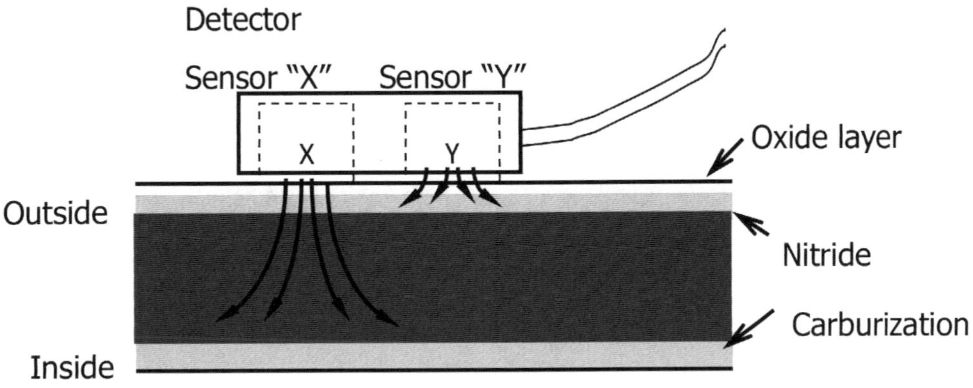

X: Measurement of oxide, nitride and carburization
Y: Measurement of oxide and nitride layer
X-Y: Measurement of carburization layer

Figure 5: Carburization detector probe, sensor X measures the magnetic pull from ID of the tube.

The following three case studies show the observations from three difference sites and coil types using ANK 400 for over 3 years. Three techniques are used to show a trend, that diffuse carburization is low over a broad area, that the inner surface is resisting the migration of carbon into the bulk alloy and that there is low carbon pick-up by the alloy below an operating temperature of 1100°C.

It must also be mentioned as a point of discussion that the temperatures described here are based on the subjective interpretation of the author. Every effort was made to relate the observed structures in field samples, to similar chemistry, isothermally aged samples. Based on the coarsening of the primary carbide structure, and the dissolution of secondary carbides at specific temperatures it is assumed a maximum operating temperature can be estimated. Regardless of the accuracy of this value, the principle is demonstrated that in a petrochemical cracking environment, a coil with an oxide barrier technology can resist carburization in a novel way as the result is superior to the expected bulk material carburization result.

Case 1

Time in service: 7 years, Ethane feed stock, Modified Split U-type coil.

Operating temperature was estimated at: (1)a highest TMT of 1100°C±10°C in the zone of highest heat flux, and (2) 1075°C±10°C in adjacent tubes.

This is a comparison of the hottest exposure locations on neighbouring passes.

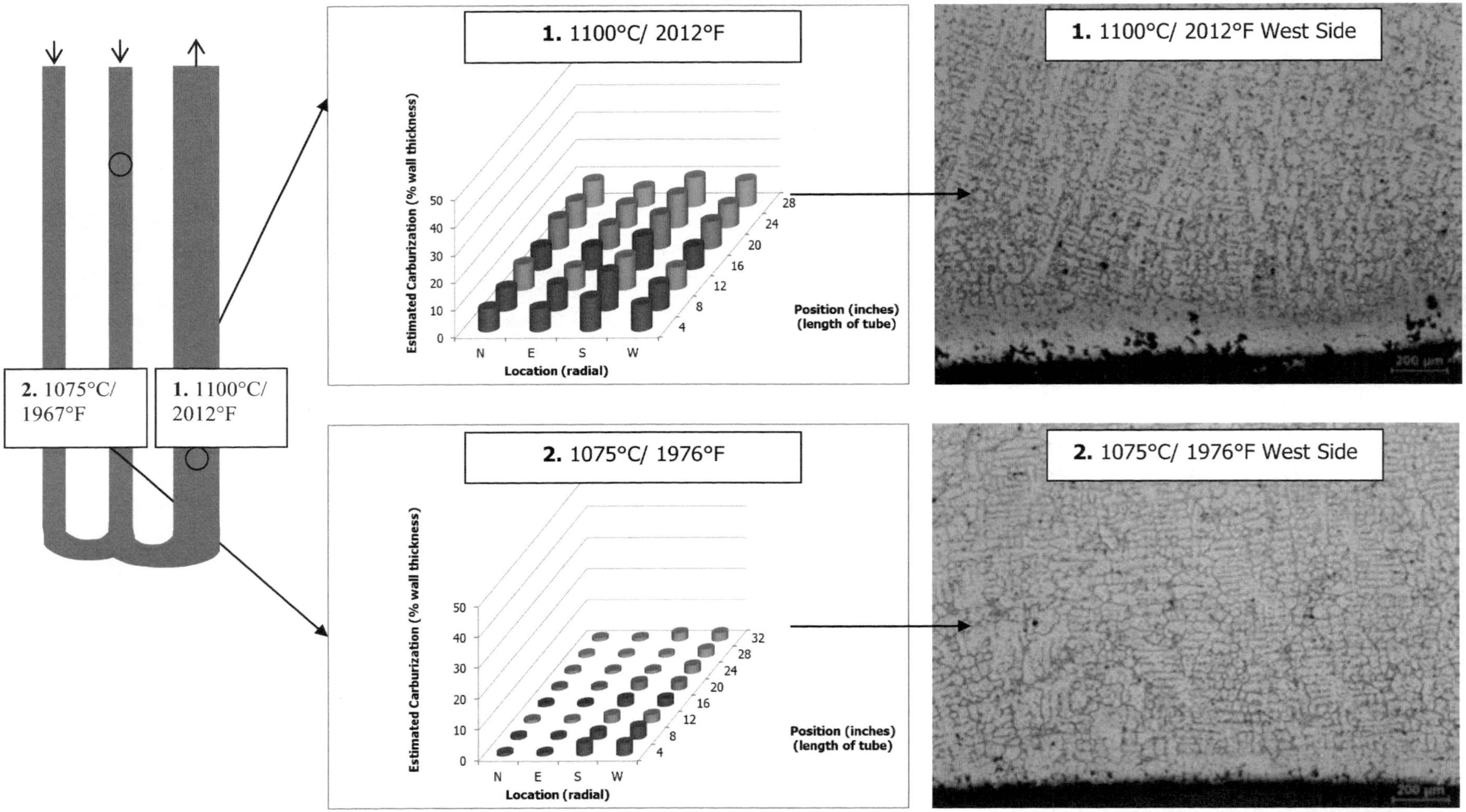

Figure 6: Seven year service inspection of ANK 400 between outlet and inlet pass for carburization.

Case 1

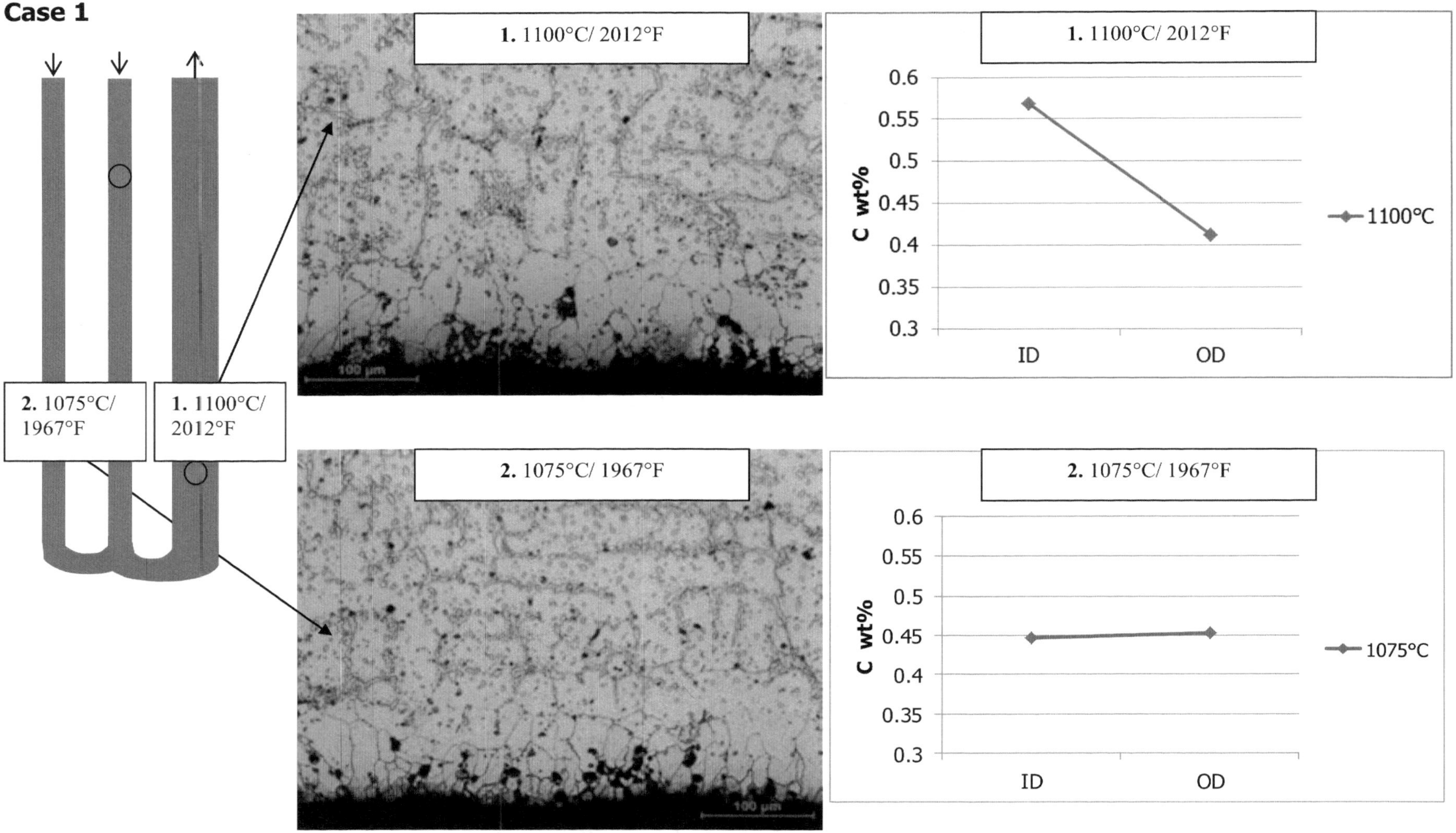

Figure 7: Seven year service inspection of ANK 400 between outlet and inlet pass for carburization.

Case 2

Time in service: 3 years, Ethane feedstock, Multipass coil

Operating temperature was estimated: (1) a highest TMT of 1100°C±10°C on the outlet tube, and (2) 850°C±10°C on the Inlet pass (2).

This is a comparison of the inlet to outlet pass.

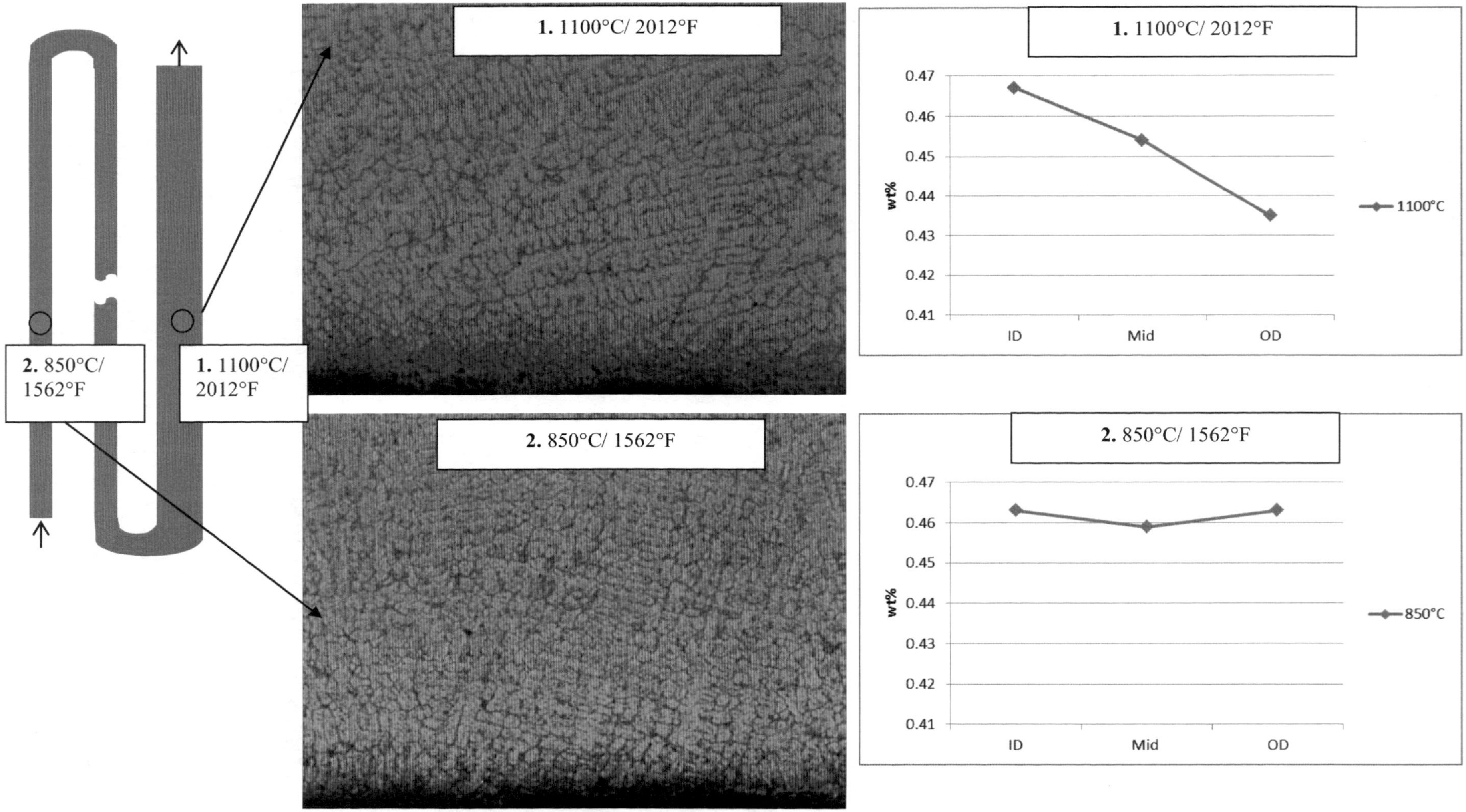

Figure 8: Three year service inspection of ANK 400 between outlet and inlet pass for carburization.

Case 3

Time in service: 6 years, ethane feedstock, serpentine coil.

Operating temperature was estimated: (1) a highest TMT of 1100°C±10°C on the outlet tube, and (2) 1050°C±10°C on the intermediate pass. This is a comparison of the inlet to outlet pass.

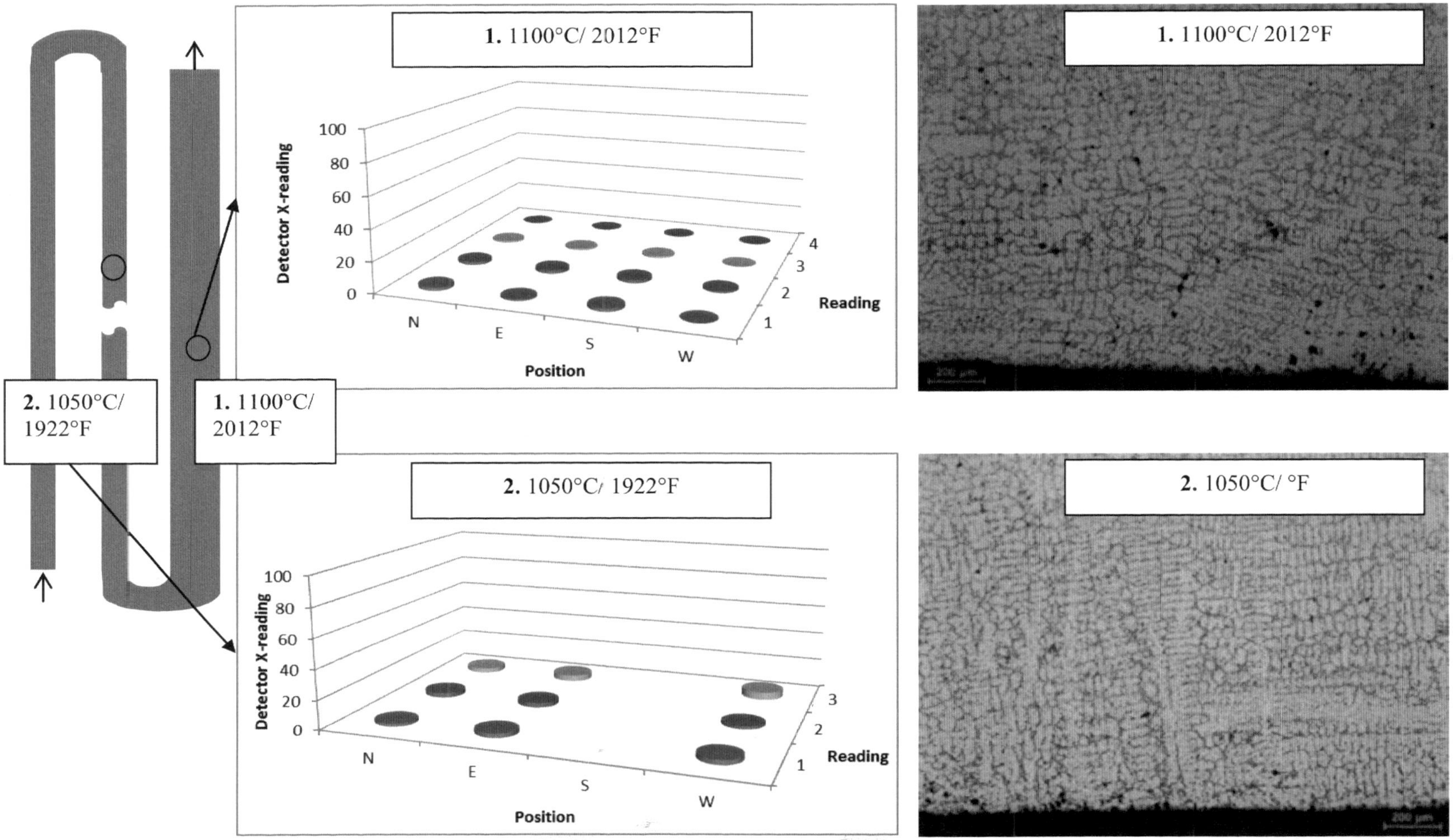

Figure 9: Six year service inspection of ANK 400 between outlet and inlet pass for carburization.

Discussion

ANK 400 is a cracking tube with a continuous oxide on its inner surface that has many beneficial effects to improve the steam cracking process and to reduce maintenance work loads. End-of-life investigations show that intact oxide is an effective barrier to carburization. The effect is not directly causal, in that lower tube metal temperatures, fewer decokes and less coke all have a contributing effect to the end-of-life condition of the tube. However to characterize a trending non-result, this surface shows a robustness to operation below 1100°C/2012°F, to which carburization is significantly reduced as summarized in Figure 10. Note that in this figure, carbon saturation is around the 3.5wt% level. Even the elevated levels indicated are far from fully carburized.

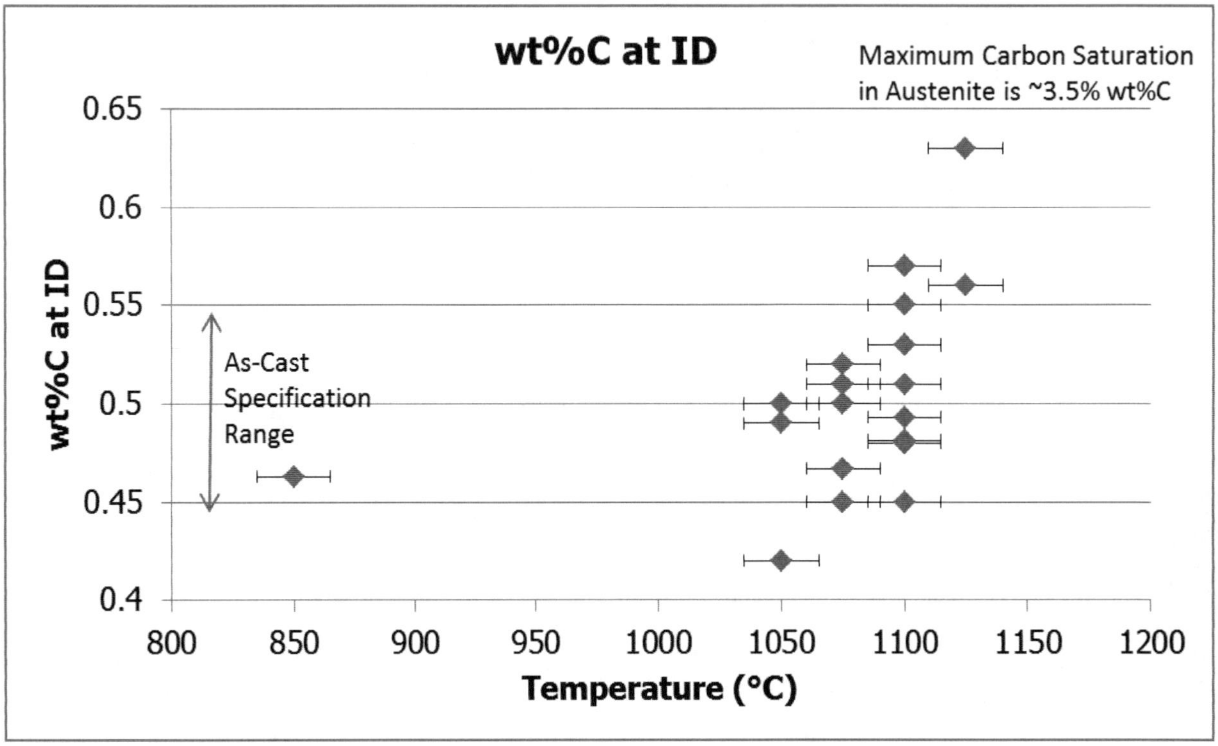

Figure 10: Plot of carbon measurement at the tube ID vs estimated tube metal temperature.

The case studies show results after up to 7 years into the life of the coil in figure 11. It is presented here that 1100°C is the limit below which carburization is effectively blocked by the spinel oxide in a way that is different to exclusively the bulk carburization resistance of the alloy. This paper does not suggest that ANK 400 will not carburize, only that there is improved resistance to carburization with the presence of a robust stable oxide surface. This is characterized by measuring the magnetic pull from any carburized layers at the ID of the tube; by measuring the carbon content in a small volume at the inner surface; and by direct observation of the inner surface with metallography. It is observed in multiple locations and on multiple coils, that a robust oxide provides protection against carburization. This phenomenon has resulted in an achieved 20% improvement in coil life for the spinel oxide users. Less carburization in the coil may result in a possible 50% extension in coil life based on continuing analysis.

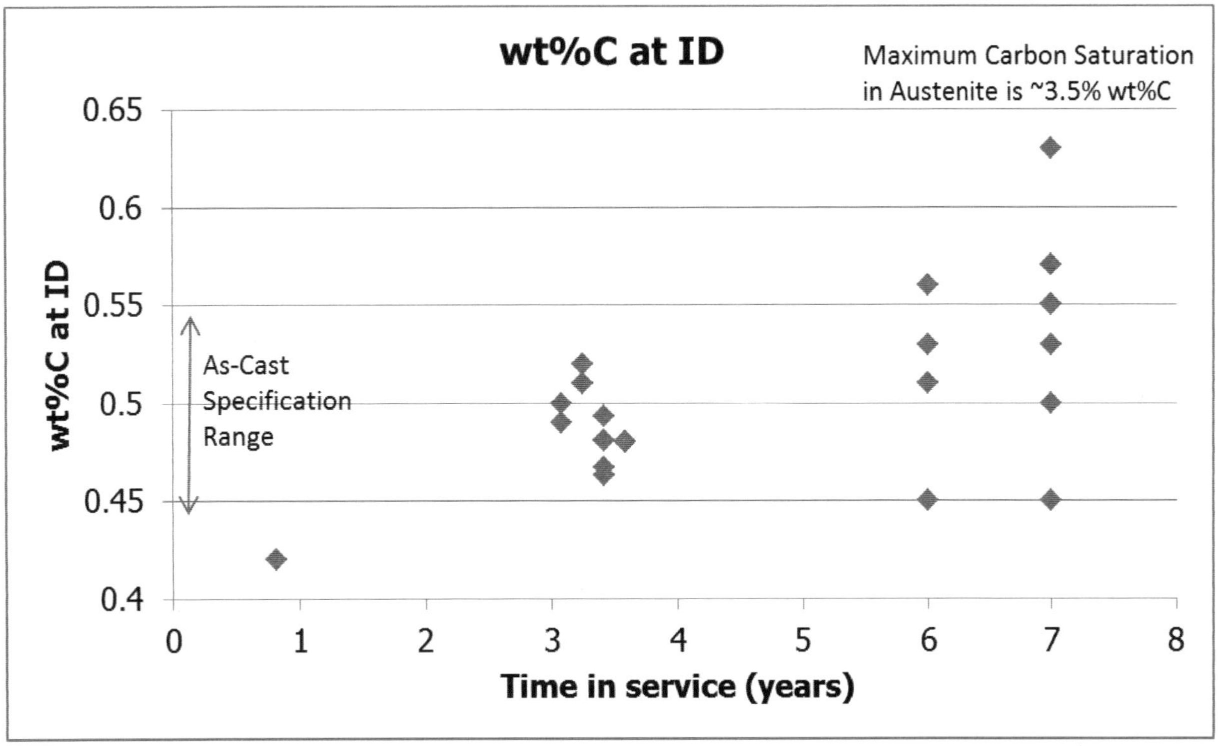

Figure 11: Plot of carbon measurement at the tube ID vs time in service.

Oxide technologies are being introduced to the cracking tube marketplace that will be able to improve reliability, availability and overall equipment life. From a maintenance perspective, if carburization can be mitigated, it means coil life will be determined by coil creep rates, and the aged ductility of the material. Additionally both of these mechanisms will benefit from a reduction in carburization rates, as the coils will creep less if not subjected to the volumetric expansion of carbide formation. Also, mechanical properties will be improved since there are fewer continuous brittle phases (carbides) in the aged tubes. This allows scheduled planning and coil performance to determine coil turn-arounds.

Conclusions

ANK 400 effectively and repeatably reduces coke deposits in ethane steam cracking coils. The spinel oxide on ANK 400 is stable up to 1100°C/2012°F. The reduction of TMTs, the improved pressure drop across the coil, the reduced number of coil decokes and the extended run lengths all indicate a significant reduction in coking with to ANK 400 tubes. It is shown that a robust oxide surface can act on a commercial level to effectively reduce carburization rates, improve coil life and reduce maintenance workloads.

References

1. T.Ren, M. Patel, and K. Blok, "Olefins from conventional and heavy feedstocks: Energy use in Steam cracking and alternative processes" *Energy*, 31, 2006, 425-451.
2. M.Gyorffy, L. Benum, N. Sakamoto, "Increased Run Length and Furnace Performance with Kubota and Nova Chemicals' ANK 400 Anticoking Technology; Data from Current Installations as Well as Technology Improvements for Higher Thermal Stability and Decoking Robustness", AiCHe 2006 Spring Meeting.
3. H. Cai, A. Krywicki, M.C. Oballa, "Coke Formation in Steam Crackers for Ethylene Production", *Chemical Engineering and Processing*, 41, 2002, 199-214.
4. G. Bach, G. Zimmermann, F. Kopinke, S. Barendregt, P. van den Oosterkamp and H. Woerde, "Transfer-Line Heat Exchanger Fouling during Pyrolysis of Hydrocarbons. 1. Deposits from Dry Cracked Gases", *Ind. Eng. Chem. Res.*, 34, 1995, 1132-1139.
5. C. Yin, and W. Tsai, "Nano-sized Carbon Filament Formation during Metal Dusting of Stainless Steel", *Materials Chemistry and Physics*, 82, 2003, 929-936.
6. A. E. Munoz Gandarillas, K. Van Geem, M. Reyniers and G. Marin, "Influence of the Reactor Material Composition on Coke Formation during Ethane Steam Cracking." I&EC Research, ACS 2014, 53, 6358-6371.
7. H. Li, W. Chen, "Stability of MnCr204 spinel and Cr2O3 in high temperature carbonaceous environments with varied oxygen partial pressure. Corrosion Science 52, 2010, 2481-2488.
8. H. Grabke, Carburization A High Temperature Corrosion Phenomenon. MTI publication No. 52., 1998.

479863 Lifecycle Analysis of Brazed Aluminum Plate Fin Heat Exchangers (BAHX)

Abstract:

Plate fin heat exchangers, commonly referred to as brazed aluminum heat exchangers (BAHX), are widely used in cryogenic fractionation and processing because of their high thermal efficiency and reliability, compared to other heat exchanger types. Applications include air separation, natural gas liquefaction, nitrogen rejection, natural gas liquids recovery, ethylene production, hydrogen recovery and propane dehydrogenation.

Typical industry expectation for BAHX lifespan is approximately 20 years, although experience shows that units that have been operated diligently, with respect to the operating and maintenance guidelines, can double their expected service life. The same experience also reveals that lifespan depends entirely on the care afforded the BAHX, particularly with respect to thermal gradients and most are retired from service due to leaks caused by some form of thermal stress damage.

BAHXs don't exhibit wear patterns during operation (like brake pads and tires) hence there has traditionally been no way of predicting change over time in BAHX performance due to operating conditions, that may eventually result in thermal fatigue. Chart has recently developed two tools to assess the impact of plant operations on the mechanical integrity of the brazed exchangers.

Distributed Control System (DCS) field data analysis provides quick proactive feedback on plant operations using Chart's Installation, Operation & Maintenance (IOM) guidelines. Advanced dynamic thermal analysis performs advanced simulations to calculate BAHX fatigue life. These methods generally highlight process control improvements for greater reliability and prevention of harmful thermal gradients.

DOW

Aging Assets: more than just Mechanical Integrity

EPC Conference,
March 28th 2017, EPC Ethylene Plant Maintenance & Reliability Session

Hydro Hc carbons

Hydro Hc carbons | DOW

Geographic locations

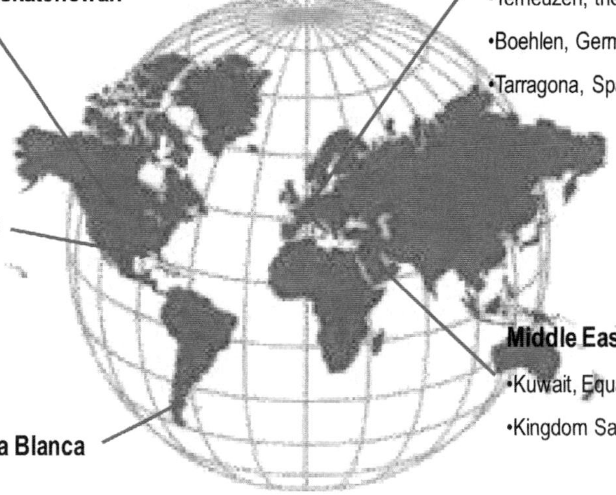

Hc

Hydro Hc carbons | DOW

Most of the (Petro) Chemical Industry in Western-Europe is build in the period 1960 – 1980

Almost all are in operation today.

The expected lifetime of these Assets is 2 – 3 times as original designed for!

This presentation provides an holistic view of external influences in combination with ageing of these Assets.

Specific answers will not be provided, this presentation can be used as input for a Business or Facility Asset Management Strategy.

Hc

Hydro Hc carbons | DOW

Ageing Must be part of Asset Management (Lifecycle Management) Strategy:

- What type of Industry do we belong to. What is the risk we face?
- A company is part of the society.
- Goals to be set for Safety, Health, Environmental, Capital/Expense etc.

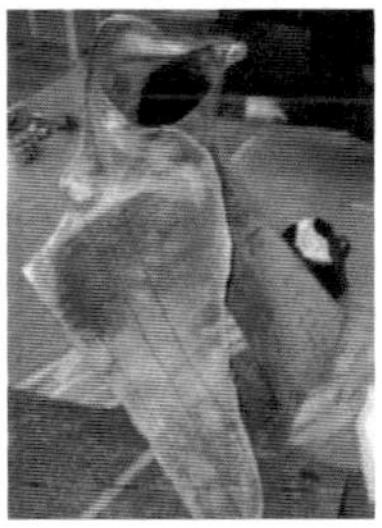

Asset Strategy at all levels:

- Corporate / Business
- Facility /plant
- Equipment / Component

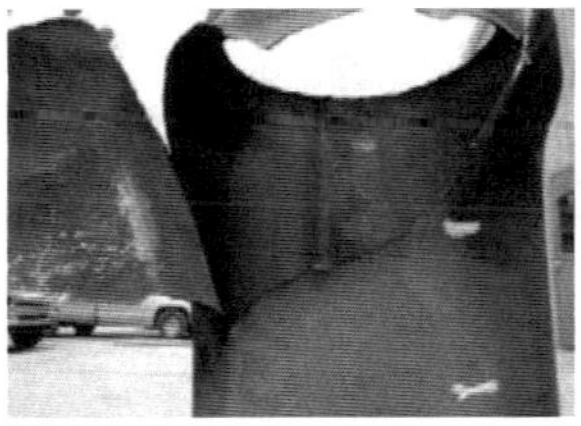

Hc

Hydro Hc carbons | DOW

Failure of the Mechanical Integrity caused by Ageing is one of the highest risks and challenges we have in the Process Industry.

Our programs and standards must minimize the risk of having these type of incidents!

The Company Future / Existence can be at DIRECT risk!

But, is it only ageing or are there more causes?

CUI accelerated by dripping condensate on pipe:
Have we the discipline in place for immediate repair?

Cyclic service will accelerate corrosion. Pipe failure is the result:
Have we identified all locations and are we inspecting?

Freezing conditions in combination with water (ice) in a dead-leg:
Have we identified them and are they included in our inspection program?

KiK | MPi

27-Sep-16

Hydro Hc carbons | DOW

Process Safety Calamities have **DIRECT influence** on the company future / existance.

- **Ageing Assets** – Failure of Mechanical Integrity is one of the major causes.
  - Is this the end for the site / plant?
  - Will it be re-constructed again?
- ***Operating Discipline***
  Deep-water Horizon Oil spill (BP)
  Ineffective cementjob (Root Cause)

- **External factors**
  - Mergers, acquisitions and divestments
    Might result in 10+ years without required investments and minimal capital and maintenance expense. Might be limited to legal must do only (*).
  - Synergy; closure of "low margin assets".
    Aged Assets have often a high ACE/RAB value.
    Aged Assets operate quit often in the lower Quartyle of Benchmark studies (e.g. Solomon).

(*): Source Zurich Insurance Company Ltd

Hc

DOW CONFIDENTIAL - Do not share without permission 2/20/2017 6

Hydro Hc carbons | DOW

Do we know the <u>remaining life</u> of our equipment and components?
Can we eliminate unexpected failure?
Do we have a program for "*Mandatory Retrofit*"?
How effective are our QA/QC program on (cooling) water treatment?

Can we prevent <u>fouling/corrosion?</u>

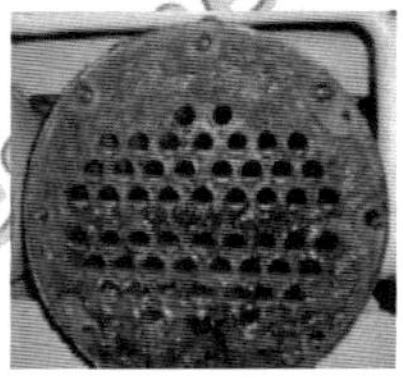

Will our current design prevent collateral damage? Have we reviewed Explosion / Fire Hazards (F&EI Circles) to update design?.
As example; are fire resistant walls installed between Trafo's?

Is there an investment plan for Outdated Technology?
Do we know when in future we cannot service our Process Automation System?

Hydro Hc carbons | DOW

Select the appropriate **RELIABILITY Strategy**!
Can we accept "End of Life"?
Shall be a risk driven decision:

- Operate to failure.
- Preventive Maintenance.
- Predictive Maintenance (big data?) / Condition Based Maintenace.

Pressure/Level measurement
3000 instruments per plant.
- Operate to failure:
  - Often 1oo2 systems.
  - Replace/repair job?

Turbine/compressor system:
Unique, failure will cause plant breakdown.
- Condition based Maintenance:
  - Machine monitoring (big data).
  - Lube Oil System.
  - T&T Valve design.

Hc

Hydro Hc carbons | DOW

Thesis: The <u>System</u> has got <u>Endless Life</u>. Will exist "forever" when maintained.

Preventive/Predictive component repair & replacement.

- Failure Mechanism must be understood!
- Risk and economic driven decision .

<u>Instrumentatie:</u> "Subcomponent" in a "system".

Focus is often "Cost management".

<u>Turbine</u> "*Reliability Strategy*": predict Failure!

Eliminate product losses by installing e.g. Machine monitoring

Turbine rotor :

- Replace broken Blades?
- Replace Turbine by complete new Spare to minimize repair duration and optimize quality?

Turbine casing (robust, thick wall):

- internal erosion -> inspect and weld repair?

Hc

DOW CONFIDENTIAL - Do not share without permission 2/20/2017 9

Hydro Hc carbons | DOW

Thesis: The <u>System</u> has got <u>Endless Life</u>. Will exist "forever" when maintained.

<u>Ethylene furnace:</u> "*Reliability Strategy*" is maximize Production:

- Radiant Coils will be replaced every 4 – 8 jaar years (operate just before failure).
- Convection Bank expected lifetime is 15 – 30 years before replacement:
    - Relative less predictive inspection, low maintainability etc.
- Refractory and Furnace "wall":
    - Visual inspection followed by on-need repair.
- De-coke piping (failure might result in SHE consequence):
    - Visual and Wall Thickness Measurement to track erosion, followed by repair or replace.

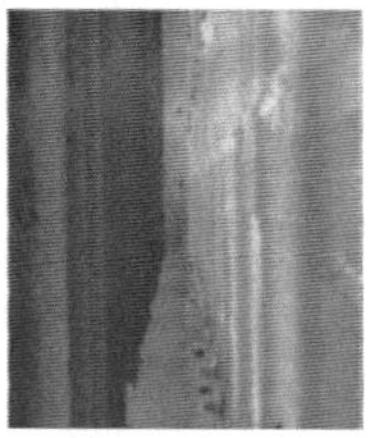

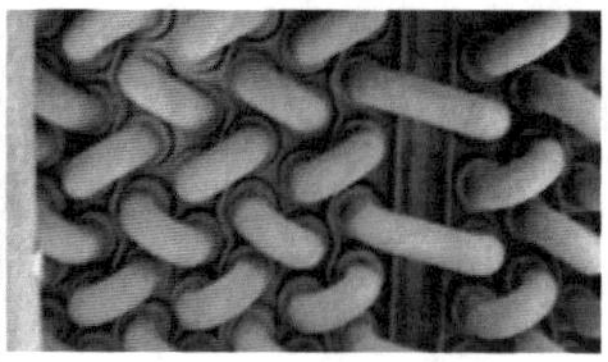

Hc

DOW CONFIDENTIAL - Do not share without permission 2/20/2017 10

Hydro Hc carbons | DOW

Hydro Hc carbons | DOW

The EEPC and EPC are organisations of ethylene producers and technology contractors in Europe and the States, with the purpose of providing a platform for the exchange of non-confidential information relating to the health, safety and environmental issues encountered in the production and handling of ethylene.

Just two EEPC examples recently shared in a Maintenance IG meeting;

Valve packing leaks in cryogenic service:

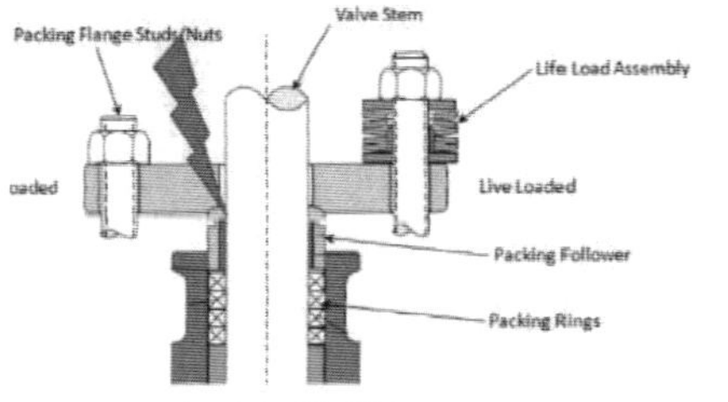

Non-Live Loaded vs Live Loaded Packing System

Company used PTFE packing.
Changed to graphite / metal.

S.I.C. failure in convection section:

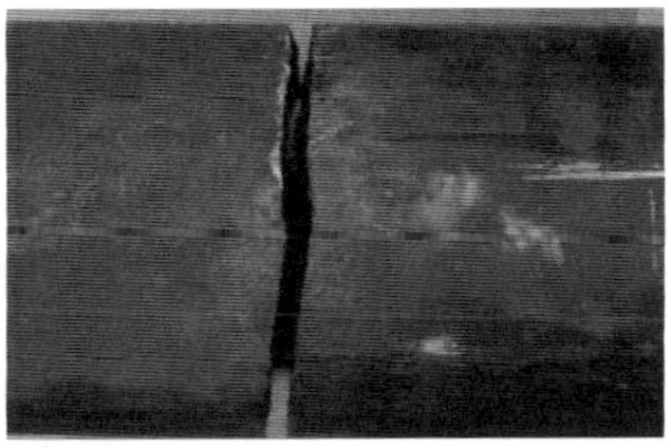

Tube base material – 800H.
Stress – Temp (500 – 700 °C), ***AGE***?

Hc

Hydro Hc carbons | DOW

The EEPC Maintenance IG – Examples of recent activities:

Feb 2016: CUI workshop:
Out if Sight, Out of Mind?

64 Participants from 23 different companies.
Growing awareness for CUI management.

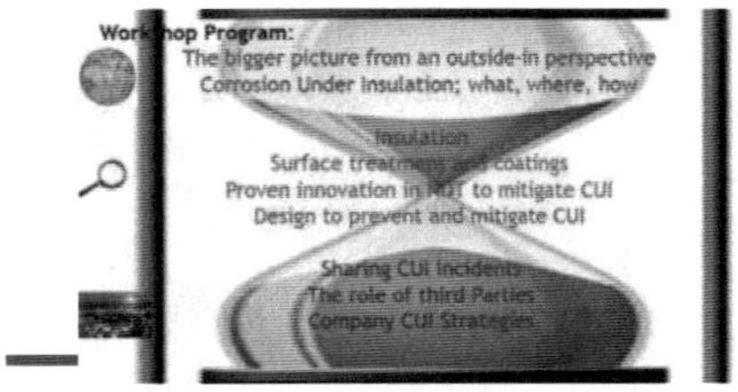

EEPC Maintenance IG meeting June 2016:
Strategic Topic - Composite wrapping

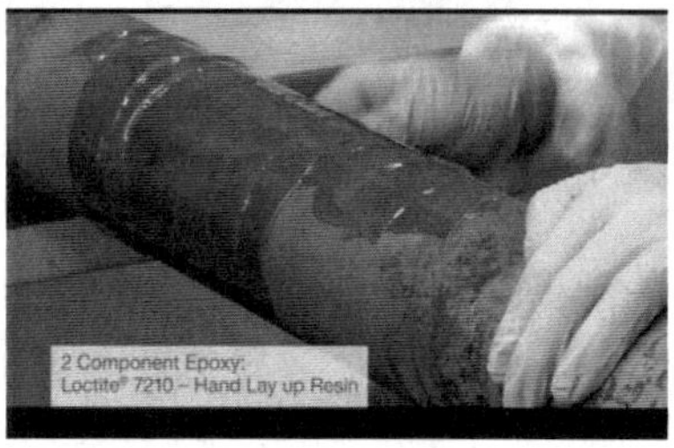

Latest Technology and Working Methods shared by Applicators (e.g TeamFurmanite & Henkel).
Temporary versus Permanent repair Discussion.

Hc

Hydro Hc carbons | DOW

Obsolescence:

Costs per [MT], Yield, Energy.
Wat is "world-scale"?

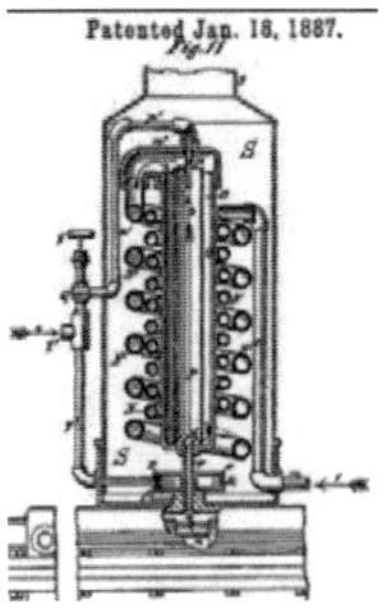

Hc

Hydro Hc carbons | DOW

In the past Standards & Norms were different compared to today!

The trend for more severe standards is good, a company is part of the society.

- Environmental – (fugitive) Emissions, Sound (fackel) or Nox (furnaces, boilers) & CO2 emissions.
- Maximum Allowed Concentration (MAC): downward trend. Is Plant Clearance designed for these new Norm?
- BREF: BAT Reference documents per EU IPPC-directive.
- BAT: Best Available Techniques is in place or is in development in Europe
- PGS29 The Netherlands, Above ground storage flammable liquids in vertical Cylindric Tanks
- CSB recommendations – Industry shall not Ignore.

Investmens are required to meet requirements.

How sound are the Economics"? Are they feasible?

Hc

DOW CONFIDENTIAL - Do not share without permission 2/20/2017 15

Hydro Hc carbons | DOW

It is a fact - the Process Industry has got Aged Assets.

An optimized maintenance strategy at all levels (company, facility and equipment) is required to:

- Minimize the risk of having (large) SHE and Process Safety Incidents
- Optimize Reliability and Costs as required at company / business level

External factors play a large role within the "lifecycle of an asset" and must be controlled.

Organizations like the EEPC, EPC or CSB can assist companies in sharing experiences to mitigate the risk of having Aged Assets.

Hc

DOW CONFIDENTIAL - Do not share without permission 2/20/2017 16

Hydro Hc carbons

Thank You!

481724 The Impact of Moving to Gas Feeds on the Maintenance & Reliability of the Ethylene Plant

Abstract:

The shale gas revolution in the United States has caused ethylene producers to push plants originally designed for heavier liquid feedstocks to crack increasingly lighter feedstocks to capitalize on the competitive advantage that they offer. The economic drivers and process modifications required to adapt to the changing market conditions are well documented, but the forecasted impacts to maintenance and reliability were educated assumptions. A few years into this shale gas revolution, those impacts are now being realized. What might operators and maintenance personnel expect to see as a result of the shift to lighter feeds? Certainly there is the potential for some process areas and pieces of equipment to see increased maintenance and reduced reliability because of the change in operating conditions or feedstock characteristics. Being alert to and prepared for these possibilities can minimize their negative impact. This paper will highlight areas of the ethylene plant affected by the shift from heavier liquid feeds to gas feeds and offer ideas of "things to watch for" based on the experience of several plant operators. Examples of monitoring points and case studies are provided to illustrate how changing feedstocks may affect a process and its equipment so that maintenance and reliability professionals may be better informed when facing the challenges imposed upon their plants by the shale gas revolution.

483751 Panel Session - Impact of Lighter Feeds and Plant Age on Maintenance & Reliability

Abstract:

Ethylene is one of the largest volume base chemical and building block for the petrochemical industry. It's of paramount importance that these units run reliably to support the downstream polymer and derivative units. High utilization rates of these assets drive the overall economics. Shale revolution has resulted in abundant supplies of lighter feedstock, particularly ethane, at cheaper price as compared to other regions of the world. Therefore, ethylene operators in North America have shifted to maximizing ethane (and lighter feeds) cracking. This shift can result in maintenance and reliability issues in certain areas of the plant and require special attention to meet high reliability expectations. In addition, the existing assets require attention to maintenance and reliability due to age of these facilities. The latest operating plant in North America was started-up more than 15 years ago and some of the older units have been operating for nearly 50 years. This panel session will focus on the maintenance and reliability issues impacted by shift to lighter feeds as well as plant age. We will hear from our panel of industry experts covering the following:

- Their experiences with shift to lighter feeds and related to plant age
- Options to consider for addressing M&R issues

Panelist Include:

Theo Knijff, The Dow Chemical Company
John Weikel, BASF
Jimmy Cleavinger, CPChem
Mike Pelton, LyondellBasell
Sanjeev Kapur, Apex PetroConsultants
Rene Kulik, ARVOS
Michael Gyorffy, Kubota
Bruce Harle, CHART

480717

Building with Natural Gas

Radaelli, Guido

Executive Vice President, Engineering

Siluria Technologies

Iyer, Rahul

Executive Vice President, Strategy & Corporate Development

Siluria Technologies

Rafique, Humera

Process Engineer II

Siluria Technologies

Prepared for Presentation at the 2017 Spring National Meeting

San Antonio, TX, March 28, 2017

Building with Natural Gas

Radaelli, Guido

Executive Vice President, Engineering

Siluria Technologies

Iyer, Rahul

Executive Vice President, Strategy & Corporate Development

Siluria Technologies

Rafique, Humera

Process Engineer II

Siluria Technologies

Abstract: Siluria Technologies has built a unique platform to commercialize direct Natural Gas to Ethylene by successfully demonstrating the operations of the Oxidative Coupling of Methane (OCM) technology for over two years at its pre-commercial scale in La Porte, Texas. The modularized demonstration unit is built to test the catalyst in a wide range of operating conditions (450-950°C and up to 16 barg) with specific campaigns aimed at process conditions expected in the first commercial projects. This milestone leads the pathway to commercialization of the OCM process with our ethylene industry partner, Linde Engineering. Siluria Technologies has also commenced further development of its first generation OCM technologies to deliver ethylene producing processes with higher carbon efficiencies and lower capital intensity.

In parallel with taking first generation OCM to the market place in a range of applications, Siluria has expanded its technological abilities to develop processes producing various petrochemical products via OCM. The company has developed solutions for ethylene to liquid fuels, natural gas to propylene (via ethylene), flexible ethylene/propylene processes, and OCM integration with refineries and syngas plants.

Introduction

Siluria Technologies has successfully developed an industrial process for converting natural gas to olefins based on its proprietary version of Oxidative Coupling of Methane (OCM). To develop of this technology from bench to commercial level, Siluria has established a scale-up platform to screen and validate at each scale-up step the critical process performances required for a commercially viable solution. This paper will describe the steps and modelling tools, validated with experimental data from bench, pilot and demonstration units, that are used to design Siluria's OCM based process solutions to produce olefins (ethylene and propylene) from natural gas.

OCM has been a long sought after route for olefin production from low-cost natural gas feedstock. Previous companies and researchers such as Sofranko et. al at ARCO Chemical and Union Carbide studied several OCM catalysts up to the pilot scale level. However, economical catalyst performances were only achieved at either very low pressures (typically atmospheric) and required unrealistic commercial designs of up to 8 reactors in series. Moreover, especially in the case of Union Carbide, yield and selectivity were hindered by catalyst stability issues. In all instances the OCM research programs were eventually discontinued(1).

After having determined the minimum catalyst performances, design parameters and operating conditions necessary for a commercial OCM process, Siluria developed a scale-up platform aimed at discovering, testing and validating the above mentioned performance metrics in successively larger and more complex setups.

The first step in the scale-up platform consists of the simultaneous rapid screening of over 250 catalyst formulae, which are obtained through several state-of the-art synthetic routes, from bio templating to nanotechnologies. At this stage, catalysts are screened for methane conversion and C2 selectivity over a range of operating temperatures and gas compositions.

Catalysts with promising performances are then placed through secondary screening in high throughput fixed bed reactors of 1-2mm diameter and the successful candidates are further tested in a 4-8mm diameter reactor. At this stage, catalysts are still in powdered form. Complex kinetic models are derived from the experimental data to fit the performance results shown by the tests.

High performance catalysts selected from the secondary screening are then validated at the pilot scale using a 50-100mm diameter reactor with larger catalyst particles. The previously developed complex kinetic models are coupled with the detailed description of the transport phenomena occurring in the heterogeneous environment inside the pilot units. At this stage, lifetime and aging mechanisms are also evaluated prior to scaling up catalyst production for testing in the demonstration unit.

Siluria's OCM demonstration unit is the largest OCM reactor ever built. The demonstration unit has been operating for over 2 years in campaign mode, testing various pre-validated OCM catalysts at a range of operating temperatures and pressures, superficial gas velocities, residence times, and inlet gas compositions. The reactor is designed and built to perfected mimic the design and operating conditions expected at the commercial scale. Complex reactor geometry and fluid dynamics are now added to the kinetic and transport models produced in the previous steps to predict the performances of the demonstration reactor.

Based on the extensive validation that the demonstration unit has provided for this comprehensive OCM reactor model over 2+ years of operation in a wide range of operating conditions, we can confidently utilize the model to design the first generation of Siluria's OCM commercial reactors and begin the development of more complicated second generation reactor designs.

Scale-up platform

At the core of the OCM technology is an adiabatic reactor that consists of two main sections: the OCM Reaction Section, which contains the axial-fixed OCM catalyst bed, and the Ethane/Propane Conversion Section, an adiabatic empty chamber designed to perform Post Bed Cracking (PBC).

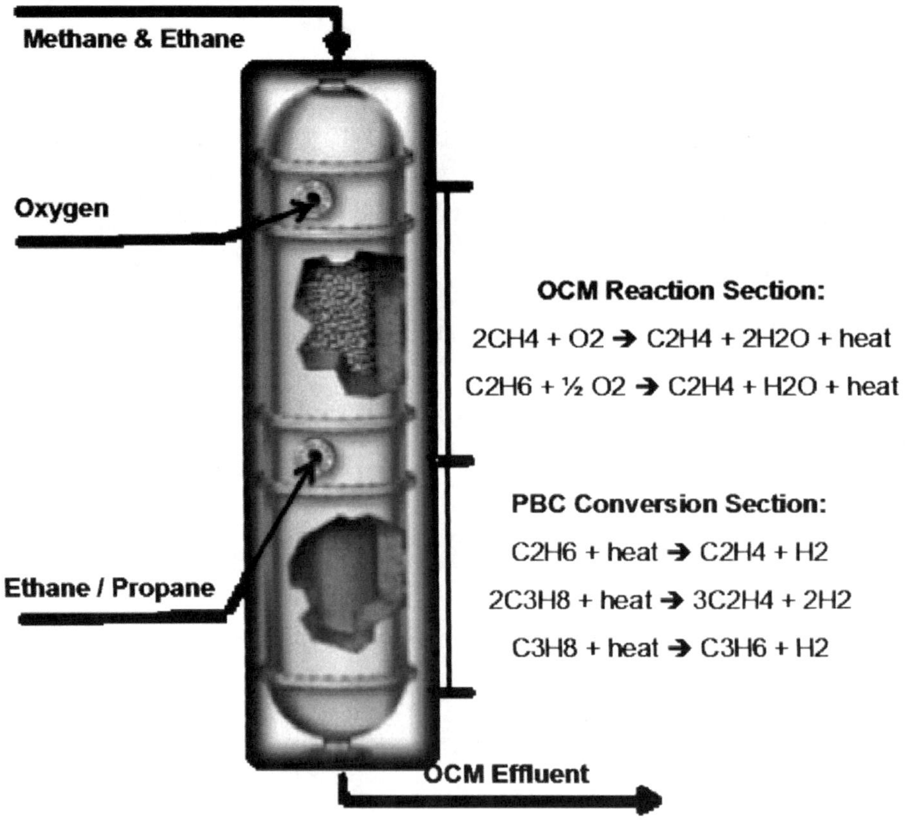

Figure 1 Illustration of OCM Reactor

Hot feed to the reactor, primarily consisting of methane with minor amounts of ethane, is mixed with oxygen in the mixer and sent to the catalyst bed. Oxidative coupling of methane reactions, alongside other side reactions begin to occur. The main reaction that takes place in the catalyst bed is shown in reaction 1.

$2CH_4 + O_2 \rightarrow C_2H_4 + 2H_2O + heat$ Reaction 1

In addition to methane oxidation, Siluria's OCM catalyst simultaneously promotes Oxy-De-Hydrogenation (ODH) of the ethane present in the feed according to the following reaction:

$C_2H_6 + 1/2O_2 \rightarrow C_2H_4 + H_2O + heat$ Reaction 2

The primary by-products generated in the OCM Reaction section are H2, CO, and CO2 produced by combustion reactions at high temperatures. The reaction heat causes a temperature increase over the OCM catalyst bed. The resulting temperature of the effluent is ideal for thermal cracking of additional hydrocarbons in the PBC section, such as ethane and propane, which produces additional ethylene and propylene. PBC reactions require a relatively short residence time and are conducted in the simple adiabatic chamber constituted by the connection between the OCM reactor and the downstream Heat Recovery Steam Generator (HRSG).

To validate this new technology on a commercial scale every aspect of the OCM reactor, from the inlet nozzle to the tubesheet of the HRSG, is studied in detail. Siluria's scale-up platform considers all these aspects and generates performance data in a wide range of operating conditions, space and superficial velocities, residence times, catalyst particle sizes and shapes, and gas compositions. All these data are used to develop and validate the comprehensive modelling and design tools adopted for commercial process designs.

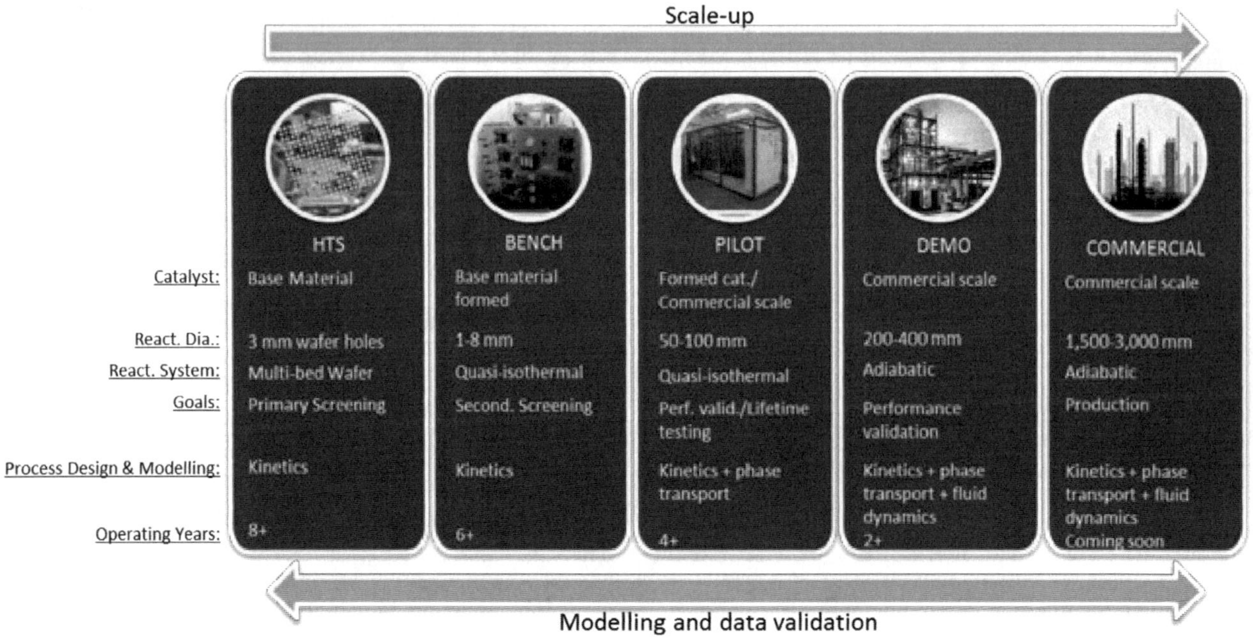

Figure 2 Illustration of Siluria's OCM scale-up platform

The first step in Siluria's scale-up process is primary screening of thousands of OCM catalyst formulae. Siluria's high throughput screening tool can screen over 250 catalysts simultaneously in real OCM reaction conditions. Siluria has, to date, screened over 30,000 catalysts for methane conversion and C2 selectivity over a wide range of temperatures to find the few catalyst formulae that meet the stringent requirements for a commercially viable OCM process.

Catalysts with promising performances are tested in a secondary screening tool, which consists of 1 to 2mm reactors. These units are filled with very small catalyst particles. Reactants – natural gas and oxygen – are mixed and fed to all reactors at constant velocity to obtain uniform and comparable data. The catalysts are tested over a wide range of operating temperatures, pressures and compositions using standard Design of Experiment techniques.

The OCM reaction network is an extremely complicated set of reactions that include not only the key product and by product species (ethylene, H2 CO, CO2) but also many reaction intermediates. Thus two separate kinetic models have been developed:

- a detailed kinetic model that includes all the thousands of elementary reactions occurring over the catalyst bed;
- a simplified activity model used to predict oxygen conversion and temperature rise as a function of a lumped composition, temperature pressure and catalyst type and aging.

Both models have been thoroughly validated over several sweeps of conditions, varying temperature, pressure, and inlet gas composition (e.g. O_2/methane ratio, ethane content, etc.). Differently aged catalysts have also been tested under these conditions to determine the effect of catalyst aging on catalyst performance and product distribution.

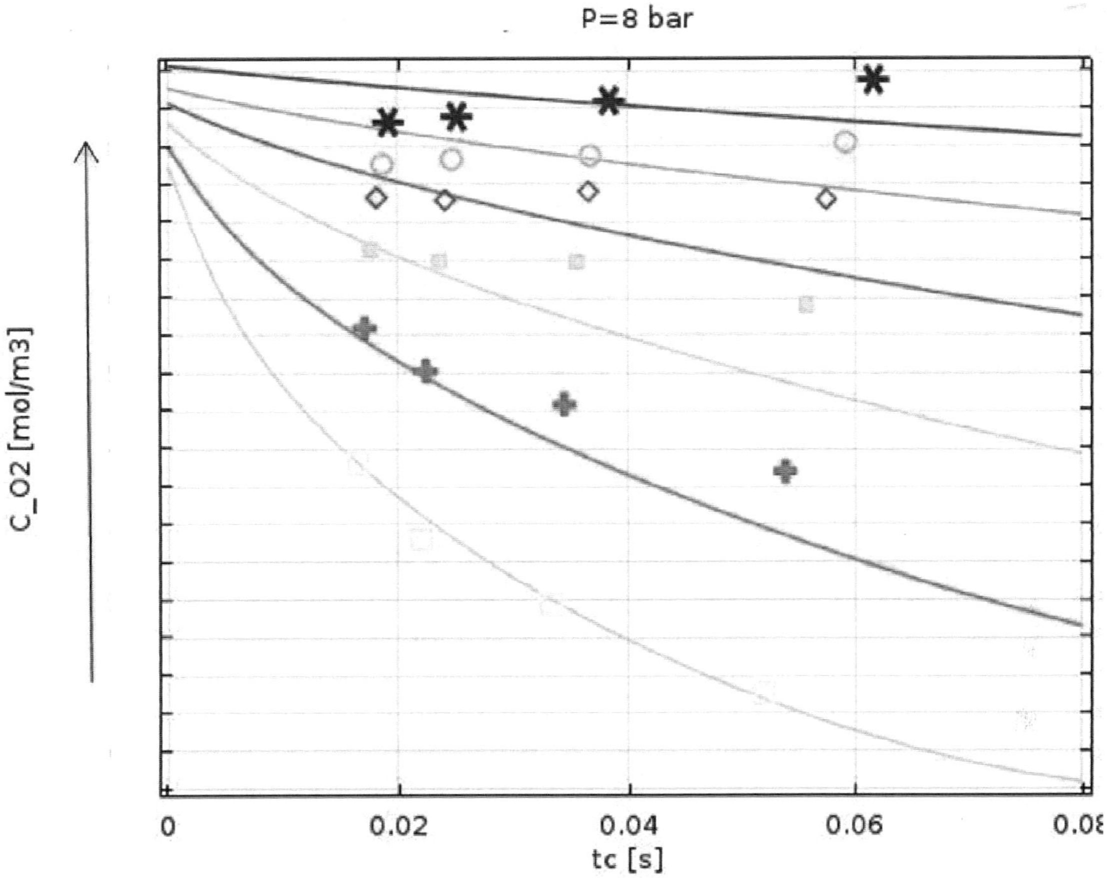

Figure 3 Evolution of the O_2 concentration profile with a contact time. Comparison between the simplified activity model and secondary screening data.

Figure 3 shows a comparison between oxygen concentration profiles calculated with the simplified activity model and the experimental data obtained with quasi-isothermal secondary screening reactor tests for different isotherms and a specific catalyst. The results show a very good agreement between model and measurements over a very wide temperature range.

The most promising catalysts are then tested at Siluria Pilot facility in Hayward, CA. Siluria' s pilot scale reactor has been in operation for over 4 years, continuously testing formed catalysts in a 50-100 mm diameter reactor to validate overall performances and lifetime and evaluate aging as a function of different operating conditions. The data obtained from the pilot facilities enable the inclusion of transport equations in the existing kinetic models derived from bench scale data as described previously.

The main goal of the pilot tests is to assess the effect of catalyst forming and shapes on process performance, in particular conversion and selectivity at different operating conditions. Pressure drop of the packed-bed and temperature profiles across the adiabatic reactor bed are also collected to validate the combined transport and kinetic models.

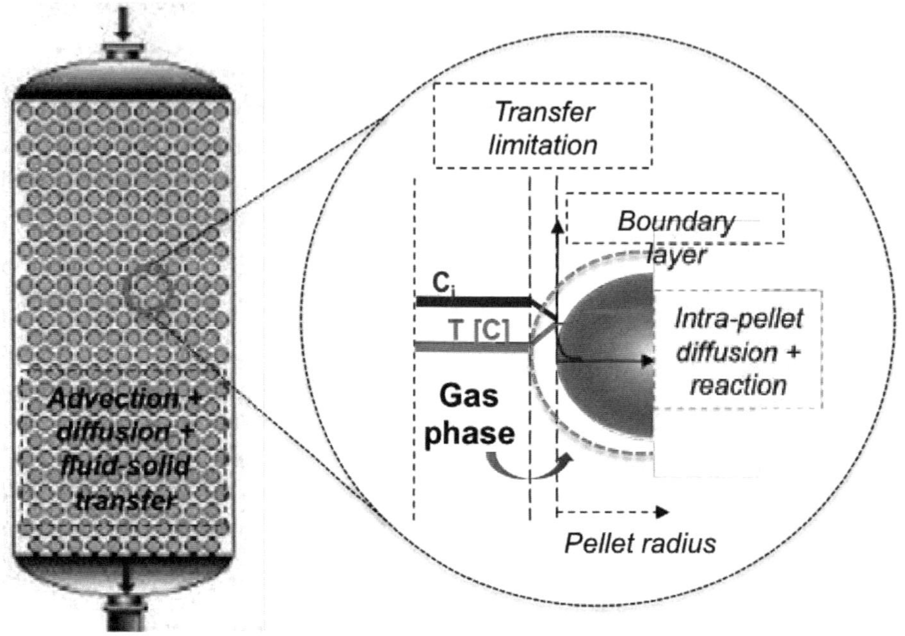

Figure 4 Fixed bed reactor model accounting for mass and heat transport between phases and inside the catalyst particle.

The model integrates various relevant bed and catalyst descriptors: bed void fraction, particle effective diameter, heat and mass transfer, catalyst BET active surface area, catalyst porosity, catalyst effective diffusivity, catalyst tortuosity and catalyst sintering kinetics. Once coupled with the simplified activity model it allows a determination of O2 consumption and temperature rise as a function of process conditions and the different catalyst descriptors. The transport model has also been integrated with the detailed kinetic model for a very detailed description of the OCM bed performance.

Catalytic performance versus operating window (inlet temperature, pressure, flow rate, feed composition, O2/methane ratio) is characterized by analyzing gas samples at different levels within the catalyst bed as well as recording catalyst and/or gas temperatures. This allows to optimize both design parameters and operating conditions in order to maximize different target functions, from C2+ selectivity to simple ethylene yield.

Separate pilots, capable of continuous unattended operation, are used to study the effect of minor species in the OCM feed and the effect of process integration with recycling of unconverted methane as done in at the commercial process.

Once the optimal operating conditions are defined, catalyst aging studies are also carried out in these pilot units in 1,000-hour long campaigns. Performance trends for conversion and selectivity are generated as well as catalyst samples for post-test analysis. Typical catalyst material characterization is used to track the different aging mechanism across different axial location within the catalyst bed.

OCM demonstration unit

Siluria's OCM demonstration unit was commissioned and started up in December 2014 in La Porte, TX. The demonstration unit was built to specifically run in campaign mode, testing a wide window of feed flows and compositions over a range of operating temperatures and pressures. The main goal of the demonstration unit is not only to validate performances and models developed during the OCM scale-up process but also to understand the intricacies of operating an OCM reactor at a commercial scale, experience not yet achieved by any other company.

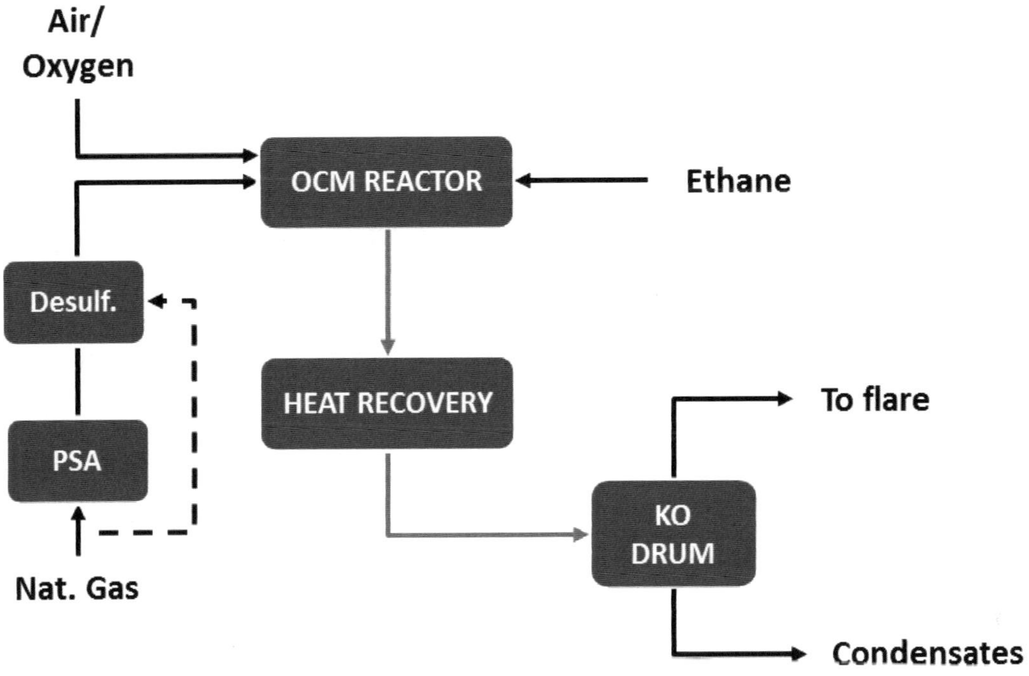

Figure 5 Block Flow diagram of the demonstration facility in La Porte, TX

In the demonstration plant natural gas from an interstate pipeline is fed to a PSA unit to adjust the concentration of heavy hydrocarbons (ethane, propane and butane) that may be present in the natural gas and enable testing over a range of feed composition. The PSA can be by-passed to test different gas compositions to the reactor.

The natural gas is heated and fed to the adiabatic OCM demo reactor, where it is mixed with air or oxygen in a proprietary low residence time mixer. As previously described, the gas then passes through the catalyst bed where the main OCM reactions occur. Ethane is also fed to the PBC section of the reactor to conduct post-bed cracking. The reactor effluent is then cooled in a steam generator and, eventually, in a cooling water exchanger. Process condensates are separated from the gaseous effluent in a simple knock-out drum.

Temperatures, pressures and gas compositions are constantly monitored through the catalyst bed and the PBC section during a campaign to validate the models derived from bench and pilot data. Figure 6 below shows a comparison between the temperature profile across the reactor and the model prediction.

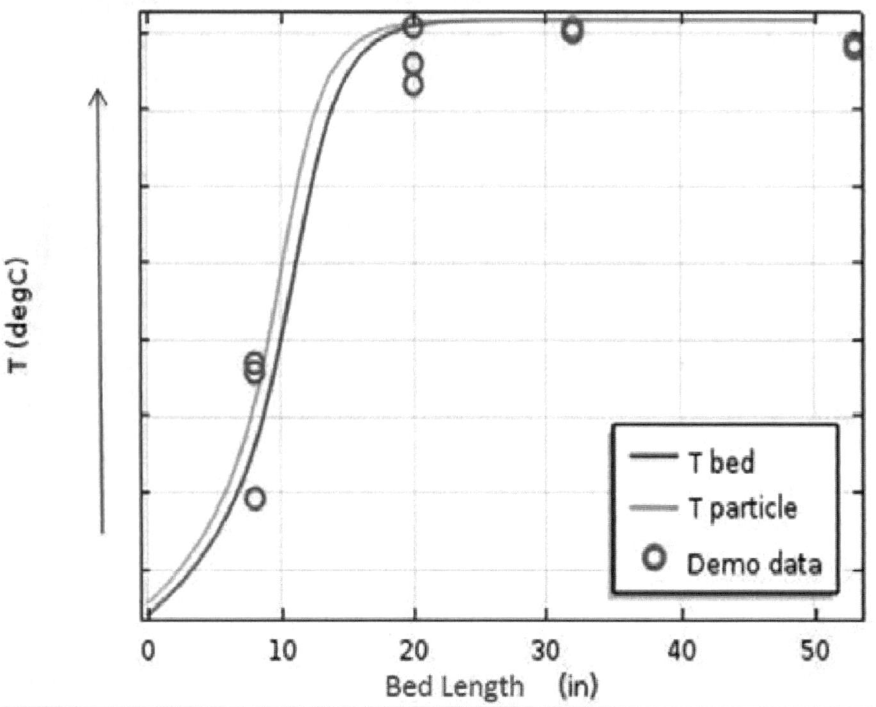

Figure 6 Actual temperature measurements compared with predicted temperature profiles

Gas compositions are closely monitored throughout the bed via appropriate online sampling and gas chromatography. Figure 7 below shows a comparison between the ethylene and ethane concentrations – as a function of the outlet temperature – predicted by the detailed kinetic and transport model and the corresponding measurements obtained at pilot and demonstration scale. The pilot reactor is not completely adiabatic and heat losses especially affect the last temperature measurement at the end of the reactor, where the reaction is completed and the temperature is the highest. Based on a determination of the temperature gradient without reaction, the temperature measurement at pilot scale is adjusted to eliminate the effect of heat losses and compared to the model predictions (that assume perfect adiabaticity of the system).

The results from the demonstration unit match the experimental data collected from the pilot units perfectly once heat losses are accounted for. Moreover, the detailed models developed by Siluria also predict catalyst performances to a great level of accuracy and enable reactor design at different scales and over a wide range of conditions.

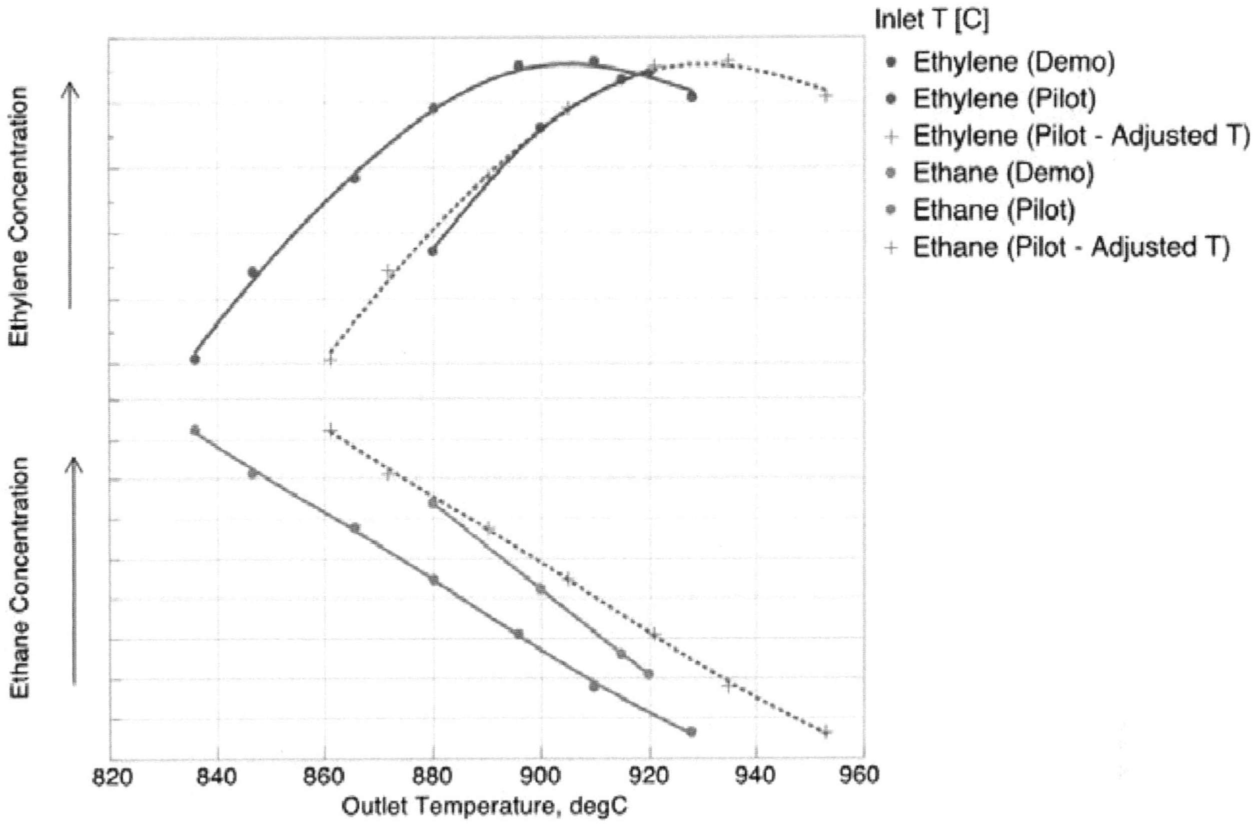

Figure 7 Ethylene and ethane concentrations at OCM reactor outlet measured at pilot and demo scale.

Ethane is also periodically fed to the demo reactor to verify PBC models produced with pilot data. Cracking is generally a well understood chemistry and existing kinetic models, coupled with fluid-dynamics, are used to predict PBC performances. As described for the bottom end of the OCM bed, heat losses through the PBC need to be properly accounted for when building the PBC model (heat losses are especially significant at pilot scale). Figure 8 below shows the validation of the PBC kinetic and fluid-dynamic model with the demo data.

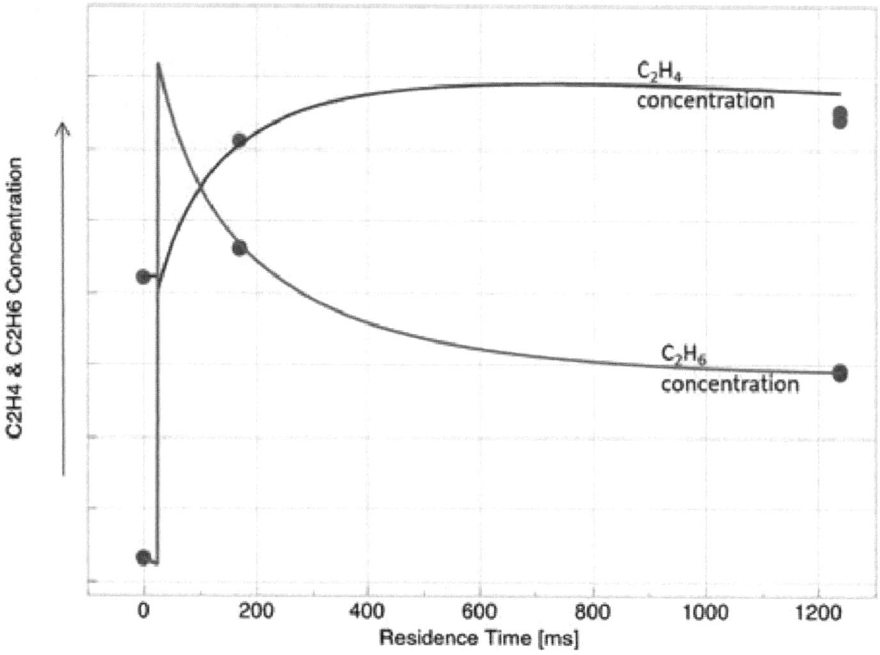

Figure 8 Ethylene and ethane concentration at PBC outlet predicted by the model and measured a

Technology readiness for commercial scale

The derivation of detailed unified kinetic, transport and fluid-dynamic models in Siluria's OCM scale-up platform and their validation at different scales enable the accurate and confident prediction of the OCM reactor performances at commercial levels. The availability of these models allow Siluria's engineers to optimize the design of all reactor features and the overall process architecture, including the window of operating conditions that the OCM reactor shall be maintained within and the appropriate process control strategy.

The operation of the demonstration plant since December 2014 also allowed Siluria to gain invaluable experience with the direct operation of an industrial OCM unit and to validate the design and operating procedures that are being adopted for the implementation of the technology at full commercial scale.

Moreover, the elucidation of the extremely complex reaction network in OCM catalyst have allowed Siluria to start the development of reactors with more advanced designs, which can, for instance, incorporate heat removal and overcome the intrinsic conversion ceiling posed by the adiabatic design.

Conclusions

Over the past eight years, Siluria has successfully established a powerful scale-up platform to screen and validate the critical process performances required for a commercially viable OCM catalyst and the relevant reactor design. Each step of the process produced data to develop modelling tools, which have been thoroughly validated with experimental data from bench, pilot and demonstration units built and operated by Siluria. The validated models are used to design commercial OCM units and for current and future generations of the technology.

References:

1. Radaelli, Guido, Kemper, Rainer. (2015) "Siluria and Linde OCM: Turning a breakthrough into commercial practice," presented at the 2015 AIChE Spring Meeting, Austin, TX

T40144 479642

IMPROOF: INTEGRATED MODEL GUIDED PROCESS OPTIMIZATION OF STEAM CRACKING FURNACES

Kevin M. Van Geem
Prof. Dr. ir.
Laboratory for Chemical Technology, Ghent University, Ghent, Belgium

Frédérique Battin-Leclerc
Prof. Dr.
Centre National de la Recherche Scientifique, Nancy, France

Georgios Bellos
Dr.
DOW Benelux B.V., Terneuzen, The Netherlands

Geraldine Heynderickx
Prof. Dr.
Laboratory for Chemical Technology, Ghent University, Ghent, Belgium

Wim Buysschaert
Mr.
CRESS B.V., Breskens, The Netherlands

Benedicte Cuenot
Dr.
European Centre for Research and Advanced Training in Scientific Computation, Toulouse, France

Marko R. Djokic
Dr. ir.
Laboratory for Chemical Technology, Ghent University, Ghent, Belgium

Tiziano Faravelli
Prof. Dr.
Politecnico di Milano, Milan, Italy

Gilles Theis
Mr.
John Zink International Luxembourg SARL, Luxembourg

Dietlinde Jakobi

Dr.
Schmidt + Clemens GmbH +CO. KG, Lindlar, Germany

Philippe Lenain
Mr.
Ayming France, Lyon, France

Andrés E. Muñoz G.
Dr.
AVGI, Ghent, Belgium

John Olver
Dr.
Emisshield Inc., Blacksburg, Virginia, USA

Jens N. Dedeyne
ir.
Laboratory for Chemical Technology, Ghent University, Ghent, Belgium

Stijn Vangaever
ir.
Laboratory for Chemical Technology, Ghent University, Ghent, Belgium

Petra Honnerová
Dr.
New Technologies Research Centre, University of West Bohemia, Plzen, Czech Republic

Zdeněk Veselý
Dr.
New Technologies Research Centre, University of West Bohemia, Plzen, Czech Republic

Prepared for Presentation at the 2017 Spring National Meeting
San Antonio, Texas, March, 28, 2017

IMPROOF: INTEGRATED MODEL GUIDED PROCESS OPTIMIZATION OF STEAM CRACKING FURNACES

Kevin M. Van Geem
Prof. Dr. ir.
Laboratory for Chemical Technology, Ghent University, Ghent, Belgium
Frédérique Battin-Leclerc
Prof. Dr.
Centre National de la Recherche Scientifique, Nancy, France
Georgios Bellos
Dr.
DOW Benelux B.V., Terneuzen, The Netherlands
Geraldine Heynderickx
Prof. Dr.
Laboratory for Chemical Technology, Ghent University, Ghent, Belgium
Wim Buysschaert
Mr.
CRESS B.V., Breskens, The Netherlands
Benedicte Cuenot
Dr.
European Centre for Research and Advanced Training in Scientific Computation, Toulouse, France
Marko R. Djokic
Dr. ir.
Laboratory for Chemical Technology, Ghent University, Ghent, Belgium
Tiziano Faravelli
Prof. Dr.
Politecnico di Milano, Milan, Italy
Gilles Theis
Mr.
John Zink International Luxembourg SARL, Luxembourg
Dietlinde Jakobi
Dr.
Schmidt + Clemens GmbH +CO. KG, Lindlar, Germany
Philippe Lenain
Mr.
Ayming France, Lyon, France
Andrés E. Muñoz G.
Dr.
AVGI, Ghent, Belgium
John Olver
Dr.
Emisshield Inc., Blacksburg, Virginia, USA
Jens N. Dedeyne
ir.

Laboratory for Chemical Technology, Ghent University, Ghent, Belgium
Stijn Vangaever
ir.
Laboratory for Chemical Technology, Ghent University, Ghent, Belgium
Petra Honnerová
Dr.
New Technologies Research Centre, University of West Bohemia, Plzen, Czech Republic
Zdeněk Veselý
Dr.
New Technologies Research Centre, University of West Bohemia, Plzen, Czech Republic

Abstract: In IMPROOF the steam cracking furnace of the 21st century will be developed and demonstrated. The ambition is to drastically improve the energy efficiency compared with the current state-of-the-art, and this in a cost effective way. Simultaneously the greenhouse gases and NOX emissions per ton ethylene produced will be reduced by 25%. This project will implement and combine several of the latest technological innovations in the field of fouling minimization and energy efficiency at pilot and industrial scale. These include the use of renewable fuels, oxy-fuel combustion, and high emissivity coatings which emit in the non-absorbent flue gas spectrum. Also, new advanced high temperature alloys that lower the coking rate will be implemented in combination with novel 3D reactor technologies leading to reduced coking and enhanced heat transfer between flue gas and the process. In 2019, the furnace will be deployed at the demonstrator at commercial scale using propane as feedstock based on the experimental and modeling data provided by the industrial partners, knowledge institutions and research organizations.

Introduction

Although steam cracking is considered a mature technology for olefin production, the complexity of the process and the harsh operating conditions allow the implementation of technology developments towards substantial heat transfer enhancement. One of the most important ways to reduce the energy input in steam cracking furnaces per ton ethylene produced is to reduce coke formation on the reactor wall of the long tubular reactors that are mounted in the furnaces [1]. Steam cracking is the most energy-consuming process in the chemical industry and globally uses approximately 8% of the sector's total primary energy [2]. Additionally, it is responsible for massive amounts of CO_2 emissions [3]. Improving the energy efficiency has an immediate payout because energy cost counts for a substantial part of the production costs in typical ethane or naphtha based olefin plants [3]. Typically, ethylene furnaces have to be decoked after 30-60 days to remove the coke that builds up in the coil. When the furnaces are decoked, production of the desired products is stopped for approximately 48 hours [4]. During the course of one run, deposited coke can reduce the heat-transfer efficiency of the firebox by 1-2 %, resulting in a 5 % increase in fuel consumption [5]. The use of advanced coil materials, combined with 3D reactor designs, improved process control, and more uniform heat transfer could increase run lengths [6], reducing simultaneously CO_2 emissions and increasing the lifetime of the furnaces. It has been proven that improved metallurgy of radiant coils can reduce catalytic coking [1], and that advanced 3D coil design such as the swirl flow reactor design [7] and the SCOPE design (by Schmidt + Clemens GMBH +CO. KG) can mitigate coke deposition leading to heat transfer improvement. Improved geometries can result in smaller radial temperature gradients inside the coil, and thus lower wall temperatures where the coke formation occurs. Advanced 3D CFD reactor simulations (see Figure 1) are extremely useful for that purpose [8].

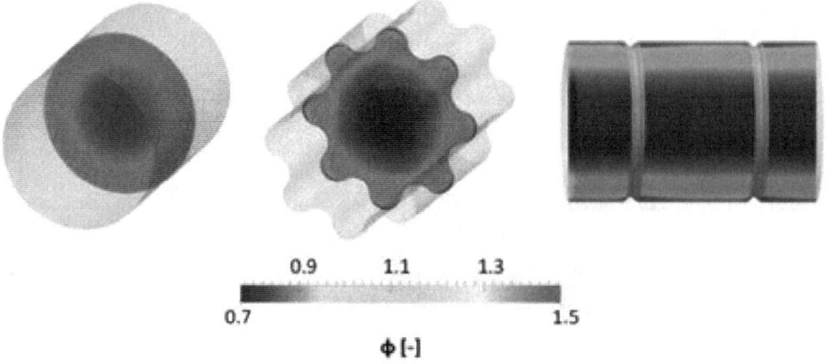

Figure 1: Three-dimensional representation of the studied reactor designs with contour plots of the corresponding local fluid age correction factors, from left to right: a bare tubular reactor, a longitudinally finned reactor and a transversally ribbed reactor [8].

An important challenge for the petrochemical industry is the upcoming stronger environmental regulations, in particular related to NO_X and CO_2. The total NO_X emissions of all the furnaces currently operating in the EU are approximately 16.3×10^3 tpa. To reduce NO_X emissions the use of advanced oxy-fuel combustion is demonstrated on pilot plant scale. The advantage is clear: because no nitrogen is added (apart from leakages), almost no NO_X is produced [9]. An additional advantage is that the produced flue gas is a concentrated CO_2 stream, that can be more easily captured, stored or used, for example, in chemical looping [10]. Bio-gas and bio-oil are renewable fuels, and hence, decrease net CO_2 production [11]. It is expected that these fuels will become available in substantial amounts in the near future in Europe and therefore can be used as fuels for steam cracking furnaces.

A final point for improvement is related to the radiant section of a steam cracking furnace, where the major part of heat transfer occurs by radiation. The radiation is emitted by the refractory walls towards the process radiant coils. Application of high emissivity coatings on the external surface of the radiant coils could improve the energy consumption [12]. Applying the improved emissivity coatings on the furnace walls [13] will decrease the required firing to reach the same process temperatures in the radiant coils. This will reduce fuel gas consumption and CO_2 emissions by an anticipated 10 to 15%. In addition to higher heat absorption, coating the external surface of radiant coils can improve the surface homogeneity and eliminate hot spots on the tube walls. An additional benefit will be extended run lengths resulting in less energy spent annually for decoking the radiant coils.

The IMPROOF project will demonstrate the advantage of combining all these technological innovations, with an anticipated reduction of emissions, and increase of the time on stream and energy efficiency.

Objectives

To answer the market need and implement novel technology for olefin production the project has set the following technical objectives:

1. Demonstrate the individual impact of novel emissive, reactor and refractory materials on pilot scale (TRL5),
2. Demonstrate the power of advanced process simulation (high performance computing and CFD) for furnace design and optimization,
3. Demonstrate the technical economic and environmental sustainability of the IMPROOF furnace at TRL6,
4. Novel combustion technology using alternative fuels and oxy-fuel combustion, and
5. Coke formation reduction and real time optimization.

In the following section the first results obtained on the first two objectives will be presented and discussed.

Results

High emissivity coatings

Emissivity is a radiometric relative property of a material that quantifies the ability of the real surface to emit radiation. It is defined as the ratio of radiances emitted by the real surface and an ideal black body at the same temperature, spectral and geometrical conditions. In the case of coatings intended for high-temperature applications in furnaces, their emissivity is among the most crucial characteristics that considerably influence thermal efficiency of the whole furnace [14].

Two novel high-emissivity coatings designed for radiant walls/floor of furnace and process coils/tubes were investigated in the experimental setup for the normal spectral emissivity at the New Technologies Research Centre within University of West Bohemia. Next to coatings provided by Emisshield company, also reference materials, namely regular stainless steel 304 and Alamo® refractory bricks (produced by HarbinsonWalker International) were inspected.

The measuring apparatus of the normal spectral emissivity measurement method has been recently described [14,15]. The experimental setup (see Figure 2) consists of a laboratory blackbody as a reference source of radiation and a FTIR spectrometer that detects emitted radiation both by the blackbody and real materials. An infrared camera is used to measure the surface temperature. The blackbody and the sample are positioned against each other. A rotary parabolic mirror is positioned in the middle of the blackbody and the sample, and is used to switch the optical paths. The radiation collected by the mirror enters the spectrometer through an entrance port. Optical and optomechanical components (translation stages, mirrors, aperture, shutter, alignment laser) define the analyzed area size and accomplish equal optical paths from both radiation sources. The optical path is enclosed inside a box; however, no evacuating or purging is used [14,15].

Steel, ceramic and coated disc substrates of 25 mm in diameter and 5 mm thickness are used (see Figure 3). The measured sample is clamped to a ceramic fiber insulation case and it is attached to the optical bench by translation stages [14].

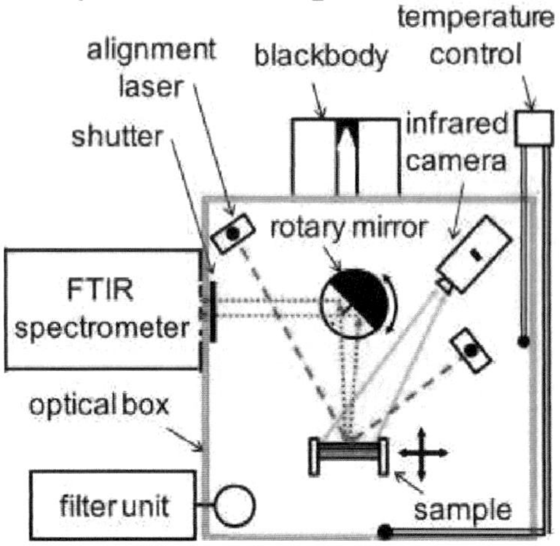

Figure 2: Schematic view of the emissivity measuring apparatus with the optical path for the radiation coming from the sample to the detector [14,15].

Figure 3: (Un)coated 304 stainless steel and (un)coated Alamo® refractory samples prepared for normal spectral emissivity measurements.

Normal spectral emissivity measurements of a coated metal sample were performed at different temperatures in order to check the temperature dependency. As can be seen from the left side of Figure 4, all spectral emissivity experiments appear to be temperature independent in the observed temperature range (700 – 850 °C). The right side of Figure 4 gives a comparison of normal spectral emissivities of coated and uncoated ceramic material obtained at 800 °C. As can be seen from the right side of Figure 4, a big difference exists in the sub 5 μm region, which is the most important for radiation in steam cracking furnaces. Thus, coating the refractory wall can significantly improve the thermal efficiency of the cracking furnaces. Nevertheless, a decrease of spectral emissivity is observed for wavelengths shorter than 5 μm. As this is not aligned with current literature data [16], the spectral emissivity measurement method will be further improved in order to evaluate this in more detail.

Figure 5 shows the comparison of normal spectral emissivity of uncoated and coated 304 stainless steel. As can be seen from Figure 5, a clear difference exists between the samples throughout the entire range of wavelengths. Emissivity of 304 stainless steel appears to be slightly higher than data found in literature [17,18]. However, the surface is rather rough compared to the smooth surfaces typically used in emissivity experiments, which as a consequence has a positive effect on emissivity values [19]. Important to note is that once the 304 stainless steel material is oxidized, spectral emissivity will increase [19] and will result in a smaller gap with coated 304 stainless steel. Figure 6 gives a comparison of the two different coatings designed for ceramic and metallic substrates. It is clear that although the composition of the two materials is different, the emissivity behavior is very similar.

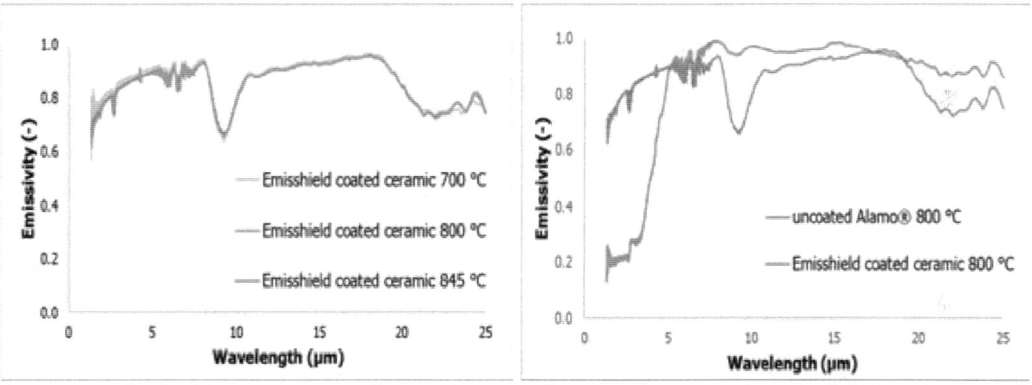

Figure 4: Left: Temperature dependency of normal spectral emissivities of coated ceramic samples and Right: Comparison of normal spectral emissivites of coated and uncoated ceramic materials at 800 °C.

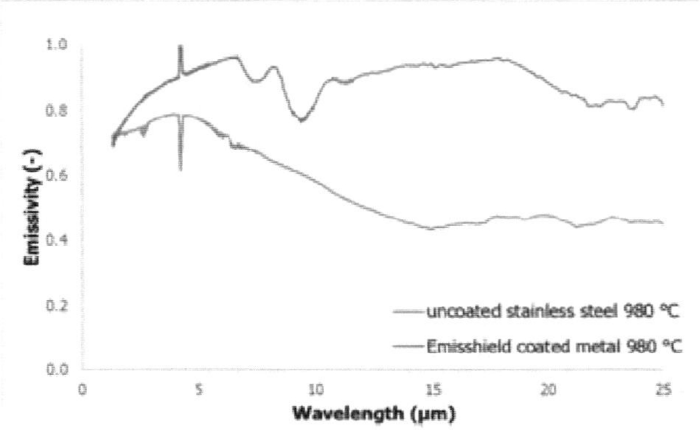

Figure 5: Comparison of normal spectral emissivity of uncoated and coated 304 stainless steel.

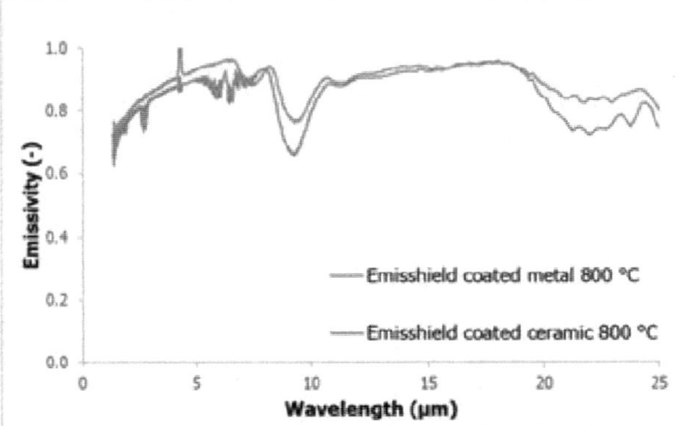

Figure 6: Normal spectral emissivities of coated metal and coated ceramic at 800 °C.

3D Reactor Technology

A lot of research focusses on the enhancement of heat transfer between the reactor coil and the process gas, as this positively affects the process in multiple ways [20]. Firstly, heat transfer enhancement at this point implies a lower temperature at the metal surface, which results in a decreased rate of coke formation and, consequently, longer run lengths. Next to this, an enhanced heat transfer allows to decrease the residence time of the process gas, resulting in a more favorable product distribution, i.e. a higher selectivity towards light olefins.

To achieve this heat transfer enhancement, different reactor designs are proposed that either increase the reactor surface, e.g. longitudinally finned reactors [21], or promote turbulence, and radial mixing, e.g. MERT [22] or SFT [7]. The increased radial mixing implies a more uniform temperature distribution of the process gas, which has additional beneficial effects on the product distribution.

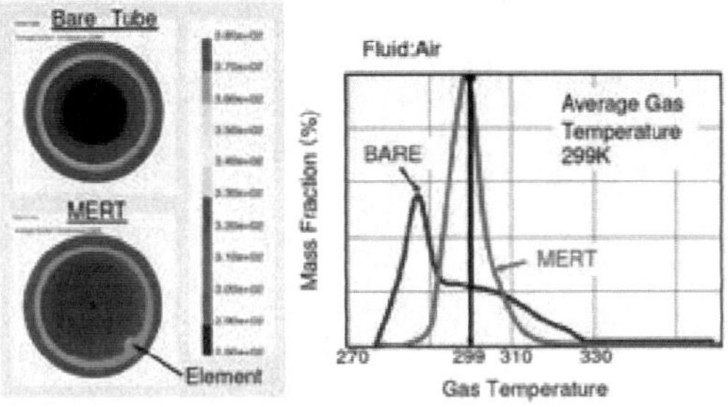

Figure 7: Gas temperature distribution in bare tube and MERT tube [22].

Next to the beneficial increase in heat transfer, the strong turbulence induced by 3D elements increases the shear stresses in the flow, resulting in an increase in pressure drop. As this is detrimental to the product selectivity, the design of 3D elements can be viewed as balancing these two effects against each other. A detailed study of the flow behavior and the overall performance of these 3D reactor designs is thus necessary.

It should be noted that small modifications to the reactor design can introduce secondary radial or swirling flow patterns, thereby completely altering the reactor performance both in terms of heat transfer and pressure drop. As these additional flow patterns are highly three dimensional in nature, simple one-dimensional simulations, which are today's industrial standard, are unable to accurately capture all flow phenomena and consequently cannot assess the impact on heat transfer enhancement. Consequently, it is necessary to assess the performance of 3D reactor technologies by means of three-dimensional CFD [23].

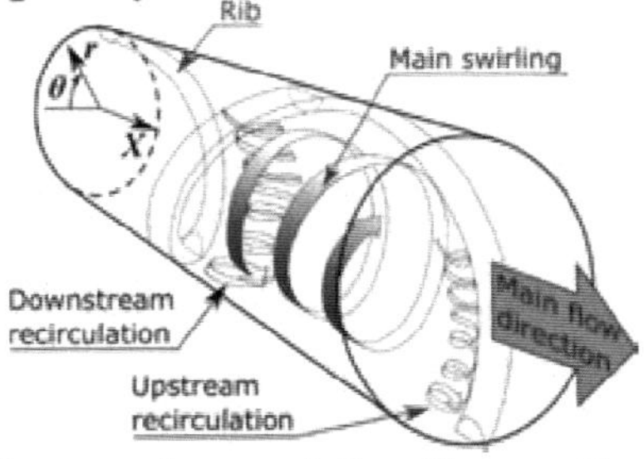

Figure 8: Three dimensional secondary swirling flow patterns and smaller recirculation zones in ribbed geometry [24].

For this performance assessment, validation of the applied CFD methodology is crucial, but only few adequately validated cases are available in the open literature. In order to have confidence in the simulation results, a cooperation with the Von Karman Institute (VKI) has been established. Their expertise in flow field measurement allows the extraction of highly detailed information of both the velocity field, by means of Stereo-PIV, as well as of the heat transfer at the wall, by means of liquid crystal thermography (LCT) for swirling flows. This experimental campaign showed that for this type of reactor design RANS (Reynolds Averaged Navier-Stokes) simulations, in which all scales of turbulence are modelled, are incapable of capturing important flow phenomena like recirculation zone size and reattachment lengths. Fully resolving all turbulence, even to the smallest scales, with Direct Numerical Simulation (DNS) is currently still impossible at industrially applied flow regimes and therefore swirling flows in 3D reactors should typically be resolved by means of Large Eddy Simulation (LES) [25]. This technique resolves the largest eddies, which contain the most turbulent kinetic energy, and models the smaller scales, thereby decreasing the computational demand.

The application of stream wise periodic boundary conditions (SPBC) [8] in the open source CFD package OpenFOAM® allows to significantly decrease the computational domain, thereby further reducing the computational demand up to a level that allows a highly accurate and detailed study of the influence of geometric parameters of the 3D design.

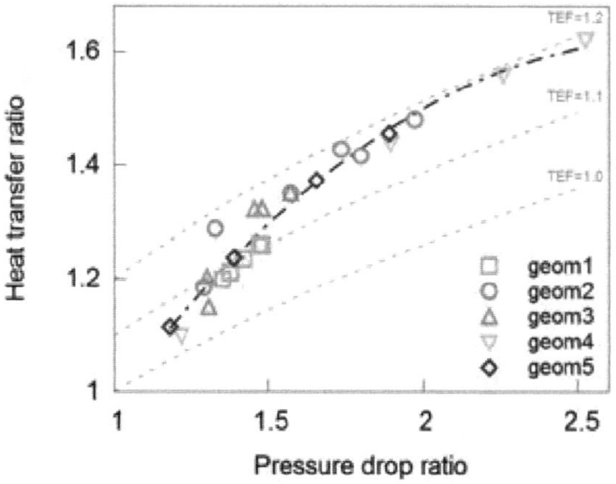

Figure 9: Parametric study of 3D reactor design showing the trade-off between heat transfer and pressure drop.

Based on this information, the performance of different designs can be compared against each other allowing to further optimize these designs to improve heat transfer with little additional pressure drop. Next to this, information from this kind of simulations can also be used to redesign the reactor in a more fundamental way, e.g. decreasing recirculation zones or enhancing mixing effects, thereby leading to novel steam cracking coil designs.

As metal temperature is less uniform due to the application of 3D elements, the coke layer will generally grow non-uniformly. This non-uniform growth will influence the reactor geometry over time which will in turn influence the flow pattern and consecutive coking behavior. In order to assess the long term performance of these 3D coils, one should not only evaluate the performance at start-of-run conditions but monitor the coke formation and overall performance throughout the run. It has been shown that there is a pressure drop penalty for the enhanced heat transfer. However, this effect is leveled out throughout the run due to the decreased rate of coke formation (see Figure 10).

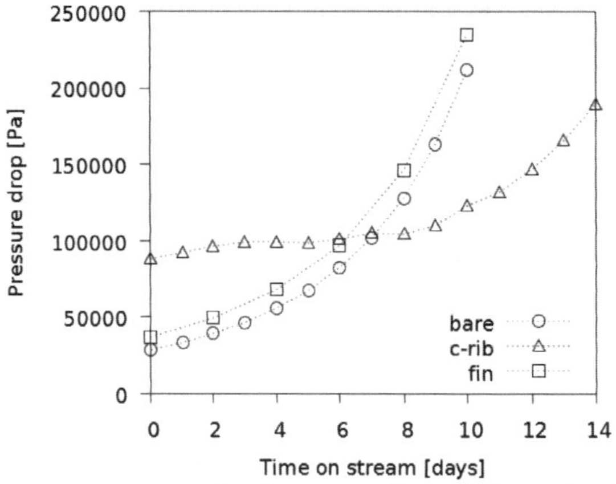

Figure 10: Pressure vs. time on stream for bare, c-rib and finned tubes for propane cracking.

Furthermore, run length simulations have shown that the on-stream time can be drastically improved by the implementation of these 3D designs.

Conclusions

The first results of the IMPROOF project show that applying high emissivity coatings on the refractory wall can substantially improve the thermal efficiency of the radiant section of a steam cracking furnace. Formulations can be optimized to fine-tune the performance. Moreover, applying the high emissivity coatings on the external surface of the reactor tubes can additionally improve the surface homogeneity and eliminate hot spots. Combining the advanced coil materials and novel 3D reactor designs can lead to more uniform heat transfer, which increases run lengths, product selectivities as well as lifetime of the furnace, while simultaneously reducing CO_2 emissions. In addition, it is believed that the current design of the 3D reactors is far from optimal, and that CFD can lead to the design of radiant coils with intensified performances. By implementing the 3D designs the on-stream time of the cracking furnaces can be drastically improved. It is key to focus not only on start of run conditions, but also on evaluating the performance over a complete run length because initial larger pressure drops can be misleading for the overall performance over the time on stream.

Acknowledgment

The work leading to this invention has received funding from the European Union Horizon H2020 Programme (H2020-SPIRE-04-2016) under grant agreement n°723706.

References

1. Andres E. Munoz G., Kevin M. Van Geem, Marie-Francoise Reyniers and Guy B. Marin (2014), "Coking Resistance of Specialized Coil Materials during Steam Cracking of Sulfur-Free Naphtha," Industrial & Engineering Chemistry Research, pp. 13644-13655.

2. Tao Ren, Martin K. Patel and Kornelius Blok (2008), "Steam cracking and methane to olefins: Energy use, CO2 emissions and production costs," Energy, May, pp. 817-833.

3. Tao Ren, Martin K. Patel and Kornelius Blok (2006), "Olefins from conventional and heavy feedstocks: Energy use in steam cracking and alternative processes," Energy, June, pp. 425-451.

4. Zhaobin Zhang and Lyle F. Albright (2010), "Pretreatments of Coils to Minimize Coke Formation in Ethylene Furnaces," Industrial & Engineering Chemistry Research, pp. 1991-1994.

5. Heinz Zimmermann and Roland Walzl (2000), "Ethylene: Ullmann's Encyclopedia of Industrial Chemistry," Wiley-VCH Verlag GmbH & Co. KGaA.

6. Yu Zhang, Feng Qian, Carl M. Schietekat, Kevin M. Van Geem and Guy B. Marin (2015), "Impact of flue gas radiative properties and burner geometry in furnace simulations," AIChE Journal, pp. 936-954.

7. Carl M. Schietekat, Marco W.M. Van Goethem, Kevin M. Van Geem and Guy B. Marin (2012), "3D swirl flow reactor technology for pyrolysis processes: hydrodynamic and computational fluid dynamic study," XX International Conference on Chemical Reactors.

8. David J. Van Cauwenberge, Laurien A. Vandewalle, Pieter A. Reyniers, Kevin M. Van Geem, Guy B. Marin and Jens Flore (2016), "Periodic reactive flow simulation: Proof of concept for steam cracking coils," AIChE Journal.

9. Abass A. Olajire (2010), "CO2 capture and separation technologies for end-of-pipe applications – A review," Energy, pp. 2610-2628.

10. Juan Adanez, Alberto Abad, Francisco Garcia-Labiano, Pilar Gayan and Luis F. De Diego (2012), "Progress in Chemical-Looping Combustion and Reforming technologies," Progress in Energy and Combustion Science, pp. 215-282.

11. Anja Oasmaa, Bert Van De Beld, Pia Saari, Douglas C. Elliott and Yrjö Solantausta (2015), "Norms, Standards, and Legislation for Fast Pyrolysis Bio-oils from Lignocellulosic Biomass," Energy & Fuels, pp. 2471-2484.

12. Georgios D. Stefanidis, Kevin M. Van Geem, Geraldine J. Heynderickx and Guy B. Marin (2008), "Evaluation of high-emissivity coatings in steam cracking furnaces using a non-grey gas radiation model," Chemical Engineering Journal, pp. 411-421.

13. Cressie E. Jr. Holocombe and Lloyd R. Chapman (1999), "High emissivity coating composition and method of use," Google Patents.

14. Petra Honnerová, Jiří Martan, Martin Kučera, Milan Honner and Jacques Hameury (2014), "New experimental device for high-temperature normal spectral emissivity measurements of coatings." Measurement Science and Technology.

15. Honner Milan, Honnerová Petra, Kučera Martin and Martan Jiří (2016), "Laser scanning heating method for high-temperature spectral emissivity analyses," Applied Thermal Engineering, pp. 76-81.

16. Bo Kong, Ting Li and Qitai Eri (2017), "Normal spectral emissivity of GH536 (HastelloyX) in three surface conditions," Applied Thermal Engineering, pp. 20-26.

17. G. Cao, S.J. Weber, S.O. Martin, K. Sridharan, M.H. Anderson and T.R. Allen (2013), "Spectral emissivity of candidate alloys for very high temperature reactors in high temperature air environment," Journal of Nuclear Materials, pp. 667-673.

18. Y.M. Wanga, H. Tiana, D.L. Quanb, L.X. Guoa, J.H. Ouyanga, Y. Zhoua and D.C. Jia (2012), "Preparation, characterization and infrared emissivity properties of polymer derived coating formed on 304 steel," Surface and Coatings Technology, pp. 3772-3776.

19. L. del Campo, R.B. Perez-Saez, L. Gonzales-Fernandez, X. Esquisabel, I. Fernandez, P. Gonzales-Martin and M.J. Tello (2010), "Emissivity measurements on aeronautical alloys," Journal of Alloys and Compounds, pp. 482-487.

20. Marco W. M. Van Goethem and Jelsma, Ellert (2014), "Numerical and experimental study of enhanced heat transfer and pressure drop for high temperature applications," Chemical Engineering Research & Design, pp. 663-671.

21. Carl M. Schietekat, David J. Van Cauwenberge, Kevin M. Van Geem and Guy B. Marin (2014), "Computational Fluid Dynamics-Based Design of Finned Steam Cracking Reactors," AIChE Journal, pp. 794-808.

22. Michael Gyorffy, Makateo Hineno, Kunihide Hashimoto, Seong-Hoan Park and Min-Soo You (2009), "In Mert performance and technology update," AIChE Spring Meeting: Ethylene Producers Conference.

23. Pieter A. Reyniers, Carl M. Schietekat, David J. Van Cauwenberge, Laurien A. Vandewalle, Kevin M. Van Geem and Guy B. Marin (2015), "Necessity and Feasibility of 3D Simulations of Steam Cracking Reactors," *Industrial & Engineering Chemistry Research*, pp. 12270-12282.

24. Ignacio Mayo, Bogdan C. Cernat, Marco Virgilio, Alessio Pappa and Tony Arts (2016), "Aerothermal Investigation on the Flow and Heat Transfer in a Helically Corrugated Cooling Channel," *ASME. Turbo Expo: Power for Land, Sea, and Air, Volume 5B: Heat Transfer*, pp. V05BT16A017.

25. David J. Van Cauwenberge, Carl M. Schietekat, Jens Flore, Kevin M. Van Geem and Guy B. Marin (2015), "CFD-based design of 3D pyrolysis reactors: RANS vs. LES," Chemical Engineering Journal, pp. 66-76.

AIChE Paper Number 124c

IMPACT OF IMPURITIES ON CARBON MOLECULAR SIEVE MEMBRANES FOR APPLICATION IN OLEFINS UNITS

Mark Brayden
R&D Fellow
The Dow Chemical Company

Liren Xu
Senior Engineer
The Dow Chemical Company

Greg Barbay
Senior Technologist
The Dow Chemical Company

William J. Koros
Professor
Georgia Institute of Technology

Prepared for Presentation at the 2017 Spring National Meeting

San Antonio, TX

March 26-30, 2017

IMPACT OF IMPURITIES ON CARBON MOLECULAR SIEVE MEMBRANES FOR APPLICATION IN OLEFINS UNITS

Mark Brayden
R&D Fellow
The Dow Chemical Company

Liren Xu
Senior Engineer
The Dow Chemical Company

Greg Barbay
Senior Technologist
The Dow Chemical Company

William J. Koros
Professor
Georgia Institute of Technology

Abstract: The separation of olefins from paraffins is important in the petrochemical industry. The current dominant technology, cryogenic distillation, is highly energy intensive. In 2013 we reported on the use of carbon molecular sieve (CMS) membranes in a hybrid membrane-distillation process to improve energy use with resulting lower energy or increased production.

The CMS membranes are produced by pyrolyzing polyimides like 6FDA:BPDA-DAM and show improved production (flux) and exceptional separation (selectivity) when compared to the precursor polymer and polymeric membranes in general.

The impact of impurities on the CMS membrane must be understood in order to use it in actual production plant applications. This paper will discuss common impurities found in the ethylene plant's purification train, specifically, the potential impact of these impurities on the membrane. These impurities include hydrogen sulfide, water, carbon dioxide, acetylenics, and carbon monoxide.

Introduction

Olefins, such as ethylene and propylene, are among the most important building blocks in chemistry. Currently, these are mainly produced by a cracking process from a wide range of hydrocarbon feedstocks (ethane, propane, naphtha, etc.). The separation of olefins from paraffins is currently dominated by distillation in the petrochemical industry. Due to the close boiling points and volatility, the purification process is very capital and energy intensive. To save the huge cost in separation processes, different separation methods have been proposed in recent years. Hybrid membrane-distillation separation has been a very promising alternative separation technology [1, 2]. Binary ethylene/ethane separation and propylene/propane separation attracted the most attention. Most previous concepts aimed to debottleneck the C2 splitter or the C3 splitter by installing a parallel membrane module with a distillation column [1-5].

Hydrocarbons separation via membranes has not yet been successful for commercial applications. Conventional polymer materials are limited by the performance upper bound between the productivity and selectivity trade-off [6-8]. For most polymer materials, the stability of membranes under aggressive hydrocarbon feeds is also a serious problem [9]. Facilitated membranes are able to deliver high selectivity; however, they are unstable under realistic operation conditions [1, 10]. Inorganic materials with rigid pore structure are able to overcome the polymer upper bound. Zeolite and Metal Organic Framework (MOF) membranes have the potential to deliver high selectivity; however, the large-scale application is inhibited by the expensive fabrication cost and their brittle natures [11, 12].

Carbon molecular sieve (CMS) membranes have emerged as promising membrane materials for the next generation membrane technology. The slit-like pore structure, which provides a unique combination of ultra-micropores (responsible for molecular sieving) and micropores (responsible for sorption and high diffusion coefficients), is ideal for gas separations [13, 14]. One particular advantage of CMS over other inorganic membranes is that CMS membranes are produced via controlled pyrolysis of polymer precursors. While polymer membranes are the state of art technology, CMS membranes can also be relatively easily fabricated into realistic hollow fiber modules. The only additional step beyond the current technology platform is the scale-up of the pyrolysis step. This feature makes CMS not only feasible for delivering high separation performance from the materials aspect, but also viable for large-scale applications.

In our previous publications, we describe ethylene/ethane separation based on CMS membranes derived from the commercially available polyimide Matrimid® (15) and also the use of other polyimides extended to propylene/propane separation, including our efforts to improve the performance beyond those of the original Matrimid®-based membranes (16).

For any CMS membrane, though, one issue that may be a potential barrier to its commercialization is the impact of impurities on the membrane. While the impurities are typically present at nominally ppm levels, they could have an impact on the long term performance and ultimate life of the membrane. This paper will describe the efforts to study these impurities, including future plans to study the long range impacts.

Background

In this study, pemeance is used to describe the membrane productivity and selectivity is used to characterize the separation efficiency. Permeance of penetrant A is defined by the trans-membrane pressure difference normalized flux:

$$\left(\frac{P}{l}\right)_A = \frac{N_A}{\Delta p_A} \tag{1}$$

Where, N_A is the steady state gas flux and Δp_A is the trans-membrane partial pressure difference of component A.

"Gas Permeation Unit" (GPU) is usually used as the unit for permeance, which is defined as:

$$1\ GPU = 1 \times 10^{-6} \frac{cm^3(STP)}{cm^2 \cdot s \cdot cmHg} \tag{2}$$

When the downstream pressure is much less than the feed pressure, membrane selectivity is defined as the ratio of gas permeances:

$$\alpha_{A/B} = \frac{(P/l)_A}{(P/l)_B}$$

Experimental

Lab scale membrane modules were fabricated using the precursor polyimides shown in Figure 1.

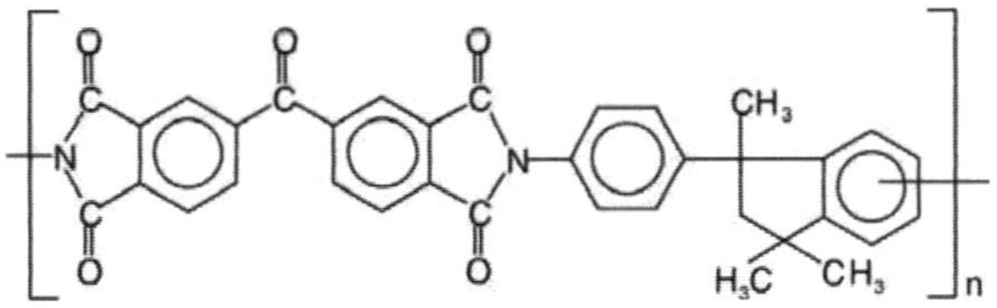

(a) Matrimid®

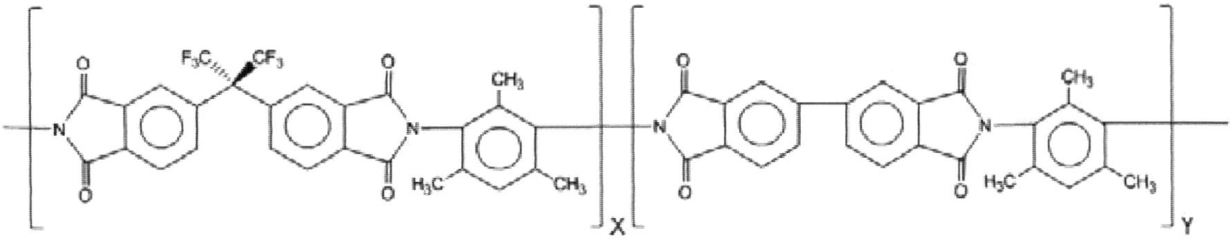

(b) 6FDA/BPDA-DAM (in this work, X:Y=1:1)

Fig. 1. Chemical structures of 2 precursor polyimides: (a) Matrimid®; (b) 6FDA/BPDA-DAM

The pyrolysis protocol and fabrication methods have been described previously. (17) For these CMS modules, typical final pyrolysis temperature is either 550 or 675 °C.

Impurities

If a membrane is to be installed in the feed to a C2 splitter or in the bottoms stream, the impurities include acetylene, carbon dioxide, hydrogen sulfide, ammonia, methanol, and water. Acetylene may be present in the ppm range, and all other impurities will be at sub ppm or even ppb levels with maybe the exception of methanol during very short intervals. In the event of a hydrate, methanol may be added to the column and may be temporarily present in the column. It is likely, though, that methanol would not come in contact with a membrane or the membrane could be bypassed during the methanol addition.

In a C3 splitter feed or bottoms stream, the most common impurities include methyl acetylene, propadiene, methanol, and carbonyl Sulfide. All of these components may be present in several ppm levels.

The effects of carbon dioxide on these membrane materials is well understood (17), and has been shown to present no problem, so it was excluded from this study. Matrimid®-based CMS membranes were used, and the following impurities were investigated

- Acetylene(3 hours at 100 ppm, 16 hours at 1000 ppm, 8 hours at 100 ppm)
- Hydrogen Sulfide (3 hours at 100 ppm, 16 hours at 1000 ppm, 8 hours at 100 ppm)
- Ammonia (3 hours at 100 ppm, 16 hours at 1000 ppm, 8 hours at 100 ppm)
- Methanol (4 hours at 20 ppm, 16 hours at 200 ppm)
- Water (16 hours with ~ 90 % humidity nitrogen at room temperature)

Another impurity that can impact membrane performance is a higher molecular weight hydrocarbon. Testing with N-alkanes (1000 ppm each of methane, ethane, propane, butane, and pentane mixture, and 1000 ppm hexane in nitrogen) and toluene (1000 ppm) were performed. The Koros lab has shown that these types of impurities could be removed with liquid propylene if necessary in a simple regeneration step (18).

Lab testing of CMS hollow fiber gas permeation

To test gas permeation properties of hollow fiber membranes, small lab-scale modules were made. Details of module making were given in previous work (19). For carbon fibers, single-fiber modules and multi-fiber modules were constructed and tested. Mixed gas permeation tests were performed at 35 °C. Gas chromatography was used to determine the downstream permeate compositions and then selectivity at typically 100 psi pressure drop (102 psig at the inlet/retentate, 2 psig on the permeate side).

The test system schematic is shown in Figure 2.

First, baseline performance was determined with lab mixture of 50:50 ethylene/ethane. Next, the membrane was exposed to the impurity, and then the membrane was re-tested with the ethylene/ethane mixture.

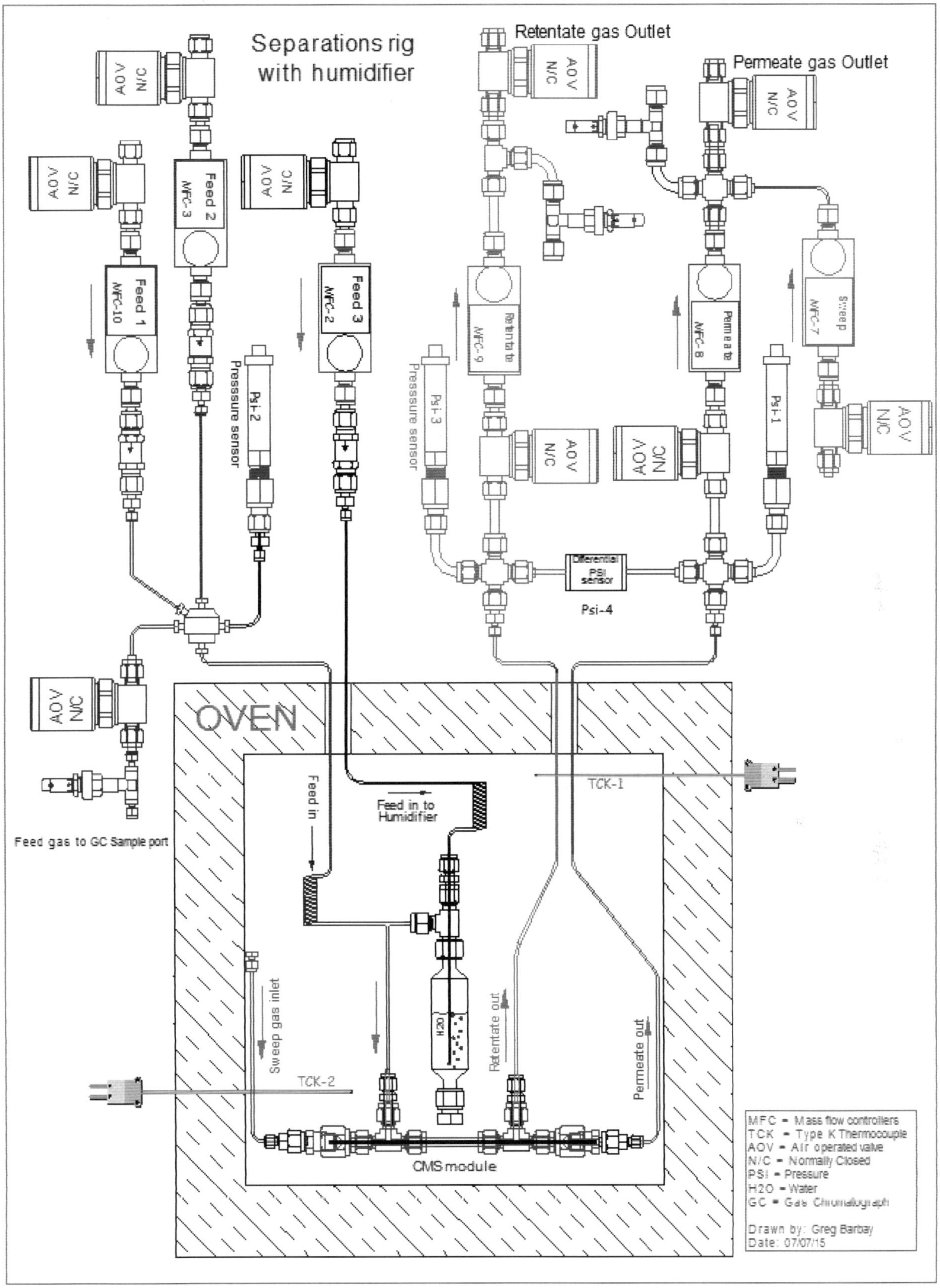

Figure 2. Laboratory Testing System

Experimental Results of Laboratory Testing

A summary of the results of the testing are shown in Table 1.

Table 1: Impact of impurities on flux and selectivity

Impurity	Impact on flux	Impact on selectivity
Hydrogen Sulfide 100 ppm	-70%	+20%
Ammonia	-10 to -20%	0%
Water	-15%	+10%
Methanol	-25%	0%
Acetylene	0%	0%

Acetylene

In this testing acetylene demonstrated no impact on the membrane performance. Subsequent testing of a mixed gas feed containing cracked gas components up to and including C3 components, the acetylenics had the highest permeance of all the components that included hydrogen, alkenes, and alkanes. Acetylene, methyl acetylene, and propadiene are all linear molecules, and even at low levels these impurities will concentrate in the permeate.

Water, ammonia, and methanol

Water is weakly adsorbed to the CMS material, but it is easily removed by heating the membrane to 60-100 °C or with liquid propylene. Testing showed a minor impact on permeance with little or no impact on selectivity.

Ammonia and methanol likely are also weakly adsorbed and show similar behavior as water. While desorption studies have not been conducted with these impurities, it is believed these can also be desorbed with low temperature regeneration.

Hydrogen Sulfide

Hydrogen sulfide is the strongest bound molecule of all of the materials tested. However, the lower concentration did not show an impact after three hours of exposure at 100 ppm H2S. This is at least two orders of magnitude higher than the typical H2S level in a C2 separation. More testing is needed to establish if low levels of H2S truly has an impact on long term membrane performance.

Our hypothesis is that the amount of H2S adsorbed will be equilibrium controlled at low concentrations of H2S and that a stable performance will be reached once the amount of H2S on the membrane is reached. Such behavior would be similar to what has been observed with low levels of O2 at elevated temperatures (20).

Higher Molecular Weight Hydrocarbons

The impact of higher molecular weight N-alkanes and toluene showed minimal impact. After a 3-hour exposure to a mixture of N-alkanes (1000 ppm each of methane, ethane, propane, butane, and pentane), and a separate test with 1000 ppm hexane, showed no change in permeate flow.

Likewise, toluene showed similar behavior after overnight exposure (17 hours) to 1000 ppm toluene.

In both of these studies the contaminant was added to the shell side of the module, so, the outer layer of the membrane with the separating layer of the asymmetric structure.

A much different result was obtained when toluene was added to the bore side of the membrane. A 19-hour exposure of the 1000 ppm toluene standard to the bore side led to lowered flow, only 20% of the baseline flow.

By exposing the bore side to liquid propylene for approximately 1 day, the permeate flow recovered to about 75% of its original flow.

Like toluene, interestingly, the feed modes had an impact on the transport properties of multiple smaller gases. For hydrogen/hydrocarbons, ethylene/ethane, propylene/propane, the performance was better when the feed was introduced from the shell side. Unlike toluene, no liquid propylene regeneration was needed after bore-side feed and the performance for shell-side feed retained after switching back. For other gas pairs such as carbon dioxide/nitrogen and carbon dioxide/methane, feeding from shell-side or bore-side also made a different.

Field testing of CMS hollow fiber gas permeation

To determine the long term effects of impurities on membranes, it is common to expose these to high levels of an impurity for a short period of time, or at actual conditions for several days. While these types of studies are valuable, the ability to extrapolate this to longer term operation is limited.

To demonstrate the impact of long term exposure may take hundreds or even thousands of hours. In order to do this, we installed a field test unit in one of our ethylene production units. This test system has been previously described. (21)

The system is basically a lab test unit in a Class 1 / Div 2 purge box. See Figures 3-5.

Figure 3. Field Testing System. Modular system in purge box for field use. Box-in-box design allows for combined electronics cabinet with z purge, and outer box for high flow experiments with all Class 1 / Division 2 rated components.

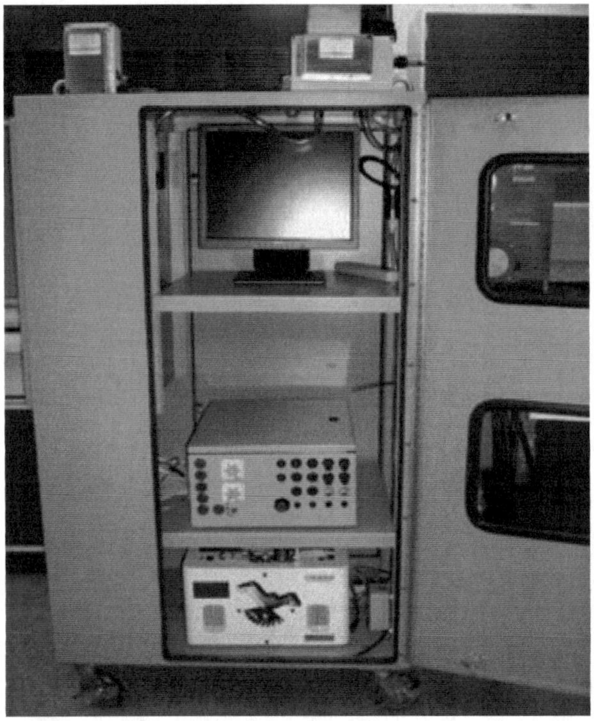

Figure 4. Purged box with electronics, including Camile control system, computer, and gas chromatograph.

Figure 5. Outer box with high flow rate membrane, including Class 1 / Div 2 rated instrumentation.

Field testing is planned to start in February, 2017, with initial testing using a mixed C2 stream prior to the tail end acetylene hydrogenation unit.

Conclusions

Laboratory testing of the impact of impurities on C2 separation indicated that

1. Acetylene has no impact on the separation of ethane and ethylene. As a matter of fact, due to the linear nature of the molecule acetylene permeates through the membrane at a higher rate than ethylene. Furthermore, methyl acetylene and propadiene permeate through the membrane at even a higher rate than acetylene.
2. Water is adsorbed onto the CMS, but it is weakly bound. While it has an impact on both permeance and selectivity, water is easily removed from the membrane with moderate heat. Ammonia and methanol have similar performance in the membrane testing.
3. Hydrogen sulfide is the most strongly adsorbed of all of the impurities tested. After 3 hours of exposure to 100 ppm H_2S, the membrane showed very little change in performance. However, after overnight exposure to 1000 ppm, the membrane showed a significant reduction in permeance accompanied by an improvement in selectivity.
4. While we hypothesize that low levels of these impurities, including hydrogen sulfide, will reach a low, equilibrium amount of the impurity adsorbed onto the CMS membrane,

long range testing is needed to verify this. This long range testing will be done in the field in an operating unit starting in February, 2017.

5. Heavier hydrocarbons like toluene have a minimal impact when fed to the shell side of the membrane. Significant permeation loss was experienced when toluene was fed to the bore side of the membrane.
6. Liquid propylene can be used to remove hydrocarbon fouling of the bore side and recover a significant amount of the flow through the membrane.

References

1. A. Motelica, O.S.L. Bruinsma, R. Kreiter, M. den Exter, J.F. Vente, Membrane retrofit option for paraffin/olefin separation—a technoeconomic evaluation, Ind. Eng. Chem. Res. 51 (2012) 6977-6986.
2. J.A. Caballero, I.E. Grossmann, M. Keyvani, E.S. Lenz, Design of hybrid distillation-vapor membrane separation systems, Ind. Eng. Chem. Res. 48 (2009) 9151-9162.
3. M. Benali, B. Aydin, Ethane/ethylene and propane/propylene separation in hybrid membrane distillation systems: Optimization and economic analysis, Sep. Purif. Technol. 73 (2010) 377-390.
4. E. Ayotte-Sauve, M. Sorin, F. Rheault, Energy requirement of a distillation/membrane parallel hybrid: a thermodynamic approach, Ind. Eng. Chem. Res. 49 (2010) 2295-2305.
5. W. Stephan, R.D. Noble, C.A. Koval, Design methodology for a membrane distillation column hybrid process, J. Membr. Sci. 99 (1995) 259-272.
6. R.L. Burns, W.J. Koros, Defining the challenges for C3H6/C3H8 separation using polymeric membranes, J. Membr. Sci. 211 (2003) 299-309.
7. L.M. Robeson, The upper bound revisited, J. Membr. Sci. 320 (2008) 390-400.
8. L.M. Robeson, Correlation of separation factor versus permeability for polymeric membranes, J. Membr. Sci. 62 (1991) 165-185.
9. W. Qiu, C.-C. Chen, L. Xu, L. Cui, D.R. Paul, W.J. Koros, Sub-Tg cross-linking of a polyimide membrane for enhanced CO2 plasticization resistance for natural gas separation, Macromolecules 44 (2011) 6046-6056.
10. R.J.B. Tim Charles Merkel, Process for regenerating facilitated-transport membranes, in: U.S. Patent (Ed.) United States Patent, Membrane Technology and Research Inc., Menlo Park, CS, United States, 2006.
11. C.M. Zimmerman, A. Singh, W.J. Koros, Tailoring mixed matrix composite membranes for gas separations, J. Membr. Sci. 137 (1997) 145-154.
12. C. Zhang, Y. Dai, J.R. Johnson, O. Karvan, W.J. Koros, High performance ZIF-8/6FDA-DAM mixed matrix membrane for propylene/propane separations, J. Membr. Sci. 389 (2012) 34-42.
13. K.M. Steel, W.J. Koros, An investigation of the effects of pyrolysis parameters on gas separation properties of carbon materials, Carbon 43 (2005) 1843-1856.
14. K.M. Steel, W.J. Koros, Investigation of porosity of carbon materials and related effects on gas separation properties, Carbon 41 (2003) 253-266.
15. L. Xu, M. Rungta, W.J. Koros, Matrimid® derived carbon molecular sieve hollow fiber membranes for ethylene/ethane separation, J. Membr. Sci. 380 (2011) 13G. Barbay, Carbon

molecular sieve hollow fiber membranes for olefin/paraffin separations, AIChE Spring Meeting, Ethylene Producers Conference, paper no. 140c, San Antonio, TX, April 28-May 2, 2013.
18. C. W. Jones and W. J. Koros, Carbon, 32, 1427-1432, 1994.
19. D.Q. Vu, W.J. Koros, S.J. Miller, High pressure CO2/CH4 separation using carbon molecular sieve hollow fiber membranes, Ind. Eng. Chem. Res. 41 (2002) 367-380.
20. M. Kiyono, PJ Williams, WJ Koros, Effect of pyrolysis atmosphere on separation performance of carbon molecular sieve membranes, J Membr Sci 2010, 359(1–2):2–10.
21. M. Brayden, G. Barbay, The Dow Chemical Company's Membrane Test Unit, North American Membrane Society annual conference, Las Vegas, Nevada, June 7, 2011.

Olefin-Paraffin Separation with Customized Amorphous Fluoropolymer (CAF) Facilitated Transport Membranes

Sudip Majumdar, Yosuke Koizumi, Ken Loprete, Ken Pennisi, Stuart Nemser, Andy Feiring, Ning Shangguan, Hannah Murnen, Dan Lousenberg

Compact Membrane Systems, Newport, DE

Background

Ethylene and propylene are the major chemical industry raw materials and consume a great deal of energy during production. The separations of these materials are some of the costliest, most energy intensive, and most technically difficult separations in the industry. The technology currently employed for the separation of ethylene and propylene from paraffins is distillation and is estimated to consume 250 trillion BTU/year of energy. A membrane based olefin-paraffin separation technology has the potential to provide substantial economic benefit to petrochemical industry and drastically reduce the energy required.

Separation of olefin-paraffin mixtures through various polymeric membranes without carriers has been studied with various different membranes such as cellulose acetate, polysulfone, PDMS, polyethylene, PPO, 6FDA–IPDA, 6FDA–6FpDA etc [Burns and Koros (2003)]. However, the permeabilities and selectivities of these membranes are generally low as olefins and paraffins usually have very similar molecular sizes and solubilities. To improve the olefin fluxes, metal-based facilitated transport membranes have been studied. Facilitating agents such as silver salts can be grafted into the polymeric membrane to achieve facilitated transport [Hsiue and Yang (1993); Teramoto et al. (2002); Park et al. (2007)].

Membranes are formed using casting solutions containing a mixture of carrier and polymer in polymer/metal salt blends [Ho and Dalrymple (1994); Peinemann et al. (1990)]. Carrier can be bound chemically or physically to a solid polymer. These membranes require quite a high loading of silver ion in order to be effective facilitated transport membranes [Pinnau and Toy (1997)].

A recent thorough study by Membrane Technology and Research [Merkel et al. (2007); Merkel et al. (2013)] on this subject focused on solid polymer electrolyte membranes, Pebax 2533 doped with silver tetrafluoroborate. The study identified carrier instability is a major problem. The lifetimes of the membranes are measured in days only. They further hypothesized a previously unrecognized phenomenon labeled as 'olefin conditioning' implying that the olefin species itself is a source of carrier instability [Merkel et al. (2007)].

Imtex Membrane Corp. [Hamza et al. (2013)] developed an olefin/paraffin separation membrane based on chitosan material and silver nitrate as a facilitation agent and reported membrane stability over 1500 hours of testing. In this approach the carrier instability problem is curbed via circulation of an aqueous silver nitrate solution over the membrane.

CMS Membrane Technology

Compact Membrane Systems (CMS) has developed a customized amorphous fluoropolymer (CAF) membrane containing silver ions that selectively transport olefin molecules from a mixture of olefin and paraffin. CAF membranes are based on a silver-containing fluoropolymer material. The technology exploits a mass-transfer mechanism known as facilitated transport. Silver incorporated into the membrane polymer acts as a binding site for the double bond in the olefins transporting them across the membrane. Saturated paraffins do not have the same interaction. This difference in affinity allows for olefins to pass through the membrane at a significantly higher rate than do paraffins.

The two key membrane properties that characterize the effects of facilitated transport are **permeance** and **selectivity**.

- **Permeance** is defined as the rate at which one component travels through a given surface area of membrane for a given driving force. This performance metric is typically expressed in GPU (gas permeation unit). CAF membranes have olefin permeances greater than 150. This translates into a propylene flux of 0.25-0.85 ton/ft^2/year under normal operating conditions (e.g., 30-200 psi, 30C, and 20-90% olefin)
- The **selectivity** of one component to another is simply the ratio of the permeance of the two components. This tells you how much olefin is passing through the membrane as compared to paraffin. CMS CAF membrane has selectivity ranging from 20-80 depending on the operating conditions and feed composition

The CAF membrane has both high permeance and high selectivity, meaning high purities can be reached with a reasonable membrane area. These characteristics yield very attractive process economics.

The membrane can perform separations of olefins from paraffins at a wide range of temperatures, pressures, and compositions. CMS has demonstrated olefin removal at temperatures up to 70°C (158°F) and pressures up to 250 psig. For an efficient separation to occur, the olefin-paraffin mixture must be in the gaseous state. Separation performance requires humidification as water vapor improves the mobility of the olefin through the membrane. The membrane can remove olefin in both concentrated and dilute streams for compositions ranging 5-95% by volume of olefin. The membrane can process mixtures of C2, C3, and C4 olefin and paraffin. Typical lab evaluations involve exposing the membrane to a stream of fixed pressure, temperature, and composition. The chart displayed on the next page shows a performance profile for a membrane separating a 20% propylene/80% propane mixture at 60 psig and room temperature. As you can see, the membrane has shown stable performance over >9 months, a vast improvement over current state of the art materials.

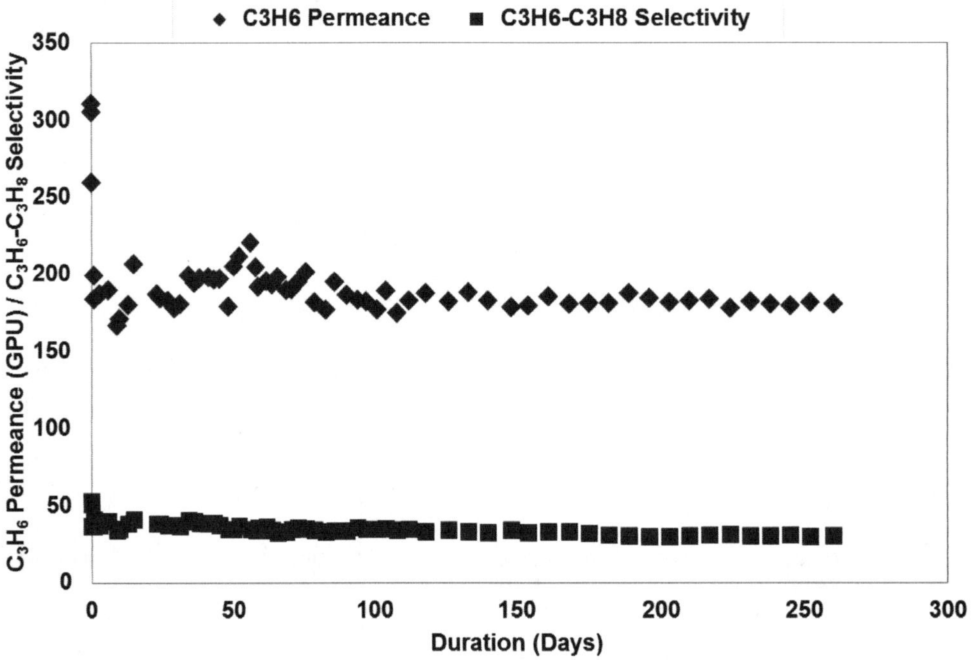

A broad range of tests have been conducted to characterize the membrane resistance to common poisons in refinery gas streams. These include testing under conditions such as 5% hydrogen or 500 ppm acetylene. We have seen good resistance to both hydrogen and acetylene from the CAF membranes. Based on our testing, we recommend process streams contain:

1. <100 ppm acetylene
2. <2% hydrogen
3. <100 ppb total sulfur

We generally include a sulfur pretreatment unit operation as part of the full membrane upgrading system to eliminate the concern that sulfur will detrimentally affect membrane lifetime.

Economics

CMS has performed multiple evaluations to assess the economic viability of the CAF membrane. These exercises rigorously examine the operational, capital, administrative, and associated auxiliary equipment (e.g. compression and drying) to return a pressurized, dry product or recycle stream. The assessments look at a representative process and use the membrane to add value by recovering olefin, purifying a paraffin rich stream and creating an olefin-rich stream.

Example Application – C3 Splitter

C3 splitters are used to separate C3 olefin (propylene) from C3 paraffin (propane). Depending on the conditions at which the column is running, olefins are often spilled out the bottom in concentrations ranging from 1-20% of the stream. Those olefins are sold at a much lower value as paraffins or used as waste heat. Using a membrane to recover the olefins can provide significant value to the producer.

The CAF membrane can be used to recover that propylene while also upgrading the propane to a higher purity. In this example, we've assumed a moderate amount of olefin in the C3 splitter bottoms (10%) and are using the membrane to generate HD5 propane and permeate a propylene rich stream which can be sold as CGP, RGP, or PGP. Alternatively, it can simply be returned to the column to expand capacity. The evaluation here evaluates the latter where the concentrated propylene is recycled to the splitter. A simple diagram of the arrangement is shown on the next page.

The proposed system augments a 40,000 ton per year C3 splitter running at 215 psig column pressure. The system with standard membranes replaced every 2 years gives

- 16 month payback time
- IRR of 80%

These economics are robust over a wide range of conditions. The sensitivity analysis for a number of variables is shown on the next page. The biggest sensitivity to the economics comes from the permeance of the membrane and the value of the stream. Higher permeance through the membrane minimizes the required area for the separation and therefore improves the capital cost for the project. CMS CAF membranes have very high permeances, contributing to positive economic outcomes across a wide variety of applications.

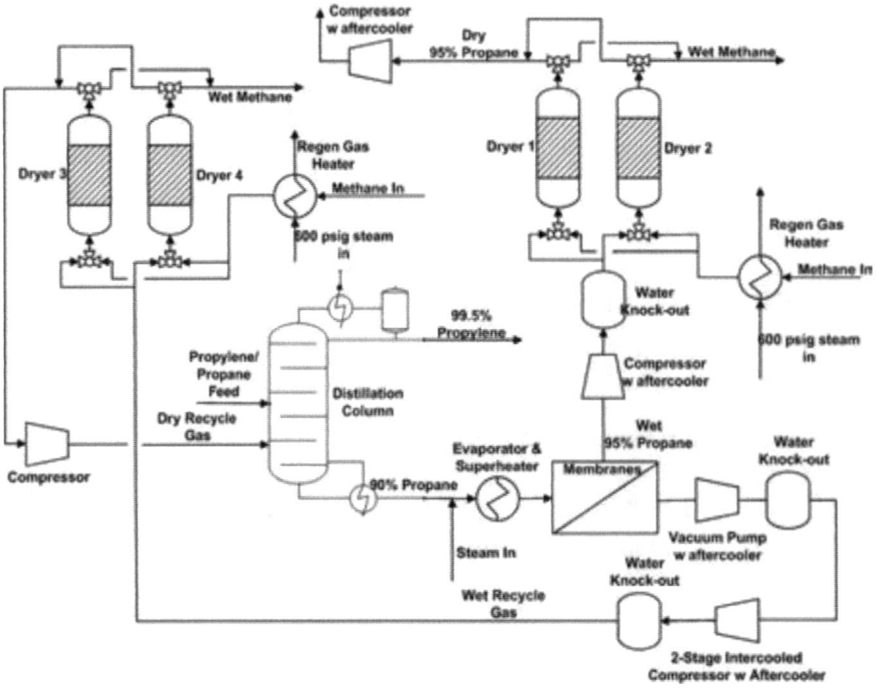

As expected, more frequent membrane replacement, lower olefin permeance, and inexpensive market prices for propane and propylene increase the payback period of the membrane system. However, in all cases evaluated across a range of likely scenarios, the investment is still strongly positive with the IRR remaining greater than 48%.

Membrane Production and Scale up

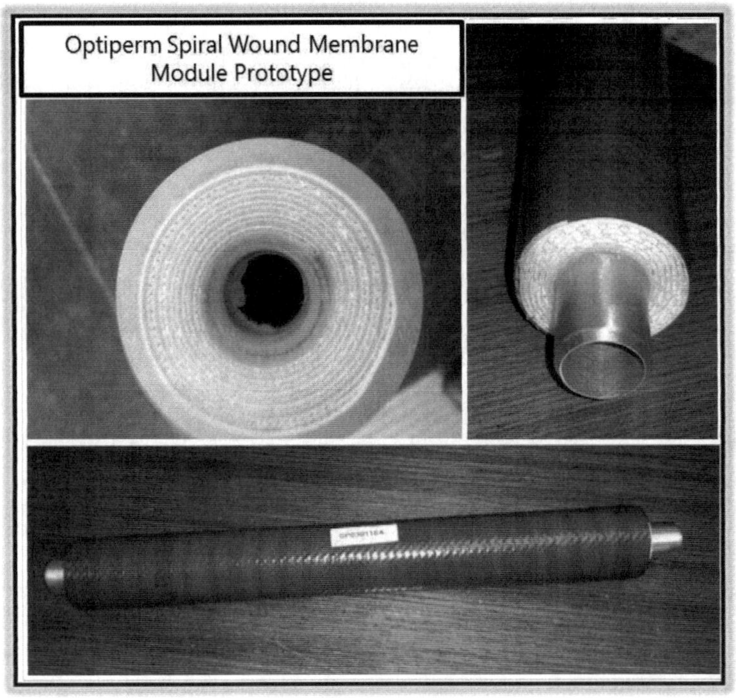

CMS is currently scaling up manufacturing to meet expected demand for the CAF membrane. These scale up activities include membrane module scale-up, supply-chain diversification, QC development, procurement of large scale coating equipment and expansion of manufacturing space. An example of a prototype CAF spiral wound membrane is shown here. This work will allow CMS to be producing its CAF membrane at significant quantities by 4Q 2017.

Pilot tests

We are working with numerous players in oil and gas and petrochemical spaces. Several interested parties have requested on-site evaluation of the CAF membrane. Our relationship with Delaware City Refining Company (DCRC) has led to a scheduled pilot system launch in May of 2017. The unit will provide real-world evidence of membrane stability in realistic conditions validating our laboratory testing. Overall the project will confirm the technical and economic viability of a full scale system. Tests will be conducted over 6 months during which time the skid will operate autonomously save for regular sampling and maintenance.

The main features of the system:

- Data acquisition system constantly monitors pressure, temperature, flow, humidity, and alarm status
- Coriolis mass flow control and measurement equipment ensures accurate evaluation membrane performance
- Glycol heating system maintains uniform temperature in unit plumbing and equipment
- Humidification vessel keeps process gas saturated with water vapor to optimize membrane performance
- Alarms for high and low temperature, flow, oxygen content, and flammable gas content
- Quick disconnect sample ports allow operators to quickly take gas samples

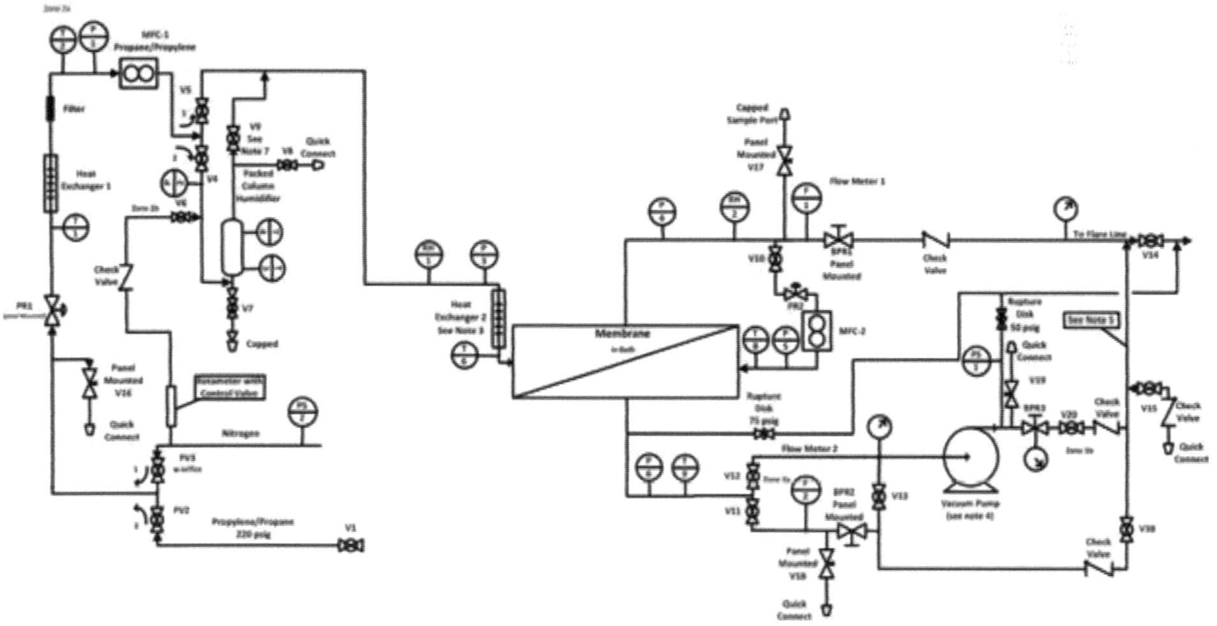

Pilot test rig for olefin-paraffin separation

Full-Scale System

As developments with interested parties progress CMS has looked forward to design full-size membrane systems for use at a refinery or polymer production operation. Some key features of a full scale system include:

- Use of steam for temperature maintenance
- Recompression of recycle stream to system pressure
- Bundling of multiple membranes (10+) into one housing

The benefit of the membrane form factor is that a large amount of material can be processed without taking up an excessive amount of refinery space. An example system was developed for a high capacity C3 splitter. The splitter generates 233,000 ton/year of paraffin rich bottoms with an average composition of 90% propane at 200 psig. A membrane system is used to recover olefin from this bottoms stream. The system recycles 58,000 ton/year of permeated material to the column, enriched in propylene to 54%. The system simultaneously generates 175,000 ton/year of HD05 propane. A proposed membrane bank contains 8 housings each holding 42 membrane cartridges. This design takes up a 28′ by 28′ area of refinery floor, or 784 ft^2. In the projected design the membranes are oriented horizontally and the housings are not stacked. Vertical orientation and stacking can both be implemented in order to reduce the footprint of the full scale system.

References

Burns, R.L., Koros, W.J., Journal of Membrane Sci., 2003, 211, 299,

Hamza, A.A., Towe, I.G., Keyvani, M., Stable Facilitated Transport Membrane for Olefin/Paraffin Separation, Paper presented at the 25th Ethylene Producers' Conference, AIChE, San Antonio, TX, April 29 – May 2, 2013

Ho, W. S., Dalrymple, D. C. Facilitated Transport of Olefins in Ag+-containing Polymer Membranes. J. Membr. Sci. 1994, 91, 13

Hsiue, G.-H., Yang, J.-S., J. Membr. Sci. 1993, 82, 117.

Merkel,T.C., Blanc, R., Ciobanu, I., Firat, B., Suwarlim, A., Zeid, J. Membr. Sci., 2013, 447, 177.

Merkel,T.C., Blanc, R., Zeid, J., Suwarlim, A., Firat, B., Wijmans, J.G., Asaro, M., Greene, M., Separation of Olefin/Paraffin Mixtures with Carrier Facilitated Membranes, Final Report for DOE ProjectDE-FC36-06GO14151,2007.

Park, J. H., Park, K. H., Kim, D. P., J. Ind. Eng. Chem., 2007, 13 (1) 27.

Peinemann, K. V., Shukla, S. K., Schossig, M., Proc ICOM 90, Chicago, IL, 1990, 792

Pinnau, I., L.G. Toy "Solid Polymer Electrolyte Membrane for Olefin/Paraffin Separation", .J Mem. Sci., Vol. 184, 39-48 March, 2001

Teramoto, M., Takeuchi, N., Maki, T., Matsuyama, H.,Sep. Purif. Technol., 2002, 28, 117.

AICHE PAPER NUMBER 124E

COMPUTATIONAL FLUID DYNAMICS BASED DESIGN OF A NOVEL REACTOR TECHNOLOGY FOR THE OXIDATIVE COUPLING OF METHANE

Laurien A. Vandewalle
Laboratory for Chemical Technology, Ghent University

Patrice Perreault
Laboratory for Chemical Technology, Ghent University

Kevin M. Van Geem
Laboratory for Chemical Technology, Ghent University

Guy B. Marin
Laboratory for Chemical Technology, Ghent University

Prepared for Presentation at the 2017 Spring National Meeting
San Antonio, Texas, March 28, 2017

COMPUTATIONAL FLUID DYNAMICS BASED DESIGN OF A NOVEL REACTOR TECHNOLOGY FOR THE OXIDATIVE COUPLING OF METHANE

Laurien A. Vandewalle
Laboratory for Chemical Technology, Ghent University

Patrice Perreault
Laboratory for Chemical Technology, Ghent University

Kevin M. Van Geem
Laboratory for Chemical Technology, Ghent University

Guy B. Marin
Laboratory for Chemical Technology, Ghent University

Abstract:

Oxidative coupling of methane (OCM) is considered as one of the most promising direct routes to convert methane to ethylene and higher hydrocarbons. However, two key challenges have to be addressed before OCM can be considered as an alternative gas-to-chemical technology, namely the low yields of ethylene and what to do with the substantial heat release of the reaction. Both these challenges can be overcome in the reactor technology that is proposed in this work, i.e. the gas-solid vortex reactor (GSVR) developed at the Laboratory for Chemical Technology. The flow characteristics and bed properties inside the GSVR have already been studied experimentally for a wide range of operating conditions. In this work, non-reactive computational fluid dynamic simulations are combined with 0D/1D reactive simulations to determine the operating conditions and reactor design necessary to achieve C2 yields of 5-15 %.

Introduction

Many innovative catalytic technologies have been developed in the past decade as a response to the world's rapidly growing demand for a more efficient and sustainable exploitation of energy and material resources [1, 2]. The decreasing crude oil reserves, the large amount of natural gas resources (shale gas) and the renewability of methane (biogas) have created a strong economic interest in developing processes that allow methane conversion into more valuable products [3, 4]. Commercial technologies to convert methane to higher hydrocarbons (e.g. Fischer-Tropsch synthesis) are mainly indirect, i.e. they require the production of syngas as a first step, which represents an inherent inefficiency. On the other hand, direct activation of methane and conversion into other useful products still remains one of the most challenging topics that the catalysis community faces nowadays [5].

One of the most promising direct routes to convert methane to ethylene and higher hydrocarbons is the oxidative coupling of methane (OCM). Ever since the pioneering work of Keller and Bhasin [6], OCM continues to attract both industrial as well as academic interest. Although the benefits of OCM have been known for over 30 years, a crucial issue among researchers still remains to find a viable catalyst with the performance needed for commercialization of the process [7]. Furthermore an appropriate reactor technology needs to be developed to address the low yields of ethylene and the strong exothermicity of the reaction [8, 9]. The main challenges to be overcome before OCM can be used as an alternative to steam cracking for the production of ethylene and higher olefins are hence twofold, with catalyst development on the one hand and novel reactor design on the other [10, 11].

Siluria Technologies recently announced the first commercial OCM process, in which a series of nanowire catalysts is used that operate at process temperatures below 600 °C [12]. Although Siluria's patent application reveals that the single-pass C_2 yield does not meet the target of 25 %, the catalysts are durable over long periods of operation at low process temperatures [13]. Catalytic packed bed reactors are used in Siluria's demonstration plant, as is the case for the majority of all laboratory-scale OCM reactors [11]. Because of the high exothermicity of the OCM reaction, thermal control of the reactor is an important issue. Most laboratory-scale experimental set-ups use small-diameter tubes and run with very dilute mixtures and low methane conversions [14-16], which is of course not practical for an actual large-scale process. The fluidized bed reactor has been designated as the optimal reactor concept for OCM by different authors [8, 9, 17, 18]. This is mainly because of its beneficial heat transfer characteristics, that cannot be achieved in any type of packed bed reactor.

From the above considerations, it may be clear that some effort is still required before OCM can be considered as an alternative to steam cracking for the production of ethylene and other olefins. Although the door toward successful commercialization has already been opened by Siluria Technologies, further investigation to enhance ethylene yields is still necessary. Catalyst design as well as the development of novel reactor technologies are important to this purpose. In the present work, the focus is on the latter. The gas-solid vortex reactor (GSVR), developed at the Laboratory for Chemical Technology (UGent) is proposed as an excellent reactor choice to demonstrate the OCM process.

Gas-solid vortex reactor technology

Gas-solid fluidized beds (FBs) are known for their enhanced heat and mass transfer characteristics and are therefore widely used for both reactive and non-reactive processes in the chemical industry. The efficiency of heat and mass transfer in these FBs is determined by the relative velocity, the so-called slip velocity, between the gas and the solid phase. As a consequence of the balance between the drag force and the gravitational force, the slip velocity in a conventional gravitational FB is limited by the terminal free-fall velocity of the particles [19]. Higher gas velocities give rise to the formation of bubbles and slugs, which negatively affect the heat and mass transfer efficiency as the extensive gas bypass decreases the gas-solid contact. If the gas flow rate is increased even more, particle entrainment is unavoidable. Higher gas throughput, more uniform fluidization, higher slip velocities, and hence better heat and mass transfer, can be achieved by using a centrifugal force instead of the gravitational force [20-22]. A centrifugally fluidized bed thus emerges as an excellent candidate for process intensification.

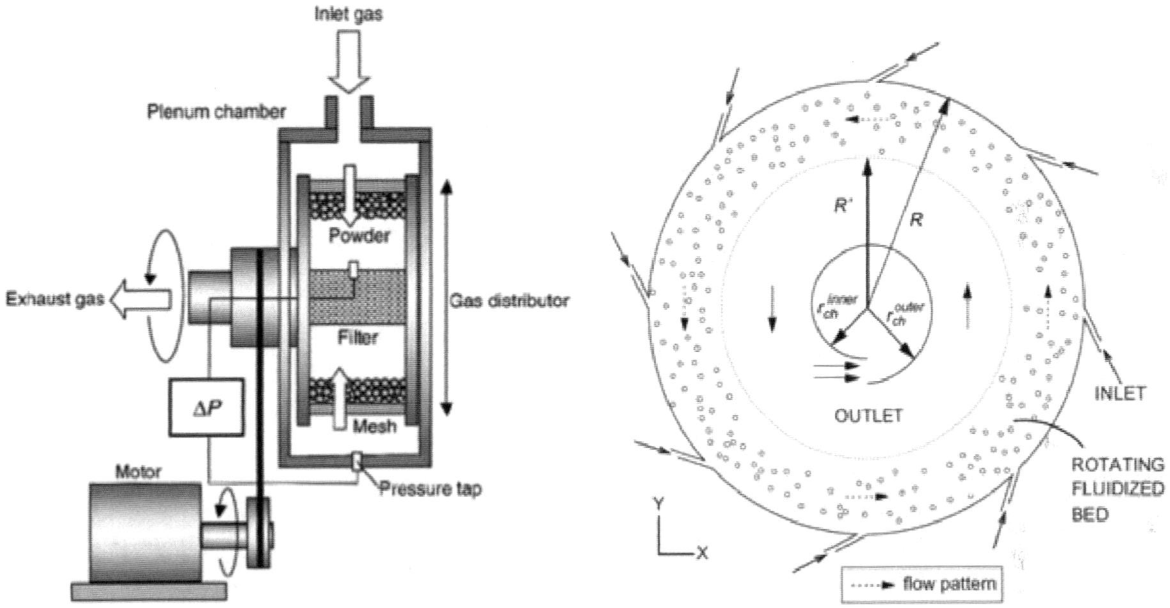

Figure 1. Left: Schematic overview of a rotating fluidized bed (RFB) [23]. Right: Schematic representation of a gas-solid vortex unit (GSVU) [24].

Figure 1 shows two possibilities to achieve a centrifugal fluidized bed: using a rotating fluidized bed (RFB), where the particles are set in motion by rotating the operating vessel itself, or using a gas-solid vortex unit (GSVU), where the particles are introduced in a swirling flow field of azimuthally injected gas in a static vessel. In a RFB, the tangential and radial velocity components can be varied independently, as the rotational velocity of the vessel and the injected gas flow rate can be controlled in a separate way. The downside of using RFBs however is the use of mechanically moving parts which imply the risk for mechanical abrasion. In a GSVU the fluidizing gas is injected tangentially through multiple gas inlet slots in the cylindrical wall of the fluidization chamber. The swirling gas transfers tangential momentum to the particles, which in turn start rotating and experience an outward centrifugal force. A fluidized state is reached when the radially inward drag force exerted by the gas overcomes the apparent weight of the solids in the centrifugal field. In contrast to the RFB, it is impossible to control the particle velocity components in a GSVU independently. Nevertheless, the GSVU is preferred to the RFB as the absence of moving parts makes the unit more appropriate for scale-up [24].

Several industrially relevant processes have been suggested for implementation in a gas-solid vortex reactor (GSVR): fluid catalytic cracking (FCC) [25], coating of cohesive particles [26], gas adsorption [27], drying [28], gasification [29], combustion [30] and pyrolysis [31]. At this moment, the first actually reactive GSVR unit for biomass fast pyrolysis is being tested at the Laboratory for Chemical Technology. Because of its excellent heat transfer characteristics and high gas throughput, this reactor also emerges as an ideal technology for the oxidative coupling of methane (OCM).

GSVR setup at the Laboratory for Chemical Technology

A schematic view of the GSVR setup at the Laboratory for Chemical Technology (LCT) is shown in Figure 2 and Figure 3. It basically consists of a cylindrical unit positioned along a vertical axis with eight gas injection slots of 1 mm width, equally distributed over the circumferential wall and tangentially inclined at a 10° angle. A reactor diameter of 80 mm and a length-to-diameter of 0.25, i.e. a reactor length of 15 mm, are defined. The diverging exhaust aims at minimizing the gas backflow and at preserving the swirling flow in the throat. The diverging exhaust wall profile follows an analytical solution for strongly swirling jets. The design of the bottom end wall and exhaust profiles was based on preliminary calculations and CFD simulations as described by Gonzalez-Quiroga et al. [32]. The design of the reactor is such that the inlet slots are easy to replace, so that the number and the angle of the inlet slots can be altered for parametric studies etc. For a given gas flow rate, the thickness of the injection slots determines the magnitude of the gas injection velocity, while the injection angle determines its radial and tangential components. The actual GSVR is operated in semi-batch mode. Solid particles are already present in the inner chamber at the beginning of an experiment. Then the hot fluidizing gas is introduced until a pseudo-steady-state swirling gas-particle flow is established in the reactor.

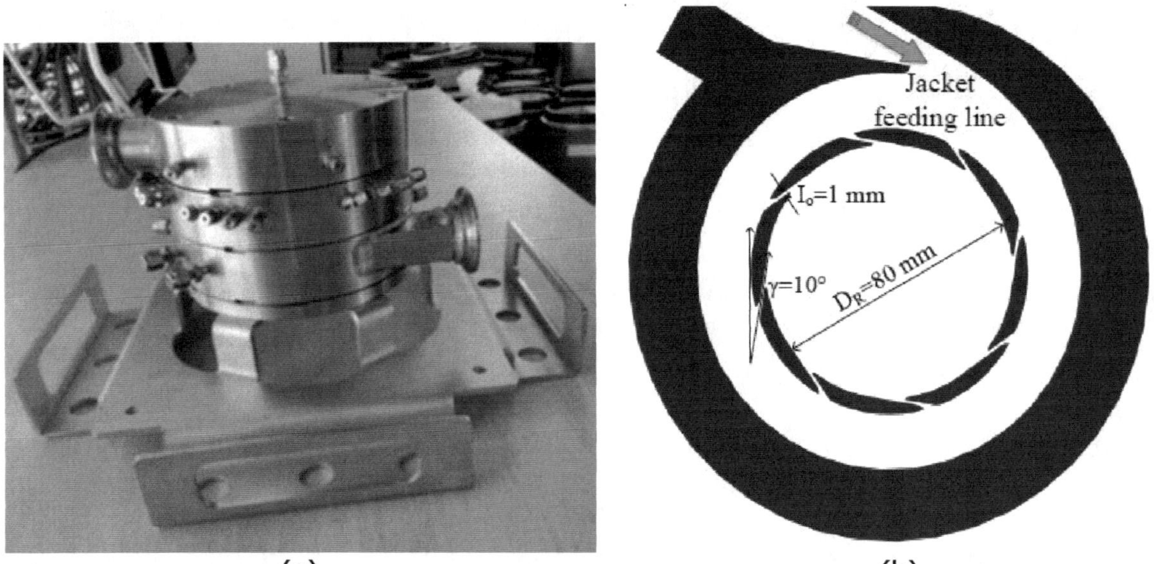

(a) (b)

Figure 2. (a) Experimental setup at the Laboratory for Chemical Technology, and (b) schematic representation of the gas feeding line, jacket and reactor chamber at a horizontal cutting plane.

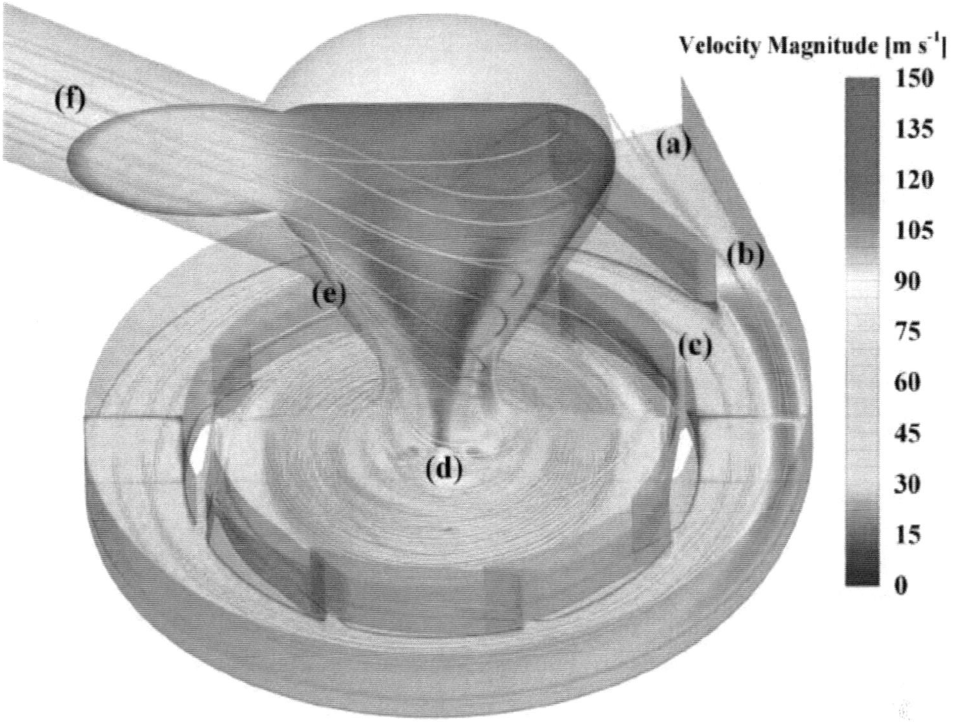

Figure 3. Flow pattern for gas-only flow in the Gas-Solid Vortex Reactor obtained via computational fluid dynamic simulations. Contours of velocity magnitude are shown in the cutting planes, and the streamlines are colored in grey. (a) Gas inlet, (b) connection of gas inlet to the jacket, (c) gas inlet slots, (d) profiled bottom end wall, (e) diverging exhaust, (f) gas outlet.

CFD simulations

Operating conditions (gas flow rates, particle density, temperature) and geometric features (number and angle of inlet slots) are found to have a large effect on the bed stability in the GSVR. Based on CFD simulations we were able to derive a range of operating conditions for which we can obtain a stable bed with the current geometry. The results of one such simulation are shown below. Also experimentally we were able to obtain stable rotating beds over a wide range of operating conditions.

For the presented case nitrogen was used as carrier gas and fed to the GSVR at a flow rate of 10 g/s at 923 K. The properties of the solid particles were chosen in agreement with the OCM catalyst particles that will be used in the GSVR: 1 mm diameter and a density of 2300 kg/m^3. Although we were able to obtain a stable bed inside the reactor (see Figure 4a,b), it is clear from Figure 4c that a lot of gas is not retained in the bed but just flows through, hence resulting in a small contact time between the gas and the particles. This is most probably not the ideal scenario if we want to use the reactor for OCM, but by finetuning the operating conditions or using an alternative design, it is possible to obtain the flow conditions required to achieve reasonable conversions and selectivities.

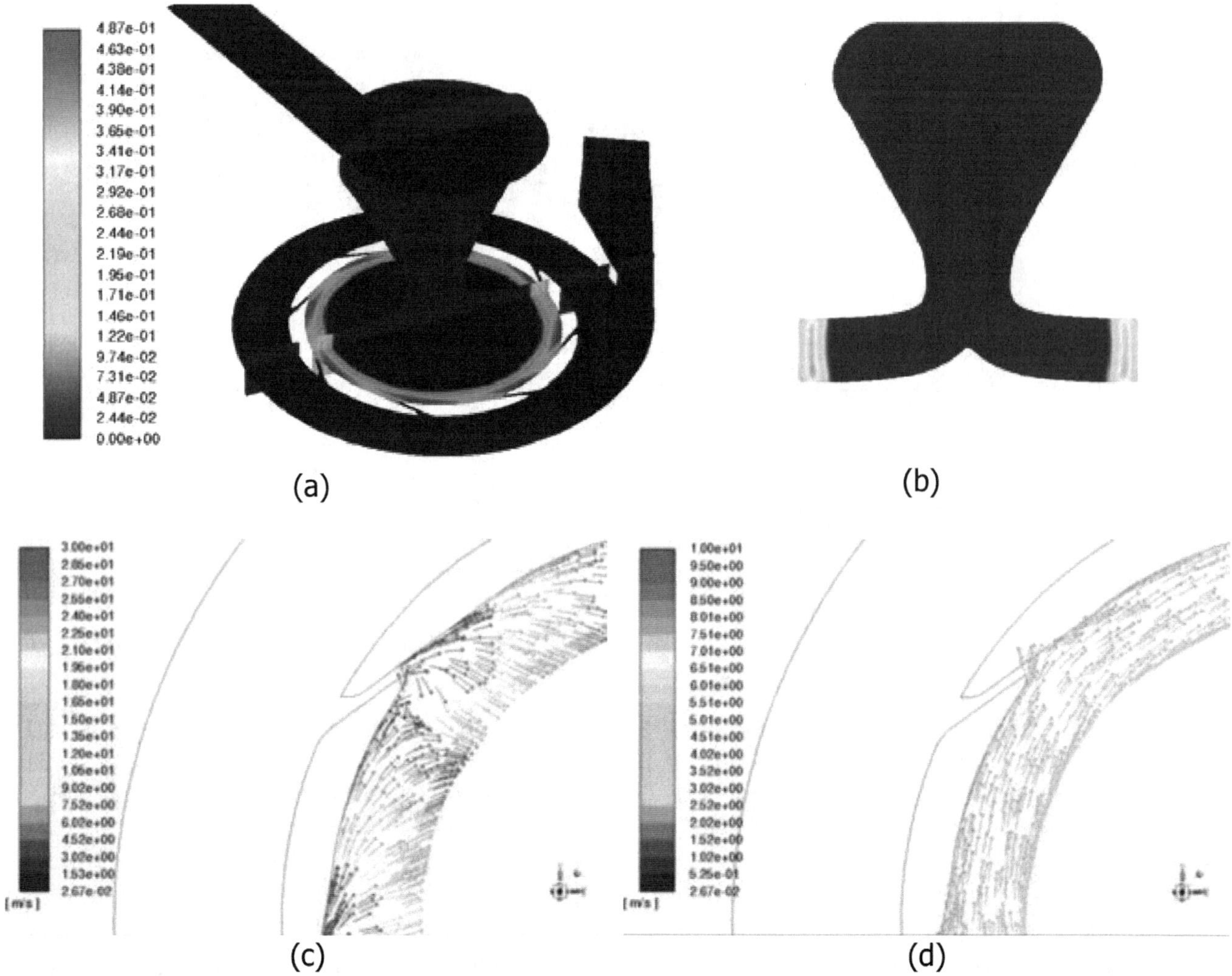

Figure 4. (a) Contours of particle volume fraction after establishing a pseudo-steady particle bed; (b) idem, on a vertical cutting plane; (c) Vectors of gas velocity, colored by gas velocity magnitude [m/s]; and (d) vectors of particle velocity magnitude, colored by particle velocity magnitude [m/s].

Oxidative coupling of methane

Essential for defining and controlling the GSVR conditions is to determine the residence times and bed properties necessary to obtain reasonable conversions and selectivities for OCM. Therefore, some 0D/1D simulations were performed for different catalysts (La_2O_3/CaO, Mn/Na_2WO_4/SiO_2, Sn-Li/MgO and Sr/La_2O_3). As the flow behavior in the GSVR is far from ideal, we can use the "tank-in-series" model to imitate the GSVR in our kinetic study. The idea behind this model is that, given the residence time distribution (RTD) in the actual reactor, it is possible to approximate it by a cascade of CSTR's. The lower the number of CSTR's in the series, the better the mixing, and in the limit of $n \to \infty$, ideal plug flow is approximated. The "tank-in-series" model comes down to solving the following equations, for every CSTR k in the series of n reactors:

$$F^k C_i^k - F^{k-1} C_i^{k-1} = V^k[\epsilon_b R_{i,g}^k + R_{i,s}^k(1-\epsilon_b)\rho_{cat} a_{cat}] \tag{1}$$

$$c_{p,g}^{k-1}(T^k - T^{k-1}) = -\sum_{i=1}^{N_g} h_i^k (Y_i^{k-1} - Y_i^k) \tag{2}$$

In these equations C_i^k is the concentration of species i in reactor k $[\mathrm{mol\ m_g^{-3}}]$; F^k is the volumetric flow rate leaving reactor k $[\mathrm{m^3\ s^{-1}}]$; V^k is the volume of reactor k $[\mathrm{m^3}]$; ϵ_b is the bed voidage $[\mathrm{m_g^3\ m^{-3}}]$; $R_{i,g}^k$ is the net rate of formation of species i in reactor k due to gas phase reactions $[\mathrm{mol\ m_g^{-3}\ s^{-1}}]$; $R_{i,s}^k$ is the net rate of formation of species i in reactor k due to surface reactions $[\mathrm{mol\ m_{cat}^{-2}\ s^{-1}}]$; ρ_{cat} is the catalyst density $[\mathrm{kg_{cat} m_{cat}^{-3}}]$; a_{cat} is the specific surface area of the catalyst $[\mathrm{m_{cat}^2\ kg_{cat}^{-1}}]$; $c_{p,g}^k$ is the specific heat capacity of the gas phase in reactor k $[\mathrm{J\ kg^{-1}\ K^{-1}}]$; T^k is the temperature in reactor k $[\mathrm{K}]$; h_i is the specific enthalpy of species i in reactor k $[\mathrm{J\ kg^{-1}}]$; Y_i^k is the mass fraction of species i in reactor k $[-]$; and N_g is the number of species $[-]$. It is assumed that the surface is in equilibrium with the gas phase. The model equations are solved using an in-house code based on Cantera [33]. A detailed microkinetic model consisting of 39 gas-phase reactions and 26 catalytic reactions [34-37] is used for Sr/La_2O_3 and Sn-Li/MgO, while a comprehensive model with 10 reactions is used for La_2O_3/CaO [38] and $Mn/Na_2WO_4/SiO_2$ [39]. The physical properties of the investigated catalysts are shown in Table 1.

Table 1. Physical properties of investigated catalysts [34].

	Sr/La_2O_3	Sn-Li/MgO	La_2O_3/CaO	$Mn/Na_2WO_4/SiO_2$
Specific surface area $[m^2{}_{cat}/kg_{cat}]$	2000	2800	2000	2000
Density $[kg_{cat}/m^3{}_{cat}]$	2300	2300	2300	2300

For a process like OCM, where gas phase and surface reactions occur simultaneously, there are two important space times, one related to the homogeneous gas phase reactions, the other related to the heterogeneous surface reactions. The results below are only shown as a function of the gas phase residence time, which also implies a proportional change in the catalytic space time.

Figure 5 shows the methane conversion and selectivity towards C2 (ethane + ethylene) as a function of mean residence time in a plug flow reactor. The four different catalysts mentioned earlier are compared. The results of Figure 5 are obtained using an isothermal reactor model, so without solving Eq. (2). It follows that the highest C2 yields, i.e. C2 selectivity times CH_4 conversion, are possible using Sr/La_2O_3, so this type of catalyst has been studied.

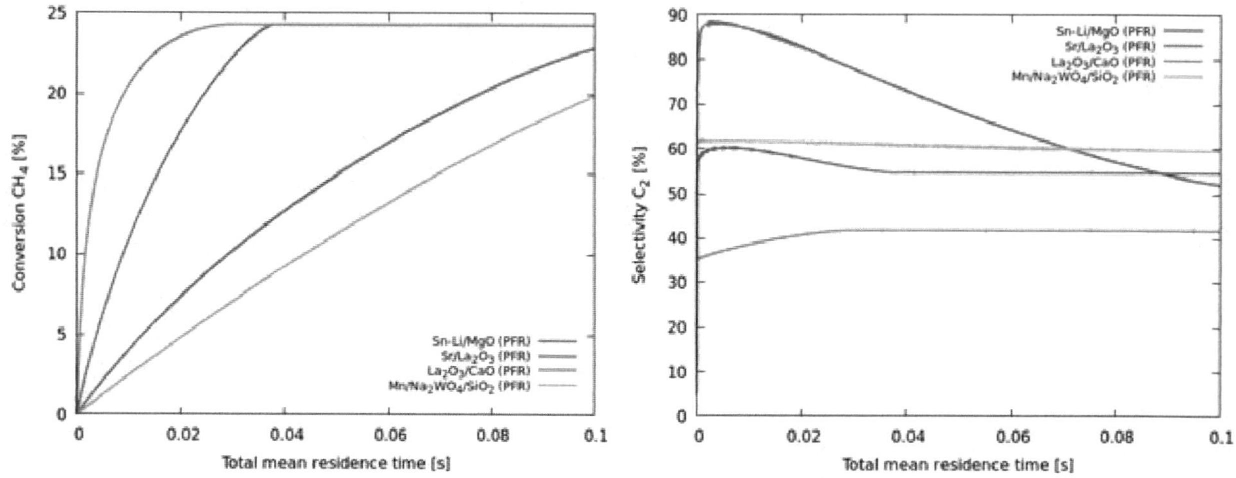

(a) (b)

Figure 5. (a) Methane conversion, and (b) C2 selectivity as a function of mean residence time in an isothermal plug flow reactor ($n \to \infty$), for different catalysts (CH_4/O_2=4, T=1023K, p=1.1bara, ϵ_b=0.6).

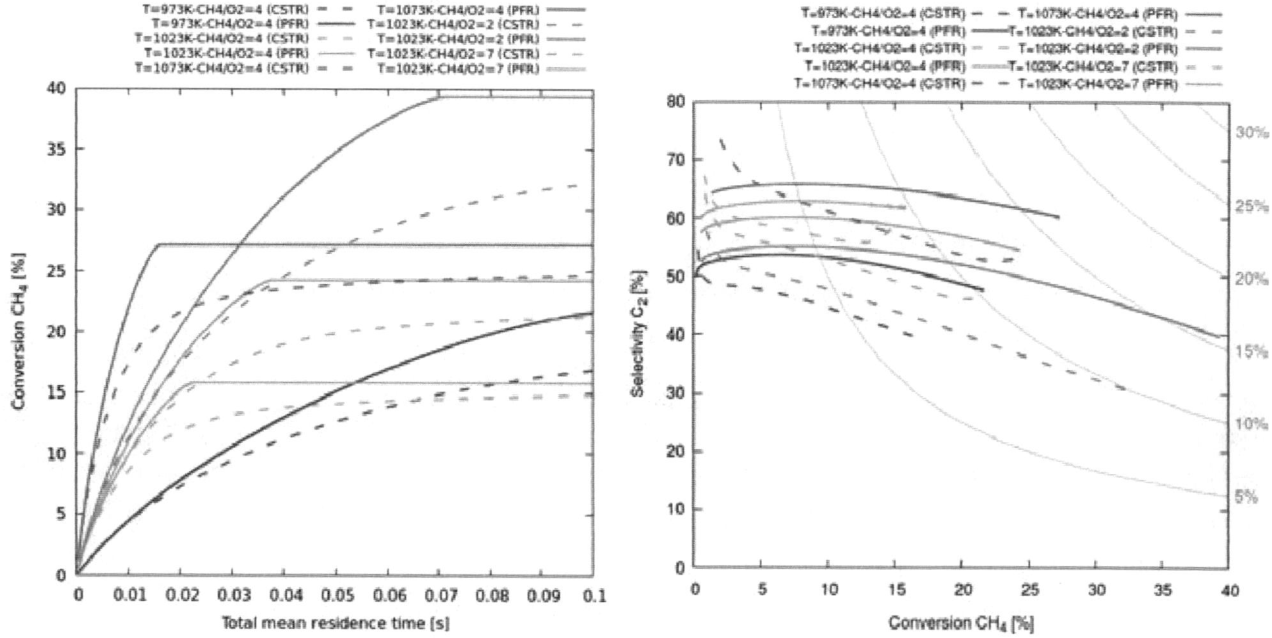

(a) (b)

Figure 6. (a) Methane conversion as a function of residence time, and (b) C2 selectivities as a function of CH_4 conversion in isothermal CSTR's and PFR's with Sr/La_2O_3 catalyst for different operating temperatures and CH_4/O_2 ratios (p=1.1bara, ϵ_b=0.6). Lines of constant C2 yield are indicated in grey in the right figure.

Figure 6 shows the methane conversion and C2 selectivity in an isothermal plug flow reactor and perfectly stirred reactor with Sr/La_2O_3 catalyst for different operating conditions. It follows from the figure that at low conversion, the C2 selectivity in a perfectly stirred reactor is higher than in a plug flow reactor, while the opposite holds for higher conversions. Conversion is always larger in a plug flow reactor. The behavior of the GSVR is in between the plug flow and perfectly stirred regime. The lower the CH_4/O_2 ratio, the higher the conversion but the lower the selectivity. For the same CH_4/O_2 ratio, higher C2 selectivities are found at higher temperatures. From the results in Figure 6 it follows that we should try to design the GSVR such that we can reach residence times of approximately 20-50 ms. In this way, it should be possible to obtain CH_4 conversions of 10-25 % and C2 yields of 5-15 %.

Although Figure 6 seems to speak in favor of the ideal plug flow regime, there are some important practical problems with this type of reactor if it is operated adiabatically instead of isothermally. Figure 7 shows the temperature profile resulting from an adiabatic simulation, i.e. where Eq. (2) is included, with the Sr/La_2O_3 catalyst. Due to the strong exothermicity of the OCM process, the temperature in the reactor rises quickly. In a packed bed reactor (as idealized by the PFR model in the figures), one should take appropriate measures to prevent reaction runaway. As reported earlier by Pak et al. [40], hot spot formation can be observed in packed bed reactors, even for small-diameter tubes. Here is where the excellent heat transfer characteristics in the GSVR come into play. The more the actual reactor resembles a CSTR, or the higher the degree of mixing (or the lower the number of CSTR's in the "tank-in-series" model), the more stable the reactor operation. The continuous mixing of the colder feed with the reacting mixture provides a way to control the temperature, so that it becomes possible to operate the GSVR almost isothermally.

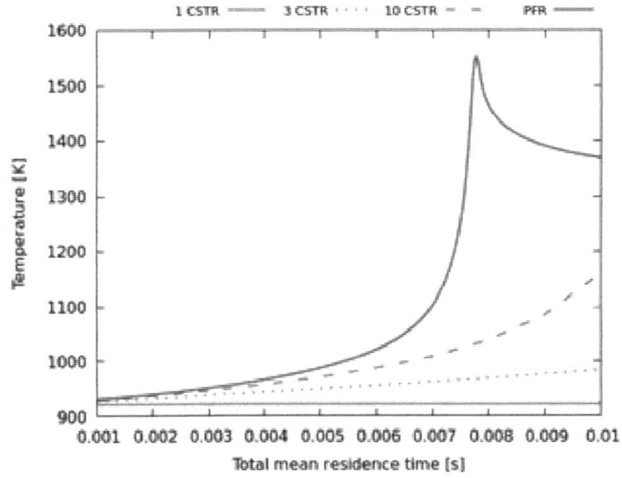

(a)

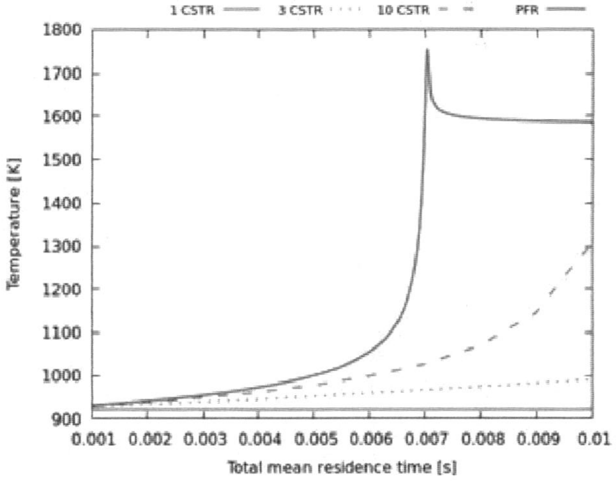

(b)

Figure 7. Temperature [K] as a function of mean residence time on Sr/La_2O_3 catalyst and with different degrees of mixing, i.e. number of CSTR's, in the *adiabatic* "tanks-in-series" model. Inlet/operating conditions: CH_4/O_2=4(a)/2(b), T=923K, p=1.1bara, ϵ_b=0.6.

Conclusions

In this work, a new reactor technology has been studied for the oxidative coupling of methane: the gas-solid vortex reactor (GSVR). Non-reactive computational fluid dynamic simulations were performed to estimate the bed characteristics and flow dynamics for a wide range of operating conditions and reactor designs. Then reactive 0D/1D simulations were performed to evaluate the reactor performance in terms of CH_4 conversions and C2 selectivities. From these reactive simulations it follows that we should design the GSVR in such way that it is possible to reach residence times of approximately 20-50 ms. For Sr/La_2O_3 catalyst, this would give CH_4 conversions of 10-25 % and C2 yields of 5-15 %.

References

[1] M. Maestri, A. Cuoci, Coupling CFD with detailed microkinetic modeling in heterogeneous catalysis, Chemical Engineering Science 96 (2013) 106-117.

[2] J.K. Norskov, T. Bligaard, J. Rossmeisl, C.H. Christensen, Towards the computational design of solid catalysts, Nat Chem 1 (2009) 37-46.

[3] E.V. Kondratenko, M. Baems, Oxidative Coupling of Methane, in: G. Ertl, F. Schüth, J. Weitkamp (Eds.) Handbook of Heterogeneous Catalysis Wiley-VCH2008, pp. 3010-3023.

[4] E.V. Kondratenko, M. Schlüter, M. Baerns, D. Linke, M. Holena, Developing catalytic materials for the oxidative coupling of methane through statistical analysis of literature data, Catal. Sci. Technol. 5 (2015) 1668-1677.

[5] A.I. Olivos-Suarez, À. Szécsényi, E.J.M. Hensen, J. Ruiz-Martinez, E.A. Pidko, J. Gascon, Strategies for the Direct Catalytic Valorization of Methane Using Heterogeneous Catalysis: Challenges and Opportunities, ACS Catalysis 6 (2016) 2965-2981.

[6] G.E. Keller, M.M. Bhasin, Synthesis of Ethylene via Oxidative Coupling of Methane, Journal of Catalysis 73 (1982) 9-19.

[7] A. Galadima, O. Muraza, Revisiting the oxidative coupling of methane to ethylene in the golden period of shale gas: A review, Journal of Industrial and Engineering Chemistry 37 (2016) 1-13.

[8] S. Jašo, H. Arellano-Garcia, G. Wozny, Oxidative coupling of methane in a fluidized bed reactor: Influence of feeding policy, hydrodynamics, and reactor geometry, Chemical Engineering Journal 171 (2011) 255-271.

[9] L. Mleczko, U. Pannek, V.M. Niemi, J. Hiltunen, Oxidative Coupling of Methane in a Fluidized-Bed Reactor over a Highly Active and Selective Catalyst, Industrial & Engineering Chemistry Research 35 (1996) 54-61.

[10] R.J. Kee, C. Karakaya, H. Zhu, Process intensification in the catalytic conversion of natural gas to fuels and chemicals, Proceedings of the Combustion Institute (2016).

[11] C. Karakaya, R.J. Kee, Progress in the direct catalytic conversion of methane to fuels and chemicals, Progress in Energy and Combustion Science 55 (2016) 60-97.

[12] E.C. Scher, J.M. Cizeron, W.P. Schammel, A. Tkachenko, J. Gamoras, D. Karshtedt, G. Nyce, F.R. Zurcher, Method for the oxidative coupling of methane in the presence of a nanowire catalyst, Google Patents, 2015.

[13] W.P. Schammel, J. Wolfenbarger, M. AJINKYA, J. MCCARTY, J.M. Cizeron, S. WEINBERGER, J.D. EDWARDS, D. SHERIDAN, E.C. Scher, J. Mccormick, Oxidative coupling of methane systems and methods, Google Patents, 2013.

[14] B. Zohour, D. Noon, S. Senkan, New Insights into the Oxidative Coupling of Methane from Spatially Resolved Concentration and Temperature Profiles, ChemCatChem 5 (2013) 2809-2812.

[15] R. Horn, K. Williams, N. Degenstein, L. Schmidt, Syngas by catalytic partial oxidation of methane on rhodium: Mechanistic conclusions from spatially resolved measurements and numerical simulations, Journal of Catalysis 242 (2006) 92-102.

[16] A. Donazzi, M. Maestri, B.C. Michael, A. Beretta, P. Forzatti, G. Groppi, E. Tronconi, L.D. Schmidt, D.G. Vlachos, Microkinetic modeling of spatially resolved autothermal CH4 catalytic partial oxidation experiments over Rh-coated foams, Journal of Catalysis 275 (2010) 270-279.

[17] L. Mleczko, M. Baerns, Catalytic oxidative coupling of methane—reaction engineering aspects and process schemes, Fuel Processing Technology 42 (1995) 217-248.

[18] F.M. Dautzenberg, J.C. Schlatter, J.M. Fox, J.R. Rostrup-Nielsen, L.J. Christiansen, Catalyst and reactor requirements for the oxidative coupling of methane, Catalysis Today 13 (1992) 503-509.

[19] G.F. Froment, K.B. Bischoff, J. De Wilde, Chemical reactor analysis and design, 3rd ed.2011.

[20] J. De Wilde, A. de Broqueville, Rotating fluidized beds in a static geometry: Experimental proof of concept, AIChE Journal 53 (2007) 793-810.

[21] J. De Wilde, A. de Broqueville, Experimental investigation of a rotating fluidized bed in a static geometry, Powder Technology 183 (2008) 426-435.

[22] K. Niyogi, M.M. Torregrosa, M.N. Pantzali, G.J. Heynderickx, G.B. Marin, Experimentally validated numerical study of gas-solid vortex unit hydrodynamics, Powder Technology 305 (2017) 794-808.

[23] H. Nakamura, S. Watano, Fundamental particle fluidization behavior and handling of nano-particles in a rotating fluidized bed, Powder Technology 183 (2008) 324-332.
[24] A. Dutta, R.P. Ekatpure, G.J. Heynderickx, A. de Broqueville, G.B. Marin, Rotating fluidized bed with a static geometry: Guidelines for design and operating conditions, Chemical Engineering Science 65 (2010) 1678-1693.
[25] W. Rosales Trujillo, J. De Wilde, Fluid catalytic cracking in a rotating fluidized bed in a static geometry: a CFD analysis accounting for the distribution of the catalyst coke content, Powder Technology 221 (2012) 36-46.
[26] P. Eliaers, A. de Broqueville, A. Poortinga, T. van Hengstum, J. De Wilde, High-G, low-temperature coating of cohesive particles in a vortex chamber, Powder Technology 258 (2014) 242-251.
[27] R.W. Ashcraft, J. Kovacevic, G.J. Heynderickx, G.B. Marin, Assessment of a Gas–Solid Vortex Reactor for SO2/NOxAdsorption from Flue Gas, Industrial & Engineering Chemistry Research 52 (2013) 861-875.
[28] L.M. Kochetov, B.S. Sazhin, E.A. Karlik, Hydrodynamics and heat exchange in vortex drying chambers, Chemical and Petroleum Engineering 5 (1969) 688-690.
[29] N. Staudt, A. De Broqueville, W. Rosales Trujillo, J. De Wilde, Low-Temperature Pyrolysis and Gasification of Biomass: Numerical Evaluation of the Process Intensification Potential of Rotating- and Circulating Rotating Fluidized Beds in a Static Fluidization Chamber, International Journal of Chemical Reactor Engineering 9 (2011) 1-31.
[30] E.P. Volchkov, N.A. Dvornikov, V.V. Lukashov, R.K. Abdrakhmanov, Investigation of the flow in the vortex chamber with centrifugal fluidizing bed with and without combustion, Thermophysics and Aeromechanics 20 (2013) 663-668.
[31] R.W. Ashcraft, G.J. Heynderickx, G.B. Marin, Modeling fast biomass pyrolysis in a gas–solid vortex reactor, Chemical Engineering Journal 207-208 (2012) 195-208.
[32] A. Gonzalez Quiroga, H.-H. Carstensen, K.M. Van Geem, G.B. Marin, Design of a gas-solid vortex reactor demonstration unit for the fast pyrolysis of lignocellulosic biomass, 21st International Symposium on Analytical and Applied Pyrolysis (PYRO 2016)Nancy (France), 2016, pp. 181.
[33] D.G. Goodwin, H.K. Moffat, R.L. Speth, Cantera: An Object-oriented Software Toolkit for Chemical Kinetics, Thermodynamics, and Transport Processes (Version 2.3.0), http://www.cantera.org, 2017.
[34] V.I. Alexiadis, J.W. Thybaut, P.N. Kechagiopoulos, M. Chaar, A.C. Van Veen, M. Muhler, G.B. Marin, Oxidative coupling of methane: catalytic behaviour assessment via comprehensive microkinetic modelling, Applied Catalysis B: Environmental 150-151 (2014) 496-505.
[35] Q. Chen, P.M. Couwenberg, G.B. Marin, Effect of pressure on the oxidative coupling of methane in the absence of catalyst, AIChE Journal 40 (1994) 521-535.
[36] P.M. Couwenberg, Q. Chen, G.B. Marin, Kinetics of a Gas-Phase Chain Reaction Catalyzed by a Solid: The Oxidative Coupling of Methane over Li/MgO-Based Catalysts, Industrial & Engineering Chemistry Research 35 (1996) 3999-4011.
[37] P.N. Kechagiopoulos, J.W. Thybaut, G.B. Marin, Oxidative Coupling of Methane: A Microkinetic Model Accounting for Intraparticle Surface-Intermediates Concentration Profiles, Industrial & Engineering Chemistry Research 53 (2014) 1825-1840.
[38] Z. Stansch, L. Mleczko, M. Baerns, Comprehensive Kinetics of Oxidative Coupling of Methane over the La2O3/CaO Catalyst, Industrial & Engineering Chemistry Research 36 (1997) 2568-2579.
[39] M. Daneshpayeh, A. Khodadadi, N. Mostoufi, Y. Mortazavi, R. Sotudeh-Gharebagh, A. Talebizadeh, Kinetic modeling of oxidative coupling of methane over Mn/Na2WO4/SiO2 catalyst, Fuel Processing Technology 90 (2009) 403-410.
[40] S. Pak, J.H. Lunsford, Thermal effects during the oxidative coupling of methane over Mn/Na2WO4/SiO2 and Mn/Na2WO4/MgO catalysts, Applied Catalysis A: General 168 (1998) 131-137.

AIChE Paper 124f 479756

Kinetic Study of Thermal Decomposition of N,N-Diethylhydroxylamine (DEHA) in Steam Crackers

Hangyao Wang, *Associate Research Scientist, Dow Chemical Company*

George Bellos, *Research Scientist, Dow Chemical Company*

Stefan Korf, *Principle Research Scientist, Dow Chemical Company*

Liwei Li, *Senior Engineer, Dow Chemical Company*

Viet Pham, *Associate Research Scientist, Dow Chemical Company*

Mark Siddoway, *Research Fellow, Dow Chemical Company*

Mark Brayden, *Research Fellow, Dow Chemical Company*

Marcos Martinez, *Associate Director, Dow Chemical Company*

Brien Stears, *Associate Director, Dow Chemical Company*

Prepared for Presentation at the 2017 Spring National Meeting

San Antonio, Texas, March 28, 2017

Kinetic Study of Thermal Decomposition of N,N-Diethylhydroxylamine (DEHA) in Steam Crackers

Hangyao Wang, *Associate Research Scientist, Dow Chemical Company*

George Bellos, *Research Scientist, Dow Chemical Company*

Stefan Korf, *Principle Research Scientist, Dow Chemical Company*

Liwei Li, *Senior Engineer, Dow Chemical Company*

Viet Pham, *Associate Research Scientist, Dow Chemical Company*

Mark Siddoway, *Research Fellow, Dow Chemical Company*

Mark Brayden, *Research Fellow, Dow Chemical Company*

Marcos Martinez, *Associate Director, Dow Chemical Company*

Brien Stears, *Associate Director, Dow Chemical Company*

Abstract: Nitrogen compounds such as N,N-Diethylhydroxylamine (DEHA) are used as antifouling additives in the distillation section of crude oil refineries and ethylene plants. These compounds are excellent radical scavengers and are proven to prevent fouling from radical reaction mechanisms. However, the DEHA might end up in ethylene plant furnaces. These are run at elevated temperatures as high as 850 °C, which may lead to the thermal decomposition of the nitrogen additives. Possible decomposition products are NO_x, which pose a serious safety concern because NO_x could lead to gums and/or ammonium nitrite and nitrate in the cold box of an ethylene plant.

In this study, we developed the thermal decomposition kinetics pathways of DEHA and diethylamine (DEA), which is a decomposition product of DEHA. Interestingly, two distinct pathways exist - one producing NO and the other producing NH_3. The kinetics, together with thermodynamics, were used to study the decomposition products of the DEHA and DEA under different temperatures by employing an in-house developed kinetic modeling platform. Temperature plays an important role in determining the major decomposition products of DEHA and DEA. While NO is one of the major decomposition products of pure DEHA, the DEHA decomposition products are dramatically different when cracking with propane. Only at lower temperatures (below 500 °C) is appreciable NO produced from DEHA decomposition with propane cracking, but NH_3 is essentially the only major product at higher temperatures (above 500 °C). NH_3 is very stable once formed and may potentially react with N_2O_3 to produce ammonium nitrite and accumulate in the cold box. If NO is fed to a furnace, the majority will be converted to N_2O, N_2O_3, and NH_3 inside the furnace, but N_2O can be converted back to NO in a quench tower and compression section if some free oxygen is available. The nitrogen compounds NO_x can possibly condense in the cold box and react with dienes such as butadiene, eventually forming so-called NO_x gum. They can also react with NH_3 to form ammonium nitrite. Both NO_x gum and ammonium nitrite pose severe process safety issues and need to be carefully monitored.

Introduction

Nitrogen compounds such as N,N-Diethylhydroxylamine (DEHA) are used as antifouling additives in the distillation section of crude oil refinery systems and ethylene plants. DEHA is an excellent radical scavenger and is proven to prevent fouling from radical reaction mechanisms. However, DEHA might end up in ethylene furnaces, and these are run at elevated temperatures as high as 850 °C, which may lead to thermal decomposition of the nitrogen additives. This poses several questions regarding the outcome of DEHA addition to the steam cracking process such as: (1) Does DEHA decompose during steam cracking? (2) Does it decompose in the radiant coils or downstream with longer residence times? (3) If so, what are the decomposition products? In particular, does it produce NO_x?

The question regarding NO_x formation is especially important due to process safety consideration. For example, the incident that happened in the BASF Port Arthur naphtha cracker was discovered to be caused by decomposition-triggered detonation of NO_x gums in the cold box[1]. NO_x gums are formed when conjugated diolefins such as butadiene react with NO_x species in the cold box.

The source of NO_x has been generally linked to refinery gases from fluid catalytic cracking (FCC) units. Nitrogen compounds in the feed, especially for naphtha crackers, could be another potential source of the NO_x species observed in the cold box. In addition, we also hypothesize that the addition of the radical inhibitor, DEHA, could be a potential source of NO_x formation. In this report, we present the predictions of a fundamental kinetic model with regard to DEHA thermal decomposition during propane steam cracking. The model simulations are based on the DEHA decomposition reaction kinetics developed in this study, together with literature kinetics on steam cracking and NO_x chemistry involving oxygen and water and small hydrocarbon molecules.

Disclaimer: the current study is a theoretical modeling and simulation effort based on estimated rate parameters and an ideal plug flow reactor. The results are used for fundamental understanding of how DEHA decomposes and how NO is produced through ethylene plants.

DEHA and NO_x Chemistries

Based on possible reactions suggested in a previous paper[2], we propose two reaction routes for DEHA decomposition. The reaction kinetic parameters are estimated using density functional theory (DFT) calculations. Reaction activation energies are calculated based on transition state theory using Gaussian 09 package[3]. Pre-exponential factors of rate constants are calculated using an in-house thermal code by using thermodynamic information such as entropy from Gaussian calculations. The details of the calculations will not be discussed here, but can be found in literature[4].

In the first route, DEHA reacts with an available radical R· through a hydrogen abstraction reaction as shown in R1 in Table 1. The hydrogen abstraction reaction has a low activation barrier, only around 6.0 kcal/mol. The resulting radical, C_2H_5-NO·-C_2H_5, is fairly stable, but possible to decompose to ethyl radical and nitrosoethane (C_2H_5-N=O) with a barrier of 45.8 kcal/mol. At relative low temperatures, e.g. below 200 °C, this high barrier would prevent this decomposition reaction of the radical. Due to the bulky two ethyl groups, this radical is also unreactive to further participate hydrogen abstraction reaction with other molecules. This is why DEHA is used as a radical inhibitor. If the temperature is high enough, nitrosoethane will be formed through reaction R2 and it will also immediately decompose to produce NO and ethyl radicals (R3), which may combine to give butane (R4). From this reaction route, we do see it is possible to generate NO from DEHA decomposition.

In case there is no available radical to react with in the beginning, DEHA itself might break the N-OH bond as illustrated in the second route in Table 1. This bond breaking requires a high energy of 62.2 kcal/mol, so it is only possible at relatively high temperatures. The resulting radical, C_2H_5-N·-C_2H_5 , can further react with hydrocarbon molecules to form diethylamine (DEA) as shown by reaction R6, which then further react to ethylamine (EA) (R7 and R8) and eventually form NH_3 (R9 and R10).

Table 1: Two reaction pathways of DEHA thermal decomposition. RH represents a generic hydrocarbon molecule.

	Elementary Reactions	Barrier (kcal/mol)
Route 1:	R1: C_2H_5-NOH-C2H5 + R· → C_2H_5-NO·-C_2H_5 + RH	6.0
	R2: C_2H_5-NO·-C_2H_5 → C_2H_5· + C_2H_5-N=O (nitrosoethane)	**45.8**
	R3: C_2H_5-N=O → NO + C_2H_5·	**40.0**
	R4: C_2H_5· + C_2H_5· → C_4H_{10}	0
Route 2:	R5: C_2H_5-NOH-C_2H_5 → C_2H_5-N·-C_2H_5 + OH·	**62.2**
	R6: C_2H_5-N·-C_2H_5 + RH → C_2H_5-NH-C_2H_5 (DEA) + R·	15~20
	R7: C_2H_5-NH-C_2H_5 → C_2H_5· + C_2H_5-NH·	**81.6**
	R8: C_2H_5-NH· + RH → C_2H_5-NH_2 (EA)+ R·	13 ~18
	R9: C_2H_5-NH_2 → C_2H_5· + NH_2·	**83.6**
	R10: NH_2· + RH → NH_3 + R·	11~16

Once NO is formed from DEHA decomposition, it is possible for it to further react and convert to other nitrogen species. NO_x chemistry is included by taking the relevant reaction kinetics from a literature paper, where the interactions between NO_x species, oxygen and C1 to C4 hydrocarbons are investigated[5]. While the reaction kinetics are largely elementary reaction based, overall reactions were included such as NO + ½ O_2 → NO_2 and NO_2 + NO → N_2O_3.

Simulation Details

Since we are interested in DEHA thermal decomposition during steam cracking, we also need to include the steam cracking chemistry. For this study, we used heptane combustion chemistry taken from the literature [6, 7] to simulate steam cracking. This detailed heptane combustion mechanism has been developed and validated by comparison to experiments in shock tubes [6, 7]. The mechanism was proven to perform well over a broad range of temperatures and pressures including steam cracking conditions.

The above mentioned DEHA and NO_x reaction kinetics are also included. The simulations were carried out in a one-dimensional (1-D) plug flow reactor (PFR) assuming ideal mixing of the reactants. We used an in-house tool similar to ChemKin [8] to solve for the governing differential equations to obtain product distributions in various simulations.

Pure DEHA Thermal Decomposition

We first study the thermal decomposition of pure DEHA under isothermal and isobaric (2.4 bara pressure, the same as the pressure used in DEHA decomposition during steam cracking) conditions. At a residence time of 0.25 seconds, which is typical for steam cracking, pure DEHA starts to decompose around 220 °C, as shown in Figure 1. At temperatures below 250 °C, there is minimal DEHA decomposition. This is because the weakest bond in DEHA is the N-O bond with a fairly strong bond energy of 62.2 kcal/mol (Table 1). Above 250 °C, DEHA decomposes with significant conversion, and the decomposition conversion increases dramatically when temperature increases. Figure 1 shows pure DEHA almost fully decomposes at 300 °C.

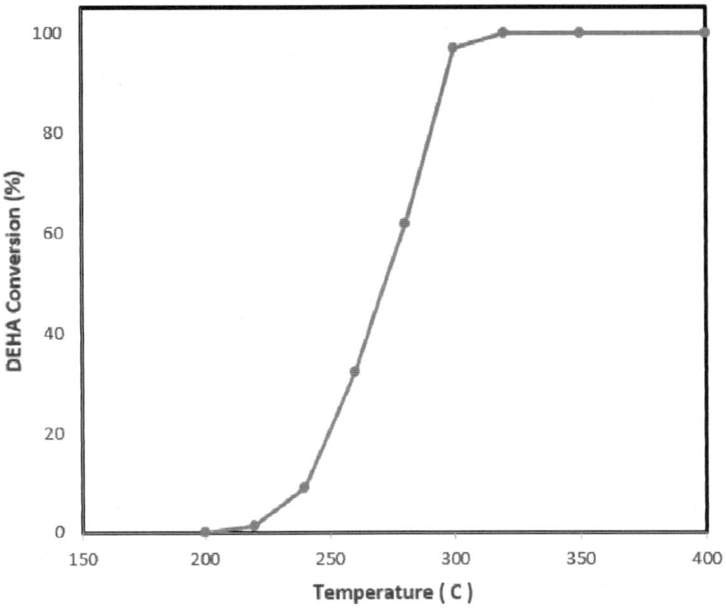

Figure 1: Pure DEHA conversion vs temperature. Residence time is around 0.25 second for each temperature. Pressure = 2.4 bara.

The DEHA thermal decomposition products include a range of small molecules such as ethylamine (EA), diethylamine (DEA), NO, and NH_3, shown in Figure 2. Water and alkanes (mostly C2 and C4) are also produced and not shown in the figure. It is particularly interesting to observe that the NO formation from DEHA decomposition increases first with temperature and peaks at around 500 °C. For temperatures above 500 °C, NO starts to decrease while more and more NH_3 is produced. There are two main possible reasons. The first reason is that at low temperatures, once the bond breaking reaction (R5 in Table 1) initiates to produce some sufficient concentration of radicals, reaction Route 1 shown in Table 1, which produces NO, dominates due to the high barriers of rate limiting reactions in Route 2. At higher temperatures, Route 2 starts to be able to compete with Route 1 and produce NH_3. The second reason is that more hydrocarbons are produced during DEHA decomposition at higher temperatures, and they can reduce NO to form NH_3 and N_2O (not shown in the figure). And at higher temperatures, the reduction reactions also take place faster. The decomposition of pure DEHA is interesting, but, in reality, it is added to steam cracking process and will have to interact with the hydrocarbons where DEHA is at very low concentrations.

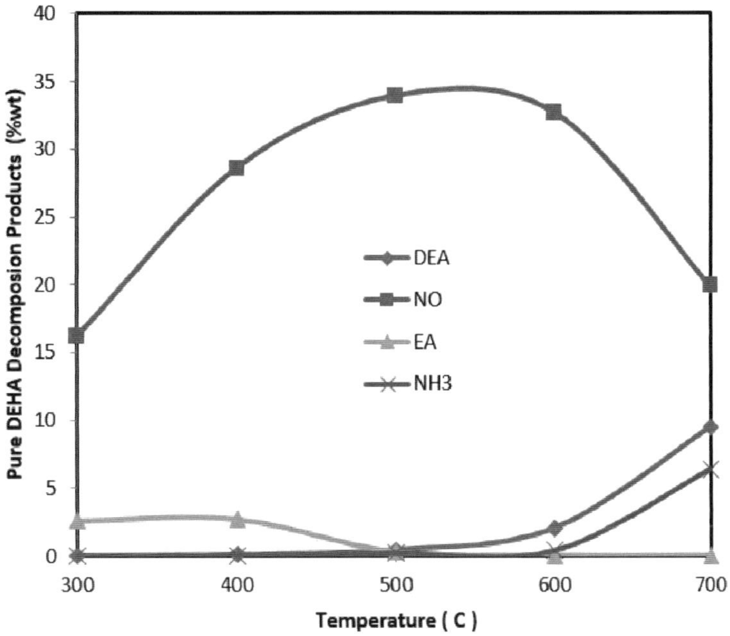

Figure 2: Major products from complete DEHA decomposition at temperatures over 300 °C. Residence time is around 0.25 second for each simulation. Pressure = 2.4 bara.

DEHA Decomposition in Propane Cracking

The next stage is to study the DEHA decomposition in typical propane steam cracking mixtures under isothermal conditions. The pressure used in the simulations are typical cracking pressure around 2.4 bara. The residence time used in the simulations is controlled to be around 0.25 seconds for comparison purpose. The concentrations of DEHA and products that will be reported in this study are all based on a dry weight basis, i.e., relative to propane feed excluding water. The DEHA concentration fed to the reactor is set to be 100 PPM. But the decomposition behavior is almost independent of this concentration as long as it is small. The propane steam cracking reactions produce many hydrocarbon intermediate radicals, which can interact with DEHA and its decomposition intermediates. Similar to pure DEHA decomposition, Route 1 of the DEHA decomposition mechanism (Table 1) dominates at low temperatures (< 500 °C) to produce DEA and NO. At higher temperatures, ethylamine (EA) and NH_3 are the major nitrogen products from DEHA decomposition. This is also similar to pure DEHA decomposition, but the temperature to produce predominantly EA and NH_3 is lower with propane steam cracking than pure DEHA thermal decomposition. This demonstrates the presence of the hydrocarbons and hydrocarbons intermediate radicals during steam cracking process that help to shift the DEHA decomposition products towards NH_3 from NO. The NH_3 formed from DEHA decomposition might also be one of the root cause of ammonia presence in ethylene products observed in industry in addition to formation from other nitrogen compounds in the feed[9].

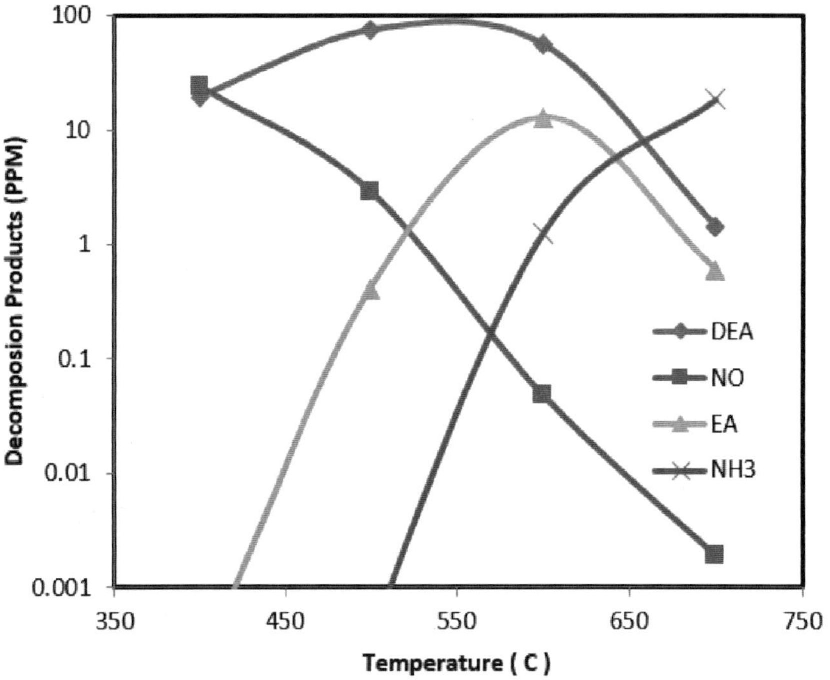

Figure 3: Major products of DEHA decomposition in propane steam cracking at hypothetical isothermal conditions. Residence time is around 0.25 second for each simulation. Pressure = 2.4 bara.

The simulation results shown in Figure 3 demonstrate that when DEHA is fed with propane to the inlet of the furnace, there will be only a minor fraction of DEHA converted to NO. This is consistent with the conclusion from a pilot plant study [10] where less than 1%wt of DEHA was found to be converted to NO when cracking DEHA together with heavy gas oil in the temperature range between 300 and 800 °C. In the same literature study [10], it was mentioned that 60~ 85% of NO can be reduced during ethane steam cracking, but almost no NO reduction was observed with heavy gas oil cracking. This could be the reason why around 1%wt conversion (relative to DEHA) of NO was observed when feeding DEHA in heavy gas oil at 800 °C in the literature study [10], while the present study revealed almost no NO formation when feeding DEHA with propane cracking at high temperatures (T > 600 °C). In the simulations, even if NO is formed from DEHA decomposition, it will be cracked/reduced in propane cracking (see the next section). Therefore, the conclusion that only a minor fraction of DEHA is converted to NO in the furnace from the simulations is consistent with the previous study.

In a separate simulation, we find that NH_3 is very stable even at very high temperature once formed. But NO can further react, which we discuss in the following section.

Fate of NO_x in Furnace

One natural question to ask is whether NO_x produced from DEHA decomposition will be able to survive through a steam cracking furnace. To answer this question, we did a simulation feeding 10 PPM NO during propane steam cracking and plotted the nitrogen products along the normalized furnace coil and transfer line exchanger (TLE) length z/L in Figure 4. The TLEs starts at around z/L = 0.7. NO is found to be largely reduced to N_2O, NH_3 and N_2O_3 inside the actual reactor (z/L from 0 to 0.7). Inside TLEs (z/L > 0.7), the temperature decreases sharply and we see that N_2O and N_2O_3 concentrations continue to increase, indicating low temperatures thermodynamically favor the formation of N_2O and N_2O_3 from other intermediate radicals containing a N atom. A previous EPC report showed that NO was reduced by 60 to 85% when cracking NO_x with ethane feed[10]. This simulation shows that NO indeed can be destroyed inside a furnace; however, the formed products are still nitrogen compounds such as N_2O and N_2O_3.

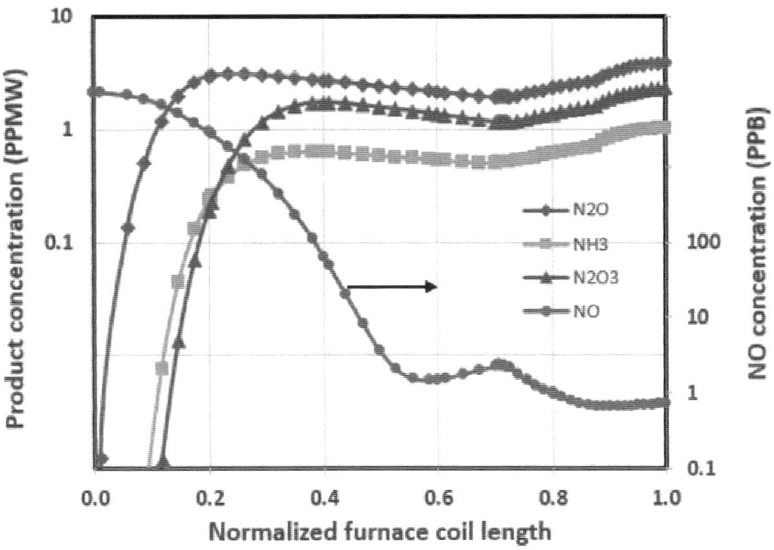

Figure 4: The fate of NO inside a typical propane steam cracking furnace. Inlet pressure is 2.4 bara.

The formed nitrogen compounds can still slowly react inside the quench and compression units between the TLEs and cold box, especially if there is some oxygen available from leakage. We did a simulation by feeding 10 PPM N_2O and oxygen and found around 5% of N_2O is converted to NO at 300 °C and 100 seconds residence time. Given longer residence times, the N_2O to NO conversion will be larger if oxygen is sufficient. Therefore, we postulate that even though only a very minor amount of NO is formed from DEHA decomposition inside the furnace it may still accumulate slowly after the furnace when a small concentration of oxygen is available.

NO goes with methane and ends up in the cold box and reacts with conjugated diolefins such as butadiene to form NO_x gums. Reaction between ammonia NH_3 and N_2O_3 to produce ammonium nitrite (NH_4NO_2) has also been observed in the cold box [10]. In one of the Dow ethylene plants, certain amount of NH_4NO_2 was found in the cold box during a turnaround. The exact amount was not known, but based on the sizing of the equipment the amount was estimated to be approximately between 1.1 and 77 kg over a period of 30 years. In order to produce this much NH_4NO_2 in 30 years, between 0.1%wt and 7.7%wt of DEHA would have to have been converted to NO based upon a typical value of DEHA in the post fractionator bottom flow of the butadiene plant (1.5 ton/h), 3 ppmv O_2 entering the cold box and the typical temperature of the critical exchanger (-127 °C). The predicted DEHA conversion to NO obtained in the time scale of seconds cannot compare directly with the plant data obtained in the time scale of years, but they qualitatively agree with each other that only a small percent (weight basis) of DEHA is converted to NO.

Both NO_x gums and ammonium nitrite pose a serious process safety risk and must be carefully monitored. Understanding of the fundamentals of their formation certainly will help to address the problem.

Conclusions

In summary, the current fundamental study helps us understand the mechanism and conditions of DEHA decomposition to produce NO_x. The kinetic parameters were estimated through functional density theory and may contain uncertainties, therefore a validation with experimental data is being considered. Nevertheless, from the model simulations, it is found that NO is the major pure DEHA decomposition product and its concentration peaks at around 500 °C. At higher temperatures, NO formation decreases while NH_3 formation increases. The decomposition behavior is quite different when DEHA is fed together with propane cracking compared to pure DEHA decomposition. Cracking together with propane, only a minor fraction of DEHA is converted to NO. Instead significant NH_3 is produced, especially at higher temperatures.

When NO is formed from the low temperature DEHA decomposition, the majority of NO is predicted to be converted to N_2O, N_2O_3 and NH_3 inside the furnace. However, at lower temperature sections such as quench tower and compression, it is possible some NO is regenerated from N_2O when there is a small concentration of oxygen available. The nitrogen compounds NO_x can condense in the cold box and react with dienes such as butadiene, eventually forming so-called NO_x gum. They can also react with NH_3 to form ammonium nitrite. Both NO_x gum and ammonium nitrite pose severe process safety issues and need to be carefully monitored.

References

1. Hack, W., R. Copeland, and D. Hines, *Investigation Of NOx Related Cold Box Incident.* Paper presented at the Ethylene Producers Conference at the 2009 Spring National AIChE Meeting, 2009.
2. Abuin, E., et al., *One the reactivity of diethyl hydroxyl amine towards free radicals.* International Journal of Chemical Kinetics, 1978. **X**: p. 677-686.
3. http://www.gaussian.com/.
4. Wang, H., et al., *Kinetic Modeling of 1, 2-Dichloropropane (PDC) Free-Radical Chlorination.* AIChE Journal, 2016. **62**: p. 1174-1191.
5. Marinov, N.M., et al., *An Experimental and Kinetic Calculation of the Promotion Effect of Hydrocarbons on the NO-NO2 Conversion in a Flow Reactor.* Proceedings of the Combustion Institute, 1998. **27**: p. 389-396.
6. Mehl, M., et al., *Kinetic Modeling of Gasoline Surrogate Components and Mixtures Under Engine Conditions.* Proceedings of the Combustion Institute, 2011. **33**: p. 193-200.
7. Curran, H.J., et al., *A Comprehensive Modeling Study of n-Heptane Oxidation.* Combustion and Flame, 1998. **114**: p. 149-177.
8. http://www.reactiondesign.com/products/chemkin/chemkin-2/.
9. Hines, D., et al., *Ammonia in the BASF FINA Petrochemicals Limited Partnership Steam Cracking Plant at Port Arthur Texas.* Paper presented at the Ethylene Producers Conference at the 2006 Spring National AIChE Meeting, 2006.
10. Halle, R.T., *Hazards of NOx Accumulations in Cryogenics Ethylene Recovery Facilities.* Paper presented at the Ethylene Producers Conference at the 1991 Spring National AIChE Meeting, 1991.

479731 Molecular Simulations for Developing New Ethane/Ethylene Separation Processes

Abstract:

Separation of ethylene from ethane is by far the most challenging separation in the chemical industry simply because these two molecules have very similar sizes, shapes, and self-interaction strengths characterized by their closing boiling points. In spite of their relative volatility ranging between only 1.2 to 3.0, cryogenic distillation has been the preferred method for their separation. To develop new separation processes with the aid of molecular simulations, there is a compelling need for developing very accurate molecular models that capture these ultra-low separation factors. In this work, we develop an improved version of the Transferable Potentials for Phase Equilibria - United Atom force field. These newly developed models for ethane and ethylene are used to probe adsorptive separation using various zeolites. Furthermore, we investigate the sensitivity of the size of the separation column to the quality of molecular models and other assumptions that are employed in simulating the phase and sorption behavior.

479772 Process Safety Training for Operators

Abstract:

It is stipulated in Belgian law for high threshold Seveso plants and by corporate guidelines to revise the process safety study on a 5 yearly frequency.

In case of the Steamcracker plant in BASF Antwerp the expectations from the business unit were defined as followed: "***Conceptual revision based on a new method of risk assessment. A method which would include all risk scenarios, risk classifications and countermeasures per process group. Additionally the information/data should be gathered as such that this can be used for review when plant is audited by the authorities (external objective) and for training purposes (internal objective).***" The presentation will not focus on the actual results of revalidation of the process safety studies but on how a training package was set up towards the operators.

Process safety studies are well documented based on which method is followed; it is an overview of scenarios translated in typical guide words such as "more", "less", "other than", ... the result of the scenario is than summarized in deviation, cause, effect and the different countermeasures. Although this type of documentation is needed as reference for the plant, it's not suitable as training material.

Setting up training material starts with defining the short and long term objectives of this training. Especially for plants running already for several decades, operators can play an important role in the evaluation and validation of some scenarios. The objectives can therefore be summarized as follows:

- Enhance knowledge on process safety concepts
- Workforce involvement by having operators participating in safety reviews

  This all will be illustrated using a distillation column. A distillation column covers many scenarios that are relevant for other process units such as loss of cooling, maximum heat input, exchanger leakage, etc. It's also very useful to explain the different safeguarding concepts (intrinsic safe, mechanical, istrumentational).

  To conclude: This presentation will show that generic training is possible - as long as a recognizable unit is taken as reference. It makes process safety studies more accessible for operators and in this way they can make a valuable contribution during revalidations.

160b

PROCESS EVENT AVOIDANCE: USING IDENTIFICATION OF PROCESS FAULTS TO AVOID ABNORMAL PROCESS EVENTS

Thomas R. Kindervater
Solution Support Manager
Center for Operator Performance

Mark Nixon
Director of Applied Research
Emerson Automation Solutions

Prepared for Presentation at the 2017 Spring National Meeting
San Antonio, Texas, March 26 - 29, 2017

PROCESS EVENT AVOIDANCE: USING IDENTIFICATION OF PROCESS FAULTS TO AVOID ABNORMAL PROCESS EVENTS

Thomas R. Kindervater
Solution Support Manager
Center for Operator Performance

Mark Nixon
Director of Applied Research
Emerson Automation Solutions

Abstract:

Process faults are a significant concern in the chemical and petrochemical industry. Process operators typically become aware of process faults only after the problem has triggered an operating alarm. This forces the operator to react to the situation after it has caused a deviation of the process from desired operating conditions instead of proactively addressing a process fault before it causes an operational deviation sufficient to trigger an alarm.

The Center for Operator Performance (COP) commissioned the University of Texas at Austin to examine new ways to identify process faults and predict events. This research developed a geometric analysis tool that beat the broadly used Tennessee Eastman Challenge Process benchmark for average detection time for twenty faults by over 90%. The Tennessee Eastman Challenge Process is a benchmark problem that is broadly used to evaluate new strategies for improving chemical plant operations. The geometric analysis tool detected three faults that were either missed or signaled far too late for corrective action by 12 other published techniques.

The geometric tool developed by the Center for Operator Performance and the University of Texas at Austin can identify process faults as leading indicators of abnormal process events with a very low rate of false positives. The technique has been successfully applied to detection of surge in a large, multistage compressor, where analysis of past surges helped identify event signatures as early as a few hours before surge onset. It has given equally impressive fault detection for distillation flooding events and for detecting precursors to flaring events, continuing to show a low false positive rate.

In addition to developing the analysis tool, the research project has provided visualization techniques for process monitoring in operator displays, abnormal event prediction, and identification of process variable(s) causing the detected process fault.

Background

Despite advances in safety interlock systems, process control instrumentation, and Human Machine Interface (HMI) data presentation in the process industry, situations still arise that operators are unable to accurately assess. These situations result in unplanned equipment outages, lost productivity, and unsafe conditions. These situations present strong incentives to provide operators with better techniques to illustrate the state of equipment and process conditions, including the ability to provide advanced warning of pending abnormal events to the operator.

The issue is not a lack of data. The amount of data being captured and made available to process operators is ever increasing. As the volume of data increases, the challenge of interpreting and analyzing the appropriate data to gain information about the state of the process has become overwhelming for operators in many cases. Process control operators rely heavily on their visual displays to monitor and stay abreast of processes and situations, anticipate future status, and proactively detect and resolve emerging problems. However, operator displays support only a subset of their tasks and promote reactive rather than proactive monitoring because of inadequate attention management. Reactive monitoring is a significant barrier to effective operations and increasing operator span of control. To improve overall operating efficiency and safety, more attention needs to made to proactive monitoring.

Prior efforts to achieve early detection of process faults have relied on several approaches: using various data analysis methods to examine process operating data to identify faults, using mathematical modeling of the process (typically first-principles models) to compare actual data with model conditions, or using expert knowledge-based rules for fault identification. Attempts to use such methods have met with mixed results, including high rates of false positives. In addition, those that are successful in identifying faults often require extensive support to maintain the accuracy of the mathematical models or do not provide sufficient lead time for pre-emptive operator action. Yet, the belief remains among process operating companies that, in the proper context, the presentation of current operating information in relation to recent historical operating data through the operators' HMI has a high potential for enabling recognition of pending events and pre-emptive decision making.

Adding to the problem is the fact that traditional visualization of process variables is limited to two and three-dimensional plots. In order for an operator to successfully monitor the condition of an operating process over time, it is necessary for them to understand and analyze data from multiple plots. The operator is largely left on their own to use the data from multiple plots to identify abnormal situations.

This issue became a topic of discussion among members of the Center for Operator Performance (COP). COP initiated a project to develop visual techniques to support the process operator in predicting, detecting, identifying and responding to abnormal situations. A key objective was to provide the indication to operators as much as 15 minutes before the event would otherwise occur. Providing operators with a better idea of the real cause would enable them to take early action which would stop an escalating incident from occurring. The COP member companies provided case studies for predicting compressor surge for a multi-stage compressor, column flooding, and flaring events. The Center for Operator Performance engaged researchers at the University of Texas at Austin to undertake an investigation into process event prediction and mitigation (EPM). Performance member company Emerson Automation Solutions also provided significant resource support to the research team.

The objective of this research was to evaluate methods of early detection of off-normal and/or abnormal operations. The research work has focused on developing a visualization-based approach to identify process faults to lead to off-normal/abnormal events, determining how to present identified faults to operators, and incorporating decision aids to assist the operator.

Approach

Since identification and visual presentation of process faults were at the core of this EPM research effort, the initial focus of the investigation was on visualization-based approaches. Parallel coordinate plots, which provide a visual representation of multi-dimensional data sets in a two-dimensional space, were used as the starting point in this effort.

Each data sample in a parallel coordinate plot is depicted by a line that intersects each variable's ordinate at the representative value for that process variable in that data sample. Figure 1 represents numerous data sets for a seven variable system. After defining a confidence region based on known good operation for each process variable, shown by the dashed black lines for upper and lower bounds, a fault is identified by data that fall outside of the confidence region.

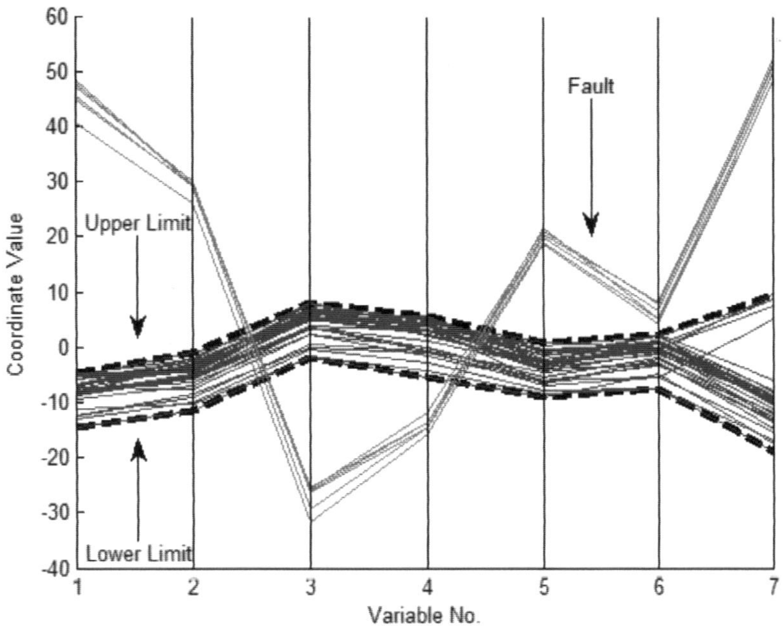

Figure 1. Fault detection using parallel coordinates. Adapted from [1,2]

While parallel coordinate plots remove the limitation of the number of dimensions that can be plotted, they do not explicitly indicate time. It became evident that the challenge of this research effort was to find a way to represent multi-dimensional data sets while also retaining the time order sequence of the data.

This time sequence aspect of data analysis is fundamentally important for analyzing operational trends and is one of three characteristics defined by the research team for good visual representation of data:

- Captures the multiple independent variables in process data,
- Retains the time order sequence of data sets,
- Provides high resolution and minimizes clutter.

At this point, visualization technique development moved to an examination of radial plots. In radial plots, the axes extend outward from a center point. Unlike parallel coordinates, radial plots represent each multi-dimensional data sample in a unique plane and a line connecting the variable values becomes a closed polygon. Additionally, the individual data samples can be time sequenced using a time axis that is perpendicular to the plane of the radial plots. When multiple data sample radial plot polygons are stacked in time order sequence, a multi-sided three-dimensional figure is formed. Figure 2 represents 7200 data samples, captured at one-minute intervals over five days, for a five-variable system.

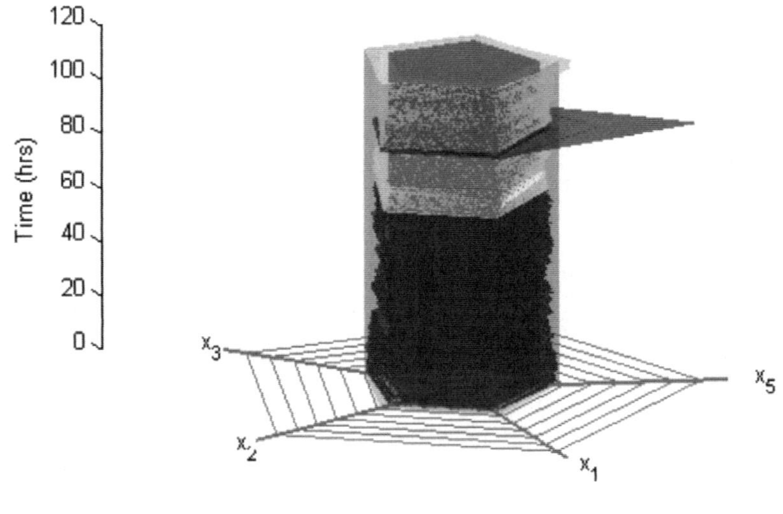

Figure 2. Data samples in time-ordered radial plots for five variables

This approach creates an operating condition visualization that can be rendered and updated in real-time, using physical process data as they are collected. The time horizon can be adjusted by setting the number of time-slice radial plots to be retained in view. First in, first out updating would drop the oldest data slice and add the newest one.

Having developed a visualization method for multi-dimensional data sets that achieved the desired characteristic high-resolution visual representation without clutter using a geometric shape to represent each data sample, the research team realized it might be possible to perform fault detection by evaluating current data against a desired operating state.

Similar to the upper and lower limits that define the confidence region for a parallel coordinates plot, the confidence limits can be determined for each process variable used in the radial plots. The maxima and minima create outer and inner hulls, respectively, which define a confidence region for operation of the process.

Figure 3 illustrates the polygonal lines (black for maxima or upper limits and red for minima or lower limits) which bound the confidence region for process operation.

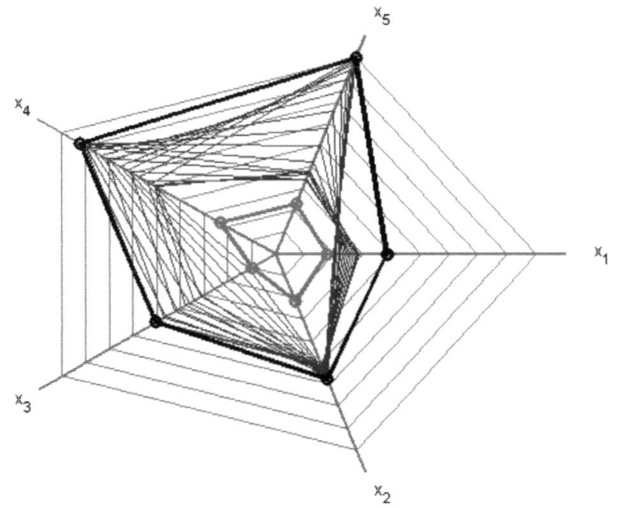

Figure 3. Inner (red) and Outer (black) Hulls bounding confidence region for process operation

It was recognized that this approach to establishing the confidence region based on individual variable confidence limits is essentially a univariate analysis. The team also recognized that process variable interactions can account for the occurrence of process faults and those interactions must be known to detect process faults. To address these concerns, the research team proposed a solution that leverages a unique feature of the radial plots. The polygon created by connecting all the data points in a specific data sample is closed and, therefore, allows calculation of the center of gravity for that corresponding data slice. That center of gravity (centroid) is a multivariate representation of the process state at that point in time.

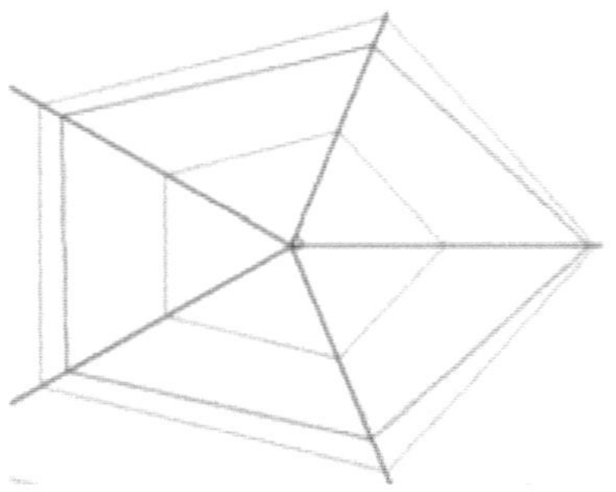

Figure 4. Radial Plot Center of Gravity (Centroid)

The data being used in this EPM method for detecting process faults are always normalized prior to plotting. This means the data for a time slice during steady-state operation will generally form a near-regular polygon and the center of gravity (centroid) will be close to the intersection of the axis of the radial plot (See Figure 4). When a fault occurs, one or more process variables will deviate from their steady-state values, causing the shape of the polygon to change for that time slice and its centroid location will move.

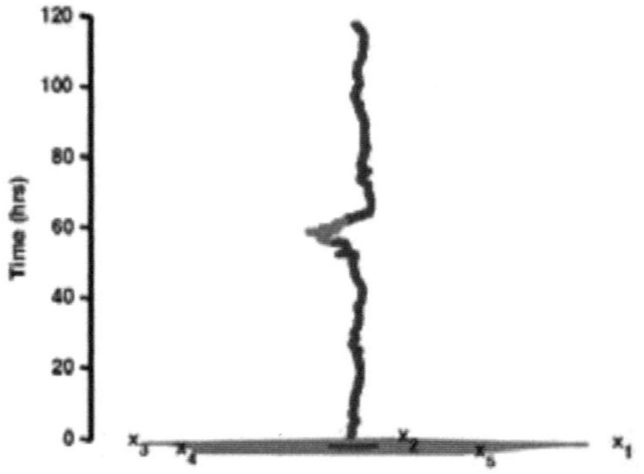

Figure 5. Time Ordered Radial Plot Centroids

As can be seen in Figure 5, a process fault can be identified by detecting a shift in position of the centroids for data sets associated with that fault condition. If the process is not oscillatory and has a single, nominal steady state, the centroids corresponding to normal operation will be closely clustered. Centroids corresponding to conditions when the operation has deviated from normal operation will lie outside of this cluster.

Based on these observations, time-ordered radial plots can be used to implement fault detection according to the following steps:

- Define a confidence region around the normal operating cluster of centroids,
- Monitor the current process operating data and provide visualization of the data on a time ordered radial plot,
- When the centroid of a newly collected data set is not inside the confidence region, a process fault should be declared.

The procedure for calculating a confidence region around the normal operation cluster of centroids is beyond the scope of this paper, as are other mathematical and analytical steps required to develop the fault detection method described in this paper. If additional detail regarding the mathematical and analytical background is desired, please refer to the paper on *Process Fault Detection Using Time-Explicit Kiviat Diagrams*, by Wang et al., published in the AIChE Journal in December 2015 [1].

Process data sets frequently consist of a large number of variables. This research effort has successfully applied principal component analysis, using linear transformation of process variables into principal components, as a means to reduce the dimensionality of the data. Usually the desired amount of steady state variance that needs to be accounted for will determine the number of principal components used to model the process.

One other benefit of using principal component analysis (PCA) to reduce the amount of data is that PCA enables the use of contribution plots to identify process variables that contribute to the process deviation away from its normal steady state. This enables the EPM approach developed by this research to provide meaningful information to the operator regarding appropriate action(s) to take to mitigate the process fault before an off normal or abnormal event occurs.

Results

The process fault identification method (EPM) described in this paper has been tested against actual plant data supplied by member companies of the Center for Operator Performance. These data sets represent compressor surge and plant flare relief events. The approach has also been compared to other publicized process fault identification methods using the Tennessee Eastman process [3]. The following results document the performance of the method developed by the University of Texas at Austin research team.

Compressor Surge

The research team was provided with 30 surge event data sets for a five-stage compressor (Figure 6). Surge events can result in machine damage, flaring, and shutdowns. More importantly, non-steady-state events are also susceptible to unsafe operations. Each data set pertained to a single surge event and contained data for 76 process variables associated with the compressor operation for four days prior to and 1 day after the surge event. The data sampling frequency was one minute and generated 7200 data samples per surge event.

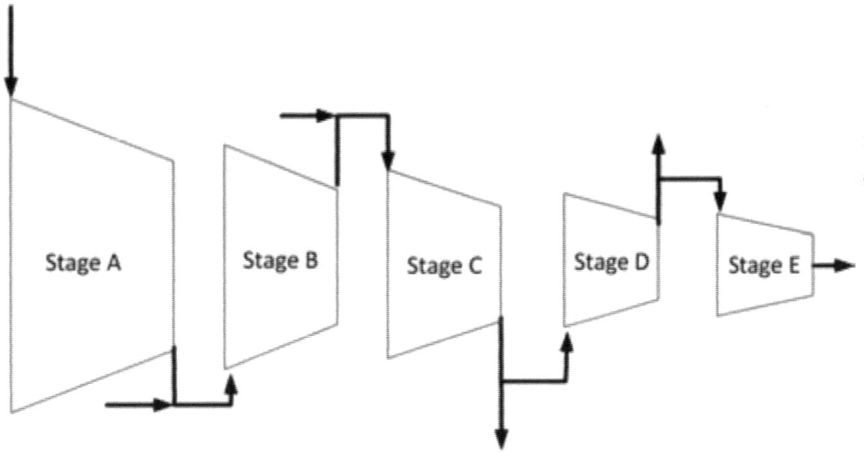

Figure 6. Compressor system simplified schematic. Adapted from [1,2].

Figure 7 illustrates a time-ordered radial plot of all process variables in one surge event data set (7200 data samples of 76 process variables). The blue data represent normal compressor operation and were collected from the four days prior to the surge event. The red data represent data collected prior to the surge event after the initial process fault had occurred. The shift in the centroids that illustrates the fault detection in advance of the surge can be seen in the red data section of the radial plot.

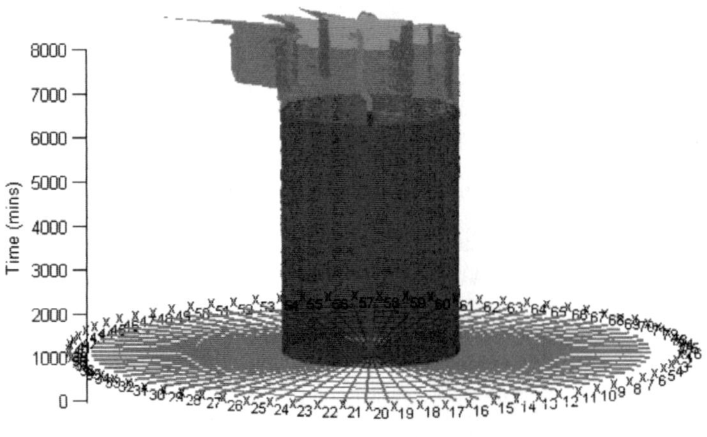

Figure 7. Time-ordered radial plot for representative compressor surge event data set

Figure 8 illustrates a radial plot top view indicating the process fault data. This view clearly shows the data deviations from the normal operation polygon cylinder once the process fault occurred.

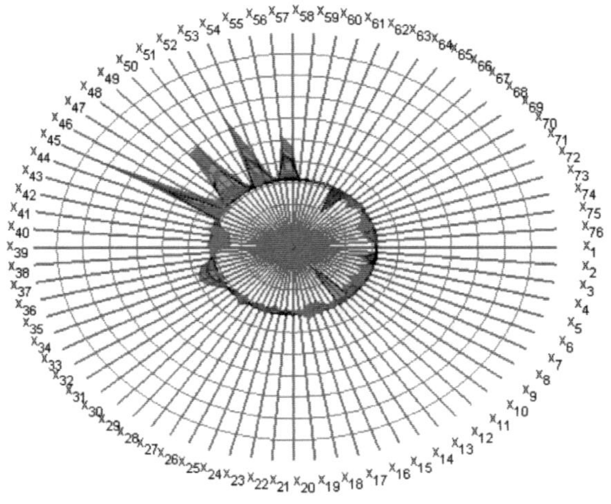

Figure 8. Compressor surge data radial plot top view showing process faults. Adapted from [1].

The steady state samples from all 30 data sets were used as training data and formed the basis for establishing a 95% confidence ellipse for the radial plot centroids. The compressor surge fault detection results are shown in Table 1. The table results are based on knowing when the surge event occurred. The early detection time (EDT) entries represent the time before (positive) or after (negative) the surge event that the process fault was detected, i.e. the time when centroids violated the confidence ellipse. A positive time indicates the fault was detected early and the operator could have taken preventative measures. A negative time indicates the fault would not have been detected prior to the surge event occurring.

Table 1. Compressor surge fault detection results				
Early Detection Times (EDT) in minutes for 30 Compressor Surge Data Sets. Results from [1].				
Data Set #	EDT PCA with Radial Plots	EDT Raw data with Radial Plots	EDT using PCA with T^2 values	EDT using PCA with Q statistics
1	31	31	31	
2	42	42	42	42
3			1768	1768
4	748	757	716	585
5	- 7	45	- 68	- 67
6	44	44	44	45
7	52	52	52	53
8	1335	1137	1336	1351
9	325	325	42	84
10	1434	878	33	33
11	721	440	70	41
12	229	143	62	33
13	1519	1335	1337	1332
14	39	43	39	36
15	877	932	263	- 13
16	17	53		- 6
17	1369	1410	1307	135
18	35	32	35	36
19	26	26	25	26
20			2463	2116
21	1079	1093	206	
22	61	34	103	114
23	1133	1109	180	180
24	561	575	143	204
25	3	37	- 15	- 28
26	- 52	- 50	- 50	- 49
27	- 27	- 64	- 28	- 62
28	33	- 43	- 59	- 41
29	25	25	25	- 25
30	178	176	58	102

In addition to using the full raw data set of 76 variables, principal component analysis was also applied to the data sets. Figure 9 provides a visual representation of the five principal component time-ordered radial plot used for fault detection, capturing 74% of the steady state variance. These times are included in Table 1, along with detection times using standard PCA T2 and Q statistics for comparison.

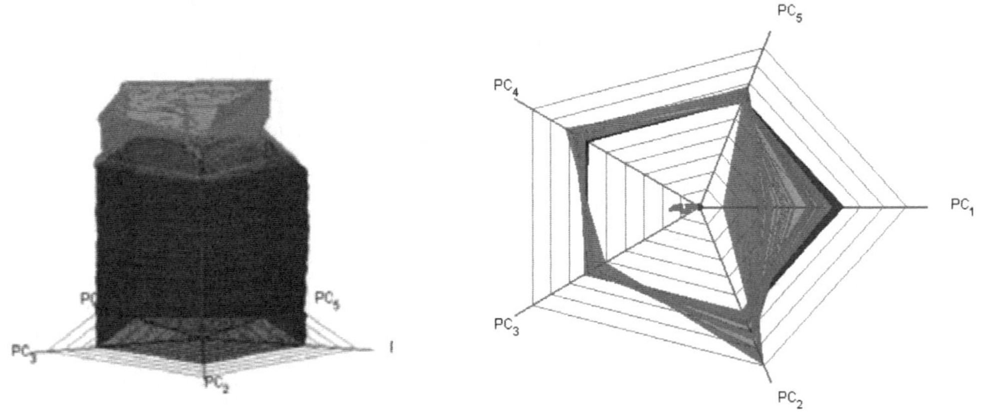

Figure 9. Principal component radial plot profile view and top view [1]

When using all 76 process variables in the data sets, a fault was detected in 28 of the 30 cases, with a positive EDT in 25 cases and a negative EDT in three cases. In two cases, no fault was detected.

When using PCA and the five principal components, a fault was detected in 28 of the 30 cases, with a positive EDT in 25 cases and a negative EDT in three cases. In two cases, no fault was detected.

When using PCA T2, a fault was detected in 29 of the 30 cases, with a positive EDT in 24 cases and a negative EDT in five cases. In one case, no fault was detected.

When using PCA Q, a fault was detected in 28 of the 30 cases, with a positive EDT in 20 cases and a negative EDT in eight cases. In two cases, no fault was detected.

EDT Raw data with Radial Plots performed as well as or better than any of the other methods. Since the actual parameters are included in the plot, it is easier to see which parameters are driving the condition.

Tennessee Eastman Process Benchmarking

The performance of the EPM fault detection method described in this paper was compared to other documented fault detection methods using the Tennessee Eastman Process (TEP). The TEP is a recognized benchmarking tool with a number of published fault detection studies that use TEP as the process simulator. The TEP was created by the Eastman Chemical Company to provide a realistic process for evaluating process control and monitoring methods. TEP simulators written in MATLAB are available for researchers to use. The TEP simulator [4] (see Figure 10 for a process diagram) includes provisions for applying 20 faults (Listed in Table 2), 15 of which are known and five of which are not.

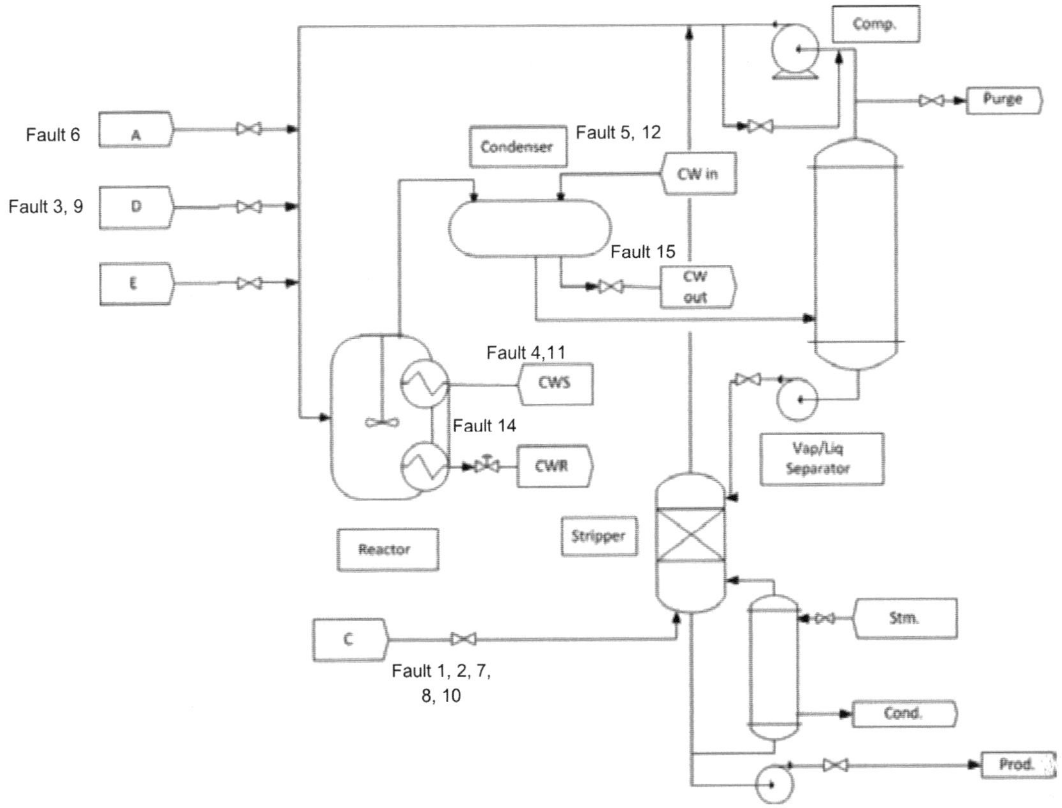

Figure 10. Diagram of the Tennessee Eastman Process [3]

Table 2. Tennessee Eastman Process Simulator Faults [3]		
Fault No.	Description	Type
1	A/C feed ratio, B Composition constant (Stream 4)	Step
2	B Composition, A/C ratio constant (Stream 4)	Step
3	D feed temperature (Stream 2)	Step
4	Reactor cooling water inlet temperature	Step
5	Condenser cooling water inlet temperature	Step
6	A feed loss (Stream 1)	Step
7	C header pressure loss – reduced availability (Stream 4)	Step
8	A, B, C feed composition (Stream 4)	Random variation
9	D feed temperature (Stream 2)	Random variation
10	C feed temperature (Stream 4)	Random variation
11	Reactor cooling water inlet temperature	Random variation
12	Condenser cooling water inlet temperature	Random variation
13	Reaction kinetics	Slow drift
14	Reactor cooling water valve	Sticking
15	Condenser cooling water valve	Sticking
16..20	Unknown	

The performance comparisons were made against published tests of nine linear PCA-based fault detection methods and three non-linear PCA-based methods. For each fault detection test, the simulator ran for a minimum of 12 hours, with a specific fault injected five hours into the simulator run. The random noise signal generator was changed for every fault test run. Nine principal components were selected based on a steady state data variance capture threshold of 70%. A 95% confidence ellipse was used for all faults.

The EPM method developed by the research team performed well compared to the other models in fault detection time (Table 3), false detection (Table 4), and missed detection rate (Table 5) with generally lower times and rates. A fault must be detected in two consecutive samples to count as a detected fault. Since the time is known when the fault is implemented, any fault detected prior to that time is considered a false detection. Any sample after that time which is not detected is considered a missed detection.

Table 3. Fault detection time comparison

	Fault Detection Time (Minutes) [1]				
Fault Number	EPM Method	Average Time of Other Methods	Median Time of Other Methods	Minimum Time of Other Methods	Maximum Time of Other Methods
1	3	13	12	6	21
2	8	41	38	27	75
3	17				
4	2	234	8	3	1386
5	2	8	3	0	48
6	2	59	3	0	633
7	2	5	3	0	15
8	46	64	62	54	75
9	201	899	899	894	903
10	52	131	116	42	303
11	39	271	63	21	912
12	14	15	8	0	66
13	124	110	119	48	147
14	8	6	3	3	18
15	32	865	27	21	2220
16	56	304	98	9	936
17	112	89	74	51	192
18	113	242	251	195	279
19		140	140	33	246
20	180	189	207	39	267

Table 4. False detection rate comparison				
EPM Method	Average for other Methods	Median for other Methods	Minimum for other Methods	Maximum for other Methods
0.032	0.0712	0.0152	0.0027	0.281

Table 5. Missed detection rate comparison					
Fault Number	EPM Method	Average Rate of Other Methods	Median Rate of Other Methods	Minimum Rate of Other Methods	Maximum Rate of Other Methods
1	0.018	0.0026	0.002	0	0.008
2	0.018	0.0175	0.0195	0.01	0.026
3	0.542	0.9742	0.9855	0.92	0.998
4	0.04	0.4876	0.439	0	0.975
5	0.002	0.5197	0.728	0	0.775
6	0.005	0.0034	0	0	0.013
7	0.007	0.1116	0	0	0.486
8	0.102	0.0714	0.0265	0.016	0.486
9	0.829	0.9796	0.9895	0.95	0.995
10	0.138	0.4394	0.575	0.099	0.666
11	0.171	0.4653	0.4355	0.18	0.801
12	0.069	0.0189	0.0225	0	0.03
13	0.283	0.0505	0.0495	0.04	0.06
14	0.04	0.0551	0	0	0.21
15	0.188	0.9551	0.957	0.903	0.988
16	0.138	0.4859	0.5645	0.084	0.834
17	0.266	0.1286	0.106	0.024	0.26
18	0.266	0.1003	0.1	0.09	0.113
19		0.6838	0.861	0.019	0.996
20	0.411	0.5312	0.5525	0.342	0.701

Flare Relief

Emissions from flare events have a significant impact on health, safety, and the environment. Flare events are the result of plant-wide abnormal operating conditions that can be predicted by analyzing the broad complement of data available from the control systems, and minimized or altogether prevented by taking appropriate control action.

The research team was provided with a year's worth of one-minute data samples from Oct. 2013 through Sept. 2014. Those 525,000 data sets each contained 130 process variables. The date and time of four flare relief events contained within the data were identified by the member company contributing the data and provided to the research team.

The goal was to see if the EPM method could be used on a large-scale system and, after building a model using the one-year dataset, could predict a more recent flare event.

Figure 11 is a simplified diagram of the fuel gas network and flare system. The fault detection model was constructed using 66 principal components (out of the 130 process variables) to capture 90% of the steady state variance. A 95% confidence ellipse for the centroids was used to identify a process fault. For the training data, process faults were detected at least one hour prior to all four flare relief events. Additionally, several other process faults and associated flare relief events were detected in the training data that had not been communicated to the research team by the contributing company.

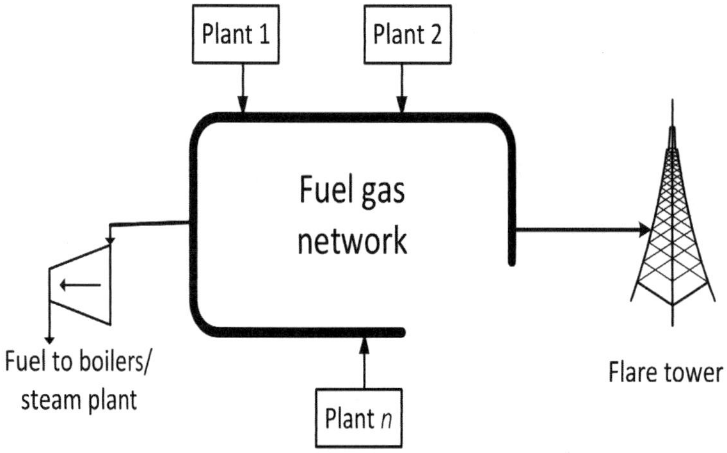

Figure 11. Fuel gas network and flare system

Using one day's worth of data from July 2015, when a known flare relief event occurred, the EPM model based on the year-long data set, detected the process fault 1.6 hours in advance of the known flare relief event. Additionally, using variable contribution from the principal component analysis, the model was able to identify specific areas of interest for determining the actual leading indicator.

The company that contributed the flare relief data has conducted additional post event assessments of other flare relief events and has concluded, "EPM would have notified the operator at least 30 minutes prior to each event and would have revealed the cause as well if it had been in real time." The company is currently implementing the EPM fault detection method in their control system.

Summary

By explicitly designing attention management into new representations of process status, a dramatic increase in effective span of control can be achieved. Better allocation of limited attentional resources during monitoring shifts the nature of monitoring from reactive to proactive. Designing for and addressing the underlying cognitive challenges of multi-variable control can result in more effective system operation, safer management of abnormal situations, and increased productivity.

The Center for Operator Performance believes that the EPM process fault detection method developed through the research efforts at the University of Texas at Austin can be a viable real-time tool to provide:

- Understandable visualization of the process conditions –
  - Capturing multiple independent variables in process data,
  - Retaining the time order sequence of data sets,
  - Providing high resolution and minimizing clutter.
- Reliable process fault detection –
  - Predicting off normal or abnormal process events with
    - Low missed detection rate
    - Low false detection rate
  - Identifying causal factors (through PCA)
- Meaningful information to operators to enhance corrective actions for minimizing or preventing the off normal or abnormal process event.

As with any such data analysis and modeling method, it is susceptible to noise and appropriate data preprocessing may be required. The selection of an appropriate steady state region is key to the quality of the confidence ellipse around the centroids of the radial plots, which serves as the model in this method.

Currently, the Center for Operator Performance, through member company Emerson Automation Solutions, is working with the research team to transition the EPM method from a research model to an application available for use by industry.

References

[1] Wang RC, Edgar TF, Baldea M, Nixon M, Wojsznis W, and Dunia R. *Process Fault Detection Using Time-Explicit Kiviat Diagrams*, by, AIChE J. 2015;61(12):4277-4293

[2] Dunia R, Edgar TF, Nixon M. Process monitoring using principal components in parallel coordinates. AIChE J. 2013;59(2):445–456

[3] Downs JJ, Vogel EF. A plant-wide industrial process control problem. Comput Chem Eng. 1993;17(3):245–255.

[4] Ricker NL. Tennessee Eastman challenge archive. Available at http://depts.washington.edu/control/LARRY/TE/download.html. Last Accessed: Sep. 28, 2015.

AIChE Paper Number: 160c

LESSONS LEARNED FROM INCIDENTS
CASE STUDY : 25 YRS OPERATION OF A CRACKER PLANT

Geert Vercruysse
Prof. Process Safety
Catholic University of Leuven

Geert Boogaerts
Visiting Professor Hazardous Processes and Products
essenscia Chair Safety Engineering KU Leuven

Prepared for Presentation at the 2017 Spring National Meeting
San Antonio, Texas, March 29th 2017

AIChE Paper Number: 160c

LESSONS LEARNED FROM INCIDENTS
CASE STUDY : 25 YRS OPERATION OF A CRACKER PLANT

Geert Vercruysse
Prof. Process Safety
Catholic University of Leuven
Geert Boogaerts
Visiting Professor Hazardous Processes and Products
essenscia Chair Safety Engineering KU Leuven

Abstract:

The process safety concept of a plant is determined during the engineering phase of a project. Once the process is started this safety concept will be validated during the life cycle of the plant, based on operational experience and lessons learned from incidents. In this paper three incidents, related to start up and shutdown, will be elaborated in detail and its impact given on the process safety concept. Further it will be illustrated that similar scenarios can/could occur in different process unit set ups.

Evaluation of process safety incidents show that for more than 50% of them the root cause are related to human and organizational factors. By using new tools provided by Industry 4.0 such as better communication and visualization software, this will certainly help in the prevention of these root causes.

Introduction

The process safety concept of a plant is determined during the engineering phase of a project. Once the process is started this safety concept will be validated during the life cycle of the plant. It is stipulated in the Seveso Directive[1] , that applies to more than 10.000 companies in Europe, transposed by Belgian legislation and associated inspection guidelines that this revalidation or review should be done at least on a 5 yearly frequency.

Three important forms of information play a crucial role in this revalidation:

- **Operational experience**. Operators or team leaders should take part in the revalidation of the process hazard analysis. Scenarios can be discussed with practical information: logics and start up sequences can be explained as well as for instance the response time of the process.
- **Lessons learned from incidents**. An incident database is an important instrument to keep a log of near misses, the activation or failure of protective devices (AFPD) and Process Safety Incidents (PSI).
- **Management of Change database (MOC)**. When MOCs have an impact on the process safety concept they should be immediately incorporated into the documentation of the process hazard analysis. Nevertheless a periodic review of these MOCs is an excellent moment to verify if they are correctly implemented and to verify that no impact on other parts of the plant/process was overseen.

Incident 1: Trip of furnace – leaking of control valve

The incident started with an unintended closure of an on/off valve in the general hydrocarbon feed distribution line to the furnaces. Thermal cracking is an endothermic process and a loss of feed scenario is similar as a loss of cooling scenario for exothermic reactions. Therefore the furnace needs to be taken out of service as quickly as possible. A furnace is schematically represented in Figure 1.

[1] Directive 2012/18/EU of the European Parliament and of the council of 4 July 2012 on the control of major-accident hazards involving dangerous substances, amending and subsequently repealing Council Directive 96/82/EC.

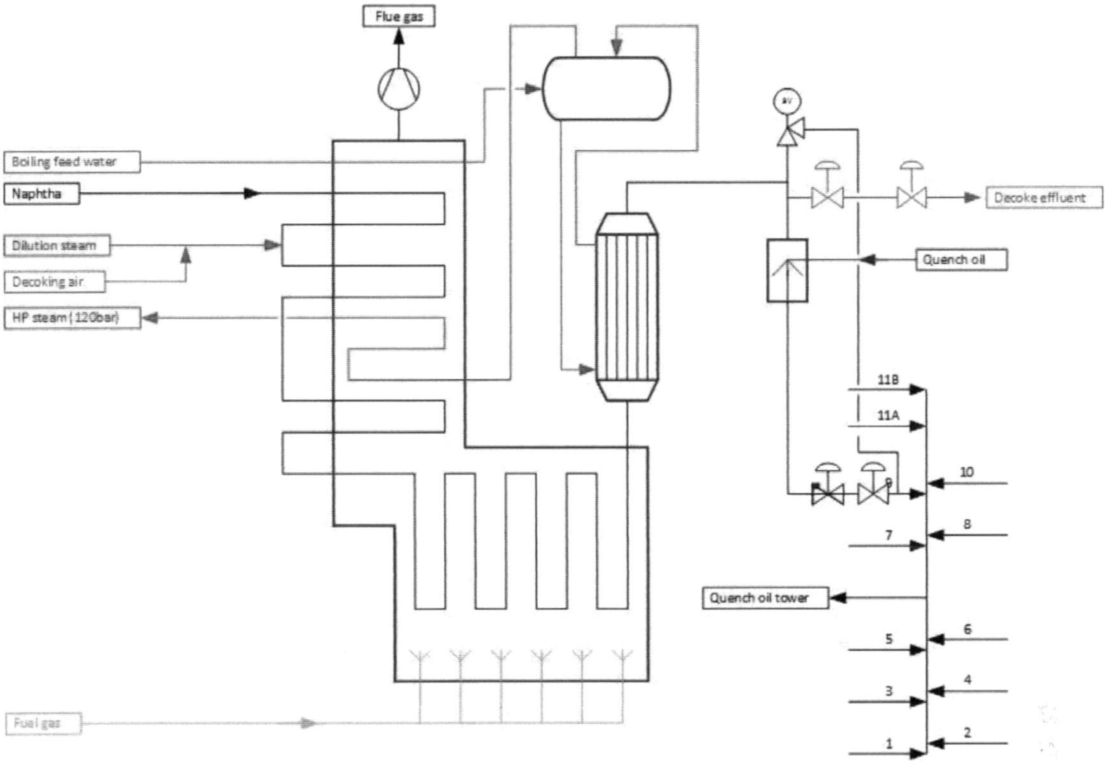

Figure 1: Schematic overview of the furnace section

A trip of a furnace initiates a sequence where all feed (with the exception of process steam) and energy sources to the furnace will be shut down (process steam will cool down the furnace coil and guarantee a positive flow towards the crack gas header).

Shortly after the trip of the furnace, a leak in a furnace coil was detected and it was decided to isolate this furnace from the crack gas header. Before this can be done a flush procedure is started where the furnace is first redirected to the decoking header and then the valve to the crack gas header is closed.

Unfortunately, the control valve of the quench oil did not close tightly enough, leading to a gradual filling of the outlet line of the furnace with quench oil. When the outlet line of the furnace was completely filled, a back flow of quench oil towards the furnace coil started. Through the leak in the coil, the quench oil entered into the furnace chamber and shortly afterwards it ignited. The furnace chamber walls were not sufficiently cooled so a spontaneous fire started. The fire could only be stopped when the manual valves in the quench oil line to the furnace were also closed. The incident investigation looked first into the functionality of the quench oil. Its primary function is to cool down the cracked gas formed in the furnace. Secondary, the quench oil loop is used to generate process steam by heat transfer in a train of heat exchangers. As stated above this process steam is necessary as a safety function in case of a trip of a furnace. So it is clear that this it is not an option to stop the quench oil pump in case of a leaking control valve as this would create other safety issues. It was decided to install an additional on/off valve.

As a result of reviewing process safety concepts, similar observations in other scenarios were made and have resulted in an increased number of on/off valves installed in series with control valves. Figure 2 shows an overview of different set ups on how a stream can be interrupted.

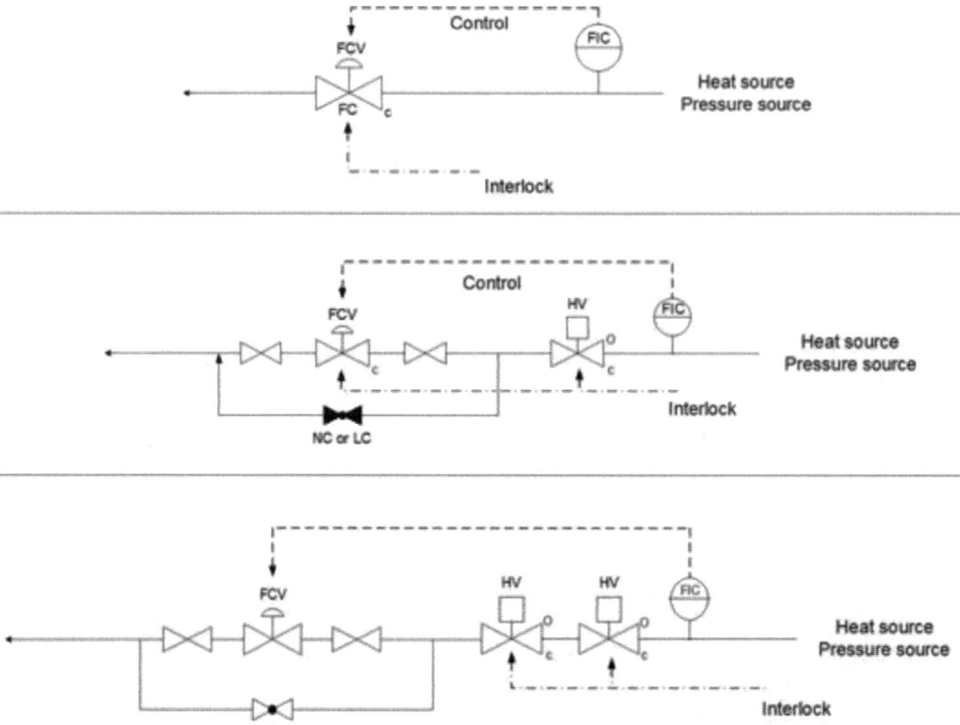

Figure 2: Different concepts to close streams by an instrumentational interlock

Nevertheless, this has led to new scenarios while a proper sequence needs to be followed to restart this stream. Additionally these on/off valves can "unexpectedly" close when they are for instance bumped.

Incident 2: Trip of furnace – coil rupture and crack gas back flow

In this case a furnace had to be taken out of service but the panel operator missed by selecting the wrong furnace number (see Figure 3).

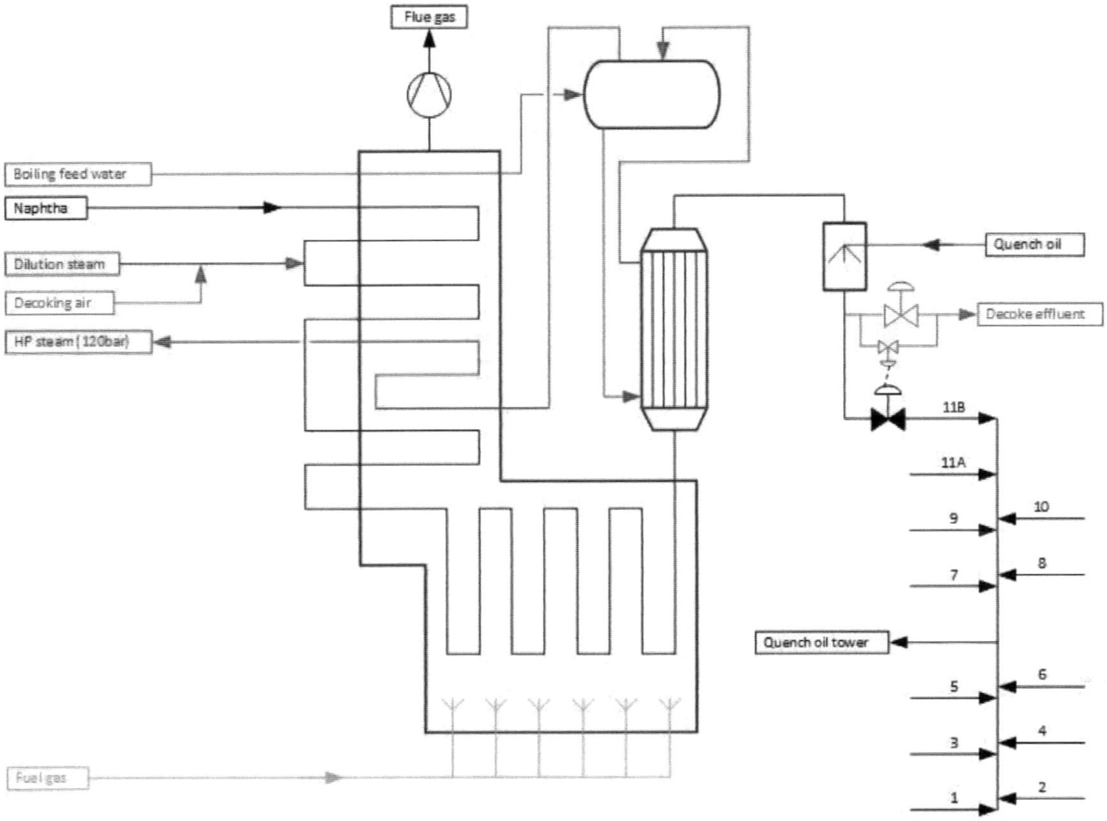

Figure 3 : Schematic overview of furnace section (furnace 11B)

Unfortunately, the selected furnace was running at the end of its cycle, which meant a thick coke layer was formed in the coil. Due to the sudden interlock – the furnace cooled in a completely uncontrolled way and the coil inside the furnace chamber ruptured. Due to the tube rupture, the process steam could not guarantee positive flow, and resulted in a back flow of cracked gas from the header towards the furnace which ignited in the furnace chamber.

For the incident investigation, a working group was formed with other steam cracker operating companies and licensors. All preventive and mitigating measures were discussed and each operating company could review its own design of what would best fit as layers of protection.

The result was that, given the actual design of the furnace, a check valve was installed to prevent effective backflow. This was selected as a better design solution than the installation of an on/off valve as this would not fit in the global safety concept of guaranteeing an open connection downstream of the furnace.

Both incidents clearly show the necessity of a periodic review and update of a process safety concept. Writing down these lessons learned, including the ones categorized as "near misses", will lead to future designs that include the above design features.

Incident 3: Startup of hydrogenation reactor filled with new catalyst

A schematic overview of the reactor unit is shown in Figure 3.

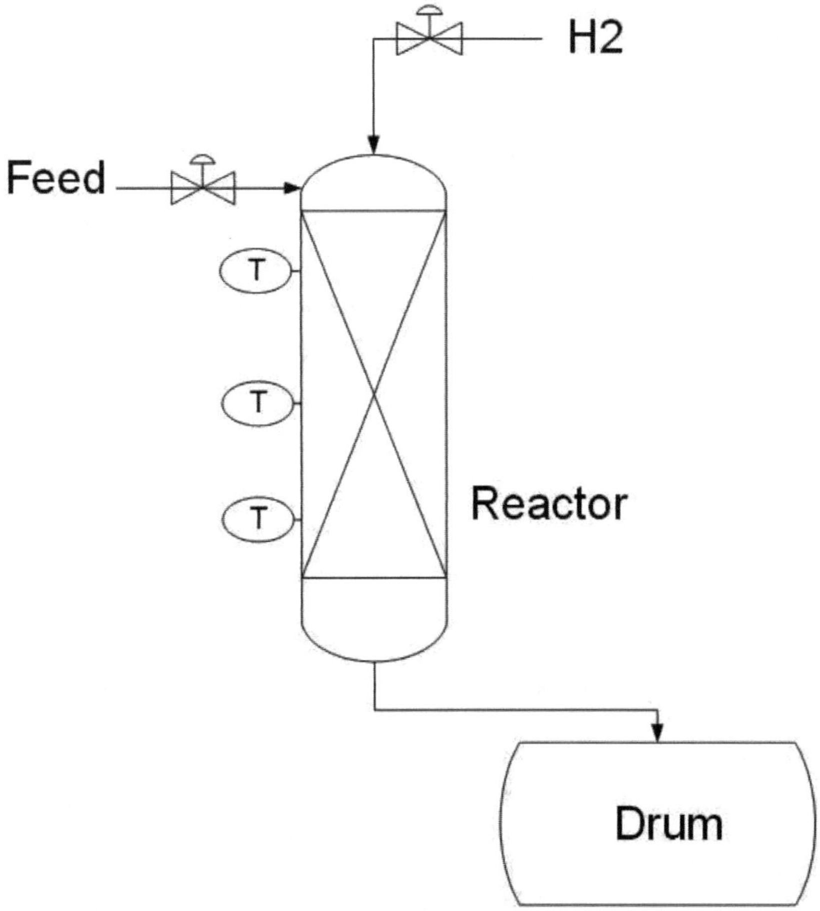

Figure 4: Schematic overview hydrogenation reactor

The reactor was filled with new catalyst during a turnaround. After startup it was noticed, during a plant tour, that the separation drum had noticeably discolored at the gas/liquid interface level (see Figure 5).

Figure 5: gas/liquid separator

This was an indication that there was still an exothermic reaction occurring at this location. Although hard proof could not be found, it was assumed that catalyst dust was entrained in the drum. With replacing hydrogen by nitrogen for the pressure control in the drum, the temperature increase at the gas/liquid interface disappeared.

Summary

Starting up a plant is one of the most critical moments within the life cycle of a plant. Three variations of startup should be studied during a process hazard analysis:

- A scheduled start up after construction
- A scheduled start up after a scheduled shutdown
- An unscheduled complete or partial start up after the activation of an interlock

Special emphasis on risk analysis during start-up and shutdown needs a continuous stress in our process safety management systems.

Process safety ... still more to be done

Still incidents are happening – only in a close region around Antwerp some severe incidents have happened over the last years. A detailed review of these incidents showed that primarily human and organizational factors are the main root cause (60%). Within Industry 4.0[2] new tools are developed such as improved communication systems between operations and maintenance, visualization software to support decision making, etc.

Conclusion

As shown with some examples, process safety concepts needs to be adjusted based on experiences observed during the life cycle of a chemical plant. Therefore an incident database is a crucial management system to provide this type of information.

Analysis of the root causes of process safety incidents over the last the three years show that many have to do with start up or non routine activities. That means that in general our process hazards analysis are correct but that more focus should be given on tools to support operations and maintenance in their cooperation and communication. Elements out of Industry 4.0 will support this further improvement and also help us to see the link between different (weak) signals.

---

[2] The term "Industrie 4.0" originates from a project in the high-tech strategy of the German government, which promotes the computerization of manufacturing. It refers to the combination of several major innovations in digital technology, all coming to maturity right now, all poised to transform the energy and manufacturing sectors.

479602 Low Temperature Safety Management Practices - an Industry Perspective

Abstract:

Cold safety considerations are a key issue for all ethylene producers. Prior industry presentations on low temperature safety have illustrated the wide diversity in safety risk management practices in this area, and there is a strong need for all producers to be as effective as possible when dealing with this critical, endemic issue. A Low Temperature Safety EPC sub-committee was formed to better understand key industry low temperature safety management elements. We will summarize the committee's understanding of these key elements concerning low temperature safety methodologies and implementation. We will also review specific elements regarding monitoring, training and stewardship while summarizing approaches to risk analysis and equipment considerations. In the end, this summary will further enhance industry low temperature safety element understanding and application.

AIChE Paper Number: 160e

Furnace Isolation for Steam Air Decoking

David Sankey

Furnace Specialist

BASF Corporation

Bert Johnson

Sales Manager

Smith Flow Control

Tim Zygula P.E

Process Engineer

BASF Corporation

Zack Givens

Account Manager

BARO Companies

Prepared for Presentation at the 2017 Spring National Meeting
Houston, Texas, March 26 - 30, 2017

Furnace Isolation for Steam Air Decoking

David Sankey

Furnace Specialist

BASF Corporation

Bert Johnson

Sales Manager

Smith Flow Control

Tim Zygula P.E

Process Engineer

BASF Corporation

Zack Givens

Account Manager

BARO Companies

Abstract:

The end of run for steam cracking furnaces is usually dictated by coke build up in the radiant coil and/or fouling of the Transfer Line Exchangers. Periodically it becomes necessary to remove this coke by "decoking". This process can be accomplished by using only steam or, more commonly, combusted with air. There are many unique methods and techniques used to accomplish this task, and there are a few operators that practice online decoking (steam/water), but most commonly the furnace piping must be isolated and cleared prior to beginning the process.

The goal of furnace preparation for decoking is to isolate the decoking air from any hydrocarbon source. In addition to various feed types, there may also be slop, quench oil and DMS/DMDS lines that must be cleared and isolated prior to the introduction of decoking air. Piping is normally isolated and drained/vented to an oily waste system or flare prior to being steamed and/or purged with nitrogen until clear of hydrocarbons. The piping is then de-pressured, drained/vented, locked and/or tagged out prior to a permit being issued for equipment blinding. Once all piping is prepared, protective clothing is donned prior to making the first break to place the blinds by the maintenance group.

Because of the number of cycles (decoke cycles * furnaces) in an ethylene facility, this task can represent a large amount of lost opportunity and an equally large opportunity for error and exposure. This paper will show one approach taken to accomplish safe isolation and sequential preparation of furnace piping system to prepare the furnace for decoking and/or maintenance activity.

Background

Operations personnel are responsible for the clearing and preparation of furnace process piping for decoking. The intent of this process is to remove hydrocarbon from the feed and utility piping and furnace prior to the introduction of decoking air.

The equipment clearing process is normally performed by steaming and/or nitrogen inerting of all piping systems to the furnace followed by the installation of blinds. Depending on the site, the maintenance group is responsible for the "blinding" task which is usually scheduled and performed as a "day" activity which somewhat limits the flexibility of the decoking plan.

Following clearing and prior to the installation of slip and/or figure eight blinds, the furnace piping must be checked, to assure the absence of hydrocarbons, locked and/or tagged, and permits issued for work. Once permitting is completed by the operations group, maintenance personnel must gather tools, gaskets, studs, nuts, and don proper PPE in preparation for a "first break", to perform the blinding task.

The use of line blinds is, arguably, an improvement over conventional blinding, making the blinding task significantly easier, allowing the operator to perform the task more economically without regard to the availability of maintenance personnel or affect/interruption to the maintenance schedule. However, there are specific items that must be closely monitored and maintained to assure this type of blind provides the positive isolation and protection from loss of containment. Latches and sealing surfaces must be checked and cleaned, "O" rings and/or gaskets must be replaced as needed or required, and actuating levers and slide gates must function to assure proper operation.

Regardless of whether slip, figure 8 or line blinds are used, bad gaskets and/or "O" rings, unlatched blinds, loose bolting, improperly prepared piping, and personnel inadvertently starting work on the wrong equipment can cause loss of containment. If these systems are left improperly prepared they can also leave hydrocarbons in the piping which can become volatile when initiating decoking air flow. To mitigate some of these issue an isolation concept was derived to avoid mechanical blinding.

Design

The design concept was intended to avoid many of the common problems encountered with blinding systems. Design considerations were to include a way of isolating piping and a sequential control method to assure piping is properly prepared and cleared. Typical isolation points for all furnaces consist of a combination of liquid and gas feeds, DMDS, slop system, quench oil and decoking air.

The isolation concept consists of a double "block and bleed" system. The "block" consists of dual "keyed" manual isolation valves for the DMDS, oily slop, quench oil and decoking air, and additional tight shut off (TSO) automated block valves for the feed supply. The "bleed" in the feed supply consists of (2oo3) pressure switches located between the "blocks" in the piping system. This assures that hydrocarbon will not leak into the furnace piping while decoking air is being used for decoking. (Figure 1)

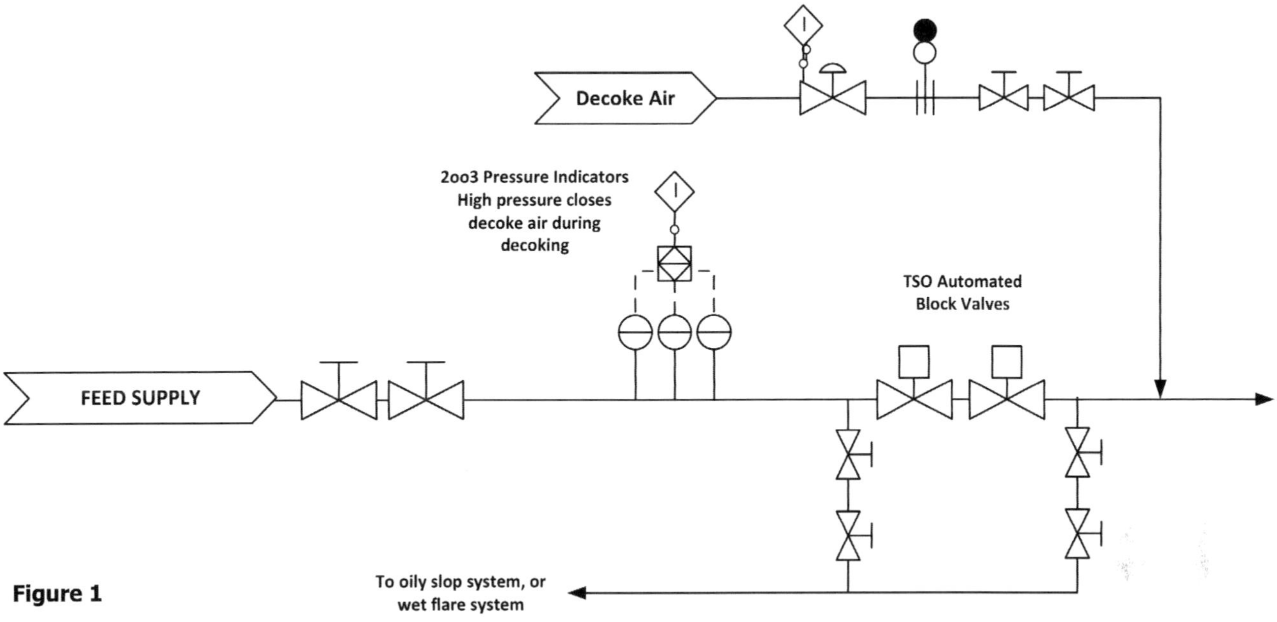

Figure 1

In place of pipe blinding, the plant utilized a system of key locked valves and the addition of (2oo3) pressure interlock system that assures hydrocarbons does not enter the furnace during the steam/air decoking process. If valves are leaking through the pressure will increase and trip the Interlock and close the decoke air automated block valve, which will stop the flow of air into the furnace. In addition to the pressure interlock the key system will have an interlock that will not allow the decoke air valve to be operated or the furnace to go into decoke mode if the preparations are not complete and the proper keys released for this operation.

All Furnaces with blind systems incorporated in the piping were replaced with the following system:

1. Two tight shut off (TSO) triple offset butterfly valves installed in the feed, DMDS, quench oil, and utility (slop and wet flare) lines. These valves are used for isolation of the feed to the furnace.
2. Two tight shut off (TSO), automated block valves were installed in the piping downstream of the manual isolation. The interlock associated with these valves will close these valves when the heater is placed in decoke mode. These valves will be released to be opened when the furnace is placed in run mode.
3. Three pressure transmitters are installed between the manual isolation valves and the automated block valves. The function of these transmitters will be to detect increased pressure from a leaking isolation valve. If a high pressure is detected, an interlock

associated with these pressure transmitters will be activated. When this happens the decoke air valve XZV valve will be closed.

4. In between the XZV valves and the triple pressure transmitters a wet flare or oily slop line is installed. These lines have two triple offset butterfly valves installed along with a swing blind. This line is used to prepare the feed and quench oil piping before the furnace is decoked.
5. Downstream of the XZV valves an oily slop line is installed. These lines will have two triple offset butterfly valves installed along with a swing blind. This line is used to prepare the feed piping before the furnace is decoked.
6. Each of the triple offset butterfly valves are installed with a key lock. The keys will be in a key control panel that allows the keys to be accessed in a particular order.
7. Figure 8 blinds will remain in the piping system for isolation and maintenance purposes only.

To assure proper preparing and clearing, specific steps had to be followed. Multiple sequential logic steps were necessary to prepare piping to slop on either side of the automated block valves. Some experience was derived from key exchange technology used for reactor switching and regeneration. This technology uses locking valves and requires a valve to be in a preset position prior to allowing another to be operated. These types of systems assure proper valve alignment to avoid process upsets.

The design team considered two options for key management and sequence control:

- Electromagnetic "key holders" which released keys when certain conditions were met, and the logic for sequencing and key control were to be managed by a PLC, or DCS in combination with the "Triconex" plant safety interlocking system.

- Mechanical sequencing and key control devices where a key is inserted and the device is indexed which releases the key(s) need to advance to the next step.

Although, both systems could adequately accomplish the task, the mechanical system was selected due to reduced cost of installation and ease of operation. The electromechanical system was complicated in design, installation, and maintenance. Another consideration was the need to minimize DCS controls traffic and the addition of points in the DCS and "Triconex" safety system.

Application

Mechanical process interlocking has been around for some time. The origin goes back to electrical switchgear. Companies which provided a positive means of electrical isolation through a unique key application were asked to provide possible solutions to similar valve issues in the North Sea. This simple request spawned an entire industry based on Safe Sequential Valve Operation.

Mechanical Process control is a simple act of positively verifying the completion of steps in a dictated order allowing for little to no variations. The need for this type of product came from growing industries working with volatile products, and/or higher pressures. A byproduct of this need is "Key Exchange Technology".

Key Exchange Technology is a permissive system that depends on the transfer of a key of sorts between each step in the operation. There have been various adaptations of this concept throughout the industry utilizing this technology in many ways. By using the single base concept, the complexity of the operation can be varied from a simple pair of valves to very complex systems involving many valves, similar to isolation and clearing of Steam Cracking Furnace operations.

The process of isolating and clearing the Steam Cracking furnace is a non-linear application so basic valve to valve isolation didn't cover all the directives given by BASF TOTAL Petrochemicals LLC (BTP). This application required the use of equipment designed to keep certain operations independent of each other.
Also, in some instances valves are to be opened then closed prior to moving to the next step in the process this operation needs to be confirmed as action taken.

Looking at the BTP furnace diagram (Figure 2) you can see that there are 4 distinct paths of operation.

1. Light naphtha feed
2. Heavy naphtha feed
3. DMDS
4. Decoke air

The first three being a function of the process. While the Decoke air is a function of cleaning the furnace. Light feed, heavy feed and DMDS each function independent of each other in the interlocking sequence. Decoke Air may only be accessed when all other processes have been blown down, cleared and isolated as required by BTP Operational guidelines. This is done through Key Exchange by requiring a representative key, verifying the completion of the appropriate steps have been completed, is inserted into a 'Key Selector Unit' (KSU). When the representative key from each process path is in the KSU only then may the key allowing access to the Decoke application be released.

Breaking the procedure down further, the Light and Heavy Feed lines must be cleared of residual product in each direction prior to beginning the Decoking procedure. The steps to accomplish this are non-linear and require isolation of an upstream block valve, opening an alternate path to clear the line, then re-isolating that alternate path before closing the downstream block valve and clearing the balance of the system through an alternate path. Then that path must be again re-isolated before moving to the Decoke process.

In a key exchange system, this can be accomplished with a Sequence Control Unit (SCU). The SCU allows for verification of operations as the progress through a series of traditional components and a machined component between them which gages which keys are in and allowing a turn which allows designed keys to be released.

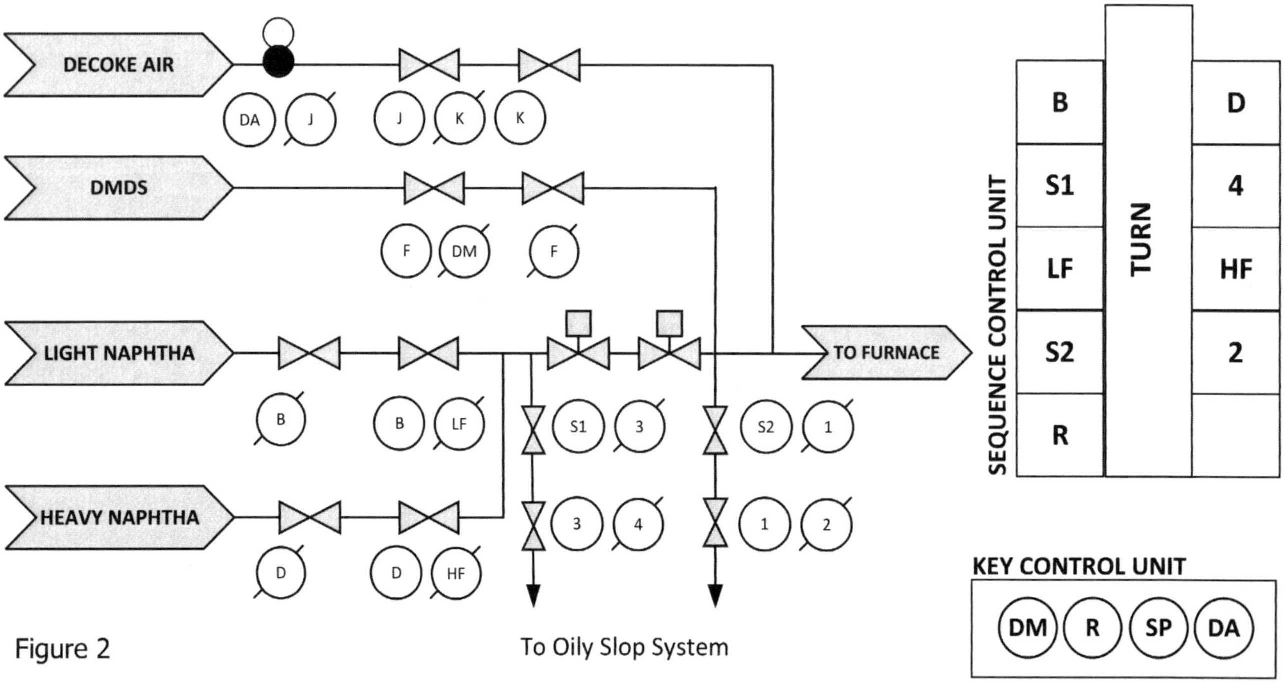

Figure 2

Installation

A successful installation is crucial to ensure that the interlock system is being used to its greatest potential. There are two different approaches that can be taken to install the interlocks onto the isolation valves. If the valves are already installed into the current piping configuration, then the locks can be installed in the field by a service technician. Since BTP was already in the process of updating piping and adding new isolation valves, a decision was made to send the new isolation valves to the installer's facility for interlock installation.

Although field installation is an option, a shop environment is almost always preferred over a field environment for several reasons. Being able to set up all the valves outside of the plant and install all of the interlocks in the correct sequence is much easier than having to work around issues that could arise in the field. We could function test all of the valves in the shop once the interlocks were installed and we were also able to easily adjust any of the valve stops/key release positions if necessary. Being able to walk through the interlock sequence in the shop prior to the valves being installed in the field was very important to BTP to ensure that the valve field installation would be as seamless as possible.

Even though the majority of the interlock installation was executed at the installer's facility, BTP required the installer company's technician to oversee the valve field installation so that any discovery issues could be quickly resolved in the field. This ensured that the overall installation was not delayed and all the furnace work could move forward as scheduled. After the valves were installed into the piping, the installation technician was able to double check that the locks/valves were sequentially correct and finalize the installation.

To ensure that the interlock system is being used correctly and piping is being isolated properly, a factory technician was scheduled to provide interlock training to anyone that would be involved in using or maintaining the system. Onsite training was provided by BARO in conjunction with Smith flow in a classroom setting to describe the purpose of the interlock system and to give a general overview of the key sequence. The classroom training was followed by a field visit to physically demonstrate how the interlock system worked and the importance of using the equipment properly. Although this may seem like a minor detail, the fact that both operations and maintenance personnel appeared to easily grasp the concept of the interlock system in the field, demonstrated the simplicity of the system and why mechanical isolation verification is so important to perform the task at hand safely and efficiently.

Preventive Maintenance

Preventive maintenance programs are common practice in industry and are necessary to reduce downtime while maintaining the longevity of the product. Typically, you would maintain a piece of equipment by using a repair kit or inserting the proper lubrication specified by the manufacturer. Valve interlocks are maintained slightly different in that the lock itself does not require much maintenance but the entire valve package needs to be maintained to ensure that the lock is able to function as designed.

The function of a gate valve is very similar to the triple offset butterfly valve in that it is also a torque seated valve. Each gate valve will also take a different number of turns depending on factors such as seat material, temperature, and pressure. Gate valves are multi turn valves and since BTP chose to have gearboxes on all of the butterfly valves, they are considered multi turn as well. Due to both gate valves and triple offset butterfly valves being multi turn and having similar seating characteristics, setting and maintaining the interlocks on each type is identical.

How the valve package is maintained is determined by the valve selection, BTP used mostly triple offset butterfly valves along with a few small gate valves. Triple offset butterfly valves are torque seated valves which means that each valve could be unique in terms of how many hand wheel turns it takes to fully operate the valve. Where a quarter turn ball valve is always going to take 90 degrees to shut, the design of the butterfly valve is slightly different. Having a torque seated butterfly valve enables you to have zero leakage by torqueing the butterfly disc into the valve seat. Understanding the design of the valve that the interlock is being placed onto is very important to ensure that the lock itself will work properly over time.

The valve directly affects the function of the interlock because the key release points are set up to either release or capture the key based on the position of the valve. Upon initial set up, the valve is fully opened and fully closed to set the key release positions. The key release positions will change over time due to the valve seats changing directly affecting where the valves fully closed position is. Over time, a combination factors will cause the valve to seat off differently. Vibration, temperature variation, process conditions, and operation will cause the valve seats to change. This means that the key release positions of the interlock could benefit from regular checks and maintenance.

Required maintenance is as simple as fully stroking the valve while tightening the corresponding set screw to adjust for the new open and closed points of the valve travel. How frequently this should be done depends on the exact process in which the valves/locks are installed. However, locks and key release positions can be checked regularly by adding this procedure as a task in the preventive maintenance program already in place. Certified technicians can come on site to assist in maintaining the locks or the technicians can train the current staff to maintain the locks internally. Due to having such a large number of interlocks installed, BTP chose to have the manufacturer come on site to train the maintenance group on proper interlock maintenance as well as having the certified technicians come on site occasionally to double check the interlocks.

Besides maintaining the key release positions, keeping all the bolting/bracketry tight is the only other maintenance point the interlocks require. As with any piece of equipment, bolting and mounting hardware has the potential to loosen over time due to temperature, vibration, and harsh environment.

One reason that the BTP interlock installation was considered a success is the fact that the internal maintenance group has a great knowledge on how the locks should work as well as keeping a close relationship with the distributor/manufacturer technicians should BTP need additional support. However, for a "self-maintenance" program to work, operators may want to consider periodic retraining at scheduled intervals. Other than the few simple tasks of maintaining the key release positions and ensuring that the entire valve package is sound, valve interlocks are typically maintenance free.

Lessons Learned

- All design ideas can be simulated to assure the logic doesn't circumvent the operation or sequencing of the valves.
- Operations should be made aware that a system of this type is a safety system. Modifying and/or circumventing the design of the equipment may be considered a violation of certain plant safety practices.
- Due to past experiences with common key storage areas for these types of valves it was imperative that the keys were to be managed in the field where the operators need them. This avoids misplacement of the keys for the system as the system needs all key in place regardless of the modes of operation.
- Operators might consider that design be done in conjunction with plant engineering, operations, maintenance, and vendors so all become familiar with the system it will be applied to and the capabilities and functionality of the hardware being installed.

- A preventative maintenance program can help maintain the operability and the reliability of the system. This may include training site personnel to install, adjust and maintain the equipment, or engaging a capable contractor or rep company that is qualified by the manufacturer to provide this service.
- Operators might consider installing hardware in the system in a manner that avoids easy disassembly. This may include tacking nuts to bolts where called for. This deters anyone from removing the mechanisms to bypass the safety system.
- Mechanical locking devices may be taken into account in a process hazards analysis and may eliminate or reduce the need for costly installations of instruments.

160f

MANAGING HAZARDOUS CHEMICAL SAFETY KNOWLEDGE

James Hillier
Process Engineer
Chevron Phillips Chemical Company LP

Ralph Price
Process Engineer
Chevron Phillips Chemical Company LP

Maruti Bhandarkar
Process Engineering Team Leader
Chevron Phillips Chemical Company LP

Prepared for Presentation at the 2017 Spring National Meeting
San Antonio, Texas, March, 26-30, 2017

MANAGING HAZARDOUS CHEMICAL SAFETY KNOWLEDGE

James Hillier
Process Engineer
Chevron Phillips Chemical Company LP
Ralph Price
Process Engineer
Chevron Phillips Chemical Company LP
Maruti Bhandarkar
Process Engineering Team Leader
Chevron Phillips Chemical Company LP

Abstract:

It is useful for chemical manufacturers to ensure that hazardous chemical safety knowledge which is accumulated over the years is properly managed, so that it can be disseminated among the employees that require it in their regular tasks. Further, as companies continue to face significant workforce changes that result in transition from veterans to new workers, it is important to find ways to bridge safety knowledge gaps. Chevron Phillips Chemical Company has been developing an in-house knowledge base for several hazardous chemicals to capture, maintain and readily access key information. Our experience in this regard will be shared using ethylene, acetylene, and butadiene as examples. We will cover the type of information contained in the knowledge base, as well as implementation and oversight of the system. We will also discuss some of the early benefits that were realized by having access to a safety knowledge base.

Introduction:

A single accident or incident in the chemical industry can cause personnel injuries, death, hundreds of millions of dollars of monetary loss, loss of profit, environmental damage, as well as tarnish a company's public image. The root causes of incidents vary, but one area of concern in a greying workforce is insufficient individual and institutional knowledge about the hazards of specific chemicals. Though many companies have engineering design standards and other hazardous chemical safety documentation, in practice employees often first rely on coworkers and their own experience and expertise to recognize risks and hazards associated with hazardous and reactive chemicals when faced with design and operational decisions. Growing, managing, transmitting, and utilizing the "tribal knowledge" associated with specific hazardous chemicals is an important activity to ensure hazardous chemical safety knowledge remains robust. Ideally, the safety knowledge about hazardous chemicals is accumulated by the next generation of technical personnel through years of working under the guidance of senior engineers and scientists. When this model is not possible, another means to capture the institutional knowledge and prepare the next generation of experts must be developed and implemented.

A possible first step to lay the foundation for future company experts is the creation and maintenance of a knowledge base about specific chemicals. Knowledge bases are common among computer programming companies and similar industries and a similar system can be beneficial to the chemical industry. The collection and evaluation of specific knowledge lays the foundation of a knowledge base and can help bridge an engineer's safety knowledge gap and starts him or her on the path to being an expert. These experts are invaluable to the company to disseminate information to others as required for their tasks. Some of the benefits/issues in creating a knowledge base and some key steps in the knowledge base creation process are discussed in this paper.

Several benefits exist for a company to develop and maintain a knowledge base related to specific chemicals, especially those chemicals found in multiple company processes. These benefits include:

1. Develop personnel with specific specialized knowledge not regularly taught in academic settings and address lack of experience in work force due to personnel changes (i.e. retirements)
2. Collect and make available relevant and reliable information for faster and better decision making
3. Gather relevant company and industry incidents and lessons learned
4. Review, update, and error check internal documents
5. Identify safety knowledge gaps and work to close those gaps

The above list is not comprehensive and other benefits will be realized throughout the process.

Key Benefits:

Develop Personnel:

One of the main benefits in collecting and compiling the information for a knowledge base is that it can be used to develop personnel to have specific knowledge. Ideally, the safety knowledge about hazardous chemicals is accumulated by the next generation of experts through years of working under the guidance of senior engineers and scientists. Unfortunately, business cycles and personnel turnovers sometimes interrupt this knowledge transfer mechanism and companies can struggle with training the next company expert(s). When a company is faced with personnel knowledge and experience gaps it needs to invest resources in developing people to fill those gaps, or expect incidents from insufficient knowledge. As an individual(s) is tasked with assembling a knowledge base about a specific chemical, the foundation is laid for their development as a company expert.

A knowledge base can also be part of a company's succession plan. An example of this situation was the planned retirement of the company expert in butadiene. Before the expert's retirement, work on a butadiene knowledge base was initiated and the expert and a mid-career employee worked together to establish the knowledge base. The mid-career employee invested many hours in collecting and reviewing internal and external documentation about this chemical and is now recognized as an expert. This employee is now involved in assisting other company personnel in decisions and questions around the safety of the chemical. In this specific situation, the realization that the company expert was approaching retirement allowed for several months of co-development and review of information to assist the mid-career employee in the endeavor.

Developing expertise by compiling a knowledge base challenges engineers to continue learning and develop skills and knowledge to enhance safety and operational discipline. Without the assignment to assemble a knowledge base, an individual is likely to only study a chemical within his or her immediate concern or job responsibility, for example when there is an incident or to prepare for an upcoming PHA (Process Hazard Analysis) or HAZOP (Hazard and Operability Study). Even then, this study is likely only cursory and information uncovered will not be documented so that it will be useful in the future. Therefore, the assignment to develop a knowledge base is anticipated to result in the developer becoming a company recognized safety expert on his/her assigned chemical.

Collect and Make Available Relevant and Reliable Information:

A quote attributed to Lew Platt, a former CEO of Hewlett-Packard, captures the motivation to collect and make information accessible, "If only HP knew what HP knows, we would be three times more productive!"[1] This relates to both the underlying information as well as the personnel that developed that knowledge. Information is vital to our jobs. A major benefit of a knowledge base is to collect information so that others within the organization can utilize it. In the information age, the availability of information has grown to nearly unmanageable quantities. A stereotypical younger engineer working in a chemical plant is just as likely to research a chemical on the internet as turning to a published source such as a chemical handbook. Significant amounts of time may be spent looking for information on a particular chemical, especially if internet sources do not agree or are incomplete.

As the knowledge base is compiled, references and digital copies (when permitted by copyright) may be linked so that employees researching a particular aspect of a chemical are able to quickly access the source material if additional information is desired. This format allows for extensive amounts of information to be referenced to supplement the information in the knowledge base. A knowledge base differs from a SDS (Safety Data Sheet) or internal standard in that the knowledge base may accommodate a wide variety of information formats and provide supplementary reference materials for further study. Additionally, it is useful to summarize important information so that it can be used for reference and training.

In addition to the physical knowledge base, it is important to recognize and utilize the individual(s) that collected, compiled, and summarized the knowledge base as a resource when he/she can add value. Since considerable effort is expended to collect and begin developing an expert, it makes sense to utilize the employee similar to how other experts are utilized and use their knowledge to help make better decisions. As an example, the the developer of the acetylene knowledge base was assigned to a project involved with mitigating lightning as a source for decomposition of acetylene in a pipeline. The employee had researched a number of relevant topics such as associated hazards and previous pipeline incidents that were relevant to the discussions.

Gather Relevant Company Incidents and Lessons Learned:

Chemical properties are not the only useful information that may reside in a knowledge base. Information about associated hazards and incidents may also be compiled and made available. In the chemical industry, when an incident happens there is almost always an investigation to determine the root cause(s), documentation explaining the root cause(s) and mitigating steps to prevent future related incidents (often called lessons learned). Incidents in the recent past (within the last ~10 years or longer depending on the severity of the event) remain in a company's (or industry's) collective institutional knowledge and the lessons learned are often fervently followed. However, as time passes and personnel move on, the lessons learned may be forgotten unless *easily* available written documentation remains. If the information about a chemical, the prior lessons learned, and other relevant information is inconvenient or difficult to access, the probability that it is read and understood by new engineers decreases. It is likely that without a strong external motivation or a specific job/task requirement, the amount of information an individual acquires is proportional to the ease in accessing the information. In the knowledge bases, there is a section that collects relevant internal and external incidents. This section includes summaries of the incidents as well as original reports and documentation for further study, if desired.

In addition to gathering previous lessons learned, the knowledge base provides a convenient platform to add new incidents and lessons learned as they become available, especially from sources outside the company. Companies commonly have a mechanism to store internal incidents and lessons learned for various purposes but incidents at *external* facilities are not the focus of these internal databases. As the knowledge base is updated and maintained, new learnings from internal and external sources can be added and made available for personnel to review.

Identify Safety Knowledge Gaps:

Every day the collective knowledge of the chemical industry is growing. While much of the positive knowledge and know-how remain company secrets, the incidents and lessons learned from industry are often publically available to help prevent similar incidents. The *Safety Incident Database* (Subscription service at http://epc.omnibooksonline.com/index.html) available through the Ethylene Producers' Conference is just one example of an industry resource for incidents and lessons learned. These incidents can provide valuable safety knowledge. As the knowledge base is being compiled and internal company and industry incidents and lessons learned are reviewed, opportunities may be revealed to improve the safety of processes within the chemical plant.

One of the mandates that may be given to the personnel compiling the knowledge base would be to evaluate current internal practices using information from the knowledge base. A major benefit would be identifying a safety gap and correcting it before an incident happens. These benefits are important even if they can never be fully appreciated. The alternate, where a safety gap is not corrected and leads to a catastrophic incident, is fully appreciated due to the high cost, both monetarily and in personnel loss. Imagine if the NASA launch team had listened to the engineers that warned them about the potential for O-ring failure due to cold weather on the Space Shuttle Challenger.

Review, Update and Error Check Internal Documents:

As the knowledge base is compiled it should include a review of both internal and external documents for correctness and completeness. Despite best efforts, occasionally information contained in internal documents may be incorrect or incomplete, whether from an unintentional error during data entry or due to outdated or inadequate information at the time a document was written. The creation of the knowledge base gives an opportunity to review internal information and cross check that information with current knowledge.

Compile Knowledge Base:

Compiling a knowledge base that will serve a company's internal personnel can be difficult. The first step is identifying which chemicals need a knowledge base within your company. The next step is identifying who will compile and maintain the knowledge base as the custodian. The custodian is likely an employee with a broad background in an area such as process operations, process safety, R&D, or someone who has advanced academic qualifications. As the compiler and custodian of the knowledge base, he/she will have to evaluate a wide range of information and ensure that the knowledge base contains relevant and accurate information. Other possible attributes of this individual are someone who has potential for longevity in the role, time available to periodically review and update the knowledge base, and the ability to communicate with others specific information contained in the knowledge base, as needed. This lifelong learning and sharing will serve to develop a company expert.

Another step, after the custodian is identified, is to gather a team to assist in reviewing and publishing the knowledge base. The team may include some key stakeholders such as plant engineers and operators, or research engineers. The team should also include IT representation. The team has the task of developing the scope of the knowledge base, establishing the timeline for completion, and reviewing the initial knowledge base. For initial deployment, IT should be involved to determine the best method to host the knowledge base and to ensure the proper IT protocols are in place. For subsequent knowledge base chemicals, it is presumed that similar IT protocols to the first would be implemented. The knowledge bases are hosted on an internal SharePoint site.

The next step is to collect and authenticate the information. This is the step that takes the most time and is the step that begins to train the next expert. It is important to keep in mind the target audience of the knowledge base and collect relevant information. Some categories might be hazards, incidents, regulatory information, toxicity, physical properties, research articles, best practices, and other relevant topics. This information should come from historical internal company information, peer reviewed research articles, handbooks, incident databases, and, in some cases, laboratory tests. The collection process will probably be a part time assignment in addition to other duties and, in Chevron Phillips Chemical's experience it may take 50-100 man-hours in team meeting discussing the knowledge base and 100 – 200 man-hours collecting and authenticating the information.

Once the information is collected and authenticated, it should be formatted to fit the previously determined system and disseminated to interested parties. A SharePoint site was created to host all documentation. This site is accessible to all company employees through the internal intranet. It was then distributed to the project and manufacturing groups to enable further review of the information. To date the initial knowledge bases for several hazardous chemicals have been completed, with additional knowledge bases in progress. Chevron Phillips Chemical has already realized several benefits from these knowledge bases, including:

1. Based on research performed for the knowledge base, several errors in internal documentation for two chemicals were discovered and corrected.
2. A new company expert was involved in analyzing a non-routine activity involving a hazardous chemical to ensure the task was performed safely.
3. The custodian of a knowledge base was utilized to help a team understand a process safety event and devise steps to prevent similar events.
4. A new subject matter expert was trained as part of a succession plan and employee development activity.

Figure 1 is a sample look at the Ethylene Knowledge Database. It includes pdf documents, sample MAT (Minimum Allowable Temperature) calculations, and informational articles. There is also a summary document in presentation form to help review the information. The knowledge base differs from a "Fact Sheet" or "Safety Assessment" document in that it is dynamic and contains multiple forms of information and resources.

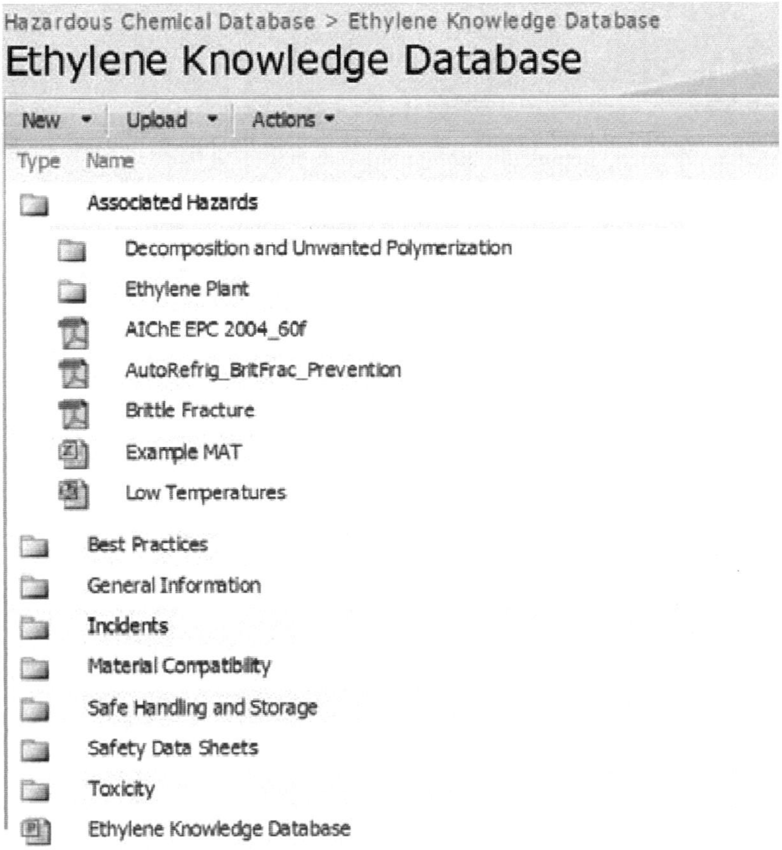

Figure 1. Example knowledge base informational categories.

Once the knowledge base is live, the custodian is tasked with maintaining it as a living database and communicating any new findings to stakeholders. If the stakeholders are invested in and use the knowledge base, they can also submit information to the custodian to be included in the knowledge base, facilitating the expansion and utilization of company knowledge.

Conclusions:

In the information age, knowledge bases are relevant to the chemical industry where knowledge is valuable for safe, reliable, and profitable operation. The effort of constructing a knowledge base about a specific hazardous chemical can serve multiple purposes including succession planning in training a company subject expert and providing a resource for interested stakeholders to gain information about a specific chemical.

References:

1. Ruggles, Rudy. "Why Knowledge? Why Not?" Web. 25 Jan. 2017. <https://www.leader-values.com/FCKfiles/File/CBI/Journal_Issue_1.pdf>.

AIChE Paper Number 187a

How big a future Cracker should you consider?

Sanjeev Kapur
Principal Consultant & President
Apex PetroConsultants, LLC

Prepared for Presentation at the 2017 Spring National Meeting
San Antonio, TX, March 29, 2017

How big a future Cracker should you consider?

Sanjeev Kapur
Principal Consultant
Apex PetroConsultants, LLC

Abstract:

The plant capacity of ethylene plants has evolved over more than last six decades. Every time businesses have struggled with the same question – how big a plant can be built? That's true even today. This overview will focus on:

- Background and trends in the industry
- Business drivers and challenges
- Factors that determine the capacity limit of the ethylene plant

2017 AICHE Ethylene Producers Conference

How big a future Cracker should you consider?

Sanjeev Kapur, Principal Consultant, Apex PetroConsultants LLC

Ethylene plants, based on the complexity of process technology and the associated controls to operate these facilities safely and efficiently, are a classical example of high technology industry. Economy of scale is one of the salient features of the high technology industries, particularly when they approach maturity in a competitive and cyclical business environment. The ethylene technology and the plant capacity have evolved over more than last six decades. The businesses have struggled with the same question about the plant capacity every time a new project is in early stages of development. The significant investment, along with long timeframe for developing and building these facilities, has inherent risks for businesses to forecast the economics accurately, specifically for the timeframe soon after these facilities will come on-stream. Businesses struggle with selection of the right project and the optimum project strategy. The success of any project, but more so for large and complex projects, depends on many factors (hard aspects including process technology, engineering design, procurement, construction, and commissioning/start-up as well as soft aspects like level of experience and skills of teams, leadership, teamwork etc.) and a constant focus on goals. The impact of hard aspects tend to be underestimated given the belief that these technologies are mature.

Industry Background

Ethylene is the largest among the seven base chemicals that form the building blocks for the petrochemical industry. Ethylene demand is driven by the end markets in automotive/ transportation, consumer products, packaging, building/construction, textiles, electrical/ electronics, health/hygiene/personal care, agricultural, medical industries etc.

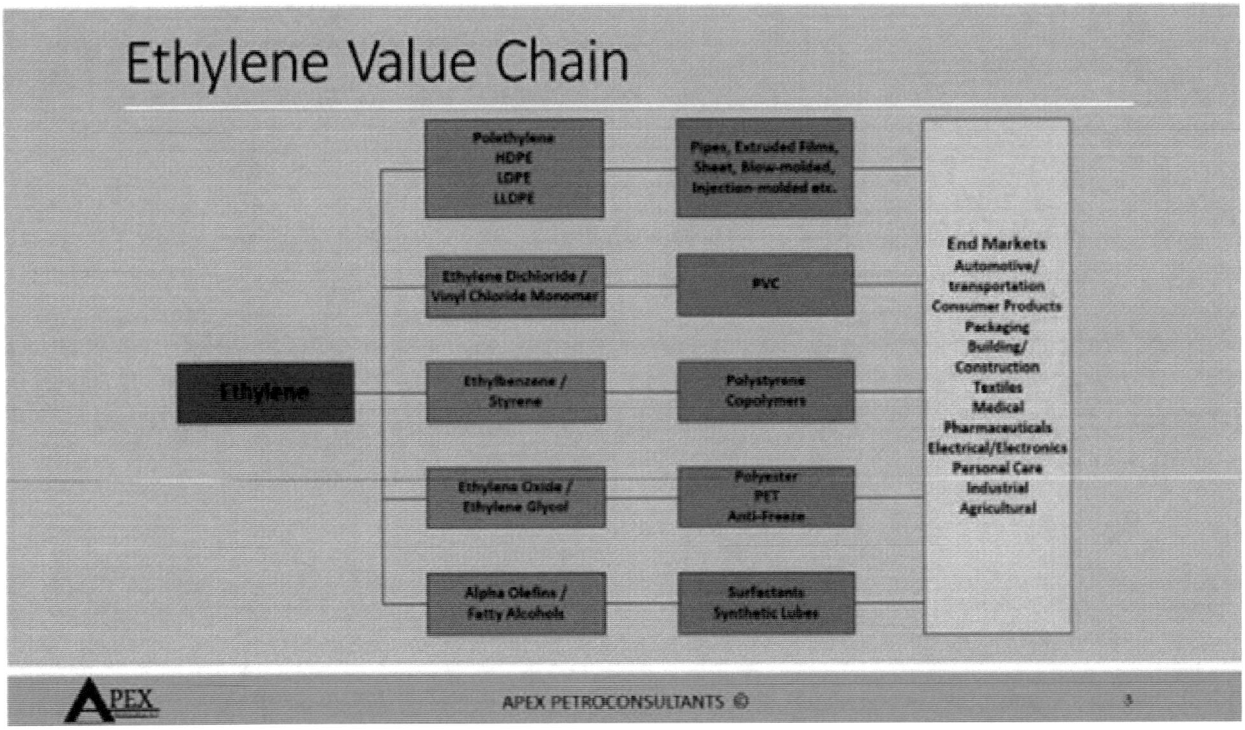

The growth in end markets is driven by the growth in GDP and population as well as trend towards urbanization. These products help improve the quality of life and make our lives easier, safer and more comfortable.

The total ethylene demand in 2015 was nearly 139 million metric tons with total name plate capacity of 160 million metric tons per annum. The overall capacity utilization rate was nearly 87% globally. Based on GDP forecasts till 2030, the ethylene demand is expected to grow by nearly 3.3% AAGR (Annual Average Growth Rate).

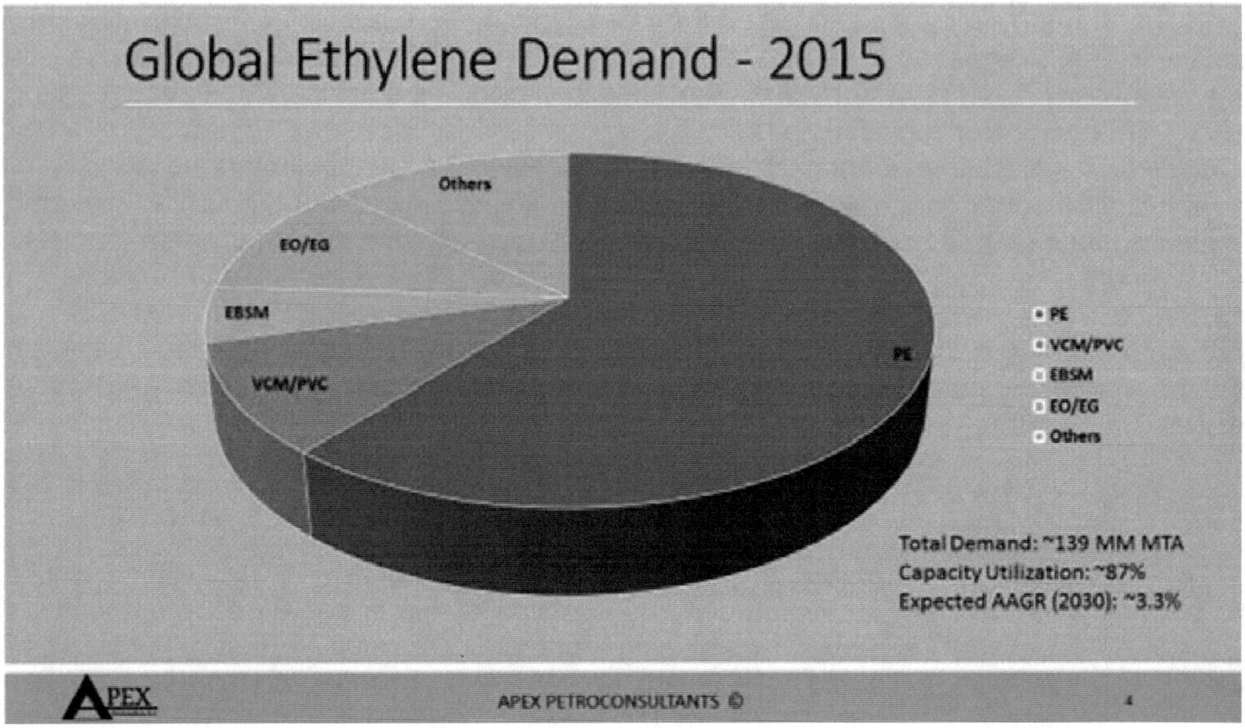

The largest consumer of ethylene is polyethylene that ends up in pipes, films, sheets and molded mechanical parts. The end applications include packaging, bags, toys and other consumer goods etc. We may not realize it but many polyethylene products come in contact with us every day. The other major derivatives of ethylene are ethylene glycol (end applications include antifreeze, PET bottles, textile etc.), vinyl chain (end applications include construction materials, PVC pipes etc.), and ethylbenzene styrene monomer (end products include polystyrene cups/packaging, synthetic rubber, appliances etc.).

The world ethylene capacity has grown by nearly 100 million metric tons over last 25 years. The capacity in 1990 was nearly 60 million metric tons per annum and in 2015 it reached close to 160 million metric tons per annum. Based on expected growth rates, another 100 million metric tons will be needed in the following 15 years (till 2030).

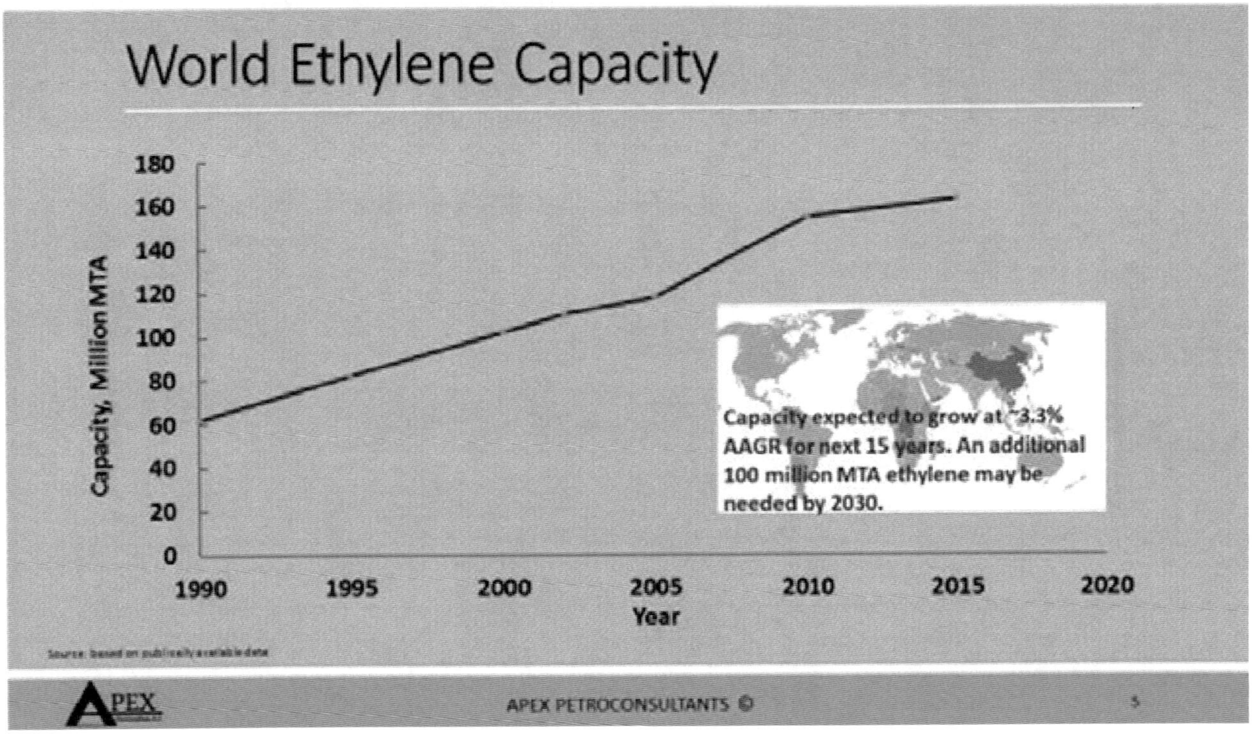

The success of petrochemical businesses depends on access to feedstock, growth markets and resources to support these high technology facilities. The economies of scale, and integration (both upstream and downstream) helps to improve competitive position through improvements in cost of production. A broader range of products, including closely held differentiating technologies improve the overall profitability and maintain competitive strength during down cycles. Global coverage to serve the customers worldwide and to provide local technical support are important factors in a highly competitive business environment.

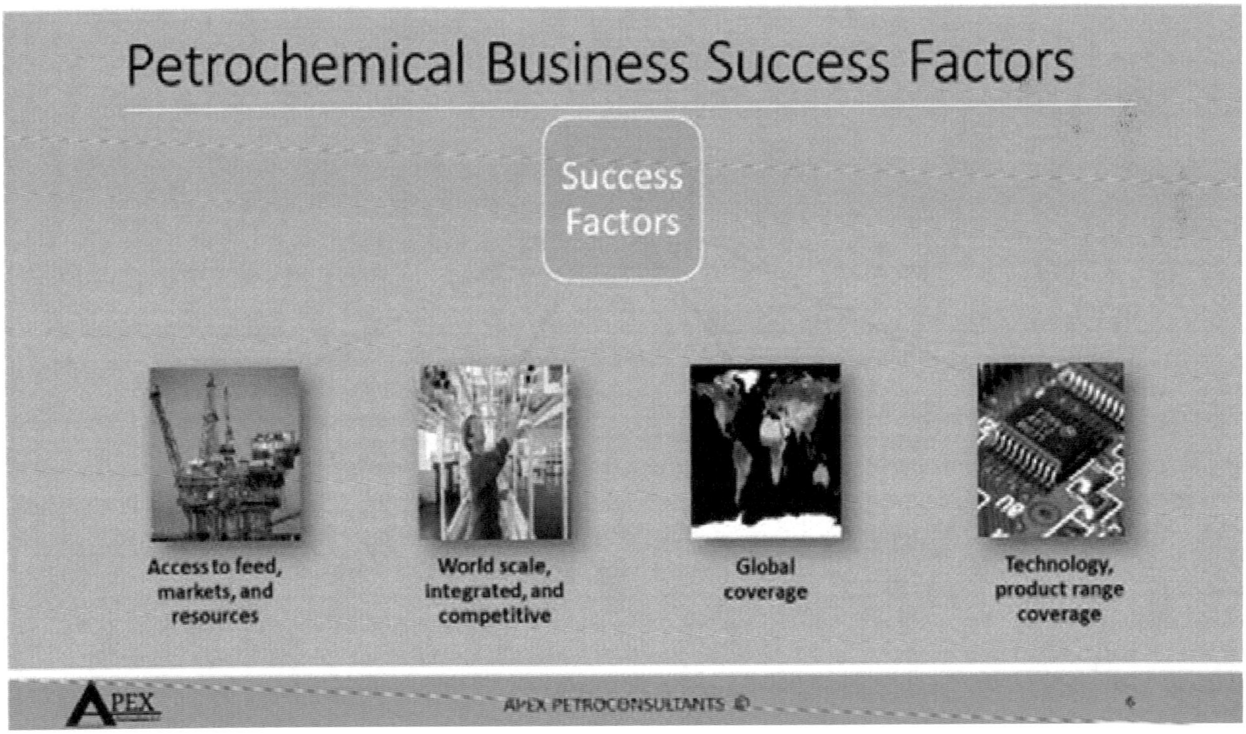

The economic viability and survival of ethylene businesses, being a competitive and mature high technology industry, depends on several contributing factors.

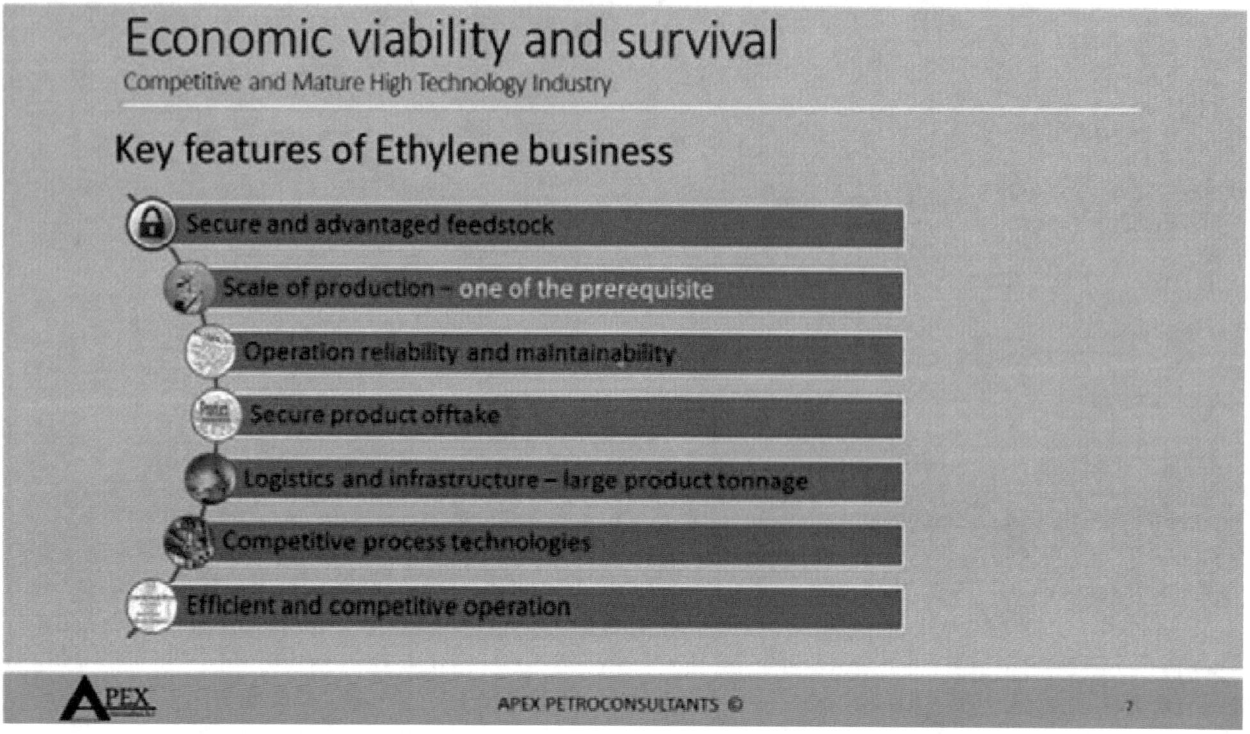

The scale of production is one of the key prerequisite for remaining competitive over a prolonged period, as the lifetime of these facilities is typically longer than 30 years.

The technology selection process can have significant impact on not only performance but also on reliability/maintainability and flexibility. The impact of proper technology choices, for site specific requirements, is often times underestimated given the belief that these technology are mature and there are no major differences in the solutions offered.

The logistics and infrastructure can have significant impact on economic viability and survival as the large capacity projects require movement of large product tonnage, potentially involving high transportation costs for delivery to growth markets.

Large Capacity Projects

The capacity of world scale ethylene plants has always trended upwards and construction of ever increasing size has been a constant theme. In the feed advantaged regions, many export oriented facilities are built significantly larger than the average world scale trend. The facilities based on heavier feedstock handle much larger amounts of hydrocarbon feed, as compared to ethane crackers, and result in high volumes of propylene/ C4 Olefins and aromatics products. These facilities have to deal with challenges related to large feed throughput as well as handling large tonnage of diverse products.

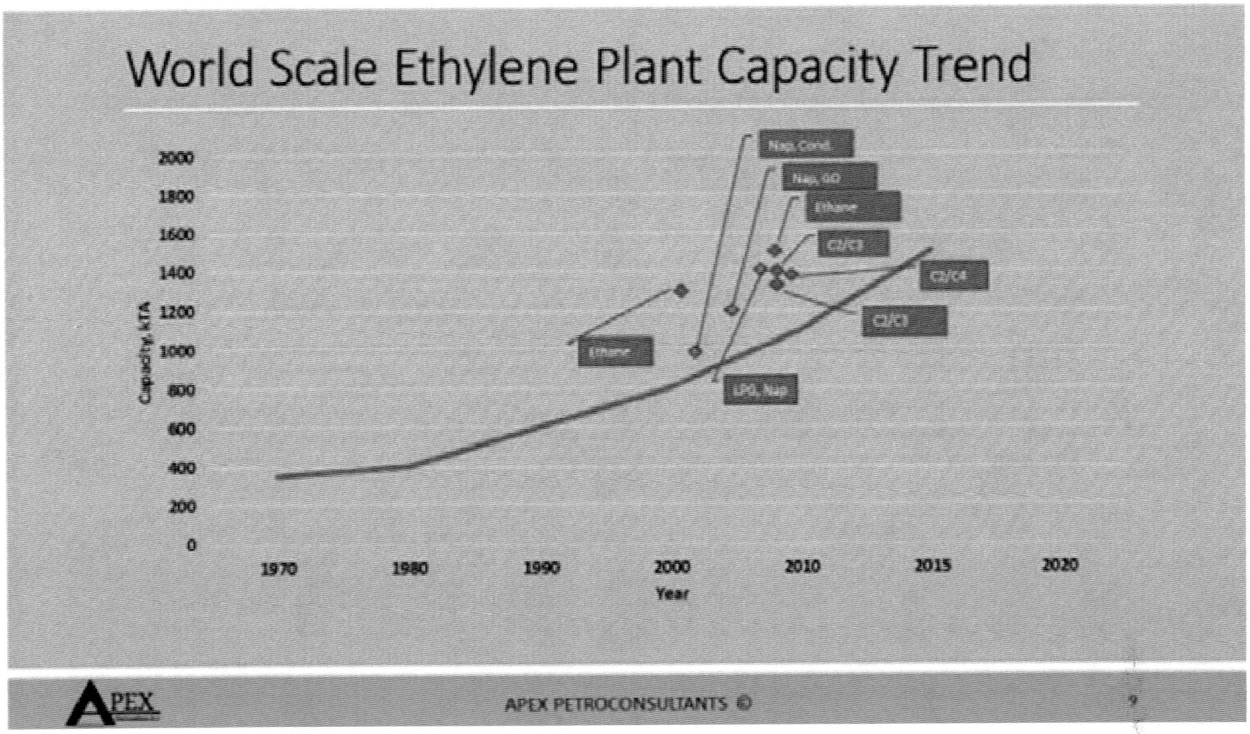

The perceived advantages of large capacity facilities are driven mainly by economics related to cost of production. The economy of scale is believed to drive the reduced investment per ton of product. The large capacity also results in lower fixed cost per ton of product.

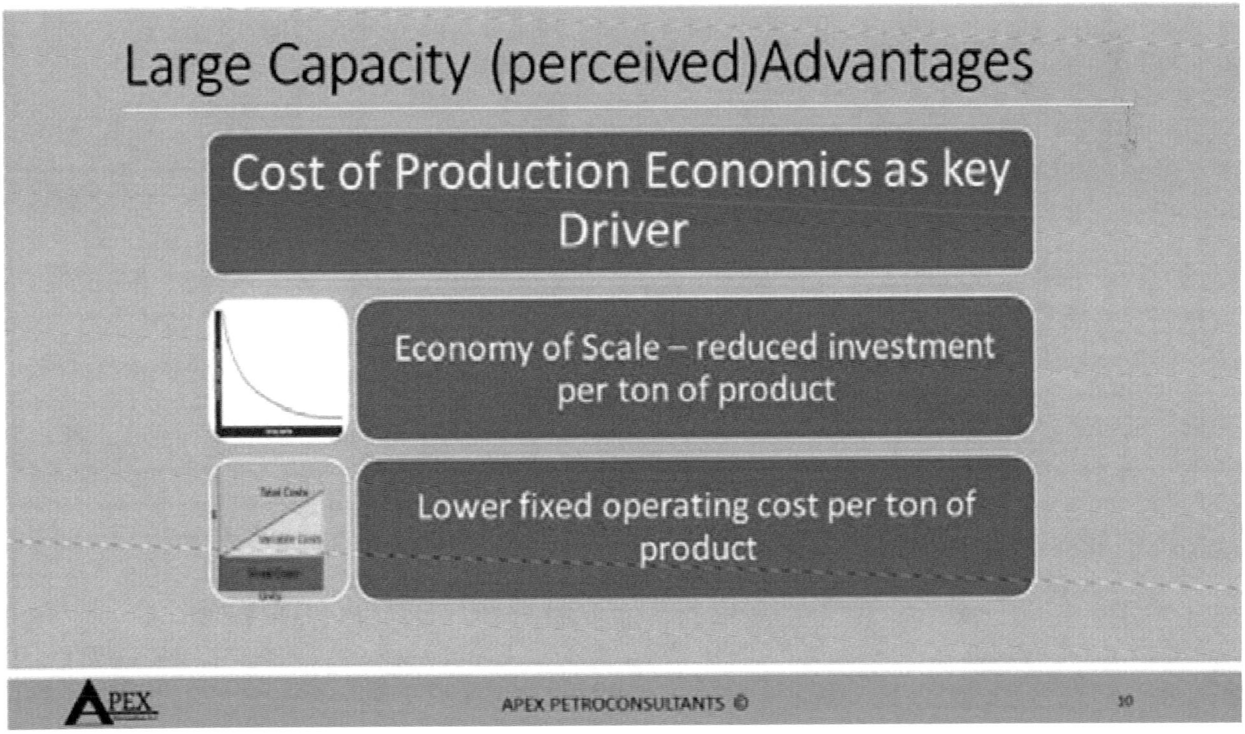

The large capacity also leads to some challenges that need to be addressed during the course of developing these facilities.

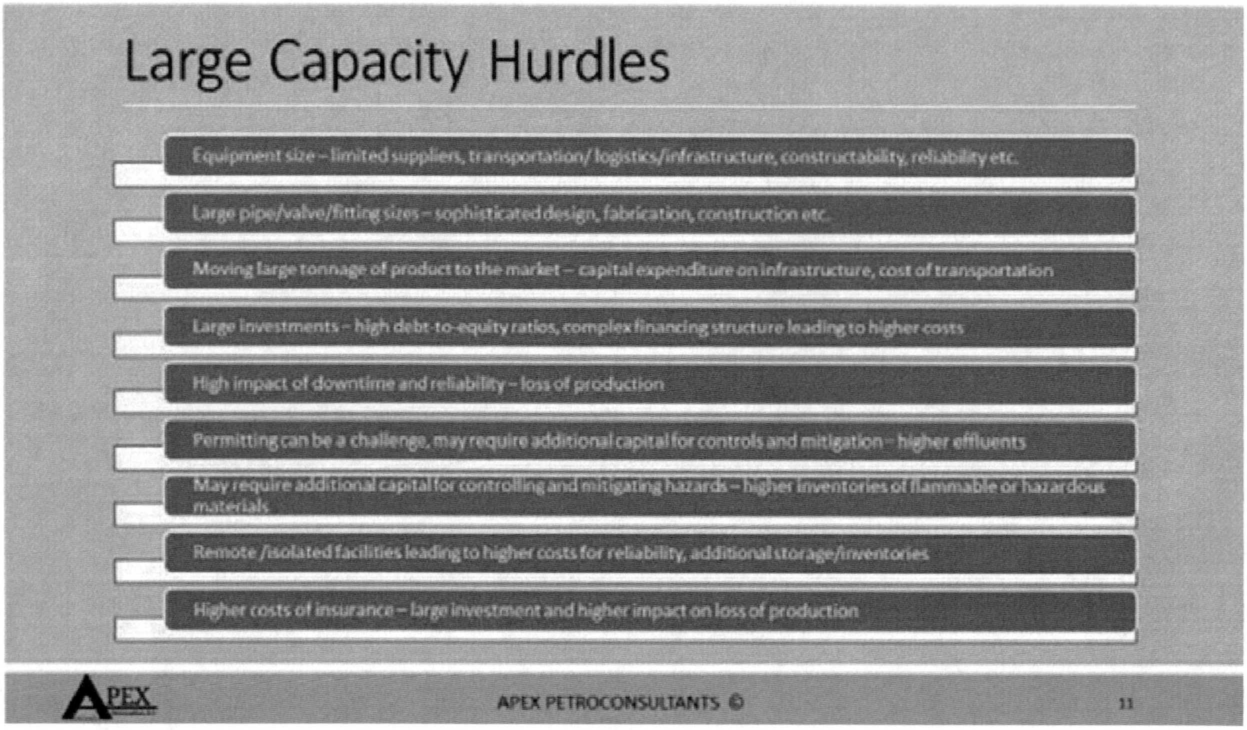

Large capacity pushes the limits of equipment sizes, potentially resulting in limiting the number of suppliers. Large sized equipment also poses challenges for transportation/ logistics/ infrastructure and constructability (impacting the sequence and heavy lifts etc.). For some of the equipment types, size may impact the reliability and that needs to be assessed early on.

Large size results in larger pipe/valves/fitting sizes and these may require more sophisticated design, fabrication and construction techniques. This impacts not only the initial investment but can have significant impact on life cycle costs.

Large capacity plants see high impact of downtime and reliability due to higher loss of production. As an example, an ethylene plant with 1500 kTA capacity will potentially lose nearly 2.5 million dollars in margin for one day production loss.

Permitting of large facilities can be a challenge, given the site location, and may require additional capital for controls and mitigation due to larger volumes of effluents.

Ethylene based petrochemical facilities are large and complex, based on most of the contributing criteria. These facilities involve high level of process complexity with all the sophisticated unit operations involved and the processing conditions. These facilities require advanced and complex controls for operating them safely while meeting high level of performance and environmental requirements. The construction, commissioning and start-up efforts pose a special challenge during planning and execution of these projects. These facilities are highly capital intensive.

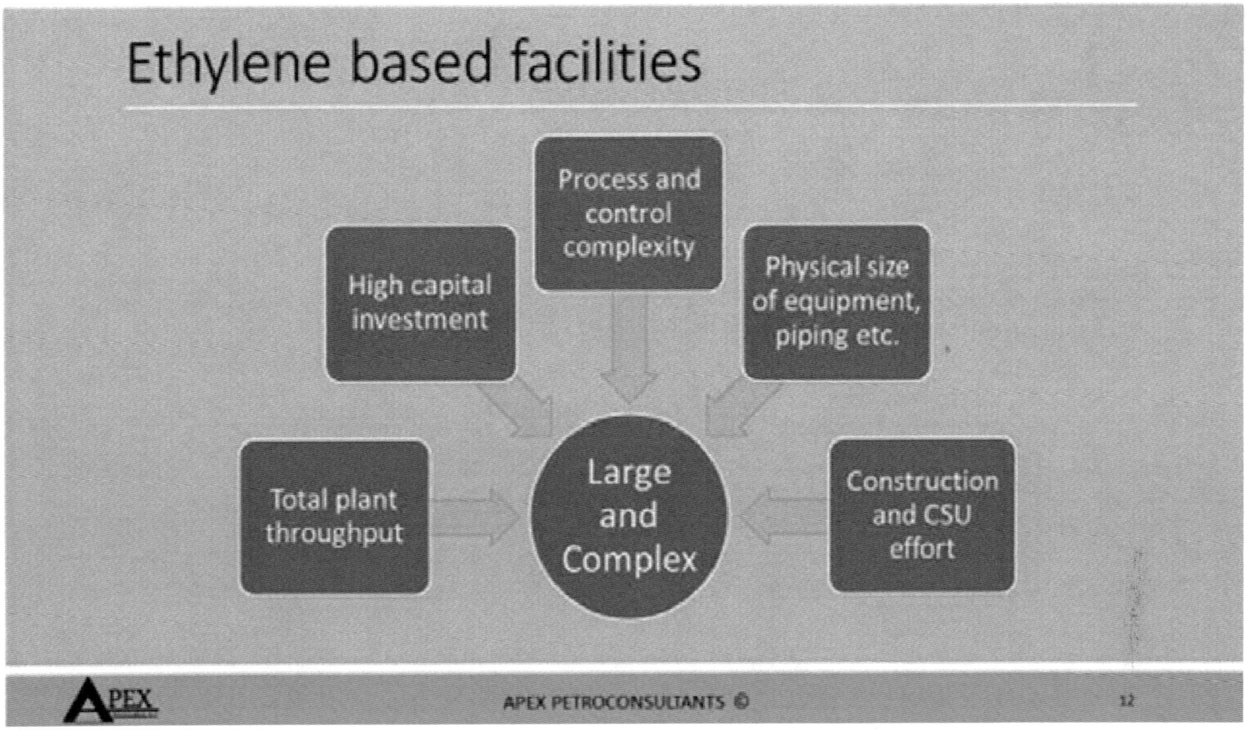

The capital cost of ethylene plants has been increasing in real terms since early 2000s. It has more than canceled out the benefits of scale. Era of escalating capital costs also makes it difficult to forecast the rates of return on capital invested.

It is nearly impossible to determine the sole impact of plant size on the capital cost due to other changes and many interactive factors occurring at the same time. As loss of production has high impact on overall economics, there is strong emphasis on reliability. The development and execution of these large facilities not only creates peak loads for equipment/ materials/ fabrication industry (particularly for items that have limited number of suppliers), it also creates a large demand for engineering resources. This peak demand often leads to higher costs and impacts the quality as well as productivity, resulting in recycles and late changes thus impacting the cost even more. Industry is losing experience and knowledge as individual retire or move-on to different roles/industry. Industry also has a tendency of lack of hiring and training during prolonged low cycles.

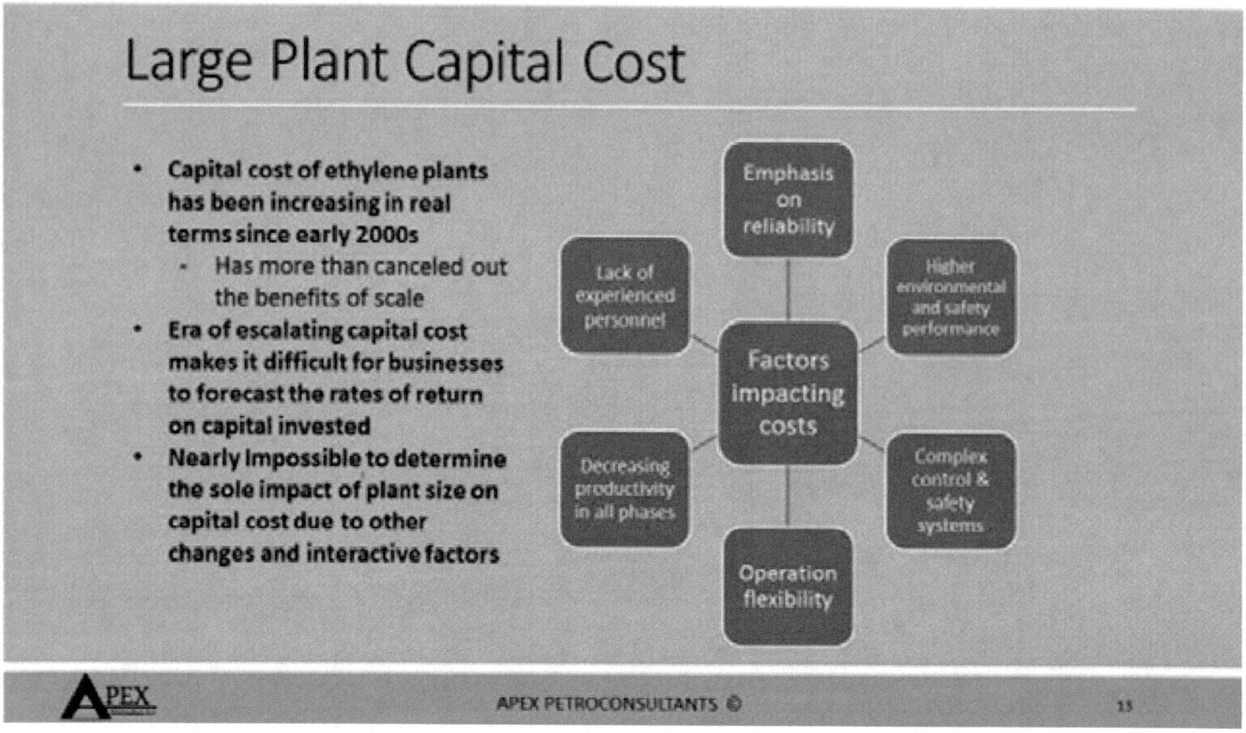

During the course of developing and executing these facilities, capital cost is estimated with different levels of accuracy during different stages. Majority of the large capital programs fail to meet their budgeted investment levels. A part of the reason is the estimation of capital cost, as it's not an exact science.

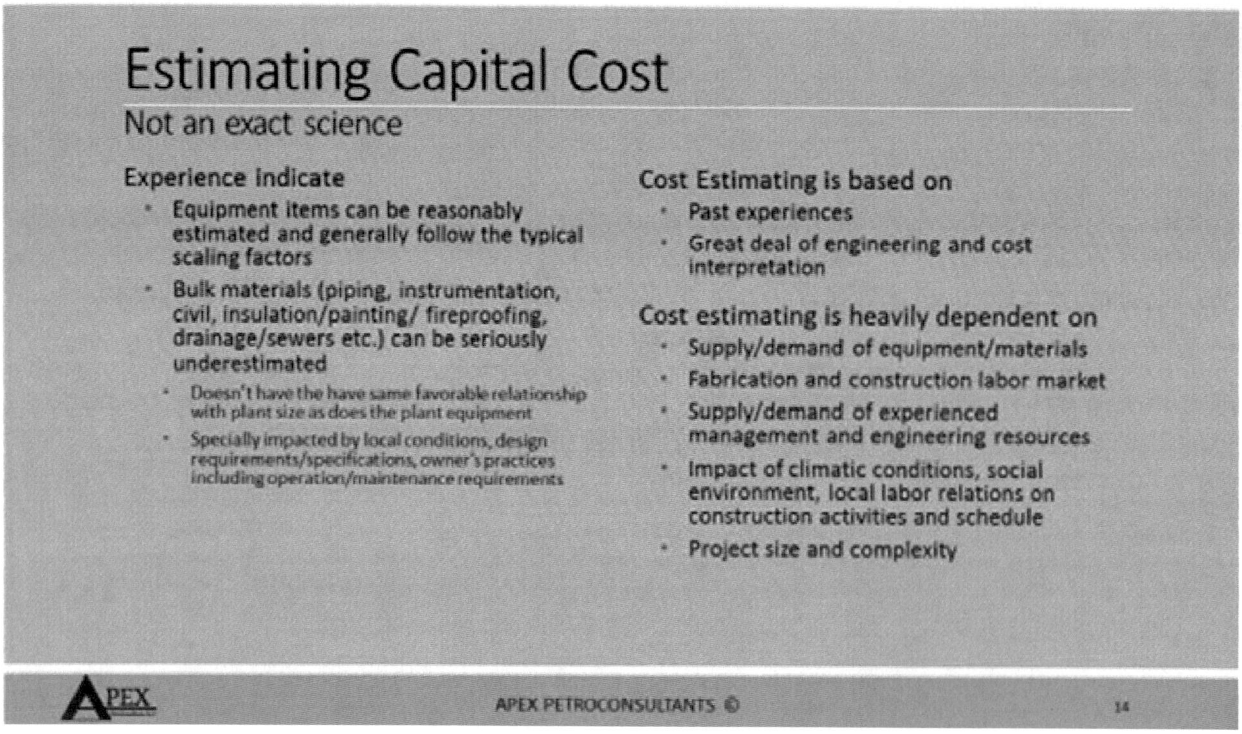

There is a high risk in estimating bulk materials, as these don't have the same favorable relationship with plant size as does the equipment. The bulk materials are specially impacted by climatic conditions, design requirements/specifications, and owner preferences. Large equipment and piping require adequate access for construction, operation and maintenance purposes resulting in larger spacing needs and plot size. This impacts directly on bulk quantities and fabrication/construction. Cost estimation requires a great deal of judgement based on past experience, interpretation of engineering data/specifications and level of details. Costs are impacted by market and macroeconomic conditions as well as local conditions. Cost is also heavily dependent on project size and complexity.

Ethylene plants (like most of the hydrocarbon industry) differ from manufacturing industry in expected productivity gains in designing/constructing these facilities over time. The plants are rarely replicated for a variety of reasons.

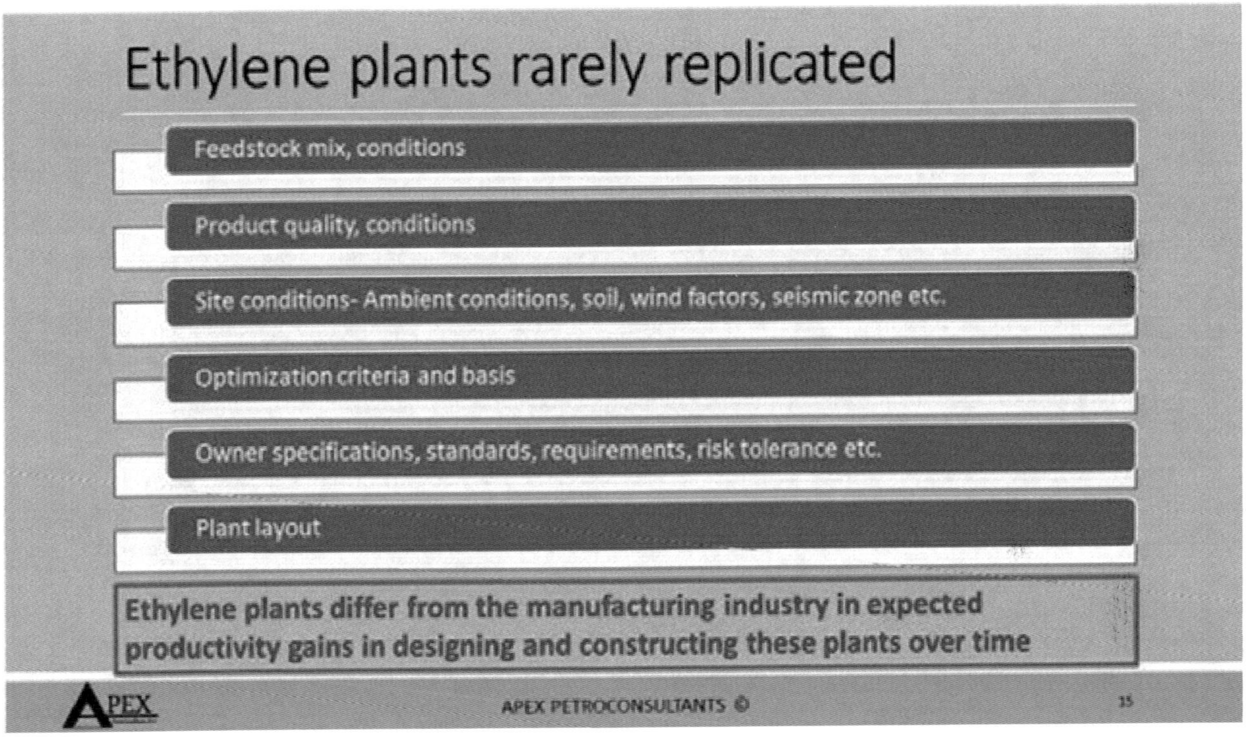

Some owners may take the approach of duplicating or replicating or introduce small changes necessitated by the business needs. This assumes that lessons learned are applied to avoid any problems experienced. The changing feedstock situation, energy costs, environmental regulations, changes in equipment design/suppliers or construction techniques may necessitate the changes.

How Large?

As explained in earlier paragraphs, the decision about plant capacity is dependent on multiple factors. These factors determine the eventual size selection basis. In addition, factors related to proven equipment technology can limit the feasible plant size (without extensive paralleling).

The economy of scale drives the large capacity plants to opt for single train machines for the cracked (or charge) gas and the refrigeration services. The capacity limit is traditionally set by the cracked gas compressor capacity. Plant reliability requirements often result in selection of proven elements for the frame, rotor assembly and all the ancillary/auxiliary systems. The steam turbine drives (along with all the ancillary systems) are also selected to meet a similar level of high reliability. Main machines have a significant impact on energy efficiency of the plant, higher efficiency is therefore essential to stay competitive. Only a few major machinery suppliers can meet these requirements for large capacity plants.

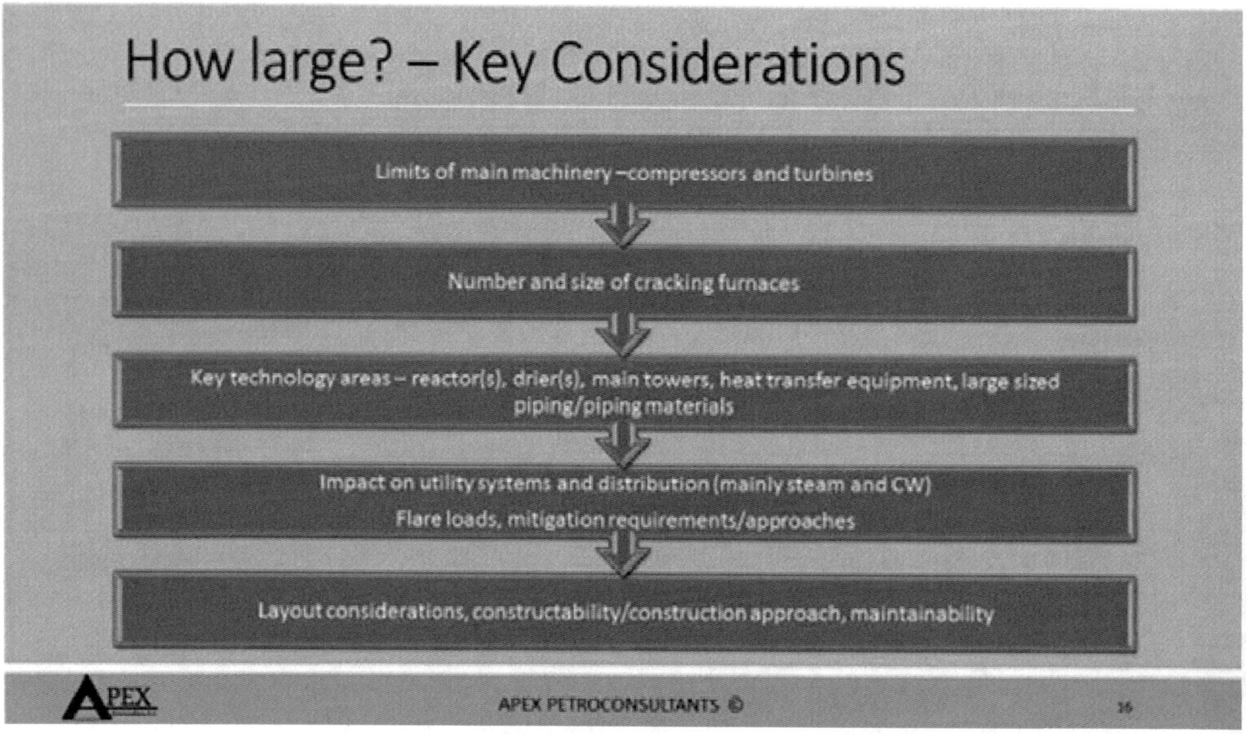

Cracking furnaces play a significant role in the overall investment, maintenance requirements, operability, flexibility and reliability. The maximum furnace size and number of furnaces are often selected based on operational considerations, reliability as well as capital investment. The size of individual furnaces has been increasing over time and generally kept up with the increase in plant size. Furnaces, being multiple for runtime limitations, do not limit the ultimate plant capacity.

Reactors and driers increase in size with the plant capacity and require sophisticated design approaches and tools to ensure high performance requirements over the longer turnaround cycles. Special care is needed to ensure proper flow distribution during all modes of operation. These equipment require advanced and complex controls/monitoring for operating them safely and meeting high level of performance.

Main fractionation towers typically use high capacity and high efficiency devices with proven operation. The diameter of these towers haven't increased as much with the plant size increases seen over time. Only a few major suppliers of fractionation trays can meet the capacity and performance requirements of the critical towers.

The improvements in heat transfer equipment have helped to minimize paralleling as the plant capacity has increased over time. The size of some of the heat exchangers have increased with capacity thus limiting the number of fabricators. The selection of design approaches can have significant impact on accessibility, maintainability and overall plant reliability.

Large size plants result in larger pipe/valves/fitting sizes and some of these may not be covered in industry standards. Very large pipe sizes typically require more sophisticated design, fabrication and construction techniques and tend have limited number of suppliers.

Large capacity plants result in very large unmitigated flare loads. Typically flare loads are mitigated to stay within the acceptable emission limits and to optimize the investment for complete flare system. Flare minimization is driven by economic and environmental pressures. The need for smokeless operation, high destruction efficiency, flame visibility, plot space, and in combination with large capacity etc. are leading to the selection of multipoint ground flares. Large size plants result in higher loads on utility systems, particularly the cooling water and the steam systems. In large facilities with multiple upstream and downstream units, integration and optimization of these systems provide opportunities for improving investment as well as overall energy efficiency. As the distribution header sizes increase, the site layout can offer significant opportunities for savings in investment and operating costs.

Large scale ethylene based investments are always challenging to develop, execute and operate. The robustness of these decisions depend on multiple interdependent factors.

High investment intensity and high economic impact of loss of production pose inherently high risk for owners/operators. Fewer decisions can be left to chance and each foreseeable occurrence tends to be covered through either monitoring, controls or safety systems, or through standardized procedures and routines. This requires a high level of

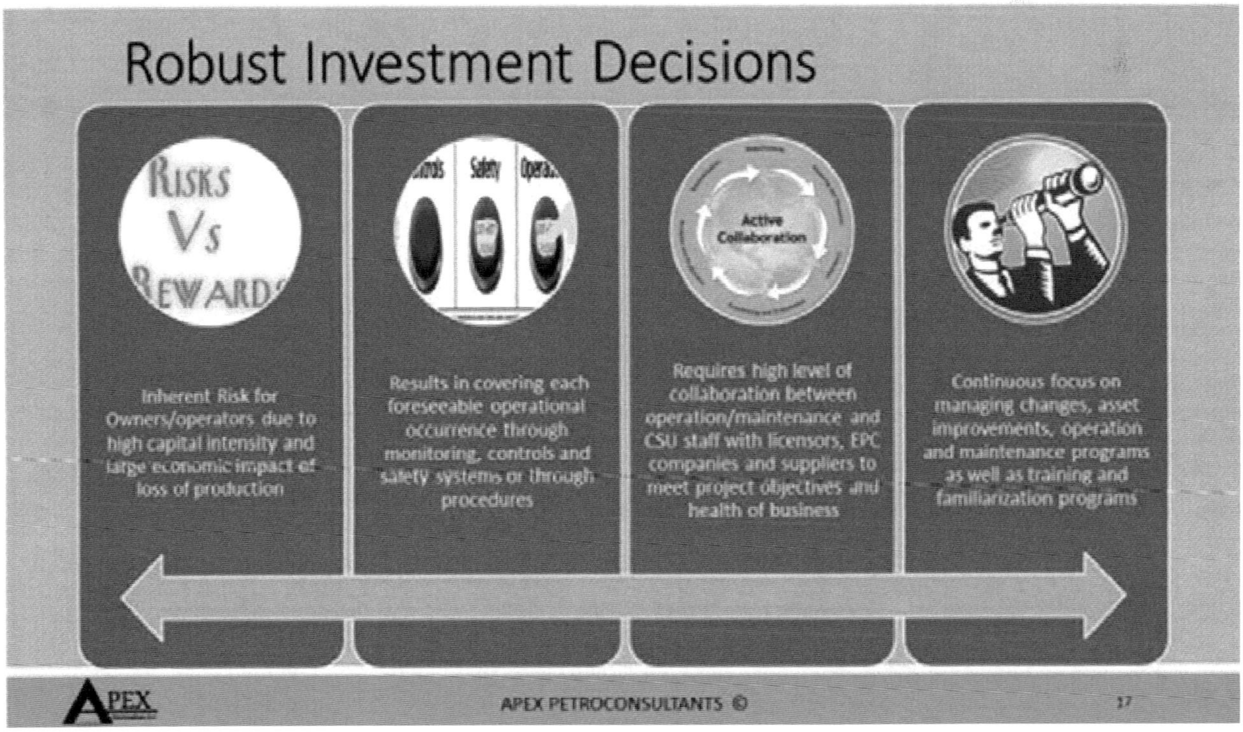

collaboration between operation/maintenance as well as commissioning/start-up teams with licensors, EPC companies and suppliers to meet the project objectives and sustainability of business over time. The long-term health of business can be assured through continuous focus on managing changes, asset improvements, operation/maintenance programs as well as training and familiarization programs. Inherent understanding of process technology, equipment/design limits and controls/safety philosophy/systems is essential for supporting these facilities effectively. It can be achieved through strong teams with right mix of experience, expertise and depth of knowledge.

Concluding thoughts

Any analysis of the consequences of plant scale selection is far from simple. There may be a tendency to carry on with simplistic approach to the economics of plant scale. Based on the feedback about the capital cost increase over the recent years, there is a need to at least challenge the perceived beliefs. A great deal will always have to be left to the informed judgement and technical expertise applied to the project specific objectives and needs. The incentive to carry out and lead this analysis lies with owners and operators due to high level of risks and rewards.

References:

1. Corn, Ron (2016), "The Extraordinary Challenges of Megaprojects," 28th Ethylene Producers Conference.
2. Panel Discussions (2016), "Innovations in Managing Capital Cost of Ethylene Based Petrochemical Investments," 28th Ethylene Producers Conference.
3. Kapur, Sanjeev et al. (2000), "Evolution of Ethylene Plant Compression Systems," 12th Ethylene Producers Conference.
4. Kapur, Sanjeev (2015), "Petrochemical Industry Overview – A Key Value Creator for the Hydrocarbon Chain," December Roundtable – Rice Global E&C Forum.
5. Weinberger, Samuel (1995), "Current & Future Challenges of the Ethylene Industry," 7th Ethylene Producers Conference

AIChE Paper Number: 187b

Building Tomorrow's Mega-Scale Ethylene Plants – General Criticality Criteria and Selected Design Aspects for Plate-fin Heat Exchangers and Cracking Furnaces

Alexander Kromp

Process Engineer Petrochemical Plants

Linde AG, Engineering Division

Gunther Kracker

General Manager Process Engineering Petrochemical Plants

Linde AG, Engineering Division

Axel Lehmacher

Senior Expert Plate Fin Heat Exchanger

Linde AG, Engineering Division

Robert Stegemann

Senior Expert Process Design Cracking Furnaces

Linde AG, Engineering Division

Mark Whitney

Principal Process Engineer Olefins

Linde Engineering North America Inc.

Prepared for Presentation at the 2017 Spring National Meeting

San Antonio, Texas, March 29, 2017

Building Tomorrow's Mega-Scale Ethylene Plants – General Criticality Criteria and Selected Design Aspects for Plate-fin Heat Exchangers and Cracking Furnaces

Alexander Kromp

Process Engineer Petrochemical Plants

Linde AG, Engineering Division

Gunther Kracker

General Manager Process Engineering Petrochemical Plants

Linde AG, Engineering Division

Axel Lehmacher

Senior Expert Plate Fin Heat Exchanger

Linde AG, Engineering Division

Robert Stegemann

Senior Expert Process Design Cracking Furnaces

Linde AG, Engineering Division

Mark Whitney

Principal Process Engineer Olefins

Linde Engineering North America Inc.

Abstract: This paper presents a general approach to determine the limits for tomorrow's mega crackers with plant capacities significantly above 1500 kta. A set of general criticality criteria for screening of mega-scale equipment and/or plant components is reported. This approach renders identification of equipment-wise breakpoints for super-sized ethylene plants, i.e. capacities, beyond which further upscale does not generate economic benefit for the specific item. By this means the economies-of-scale optimized overall plant capacity can be determined for project specific boundary conditions.

The reliability of the presented method strongly depends on the underlying knowledge of equipment specifics for mega cracker services. This paper further provides detailed insights on mega-scale design challenges for two equipment types that exhibit untypical mega-scale characteristics: (i) Plate-fin heat exchangers (PFHE) are, despite their lightweight properties and their compactness, subject to parallelization already in certain services of state of the art mega crackers. Herein, the breakpoints that limit design of single PFHE blocks are reported, an overview of design and modularization concepts for further scale-up is given and associated risks and mitigations are discussed. (ii) Also cracking furnaces are highly modularized and already parallelized equipment in ethylene plants, which do not represent any limitation for achieving super-sized plant capacities. However, optimization of size and number of cracking furnaces becomes a challenging task for mega-scale crackers. In this context, the interaction of availability, capacity utilization and component costs is discussed in detail.

1. Introduction

Ethylene plants have experienced a distinct step up in capacity during the past two decades. The development of grassroots ethylene plant capacities is captured in Fig. 1. While in the early 1990s, there were few ethylene plants with an annual plant capacity above 500 kta, the average plant capacity of grassroots plants installed after 1995 showed a steep increase [1]. This trend proceeded steadily until 2010, when Borouge took its EU2 (Ethylene Unit 2) ethane cracker in Ruwais, UAE, into operation. With an annual (grass root) ethylene production rate of 1500 kta, this cracker has marked a new mega-scale capacity class, which has set the benchmark for mega-scale ethylene plants until today.

Already during the design phase of the EU2 mega cracker, the plant design was immediately copied for a second 1500 kta plant on the same site, which was then successfully started up in 2013. Based on this experience in design, execution and operation, a 1500 kta ethane cracker standard design has been developed and applied in several projects that are currently in the construction phase, also including adaptations of the design for mixed feed cracking. Today, several ethylene technology licensors are offering ethylene plants in the 1500 kta capacity class, with several plants started up or under construction.

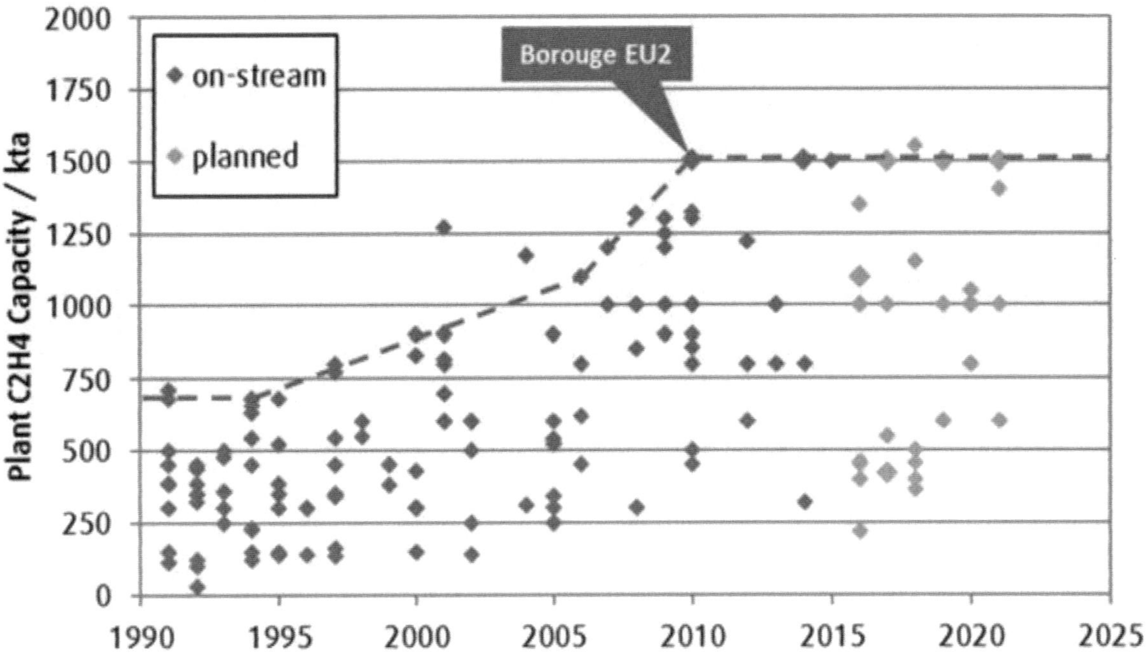

Figure 1: Global capacity trend for grassroots ethylene plants. Revamped plant capacities are not included. Planned ethylene plants represent awarded EPC projects (source: IHS, status: spring 2016).

The fundamental reason for the continuous increase in ethylene plant capacity are economies of scale, providing decreased specific capital invest and specific operation costs for increased plant capacities. With the benchmark plant capacity settled at 1500 kta, the olefins industry has now gathered several years of experience in designing, erecting and operating mega crackers. In the meantime, also process and equipment design methods, fabrication methods and testing methods have undergone a steady enhancement [2-4]. To the present day, it is hence undoubted that building ethylene plants greater than 1500 kta will generate further margin benefits.

For the players in the olefins business willing to shape the next step up in capacity, it is essential to understand in detail the scale economies in the mega cracker range. This paper reports a general methodology to determine economies-of-scale optimized plant capacities on the basis of a criticality screening on equipment level. As practical examples of mega-scale challenges for equipment design, insights on plate-fin heat exchangers and cracking furnaces are given. Both items represent equipment of LINDE's proprietary development, design and (for PFHE) manufacturing.

2. General Breakpoint Determination Methodology

In theory, the capital efficiency of a chemical plant increases with the size of the plant. This economies-of-scale effect is commonly described by a power law relation between plant costs and plant capacity [5]. While in literature a large variety of exponents of the power law relation can be found for ethylene plants [6], values close to 0.7 are most commonly assumed in the industry. This means that doubling the capacity of an ethylene plant can be realized by approx. 60% more investment. Needless to say, in practice this relation is only valid for a certain capacity range. Various factors that set barriers to the economies of scale result in certain breakpoint capacities, beyond which further scale up does not generate additional benefit.

The following paragraphs report the major criteria that need to be considered thoroughly for mega crackers. Each criterion is critical to mega-scale plant design and might turn economies of scale into diseconomies of scale. The criticality criteria have been elaborated based on the experience gained in the execution and operation of several mega crackers and in continuous investigations of the mega-cracker limitations.

Constructability / Availability (manufacturing limitations)
Besides the maximum constructible equipment size, it needs to be considered that increasing the size of the equipment is limiting the number of potential vendors. Availability of materials and fabrication capabilities can become limiting factors. The availability of suppliers is a strong decision criterion in the evaluation of mega cracker concepts.

Transport & Construction Requirements
This criterion is often the main criticality which is influencing the final decision of equipment sizes. If e.g. transport requirements are limiting the maximum dimensions of transport modules, the cost for onsite construction is increasing drastically. In this case the maximum selected equipment sizes and the degree of parallelization has to be adjusted to smaller sizes.

Mechanical / Stress Calculation
Consideration of earth quake factor or mechanical stress calculations results in equipment or piping dimensions which are difficult to handle. Therefore, especially in mega cracker design this criterion often is limiting the size of equipment. As far as the piping is concerned, mechanical stress calculation has a large impact on overall plot requirements.

Maintenance
Required plot space for accessibility, available cranes and other tools or limits when pulling heat exchanger bundles or furnace convection tubes are important aspects which have to be taken into account. Especially for maintenance also the input of the operating company of the plant has to be considered in the design.

Time Schedule
A short time schedule is one of the most important criteria for the decision if a plant concept is viable. With increasing equipment size, particularly for long lead items, considerable increase of delivery time has to be expected. Further, increased time for installation needs to be taken into account. Therefore, concepts have to consider this aspect when delivery times of equipment or installation schedules are verified.

Safety
Increased inventory of hydrocarbons and increasing flare loads have to be considered. Safety concepts have to take into account this aspect when safety instrumented systems and depressurization times are defined.

Risk and References
Large plants bear extraordinarily high risks. Therefore, mega-crackers need to rely on referenced and fully proven concepts as far as possible. When using new concepts, the factor "risk" is very important. Especially in the mega cracker design it could become critical by using concepts having no reference or using unproven equipment types. Therefore, it is very important for every new concept to perform a risk analysis considering consequences and probability of a failure.

Reliability and Plant Availability
Plant equipment, which is designed close to the constructability limits, bears a higher risk of failure. With increasing plant capacity, the importance of operation reliability becomes more important. In case of an upset of an individual unit (e.g. problems with the hydrogenation), the full production can be lost. Therefore, it needs to be checked thoroughly if a choice of equipment or a mitigation might affect the envisaged yearly operating time.

With the above mentioned criticality criteria, mega-scale breakpoints can be identified for project specific boundary conditions. To this end every critical equipment and/or plant component needs to be checked against the above listed criticality criteria.

As a starting point for the mega-scale criticality screening, it is recommended to employ a grassroots ethylene plant design that is already referenced, i.e. fully engineered, constructed and operated, in the scale of 1500 kta. The plant capacity is then continuously scaled up. A breakpoint is reached whenever, for a certain plant capacity, one of the above mentioned criteria becomes critical for a specific equipment and/or plant component. The results from the criticality check are transferred to a breakpoint evaluation matrix, as illustrated schematically in Fig. 2.

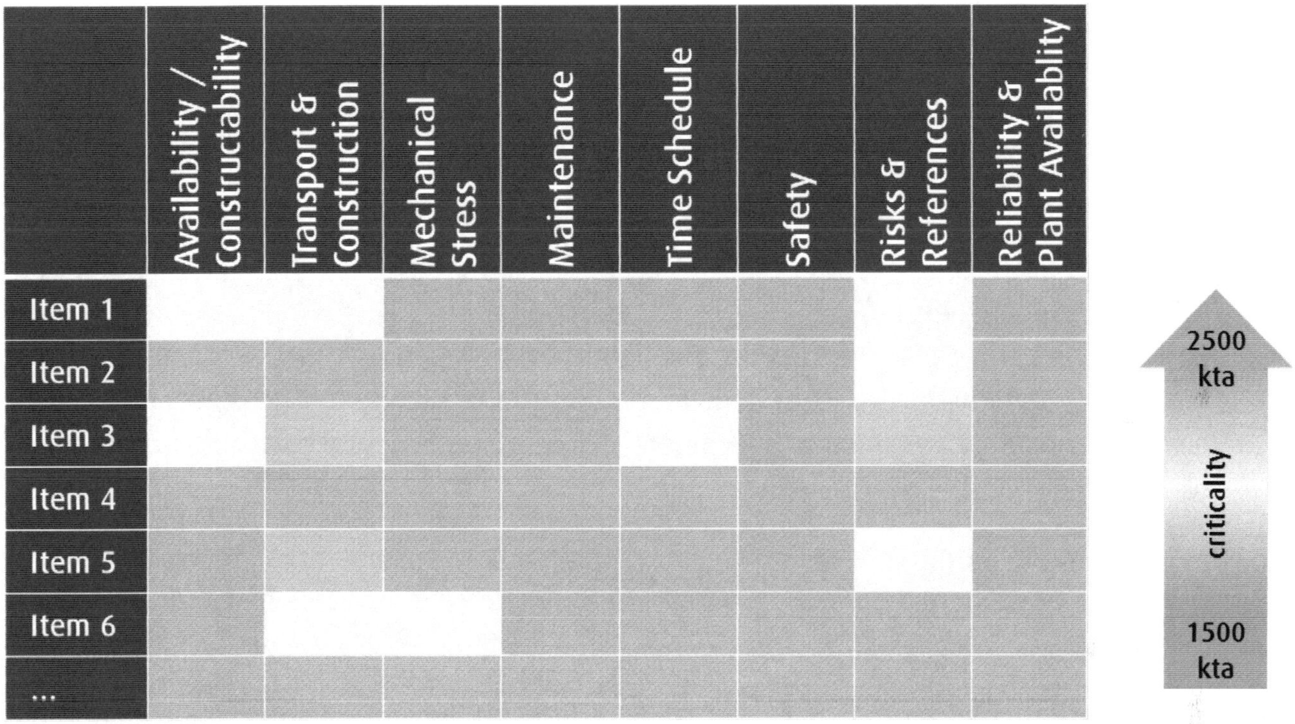

Figure 2: Schematic setup of the breakpoint evaluation matrix: Equipment, piping, plant components (items) vs. criticality criteria.

For each equipment, the lowest determined breakpoint represents the plant capacity from which a further capacity increase does not generate scale benefits for the specific equipment. It can be distinguished between major and minor breakpoints. Major breakpoints represent equipment breakpoints that have a major impact on the overall scale economies of the plant, e.g. the necessity for parallelization of main cost items like major columns or compressor trains. Equipment limitations that can be handled with moderate effort are considered as minor breakpoints. With all breakpoints obtained for any critical equipment or components of a whole plant, the plant capacity which makes optimized use of scale economies can be determined.

In case a target capacity is defined for the investigated plant scenario, the breakpoints which are below the target capacity need to be considered in more detail. These breakpoints identify the equipment items and/or plant areas, which need to be subject to further investigation in order to reach the target capacity.

The described method to determine mega-scale breakpoints relies on a consolidated database that includes comprehensive knowledge ranging from specific equipment manufacturing limits to current equipment references to highly accurate equipment and plant component cost curves. In order to keep the evaluation tool up to date, the database needs to be updated continuously.

It is important to note that the evaluation of the criticality criteria has to be executed under consideration of project specific boundary conditions (e.g. plant location, owner and authority requirements). Depending on the project specific boundary conditions, breakpoints might shift to lower or higher plant capacities and/or the impact of specific criticality criteria on the overall breakpoint evaluation might be altered. The following two examples illustrate the impact of project specific boundary conditions:

- A customer strictly requests equipment within referenced sizes. In this case, the criteria with respect to references have a high criticality. Breakpoints resulting from equipment references might thus override other breakpoints like the sheer constructability limits of the specific equipment.
- For regions with high construction and labor cost, like e.g. the U.S. gulf coast, the criticality criteria with respect to construction and time schedule accordingly need to be highly weighted, whereas other criteria like the number of available manufacturers might play a minor role.

The breakpoint evaluation matrix represents a comprehensive, reliable and convenient tool for the identification of economies-of-scale optimized mega cracker plant capacities under specific project boundary conditions. It has been successfully applied in several studies on mega-scale economies of ethylene plants far beyond today's benchmark capacities. It is important to note that the reliability of this tool strongly depends on the underlying knowledge of equipment specifics for mega cracker services. Illustrating the practical challenges that need to be considered for mega-scale cracker equipment, the following sections provide detailed insights on two equipment types that exhibit untypical mega-scale characteristics.

3. Plate-fin Heat Exchanger

Brazed aluminium plate-fin heat exchangers (PFHEs) are key components in ethylene plants. They have been implemented and operated in different services within this plant type since decades. Combining low-cost and high performance density, PFHEs offer several advantages, which make them particularly attractive for mega-scale applications:

- The heat exchangers combine a very high compactness with the possibility of integrating numerous process streams into one equipment. This leads to a considerably low plot demand as well as reduced transport limitations.
- The light weight further results in reduced demand with respect to foundation and steel structure, as well as low lifting weights.

- The application of PFHEs can save capital investment. Through the integration into other equipment, like e.g. column heads, capital equipment like reflux pumps can be saved. Also the integration into vessels as block-in-shell design is a highly efficient and cost effective alternative to standard shell-and-tube heat exchangers.

However, for the application of PFHEs in mega-scale ethylene plants, there are several limitations that have to be considered. The fabrication in brazing furnaces sets limits to the dimensions of single PFHE blocks. If a higher heat exchanger area is required, two or more individually brazed PFHE blocks may by welded together in parallel to form one composite block. Mainly depending on applied material and design pressure, composite blocks currently can reach a height up to the range of 3000 mm and a total volume with up to 25 m^3.

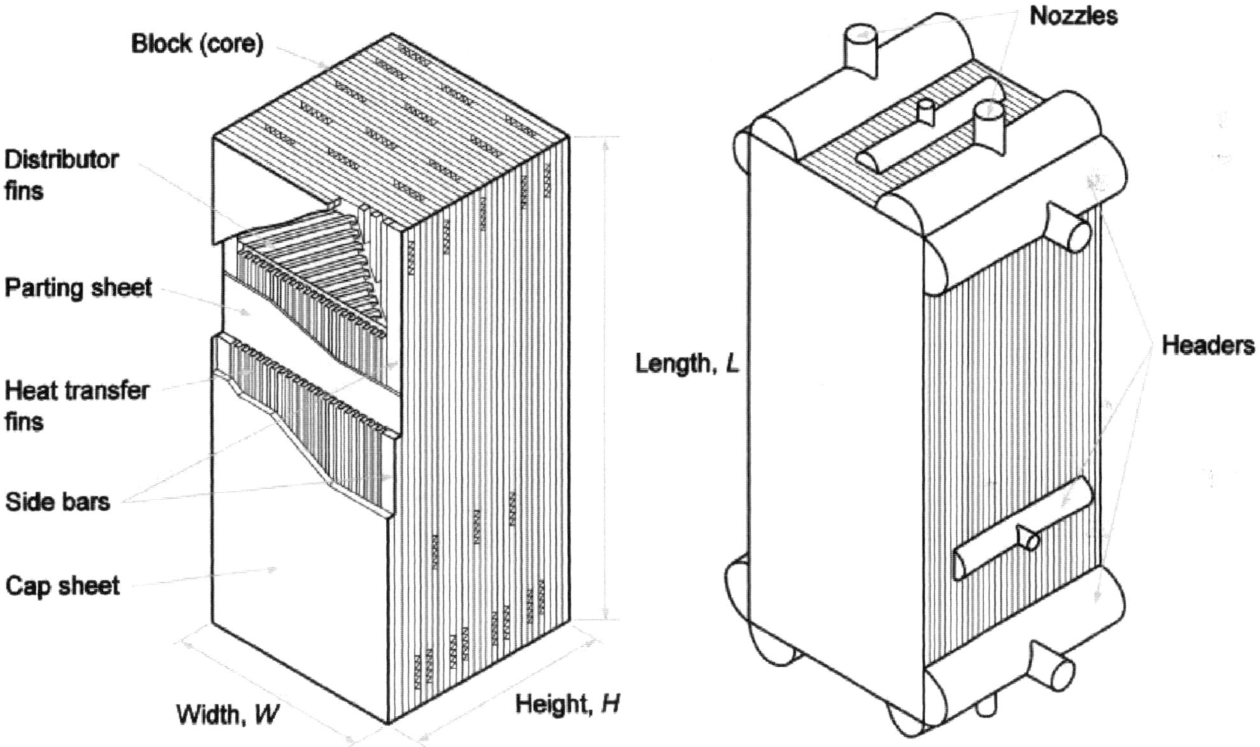

Figure 3: Components of a brazed aluminium plate-fin heat exchanger [7]

The increased cross-surface area obtained with welded blocks may come at the cost of increased thermomechanical stress, depending on the specific service. The higher the block height, the larger the gradient of thermal expansion within the PFHE block. Since the thermal expansion gradients are commonly highest for transient operation conditions (during start-up, upset, etc.), it is crucial to execute detailed integrated design reviews on PFHEs in mega-scale services [8,9]. The reviews have to cover all critical operation scenarios and should be executed in an early phase parallel to the normal engineering workflow. Based on thorough investigation by means of both steady state and dynamic thermohydraulic [10], as well as FEM-based thermomechanical simulations [11], a lifetime forecast and the definition of appropriate mitigation measures can be elaborated. Typical measures are tailored PFHE robustification, instrumentation, specific procedures, start-up lines, switches and more. The procedure of the integrated design reviews requires close collaboration of process design, systems design, safety engineering and manufacturing.

For realization of mega-scale heat duties beyond the limitations of welded composite block dimensions, the blocks can be further modularized. Figure 4 illustrates assemblies of multiple PFHE blocks connected in parallel. In such a setup, each individual stream enters the assembly through a manifold, is distributed to the inlet nozzles on each of the heat exchangers, flows through the heat exchanger and leaves the assembly through the outlet manifold.

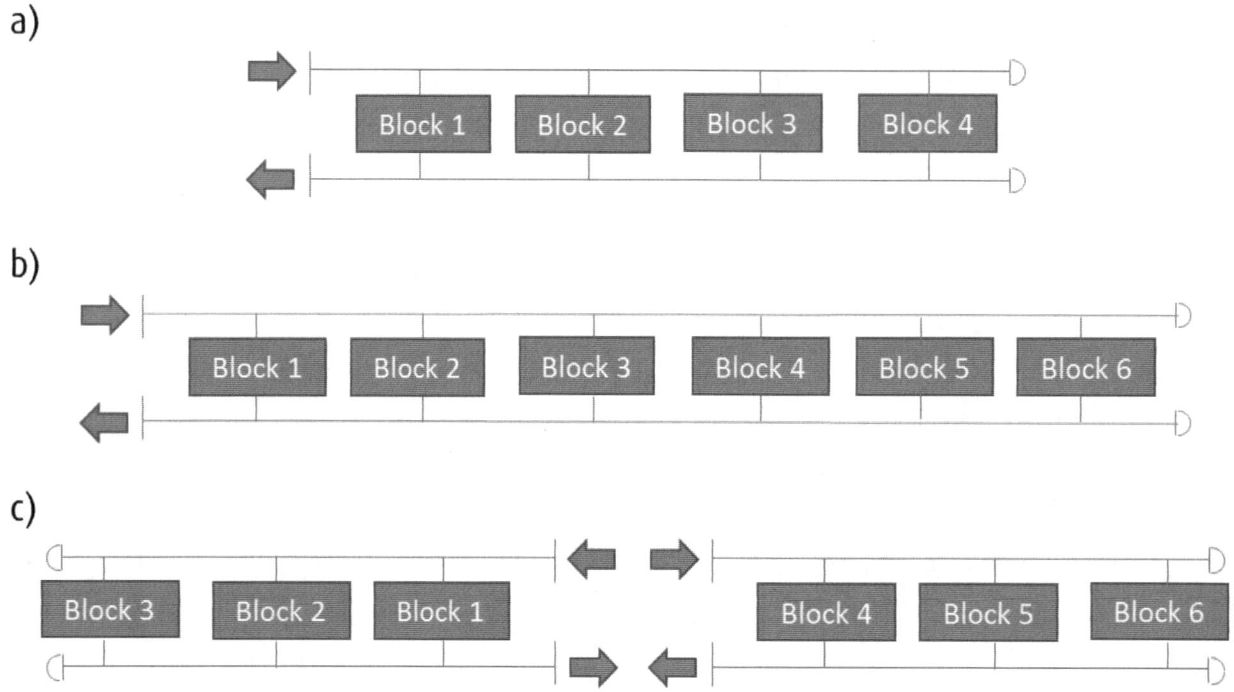

Figure 4: Concepts for modularization of plate-fin heat exchanger blocks: a) reference, b) scale up as manifolded assembly, c) scale up with further level of parallelization.

In current ethylene plant references, assemblies with up to four blocks have been implemented. For other plant types even higher numbers of blocks per assembly were realized. However, when increasing the number of parallel blocks fed from one manifold, an increasing risk of uneven flow distribution has to be considered. Caused by deviating pressure drops mainly in the manifolds, a fluid maldistribution poses the risk of reduced performance of the entire assembly.

The risk can be reduced by increasing the ratio of pressure drop over the single blocks, to total pressure drop. However, this measure implies either decreased energy efficiency due to increased total pressure drops or higher capital investment due to increased manifold diameters. Particularly in the latter case, flow maldistribution and the risk of reduced performance needs to be considered for transient operation. Therefore, simulation-based investigations of flow distribution considering the interaction of all streams needs to be executed.

Another option to overcome this hurdle is to introduce a further level of parallelization (see Fig. 4c). By dividing the unit into two parallel U-type manifold assemblies, the total flows can be distributed evenly into two sub-streams. The distribution can be performed either by geometrical symmetrical distribution (T-piece) and/or the application of flaps.

Concluding, plate-fin heat exchangers do not constitute major breakpoints that would set the barrier to an overall mega-scale plant capacity. While for certain services, the need for parallelization is reached already at today's ethylene plant capacities, smart solutions exist which enable efficient parallelization and further modularization. Nevertheless, in order to guarantee safe and reliable operation throughout a long lifetime, it is essential to perform detailed analyses by means of advanced numerical methods. This unconditionally requires a close collaboration of all involved disciplines, from process design to manufacturing.

4. Cracking Furnaces

Despite their relatively high contribution to the total installed cost of an ethylene plant, cracking furnaces do not constitute major breakpoints to feasibility and economics of mega crackers. This is mainly due to the reason that the furnaces are parallelized equipment by default already. Hence virtually any plant capacity can theoretically be achieved by simply adding furnaces of proven capacity. Another option for mega-scale cracking furnaces is to increase the capacity of the individual furnaces while keeping the number of installed furnaces constant.

Increase of Individual Cracking Furnace Capacity

While the implications of the first option shall be investigated in more detail in the present paper, the economics and implications for increasing furnace capacities shall also be discussed briefly. In general, scale economies apply also to cracking furnaces. However, the scale exponent is higher than for the overall ethylene plant, which is mainly due to the fact that multiple parts of the furnaces are modular, in particular the radiant coils, the quench exchangers and the burners. For modularized equipment, the scale economies behave rather linear than exponential.

Furthermore, when increasing cracking furnace capacities, certain incremental steps of CAPEX increase have to be considered. As an example, in a high capacity twin cell furnace with coils arranged in groups of four coils each, increasing the furnace capacity is realized by adding stepwise eight coils (i.e. steps of four coils per cell), while small capacity increases are accomplished by tuning the radiant coil dimensions (i.e. length, diameter). Another example for a typical CAPEX step is the necessity of switching from flue gas fans with variable speed drive to a high voltage fan drive, which will be reached at a certain furnace capacity.

Increasing the Number of Installed Cracking Furnaces

The following considerations for increasing the number of installed cracking furnaces are based on a state of the art furnace design for gas cracking service with an ethylene capacity of 250 kta. This capacity represents a fully proven and tested furnace design, hence no bottleneck investigations and/or conceptual adaptions are required. The following discussion considers a virtual scale up of overall plant capacities to 2750 kta by adding single furnaces. All considerations are based on a full spare furnace to minimize loss of production.

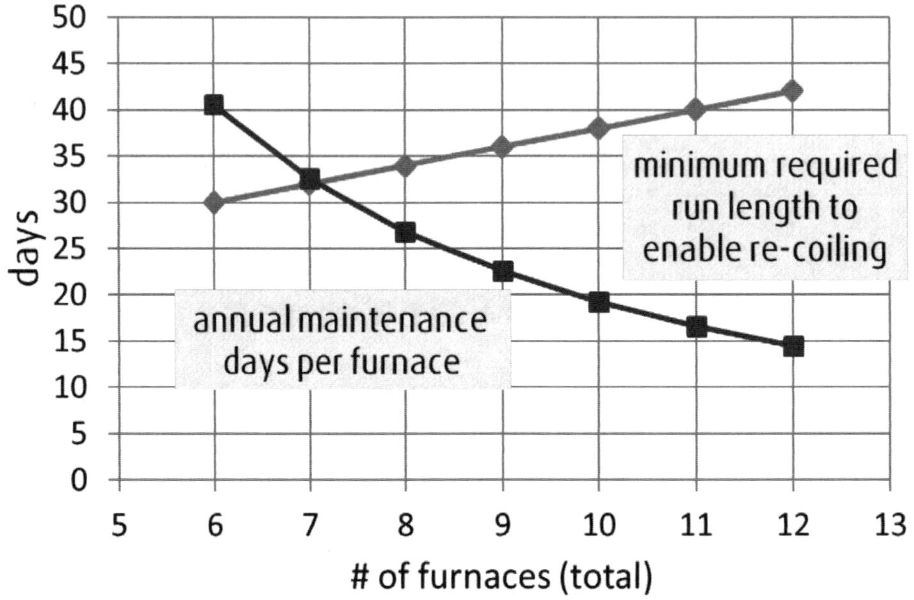

Figure 5: Consequences for increasing the number of cracking furnaces – increase of the minimum required run length and reduction of available maintenance days per furnace (based on assumption of two days for decoking per furnace and 20 days for recoiling).

For a nameplate plant capacity of 2500 kta eleven (10+1) furnaces are needed as base concept. The large size and such a high number of furnaces makes it almost impossible to perform regular recoiling of all furnaces during the plant turnaround, hence recoiling should be performed on a one at a time basis during normal operation. In order to enable recoiling during normal plant operation without loss of production, the minimum furnace run length must provide enough time for sequential decoking of all the remaining furnaces and subsequent recoiling. For state of the art furnaces equipped with a special radiant coil removal mechanism for quick replacement, a recoiling time frame of 20 days can safely be considered feasible. Accordingly, the minimum required run length of each furnace increases with the total number of installed furnaces, as displayed in Fig. 5. For a mega cracker of 2500 kta equipped with eleven (10+1) furnaces, the minimum run length must be set to 40 days.

A further consequence for increased numbers of furnaces is the reduced amount of available maintenance days per furnace. The run length, increased according to the requirements discussed above, still provides less days for maintenance on a single furnace as for a plant equipped with fewer furnaces that have a lower run length. If the furnaces are designed in a way that the run length of one furnace provides for just 20 additional days after decoking of all remaining furnaces, these 20 days represent the available standby time for the remaining furnaces. Accordingly, the standby time available for maintenance of a single furnace is determined by 20 days divided by the number of furnaces in operation. With increasing number of installed furnaces, the available time for maintenance per furnace hence decreases considerably (see Fig. 5). For the upscale from 1500 kta (6+1 furnaces) to 2500 kta (10+1 furnaces), a reduction of the available days per furnace and year from approx. 32 days to approx. 16 days needs to be considered. If this figure is considered insufficient for a specific project, the minimum furnace run length has to be even higher for a high furnace count.

In order to provide furnaces with enlarged run lengths, one possibility is to increase the surface area of the coils, resulting in a lower radiant section heat flux. This means that if mega-scale plant capacities shall be reached by increasing the number of installed furnaces, enlargement of the furnaces should be considered to a certain extent, at least for the radiant section. This effect needs to be considered when comparing the scale economies for cracking furnaces.

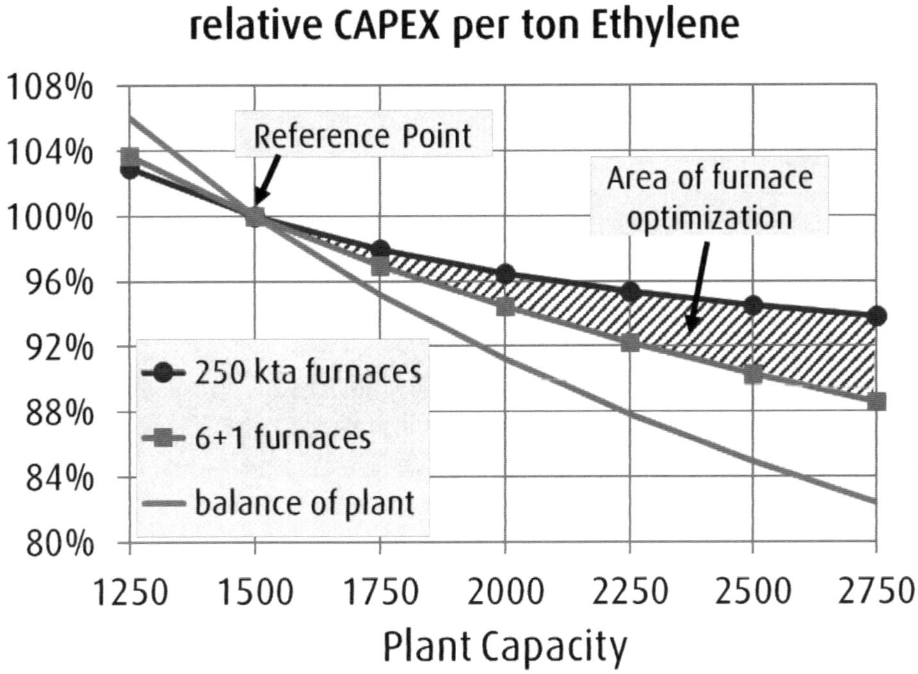

Figure 6: Mega-scale economies for ethylene cracking furnaces – comparison of increasing number of furnaces and furnace capacity upscale with the overall ethylene plant.

Increased Furnace Count vs. Increased Furnace Capacity

The cracking furnace scale economies for the two options of increasing plant capacities are compared in Fig. 6. Increasing the number of furnaces generates savings mainly due to the decreased spare capacity. The higher the total number of furnaces, the lower the share of the single spare furnace. The same effect holds for engineering, since all of the furnaces are identical. As the number of furnaces goes up, the ratio of engineering effort per total furnaces goes down. However, the demand for larger furnaces in order to provide appropriate run lengths partly compensates the benefits.

Scaling up the capacity of the single cracking furnaces while keeping the number of furnaces leads to further improved scale economies. As discussed above, also here the specific benefits are still somewhat lower than for the overall plant. It needs to be considered that this approach exceeds the proven furnace capacity range by far. For achieving a nameplate plant capacity of 2500 kta with 6+1 furnaces, single furnace capacities of almost 420 kta would be necessary. Apart from the high risk such large furnaces pose, the high uncertainty of scale economies in this capacity range should also be considered.

Both approaches to mega-scale plant capacities can hence be regarded as the two extremes, which mark up the field of action. An optimized solution for mega cracker furnaces will most probably represent a combination of both approaches. The scale economies for mega-scale cracking furnaces consequently lies in the field between both curves.

For realization of a specific mega cracker project, the final investment optimization of furnace size and count should be done once the specific plant size and cracking conditions are selected. From the discussion above, it is clear that a proper definition of requirements for scheduled / unscheduled maintenance times and determination of required minimum furnace run length between decoking is essential.

5. Conclusion

After six years of experience in operating 1500 kta ethylene plants the olefins industry seems poised to take the next step up in capacity. For the design of super-sized plants, two factors become essential: (i) exact understanding of the limitations to the economies of scale and (ii) the ability to provide tailor-made solutions for mega-scale services.

To understand the overall plant scale economies and to provide economies of scale optimized plant capacities for project specific boundary conditions, the mega-scale characteristics of the plant need to be investigated on equipment level. Besides sheer fabrication limits, construction and transport constraints, there are several further criticality criteria against which any equipment and/or plant component needs to be screened thoroughly. This breakpoint evaluation relies on a consolidated and continuously updated database that includes comprehensive knowledge ranging from specific equipment manufacturing limits to current equipment references to highly accurate equipment and plant component cost curves.

Of particular interest for investigation are equipment types that exhibit untypical mega-scale characteristics. Plate-fin heat exchangers (PFHEs) as well as cracking furnaces represent highly modularized and parallelized equipment that does not constitute major breakpoints to an overall mega-scale plant capacity. For PFHEs, smart solutions exist which enable efficient parallelization and further modularization. In order to guarantee safe and reliable operation in mega cracker services, integrated design reviews with close collaboration of all involved disciplines, from process design to manufacturing are crucial. For cracking furnaces, both theoretical approaches to achieve super-sized plant capacities exhibit particularities: While for scaling up single furnace capacities, reduced scale economies and incremental cost steps need to be considered, increasing the number of installed furnaces still implies an enlargement of parts of the furnaces. An optimized solution for mega cracker furnaces will most probably represent a combination of both approaches.

6. References

1. Schmigalle, H. (2006), "Challenges of Mega Cracker Design", in *Petchem Arabia*, Abu Dhabi, UAE.
2. Whang, Y., Beck, B. (2017), "Safely Overcoming Capacity Constraints Challenges Using UOP Process Equipment and Callidus Flares", in *Ethylene Producers Conference, AIChE Spring National Meeting*, San Antonio, USA.
3. Thomas, I. (2011), "A Process Unit Modeling Framework within a Heterogeneous Simulation Environment", in *Proc. 21st Europ. Symp. Comp. Aided Proc. Eng. (ESCAPE-21)*, Chalkidiki, Greece.
4. Petrie, D. (2013), "Mechanical Challenges in Very Large Compressors and Mechanical Drive Steam Turbines", in *Ethylene Producers Conference, AIChE Spring National Meeting*, San Antonio, USA.
5. Moore, F. D. (1959), "Economies of Scale: Some Statistical Evidence", in Quarterly Journal of Economics, 73 (2), pp. 232-245.
6. Remer, D. S., Chai, L. H. (1993), "Estimate Costs of Scaled-up Process Plants ", in Encyclopedia of Chemical Processing and Design, ed., McKetta, J. J., Marcel Dekker, Inc., 44, pp. 14-39.
7. The Brazed Aluminium Plate-fin Heat Exchanger Manufacturers' Association (2016), "The Standards of the Brazed Aluminium Plate-fin Heat Exchanger Manufacturers' Association", www.alpema.org
8. Hölzl, R. (2012), "Lifetime Estimation of Aluminium Plate Fin Heat Exchanger", in *Proc. ASME 2012 Pressure Vessels & Piping Division Conf.*, Toronto, Canada.
9. Freko, P., Thomas, I., Hölzl, R., Lehmacher, A., Woitalka, A. (2015), "Optimization of Lifetime Expectation for Heat Exchangers with Special Requirements", in *Proc. 15th Int. Heat Transfer Conf.*, Kyoto, Japan.
10. Woitalka, A., Thomas, I., Freko, P., Lehmacher, A. (2015), "Dynamic Simulation of Heat Exchangers Using LINDE's in-house Process Simulator OPTISIM®", in *Proc. CHT-15, ICHMT Internat. Symp. Advances in Computational Heat Transfer*, Piscataway, USA.
11. Hölzl, R., Flüggen, R. (2013), "Experimental Fatigue Evaluation of Aluminium Brazed Plate Fin Heat Exchangers", in *Proc. ASME 2013 Pressure Vessels & Piping Division Conf.*, Paris, France.

AIChE PAPER NUMBER: 187c

KEY TECHNOLOGY CHALLENGES FOR MEGACRACKERS

Dennis Maifield

Vice President, Process Engineering

TechnipFMC, Process Technology, Claremont, CA

Prepared for Presentation at the 2017 Spring National Meeting
San Antonio, TX, March 26-30, 2017

KEY TECHNOLOGY CHALLENGES FOR MEGACRACKERS

Dennis Maifield

Vice President, Process Engineering

TechnipFMC, Process Technology, Claremont, CA

Abstract:

The purpose of this paper is to identify key technology challenges in equipment design, utility systems and piping design that have arisen during the engineering phase of large capacity ethylene plants. The paper will specifically review the key technology aspects for reactors, Dryers and drums design, separation towers design and heat transfer equipment. The paper will also examine the impact on utility systems design – particularly CW and steam systems and the design criteria for large sized specialized piping/piping materials etc. TechnipFMC has recently designed 15 ethylene plants that can be classified as megacrackers and this paper will concentrate on the technology challenges faced in the largest capacity plants. These technology challenges while primarily impacted by plant capacity are also affected by plant location and owner requirements so these features are also discussed. As successful solutions for the key technology challenges in equipment design, utility systems and piping design are achieved, then a forecast of the next hurdles for future megacrackers is presented.

Introduction

The purpose of this paper is to identify key technology challenges in equipment design, utility systems and piping design that have arisen during the engineering phase of large capacity ethylene plants. The paper will specifically review the key technology aspects for reactors, Dryers and drums design, separation towers design and heat transfer equipment. The paper will also examine the impact on utility systems design – particularly CW and steam systems and the design criteria for large sized specialized piping/piping materials etc. TechnipFMC has recently designed 15 ethylene plants that can be classified as megacrackers and this paper will concentrate on the technology challenges faced in the largest capacity plants. This megacracker experience includes ethylene plant capacities that range from 1,000 KTA to 2,000 KTA of ethylene production. Most of these megacrackers have gone through plant startup and are currently in production. Several of the recent USGC megacrackers are under construction and scheduled to start up within the next year.

As the steam cracker plant capacities have increased, we have developed practical solutions for larger equipment item sizing as well as process and utility piping systems for these megacrackers. In conjunction with increased individual equipment sizes, equipment manufacturers, transportation companies and construction contractors have also supported these project demands associated with megacrackers.

Reactors

In the steam cracking process, the most important reactions are those associated with the thermal cracking of the hydrocarbon feedstock into ethylene, propylene and other byproducts in the radiant section of the pyrolysis cracking furnaces. TechnipFMC's proprietary furnace technologies contain a wide range of design options for reliable, flexible and highly selective solutions that are applicable to megacracker furnace operation. However, furnace design applications have already been addressed in another presentation so they will not be covered in this paper.

There are several other unit operations in the steam cracking process that require catalyst to promote the desired reactions. A list of the reactions and catalyst are given below:

Processes / Catalysts used in ethylene plant

• Acetylene Hydrogenation	Pd on Alumina
• MAPD Hydrogenation	Pd on Alumina
• Butadiene Hydrogenation	Pd on Alumina
• Pyrolysis Gasoline Hydrogenation	Ni or Pd on Alumina
• Hydrogenation for C3, C4 or C5 recycle	Pd or HTC

The primary focus for this paper will be the acetylene hydrogenation reactors because these reactors are the largest in the megacracker flow scheme since the process stream is a much larger flow rate than any of the other hydrogenation reaction unit operations. The predominant method for acetylene removal is through selective hydrogenation. Most acetylene hydrogenation is carried out in vapor phase, adiabatic fixed bed reactors. The number of reactor beds ranges from one to four with intercooling between the beds. One ethylene licensor offers acetylene hydrogenation carried out in isothermal tubular reactors and this proprietary design alternative will not be addressed.

An acetylene hydrogenation unit can be located at one of several locations within the ethylene plant sequence. There are several possibilities for these locations as follows:

<u>**Acetylene Reactor Location**</u>	<u>**TechnipFMC Megacrackers**</u>
• Front end	
➢ Deethanizer Overhead in a DeC2 1st process sequence	8
➢ Depropanizer Overhead in a DeC3 1st process sequence	2
➢ Caustic Tower Overhead in a Raw Gas application	-
• Tail end or Back end in a DeC1 1st process sequence	<u>5</u>
Total	15

As noted in the table above, TechnipFMC has experience in megacracker designs for almost all process configurations for the acetylene reactor. There are several catalyst manufacturers that provide tailored catalyst formulations for each of these process sequence applications. TechnipFMC works closely with the catalyst vendors to select the design conditions and the number of catalyst beds required. The detailed sizing of a catalyst bed is a complex kinetics problem and depends on flow rates, partial pressures of reactants, temperatures, pressure, catalysts parameters, and a number of other factors to ensure uniform flow distribution and reasonable pressure drop. The reactor bed volume is determined primarily based on recommended space velocities data received from the catalyst vendors.

The acetylene reactor sizing for some recent megacrackers is provided below:

Megacracker Acetylene Reactors – Table 1

Plant type	Capacity, KTA	Reactor Location	Number of beds Total/Oper./Spare	Reactor Size ID x T-T, m	Reactor Weight, Ton
Gas Cracker	1700	Back end	2/1/1	4.6 x 6.2	75
Gas Cracker	1700	Front end	2/2/0	4.3 x 6.4	80
Gas Cracker	2000	Front end	2/2/0	4.9 x 5.8	100

The reactor dimensions and weights are well within vessel manufacturers fabrication capabilities and are on the low side of consideration for heavy haul or heavy lift criteria by the construction contractor. Therefore, acetylene hydrogenation reactors are not a design constraint for today's or future larger capacity megacrackers.

Dryers

The steam cracking process utilizes the addition of dilution steam in the thermal cracking of hydrocarbons to reduce the partial pressure of the hydrocarbon reactions and improve the yield of olefin products. While most of the condensed water is separated from the cracked gas effluent in the Quench Tower and Cracked Gas Compressor inter-stage knockout vessels, the final water removal is accomplished in the Primary Gas Drying unit operation. The functional requirements of the cracked gas dryers are described below:

Primary Gas Dryers – Functional Requirements

- Compressed cracked gas is dried in the Primary Gas Dryers to prevent freezing and hydrate formation in the subsequent low temperature operations.
- The primary dryer system is provided with an operating spare.
- The dryer system operates cyclically, with one on-line and the other on regeneration or on standby.
- The dryers are designed with enough cycle time at EOR condition to accommodate the regeneration requirements with the available regeneration gas.
- Each dryer is provided with sufficient zeolitic molecular sieves to achieve a moisture content of less than 1 ppm by weight in the effluent at the end of the desiccant life.
- Each dryer desiccant bed also includes a guard bed to protect from moisture break-through.

While the Primary Gas Dryer is located in the process sequence after the Caustic Tower (removal of acid gases; H_2S and CO_2), the Primary Gas Dryer could be located at the 3rd, 4th or 5th stage discharge of the cracked gas compressor. TechnipFMC designs for megacrackers have utilized all of these process sequence options for the Primary Gas Dryer location. There are several suppliers that offer molecular sieve adsorbent applications for the Primary Gas Dryer. The ethylene plant owner also provides specific operational requirements in regards to cycle times and reserve guard bed allowance before moisture breakthrough. The cyclical operation of the Primary Gas Dryers also impacts the expected life of the adsorbent which for cracked gas service is about 5 years. The turnaround cycle of megacrackers is often targeted at 7 or 8 years because they feed multiple derivative petrochemical units. This planned turnaround cycle could be constrained by Primary Gas Dryer molecular sieve replacement unless the Dryers are designed to facilitate adsorbent replacement while the ethylene plant remains in operation.

The Primary Gas Dryer sizing for some recent megacrackers is provided below:

Megacracker Primary Dryers – Table 2

Plant type	Capacity, KTA	Dryer Design pressure, Barg	Number of Dryers Operating/Spare	Cycle Time @ EOR,Hr	Dryer Size ID x T-T, m	Dryer Weight, Ton
Gas Cracker	1700	40	2/1	72	4.6 x 11	125
Gas Cracker	1700	44	2/1	48	5.7 x 8.5	234
Gas Cracker	2000	16	1/1	30	7.9 x 7.2	162

The vessel dimensions and weights are also well within vessel manufacturers fabrication capabilities and are candidates for heavy haul or heavy lift criteria by the construction contractor. While Primary Gas Dryers are not a vessel design constraint for today's megacrackers, the molecular sieve replacement cycle must be accounted for in achieving the overall plant turnaround cycle.

Drums

An ethylene plant has 50 to 70 pressure vessel services that can be classified as drums.

Ethylene Plant Drums

- ~25% have specific design features (Oily water separator, Spent caustic degassing drum, Deaerator, etc.)
- ~45% are Vapor-Liquid separators
- ~30% are storage vessels

A vapor-liquid separator drum is a vertical vessel into which a liquid and vapor mixture (or a flashing liquid) is fed and wherein the liquid is separated from the gas phase by gravity, falls to the bottom of the vessel, and is withdrawn. The vapor travels upward at a design velocity which minimizes the entrainment of any liquid droplets in the vapor as it exits the top of the vessel. The size a vapor-liquid separator drum (or knock-out pot, or flash drum, or compressor suction drum) should be dictated by the anticipated flow rate of vapor and liquid from the drum.

Ethylene Plant Vapor-Liquid Separator Sizing

- Design vapor velocity is based on:

$$V_d = k \times \sqrt{\frac{\rho_L - \rho_G}{\rho_G}} \qquad \begin{cases} V_d = Design\ Velocity \\ \rho_L = Liquid\ phase\ Density \\ \rho_G = Gas\ phase\ Density \\ k = Capacity\ Factor \end{cases}$$

- Capacity Factor "k" depends on the internal mesh pad or vane element
- GPSA Engineering Data Book recommends k values for vertical drums with horizontal mesh pads based on the operating pressure. GPSA recommends to use one-half of the K value for approximate sizing of vertical separators without mesh pads.

Since there are a large number of Vapor-Liquid Separator service applications in an ethylene plant, it is often advantageous to use internal devices to improve the separation efficiency and reduce the diameter of the drum. Inlet diffusers models like the Schoepentoeter™ or Vapor Flute™ could also reduce the diameter of drums but are a better design application for large diameter vessels such as the Water Quench Tower or Oil Quench Tower.

Drum Internals – Mesh Pad

- Characteristics:
  - ⇨ Cost Effective
  - ⇨ Low Pressure drop
  - ⇨ Simple installation (and replacement)
  - ⇨ "k" values from 0.07 $m.s^{-1}$ to 0.15 $m.s^{-1}$
- Theoretical 15% to 20% diameter reduction
- Considerations
  - ⇨ Efficiency based on Mesh filament diameter
  - ⇨ Thicker diameter increase pressure drop

Drum Internals – Vane Element

- Characteristics:
  - ⇨ Efficiency/Pressure Drop optimization with Vane profile
  - ⇨ Good liquid load capacity if coupled with drain channel
  - ⇨ "k" values from 0.1 m.s-1 up to 0.4 m. s^{-1}
- Theoretical 10% to 50% diameter reduction
- Considerations
  - ⇨ Higher pressure drop than Mesh
  - ⇨ More expensive
  - ⇨ Clean service preferred (Refrigerant loop, Steam …)

Today's ethylene plant designs utilize mesh pads or vane elements for most compressor suction drum applications. High efficiency vane elements are often selected for Primary Gas Dryer feed separators.

Ethylene plant drums serve a wide variety of process applications. Some of the largest drum sizes are required for the special applications and storage drums. Several process services for megacrackers are listed below:

Megacracker Drums – Table 3

Plant Capacity, KTA Equipment service	Design pressure, Barg	Drum Size ID x T-T, m	Orientation Horizontal or Vertical	Empty Weight, Ton
1700 Plant				
Oily Water Separator	4.5/FV	4.3 x 17.2	Horizontal	53
Flare KO Drum	5/FV	6.6 x 20	Horizontal	91
C3R Suction Drum	4	5.2 x 9	Vertical	71
C3R Accumulator	26	6.2 x 17.2	Horizontal	137
2000 Plant				
Oily Water Separator	4/FV	5.9 x 17.5	Horizontal	101
Flare KO Drum	3.5/FV	11.4 x 21.3	Horizontal	119
C3R Suction Drum	14/FV	6.4 x 10	Vertical	100
C3R Accumulator	21	5 x 22.5	Horizontal	146

The drum dimensions and weights are well within vessel manufacturers fabrication capabilities and several drums are classified for heavy haul or heavy lift criteria by the construction contractor. Since there are many drum services it is important to apply efficient and cost effective sizing criteria to reduce the drum footprint in ethylene plant layout. Ethylene plant drum sizing is not a design constraint for today's or future larger capacity megacrackers.

Towers

The separation of the pyrolysis furnaces cracked gas effluent into high purity, polymer grade ethylene product and other valuable byproducts is accomplished by the separation towers in the recovery section of ethylene plants. As previously mentioned in the acetylene hydrogenation reactor section of this paper, TechnipFMC has experience in megacracker design for three process sequence arrangements. It is not the intent of this paper to elaborate on the process design features of each of TechnipFMC's ethylene plant design alternatives. This paper is also not intended to serve as a design guideline for separation tower design methodology. For those interested in these subjects, we would recommend attending Henry Kister's sessions on Distillation design at the AIChE conference. The focus of this paper is the tower size constraints that may arise in megacrackers.

Megacrackers include the design of several major towers. Most of the examples referenced in the prior equipment discussion were based on plant designs for gas crackers (i.e. ethane and propane feedstocks) and a list of those major towers follows:

Gas Cracker – Main Towers

- Water Quench Tower
- Caustic Tower
- Deethanizer
- Demethanizer
- C2 Splitter
- Depropanizer
- C3 Splitter ?

The C3 Splitter is listed as a question mark since most of the recent gas crackers based on ethane feeds in the USGC have elected to export the mixed C3 product to other ethylene plants with available C3 Splitter capacity. For liquids crackers or mixed feed crackers, the Primary Fractionator (Oil Quench Tower) is the first tower and will introduce numerous design considerations for megacrackers. An Oil Quench Tower in a 1000 KTA liquid megacracker will be roughly 13 m in diameter and up to 45 m tall. Erection weight for such a tower would exceed 1100 tons. Some of the first megacrackers designs often involved sectional tower delivery and onsite tower ring welding for these large towers. Today, entire towers can be delivered to the jobsite without such construction challenges.

The main tower sizing for some recent megacrackers is provided below:

Megacracker Towers – Table 4

	1700 KTA Cracker DeC1 1st	**1700 KTA Cracker DeC2 1st**	**2000 KTA Cracker DeC2 1st**
Quench Tower	8.2 m ID x 37 m T-T 3.5 Barg/FV 340 ton	9.8 m ID x 37 m T-T 4 Barg/1/2 V 520 ton	9.2 m ID x 38 m T-T 3.5 Barg/FV 440 ton
Caustic Tower	5.4 m ID x 40 m T-T 16 Barg 280 ton	4.9 m ID x 56 m T-T 28 Barg 503 ton	5.5 m x 42 m T-T 16 Barg 319 Ton
Deethanizer	4.9 m ID x 38 m T-T 29 Barg 290 ton	6.3 m ID x 34 m T-T 40 Barg 527 ton	5.8 /3 m ID x 33 m T-T 16 Barg 209 Ton
Demethanizer	5 m ID x 42 m T-T 13 Barg	5.7 m ID x 64 m T-T 35 Barg	4.3 m ID x 36 m T-T 15 Barg

	188 ton	573 ton	165 Ton
C2 Splitter	7.1 m ID x 88 m T-T 24 Barg 1000 ton (LTCS)	6 m ID x 65 m T-T 11 Barg 463 ton (304 SS)	7 m ID x 58 m T-T 8 Barg 408 Ton (SA 203)
Depropanizer	3.4 m ID x 29 m T-T 7 Barg 70 ton	4.4 m ID x 54 m T-T 23 Barg 325 ton	2.3 m ID x 37 m T-T 7 Barg 30 ton

Most of the main towers in today's megacrackers require special transportation to the jobsite. The availability of barge transportation from the vessel fabrication facility to the jobsite is an important criterion in main tower vendor selection. The Water Quench Tower has the largest tower diameters in the Main Towers listed above and photos of Water Quench Tower barge transport, hauling to jobsite and lifting are shown below.

Water Quench Tower Logistics – Figure 1

The C2 Splitter also warrants special transportation and lifting requirements due to the tower dimensions and weight requirements.

C2 Splitter Logistics - Figure 2

Freight transport systems adjacent to ports can handle towers up to 2000 tons and equivalent major jobsite erection facilities can also handle such tower weights. As plants become larger the tower diameters increase on a square root ratio. In certain tower applications, the diameters and corresponding weight can be reduced by reducing the reflux ratio and increasing the tray count. Such tower design optimization can apply to the C2 Splitters and C3 Splitters for megacracker applications. The largest tower(s) in most liquid or mixed feed crackers is the C3 Splitter. A C3 Splitter typical reflux ratio is on the order of 15:1 or greater and therefore both tray count and diameter are extended due to the tower liquid traffic rates. There is also a corresponding impact on the C3 Splitter condenser and reboiler duties. For reference, tower sizing for some recent C3 Splitters in TechnipFMC megacrackers is provided below:

Megacracker C3 Splitters – Table 5

	1500 KTA Cracker **400 KTA C3=**	**1700 KTA Cracker** **400 KTA C3=**
Tower Size, m	6.8 m ID x 90 m T-T 4.9 m ID x 47 T-T	7.8 m ID x 88 m T-T
Design Pressure	24 Barg / 24 Barg	13 Barg
Shipping Weight	500 ton / 144 ton	835 ton

The design, fabrication, transportation and erection of C3 Splitters can often be a complex project on their own. TechnipFMC has experience with one, two and three tower C3 Splitter designs. Our largest capacity C3 Splitter is 950 KTA of propylene. In liquid crackers, C3 Splitters with diameters of 10 m and weighing more than 1400 tons are commonplace. There are a limited number of manufacturers in the world who are capable of building these towers with the requisite ocean freight port facilities so the bidding process must consider these unique requirements.

Heat Exchangers

An ethylene plant has a wide variety of heat exchanger applications. Roughly one third of all equipment items in an ethylene plant are heat exchangers. The equipment cost for heat exchangers is about 20% of the total equipment investment which ranks it in third place after furnaces and compressors/drivers equipment costs. A list of common heat exchanger applications in ethylene plants are as follows:

Heat Exchangers in ethylene plants

- Double Pipe
- Shell and Tube
- Air Cooled Exchanger
- Plate Fin
- Plate & Frame

Modern transfer line exchangers for ethylene cracking furnaces are either of multiple tube or linear exchanger design. Transfer line exchangers are highly thermally loaded and require special design concepts to meet the process requirements. These special double pipe exchangers and plate fin exchangers are the subjects of other presentations.

Shell and Tube heat exchangers is the most prevalent exchanger application in ethylene plants. There are many variations on the shell and tube heat exchanger design and they are custom designed for virtually any capacity and operating conditions, from cryogenics to high temperatures and any temperature and pressure differences between the fluids, limited only by the materials of construction. Shell-and-tube heat exchangers are classified and constructed in accordance with the widely used Tubular Exchanger Manufacturers Association (TEMA) standards, DIN and other standards in Europe and elsewhere, and ASME Boiler and Pressure Vessel Codes. TEMA has developed a notation system to designate the main types of shell-and-tube exchangers.

TEMA recommends to use a 2.54 meter (100") heat exchanger diameter maximum however this does not restrict the design selection as long as it's within the capability of vendor and the economics. Shell and tube exchangers for megacracker have often been extended beyond their traditional dimensions. Several of our recent megacracker designs have incorporated single mega exchanger in high capacity services such as cracked gas aftercoolers, refrigerant condensers and tower condensers. Examples of large TEMA heat exchanger designs are provided in the following table.

Megacracker Shell and Tube Exchangers – Table 6

Service	TEMA	Area	Size (ID x TL)	Shipping Weight
	Type	m^2	mm	ton
Quench Water Chiller	BKM	5,590	1590/2360 x 13400	72
Primary DeC1 Feed Chiller	BKM	7,060	1650/2600 x 11000	36
Dryer Feed Chiller	BKM	3,420	2000/3100 x 12800	50
HP DeC3 Reflux Chiller #3	BKU	5,190	2120/3040 x 6100	34
C.G. 2nd Stage Discharge Cooler	BEM	4,460	2580 x 12200	53
C3 Refrigerant Condensers	AJL	5 x 14700	2300 x 13400	66 each
Ethane Feed Vaporizer	NKU	11,100	2160/3350 x 11600	120
Ethane Recycle Vaporizer	BKM	5,240	1960/3000 x 9100	71
Deethanizer Reflux Chiller #1	BKM	3,180	1880/3050 x 14000	93
Cracked Gas 1st Stage Discharge Cooler	BEM	4,160	2900 x 9800	104
C3 Refrigerant Condenser	AJL	4 X 9,950	1980 x 14000	84 each

Shell and tube heat exchangers are used for the majority of ethylene plant service applications as is shown in the following pie chart comparing the total duty for each exchanger type. When the distribution is presented based on exchanger area for each exchanger type, it is evident that total surface area for plate fin services like core in kettle and cold box exchangers are the largest.

Heat Exchanger Summaries – Figure 3

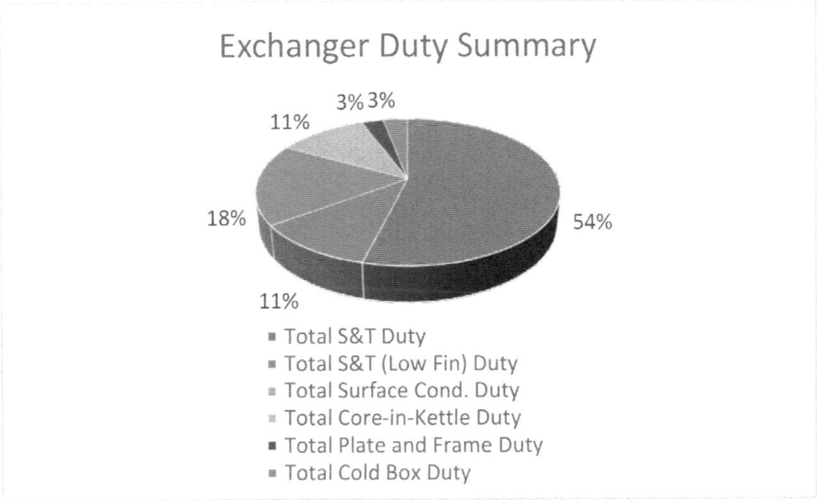

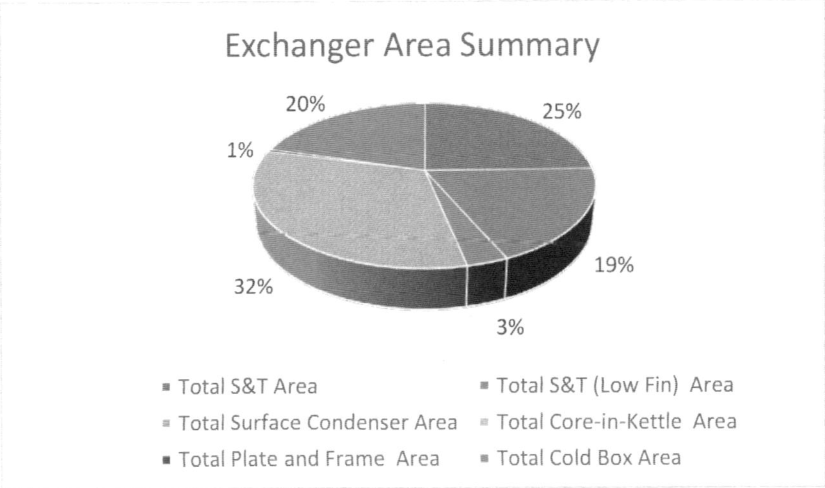

Air coolers are generally used where a process system generates heat which must be removed, but for which there is no local use. One of the most common applications for air coolers in ethylene plants is for cooling the circulating quench water. The principal benefit is that it reduces the total cooling water circulation requirements. A recent USGC ethylene plant provided the following application.

Megacracker Air Cooler – Table 7

Service	Number of bays	Duty Mw	Bare Area m^2	Empty Weight ton
Quench Water Air Cooler	16	39	5,000	426

The air coolers are installed above the pipe rack at top elevation so no additional plot space is required.

Piping Systems

An ethylene plant is connected by a vast network of piping systems. The plot layout and hydraulic analyses of the piping are key elements of megacracker designs. This is of particular importance because the proportional cost of piping systems in mega plants can be between 35% to 40% of the total plant investment. These incremental cost components include design, fabrication, associated valving, installation, insulation, testing, pre-commissioning and start-up.

The large bore piping systems for a megacracker are listed below:

Megacracker Large Bore Piping – Table 8

	Material	2000 KTA Plant
Furnace Transfer Line(s)	150# CS	2 – 66″
CGC 1st Stage Suction	150# CS	72″
CGC 1st Stage Discharge	150# CS	66″
Deethanizer OVHD	150#SS	42″
C2 Splitter OVHD	150# SS	42″
C3 Refrigerant 1st Stage Suct.	150# CS	48″
C3 Refrigerant Compr. Disch.	300# CS	42″
VHP Steam Header	2500# 2¼ Cr-1Mo or 9 Cr-1Mo-V	30″
CW Supply/Return HDRs	150# CS	84″
Wet Flare Header	150# CS	60″
Dry Flare Header	150# SS	48″
Main/Combined Flare HDR	150# CS	60″

While the cracked gas headers, cold section alloy pipe and refrigeration compressor piping have important design requirements, this paper will focus on the challenges for megacrackers with the large utility piping for the steam system and cooling water system.

Ethylene plants and power plants built in the past, most commonly used 2 1/4Cr-1Mo steel (P22) for very high pressure (100 Barg and above) steam piping applications. Today's designs increasingly use a modified 9 Cr-1Mo-V (P91) for these applications. P91 is significantly stronger than P22 at elevated temperatures. P91 typically allows a wall thickness reduction by at least a 2-to-1 ratio. If the wall is thinner, pipe rack loads are lighter. A pipe wall that is 50 percent thinner takes less time to weld and uses less weld filler metal so from a designer standpoint, it's a wonderful material. However, from a fabrication standpoint, P91 presents some challenges. Preheat and postweld heat treatment are mandatory and absolutely critical to the service life of the alloy. It is incumbent on the spool fabricators and construction welders to learn how to work with it because the material cost savings and weight reduction are substantial. The selection of P91 material for VHP piping will also accommodate larger bore sizes and higher pressure VHP (>120 Barg) design so VHP steam piping is not a design constraint for today's or future larger capacity megacrackers.

The design of the cooling water supply and return system is an integral part of an ethylene plant. The hydraulic analysis of the overall network must include all the heat exchangers in CW service as well as their location on the plot plan. These large diameter underground piping components will be one of the first priorities of the construction plan. The selection of piping materials for this service must also be decided in the initial project stages. The current materials for large bore CW piping are as follows:

CW Piping Material

- CS (coated and wrapped)
  - Corrosion resistance
  - Resistance to shock
  - High strength
- Fiberglass Reinforced Plastic – FRP
  - Corrosion resistance
  - Lightweight
  - High Strength
- Prestressed Concrete Cylinder Pipe – PCCP
  - Corrosion resistance
  - Ease of field assembly
  - Structural integrity

TechnipFMC has recently used FRP in 120" ID CW supply and return headers and has selected CS in 84" OD CW supply and return headers. Each megacracker application requires a careful analysis of the project requirements to determine the best solution for each plant. The availability of hydrotest water at the jobsite will play a significant role in commissioning strategy for this system.

Summary

The purpose of this paper was to identify key technology challenges in equipment design, utility systems and piping design that have arisen during the engineering phase of large capacity ethylene plants. World-scale Ethylene Plants have been increasing in capacity and size. As discussed in this paper, equipment size constraints for major equipment including reactors, Dryers, drums, separation towers and heat exchangers have been addressed with practical engineering solutions for the megacrackers in the 1000 KTA to 2000 KTA size range. In coordination with these larger equipment sizes, equipment manufacturers, transportation companies and construction contractors have supported these project demands associated with megacrackers. The site locations for future megacrackers will be strongly influenced by ease of access for these large equipment shipments. The industry has responded well to the increasing plant capacity demands and we must further challenge our past design limits to achieve cost effective designs for tomorrow's megacrackers.

480980 Turbomachinery Capacity for Super Mega Ethylene Plant

Abstract:

Abstract:

One of the major bottlenecks relating to mega ethylene plant capacity is turbomachinery. Turbomachinery OEM's have been moving forward to increase capacity through the application of proven technologies for critical components in order to provide more efficient & reliable turbomachinery in increasingly larger scale ethylene plants.

In regards to targeting 2.5 MTPA capacity, the authors of this abstract studied and summarized the required challenges for the design, manufacturing and testing of turbomachinery within such large scale crackers.

In addition to sumamrizing the challenges of greenfield turbomachinery mega-ethylene technology, this presentation will also focuse on the debottlenecking of brownfield ethylene crackers via footprint replacement or revamping of existing turbomachinery casings to increase plant throughput.

480987 Safely Overcoming Capacity Constraints Challenges Using UOP Process Equipment and Callidus Flares

Abstract:

Distillation column diameters have increased significantly over the past 20 years as plant nameplate capacities have grown at an accelerated pace. Column diameters exceeding 10 meters in new propylene and aromatics separations are becoming more commonplace. Large column diameters pose new challenges for the fabrication, transportation and erection of the column shell as well as the design, supply, and installation of the associated column internals. This paper will discuss how UOP has addressed these challenges through continuous evolution of mechanical improvements to UOP trays over the past forty (40) years. These design improvements are utilized today to ensure mechanical integrity and address isometric concerns with columns employing multiple reboiler and condenser shells. UOP tray designs able the operator to intuitively and easily install the trays safely even under constrained turnaround schedules.

Multi-point Ground Flares (MPGF) are the preferred technology for pressure relief in many ethylene plants. Like elevated flares, they provide a safe means of depressurizing the plant should an upset occur. Unlike elevated flares, proper design criteria have not been widely disseminated or discussed within the industry. Based on a series of performance studies conducted at the UOP Callidus multi-point flare test facility, engineering calculations, and computational fluid dynamics (CFD) models have been developed to provide greater insight into the design of MPGF. Various design inputs are discussed including physical test data and calculations, burner head design, plot space requirements, fence height requirements, and the number of stages required for correct operation. The resulting impact of the MPGF design on capital equipment cost, operating cost, environmental impact, and safety are highlighted.

485700 Superfractionator Tower Internals Designs for Megacrackers

Abstract:

This paper will review the current state of the art in large superfractionator distillation tray designs. Koch-Glitsch has specific experience of designing very large diameter fractionation trays for olefins plants for more than 40 years. We will detail some of the process and mechanical design improvements that have been implemented in these designs and successfully in operation in a few splitters. The current technologies are readily scalable to the next generation of even larger megacrackers that will require an evolution of the current distillation technology rather than a complete revolution. We will also examine potential revamp paths for future incremental capacity increases of today's largest crackers.

2017 AIChE Ethylene Producers' Conference,
San Antonio, TX, March 29, 2017

SIS Fundamentals and Implementation in Ethylene Plants

Steve Gandy, CFSP, MBA, DipM, MIET
Denise Chastain-Knight, PE, CFSE, CCPSC

1

Steve Gandy CFSP, MBA, DipM, MIET, AMBA

- VP Global Business Development for exida with 39 years industrial experience in safety and controls
- Responsible for exida's end user business
- Functional Safety Engineering (FSE100) trainer
- Certifications
  - CFSP, Certified Functional Safety Professional
- Industry Associations
  - ISA Member
  - Association of MBAs
- Publications
  - Author of Managing Risky Projects
  - Concurrent Engineering – Reducing Time to Market
  - Author of Conforming to IEC 61511: Operation and Maintenance Requirements

2

exida

Denise Chastain-Knight, PE, CFSE, CCPSP

- BS ChE, Georgia Institute of Technology
- Senior Functional Safety Engineer
- More than 31 years experience in engineering design, process improvements, process risk analysis and safety instrumented system implementation
- Experienced in the chemicals, resins, coatings, monomers, polymers, pharmaceutical, silicate, bio-fuels, petroleum, power…
- Affiliations
  - CCPS Technical Steering Committee
  - Past Chair BOK Development Committee
  - Past Chair NCEES PE (ChE) Exam Development Committee
  - Fellow AIChE

exida

Session Objectives

- Functional Safety Lifecycle Overview
- Starting with a firm Foundation
- Risk Identification and Safety Integrity Level (SIL) Targeting
- Safety Requirements Specification (SRS) Development
- Getting Good Data
- Verification Hurdles
- Personnel Competency

Industrial Accident Primary Causes

HSE study of accident causes involving control systems:

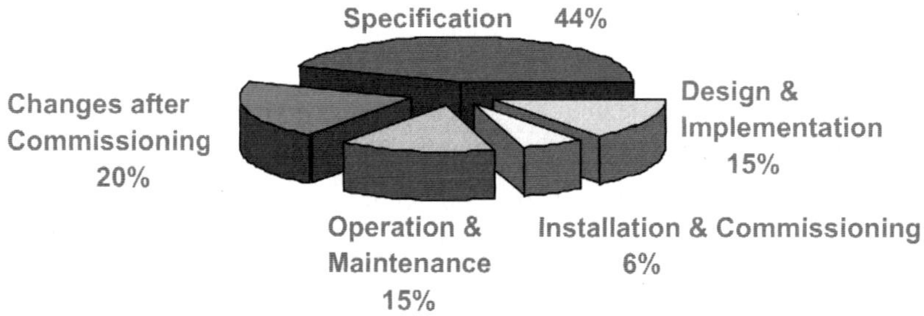

"Out of Control: Why Control Systems go Wrong and How to Prevent Failure," U.K.: Sheffield, Health and Safety Executive, 1995 (Ed 2, 2003)

IEC 61511 SIS Safety Lifecycle

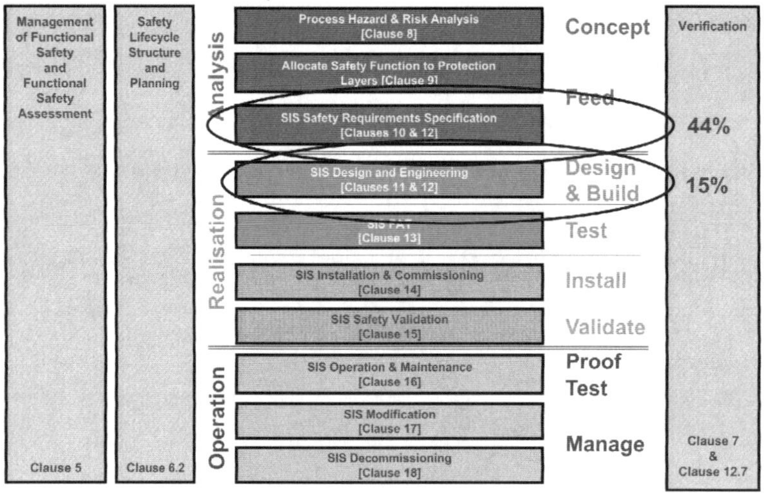

Frequent Implementation Barriers

- Poor Functional Safety Management Planning
- Misleading 'certificates'
- Complex trips
- Large package equipment
- Personnel competency

FSMP Issues

- No written plan therefore:
  - Functional Safety lifecycle activities not completed
  - Deliverables non-compliant /incomplete /missing
  - Design fails performance goals
- Procured ≠ Specified
- Cost overruns
- Schedule delays

Functional Safety Planning

- Prepare a Functional Safety Lifecycle Plan
  - What activities are to be done?
  - What deliverables are to be produced?
  - Who is responsible?
  - What are competency requirements?
- Schedule
  - List key activities and deliverables
  - Integrate with project inputs and outputs

Preparation

- Prepare philosophy and technical standards
  - Functional Safety Management guidance
    - Methods (HAZOP, LOPA, Markov)
    - Constraints (turnaround cycle)
    - Targets (MTFS, MT)
  - Approved vendors / equipment list
  - Procurement plan / sole source approvals

What is Risk?

- Risk is a measure of the ***LIKELIHOOD*** and ***CONSEQUENCE*** of an adverse effect
  (i.e., How often can it happen and what will be the effects if it does?)

- Risk receptors:
  - Personnel
  - Environment
  - Financial
  - Equipment/Property Damage
  - Business Interruption
  - Business Liability
  - Company Image
  - Lost Market Share

Basis for Risk Tolerance

- Risky activities are tolerated because they provide **benefits** and are always traded against other risks
  - Zero risk does not exist in the real world
  - Understanding the various risk and benefit options is critical to understanding what kind of risk can be tolerated in trade for what kind of benefit

SIF & Risk Reduction

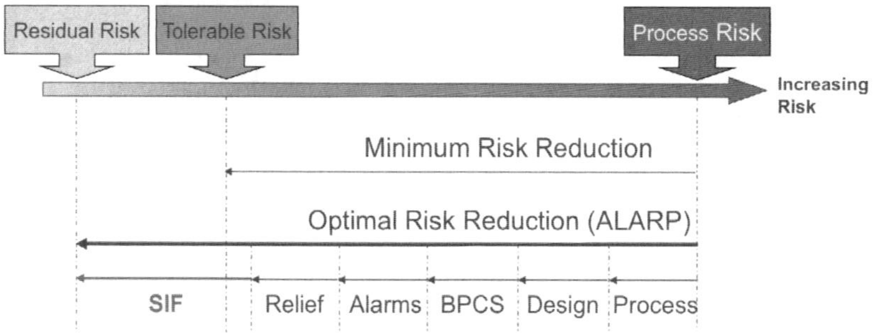

Process Hazard & Risk Assessment Objectives per IEC 61511 Clause 8

- Determine
  - the hazards and hazardous events of the process & associated equipment
  - the sequence of events leading to the hazardous event
  - the Process Risks associated with the hazardous event
  - any requirements for risk reduction
  - the safety functions to achieve the necessary risk reduction
  - if any of the safety functions are SIFs

exida

Identifying Hazard Scenarios for further evaluation from PHA reports

- Information needed
  - Description of the safety instrumented functions from previous studies
  - Hazard being prevented and its consequences
  - Initiating Events which cause the consequence
  - Safeguards (both SIS and non-SIS) used to prevent the consequence from occurring

15

exida

Typical Layers of Protection

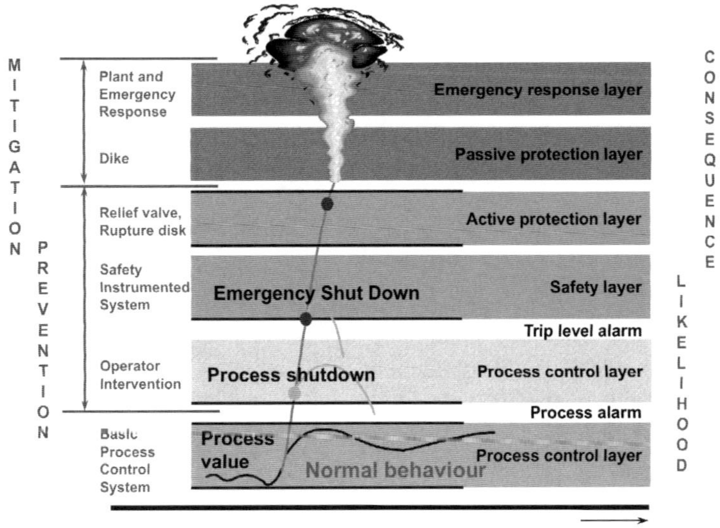

16

exida

Event Tree

- An event tree provides the analyst with a method to look at a specific cause and determine multiple outcomes
- Can be useful to understand a single scenario that leads to an outcome severity defined by a LOPA Hazard Scenario

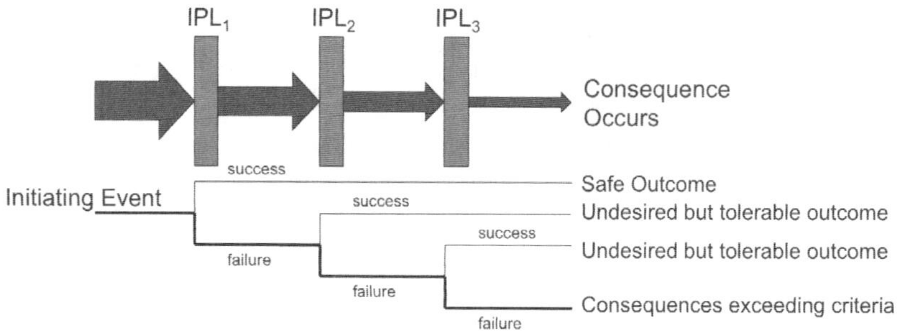

exida

LOPA

- LOPA provides the analyst with a method to look at one or more cause consequence pairs in the same analysis
- If multiple specific causes lead to the same outcome severity defined by a LOPA Hazard Scenario can model one or all in a single analysis

Individual | Multiple | Safety | Load From PHA

	Target Frequency	Actual Frequency	RRF
S	1.00E-5	1.10E-2	1100.55

2.7 demands/year		ECs	IPLs					CMs	
Initiating Events	Frequency [per year]	Continuous Operation	Alarm and operator response	Process trip	Independent PLC low temperature trip	Probability of pipe rupture	Operator response to weather	Ignition Probability	Intermediate Frequency [per year]
Flow imbalance between streams	2.5	1	0.05	NA	NA	0.33	NA	0.23	9.49E-3
Weather extreme	0.2	1	NA	NA	NA	0.33	0.1	0.23	1.52E-3

Design Requirement PFDavg/PFH

- PFDavg (low demand mode applications)

SAFETY INTEGRITY LEVEL	LOW DEMAND MODE OF OPERATION (AVERAGE PROBABILITY OF FAILURE TO PERFORM ITS DESIGN FUNCTION ON DEMAND)
4	$\geq 10^{-5}$ to $< 10^{-4}$
3	$\geq 10^{-4}$ to $< 10^{-3}$
2	$\geq 10^{-3}$ to $< 10^{-2}$
1	$\geq 10^{-2}$ to $< 10^{-1}$

- PFH (high or continuous demand mode applications)

SAFETY INTEGRITY LEVEL	HIGH OR CONTINUOUS DEMAND MODE OF OPERATION
4	$\geq 10^{-9}$ to $< 10^{-8}$
3	$\geq 10^{-8}$ to $< 10^{-7}$
2	$\geq 10^{-7}$ to $< 10^{-6}$
1	$\geq 10^{-6}$ to $< 10^{-5}$

SIF Mode

- Consideration of multiple initiating events vs Proof Test interval can drive SIF to High Demand mode
- High demand modeled using continuous mode criteria
- Continuous mode negates any benefit of Proof Testing since it will not be able to detect dangerous faults before an event occurs

High Demand vs Low Demand

- Low Demand
  - operation where the SIF is only performed on demand
  - demand is no greater than one per year
- High Demand
  - operation where the SIF is only performed on demand
  - where the frequency of demand is greater than one per year
  - Demand more frequent than twice the PTI

For 4 year PTI, high demand if f > (1/8) or 0.125/yr

SRS – The Source of Knowledge

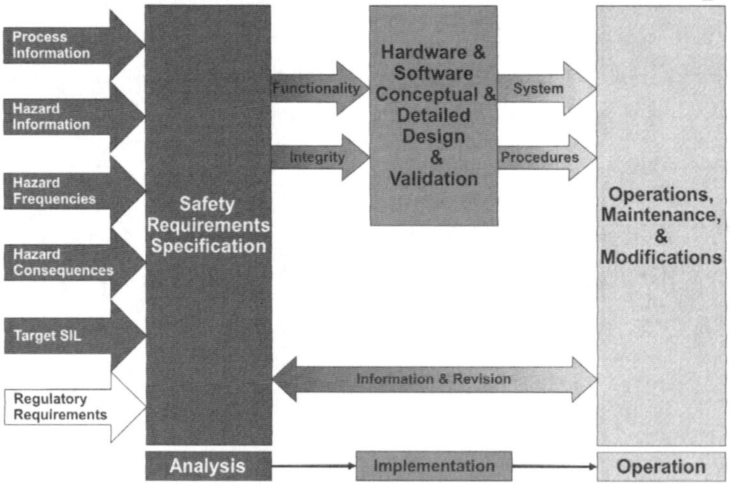

Specification = Communication

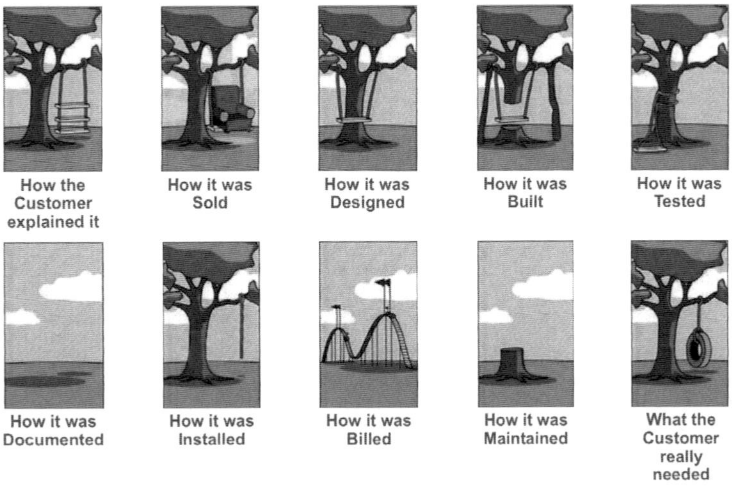

 23

SRS Functional Requirements

- Definition of the safe state
- Process Inputs and their trip points
- Process parameter normal operating range
- Process outputs and their actions
- Relationship between inputs and outputs
- Selection of energize-to-trip or de-energize-to-trip

 24

SRS Functional Requirements

- Consideration for manual shutdown
- Consideration for bypass
- Actions on loss of power to the SIS
- Response time requirements for the SIS to bring the process to a safe state
- Response actions for overt fault
- Operator Interface requirements
- Reset functions

SRS Integrity Requirements

- The required SIL for each SIF
- Requirements for diagnostics to achieve the required SIL
- Requirements for maintenance and testing to achieve and maintain the required SIL
- Reliability requirements if spurious trips may be hazardous (or costly)

Failure Rate Data

- Documentation for certified devices addresses random failure rate and systematic capability
- Systematic capability is assessed via thorough review of the design and manufacturing process
- Failure rate data must:
  - Represent the useful life of a product
  - Consider the entire population
  - Be based on a statistically valid dataset

 27

Safety Lifecycle (SLC) Objectives

- Build **safer systems** that do not experience as many of the problems of the past
- Build **more cost effective systems** that match design with risk
- **Eliminate "weak link" designs** that cost much but provide little
- Provide a global framework for consistent designs

Avoid Systematic Faults

Reduce the Potential for Random Faults

exida

Useful Life

- Useful life is pessimistically approximated by a constant failure rate based on early life failure data
- Infant mortality failures are ignored
- Wear-out is excluded

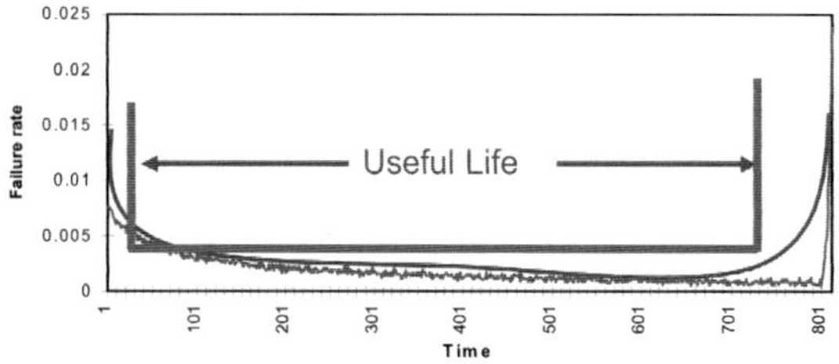

29

exida

Vet the Certificate

- Is the certification for the appropriate standard?

Has been assessed per the relevant requirements of:

IEC 61508 : 2010 Parts 1-7 ✓ **IEC 61508-2:2010** ?

- Obtain and review supporting reports

The following documents are a mandatory part of certification:

Assessment Report: BIF 11/02-075 R002 V2R1

Safety Manual: SM.SPR or SM.SPR-P16 3/2

30

Vet the Certificate

- What is source of data?

Dangerous undiagnosed failures	λ_{DU}	From return data (Route 2_H)	5.04E-09

5 FITs?

- Is the data reasonable?
- http://silsafedata.com/

60-400 FITs?

			Failures per 10^9 hours (FITs)	
Category	Type	Other	Low Bound	High Bound
Solenoid Valve	Poppet 3 way		60	400

Field Return Data Studies

- PLUS: Real Data
- MINUS: Calculation methods vary widely
- Cannot know what percentage of actual failures are returned
- Different definitions of "FAILURE"
- Operational hour basis

B10 Failure Rate Data

- The B10 method assumes that the constant failure rate during the useful life is due entirely to premature wear-out AND all other failure modes are insignificant.
- Research shows other failure modes become significant when products do not move frequently. Stiction, cold-welding, corrosion binding, etc. are dangerous failures.

Therefore B10 data is NOT applicable to most process industry applications

Vet the Certificate

- Look for evidence that certificate was developed by a body with credentials to certify safety devices
- House certificate red flags:
  - Certificate of conformity
  - Produced by device manufacturer or non-accredited third party

exida

Accreditation

- An **Accreditation Body (AB)** will audit and accredit a **Certification Body (CB)**.
- Certification Bodies must operate any product certification program under ISO/IEC 17025 and ISO/IEC 17065 requirements.

35

exida

Accreditation Confirmation

- A Certification Body will show the Accreditation Body (AB) logo on the certificate for all work done under the accredited procedures.

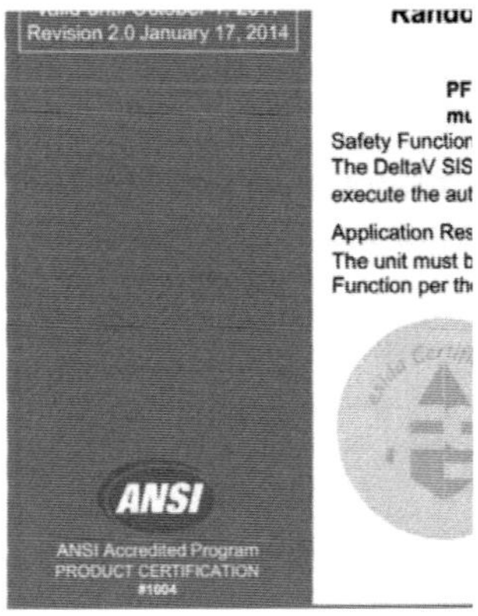

36

"House" Certificate

- When product certifications are done without using accredited procedures or done by non-accredited subsidiaries, the CB is not allowed to use the AB logo.
- This is done by some CBs and called a "House Certificate."
- There is no AB logo.

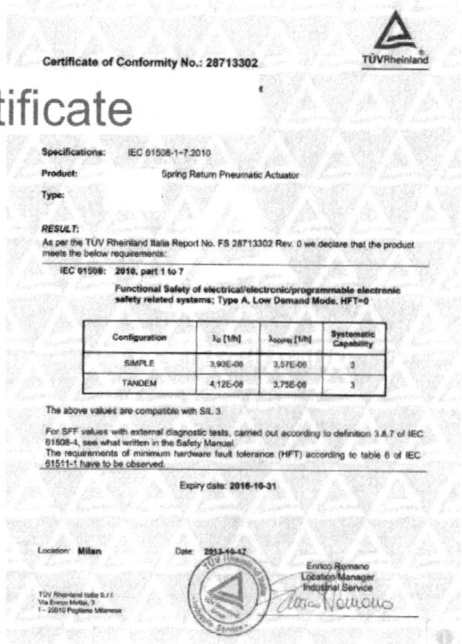

Certificate of Conformity No.: 28713302

TÜVRheinland

Specifications: IEC 61508-1-7:2010

Product: Spring Return Pneumatic Actuator

Type:

RESULT:

As per the TUV Rheinland Italia Report No. FS 28713302 Rev. 0 we declare that the product meets the below requirements:

IEC 61508: 2010, part 1 to 7

Functional Safety of electrical/electronic/programmable electronic safety related systems; Type A, Low Demand Mode, HFT=0

Configuration	λ_{D} [1/h]	λ_{DU} [1/h]	Systematic Capability
SIMPLE	3,93E-08	3,57E-08	3
TANDEM	4,12E-08	3,75E-08	3

The above values are compatible with SIL 3

For SFF values with external diagnostic tests, carried out according to definition 3.8.7 of IEC 61508-4, see what written in the Safety Manual.
The requirements of minimum hardware fault tolerance (HFT) according to table 6 of IEC 61511-1 have to be observed.

Expiry date: **2016-10-31**

Location: **Milan** Date: **2013-10-17**

Enrico Romano
Location Manager
Industrial Service

TÜV Rheinland Italia S.r.l.
Via Enrico Mattei, 3
I – 20010 Pogliano Milanese

Prior Use/Proven in Use

- Required for non-certified devices
  - Manufacturer's Quality System Evaluation
  - Adequate identification and specification of the components or sub-systems
  - Demonstration of the performance
  - Monitor field performance with no dangerous failures
- Requirements escalate as SIL level increases
- Requires OEM input
- Can be time consuming and costly

PRACTICAL IMPLEMENTATIONS

Oxygen Mixing Shutdown

- SIF is initiated by a large number of interlocks
- SIL2 with RRF sometimes required
- Multiple final elements, complex architecture
- Fast SIF response time (Process Safety Time?)
- Proven in use requirements

exida

Complex Architecture

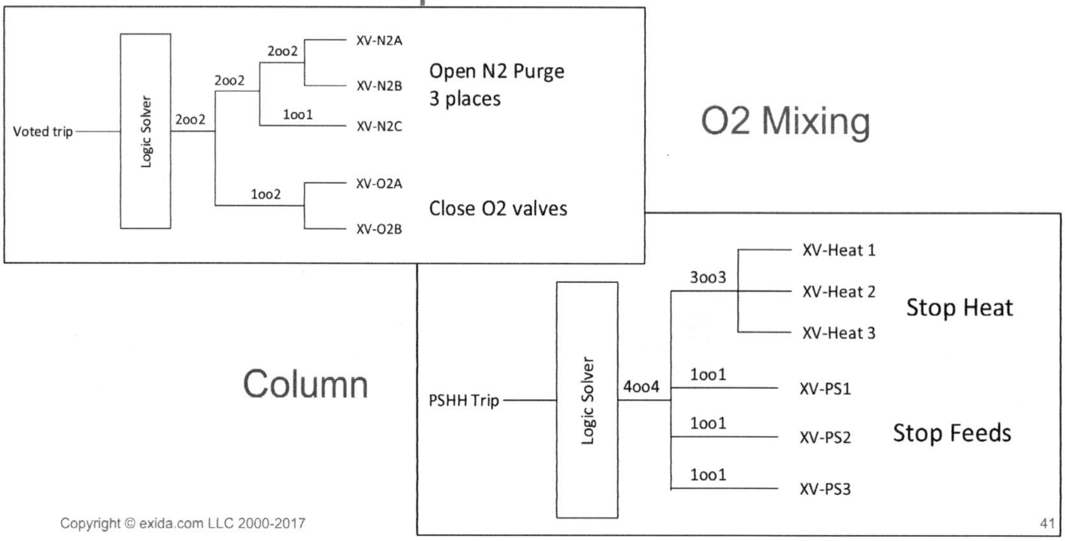

41

exida

Oxygen Mixing Shutdown DOs

- Perform LOPA to thoroughly understand what initiating events apply to the scenario and set RRF appropriately
- Definition of safe state to include what is required and exclude ancillary devices
- Calculate Process Safety time to provide foundation for SIF response time
- Use fully certified devices where possible

42

exida

Compressor Shutdown

- "Black box" package - not likely SIL certified
- Compressor package Trip design is complex
- Shutdown signal to package logic solver for trip defeats independence
- Functional Safety requirements can be contradictory to performance guarantee request

exida

Typical Arrangement

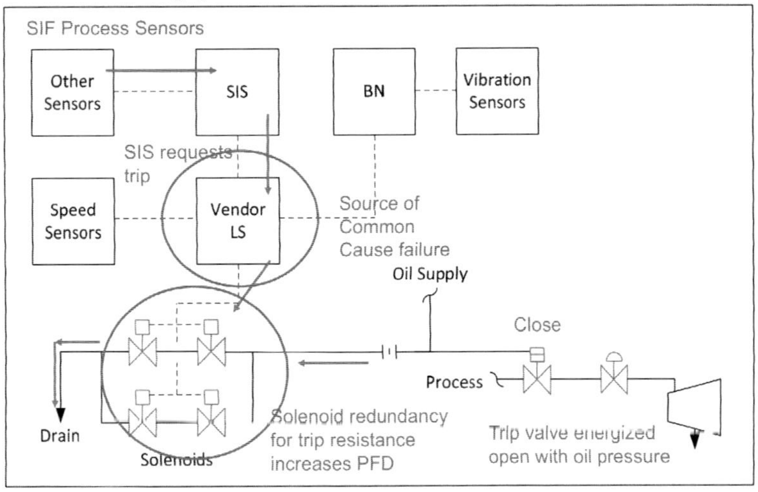

Custom Arrangement

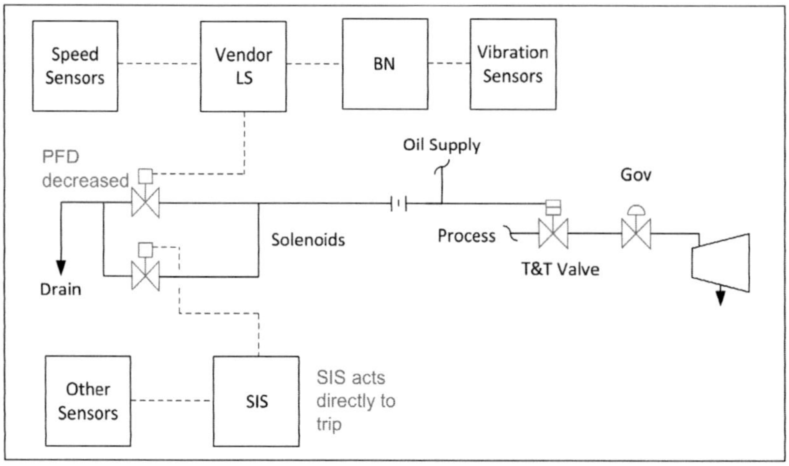

Compressor Shutdown DOs

- Where SIF trip is via independent means, take advantage of independent logic solver in LOPA by citing applicable machine protection trips as IPL
- Require vendor to provide FMEDA/Certification on trip valves
- Make timely delivery of documentation and design modifications a contract deliverable

Personnel Competency

- 5.2.2.2 Persons, departments or organizations involved in SIS safety lifecycle activities shall be competent to carry out the activities for which they are accountable.
- 5.2.2.3 Procedures must be in place to manage competence of all those involved in the SIS lifecycle.
- Periodic assessments shall be carried out to document the competence of individuals against the activities they are performing and on change of individual within a role.

IEC61511:2016

Competency Requirements

- Knowledge, training, experience appropriate to application, technology, devices
- Knowledge of safety, legal and regulatory requirements
- Management and leadership skills
- Understanding of potential consequences
- SIL of the SIF
- Novelty and complexity of the application of the technology

Thank You

sgandy@exida.com
dchastainknight@exida.com

Acronyms

- ALARP – As Low As Reasonably Practicable
- BPCS – Basic Process Control System
- FMEDA – Failure Modes, Effects and Diagnostic Analysis
- IPL – Independent Protection Layer
- LOPA – Layer of Protection Analysis
- OEM – Original equipment Manufacturer
- PFDavg – average Probability of Failure on Demand
- PFH – Probability of Failure per Hour
- PTI – Proof Test Interval
- RRF – Risk Reduction Factor
- SIF – Safety Instrumented Function
- SIL – Safety Integrity Level
- SIS – Safety Instrumented System
- SLC – Safety Life Cycle
- SRS – Safety Requirements Specification

exida

exida®

AIChE Paper Number 188c

Impact of Testing on the Application of SIS for Flare Load Mitigation

Brian K. Sullivan (Presenter)
Manager of Aftermarket Services
CB&I
Petrochemicals Business

Peter Kuzma Jr
Technology Manager
CB&I
Petrochemicals Business

Stephen De Haan
Distinguished Technologist
CB&I
Petrochemicals Business

Prepared for Presentation at the 2017 Spring National Meeting

San Antonio, Texas, March 26-30, 2017

Impact of Testing on Use of SIS for Flare Load Mitigation

Brian K. Sullivan (Presenter)
Manager of Aftermarket Services
CB&I
Petrochemicals Business

Peter Kuzma Jr
Technology Manager
CB&I
Petrochemicals Business

Stephen De Haan
Distinguished Technologist
CB&I
Petrochemicals Business

Abstract: Many plants use safety instrumented systems (SIS) to reduce their flare load. This has been done for a variety of reasons including meeting emissions permit limitations, reducing flare equipment investment cost, reducing flaring losses, and accommodating capacity expansion. To ensure that these systems are capable of limiting the flare load during all feasible scenarios, they must be designed to, and maintained at, the appropriate Safety Integrity Level (SIL). A fault tolerance is built into the system design based upon an assumed reliability for each of the components. Inherent in this assumed reliability is the requirement that the components be tested at, or more frequently than, specified intervals. It is essential that each plant be aware of the testing frequency that was assumed when calculating the basis for the reliability of their system and that plants adhere to that testing frequency schedule. This paper will present an overview of the principles related to flare load reduction design and provide simplified examples of the impact that testing frequency has upon the reliability of the system to emphasize the importance of testing frequency. Suggestions will be presented to assist operating plants with confirming the expected reliability of their system.

Introduction

The flare system in a petrochemical plant is designed to provide a safe outlet for combustible hydrocarbon gases that could otherwise give rise to over-pressuring of plant equipment during atypical operation such as startup, shutdown, process excursions, and emergency situations. The combustible hydrocarbon gases are sent to a flare stack or ground flare, where they are burned. Refer to Figure 1 for a simplified sketch of a flare system in an ethylene plant.

The flare system is typically designed to handle the load created by the maximum relief case. That case is determined by calculating the estimated loads produced by all potential over-pressure gas release scenarios and selecting the scenario with the largest relief load. Typically, this maximum relief load is determined by a common cause failure, such as a site-wide loss of cooling water supply or electrical power. Such failures cause pressure relief releases in many plant sub-systems concurrently and therefore produce the largest relief loads.

If the relief load created by a particular release event exceeds the capacity of the flare system, then excessive back-pressure in the flare headers can result in over-pressuring of the process equipment or flare system, which can potentially result in loss of containment and catastrophic failure. Theoretically, it is also possible to overload the flare tip and put the flame out with a flare load that greatly exceeds the design rate. Potentially, this could also have dire consequences. Therefore, to minimize the risk of fire and explosion, it is essential to ensure that design relief load of the flare system is not exceeded.

Many modern plants (i.e. built/expanded within the last 25 years) use Safety Instrumented Systems (SIS) to reduce their flare load. These SIS applications, which are called High Integrity Pressure Protection Systems (HIPPS), sense impending overpressure of the equipment and cut off the energy supply to the affected equipment to prevent further pressure rise and subsequent release to the flare system. They have been implemented for a variety of reasons including reducing flaring losses, meeting emissions permit limitations, reducing flare equipment investment cost, and accommodating capacity expansion with limited plot space. To ensure that these systems are capable of limiting the flare load during all feasible scenarios, they must be designed according to the regulatory requirements set forth in the applicable codes and maintained at the appropriate safety integrity level (SIL). A discussion of those codes is beyond the scope of this paper.

A fault tolerance is built into the HIPPS design based upon an assumed reliability for each of the components. Inherent in this assumed reliability is the requirement that the components be tested at, or more frequently than, specified intervals. It is essential that each plant be aware of the testing frequency that was assumed when calculating the basis for the reliability of their system and that plants adhere to that testing frequency schedule. Unfortunately, a recent industry survey of more than 50 ethylene and propylene plants with flare load reduction systems indicated that many of them may not be testing their system components at the designed frequency.

This paper will present an overview of the principles related to flare load reduction design and provide simplified examples of the impact that testing frequency has upon the reliability of the system to emphasize the importance of this issue. Suggestions will be presented to assist operating plants with confirming the expected reliability of their system.

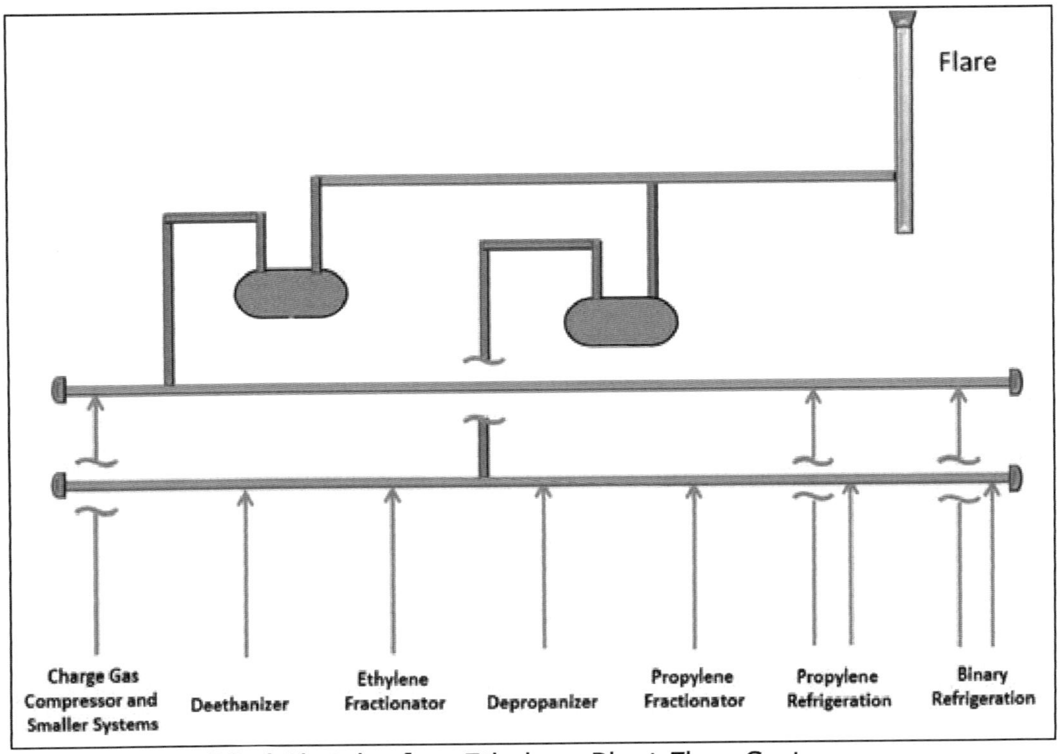

Figure 1. Simplified Sketch of an Ethylene Plant Flare System

Refresher on Safety Instrumented Systems (SIS)

This paper is not intended to be a primer on the fundamentals of SIS/HIPPS applications and the industry standards that govern the design process. However, a few basic facts are presented here to ensure that the key features of these systems are well understood within the context of this paper.

An SIS is an instrumented system that is composed of sensors, actuators, logic solvers, and final elements that are designed to automatically take the process to a safe state when predefined conditions are reached and before unsafe operating conditions occur. Other terms that have been used for an SIS include emergency shutdown system (ESD), safety shutdown system (SSD), and safety interlock system. The SIS acts independently from the process control system that is normally implemented in a distributed control system.

A HIPPS is a high integrity SIS for the purpose of flare reduction. There are a number of design standards and guidelines that address the use of HIPPS, which include:

- API Standard 521, Section 5.4.3
- ASME Section VIII, UG-140 (formerly code case 2211)

- IEC Standard 61511
- ISA Standard 84.00.01
- WRC Bulletin 498

As defined by the International Electrotechnical Commission (IEC) and the Instrumentation, Systems and Automation Society (ISA), the safety performance of an SIS is essentially measured by its average Probability of Failure on Demand (PFD). A PFD of 10^{-2} means that the system is likely to fail once per every 100 times that it is called upon to act. The PFD for each of the components is a function of the fail dangerous rate (λ or lambda) that is determined by the component manufacturer, the testing interval (T), and the beta factor (B), which is a fraction that indicates the potential for common cause failures of similar components, as shown in the equation below:

$$PFD = (lambda^2 \times T^2)/3 + (beta \times lambda \times T)/2$$

Per IEC and ISA, SIS safety performance is ranked according to safety integrity level (SIL), which is related to the PFD as shown in Table 1.

SIL	PFD
1	10^{-2} to 10^{-1} (once every 10 to 100 times)
2	10^{-3} to 10^{-2} (once every 100 to 1,000 times)
3	10^{-4} to 10^{-3} (once every 1,000 to 10,000 times)
4*	10^{-5} to 10^{-4} (once every 10,000 to 100,000 times)

Table 1. SIL vs PFD Relationship

* SIL 4 SISs are rare in the process industries and some organizations do not accept that they are achievable.

The range of PFDs and SILs is shown in Figure 2. Given the consequences of failure of the flare load reduction system, an overall SIL 3 performance is commonly selected. This requires a system that is at the high end of the estimated reliability range for PSVs that have been traditionally employed to protect systems from overpressure (see Figure 2). However, the actual SIL level is selected based on the risk tolerance set by the owner and the safety regulatory authority.

The designer then develops an SIS to meet this requirement. Once the SIL is selected, the designer ensures that the selected components, the component configuration (e.g., redundancy, voting, etc.) and component testing intervals result in an SIS safety performance that meets the SIL requirement. The SIL design calculations are based on plant-specific component failure rate data in the same or similar service and assumed component testing intervals. If the calculations do not verify the required SIL, then the design must be modified or the testing interval shortened, until the require SIL for the system is met. The final calculations are typically prepared by the detailed engineering contractor (DEC) based upon the actual equipment/instrumentation selection and final flare system layout.

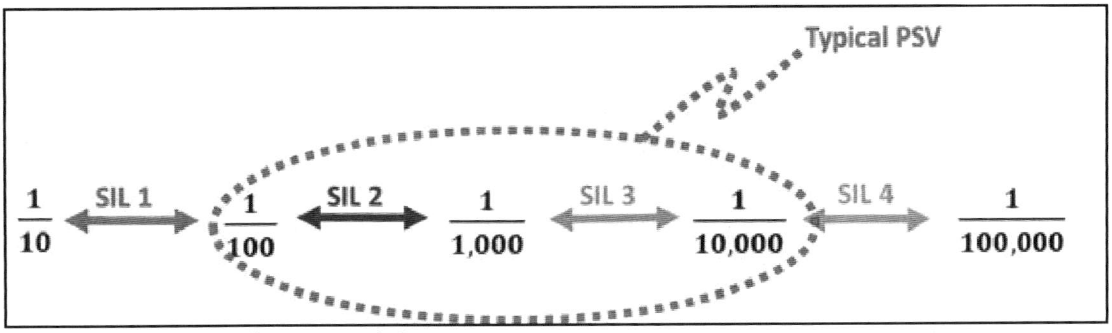

Figure 2. Range of PFD and SIL

Flare Load Reduction

As previously noted, the maximum relief load is typically caused by a common cause failure, such as a site-wide (i.e. global) loss of cooling water supply or electrical power. Such failures cause pressure relief releases in many plant sub-systems concurrently and therefore produce the largest relief loads and maximum back pressure in the flare system.

A HIPPS application can be used to reduce the maximum load on the flare system. In modern petrochemical plants, HIPPS are often used to reduce the maximum flare load that occurs during site-wide failures, such as loss of cooling water or outage of electrical power.

As shown in the simplified illustration in Figure 3 for an ethylene plant, safety instrumented functions (SIFs) are used to manage individual relief loads from compressors, fractionation systems, and in some cases cracking heaters. For example, an independent SIF on a fractionation tower will automatically cut the heat source to the tower reboiler to reduce or eliminate the vapor relief load from the tower. Alternatively, an independent SIF may trip a refrigeration compressor to prevent a release to flare due to high discharge pressure upon loss of cooling water. There are typically several managed systems within an ethylene plant flare load reduction system, each with its own SIF.

The PFD of the overall HIPPS is the sum of the PFDs of the individual SIFs. Based upon the simple example shown in Figure 3, if each individual SIF is low SIL 3 (1/1000) in a HIPPS application with 6 individual SIFs, then the PFD of the overall HIPPS will be less than SIL 3, per the following:

$PFD_S = PFD_1 + PFD_2 + PFD_3 + PFD_4 + PFD_5 + PFD_6 = 6 \times 10^{-3} > 1 \times 10^{-3} \rightarrow$ SIL 2

So, with such a system, there are two options to achieve an overall SIL 3 level of reliability:

1. Redesign all individual SIFs for a SIL 3 rating higher than the minimum (i.e. 1/6,000 instead of 1/1000) to ensure that the sum of the PFD of the individual SIFs is still in the SIL 3 range. Achieving this high level of reliability may be impractical or uneconomic, especially if there are many individual SIFs. In addition, expanding the number of components may also increase the likelihood of a false trip.

2. Assume that one SIF (the one with the largest relief load) will fail and size the flare accordingly. The probability of exceeding this relief load now becomes the sum of the probabilities that a second SIF will fail at the same time as the largest SIF, as shown in Figure 4. The simplified calculation for this option is shown below, but use of this option requires that the flare system be designed for a larger load than the approach shown in option 1. Note that in the equation, PFD_L represents the PFD of the system with the largest relief load and PFD_X represents the PFD of the other individual systems (i.e. x = 1, 2, 3, etc.), where each PFD is 1×10^{-3}

$$PFD_S = (PFD_1 \times PFD_L) + (PFD_2 \times PFD_L) + \ldots + (PFD_6 \times PFD_L) = 6 \times 10^{-6} < 1 \times 10^{-4}$$

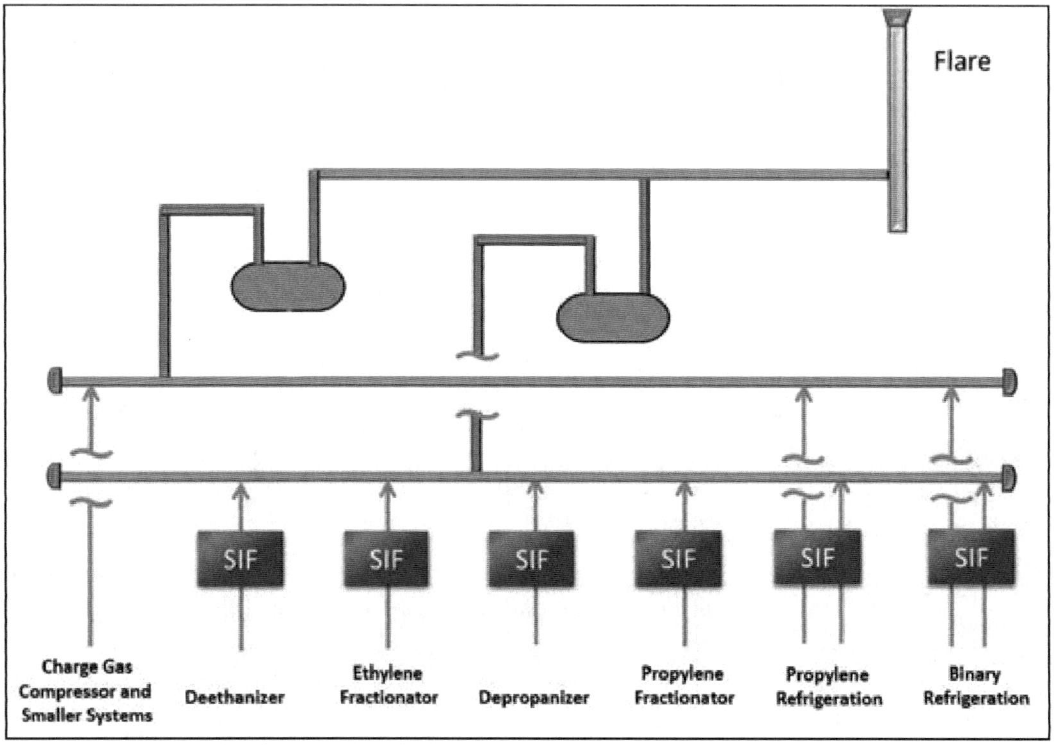

Figure 3. Simplified Sketch of Flare Load Reduction System

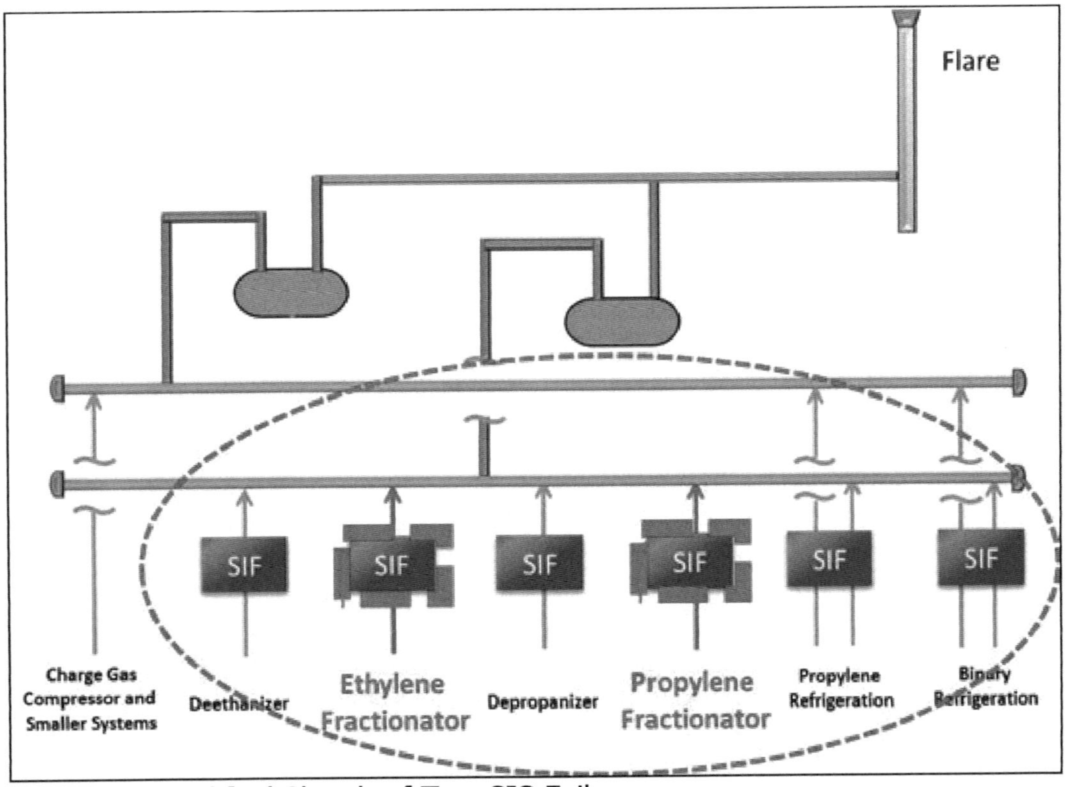

Figure 4. Simplified Sketch of Two SIS Failures

Examples with Implications for the Impact of the Testing Interval on Reliability

The following examples demonstrate the impact of the testing interval on the probability of failure upon demand (PFD) calculation and therefore the level of risk to which the plant is exposed. Please note that a simplified approach has been used to illustrate these examples and focus the reader's attention on the sensitivity of the relationship between these variables. The actual relationship between the testing frequency and the SIL of the HIPPS application for the flare system must be calculated using the specific design parameters for a particular plant.

In the first example, let's look at an individual SIF for a propylene fractionator, as shown in Figure 5. In this example, tower overhead pressure is monitored and activates isolation valves on the heating medium to the reboiler to manage the relief load from the fractionator system. This system has the following components:

- Three (3) pressure sensors, with 2 out of 3 voting to detect a high pressure event
- Redundant (2) logic solvers, with independent processing
- Redundant (2) final elements/trip valves, each provided with a single "de-energize to trip" solenoid

The PFD for the overall system is calculated as the sum of the individual probabilities:

$PFD_{Overall} = PFD_{Sensors} + PFD_{Logic\ Solver} + PFD_{Valves}$

For this system, which has redundancy for each component, using typical beta factors and fail dangerous rates for commonly used components, the relationship between the component testing interval, the PFD and the SIL for this single SIF is shown in Table 2. The PFD and SIL for a simple HIPPS that consists of 6 of these independent SIFs (in which failure of any one of them can potentially produce a flare overload) is also shown in Table 2. As previously discussed, this combined simple system does not achieve SIL3 level of reliability and quickly degrades to a lower rating if the testing frequency is reduced.

Test Interval (months)	Individual Probability of Failure upon Demand (PFD)	Individual Safety Integrity Level (SIL)	Simple System PFD with 6 SIFs in the HIPPS	Simple System SIL with 6 SIFs in the HIPPS
4	0.4×10^{-3}	3	2.5×10^{-3}	2
8	1.0×10^{-3}	3	5.8×10^{-3}	2
12	1.6×10^{-3}	2	9.8×10^{-3}	2
24	4.3×10^{-3}	2	2.6×10^{-2}	1
36	8.1×10^{-3}	2	4.9×10^{-2}	1

Table 2. Relationship between Test Interval, PFD, and SIL for Individual SIS Example

For a flare load reduction system that includes the same 6 individual SIFs, but has been designed to handle a maximum relief load that includes the assumption that the largest SIF will fail, such as the one shown in Figure 4, the relationship between the component testing interval, the PFD, and the SIL for the total system is shown in Table 3.

Test Interval (years)	Probability of Failure upon Demand (PFD)	Safety Integrity Level (SIL)
1	1.6×10^{-5}	3 (4)
2	1.1×10^{-4}	3 (4)
3	0.4×10^{-3}	3
4	1.1×10^{-3}	2/3
5	2.1×10^{-3}	2

Table 3. Relationship between Test Interval, PFD, and SIL for Combined SIS Example

For this overall flare load reduction system case, a general test interval of 3 years will still provide an acceptable safety integrity level. However, if the test interval is extended to 4 years, which could be equivalent to only testing during plant turnarounds, the safety integrity level approaches SIL 2.

Per the earlier explanation of flare load reduction systems, if the test interval of all of the individual systems was extended beyond their design value, eventually the cumulative PFD of the combined system will exceed 1×10^{-3}, which will reduce to reliability of the overall system to SIL 2. In other words, as the testing interval of the components increases, the probability of overloading the flare during a relief release event will increase.

In our final example, let's consider the case in which the flare system was not designed to handle a failure of the largest SIF load (i.e. a case in which there was no available plot space for additional flaring equipment in a revamp situation). This can be accomplished by installing dual independent SIFs on the individual system with the largest relief load, as shown in Figure 6. In that case, it is possible to design for a PFD less than 1×10^{-4}, which the owner and regulatory authority may accept as a non-credible failure case (i.e. the probability of failure is so low that the outcome is not considered to be credible). As a non-credible case, this relief load can be removed from the flare load.

Such a system, using 2 independent SIFs, is shown in Figure 7 for the example of the propylene fractionator. In addition to the SIF system shown in the initial example, a second independent system (i.e. based upon separate and independent measurements, logic solvers, and valves) is added that has the following components:

- Three (3) pressure drop sensors, with 2 out of 3 voting, to detect a low cooling water pressure drop (i.e. low supply flow) event
- Redundant (2) logic solvers, with independent processing
- A single final elements/trip valve provided with a single "de-energize to trip" solenoid

For this dual system, using typical beta factors and fail dangerous rates for commonly used components, the relationship between the component testing interval, the PFD and the SIL for the system is shown in Table 4.

Test Interval (years)	Probability of Failure upon Demand (PFD)	Safety Integrity Level (SIL)
1	0.2×10^{-5}	3 (4)
2	0.8×10^{-5}	3 (4)
3	0.2×10^{-4}	3 (4)
4	0.4×10^{-4}	3 (4)
5	0.6×10^{-4}	3 (4)

Table 4. Relationship between Test Interval, PFD, and SIL for Combined SIS with Dual System

As a consequence, it can be seen that installation of this type of dual SIF for an individual relief load can maintain a very high SIL rating and allow the owner and the safety regulatory authority to accept the relief of that individual system as a non-credible case. Therefore that load can be removed from the flare system and the SIL rating for the remainder of the HIPPS would be calculated per the above examples, depending upon the design.

It is important to emphasize again that these are hypothetical examples that are based upon typical assumptions and simplified systems. The tables presented in this section are only intended to demonstrate the sensitivity of the safety integrity level of the HIPPS that protects the flare system to the component testing frequency. The relationship between testing frequency and the SIL of the HIPPS application for the flare system in a particular plant must be calculated using the specific design parameters for that plant.

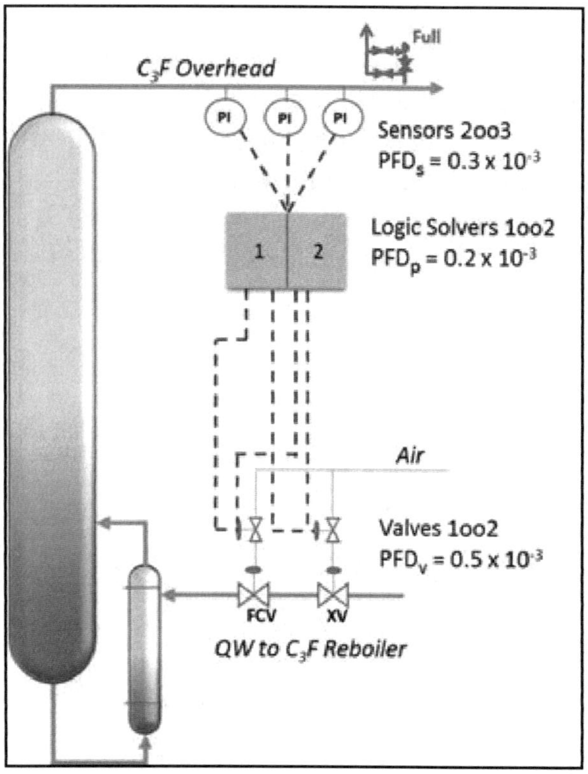

Figure 5. Example of a Single SIF for a Propylene Fractionator

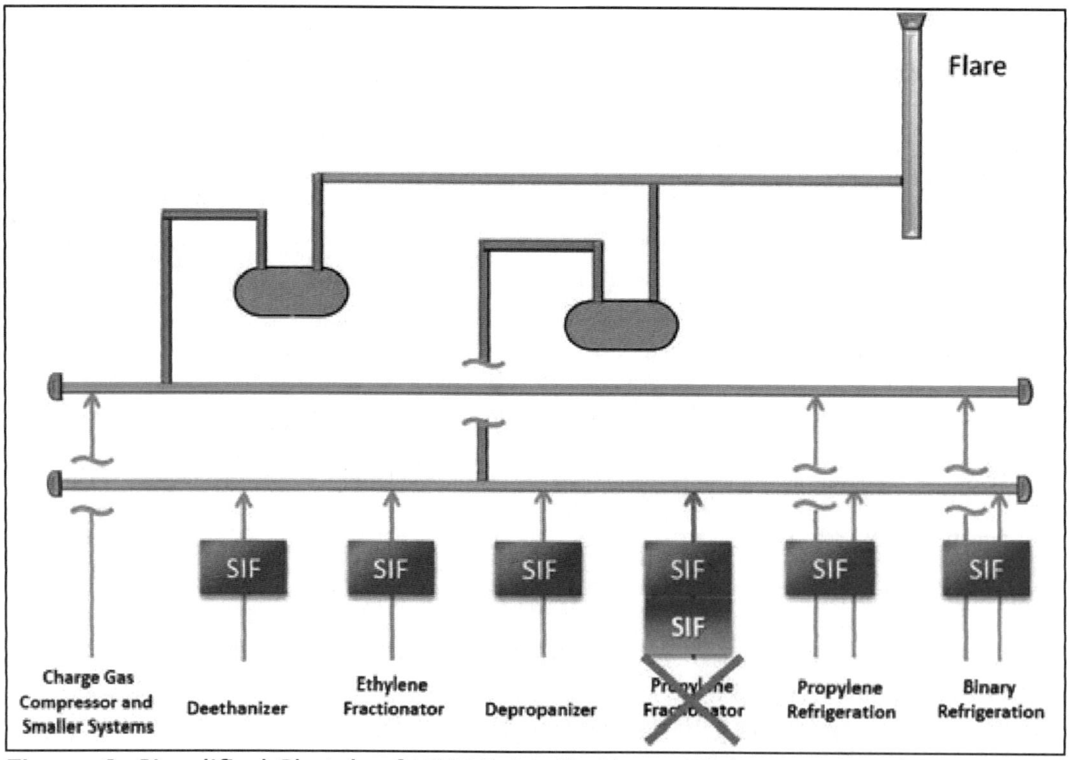

Figure 6. Simplified Sketch of HIPPS Application with Largest Load Removed from Flare Load

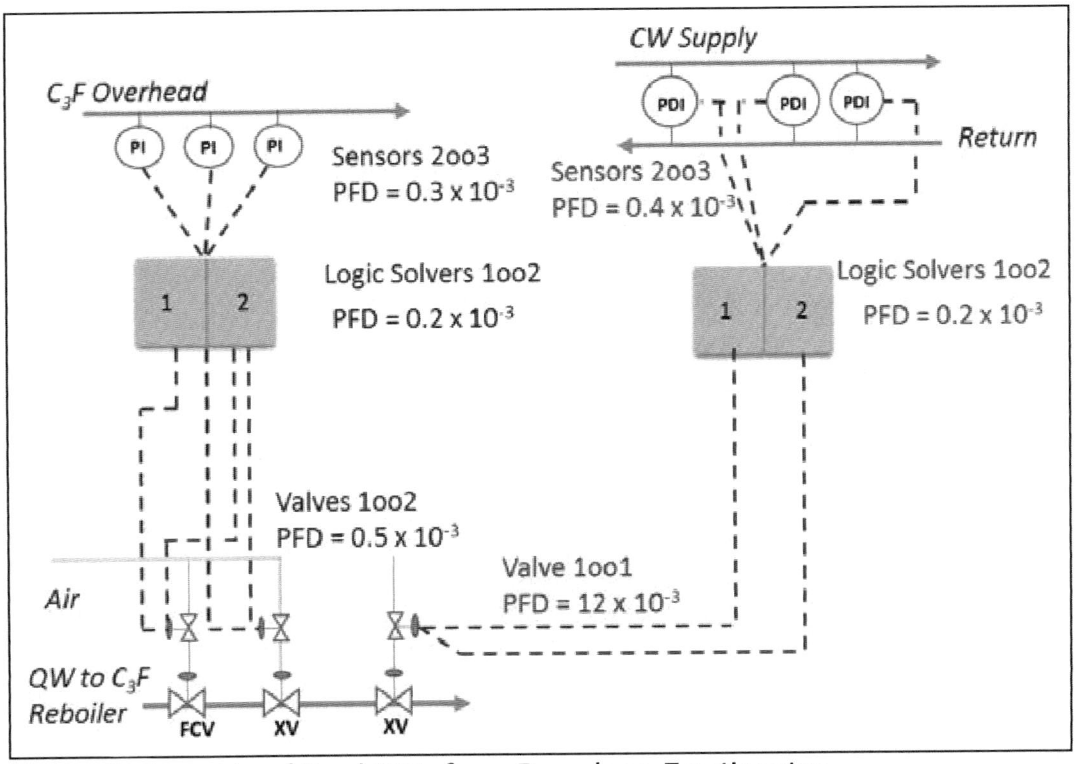

Figure 7. Example of Dual SIFs for a Propylene Fractionator

Why is It Important to Know This?

As a licensor of process technology, we periodically survey our licensees on a variety of operational issues for the purpose of improving the performance and safety of the plants. This past year, we surveyed 76 ethylene and on-purpose propylene plants to ask the following questions:

- Do you use a safety instrumented system (i.e. SIS) to reduce flare load?
- Does your plant have a testing program for your flare load reduction SIS trip sensors (i.e. measurements, etc.), actuators (i.e. solenoids, etc.), logic solvers (i.e. PLCs, etc.), and pressure relief devices (i.e. XVs, PSVs, etc.) between turnarounds?
- How often are the components in your flare load reduction safety instrumented system tested?

Of those 76 plants, 57 reported that they used a flare reduction safety instrumented system. (Note that the 19 plants that did not have this system were mostly older plants that had not been expanded, although they may have had other SISs that were not used in a HIPPS flare load reduction application.) The responses from the 57 plants that do have an SIS for flare load reduction revealed the general information regarding their HIPPS flare load reduction testing programs that is shown in Table 5:

System Component	Report Having a Test Program	Actual Test Between Turnarounds	Average/Median Test Interval
Sensors/Measurements	77%	49%	Every 32/27 months
Actuators/Solenoids	76%	51%	Every 31/24 months
Logic Solvers/PLCs	45%	20%	Every 43/48 months
Valves/XVs/PSVs	84%	49%	Every 38/36 months

Table 5. Survey Results

Viewed a different way, a plot of the testing interval used by the plants who responded to the survey (shown in Figure 8) indicates that approximately 40% of plants test their SIF components an average of every 3 years or more.

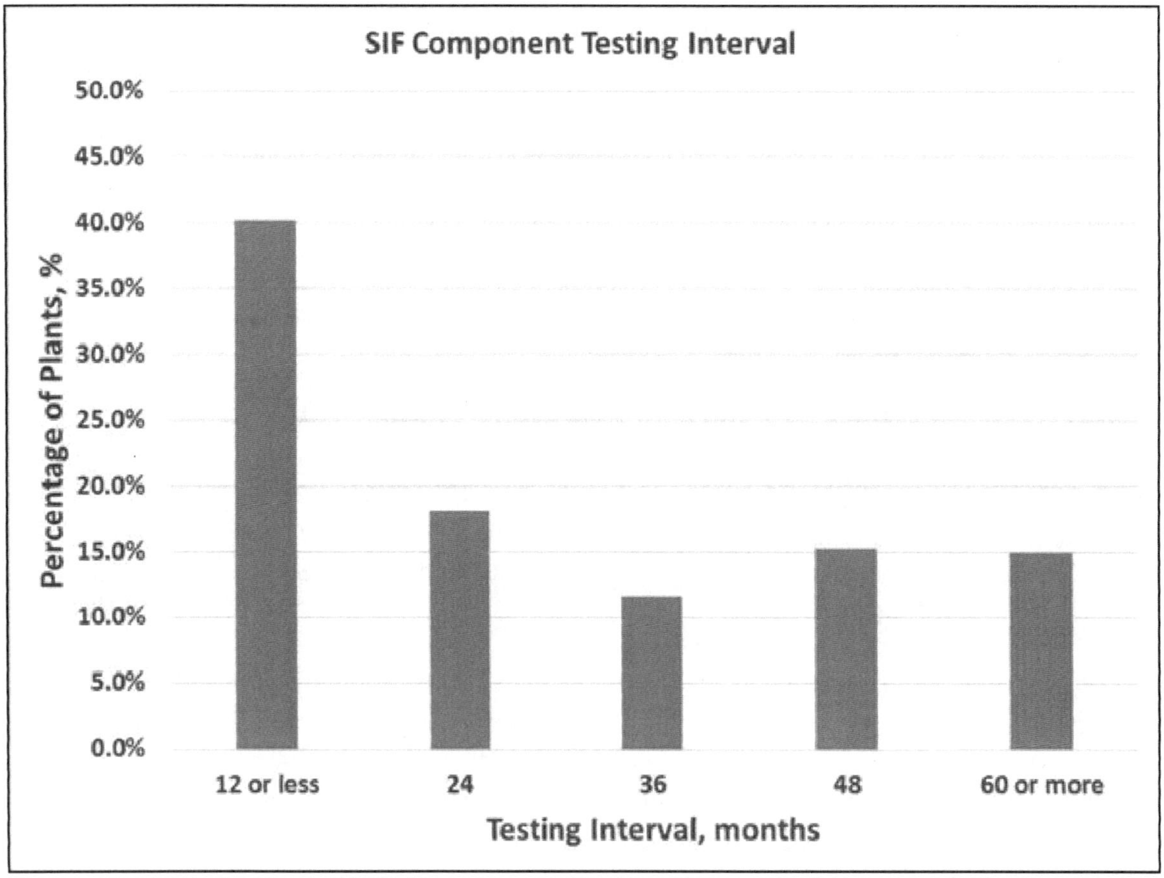

Figure 8. Distribution of SIF Component Testing Interval of Surveyed Plants

Per the previous discussion, depending upon the system design, the testing interval required to maintain a SIL 3 integrity level for the flare reduction system can vary significantly, but typically may require testing every 2 to 3 years. This data shows that a large percentage of the plants surveyed that have flare reduction safety instrumented systems may not test their system components frequently enough to maintain a SIL 3 integrity level.

The lack of a formal testing and maintenance program at some of these plants greatly increases the risk that one or more components may be in a failed state when called upon to act. Not all system failures will result in a catastrophic incident, but the plant's risk of exposure to such a failure is increased as the testing interval for the components in the HIPPS is extended. When an SIS/HIPPS is regularly tested and maintained according to the SIL selected, the risk is properly managed.

Suggestions for Operating Plants

When the survey responses are viewed in light of the design concepts used for the flare reduction safety instrumented system, it is apparent that more attention may need to be paid to the testing interval of the SIF components to ensure that the integrity level of the system is maintained high enough to sufficiently minimize the risk of a failure event and maintain the SIL selected for the plant.

To accomplish this goal and mitigate potential incidents due to failure of SIFs, operating plants should consider the following practices, if they have not already done so:

- Identify where SIFs are used in the application of SIS/HIPPS in the plant and understand the consequences of failures of these systems.
- Confirm and document the design safety integrity level of the plant's flare reduction system (and other safety systems) and the testing interval that was assumed by the detailed engineering contractor to achieve that integrity level for all the components in the SIS/HIPPS. This is typically available in the SIL Verification Report.
- Maintain a database of the SIS components that includes their required test interval, date of last test and maintenance, and their next required testing date.
- Maintain a set of testing procedures for each component that ensures the test is performed correctly and that the component is returned to its fully operational state after the test/maintenance.
- Plan testing schedules for each of the components in the SIS/HIPPS and perform periodic testing of the SIF components according to that schedule.
- In cases where testing during major turnarounds alone does not allow the SIL to be achieved, then it will be necessary to consider alternatives, such as the following:
    - Provide a means for on-line testing. This is most difficult for the final control elements (i.e. valves, etc.). Consider use of:
        - Partial-stroke testing while the plant is running to reduce the full-stroke testing interval that is required to maintain the target SIL.
        - Install bypasses to allow full-stroke testing while the plant is in operation.
    - Maintain a procedure in place to allow the critical testing (i.e. high testing frequency components) to be addressed quickly and efficiently during unplanned outages, based upon the documentation and database recommended above.

While this paper has focused upon the impact of the testing frequency on the reliability of a flare load reduction safety instrumented system, safety instrumented systems are used for other purposes in petrochemical plants, such as reactor safety systems. The above recommendations should be followed for those systems as well.

Summary

In conclusion, safety instrumented systems are important protection layers in the plant's overall safeguarding strategy and in particular, in the plant's flare load reduction scheme. It is essential that all the elements of the HIPPS flare load reduction system operate in the manner in which they were designed to ensure that the plant is brought to a safe condition during an emergency event. As the testing interval of the components increases, the probability of overloading the flare during a relief release event will increase. Failure to maintain the required test intervals of the SIF components will result in a lower level of safety integrity than required/expected, as was illustrated in the examples provided in this paper. If failure of a component occurs and goes undetected, then an emergency event in the plant could lead to operation at conditions beyond the design limits of the equipment and potential damage to equipment and risk to operating personnel. In cases where the design flare system load has been reduced as a result of an SIS/HIPPS, failure of SIF components could overload the flare system during an emergency. This could result in piping and/or equipment operating above their design pressure and potential loss of containment.

AICHE Paper Number: 188d

MODELING ETHYLENE PLANT CRACKING HEATER BURNER LIGHT-OFF SAFETY INSTRUMENTED FUNCTIONS

David K. Lewis

Functional Safety Engineer

NOVA Chemicals Corporation

Prepared for Presentation at the 2017 Spring National meeting

San Antonio, Texas, March 26-30, 2017

MODELING ETHYLENE PLANT CRACKING HEATER BURNER LIGHT-OFF SAFETY INSTRUMENTED FUNCTIONS

David K. Lewis

Functional Safety Engineer

NOVA Chemicals Corporation

Abstract:

Ethylene Plant Cracking Heaters require the design of Safety Instrumented Functions to help manage the risk of startup and operation of the Heater. Lighting off the burners on multiple burner Heaters can represent a high risk which may require a SIL 2 Safety Instrumented Function (SIF) especially if the burners are manually lit by an operator. Designing and modeling this SIF can present challenges where multiple fuel isolation valves are required on multiple burner heaters and where there is limited opportunities to proof test the SIF.

This paper will discuss the challenges of designing and modeling the safety reliability of multiple burner Cracking Heater burner light-off SIFs and provide an example of how this can be accomplished.

1. Introduction

Lighting off burners on a fired heater is likely the highest risk in its operation. In an Ethylene plant this can also represent the highest aggregate risk per system due to the multiple cracking heaters and boilers in one plant. The degree of risk during a burner light off can vary due to many factors such as:

- Size and type of firebox
- Local or remote burner ignition
- Frequency of burner light offs
- Number of burners
- Number of combustible gas sources

Cracking Furnaces (or Heaters) can be higher risk for process fires than other process fired heaters due to larger fireboxes, multiple combustible gas sources and requirements for manual local ignition of multiple burners. Managing the risk of a Cracking Furnace during burner light off will require high integrity protective layers to ensure the residual risk is acceptable.

This paper is written for an intended audience with basic knowledge of BMS, LOPA, NFPA 85/86, API RP556, CSA B149.3, IEC 61511 and Cracking Furnace operation.

2. Risk Assessment

The LOPA (Layer of Protection Analysis) risk assessment method can be used to evaluate the risk of a firebox explosion during burner light off in a Cracking Furnace. To account for the effect of a purge control system, the burner light off hazard scenario can be evaluated in two scenarios. One for "Low Leakage" and one for "High Leakage". "Low Leakage" is defined as the maximum amount of combustible gas leakage where the normal controlled minimum purge rate will keep the gas mixture below the lower explosion limit (LEL). "High Leakage" is defined as the minimum amount of combustible gas leakage which will result in a mixture above the LEL with the normal controlled minimum purge rate. This accounts for failure situations where the leakage rate overwhelms the purge rate (e.g. wide open fuel gas valves).

Other burner related typical LOPA scenarios to develop would be:

- Fired heater burner light-off delayed with open burner valves.
- Fired heater burner light-off during high combustible gas leakage due to coil rupture.
- Introducing combustible fluid into the fire box (for incineration, not using a burner), below auto-ignition temperature.
- Burner Flameout due to fuel pressure control failure
- Various Air/Fuel Imbalance issues due to equipment failures or operator error

This paper will focus on the low and high combustible gas leakage hazard scenarios in a typical Ethane Cracking Furnace. In the following simplified Cracking Furnace diagram there are four combustible gas streams which could leak into the firebox before or during a burner light off:

- Fuel Gas though a burner. Note the BAT (Before Auto-ignition Temperature) isolation valves remain closed until the first deck burners bring the firebox up to auto ignition temperature.
- Product Feed (e.g. Ethane) path through the furnace coils and the open decoke valve (normally open for the furnace startup) and into the firebox
- Acid Gas (e.g. CO2, H2S and Ethane in certain upset conditions)
- Cracked Gas backflow from the Cracked gas header through the decoke valve.

Simplified Cracking Furnace diagram:

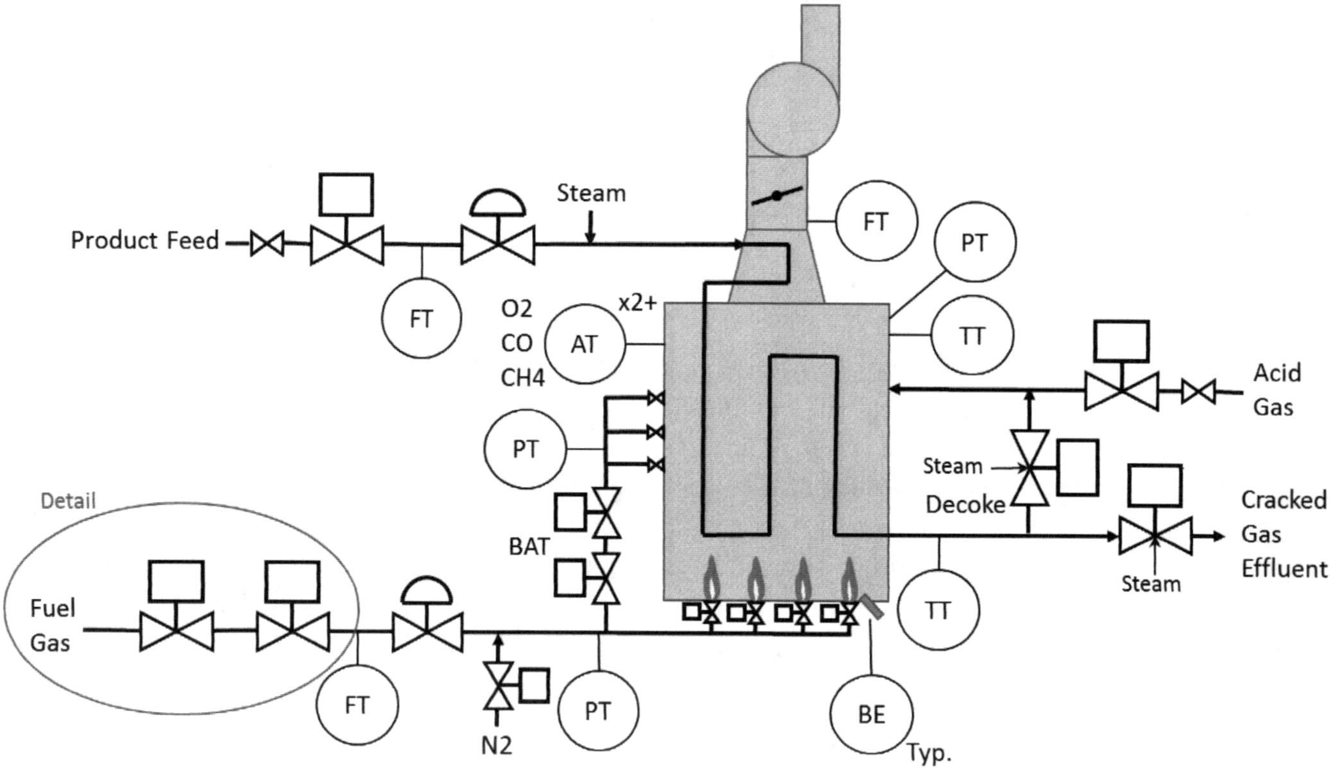

The following LOPA event frequency analysis evaluates the Low Leakage and High Leakage scenarios from all sources assuming only the normal fuel gas flow control, firebox purge flow control and operating procedures for startup manual isolation are in place. The Independent Protection Layer requirements are then developed and evaluated for effectiveness and degree of risk reduction to ensure the residual risk is acceptable.

Hazard Scenario for the Low Leakage scenario:

LOPA Hazardous Event Description		Event Frequency Analysis					
		Initiating Event		Enabling Events		Unmit. Event Freq (Events /yr)	Min Reqd RRF
Hazard Scenario Description	TRC/ y	Initiating Event	Events /yr	Enabling Event	Prob.		
1. Fired heater burner light-off during a small combustible gas leakage and low purge flow and/or low purge time, explosion and fatality. "Small leak" defined as the maximum amount where the normal minimum purge rate will keep the gas mixture below the LEL.	1E-04	Operator lighting burners once a year during a small gas leakage. (multiple equipment or operator failure modes)	1.E-01	1. Loss of purge control causing low air flow and combustible mixture accumulation in firebox.	0.1	9.E-04	10
				2. Ignition	1		
				3. Explosion - Probability of significant explosion is less then scenario # 2 due to less possible gas flow.	0.3		
				4. People presence – operator manually lighting burner	1		
				5. Fatality	0.33		

TRC – Tolerable Risk Criteria

RRF – Risk Reduction Factor

Low Leakage Scenario Protection Layer requirements:

Independent Protection Layers				LOPA	
Independent Protection Layer Description	IPL Type	IPL RRF	IPL Total RRF	Mitigat. Event Freq (Events /yr)	RRF Gap
1.1. Startup purge interlock sequence to allow the burner light off sequence to begin which includes opening of burner fuel gas valves. Safety functions include: - Verification of safe minimum purge flow rate and adequate air distribution (CSA B149.3-15 - 60% of maximum input) -Time to complete startup purge cycle to ensure no large accumulation of combustible gas before burner light-off, (CSA B149.3-15 and NFPA 86 (2015) – at least 4 volume changes) - Limit the maximum fuel source leakage rate by isolation & verification of isolation.	SIF	10	10	9.E-05	1

Hazard Scenario for High Leakage:

LOPA Hazardous Event Description		Event Frequency Analysis					
		Initiating Event		Enabling Events		Unmit. Event Freq (Events/yr)	Min Reqd RRF
Hazard Scenario Description	TRC/y	Initiating Event	Events/yr	Enabling Event	Prob.		
2. Fired heater burner light-off during **high** combustible gas leakage and normal purge rate, explosion and fatality. **"High"** gas flow defined as minimum amount which will result in a mixture above LEL with normal minimum purge flow rate.	1E-04	Operator lighting burners once a year during a large combustible fluid leak and normal purge rate. Large combustible fluid leak caused by loss of fuel source (e.g. Fuel gas, Acid Gas, HC Feed, Cracked Gas) control or isolation (multiple equipment or operator failure modes).	1.E-01	1. Ignition	1	1.1E-02	109
				2. Significant Explosion	0.33		
				3. People presence–operator manually lighting burner	1		
				4. Fatality	0.33		

High Leakage Protection Layer requirements:

Independent Protection Layers				LOPA	
Independent Protection Layer Description	IPL Type	IPL RRF	IPL Total RRF	Mitigat. Event Freq (Events/ yr)	RRF Gap
2.1. Startup sequence interlocks for combustible gas isolation. Safety Functions include: - Adequate isolation and verification of isolation of all combustible fuel sources whose capacity could exceed the safe limit with the minimum purge rate. - Verification of valve isolation to include closed limit switches on valves and pressure or gas sensing to prove below safe limit of leakage. - Interlock logic to prevent fuel source valves from opening and burner ignition to occur on failed isolation verification.	SIF	110	110	9.9E-05	1

3. Reliability Models

The objective of the identified Safety Instrumented Functions (SIFs) is to prevent and detect any leakage of combustible gas into the firebox and to prevent the ignition stage if there is a failure in either combustible gas isolation prevention or detection of combustible gas leakage. Below is the overall safety reliability model to be used for reliability calculations.

Overall reliability model for Combustible Gas Leakage Prevention or Detection:

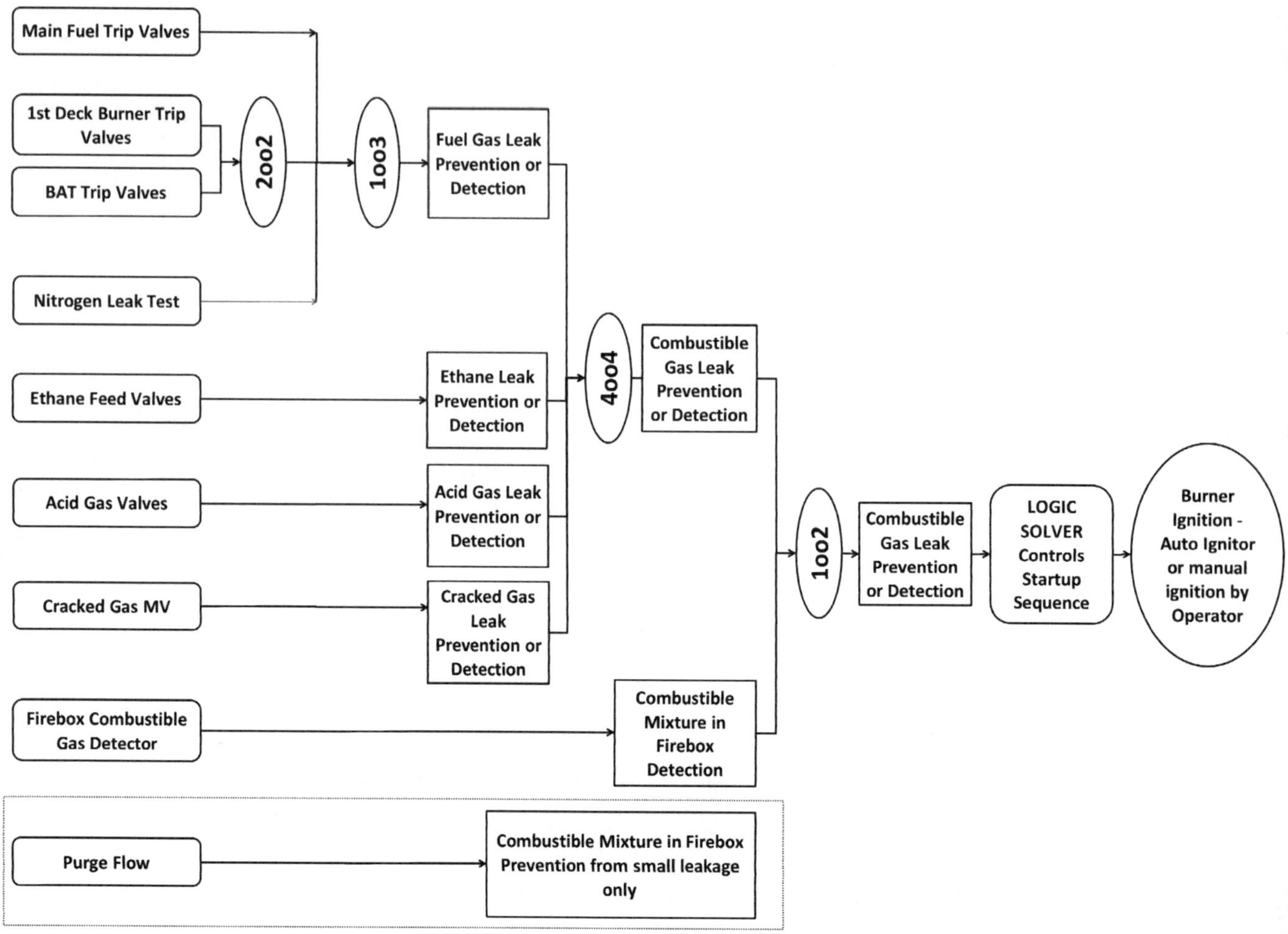

The first column of inputs to the model represent a summary of all the components for that sub-function. The model diagram is separated in to sub-functions to allow easier display of the model. To show every component of the reliability model on one diagram would create a very large diagram. The sub-function inputs are described as follows.

Main Fuel valve system model for Fuel Gas Leak Prevention or Detection:

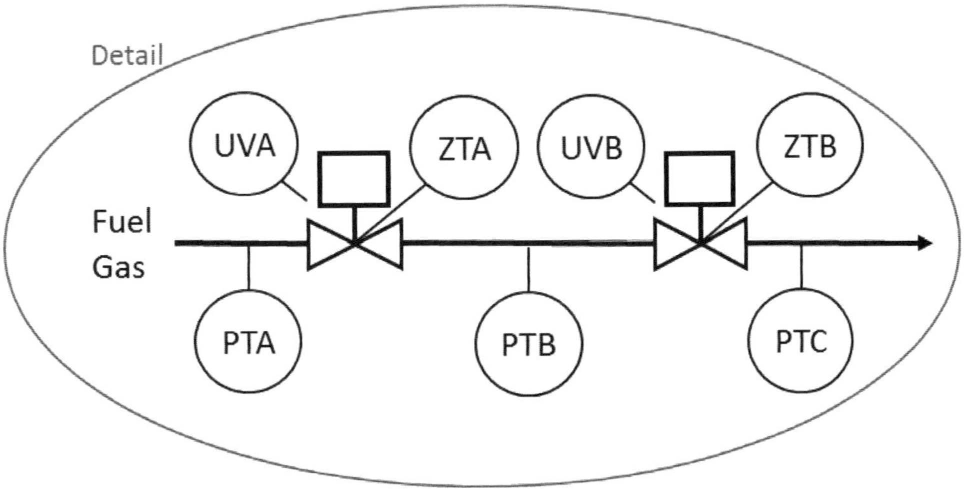

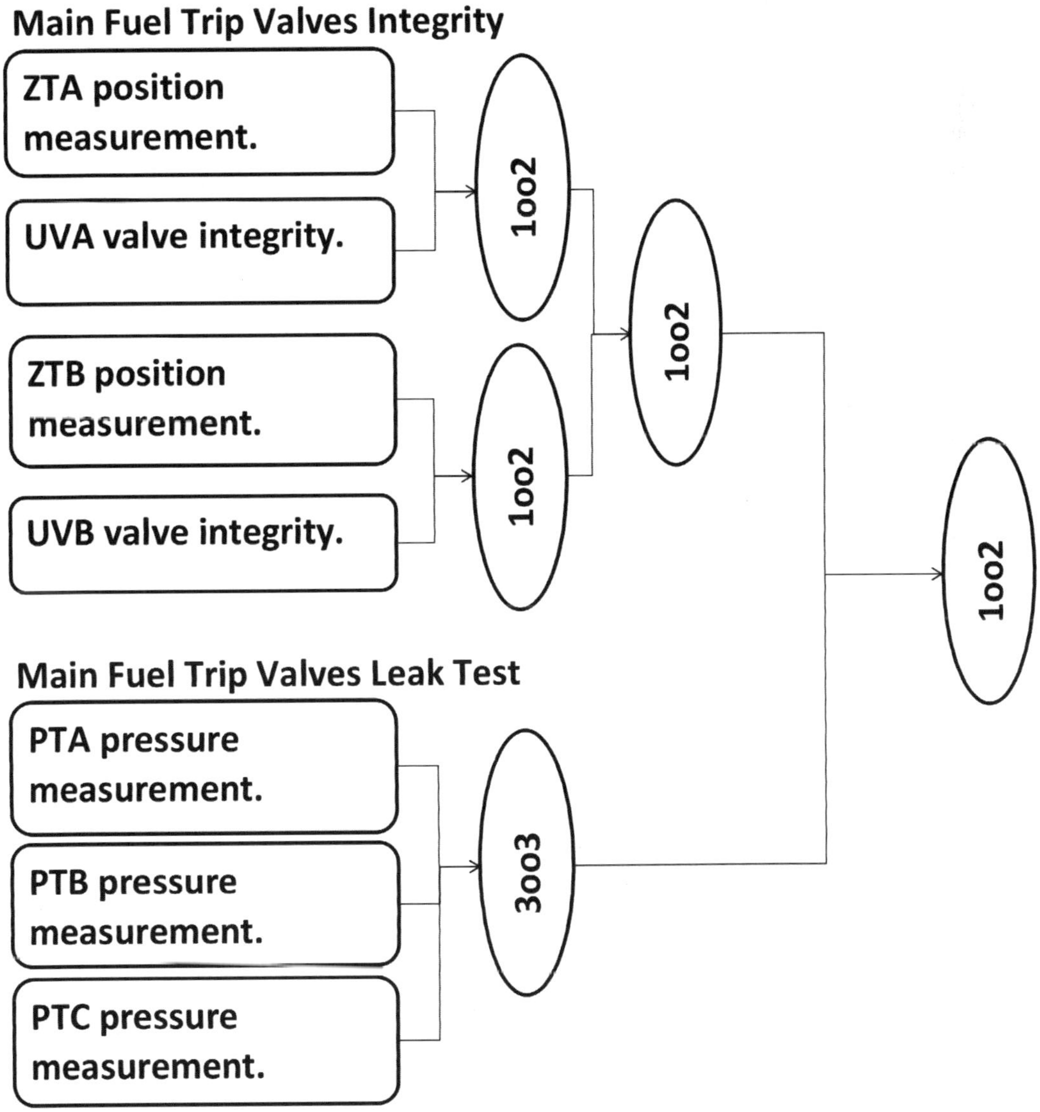

Bottom Burner and BAT Valves:

The first deck burner valves and the BAT isolation valves all have closed limit switches. The reliability model includes 1oo2 voting on each burner valve for valve integrity and reliability of the closed limit switches. Each 1oo2 voting are then inputs to NooN voting to combine all the burner valves. The BAT valves are also 1oo2 voting for valve integrity and reliability of the closed limit switches with the 1oo2 voting inputting to another 1oo2 voting to account for the two BAT valve in series.

Nitrogen Leak Test:

The Nitrogen Leak Test reliability modeling includes the Nitrogen valve integrity and closed limit switch plus all the pressure transmitters required to perform the nitrogen leak test of all burner isolation valves and the BAT valves.

Product Feed (Ethane) or Acid Gas Valve Reliability Model for leakage prevention and verification of isolation:

The product Feed and Acid Gas valves are not opened until after the burner light off sequence and a stable normal furnace temperature is achieved above auto ignition temperature. Therefore in this example a manual isolation valve is used to provide redundant isolation. If a practical operating procedure can be developed to manually open (and close) this valve when required then this can be a cost effective means to provide redundant isolation and verification of isolation to improve the safety reliability.

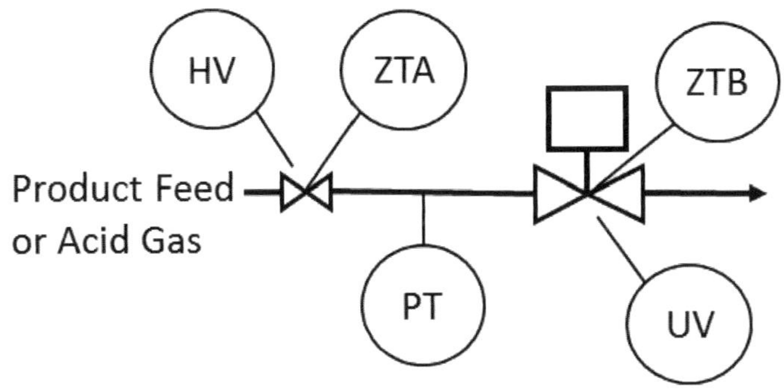

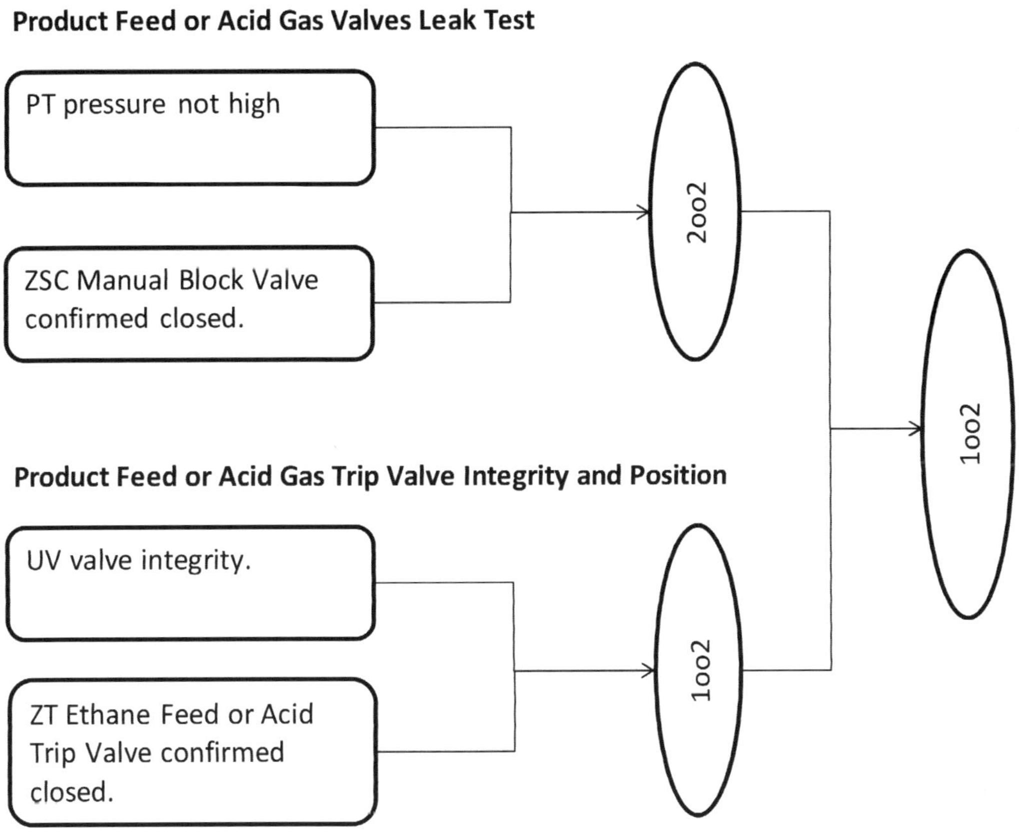

Cracked Gas Effluent Valve Reliability Model for leakage prevention and verification of isolation:

The Cracked Gas Effluent and Decoke valve in this example are large wedge plug valves. They require a steam purge to minimize coke build up on the trim seating surfaces when the valve is being operated and to provide a positive pressure barrier fluid when the valve is in the closed position. The steam purge flow and pressure is measured to indicate valve seat integrity together with the valve closed limit switch. Redundant valves could also be considered to improve safety reliability.

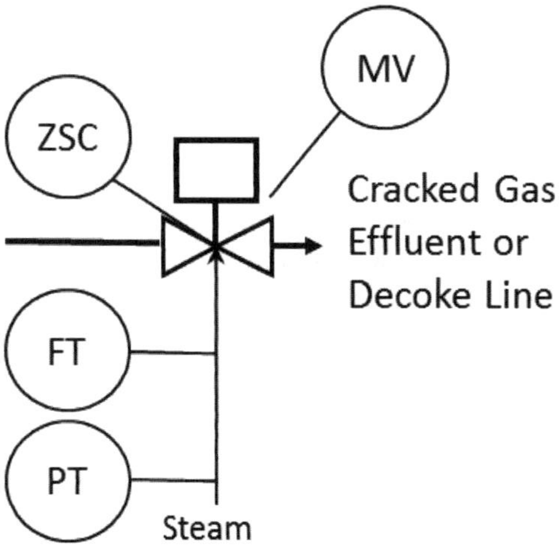

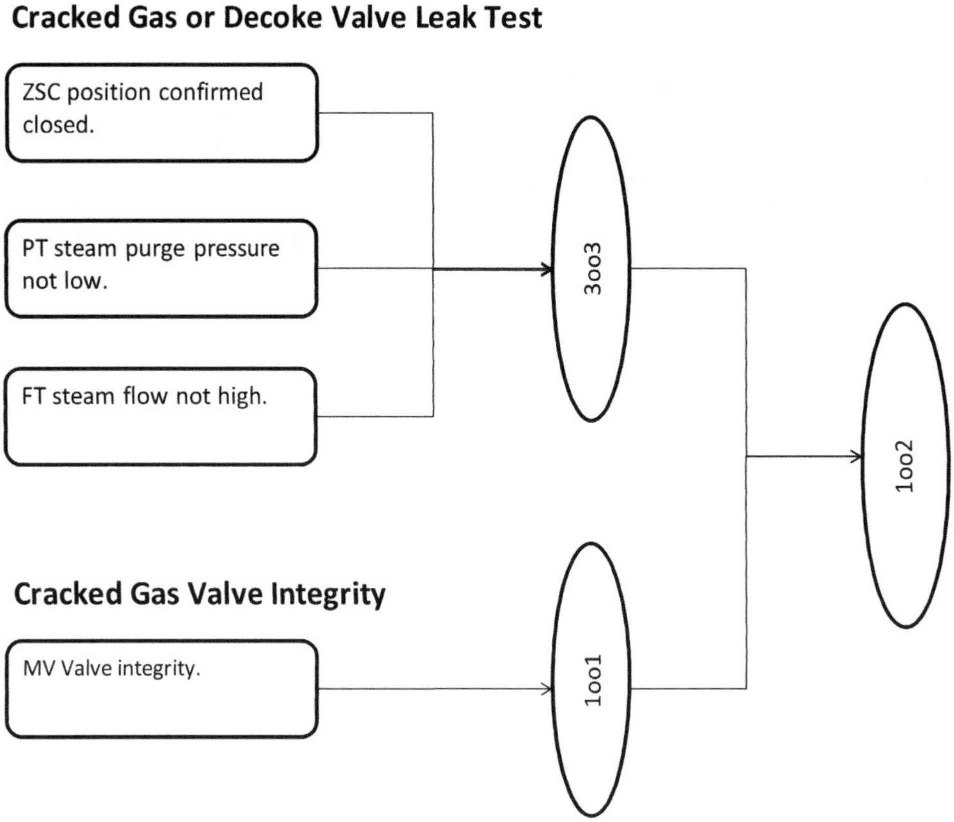

Firebox Combustible Gas Detector:

The Firebox Combustible Gas Detector can be a very effective way of improving the safety reliability of the SIF. The challenge however is designing a gas detection system that can measure a small leak from any of the combustible gas sources from many locations in a large cracking furnace firebox. The detector system also has to be reliable and proven in this application. Tunable Diode Laser (TDL) Analyzer systems are promising new technology that are being used and tested in this application to detect ppm levels of methane.

Purge Flow:

The Purge Flow input is only used for the SIF protecting the "Low Leakage" hazard scenario where the purge flow is effective. The final voting would differ from the overall reliability diagram by using a 1oo3 final voting to include the purge flow components. The Purge flow sub-function components would include voting on the purge flow measurements, any damper or louvre positions and fan status inputs.

Probability of operator error during manual ignition:

In this cracking furnace example the ignition of the burners is performed at the burner by the field operator. The operating procedure would be for field operator to be informed of which burner to light when the startup sequence logic determines it is safe to do so. When the field operator is informed that a burner is ready to be lit by the panel operator, the panel operator will remotely open the burner trip valve. The field operator will then open the manual burner valve and insert the burner light-off torch into the burner when proper communication is received and upon witnessing the burner trip valve open. The failure mechanism of this step would be the operator choosing to insert the torch into a burner during a detected or undetected fault. Stringent thorough Operating Procedures are required, plus proper field communication, signage and clear startup sequence panel graphics to ensure this step in the safety reliability model does not significantly influence to total safety reliability of the safety function.

4. Safety Reliability Calculations

Using the above safety reliability models and an assigned proof test plan including yearly on-line inspections of the field equipment and near full SIF proof testing at a Furnace outage, the total Risk Reduction Factor (RRF or PFDavg) can be calculated. The calculated RRF values for both SIF's can exceed the required LOPA IPL RRF values with proper design and selection of reliable components and development of thorough proof test plans.

5. Conclusions

1. A SIL 2 SIF can be designed and modeled to provide protection of a cracking furnace firebox explosion with multiple combustible gas streams, during a burner light off.
2. TDL Flue gas analyzers are a potential enhancement to improve the burner light-off sequence SIF safety reliability if proven reliable and effective.
3. Remote burner light off and restricting access to the heater during burner light off will inherently reduce the risk and should be considered.
4. Valve proving systems and auto leak testing can be modeling with other SIF components to provide greater safety reliability then conventional double block and bleed valve systems.

6. References

1. "Layer Of Protection Analysis – simplified risk assessment – A CCPS concept book"; Copyright Published by The Center for Chemical Process Safety of the American Institute of Chemical Engineers,
2. IEC 61511: 'Functional safety instrumented systems for the process industry sector'.
3. "Guidelines for Safe and Reliable Instrumented Protective Systems", CCPS, Copyright Published by The Center for Chemical Process Safety of the American Institute of Chemical Engineers and John Wiley & Sons, Inc.
4. NFPA 85 Boiler and Combustion Systems Hazards Code
5. NFPA 86 Standard for Ovens and Furnaces
6. API RP556 Instrumentation and Control Systems for Fired Heaters and Steam Generators
7. CSA B149.3 Field Approval of Fuel-Related Components on Appliances and Equipment

AICHE PAPER NUMBER: 188E

SUCCESSFUL AUTOMATION STRATEGIES TO PREVENT BRITTLE FRACTURE IN A BIMETALLIC DEMETHANIZER

Erin Daly
Process Control Engineer
Flint Hills Resources, Port Arthur Chemicals

Prepared for Presentation at the 2017 Spring National Meeting
San Antonio, TX, March 26 – 30, 2017

SUCCESSFUL AUTOMATION STRATEGIES TO PREVENT BRITTLE FRACTURE IN A BIMETALLIC DEMETHANIZER

Erin Daly

Process Control Engineer

Flint Hills Resources, Port Arthur Chemicals

ABSTRACT

Older demethanizer towers are commonly constructed of two different types of metals, sufficient to handle normal operating conditions: Stainless steel in the colder top section and carbon steel in the warmer bottom section. Risk of cold temperature embrittlement and brittle fracture can be initiated if the cold liquid from the top section reaches the bottom section or if there is a sudden loss of pressure in the column. The process is more likely to reach those conditions during upsets or transitions between operating states, where automated process controls can be used to help provide safeguards and reestablish stability. In accordance with the industry practice in 1975, at the time of the original Light Olefin Unit design at what would eventually become the Flint Hills Resources Port Arthur facility, the embrittlement risk was mitigated by inclusion of a "warm up" line that would route 60 degree Fahrenheit vapor from cracked gas dryers to the demethanizer tower to heat it when needed. In 2006, a site incident related to corrosion under insulation on that process line initiated the development of alternative strategies. The automated control was built up over time from the simple warming vapor injection to a complex multi-prong approach that successfully avoided cold temperature embrittlement operating regimes and incidents. In 2015, the bimetallic tower was replaced by one entirely constructed of stainless steel as the final iteration of risk reduction efforts.

Introduction

Cold temperature embrittlement and brittle fracture have been studied since around the time of World War II, yet persist as significant process hazards today. Brittle fracture has been described as "the sudden and often catastrophic fracture of a material with little or no plastic deformation."[10] Some of the earliest examples of brittle fracture occurred on WWII ships that had been welded together and made of large grain steel, which failed when operated in the cold waters of Northern climates.[3,9] Since then, the history of the chemical process industry has been spattered with similar, sometimes fatal, failure events such as a 1965 flare line fracture, a 1975 naphtha cracker explosion, a 1989 cracker gas leak and fire, and a 2008 olefins facility flare base fracture.[9,6,10] One contributing factor to many of the recorded industrial incidents is that most plants were designed for steady-state operations, before a thorough understanding of brittle fracture was established.[3] Today, the ASME boiler and pressure vessel code requires material and steel impact testing, but the requirement was not incorporated until 1987, so equipment built before then was not tested to a common standard.[2] There was also no guidance for how to evaluate older equipment for safe operation until 2000, when API 579 formulated the fitness-for-service analysis.[1]

Brittle fracture requires three concurrent conditions, often referred to as a triangle: 1) susceptible material properties at a given temperature 2) sufficient applied stress 3) material flaw of a certain size. A metal can become brittle, or lose ductility and the ability to absorb energy before failure, below its ductile-brittle transition temperature, which varies based on the material properties of the metal. The minimum design metal temperature (MDMT), equivalent to the minimum allowable temperature (MAT) at the maximum allowable working pressure (MAWP), approximates to a single transition temperature that equipment should not be operated below.[10] Since most vessels operate above the stress threshold required for fracture and have some intrinsic fabrication or use-generated flaw, the best way to prevent completing the brittle fracture triangle, is to prevent the temperature from falling below the brittle threshold.[4]

Three typical scenarios that lead to brittle facture exposure include loss of heating medium, loss of pressure and subsequent auto-refrigeration, and loss of pressure and auto-refrigeration followed by re-pressurization while the metal is still cold.[3,10] These scenarios are most likely to be reached during dynamic upset conditions or changes of operating state. While the brittle fracture hazard can be eliminated from new equipment by design that takes dynamic behavior into account, the strategy to mitigate the hazard in older equipment must be a combination of engineering and administrative controls.[15] The process controls system is a critical component of the engineering controls that are available to prevent cold temperature embrittlement and brittle fracture, especially in older equipment that may be at risk, such as bimetallic demethanizers.

Basic Process Overview

The Flint Hills Resources (FHR) demethanizer has four feed separation drums and four resulting feed streams. Liquefied cracked gas in the first feed to the column is the warmest at a nominal -30°F. The vapor is chilled and separated again, and this process is repeated to form three other feeds into the column, nominally -90°F, -150°F, and -155°F. Feeds to the tower are controlled by level in the separator drums. The overhead condenser is cooled by ethylene refrigerant (C2R) and the bottoms reboiler is heated by propylene refrigerant (C3R). The bottoms temperature cascades to the heating medium flow through the reboiler, but there is also an override to ensure vapor does not blow through the refrigeration system. The liquid flowing from the tower bottom is the feed stream for the deethanizer tower. The overhead vapor out of the demethanizer is sent into the fuel gas system.

Original Design

The originally installed Light Olefin Unit (LOU) Demethanizer tower was a bimetallic tower constructed of stainless steel in the colder top section, and carbon steel in the warmer bottom section.[14] Under normal steady state conditions, embrittlement exposure temperatures would not be violated, however, operating temperatures were routinely close to those limits. The original 1975 design operating manual by Stone & Webster Engineering Corp. recognized the risk and advised in the low temperature protection section that "caution should be taken." To prevent brittle fracture, several manual protective measures were advised, and a "warm up" process line was installed to route 60°F material from the cracked gas (CG) dryers outlet to the demethanizer tower reboiler return line (Appendix B, Drawing 1). Flow through the line could be managed by a flow controller, but it was rarely used.

On April 29, 2006, approximately 30 years after the plant startup, the warming line ruptured, releasing hydrocarbons and fueling a jet fire in the plant.[8] Investigation revealed the root cause was related to corrosion under insulation and pressure spikes and surges during an upset.[8] The incident resulted in over 14 months of down time and $300 million for repairs.[8] After the rupture, the warming line, one of the main preventative measures against embrittlement in the column, was now absent. Though it was among the alternatives considered, the final decision not to rebuild that line was swayed by concern over relying on a process line that mainly sat idle, and perhaps by recency bias as well, which occurs when people are "overly influenced by dramatic, memorable, or recent events...".[5] When the plant was started again in 2007 without the warming line, alternative methods were needed to avoid brittle fracture.

Permissive and Basic Controls Automation

Alternative and more advanced control schemes that were developed to avoid embrittlement temperatures in lieu of the warming line included a permissive and advanced control logic. Before the plant was re-started in 2007, a Trisen programmable logic controller (PLC) safety instrumented system (SIS) permissive was added. It was designed to address a CG compressor shutdown, one of the most likely causes of a depressure and auto-refrigeration scenario within the demethanizer. Skin temperature elements were added to the tower and tied to a 1oo3 voting low temperature interlock that would not allow the compressor to be restarted (Appendix A, Chart 1). This prohibited system pressure from being increased too quickly while the demethanizer was cold. Aside from this permissive, risk mitigation still relied on manual operator intervention, as it had in the original design.

Table 1. Permissive Summary

Permissive	Action	Effect
Low Temperature	Cannot reset tripped valves or compressor shutdown signals	Prevent demethanizer from being pressurized while it is cold

Now under FHR ownership, a demethanizer tower hazard analysis completed in 2010 confirmed the risk for brittle fracture was still present and there was a need for supplementary safeguards. With the next turnaround scheduled for 2012, there was not enough time to eliminate the hazard with significant equipment changes, so focus was placed on automated systems upgrades. While studies were being completed and plans were being organized for more robust field and control upgrades, immediate action was taken with customizable control logic and points within the distributed control system (Appendix B, Drawing 2).

The first code that was implemented would compare the tray temperature in the bottom section that was closest to the top section to a parametrized low trip point. If the tray temperature low alarm was not equal to the parametrized value, it was reset to be equal. This adjustment ensured the correct value was used, even if the alarm on the point had been inadvertently changed. If an alarm flag became active because a low value was matched or exceeded, a message was sent to the operator station, which served to increase awareness of critical conditions.

The next code automated one of the originally advised manual protective measures. It would take control of the first feed separator bottoms level valve and set it to manual mode and 80% open to add relatively warm fluid to the tower without overwhelming the overhead pressure control. It would send a message to the operator to alert him that corrective action had been taken and what the cause had been. The code would then release control so the operator could adjust the output as needed. It would be activated if there was a loss of reboiler capability in the tower, as indicated by low C3R compressor speed, low transition tray temperature, or low reboiler propylene refrigerant flow. The opened valve would send -30F warm vapor and liquid feed to the tower to counteract the loss of heating from the bottom (Appendix A, Chart 2, 3, 4).

The final code addressed the re-pressure risk by enabling pressure rate of change alarms whenever the transition tray temperature was below its embrittlement point. The purpose of the code was to alert an operator that the pressure in the tower may be rising too fast while the carbon steel section was below its ductile-brittle transition temperature.

Table 2. Code Summary

Code	Cause	Action	Effect
1	Transition temperature low	Set alarm equal to parameter, Send message	Alarms at correct values Increase awareness
2	Transition temperature low, Compressor speed low, Refrigerant flow to reboiler low	Open valve to send warm fluid from the first feed separator, Send message	Increase tower bottom temperature Increase awareness
3	Transition temperature low	Enable rate of change alarms	Awareness of rising pressure while temperature is low

This combination of alarm points, control schemes, and customized codes greatly improved the overall system protection against embrittlement after the loss of the warming line by automating corrective process action and improving operator awareness of critical conditions with messages to the console. However, the effectiveness of these controls in detecting and counteracting upset conditions was largely limited to single sensors, transmitters, and one final control element, the first feed separator bottoms level control valve, and still partially relied on supplemental operator response to prevent a brittle fracture event from occurring.

Extended Independent and Redundant Control

The studies that had started in 2010 and the successes of the coded automation paved the way for more robust automated safeguards to avert cold temperature embrittlement to be put in place during the September 2012 Turnaround in the form of SIS. A 2010 Layer of Protection Analysis (LOPA) recognized that the best solution was to upgrade metallurgy, but then considered two common failure scenarios to determine risk reduction required from automated functions, given severity and likelihood, and then suggested possible layers of protection.[13,11] Further analysis was needed to verify the LOPA recommendations were feasible and therefore credible protection layers. Thus, another study was completed in 2011, with detailed simulations of the proposed safety instrumented functions (SIF) and their resultant impacts.[14] The analyzed functions included warm propylene (C3E) injection, the addition of warm fluid via the first feed separator bottoms valve, and the stoppage of cooling fluids to the condenser and precooler. The cold re-pressure scenario analysis highlighted the need for procedure revisions and adequate operator training, but most effort was centered around assessing the loss of reboiler duty scenario. As noted in a 2001 demethanizer study, "reboiler failure with the feeds and reflux continuing to feed the demethanizer resulted in the coldest excursion temperatures."[7]

For the loss of reboiler duty scenario, the mitigating action of using the warm feed from the first stage separator resulted in a temperature that crossed slightly below the risk threshold. The transition tray would only reach a safe steady state temperature with the assistance of multiple cooling stream trips. The next scenario considered was a worst-case scenario of reboiler failure with all feeds and reflux continuing to flow at normal rates. The results were that the bottom section temperature would fall well below the brittle limit, and could not be corrected solely by the addition of available warm C3E feed. This confirmed that to have broader protection against embrittlement, a combination of the cooling trips, warm feed supply, and warming injection was required.[14]

Based on the above studies, a Triconex triple redundant safety system was purchased to perform the multi-stream trip and ambient C3E injection. The transition tray temperature instrumentation was triplicated and 2oo3 voting was required for any SIF to activate. A bypass around the bottom control valve of the first feed separator was installed to add warm fluid, shutoff valves on the cooling media supply to the demethanizer overhead condenser and precooler were installed to stop cooling and reduce the amount of cold liquids flowing down the column, and a line from the propylene pipeline to the second demethanizer feed line was installed to inject warm C3E (Appendix A, Chart 5). The new propylene line served the same function as the demolished warming line, but was used more frequently because it also served other purposes for the process. Three different temperature measurement values were used for progressively more aggressive corrective actions:

Table 4. Staged SIS Responses

Trip Temperature	Action	Effect
Low (L)	XVS-1 Open	Warm fluid from first feed separator added through the valve bypass into the tower
Low Low (LL)	XVS-2 Close XVS-3 Close	Stop flow of C2R to overhead condenser Stop flow of C3R to precooler
Low Low Low (LLL)	XVS-4 Open	Warm C3E added to the second feed separator liquid feed point into the tower

In addition, the operator was given a button which would trigger all the actions above, and a flow controller that would allow C3E to be proactively metered into the system when needed. (Appendix B, Drawing 3). Though this SIS provided a high level of protection against cold temperature embrittlement, the residual risk would be engineered out at the next opportunity.

Metallurgy Upgrade and Downstream SIS

Elimination of the brittle fracture risk from the demethanizer was finally possible during the 2015 outage after finalization of the project work that had begun after the 2010 process hazard analysis. The project included a tower with fully stainless steel metallurgy, as well as updates to the related condenser, reflux drum, reboiler, and condensate pot. The stainless steel metallurgy removed concern of low material toughness or embrittlement. This allowed for the extraction of most prior safeguards that had been added, as outlined above, but also created potential for new issues to arise.

The brittle fracture concern now had to be addressed downstream at the deethanizer, which receives feed flow from the demethanizer bottoms. The deethanizer tower temperature limits based on metallurgy are higher than they were in the old demethanizer, but could still potentially reach exposure temperatures. To prevent the problem from being passed to the next equipment in line, a new SIS that would stop the demethanizer bottoms flow if triggered by low bottoms temperature was also installed during the 2015 outage.[12] Deethanizer feed temperature transmitters were triplicated and 2oo3 voting was required to activate the safety functions (Appendix A, Chart 6).

Table 5. Downstream SIS

Trip Temperature	Action	Effect
Low Low (LL)	XVS-5 Close FV-3 Close	Stop demethanizer bottoms flow to deethanizer

The operator was given a button that would initiate the actions above, and the ability to proactively flow warm C3E to the demethanizer had been retained for use when needed (Appendix B, Drawing 4). These new SIS trips, along with basic controls system capability and operating procedure revisions, met the safeguard requirements needed to minimize the brittle fracture risk in the process area downstream of the new demethanizer.

Conclusion

Metallurgy replacement is the surest way to eliminate the risk of cold temperature embrittlement and brittle fracture in older equipment, such as bimetallic demethanizers, but well-designed basic process controls and safety instrumented systems along with administrative controls can also provide a defense that is robust enough to mitigate the hazard in most situations. An industry survey completed in 1996 found most brittle fracture failures occurred because of material susceptibility and cold temperature upsets followed by re-pressurizing too quickly, but it also found that upsets did not result in failures where correct materials or engineering or administrative controls were present.[2] The final iteration of controls that had been used at the FHR Port Arthur site is one example of a controls strategy that worked. The failure of the initial warming line from the CG dryers lead to the development of much stronger controls schemes that used customizable advanced control language, highly reliable instrumentation and logic-based trip systems, and a more dependable new warming line. When it is decided that engineering and administrative controls are not enough and full tower replacements are going to be pursued, it is critical to reassess the whole process and be sure to take downstream and cross stream impacts into account because in a system where one susceptible material has been identified, it is likely there are more. Just as was the case at this site, additional controls may be required to prevent simply pushing the brittle fracture hazard elsewhere in the process.

References:

1. Bernard, André. "Cold Brittle Fracture Potential on Cold Box Outlet Piping". Paper presented at Ethylene Producers Conference, AIChE Spring National Meeting, Orlando, Florida, April 23-27, 2006.
2. Buchheim, Gerrit M. and Clay White. "Brittle Fracture Assessment of Ethylene Unit Equipment." Paper presented at Ethylene Producers Conference, AIChE Spring National Meeting, Houston, Texas, April 22-26, 2001.
3. King, Ralph E. "Auto-refrigeration /Brittle Fracture Prevention Overview Training". EPC Learning Module.
4. King, Ralph E. "Auto-Refrigeration / Brittle Fracture Analysis of Existing Olefins Plants – Identification of Potential Excursions". Paper presented at Ethylene Producers Conference, AIChE Spring National Meeting, New Orleans, Louisiana, April 25-29, 2004.
5. Koch, Charles. *Good Profit.* Crown Business, 2015, 186.
6. Melching, Jeff S. Brittle Fracture. Paper presented at Ethylene Producers Conference, AIChE Spring National Meeting, Houston, Texas, April 22-27, 2007.
7. Messinger, John and Matt Findlay. "Cold Safety Risk Assessment of Existing Bimetallic Demethanizers." Paper presented at Ethylene Producers Conference, AIChE Spring National Meeting, Houston, Texas, April 22-26, 2001.
8. Prows, John B. and Robert Simon. "Killing Rattlesnakes Before They Bite You!" Paper presented at Ethylene Producers Conference, AIChE Spring National Meeting, Tampa, Florida, April 29, 2009.
9. Reid, J.A. An Overview of Olefin Unit Accidents and Safety Lessons Learned. Paper presented at Ethylene Producers Conference, AIChE Spring National Meeting, Houston, Texas, 1999.

10. Verdurmen, Werner A. M. Prevention of Brittle Fracture Scenarios in Olefins Plants. Paper presented at Ethylene Producers Conference, AIChE Spring National Meeting, New Orleans, Louisiana, March 30- April 3,2014.
11. Williams, Robert I. "Understanding SIS Industry Standards." Controleng.com. May 14, 2013. Accessed December 27, 2016. http://www.controleng.com/industry-news/single-article/understanding-sis-industry-standards/90db923905bf9f60271f67a82ad3592c.html.
12. Baker Engineering and Risk Consultants, Inc. "Ethylene Feed Flexibility Project (EFFP) and Demethanizer Tower Revamp Project (DTRP)." Study date March 3, 2014. Draft report May 5, 2014.
13. Kenexis. "Ethylene Cracking Furnace SIL Selection and Demethanizer Tower LOPA." Feb. 2010.
14. The Shaw Group, Inc. "LOU Demethanizer Cold Embrittlement Study." May 9, 2011.
15. "Hazard Prevention and Control." Osha.gov. Accessed December 11, 2016. https://www.osha.gov/shpguidelines/hazard-prevention.html.

Acknowledgements:
The author would like to thank Sasha Vragolic for his encouragement to write this paper, his knowledge-sharing on the subject, and his eloquent editing suggestions. Thanks are also owed to Bill Shelden, Mark Bergeron, Tracey Hanna, Keith Moreau, and Jadene Hathaway for their sharing and help in finding pertinent information.

APPENDIX A: PROCESS DATA AND CHARTS

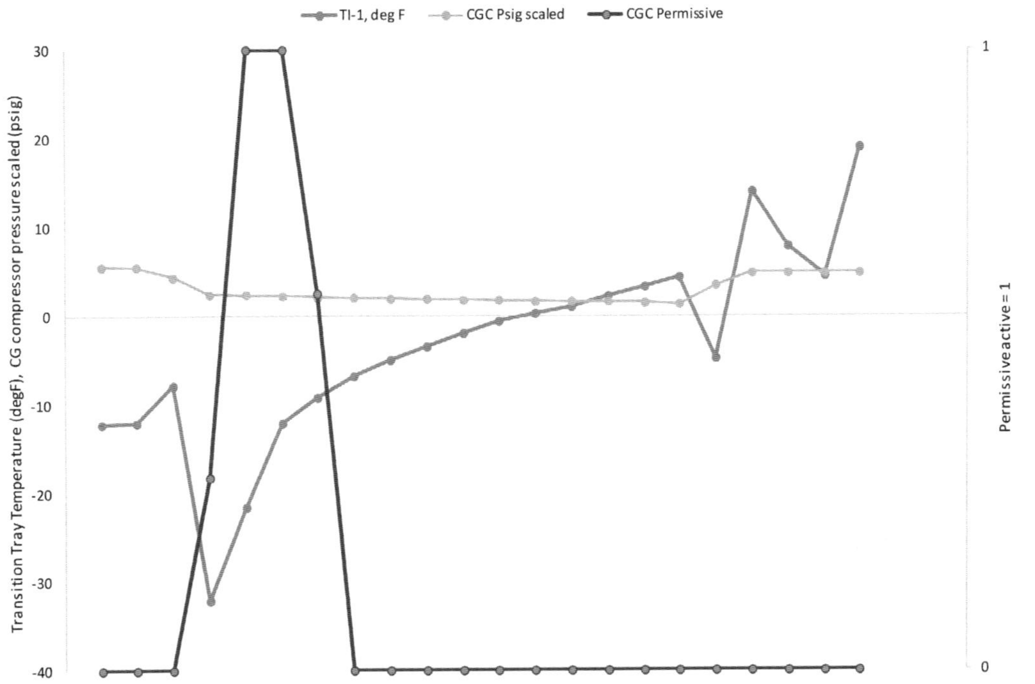

Chart 1. Example of CGC permissive flag activating when the transition tray temperature, TI-1, is low and CG compressor discharge pressure has fallen.

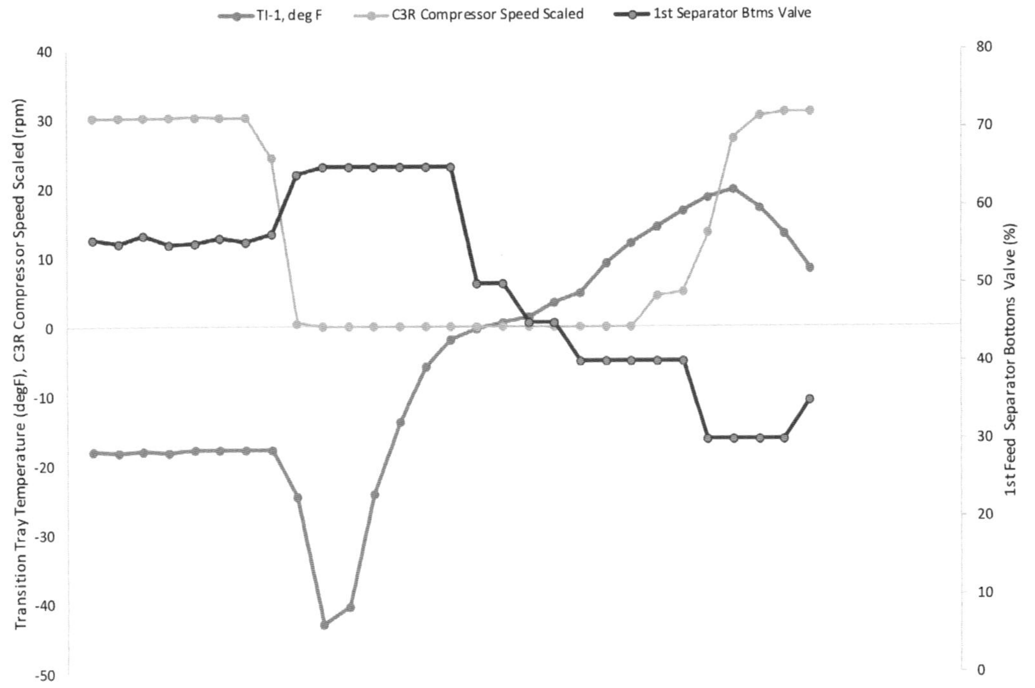

Chart 2. 2010 example of coded response, LV-1 opened to quickly restore temperature after compressor speed falls low.

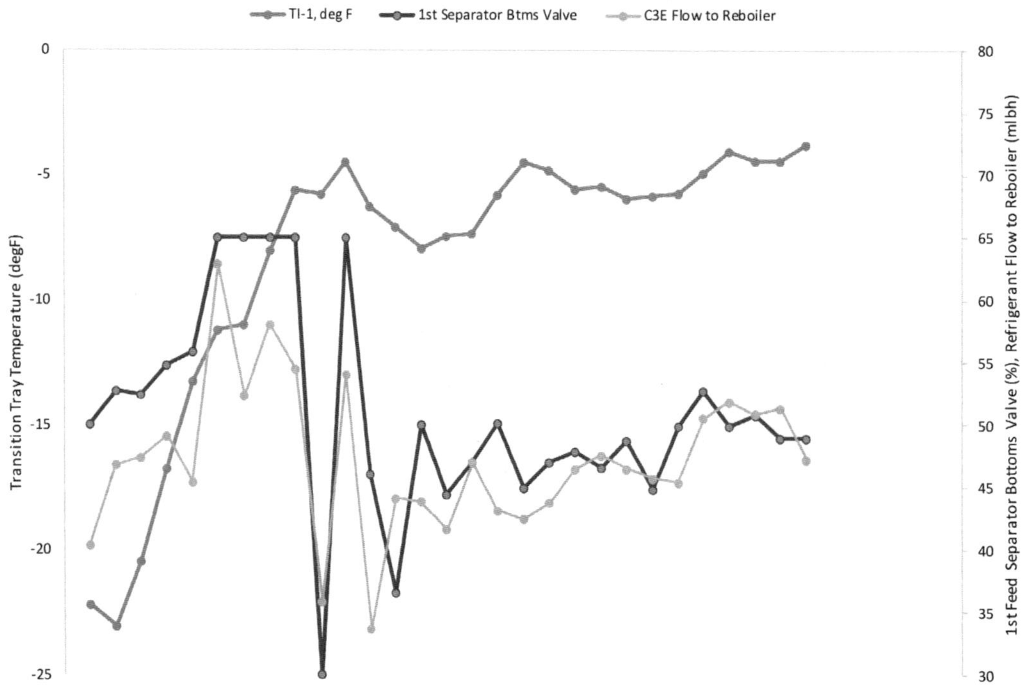

Chart 3. 2010 example of coded response, LV-1 opened to maintain TI-1 above brittle limit when heating medium flow to reboiler, FV-2, is low.

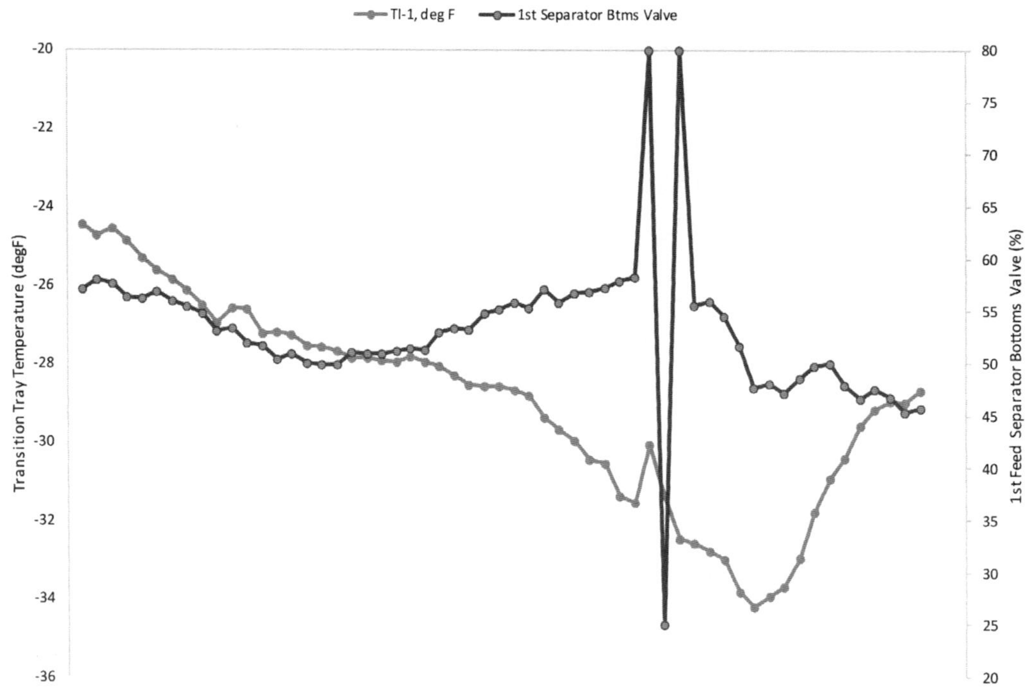

Chart 4. 2010 example of coded response, LV-1 opened to 80% when TI-1 falls low, then placed back in automatic.

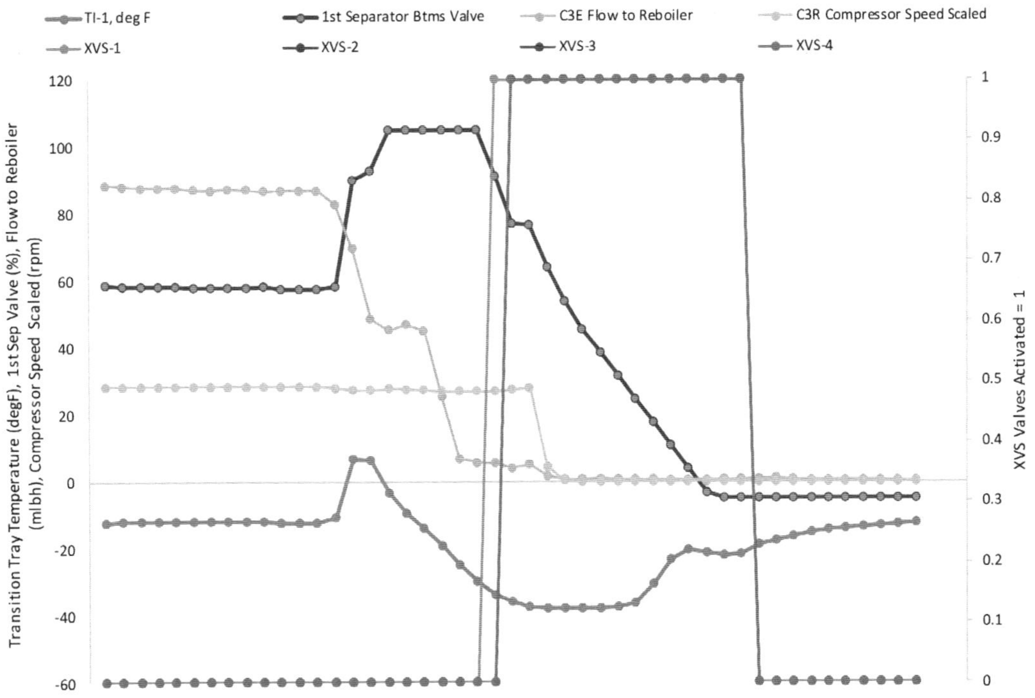

Chart 5. 2013 example of low compressor speed, low FV-2, and low temperature TI-1 causing all SIS responses to activate.

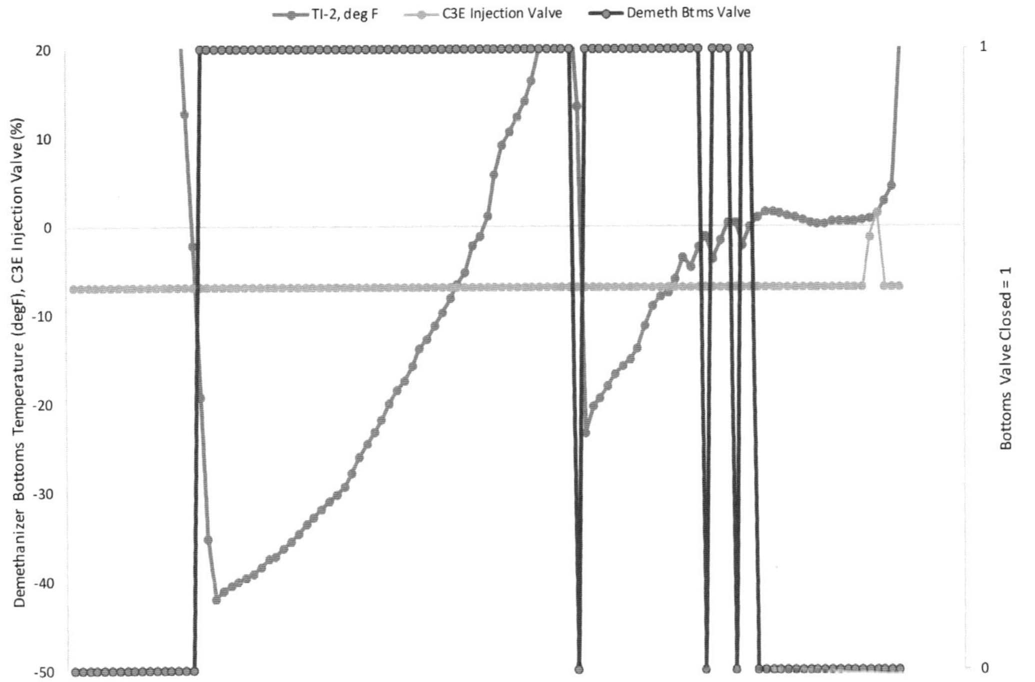

Chart 6. 2016 example of demethanizer bottoms flow valve XVS-5 closing due to low demethanizer bottoms temperature. Some use of C3E injection valve to help warming.

APPENDIX B: DRAWINGS

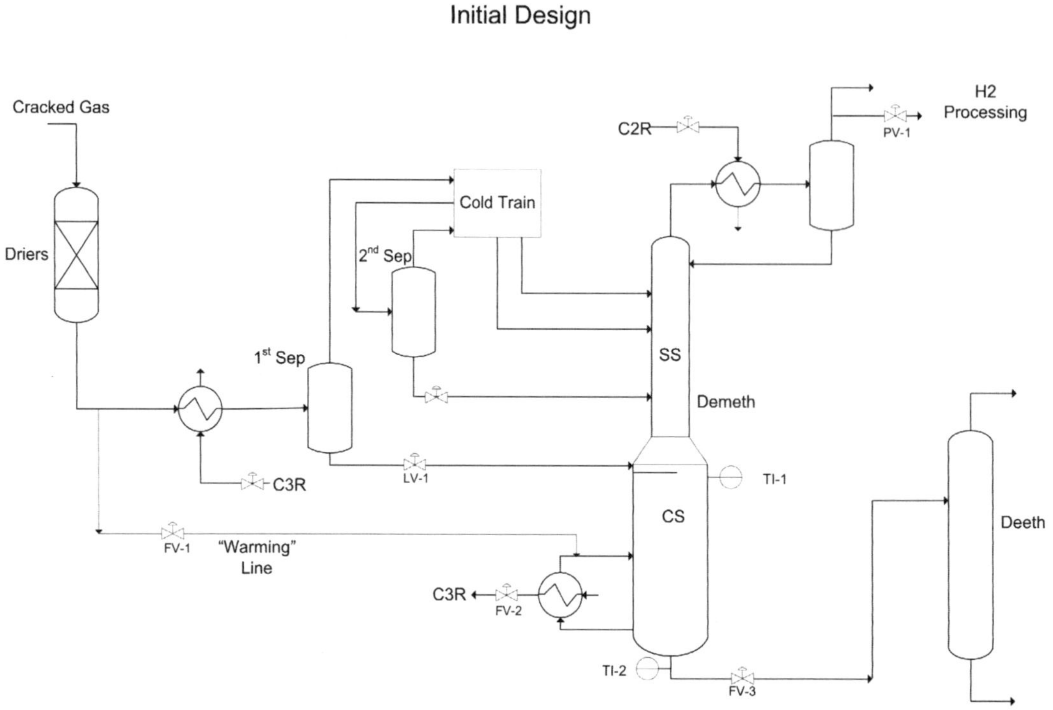

Drawing 1. Initial design with warmup line.

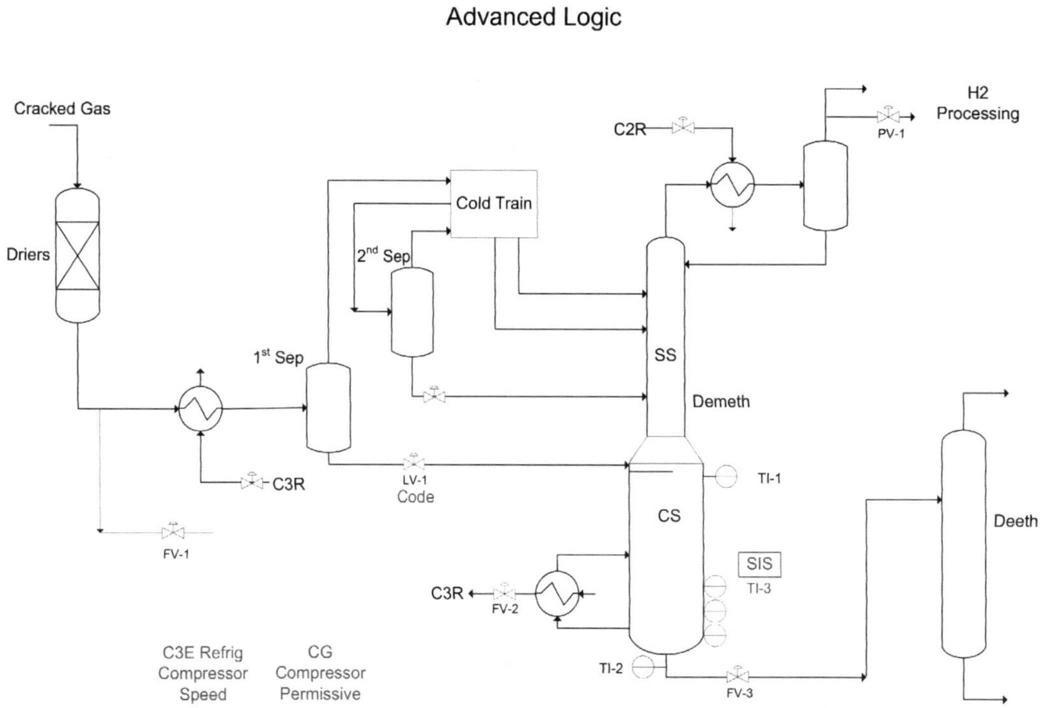

Drawing 2. Addition of low temperature permissive and logic.

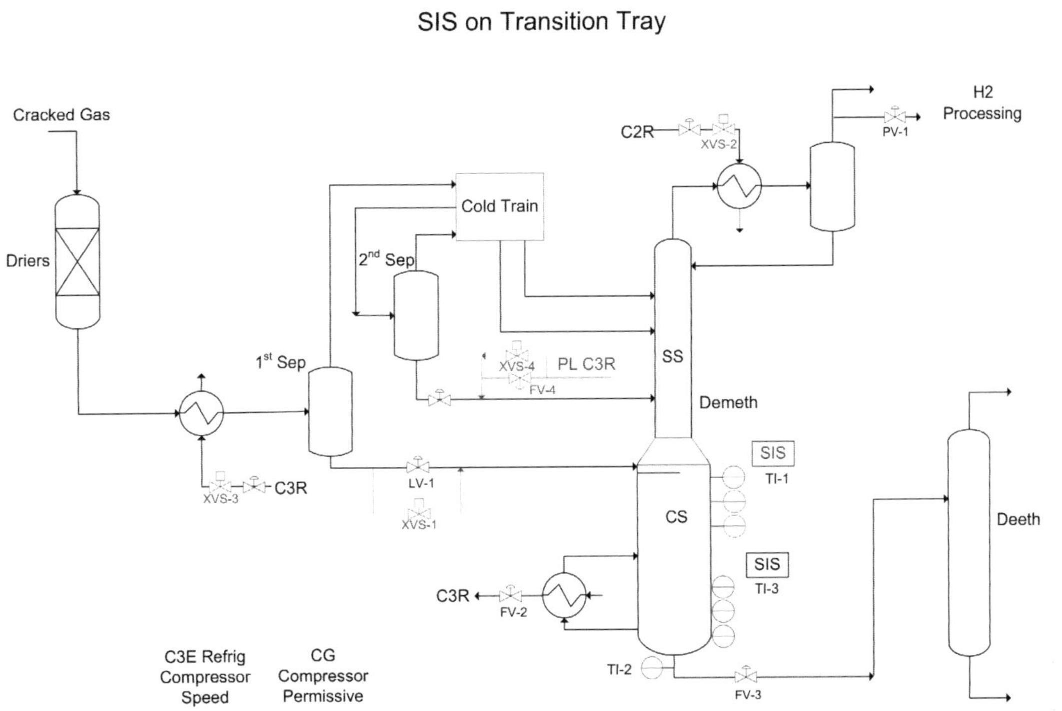

Drawing 3. Addition of low transition tray temperature SIS.

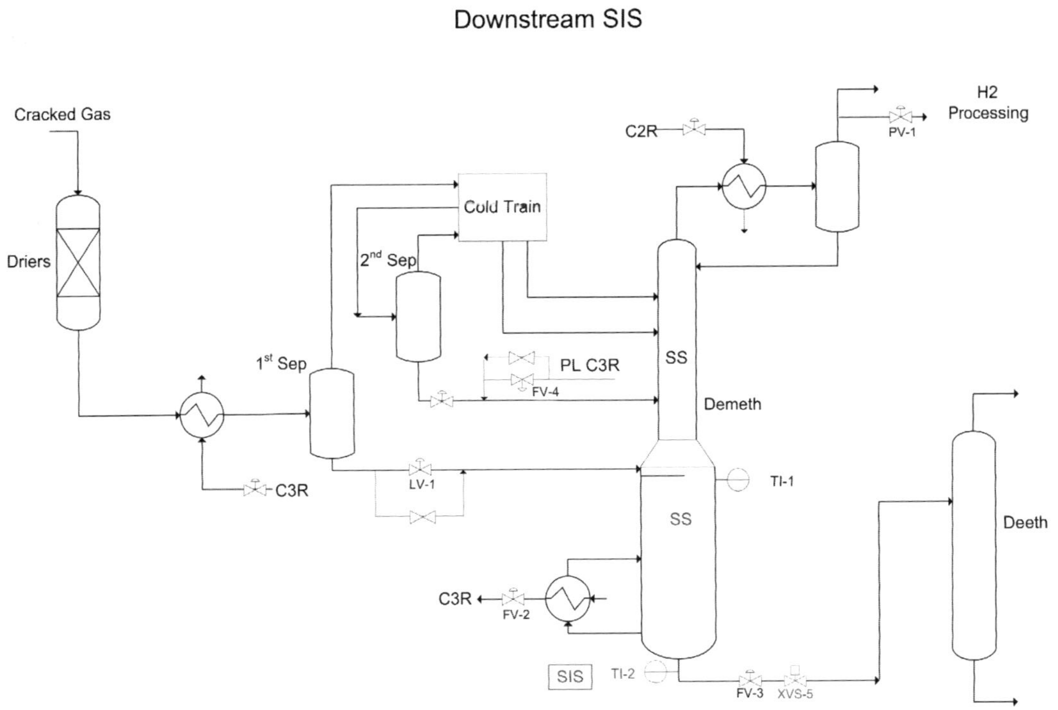

Drawing 4. Tower replacement and downstream SIS.

AIChE Paper Number: 194a

Current and Future Environmental Challenges for Ethylene Producers

Amy M. Marshall, PE

Air Quality Practice Director

AECOM

Prepared for Presentation at the 2017 Spring National Meeting

San Antonio, TX, March 27-30, 2017

CURRENT AND FUTURE ENVIRONMENTAL CHALLENGES FOR ETHYLENE PRODUCERS

Amy M. Marshall, PE
Air Quality Practice Director
AECOM

Abstract: EPA has recently made several changes to air regulatory programs that impact the ethylene industry and is contemplating several more. These actions include ever-tightening ambient air quality standards, changes to permitting and modeling requirements, new greenhouse gas emissions requirements, removal of startup/shutdown/malfunction exemptions, and MACT risk and technology reviews that impose burdensome information collection activities and more stringent regulatory requirements. This presentation will discuss recent and potential future EPA air regulatory actions that will challenge the ethylene industry and provide insight on how to prepare for those challenges.

Current and Future Environmental Challenges for Ethylene Producers

Introduction

The Obama EPA developed, proposed, and finalized many environmental regulations impacting industrial facilities. Most of the new or revised regulations were air quality related and many had cross-cutting impacts. These regulatory actions include ever-tightening ambient air quality standards, changes to permitting and modeling requirements, new greenhouse gas (GHG) emissions requirements, removal of startup/shutdown/malfunction exemptions, electronic reporting requirements, and MACT risk and technology reviews that impose burdensome information collection activities and more stringent regulatory requirements. Some of these actions were as a result of existing regulatory obligations to develop or review rules or standards and some actions were initiated in response to a desire to address court decisions or newly identified environmental challenges. Many of the regulatory actions affect ethylene producers either directly or indirectly. For example, actions that revise NAAQS and permitting requirements affect ethylene producers directly, while actions that affect the power industry may indirectly affect ethylene producers. Other regulatory actions are precedent setting and provide an indication of what requirements could be included in future regulations. For example, EPA is working on a risk and technology review (RTR) of the air toxics standards that apply to ethylene producers, and we can infer what new requirements may be included from a review of recent similar RTR actions.

The priorities of the new EPA administrator are uncertain at the time this paper is being written. However, recent executive orders (EOs) can provide some information on the upcoming regulatory landscape and possible actions by the Trump EPA. 2017 will likely be a period of proposals for regulatory rollback and policy changes that could result in a more favorable environment for industrial permitting and expansion than producers experienced over the last 8 years.

Air Quality Regulatory Background

The Clean Air Act (CAA) was originally enacted in 1970 and was amended in 1977 and 1990. EPA developed several air quality regulatory programs as a result of the CAA. These programs include national ambient air quality standards (NAAQS), new source review (NSR) air quality permitting regulations, new source performance standards (NSPS), and air toxics regulations (e.g., MACT standards). Each state also has their own set of regulatory programs to implement CAA requirements and protect the NAAQS. Recent regulatory activity has occurred in each of these program areas that affects ethylene producers. New and historical policy and guidance documents also play a role in influencing how regulatory agencies make air quality permitting and compliance decisions. The following sections describe recent regulatory challenges and upcoming activities in each area.

NAAQS, NSR, and Air Permitting

Recent EPA activity related to NAAQS and air permitting with impacts to ethylene producers includes the following:

- Lowering the ozone NAAQS to 70 ppb;
- Proposed changes to GHG permitting regulations;
- Revisions to air quality modeling guidelines and draft guidance on ozone and $PM_{2.5}$; modeling requirements;
- Startup, shutdown, malfunction (SSM) SIP call; and
- EPA control cost manual updates.

The ozone NAAQS was lowered in 2015 to 70 ppb from 75 ppb. Due to improved air quality in many regions of the country from 2014 to 2016, this action is not projected to have as large an impact on industry as previously thought; however, it could still result in either a change in status of existing ozone nonattainment areas or more stringent requirements for sources located in those areas, since the classification thresholds also change with a lower standard. EPA and states will also look at possible revisions to transport rules or SIP requirements to prevent deterioration of ambient air quality. Currently, the Cross State Air Pollution Rule (CSAPR) applies only to electric generating units, but as ambient air quality standards are lowered, applicability could eventually extend to large industrial sources of NOx, such as heaters, boilers, and furnaces. The revised ozone NAAQS is currently the subject of litigation. Additional upcoming regulatory activity includes NAAQS implementation and ongoing NAAQS reviews (including the NO_2 NAAQS).

Revisions to the 40 CFR 51, Appendix W air quality modeling guidelines were finalized on December 20, 2016 and published in the Federal Register on January 17, 2017, but the effective date has been deferred to March 21, 2017 by the new administration. For the most part, the changes to the models and establishment of certain modeling techniques as approved approaches were wanted and needed. However, some of the final changes were a result of a Sierra Club suit related to ozone precursors and modeling requirements to address formation of ozone as a result of PSD modifications. EPA did not finalize specific procedures for ozone-related modeling but instead has proposed guidance on how states should develop site-specific procedures for ozone impacts analyses and has also proposed significant impact levels (SILs) for ozone and $PM_{2.5}$. The fact that EPA is attempting to establish new modeling requirements through guidance and also suggesting that the requirements will differ by geographic region creates some uncertainty for facilities that desire to permit and implement projects that trigger PSD review for VOC or ozone precursors. Permits may be delayed because agencies must wait for EPA to provide guidance on what analyses must be performed, and concern the required analyses may be overly stringent or result in revisions to project plans that could lead to an increase in capital and operating costs.

EPA is also in the process of updating its Control Cost Manual, which is used by both permit applicants and regulatory agencies to evaluate the cost of air pollution controls. Sections on SCR and SNCR controls were recently updated, but with a focus on cost estimates based on power industry sources, which do not accurately reflect costs for smaller industrial combustion sources. In 2016, EPA proposed updates to the cost methodology chapter and VOC control chapters. The updates to the control cost chapters contained little to no new information on control equipment cost, and as a result, the Control Cost Manual continues to underestimate the expected control costs for industrial sources. The lack of updated cost information based on actual recent vendor estimates results in a manual that is not beneficial to industry, since its use could be misleading that specific air pollution control devices are cost effective in certain regulatory or permitting settings.

Currently, only projects that trigger PSD for some other pollutant can trigger PSD review for greenhouse gases. In response to the court decision that ruled the PSD tailoring rule illegal, EPA proposed to codify PSD permitting requirements for GHG in 2016. EPA requested comment on whether the significant emission rate (SER) of 75,000 tpy CO2e should be retained and did not propose to require modeling for GHG. The rule has not been finalized and it is unclear what direction the new EPA administration might take with this rule; the fact that the incoming administration does not support the Obama EPA's climate regulations could mean that the SER for GHGs could be finalized at a higher mass emission rate and result in even less projects triggering PSD review for GHGs.

As a result of a Sierra Club lawsuit, EPA has been making changes to federal regulations and has required states to make changes to their SIPs that eliminate the SSM exemption. These changes have or will result in additional requirements that must be followed during startup and shutdown (emission limits, work practices, monitoring, and/or recordkeeping), and essentially eliminate any allowance for exceedances or deviations during malfunction events. Litigation is pending on the SSM SIP call, but EPA has finalized many changes to MACT and NSPS regulations that remove the SSM plan requirement and add requirements for startup and shutdown periods and the regulatory agenda indicates that revisions to the Part 63 general provisions regarding SSM are planned.

The White House has recently issued presidential memos and executive orders meant to affect future regulatory activity and policy related to permitting and modeling. Following the inauguration of President Trump, White House Chief of Staff Reince Priebus issued a memorandum to the heads of EPA and other agencies detailing a series of regulatory prohibitions and requirements for recently-issued and pending rules. Generally, the memo bars agency officials from sending any new or revised regulations to the Federal Register for publication until they have been approved by a Trump administration official. Regulations that had already been sent to the Federal Register but not yet published were required to be withdrawn for review and approval. For rules that have already been published but were not yet effective, Priebus required agencies to postpone their effective dates for 60 days "for the purpose of reviewing questions of fact, law, and policy they raise." He also stated that agencies should consider "further notice-and-comment rulemaking." A January 24, 2017 federal register notice implementing this memo delayed the effective date of 30 regulations, including the Appendix W modeling guidelines revisions, until March 21, 2017.

A presidential memorandum with subject line Streamlining Permitting and Reducing Regulatory Burdens for Domestic Manufacturing was issued on January 24, 2017. This memorandum directs agencies to support the expansion of manufacturing in the United States through expedited reviews of and approvals for proposals to construct or expand manufacturing facilities and through reductions in regulatory burdens affecting domestic manufacturing. The order directs the Secretary of Commerce to conduct outreach to stakeholders concerning the impact of Federal regulations on domestic manufacturing and solicit comments from the public for a period not to exceed 60 days concerning Federal actions to streamline permitting and reduce regulatory burdens for domestic manufacturers. Within 60 days after completion of that process, Commerce is required to submit a plan to streamline Federal permitting processes for manufacturing and reduce regulatory burdens affecting manufacturers. The report will identify priority actions and recommended deadlines. The report also may include recommendations for any necessary changes to existing regulations or statutes, as well as actions to change policies, practices, or procedures that can be taken immediately under existing authority. Several trade associations are preparing recommendations to Commerce on regulatory actions and policy changes that will accomplish the goal of the EO. If these recommendations gain traction, it could mean a period of improved regulatory certainty and flexibility for industry.

Air Toxics Regulations

In the 1990s, EPA began establishing source category-specific national emission standards for hazardous air pollutants (NESHAP) under 40 CFR 63. These standards are based on application of maximum achievable control technology, or MACT. The ethylene production source category includes chemical manufacturing process units in which ethylene and/or propylene are produced by separation from petroleum refining process streams or by subjecting hydrocarbons to high temperatures in the presence of steam. Ethylene production operations are covered under a rule known as the "Generic MACT" at 40 CFR 63, Subpart YY. Ethylene production waste and heat exchange systems are covered under Subpart XX. Existing affected sources were required to be in compliance with Subpart YY by July 12, 2005.

The Clean Air Act (CAA), Section 112, requires EPA to review each MACT standard periodically. The first periodic review is required to be a risk and technology review (RTR). The purpose of the risk review is to evaluate whether there is remaining risk to the public following implementation of the rule and whether the standards should be revised to reduce the residual risk. The technology review is meant to determine if there have been changes in practices, processes, and control technologies that warrant a revision to the MACT standards. Although the CAA specifies that EPA is to conduct the RTR within 8 years of promulgation, EPA is behind schedule on this work and is just now conducting many RTRs for rules promulgated over 10 years ago, including the ethylene MACT standard.

As they conduct their RTR work, EPA is also covering other issues that they believe necessary to address as a result of several court decisions over the last several years. These court decisions have changed the way EPA is conducting their MACT work and increased the coverage and stringency of MACT standards.

Ethylene MACT RTR Information Collection Request (ICR)

EPA started gathering information to perform their ethylene MACT RTR over 2 years ago. They issued a component I questionnaire in 2014, and in 2016 sent component II of the ICR to 9 companies. The first ICR (Component I) focused on information related toethylene plant configurations, process information, and emissions. The information collected for component I allows EPA to construct a model input file to conduct the risk analysis portion of its RTR. However, because HAP emissions from ethylene furnaces and decoking operations have not been regulated or measured previously, EPA issued component II of the ICR, which requires extensive air emissions testing for furnaces and decoking operations at 21 facilities, to augment its information and determine how MACT standards for these operations can be set. Although the first component of the ethylene ICR required significant time and personnel resources to complete, the second component was much more burdensome in that it mandated on the order of a half million dollars in testing costs for each facility. The ICR required 7 test runs of sampling for multiple criteria and hazardous air pollutants on ethylene cracking furnace exhaust and 3 test runs (immediately after air is introduced, mid-way through the cycle, and at the end of the cycle) during each of 3 decoking events. For facilities that route decoking exhaust through a decoking pot stack rather than into another ethylene furnace, the ICR required simultaneous testing of the furnace exhaust and the decoking pot stack exhaust during the decoke process. Various process parameters were recorded and reported with the test results via the EPA's electronic reporting tool (ERT). Heat exchanger sampling was also required. Most of the sampling has been completed and submitted by the end of 2016.

Challenges to Successfully Completing the Ethylene MACT ICR Testing Requirements

Along with the significant testing costs come challenges to successfully completing the required testing program. The first challenge to successfully sampling during decoking events relates to scheduling the test. The interval between decoking activities varies significantly from site to site and is based on furnace characteristics (e.g., shape, time in service, feedstock). In addition, furnace decoking is not scheduled based on the calendar days since the last decoking event, but is targeted 7 to 10 days in advance based on various monitored furnace operating parameters. Extending a furnace run beyond the targeted decoke date is detrimental to the furnace radiant tubes and may lead to damage to the radiant tubes. Therefore, this short advance notice and the need to avoid going beyond the decoke target date presented logistical challenges when contracting with stack sampling firms. The time between furnace decoking events can range from 17 days to 4 months. Therefore, given that the window for beginning decoking can span days and it is not advantageous to the facility to begin decoking early; scheduling testing crews to complete sampling during 3 separate decoking cycles can require standby time and its associated costs.

In addition, decoking cycle length can vary tremendously. Based on facility histories, these cycles can be anywhere from 12 to 90 hours. The ICR requirement to sample during the beginning, middle, and end of the decoking cycle determines the number of field personnel that are needed to capture the data at the appropriate time. If a cycle is less than 30 hours, it requires double the personnel and 24-hour test coverage to complete the ICR requirements (as many as 20 people for a test program).

Because the ICR sampling is being conducted at facilities predominantly in the same geographic area, can require multiple mobilizations, requires simultaneous sampling of multiple pollutants using multiple methods, and requires testing firms with specialized expertise (e.g., FTIR), completing the testing and submitting the required data by EPA's deadline of November 30, 2016 was not achieved by all facilities. Although industry indicated in comments on the January 2016 draft ICR that logistics would prevent collection and submission of all data by this deadline, the final ICR issued in April 2016 retained the deadline of November 30, 2016.

Based on experience during the Refinery RTR ICR testing on delayed coker vents, industry anticipated that there would be challenges associated with obtaining valid sampling data from ethylene furnace decoking pot stacks. The American Chemistry Council (ACC) and American Fuel and Petrochemical Manufacturers (AFPM) had AECOM perform a pretest on a decoking pot stack to identify optimum equipment configurations, communicate lessons learned, and provide guidance to facilities that received the ICR so that they would have the best chance of obtaining valid data. The difficulties in accurately sampling decoking pot stacks arise due to the extremely high moisture content in the stack. The stack gas from a decoking pot typically contains in excess of 95% moisture due to the amount of steam necessary to complete the decoking process. The high moisture content requires the use of a dilution probe to deliver the flue gas to the CEMS and FTIR. A dilution ratio of 20:1 was used for the pretest location that had 95-98% stack moisture. Depending on the source and the steam to air ratio in the decoking pot stack gas, the 20:1 dilution may be reduced to provide better detection limits for the instrumental test methods. Dilution ratios should be evaluated based on the measured or monitored moisture levels throughout the decoke cycle. Concentrations of the constituents being measured using FTIR may require low level standards since detection limits will be elevated due to use of the dilution probe. High detection limits will make it difficult to interpret the test results and determine if work practices or numeric emission limits are warranted for decoking.

Moisture in the sampled stack gas must be condensed and collected during sampling – for both the CARB Method 429 and the EPA Methods 5/29 sampling trains – prior to the typical cooling and collection components of the respective sampling trains. Because of the high moisture content of the decoking pot stack gases, large volumes of condensate will be collected. The volume of condensate (water) collected is higher than conventional source sampling systems are designed to handle, and because of the moisture content of the sampled stream more cooling capacity is required. Additional cooling capacity and additional capacity for the collection of condensate must be included in the sampling train. In addition, with the high moisture content and associated temperature of the decoking pot stack gas, hundreds of pounds of ice may be used for each 1-hour test run. Accommodations must be made to have an adequate amount of ice available during the test program, and safe means to get the ice to the sampling platform level.

A requirement of EPA Methods 5 and 29 and CARB Method 429 is that the sampling must be performed within ±10% of 100% isokinetic conditions. As moisture concentrations in the sample gas increase, errors in the assumed moisture fraction have a greater impact on the sampler's ability to maintain an isokinetic sampling rate within ±10%. For example, if a gas stream is assumed to be 99% moisture during sampling and the actual moisture content determined at the conclusion of sampling indicates that the moisture fraction in the gas stream was actually 98%, then the sampling rate will be 50% lower than 100% isokinetic sampling (e.g., 50% isokinetic) which does not meet the requirements of the EPA and CARB methods. Due to the high stack gas moisture, very little dry gas is extracted during the isokinetic testing of the decoking pot stack. EPA has addressed this – to a degree – in the Section 114 requests by requiring a "nominal actual cubic meters per run for decoking operations" and a "dry cubic meters per run for ethylene cracking furnaces." The requirement for decoking operations requires a minimum volume of the wet stack gas be collected. However, EPA's Section 114 Request does not revise the requirement for the isokinetic sampling rate that would still be 100 ± 10%.

Finally, the capabilities of the test methods to detect HAP emissions are challenged when they are applied to ethylene cracking furnace exhausts. All cracking furnaces used in ethylene production are fired with clean gas (e.g., natural gas, methane, hydrogen, refinery fuel gas, olefin unit process gas, or mixtures of these gases). Emissions from ethylene cracking furnaces are from combustion of clean gas at high temperatures to heat the tubes in the furnace, not from the cracking of gas or liquid feedstocks. The cracking of the feeds takes place inside the radiant tubes. The cracked gas exits the radiant tubes and does not come in direct contact with the combustion gas.

EPA gathered data on emissions from combustion of clean gas in industrial furnaces, heaters, and boilers as part of the development of the Industrial Boiler and Process Heater (BPH) MACT (Subpart DDDDD) and determined that subjecting units firing clean gas to a work practice under section 112(h) of the CAA rather than a numeric emissions standard under sections 112(d)(2) and (3) is appropriate. The following is EPA's rationale, as presented in the preamble to the final rule.

> EPA has determined that it is not feasible to prescribe numerical emissions standards for Gas 1[1] units because the application of measurement methodology is not practicable due to technological and economic limitations. Therefore, EPA is finalizing the work practice standards for Gas 1 units. The commenters correctly point out that the measured emissions from these units are routinely below the detection limits of EPA test methods, and, as such, EPA considers it impracticable to reliably measure emissions from these units. Even CO, which commenters correctly point out was tested at many natural gas and refinery gas-fired units, was below the level EPA considers to be a reliable measurement for more than 80 percent of the test runs that were conducted on Gas 1 units. The case for other pollutants is even more compelling as the

[1] Units in the gas 1 subcategory include any boiler or process heater that burns only natural gas, refinery gas, and/or other gas 1 fuels. Other gas 1 fuel means a gaseous fuel that is not natural gas or refinery gas and does not exceed a maximum concentration of 40 micrograms/cubic meters of mercury. 40 CFR 63.7575

majority of measurements are so low as to cast doubt on the true levels of emissions that were measured during the tests. Of the 48 test runs for HCl, 98 percent were below three times the maximum reported measurement detection level; similarly, 100 percent of the Hg runs, and 45 percent of the PM data were below three times the maximum reported measurement detection level. It is unusual to see numbers near the detection limit for PM since the "detection" involves a comparatively simple (compared to other test methods) weighing procedure, and the overall result indicates that the emissions are very close to zero. All of the dioxin tests had multiple non-detect isomers. Overall, the available test methods are greatly challenged, to the point of providing results that are questionable for all of the pollutants, when testing natural gas units. Because of these technological limitations that render it impracticable to measure emissions from Gas 1 units, EPA is also unable to establish the actual performance of the best performers as well as sources outside of the top performing 12 percent. The inability to accurately measure emissions from Gas 1 units and the related economic impracticability associated with measuring levels that are so low that even carefully conducted tests do not accurately measure emissions warrant setting a work practice standard under CAA section 112(h). EPA is establishing a requirement to implement a tune-up program as described in Section III.D of this preamble. As noted by many commenters, the tune-up program is an effective HAP emissions limitation technology. The requirement of an annual tune-up will allow these units to continue to combust the cleanest fuels available for boilers while minimizing emissions to the same degree that is consistent with the operating practices of the best performing units in the subcategory. [76 Fed. Reg. 15638, March 21, 2011]

Industry believes that BPH clean gas data are representative of emissions from ethylene furnaces during ethylene production operations. Ethylene furnaces operate at temperatures above 2000°F and at this temperature, no measurable organic HAP emissions are expected. EPA acknowledged as much in the original preamble to the proposed Ethylene MACT: "Based on information provided in survey responses, the furnaces are fired with natural gas, refinery gas, off-gas from the production process, or a combination of the three. Ethylene cracking furnaces are expected to have relatively low HAP emissions. The fuels burned in cracking furnaces contain relatively little HAP, and most organic HAP are destroyed in the combustion process. In fact, process heaters are used as control devices for process vents containing HAP." [65 Fed. Reg. 76431, December 6, 2000] However, EPA has not acknowledged the available BPH clean gas emissions data in the ethylene RTR rulemaking and the ICR required collection of 7 test runs of HAP and criteria pollutant emissions data.

Possible Outcomes of an Ethylene RTR Proposal

Based on the information EPA has requested to date and a review of recent RTR rulemakings, we can anticipate some possible outcomes of an EPA RTR proposal. For example, although the residual risk from the refinery sector was determined by EPA to be acceptable, EPA deemed that additional requirements were necessary to provide an ample margin of safety. In the refinery RTR rulemaking, EPA finalized new requirements for flare combustion efficiency, fenceline monitoring, delayed cokers, storage vessels, startup/shutdown/malfunction periods, and testing, monitoring, and reporting. Although EPA has not been given a deadline for the ethylene RTR proposal as of the time this paper was written, we can hypothesize that they will receive a deadline that requires them to propose a rule by the end of 2017 and finalize a rule by the end of 2018, based on other recent court-ordered deadlines. The following are issues that will likely require the ethylene industry's attention as the proposed rule is reviewed and comments are developed and submitted to EPA.

Additional standards. EPA could regulate additional HAPs and develop standards for additional sources. As previously mentioned, ethylene cracking furnaces and decoking operations are not currently subject to requirements under the Ethylene MACT. However, due to recent court decisions, EPA has determined that it must develop either an emission standard or work practice to cover all HAP emission sources and any surrogates must be adequate to cover all HAPs. The Component II ICR information will be used by EPA to determine whether they should develop either emission limits or work practice standards for furnace and decoking emissions. Under the recent BPH MACT, process heaters fired with clean gas are subject only to periodic tune-up requirements. Under the recent Refinery RTR, work practice standards (instead of numeric emission limits) were added for delayed cokers. However, if a significant amount of the HAP data for ethylene furnaces and decoking operations are above detection limits, EPA will likely develop numeric emission limits, based on a 99 percent upper prediction limit (UPL) statistical calculation. It is possible that subcategories based on size, type, or class of operation could be developed that result in different requirements for different types or sizes of units. Industry will need to consider whether subcategorization can be justified and also the proper form of the standard.

EPA's risk analysis and technology review will also guide their determination of whether additional standards need to be proposed. If their risk modeling analysis indicates that the risk is unacceptably high, then they could propose measures to reduce HAP emissions from facilities further, either by making existing standards more stringent or developing new requirements. Under the refinery RTR, although risk was acceptable, EPA set out additional requirements for storage vessels and flares in order to provide an ample margin of safety. Similar requirements could be established for ethylene production facilities, for example, development and implementation of enhanced flare operational requirements and a flare management plan to minimize emissions from flaring.

Under some recent RTR rules, EPA has developed additional standards for HAPs not due to unacceptable risk but to fill gaps in coverage that they perceived to exist. For example, phosphate rock calciners now have a mercury emission limit as a result of their RTR. Based on their technology review, EPA could determine that standards need to be adjusted based on how technology and operations have changed since the original MACT standards were set. Based on experience with other RTR processes, it is advantageous for industry to open lines of communication with EPA about what types of changes they are considering as a part of this component of their review because sometimes an uninformed analysis can result in an outcome in a proposed rule that requires significant effort to refute (e.g., an assumption that industry can comply with a more stringent standard can be faulty because all facilities operate with a certain margin of compliance and do not typically operate right at the level of the standard). EPA is allowed to consider cost in its review of existing standards, so industry could also provide data that would show that requiring a certain technology across the industry would be too costly.

Startup, shutdown, and malfunction (SSM). Recent court decisions have resulted in EPA determining that they must have some standard that applies at all times, including startup, shutdown, and malfunction. All recent MACT and NSPS rulemaking actions have removed SSM exemptions, removed the requirement for SSM plans, added requirements that apply during startup and shutdown, and have not set separate standards for malfunction periods. Some standards have work practice requirements for periods of startup and shutdown, which could be added to the ethylene MACT. But EPA could also establish standards that apply at all times and perhaps lengthen the averaging period to accommodate any emissions during startup and shutdown that might be higher than those during normal operations.

Also related to this issue is treatment of pressure relief devices (PRD). EPA is addressing PRD releases under their various RTR rulemakings and in the refinery RTR set out work practices for PRDs with emissions potential above a certain threshold that are meant to limit releases from PRDs that vent to atmosphere and determine root causes of any releases. This treatment is different than some other fairly recent RTR rules for the chemical industry (e.g., polymers and resins) where EPA added PRD monitoring requirements and indicated that PRD releases to the atmosphere are a violation of the emission standard. Industry will want to provide justification to EPA that their PRD releases to atmosphere can be treated similarly to the way they were treated in the refinery RTR rulemaking, and not simply treated as violations. Under the refinery RTR, PRD requirements are based on two California rules that were determined to be the requirements that applied to best performers. EPA's analysis showed that work practice standards for certain releases would incentivize facilities to make changes in operations and practices to prevent PRD releases and emergency flaring, rather than to build more flares to control the releases.

Additional monitoring. Any new standards EPA sets will also require compliance monitoring. For example, if a numeric emission limit is set for furnaces or decoking operations, EPA will either require parametric or emissions monitoring over a specified averaging period to demonstrate ongoing compliance with the limit. Based on experience with the BPH MACT, one of the more challenging implementation steps was setting up electronic systems for compliance monitoring, performing data validation, calculating 30-day rolling averages, and differentiating between startup and shutdown and normal operation. However, the required compliance monitoring for ethylene production facilities could be expanded similar to the new fenceline monitoring requirements that were established in the 2015 Refinery RTR rule. Due to concern over the magnitude of fugitive emissions, EPA established requirements for refineries to install fenceline benzene monitors and report the monitoring results quarterly. This type of monitoring could be something EPA begins to require for all petrochemical facilities as they complete their RTR rules.

Additional testing and reporting. EPA has been systematically adding requirements to submit test data and other compliance data electronically into their Compliance and Emissions Data

Reporting Interface (CEDRI). The reported data is easier to review and process by agency compliance personnel and easier to make available to the public when it is collected in this manner. This process makes desktop enforcement easier for the agency and provides the public instant access to a facility's emissions and compliance data. For facilities, it represents an additional layer of reporting, since paper reports are often still prepared, as well as electronic reporting files. With respect to emissions testing for existing requirements, EPA has been adding requirements for repeat testing to many of the MACT rules that initially only required a one time performance test. These repeat tests will have to be reported via CEDRI and the data will be ported to EPA's Webfire tool after they are reviewed. The Webfire data can be used to view facility-specific data or to develop industry or process-specific average emission factors since AP-42 is no longer being updated.

Other Recent Relevant Regulatory Activity

Other recent regulatory activity with relevance to ethylene producers includes the Clean Power Plan, RMP rule revisions, the Boiler MACT remand, recent changes to ODS rule requirements, and limitations on emissions of methane from the oil and gas industry. If the Clean Power Plan stands, it could result in increased electricity costs to ethylene producers. However, the new EPA administrator has indicated that he plans to withdraw the regulation. He may also stop work on other actions related to regulation of CO_2 and methane. Absent an action that overturns the GHG endangerment finding, however, EPA must still address emissions of GHG under the Clean Air Act. EPA also recently finalized revisions to the RMP rule that include numerous new requirements for facilities. The effective date of the revisions has been delayed until March 21, 2017, however, and industry is fighting to have the revisions withdrawn as part of the Congressional Review Act. The Boiler MACT remand and remaining litigation could have implications for how MACT floors are developed, whether work practices are appropriate for startup and shutdown, and whether surrogates are appropriate. ODS rule changes effective in January 2017 provide additional opportunities for non-compliance with the regulations based on failure to keep required records.

Conclusion

The ethylene MACT RTR activities are part of a succession of challenges for the industry related to environmental regulatory activity. It is yet to be seen how the priorities and policies of the new administration will affect ethylene producers or EPA's RTR process. However, recent executive orders are promising a regulatory climate that will be more favorable to manufacturing than the Obama EPA. One January 2017 executive order called for a net zero cost of regulation and stated that for every new regulation an agency wants to propose or finalize, it must identify 2 regulations for repeal. However, increased litigation by environmental groups is also to be expected if they are unhappy with how regulations are being developed and enforced. In addition, if EPA's priorities shift, it could fall to states to address regional environmental issues, which could result in a patchwork of environmental requirements for different facilities across one company. As EPA completes the proposed and final ethylene MACT RTR rulemakings, there will be a significant amount of effort expended by industry to provide useful information to EPA that they can use to craft a rule that meets statutory requirements and is workable for industry. Communication between all parties and careful analysis of the information collected are essential to a favorable outcome for both the testing effort and the associated rulemaking. Participation in trade association activities and outreach efforts is also key to effecting regulatory and policy changes that favor manufacturers.

AIChE PAPER NUMBER: 194b

ENVIRONMENTAL CHALLENGES IN MEETING AMBIENT AIR QUALITY STANDARDS

Arijit Pakrasi
Ph.D., P.E., BCEE
CB&I

Prepared for Presentation at the 2017 Spring National Meeting
San Antonio, Texas
March 29, 2017

Environmental Challenges in Meeting Ambient Air Quality Standards

Arijit Pakrasi
Ph.D., P.E., BCEE
CB&I

Abstract:

In the last five years, the United States Environmental Protection Agency (USEPA) has revised three national ambient air quality standards (NAAQS) which became more stringent. USEPA is currently reviewing several other NAAQS, and the outcomes are expected in the next couple of years.

Meeting NAAQS is a key requirement for permitting any major facility, including ethylene plants, in an attainment area. The permitting process requires estimating the air quality impact of the project's emissions using specialized air dispersion modeling techniques. The modeling requires numerous input data, site-specific assumptions, and model defaults to ensure compliance with USEPA-approved modeling methodologies, which are continuously evolving based on the latest scientific data. Special modeling demonstrations are required for Class I areas.

Ethylene crackers are typically subjected to nitrogen oxides (NOx) and particulate ($PM_{10}/PM_{2.5}$) NAAQS. Compliance is required with both short-term and annual NAAQS for these pollutants. Higher hydrogen content in fuel gas more commonly used in the current and future ethane crackers typically increases NOx formation and poses challenges in meeting the NOx NAAQS. Particulate emissions during decoking pose challenges in meeting the short-term and annual $PM_{10}/PM_{2.5}$ NAAQS.

This presentation will provide an overview of the various NAAQS applicable to ethylene plants, discuss the air dispersion modeling techniques available, and bring awareness to the challenges of meeting the NAAQS. The challenges will be explained through a few case studies. The case studies will be based on published permitted data on emissions from typically sized ethylene plants and publicly available land use and meteorological data. The presentation will conclude by providing recommendations for meeting the challenges of demonstrating compliance with the NAAQS.

Introduction

The Clean Air Act (CAA) as amended in 1990 requires the United States Environmental Protection Agency (USEPA) to protect public health and welfare by setting and managing national ambient air quality standards (NAAQS) for pollutants considered harmful to public health and the environment. CAA identifies two types of NAAQS: (i) primary standards, which provide public health protection, including protecting the health of "sensitive" populations such as asthmatics, children, and the elderly; and (ii) secondary standards, which protect public welfare, including decreased visibility and damage to animals, crops, vegetation, and buildings. The NAAQS represent the maximum level of air quality that is expected to provide the protection of public health and protection called for in the CAA.

In the Code of Federal Regulations, 40 CFR 50, USEPA has set NAAQS for six principal pollutants, which are called "criteria" air pollutants. USEPA is mandated to review these ambient air quality standards every five (5) years and make revisions if justified for the protection of public health and welfare. The current NAAQS relevant to Ethylene plants are listed in **Table 1**.

Table 1. Current NAAQS Relevant to Ethylene Plants

Pollutant	Averaging Period	SIL	NAAQS	Form of NAAQS
		μg/m³	μg/m³ (ppb)	
NO_2	1-hour (primary)	10	188 (100)	H8H
	Annual (primary and Secondary)	1	100 (53)	Annual arithmetic mean
SO_2	1-hour (primary)	10	196 (75)	H4H
	3-hour (secondary)	25	1309 (500)	NTBE more than once per year
$PM_{2.5}$	24-hour (primary)	1.2	35	H1H
	Annual (primary)	0.3	12	H1H
PM_{10}	24-hour (primary and secondary)	5	150	NTBE more than once per year on average over 5 years
CO	1-hour (primary)	2000	40,000 (35,000)	NTBE more than once per year
	8-hour (primary)	500	10,000 (9,000)	NTBE more than once per year

In general, when the ambient air concentration in a specific geographical area is below a specific NAAQS, the area is designated as having "attainment" status for that NAAQS. When the area is above the NAAQS, the area is designated as having "nonattainment" status. There are many geographical areas in which adequate data are not available to make the designation; these areas are designated "unclassified" pending availability of data.

In order to manage the air quality below the NAAQS nationwide, USEPA issues federal regulations limiting emissions of these criteria pollutants from emission sources in both existing and new facilities. States and local air pollution control agencies may also issue separate and additional regulations for specific sources and specific geographical areas.

New Source Review Regulation

For new "major sources" of emissions of criteria pollutants and "major modifications" undertaken in existing major sources, USEPA has promulgated New Source Review (NSR) regulations (40 CFR 51 and 40 CFR 52) to manage their impacts on air quality. The terms "major source" and "major modifications" are based on annual emissions of criteria pollutants, but the thresholds vary depending on: (i) the criteria pollutant in question, (ii) the industrial category of the facility, and (iii) the existing designation status of the area where the "major source" or "major modification" will be located. For example, a new chemical process plant in an attainment area for nitrogen oxide (NOx) will have an NSR threshold of 100 tons per year (tpy) to be a "major source" whereas a wood-board manufacturing plant in the same area will have the threshold of 250 tpy. The thresholds for "major modifications" for existing sources, however, are fixed for each pollutant and independent of the industrial category. Ethylene plants are in the industrial category of chemical process plants, and the thresholds for these plants are shown in **Table 2**.

Table 2. Major Source and Major Modification Thresholds for Ethylene Plants

Regulated Pollutant	Major Source Threshold (tpy)	Major Modification Threshold (tpy)
NOx	100	40
SO_2	100	40
PM_{10}	100	15
$PM_{2.5}$	100	10
CO	100	100
Ozone (as VOC or NOx)	100	40

For new major sources and major modifications, NSR regulations require a different permitting path based on whether the ethylene facility is located in an attainment area or nonattainment area as shown in **Figure 1**.

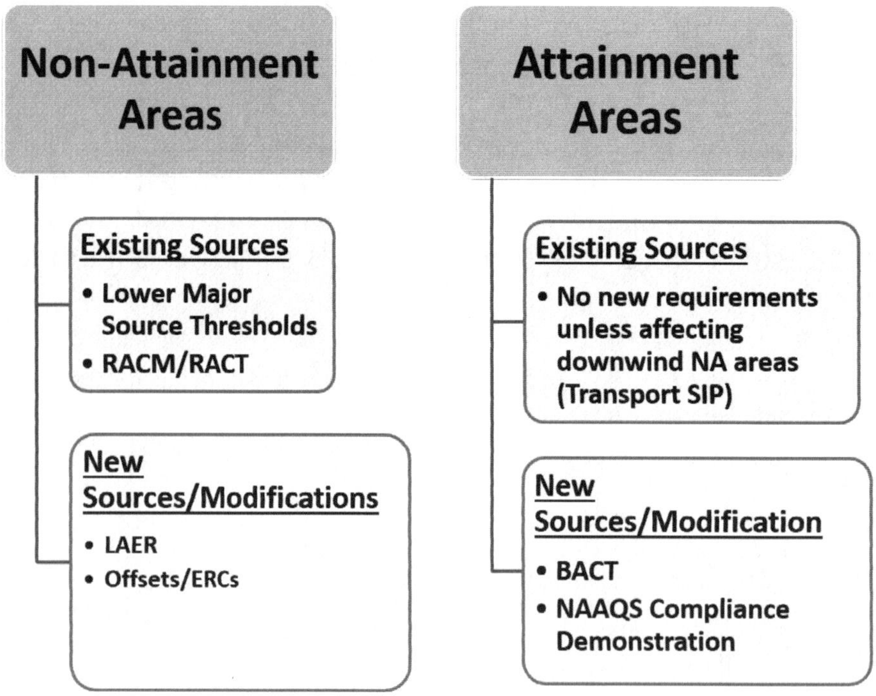

Figure 1. Permitting Paths for Ethylene Plants in Attainment and Nonattainment Areas

For attainment areas, the permitting path ensures prevention of significant deterioration (PSD) in air quality even when the new or modified emission sources will be operational. This permitting path is often called PSD permitting. There are many elements that are reviewed by the regulatory agency in a PSD permit application for a new source or modification as shown in **Figure 2**. They are:

- Potential to Emit Calculations: These are the maximum emissions from the new or modified source taking into consideration physical and operational design limits.
- Regulatory Analysis: A comprehensive analysis of all federal, state, and local air regulations that are applicable for the emission source.
- Best Available Control technology (BACT) Review: An analysis of commercially available technologies for ensuring the maximum degree of control of the pollutant taking into account energy, environmental, and economic costs and other costs.
- Air Quality Impact Analysis (AQIA): A determination of anticipated impacts of the emissions on air quality in the surrounding area.
- Additional Environmental Impact Analysis: An analysis of the impact of emissions on the soil, vegetation, and visibility in the area surrounding the project.
- State or local air pollution control agency requirements: A list of specific requirements for the projects at the state and/or local level.

It is important to note that the PSD permit is a pre-construction requirement and, therefore, a facility cannot start the construction of a new source or modification without having obtained the approval of the PSD permit application in the form of a PSD permit.

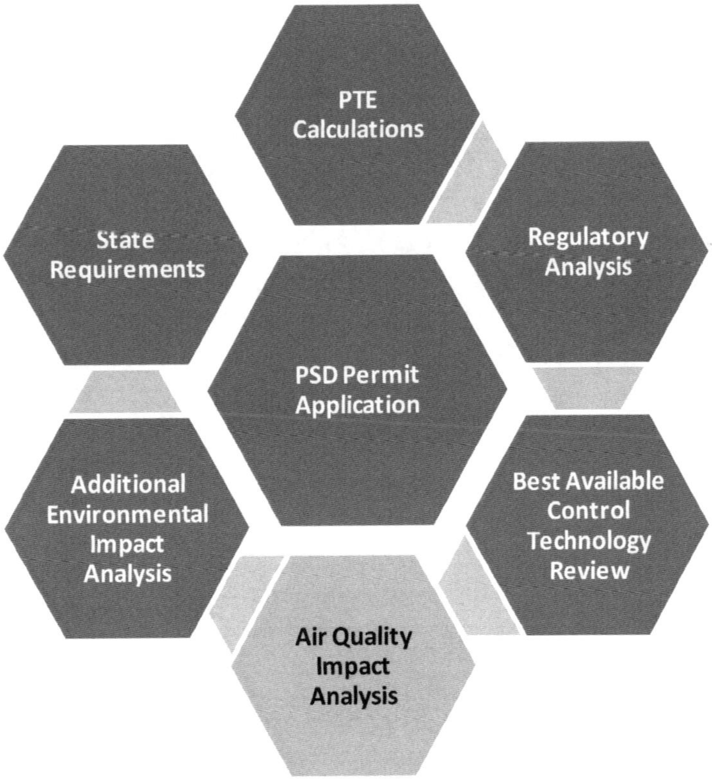

Figure 2. PSD Permit Application Elements

Air Quality Impact Analysis

Ethylene plants built in the United States are typically major sources of air emissions. Also, in most cases, modifications to ethylene plants such as capacity upgrades or the addition of downstream product manufacturing results in major modifications. In attainment areas, therefore, ethylene plants typically are required to obtain PSD permits and conduct required AQIA.

The main objective of AQIA is to demonstrate that the project will not result in the exceedance of any NAAQS. In addition, in order to ensure future growth, the project must not result in significant deterioration of air quality over the baseline level, typically quantified in terms of "PSD increments." PSD increments vary based on the criteria pollutant and the concentration averaging time for the pollutant. The focus of this paper is to discuss the challenges in demonstrating compliance with the NAAQS only.

There are two steps in the compliance demonstration for NAAQS. First, in a screening level analysis, the maximum air quality impacts of only the project emissions for all affected criteria pollutants are compared to respective screening thresholds termed as "Significant Impact Levels" or SIL. The concept of SIL is that if the project's maximum air quality impacts are below this level for all affected criteria pollutants, the project is not expected to cause or contribute to the exceedance of the NAAQS, and no further analysis is necessary. In other words, the project's air quality impact is insignificant. On the other hand, if the project's maximum air quality impact for any affected criteria pollutant is either equal to or greater than the SIL, then the project is considered significant and a refined analysis is required.

The refined analysis requires the estimation of cumulative impact of the following: (i) the project's emissions; (ii) emissions from all other existing sources in the facility (if any); and (iii) impact from off-site emission sources within a certain geographical area, the extent of which is dictated by the level of the project's impact over the SIL. This cumulative impact is then added to a representative background concentration in the vicinity of the project and the sum is compared to the NAAQS to determine if there could be a potential for exceedance of NAAQS. The process is shown in **Figure 3**.

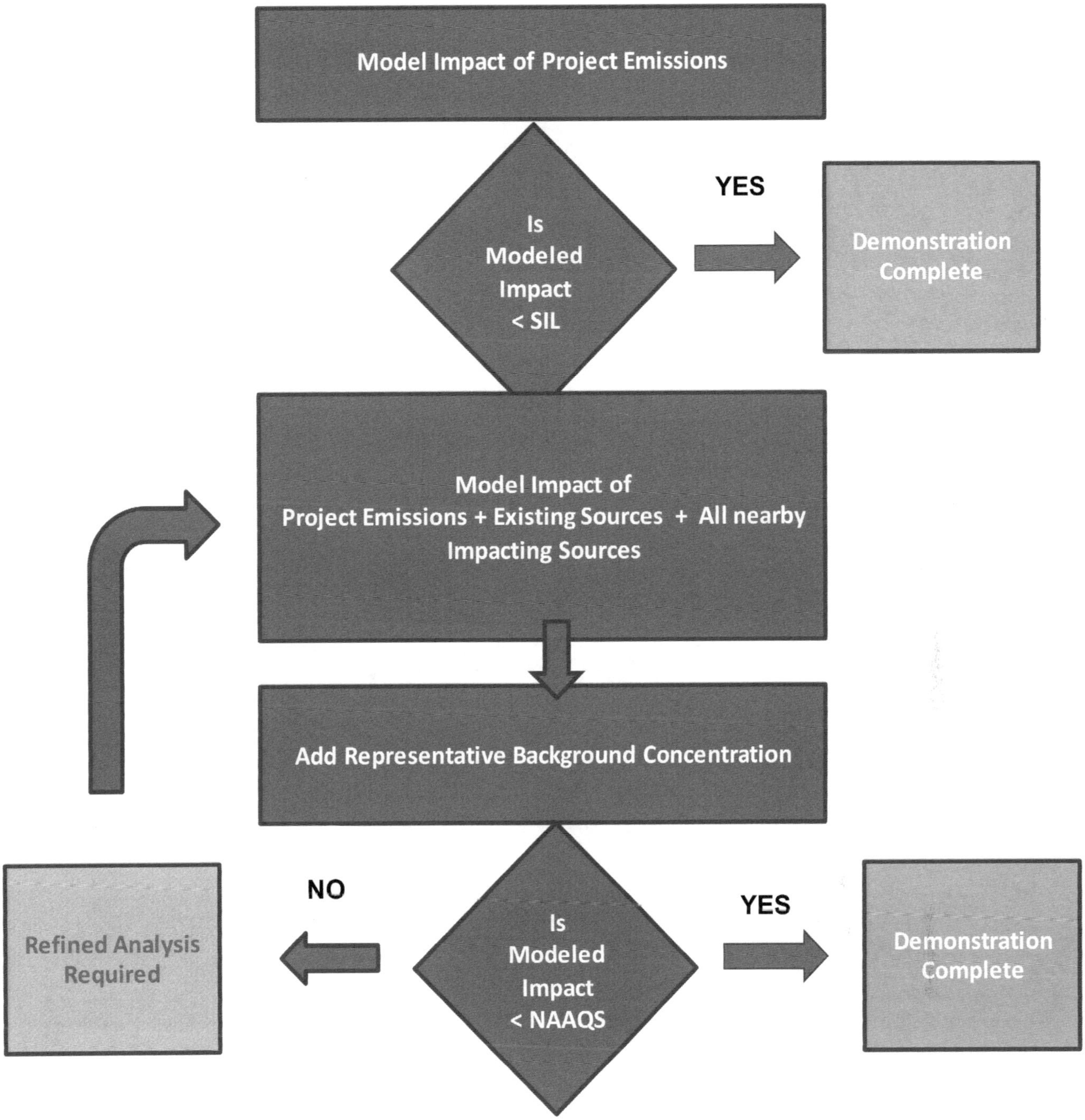

Figure 3. NAAQS Compliance Demonstration

If the sum is below NAAQS, the project is deemed to have demonstrated compliance with NAAQS. If not, further review is required to satisfy this precondition for issuance of a PSD permit.

For most new ethylene plants or major modifications, refined analysis is typically required for the criteria pollutants nitrogen dioxide (NO_2) and particulate matter with aerodynamic diameter of 2.5 microns or less ($PM_{2.5}$). These criteria pollutants will be the subject of this paper.

Air Dispersion Modeling Basics

AQIA is conducted using air dispersion models, which simulate the fate of the emissions exiting from the stacks into the atmosphere to locations where the plume reaches ground levels. This extremely complex atmospheric process is quantified using simplified equations representing the rise of the plume due to buoyancy and momentum, entrainment of ambient air due to wind shear, and both molecular and turbulent mixing of the ambient air with the emissions within the plume. **Figure 4** shows a simplified representation of the process. Building downwash, terrain elevations, and land-use are important variables affecting the dispersion of plume. USEPA has developed and approved several air dispersion models such as AERMOD and CALPUFF, which are typically used in the AQIA for PSD permit applications.

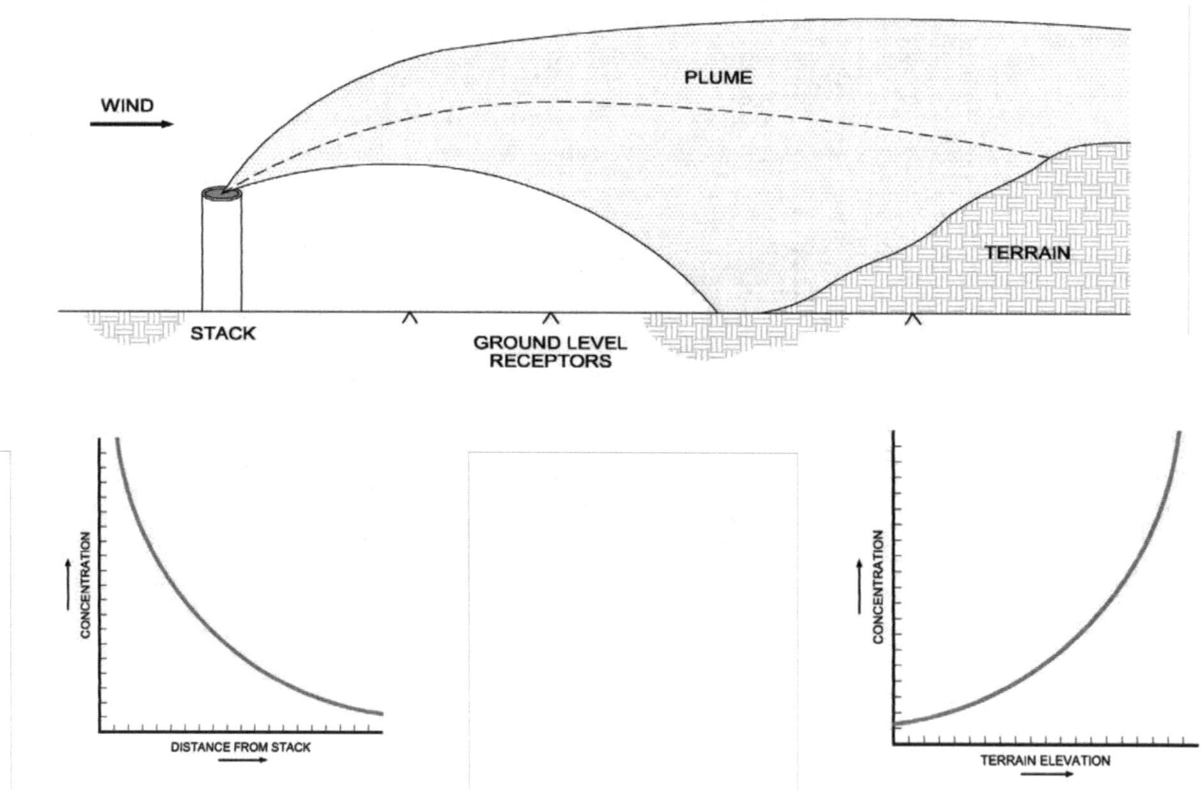

Figure 4. Air Dispersion from Stack

Ethylene Plant Emission Sources

In the United States, ethane is typically used as the raw material for production of ethylene. Ethane is thermally cracked into ethylene, propylene, methane, hydrogen, and other byproducts in the pyrolysis furnaces at temperatures up to approximately 1560°F in the presence of steam, which acts as a diluent. The furnaces act as pyrolysis reactors in which a wide range of mainly light hydrocarbons are produced. The general chemical reaction scheme is:

$$C_2H_6 + \text{heat} \rightarrow C_2H4 + H_2$$

Several byproducts are produced as part of the ethane cracking process including carbon, which is deposited as coke in the furnace tubes and is periodically removed using decoking operations. The cracking process is highly exothermic, and heat is supplied by the combustion of natural gas fuel and recycled tail gas, which is typically rich in hydrogen. If the economics are favorable, plants may separate hydrogen via a pressure swing adsorption system and sell the recovered hydrogen commercially or supply it to a downstream operation. Steam for the furnaces is provided by cogeneration units where electric power is also generated for internal use or for supply to the local electrical grid.

Figure 5 shows the major emission units for ethylene production plants. The major emissions are from combustion of natural gas and tail gas in furnaces and in cogeneration units, and emissions from the cooling towers used for cogeneration units.

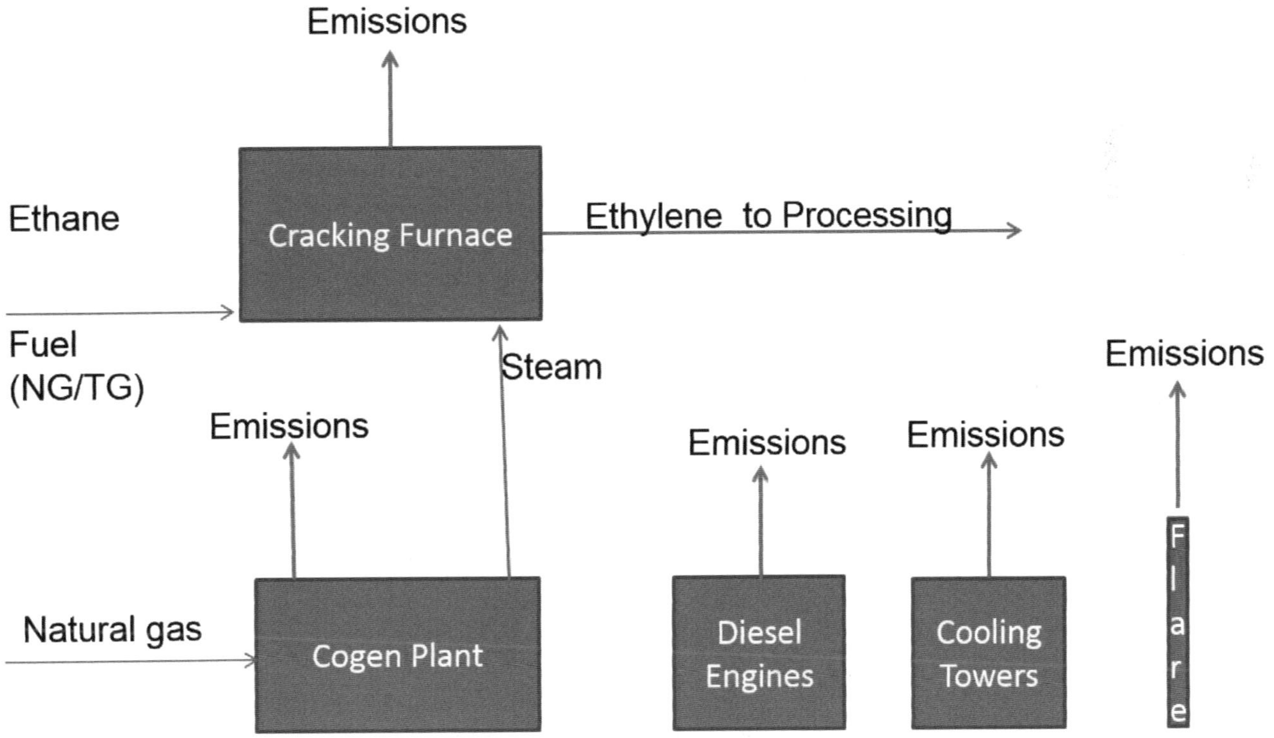

Figure 5. Major Source of Emissions for Ethylene Production Plants

Challenges in Meeting NAAQS

Incomplete Off-site Emission Inventory Data

As described earlier, for new "major" ethylene plants or for "major modifications" to existing ethylene plants, demonstration of compliance with NAAQS is necessary to obtain a PSD permit. In many cases, the screening level AQIA for these projects show exceedance of the SILs and refined modeling is necessary. As shown in **Figure 3**, estimation of impacts from off-site sources is a part of the cumulative impact and, in some cases, could be the most significant part of the cumulative impact. A complete and accurate off-site inventory is therefore critical in these cases. Often, the availability of a complete inventory of off-site emission sources in the area surrounding the project poses severe challenges in developing a robust AQIA for the project.

Air dispersion modeling requires numerous inputs for off-site (and project) emission sources. For stack releases, these are hourly and annual emission rates, stack locations coordinates, stack base elevation, stack height, stack diameter, stack gas flow rate, and stack gas temperature. For area sources, data are required for hourly and annual emission rates, horizontal dimensions of the source, the emission flux, orientation of the source, and elevation of the source. The off-site source data are typically obtained from the regulating state agency or the local air pollution control agency. The data are collected by the agency through permitting actions, annual emission inventories, and periodically via information collection requests. The quality of the data is seldom verified, and over time, the data become less and less representative of the actual operations for these sources. In many cases, there are numerous gaps in the data.

It is the permitee's responsibility to ensure that the off-site inventory data are representative of operation at the time of the PSD permit submittal. This often requires extensive research in the agency databases and field verification of the off-site emission sources, if allowed. In spite of these actions, the permitee is often faced with filling the data gaps using engineering judgments based on experience for similar emission units. To satisfy the agency, some of these judgments are necessarily conservative, thus over-predicting the contribution of the off-site emission sources in the NAAQS compliance demonstration process. Any over-prediction of the impacts of the off-site emission sources results in less room for the project's impact and may trigger stricter and often expensive control of the project emissions.

Representative Background Concentration

As shown in **Figure 3**, background concentrations are another critical part of the demonstration of compliance with NAAQS. Background includes pollutant concentrations due to: (i) natural sources, (ii) nearby sources other than the one(s) currently under consideration, and (iii) unidentified sources. The background concentration must be representative of the area of impact by the project's emissions.

Ideally, the project site will have on-site monitoring data representative of the air quality, but this is seldom the case. Project applicants mostly have to rely on the state and federal ambient monitoring network to determine a representative background concentration. Unfortunately, there are not adequate monitoring networks in place and some of the monitors do not have adequate data. This is especially true for the short-term NAAQs such as 1-hour NO_2, 1-hour SO_2, and 24-hour $PM_{2.5}$. **Figure 6** shows the national network of monitors for NOx, which includes NO_2. Although there are several monitors near the large metropolitan areas and industrial clusters, there are few beyond these areas.

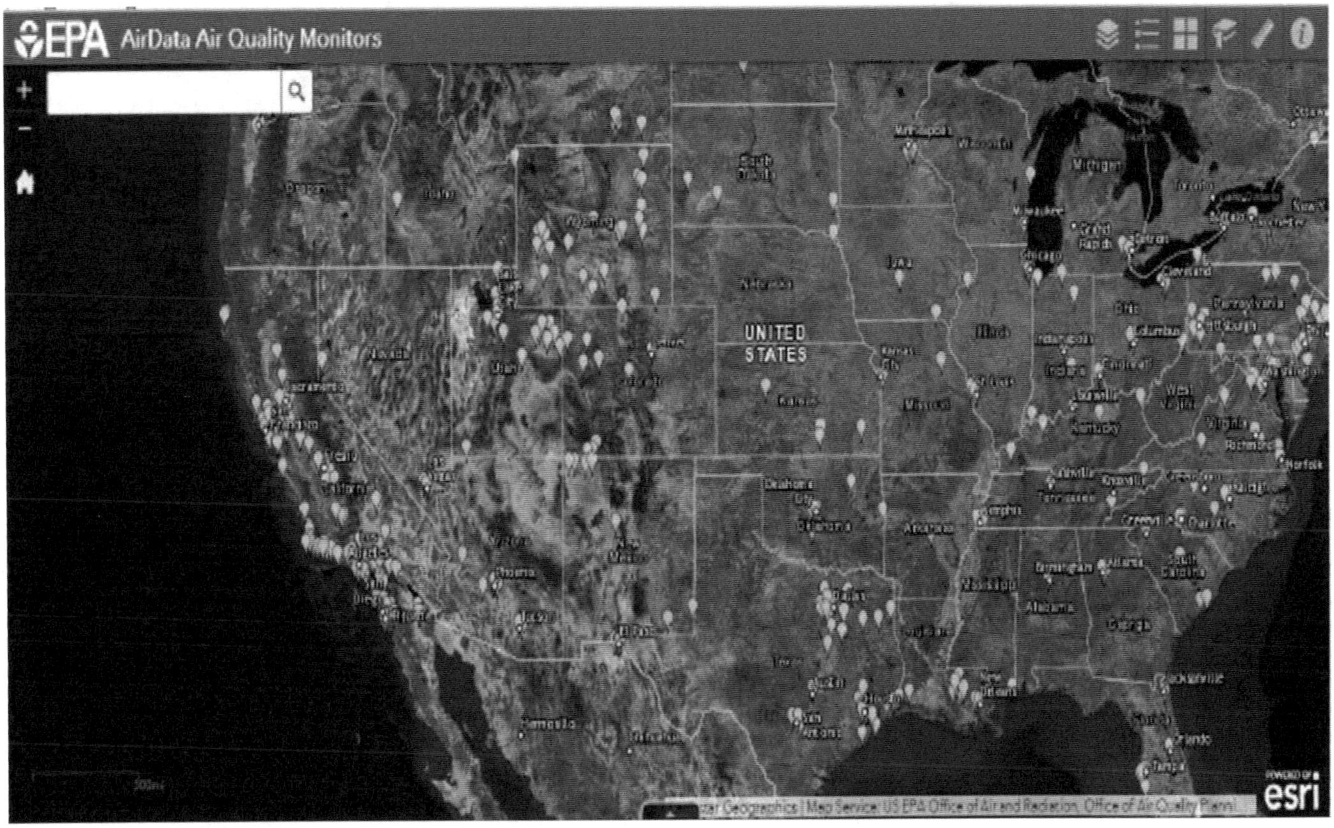

Source: USEPA AIRDATA :
https://www.epa.gov/outdoor-air-quality-data/interactive-map-air-quality-monitors

Figure 6. NOx Ambient Monitoring Network in the United States

Lack of nearby monitoring data forces the project applicant to evaluate data from other monitors in the state (or even in other states) based on similarity of emissions surrounding the monitor, the wind pattern, the potential mobile source emissions, and emissions from smaller (area) sources. The analysis is complicated and often hampered by the fact that adequate data for comparison do not exist. In that case, one option is to select data in demonstrably more industrialized areas in order to err on the conservative side. Because a conservative background concentration takes away the flexibility of allowable higher impacts from the project emission sources, this could lead to more stringent and expensive control requirements.

For monitors in industrial areas where the nearby industrial sources are also included in the refined modeling, there is another issue to be considered. The monitoring data will be most likely impacted by the industrial sources which are being modeled, leading to "double counting" of the impact of these nearby sources and making the analysis overly conservative. A detailed analysis is required in these cases to separate the impact of the source being modeled explicitly from the monitoring data from sources that are not—often a very tedious process. However, such an analysis restores some of the flexibility to the project emissions because the background concentration is not overly conservative.

1-Hour NO_2 NAAQS – The Hydrogen Issue

Hydrogen is generated in the cracking process, and the generation rate is dependent on the hydrogen content of the raw material (e.g., ethane) used. In some cases, the hydrogen is separated from the cracked gases via a pressure swing adsorber and is used elsewhere in the plant as fuel or a raw material. However, in some projects in which this opportunity may not exist, the hydrogen-rich tail gas is recycled in the cracking furnace as a fuel, thereby reducing the consumption of natural gas fuel. While this is an efficient approach for energy management, it may result in some difficulties in demonstrating compliance with the 1-hour NO_2 NAAQS due to increased generation of NOx.

Thermal NOx generation is very sensitive to the flame temperature. The adiabatic flame temperature (AFT) for methane (single largest constituent of natural gas) is approximately 3260°F. In contrast, AFT for 80%/20% H_2/Methane (e.g., a recycled tail gas) is approximately 3450°F. Thermal NOx formation increases exponentially above 2800°F. The higher flame temperature for a H_2-rich fuel generates much higher NOx than natural gas fuel; in some cases, this could result in up to 60% higher NOx formation. The higher NOx will either require higher efficiency from the NOx control technology (selective catalytic reduction), or it may result in significantly higher impacts, thereby triggering cumulative refined modeling and difficulties in meeting the NO_2 NAAQS. Because the 1-hour NO_2 NAAQS is more stringent than the annual NO_2 NAAQS, the impact of H_2-rich tail gas fuel in the cracking furnace could be more significant in demonstrating compliance with this short-term NAAQS.

There are two other challenges in demonstrating compliance with 1-hour NO_2 NAAQS. In many projects, refined modeling may be required to account for ambient conversion of NO to NO_2 in the presence of ozone (Tier 3 analysis), based on the following basic chemical mechanism, known as titration:

$$NO + O_3 \rightarrow NO_2 + O_2$$

Large quantities of ozone in the background thus results in complete conversion of NO to NO_2, resulting in higher impact. But if the atmospheric ambient ozone concentration is limiting, the conversion is limited resulting in lower NO_2 ambient impacts. This approach (known as the Tier 3 method) requires accurate data on the initial NO/NO_2 ratio in the exhaust gas from the stack, also known as in-stack ratio (ISR), and hourly data on ozone ambient concentration in the vicinity of the project, both of which pose challenges.

USEPA allows a default ISR of 0.5 for combustion sources and maintains an ISR database, but the dataset is quite limited. Currently, there are no data in the ISR database for ethylene cracking furnaces. For several combustion sources, stack tests in the past have shown ISR values as low as 0.1. If ozone background concentration is limiting, lower ISR results in lower NO_2. Therefore, it is important that a representative ISR value is obtained for the ethylene cracking furnaces; otherwise, conservative estimates of 1-hour NO_2 impact may result in some projects.

The challenge of obtaining a background ozone concentration representative of the vicinity of the project is same as that for other pollutants discussed above—the lack of an adequate monitoring network. Ozone is a secondary pollutant formed in the atmosphere primarily from precursors NOx and VOC in the presence of sunlight and the kinetics are quite complex. Due to this complexity, if ozone background concentrations are not available, it is more challenging to identify a monitor representative of the area surrounding the project. USEPA and several agencies allow a conservative default ozone concentration which, in most cases, results in complete transformation of NO to NO_2, thus defeating the very purpose of the Tier 3 method.

24-Hour Average $PM_{2.5}$ NAAQS

There are two major challenges in demonstrating compliance with the $PM_{2.5}$ NAAQS. The first challenge is based on the fact that $PM_{2.5}$ is a fairly new regulated air pollutant (the first NAAQS being promulgated in 2011) and as such, there are not many emission or background concentration data available. Much more data are available on PM_{10}, which is another older particulate standard. In many cases, where $PM_{2.5}$ data are not available, the only option is to consider that all PM_{10} emissions are $PM_{2.5}$. While this is a reasonable assumption for some types of emission sources, for many other cases, it is not. In the case of ethylene plants, it is fairly reasonable to assume that the majority of the particulates generated in the combustion of natural gas or tail gas in the cracking furnace will be $PM_{2.5}$. However, this is possibly not accurate for decoking operations when the solid carbon deposits on the furnace tubes are combusted.

The problem is compounded by the fact that the $PM_{2.5}$ NAAQS are more stringent than the PM_{10} NAAQS as shown in **Table 1** above. Although the method of estimating the impacts are slightly different, the 24-hour average NAAQS for $PM_{2.5}$ is 35 micrograms per cubic meter ($\mu g/m^3$) compared with the 24-hour average PM_{10} NAAQS of 150 $\mu g/m^3$. Speciation of the $PM_{2.5}$ in the PM_{10} is therefore more important for this short-term $PM_{2.5}$ NAAQS compliance demonstration.

The second issue is that $PM_{2.5}$ is also generated beyond the combustion chamber. Most ethylene cracking furnaces use selective catalytic reduction (SCR) for NOx control and an oxidation catalyst for CO control. The SCR typically uses ammonia to remove NOx by reaction, but there is always some residual ammonia that slips through the SCR (ammonia slip). These catalysts oxidize the sulfur dioxide (SO_2) in the gas stream to sulfur trioxide (SO_3), which in the presence of moisture reacts with residual ammonia in the stack to form ammonium bisulfite and ammonium sulfate, which add to the emissions of $PM_{2.5}$. It is a classic case where control of one regulated pollutant results in generation of another regulated pollutant.

$PM_{2.5}$ is also formed in the atmosphere as aerosols via complex atmospheric reactions. Current USEPA regulations for AQIA require qualitative or quantitative accounting of this phenomenon. Accounting for the secondary formation of $PM_{2.5}$ via photochemical reactions in the atmosphere is very complex and requires the use of specialized photochemical models. Even then, many assumptions have to be made in these models due to the lack of available data, which may lead to conservative estimates of secondary $PM_{2.5}$ impacts. USEPA has provided guidance on equivalency of $PM_{2.5}$ emissions (pollutant offset ratio) for two of the precursors, namely NOx and SO_2. In this method, 200/100 tons of NOx emissions (eastern/western US) and 40 tons of SO_2 emissions (nationwide) each are considered equivalent to 1 ton of secondary $PM_{2.5}$. It is a simpler, albeit conservative, method for estimating the secondary $PM_{2.5}$ impacts. This additional $PM_{2.5}$ also creates more challenges in the demonstration of compliance with the stringent short-term NAAQS for $PM_{2.5}$.

Conclusions

Most new ethylene plants and major modifications in ethylene plants located in attainment areas will require an AQIA as part of a PSD permit application. Demonstration of compliance with NAAQS is a key part of the AQIA. Over the last five years, USEPA has promulgated three stringent short-term NAAQS (1-hour NO_2, 1-hour SO_2, and 24-hour $PM_{2.5}$), which has made the NAAQS compliance demonstration challenging in many projects. The main issues are: (i) lack of off-site emission inventory data, (ii) lack of ambient (background) monitoring data, (iii) lack of speciation data for $PM_{2.5}$, and (iv) addressing secondary impacts of $PM_{2.5}$.

Each project is unique and may face none or all of the above challenges. The goal of the permit applicant is to strive for the most accurate depiction of impacts of other sources and background concentrations so that project emissions could be maximized within regulatory limits. To meet this objective, some recommendations prior to undertaking the AQIA are listed below.

- Evaluate the air monitoring data in the surrounding area of project to develop a representative background concentration for the project. If such data are unavailable, discuss with the permitting authority to determine how to obtain such data without being too conservative.
- Obtain and analyze emissions and release parameters data for large emission sources within 50 to 70 kilometers of the project. If refined modeling is required, these sources may be affecting the total impact significantly. For all large sources, ensure that the emissions data are current and representative.
- Ensure representative emissions and stack gas data are available for operating scenarios other than normal operation such as startup, shutdown, and decoking. These operations may be the controlling operating scenarios for the short-term NAAQS.
- Use project-specific in-stack ratios for NO/NO_2 if available from test data on similar furnaces.
- Develop speciation of $PM_{2.5}$ for decoking operation if possible.

While the NAAQS compliance demonstration is going to be challenging in many projects, with an early start, prudent decision making, and detailed data analyses, the challenges could be significantly mitigated.

AIChE Paper Number:

Flare Emissions Regulatory Update

Kevin Leary
Applications Engineering Director – Flare Systems
John Zink Hamworthy Combustion

Prepared for Presentation at the 2017 Spring National Meeting
San Antonio, Texas, March 30, 2017

Flare Emissions Regulatory Update

Kevin Leary
Applications Engineering Director – Flare Systems
John Zink Hamworthy Combustion

Abstract:

Several new regulations have been recently published by the US EPA that will have significant effects on the design and operation of industrial flares. This paper will provide a brief summary of requirements that may impact the design and operation of flare systems typically used in the petrochemical industry.

Proper control of assist steam, assist air and enrichment fuel gas is required to ensure smokeless flaring of hydrocarbons with good destruction efficiency. Historically the requirements to comply with EPA air emissions standards were focused on designing for a maximum exit velocity based on the net heating value of the gas stream and controlling the steam or air assist medium to prevent smoke formation. Testing has shown that improper control of the steam or air assist rates can cause emissions outside the EPA standards for flares. Based on this testing the EPA has issued revision of the standards required for flare operation.

In December 2015, the EPA published a final rule entitled "Petroleum Refinery Sector Risk and Technology Review and New Source Performance Standards" under 40CFR Parts 60 and 63. This standard adopted several new requirements for the operation of flares for petroleum refineries. These new requirements require refinery operators to significantly change the way they monitor and control their flares. Although these regulations only currently apply to petroleum refineries, many within the petrochemical industry believe that similar requirements will be implemented in their sector in the near future. This paper will summarize the new measurement and control requirements to meet these new standards.

In addition, Multi-Point Ground Flares (MPGFs) have been used for many years as an effective flare design to minimize the impact of flaring and to allow design for very high smokeless capacities that are not feasible with conventional elevated flares. The EPA requirements for flare design contained in 40CFR60.18 and 40CFR63.11 are focused on steam-assisted, air-assisted and nonassisted flares. These requirements do not allow for the use of high pressure MPGF's as a method to comply with EPA air emissions standards since these flares operate at much higher exit velocity than is allowed in these regulations. However, recent testing has shown that these flares can provide high destruction efficiency and very large smokeless capacity.

To facilitate use of this technology, in April 2016, the EPA published a notice entitled "Notice of Final Approval for the Operation of Pressure-Assisted Multi-Point Ground Flare at Occidental Chemical Corporation". In this notice the EPA published section IV, "Final Framework for Streamlining Approval of Future Pressure-Assisted MPGF AMEL requests". Several petrochemical sites located in the U.S. have received AMELs (Alternate Means of Emissions Limitation) for operation of MPGFs. Several additional operators in the petrochemical industry are proposing using similar MPGFs for their new sources. This paper will discuss the benefits of using MPGFs in new petrochemical facilities and summarize the requirements of the EPA streamlined MPGF AMEL approval process.

Background

Flares are used for the safe and effective relief of hydrocarbon waste gases as part of the pressure relief system in an industrial plant. The flare system should be designed to safely combust all the waste gases while maintaining the back pressure in the relief system within the allowable design while producing safe levels of flare radiation and gas dispersion.

The effectiveness of the flare system relies on providing stable combustion of the flare gases with acceptable levels of smoke production and high combustion efficiency. Typically, most elevated flares utilize either steam-assisted or air-assisted flare tips to prevent the formation of smoke from the flare. Proper control of steam injection or assist air is required to ensure smokeless flaring of hydrocarbons. The amount of steam or air assist medium required depends on many factors including, the flare gas flow rate, gas composition, design of flare tip, size of flare tip, wind conditions, etc.

The stability of the flare flame must also be maintained to ensure good destruction efficiency. The net heating value of the flare gas must be maintained at some minimum value to ensure a stable flare flame. Gas streams that are highly diluted with inert gases can often have insufficient heating value and require enrichment with fuel gas to maintain a stable, efficient flame.

Historically most flares operated in hydrocarbon and petrochemical industries in the US have been required to be designed and operated within the requirements listed in the US EPA 40CFR60.18 regulations. 40CFR60.18 requires that all flares be operated with no visible emissions for more than 5 minutes in any 2-hour period. The regulation establishes a minimum heating value that must be maintained for all flares regulated by this rule (200 BTU/SCF for nonassisted flares and 300 BTU/SCF for steam-assisted and air-assisted flares).

This regulation also establishes a maximum allowable exit velocity at which a flare can operate at that is a function of the net heating value of the flare gas. There are two separate equations in the regulation that define this maximum velocity; one equation that is applicable for air-assisted flares and another equation for steam-assisted flares or nonassisted flares. The basis for these maximum velocity equations is flare emissions data collected during the EPA and Chemical Manufacturers Association (CMA) tests in the 1980s.

Recent testing by the EPA and other regulators has resulted in a reevaluation of the rules governing flare operation. This testing included the 2010 TECQ Flare Efficiency Test which was conducted on full size steam and air-assisted flares at the John Zink Company research facilities in Tulsa, OK. This test focused on measurement of hydrocarbon destruction efficiency at turndown rates at varying steam assist and air assist injection rates. This test concluded that flares can produce relative low destruction removal efficiency, or DRE, (under 98%) under conditions of over-steaming or excessive aeration.

2010 TECQ Flare Efficiency Study Testing at John Zink Company in Tulsa, OK

There were also numerous field studies conducted on actual operating flares in refineries using various imaging technologies and remote optical monitoring techniques that concluded that over-steamed flares could reduce DRE in operating flares. The EPA issued an enforcement alert in August 2012 which summarized recommended operating practices to prevent excessive emissions that can be created by over-steaming or over-aerating of assisted flares.

Smoking Flare

Well Operated Steam Flare at Incipient Smoke Point

Over-Steamed Flare

New Refinery Rule

On December 1, 2015, the EPA published a new ruling in 40CFR63.670 that establishes new MACT requirements for the operation of flares in all affected refineries in the US. The new refinery rule supersedes the requirements of 40CFR60.18 and establishes new, more stringent design and operating requirements for refinery flares that must be implemented by January 30, 2019. This new rule imposes significant new requirements for the operation, monitoring, reporting and record keeping associated with flare control devices.

Although the new MACT requirements in 40CFR63.670 currently only apply to flares operated in refineries, it is widely believed within the industry that very similar regulations will be promulgated in the near future to apply to other industries including the petrochemical industry.

Net Combustion Zone Heating Value

The new refinery rule outlined in 40CFR63.670 introduces two new operating parameters associated with the operation of all flares. The two new operating parameters are the Net Combustion Zone Heating Value or NHV_{cz}, and the Net Heating Value Dilution Parameter or NHV_{dil}. The NHV_{cz} and the NHV_{dil} are intended to define the heating value of the flare gas just prior to combustion including the dilution effect of assist steam or assist air that is injected into the flare tip.

This new rule abandons the minimum heating value rules of 40CFR60.18 which were based solely in the heating value of the flare gas entering the flare tip. The new rule requires that the two new parameters, the NHV_{cz} and the NHV_{dil} be monitored and controlled. The NHV_{cz} parameter must always be maintained above 270 BTU/SCF regardless of the type of flare used. The NHV_{dil} parameter must always be maintained above 22 BTU/ft2 during all periods that perimeter assist air is used.

This will require the operator to monitor and control the flare gas flow rate, flare gas net heating value, and assist steam/air rates to ensure compliance with this new operating parameter. Since all steam and air-assist flares require some minimum cooling steam or minimum air flow rate to prevent overheating and damage to the flare tip it is important that operators maintain these minimum cooling rates without dropping below the minimum 270 BTU/SCF NHV_{cz}. The measurement of NHV_{cz} is based on a 15-minute block average whenever regulated material is flared for at least 15 minutes. The measurement of NHV_{dil} is based on a 15-minute block average whenever regulated material is flared and perimeter air assist is used.

It is worth noting that hydrogen receives special treatment in the new regulations. The existing 60.18 rules allow for an alternate (potentially higher) maximum vent gas velocity for nonassisted flares containing greater than 8% by volume hydrogen. This acknowledges the fact that, although hydrogen has a relatively low net heating value (274 BTU/SCF), it does an excellent job in contributing to the stability and combustion efficiency of hydrocarbon flames. The new refinery rule assigns hydrogen a pseudo net heating value of 1212 BTU/SCF to be used in the calculation of NHV_{cz} and the maximum velocity.

New Monitoring Requirements

The new refinery rule outlined in 40CFR63.670 establishes a minimum amount of monitoring that all refinery flare operators must maintain when operating a flare. A summary of these new monitoring requirements includes:

The flare vent gas volumetric flow rate must be measured, calculated and recorded continuously as well as any supplemental natural gas used. The steam and air-assist media volumetric flow rate must be measured, calculated and recorded continuously. Note that the rule does allow measurement of air blower speed or power along with fan curves as a suitable method of measuring air assist flow rate.

The flare vent gas composition and/or net heating value must also be measured. If vent gas composition is used it is required to continuously measure (at least once every 15 minutes), calculate and record the individual component concentrations present in the vent gas. This is typically done by using continuous gas chromatography. The ruling also allows for use of grab-samples every 8 hours in lieu of continuous measurement. But if a grab sample is used this sample must be assumed constant for all of the 15-minute intervals between each 8-hour grab samples. If a grab sample is taken when a particularly low BTU flare gas mixture is present, this could require the operator to use large amounts of supplement fuel gas for the subsequent 8-hour period to meet the minimum NHV_{cz} requirement. The ruling does include exceptions to gas composition measurement when the operator can document that flare only operates with "purchase pipeline quality" gas and other consistent composition vent gas streams

Alternately the operator can measure the flare vent gas net heating value in lieu of gas composition. In this case the operator must operate, calibrate and maintain a calorimeter capable of continuously measuring, calculating and recording the NHV of the flare gas at standard conditions. Since heating value measurement does not allow correction of the NHV_{cz} for hydrogen content, the operator may choose to operate, calibrate and maintain an analyzer capable of continuously measuring, calculating and recording the hydrogen concentration in the vent gas in order to take advantage of the pseudo hydrogen heating value.

The new rule also includes new visible emissions (smoke) monitoring requirements. Under this rule the operator must perform, at least once per day, a Method 22 visual emissions monitoring for 5-minute period and immediately any time an operator sees visual emissions. Alternately the operator may use a video surveillance camera to continuously record (at least one frame per 15 seconds with time and date stamp) images of flare flame suitable for visual emissions observations and provide continuous real time video to control room. This data must be saved for a minimum 5-year period.

There are flare pilot flame monitoring requirements in the new rule that are similar to the current 40CFR60.18 requirement, however, the new rule states that it is a violation for each 15-minute block period in which no pilot flame is present for at least 1-minute. This increases the need for the operator to ensure that very reliable pilot flame monitoring is maintained.

Flare Tip Exit Velocity

Another significant change that is included in the new refinery rule compared to the existing 40CFR60.18 requirements is the limitations of flare tip velocity. There is only one equation for the calculation of Vmax. This equation matches the velocities for the steam/unassisted equation from 60.18 (300 btu/scf is 60 ft/sec, 1000 btu/scf is 400 ft/sec). While the equation for calculation of the maximum allowable velocity are similar, the gas conditions under which the velocity is calculated have changed. 40CFR60.18 requires that the exit velocity be calculated using the volumetric flow rate in units of standard temperature (68°F) and pressure (29.92" Hg). The refinery rule requires that this velocity be calculated using the volumetric flow rate at actual conditions. For flares with elevated temperatures, which is common at emergency fire case flaring scenarios, the calculated exit velocity is higher under the refinery rule requirements versus 40CFR60.18 requirements. This means that it is easier to exceed Vmax under the refinery rule.

Flare Management Plan

The requirements for smokeless flaring and maximum exit velocity under the existing 40CFR60.18 requirements for flares have generally been assumed to be required only under normal smokeless operating conditions. The rule has allowed for exceeding these limits during periods of startup, shutdown and maintenance (SSM exclusion). The new refinery rule does not include this SSM exclusion. Instead the refinery rule requires operators to declare the maximum smokeless capacity of their flare. If the flare has the potential to operate above this smokeless capacity the operator must develop a flare management plan to minimize the possibility of flaring during startup, shutdown or emergencies.

Flare Management Plan requires specifying maximum smokeless capacity for all refinery flares

The flare management plan must include a detailed assessment of all process units and equipment feeding each flare with a detailed plan for minimizing flaring, including cost analysis to install flare gas recovery, modification of startup/shutdown procedures, modification of pressure relief valves feeding the flare, etc. Additionally, the flare management plan requires a detailed analysis of every affected flare including detailed process flow diagrams, detailed description of all flare components, detailed specifications of all the instruments used to measure the flare gas flow rate, composition, steam assist rates, air assist rates etc.

Violations of the Standard

The new refinery rules define new conditions that are considered a violation of the standard for which penalties will apply. It is important to note that the flare must comply with the various performance criteria only when regulated material is routed to the flare. For operation below the declared smokeless capacity of the flare, any smoking event (>5 minutes in a 2-hour period), operation below the required NHV_{cz} (15-minute block average), operation below the required NHV_{dil} (if applicable), or operation above the allowable exit velocity (15-minute block average) is considered a violation of the standard.

For operation above the declared smokeless capacity of the flare, any smoking event (>5 minutes in a 2-hour period) or velocity exceedance triggers a requirement to perform a root cause analysis and a corrective action analysis of the exceedance event. The event is considered a violation if it meets any one of the following criteria:

- The root cause analysis determines operator error or poor maintenance.
- Any two smoking exceedances that are not caused by a force majeure event in a 3-year period on the same flare for the same root cause
- Any two velocity exceedances that are not caused by a force majeure event in a 3-year period on the same flare for the same root cause
- Any three smoking exceedances that are not caused by a force majeure event in a 3-year period on the same flare for any reason

- Any three velocity exceedances that are not caused by a force majeure event in a 3-year period on the same flare for any reason.

These new requirements for establishing a flare minimization plan, corrective actions for exceedances, and violations for specific root causes will greatly increase the complexity and the cost of operating a flare system in compliance with the new refinery rule. As noted earlier, it is expected that similar rules will be applied in the petrochemical sector and other sectors which widely use flares as control devices.

Recent AMEL Approvals for the Use of MPGF Flares

Background

In recent years, there has been an increased interest within the industry to use pressure assisted multi point ground flare (MPGF) systems in new and expansion facilities in the Oil and Gas and Petrochemical industries. These flares use arrays of small capacity, pressure assisted burners to produce short, stable, smokeless flames. The burners are aligned in staged rows and can consist of hundreds of burners, allowing very high capacity with short flames that can be shielded from view with a perimeter fence.

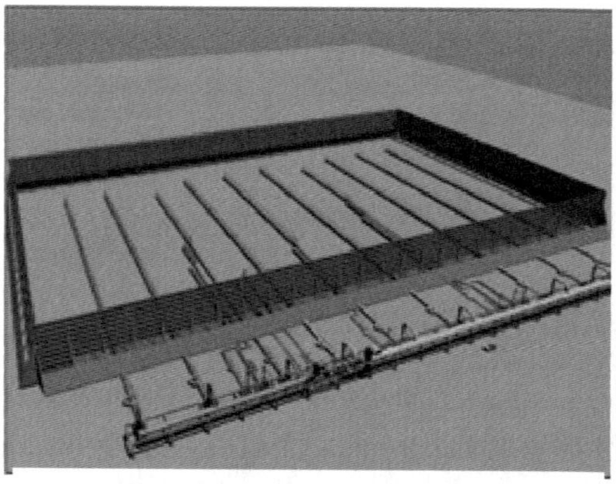

Typical MPGF Layout Sketch

One of the drivers for use these MPGF installations is to provide an effective flare system that does not require the complex and costly control, reporting and operating costs associated with operating steam or air-assisted flares under the new refinery rule. MPGFs can have many significant benefits over conventional elevated flares including:

- Provides for hidden flame that can be contained within a fenced perimeter.
- Provides 100% smokeless capacity including emergency rates.
- Little or no assist medium (steam or air) is required for smoke control.
- High hydrocarbon destruction efficiency can be assured (> 99.5% DRE)
- No steam or air controls required to maintain high DRE

Dow Chemical / ExxonMobil AMEL for Use of MPGF

In 2014 Dow Chemical Company and ExxonMobil Chemical Company submitted applications for AMEL to use MPGF's as an acceptable control device in several new petrochemical facilities planed in the Texas Gulf. As technical justification for the AMEL Dow and ExxonMobil submitted publicly available data from previous MPGF burner emissions tests that were conducted by John Zink for Dow and Marathon on LRGO high pressure burners used in both new and existing facilities. On August 31, 2015, the EPA published a ruling granting approval of the AMEL for use at the facilities requested. Based on the test data that was submitted their flares were approved for operation under specific operating conditions.

Within the Dow / ExxonMobil AMEL the EPA acknowledged that several additional facilities had submitted applications for similar MPGF AMELs and more were in the planning stages. In this ruling the EPA therefore published a proposed framework for a streamlined process for approval of future MPGF AMELs. This framework for streamlined approval was finalized April 21, 2016.

Framework for Streamlined Approval Process for Future MPGF AMELs

On April 21, 2016 AMEL approval for Occidental Chemical, the EPA published a streamlined process for future applicants to submit MPGF AMEL requests. This process will help new applicants to submit sufficient data with their requests to achieve expedited approval.

The streamlined process highlights the testing data that must be included with the AMEL application and highlights design / operational features that the MPGF must have in order to be eligible for streamlined AMEL approval.

Burner Equivalency, Burner Stability and Cross light Testing

To have confidence that the MPGF burners will meet the minimum stability and DRE requirements of the standards, the EPA has specified several sets of test data that must be submitted with the AMEL application. The first set of data relates to burner emissions. Burner emissions can be demonstrated by testing the burner and measuring the emissions or by an engineering assessment. If an engineering assessment is done, the proposed burner must be shown to be equivalent to a previously tested burner. An engineering assessment can be used if the operator intends to use a burner design that is equivalent to that used in previously published burner emissions data that has been EPA approved and is in the public record. The engineering assessment must make the case as to why the proposed burner is equivalent to a previously emissions tested burner. If the proposed MPGF burner is not

DRE Testing of high pressure LRGO Burner (MPGF Burner) for AMEL

equivalent to a previously testing model, then the operator must submit detailed emissions testing data to the EPA for that specific burner. The test must be conducted per the requirements detailed in the framework. It is best if the test protocol is submitted for EPA review and approval prior to testing.

The second set of test data that must be compiled is related to burner stability. One or more burners of the same design to be used in the proposed MPGF must be tested over the range pressures and at the lowest NHVcz for which the operator intends to permit. These burner stability tests are simple tests that are normally conducted at the flare manufacturer's test facility. This is simply a visual test to verify that the burner flame remains lit and stable over the range of operation tested.

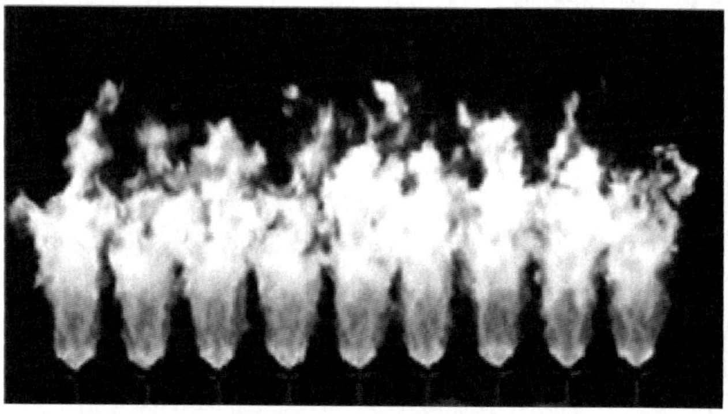

Cross Light Testing of a Large Row of Burners

The third set of test data that must be compiled is cross light reliability testing. A critical feature of MPGF design is the requirement that the burners along a row effectively cross light off a single piloted burner at the beginning of the row. The minimum cross light distance is dependent on many variables including burner style, burner capacity, gas pressure and gas composition. A cross light test verification report is therefore required as part of the AMEL application. A small row of burners (3 minimum) of the same design and spacing to be used in the proposed MPGF must be tested for cross light reliability at the minimum NHVcz for which the operator intends to permit. These cross-light tests are relatively simple tests that are normally conducted at the flare manufacturer's test facility. This is a visual test to verify that the burners cross light and remain stable over the range of operation tested.

These three test components, are submitted as part of the AMEL application. The EPA suggests that following this process will result in streamlined approval of future MPGF AMELs but notes that all AMEL requests are subject to public notice and comment.

New Operating and Design Parameters for MPGFs

Within the Dow / ExxonMobil AMEL approval and the Occidental AMEL approval the EPA established several other minimum requirements for the design and operation of MPGFs. These will likely be similar requirements for all new MPGF's going forward. The following is a list of the significant operating requirements identified:

- The high pressure MPGF burners must operate with a flare gas LHV of 800 BTU/SCF or greater, and a flare gas lower flammability limit (LFL) of 6.5% or lower.
- The MPGF must be operated with no visible emissions above the fence line except for 5 minutes in any 2-hour period.
- The operator must provide visual indication (CCTV) and record of burner operation continuously to document smokeless operation.
- The operator must continuously measure and record pressure in the flare header.
- The operator must continuously measure and record vent gas flow rate.
- The operator must continuously measure, calculate and record vent gas NHV via continuous GC or BTU analyzer.
- The operator must continuously measure and record each staging valve position.
- The operator must provide a minimum of two flare pilots per stage with continuous pilot flame monitoring. Loss of both pilots on an active stage is violation.

Summary

The EPA recently promulgated new rules for the design and operation of flares in refineries. It is anticipated that these rules we soon become a requirement in other industries including the petrochemical industry. These new rules require detailed flare management plans to be established by 2019. The rules also impose much stricter minimum BTU levels and smokeless operation that a flare must be operated under. These rules will require a significant increase in the measurement, control, recording and reporting associated with elevated flares. This could result is increased operating costs to meet the requirements and increased chance for violations when a flare is not operated within the guidelines.

The US Oil & Gas and Petrochemical industries have significantly increased the use of MPGF's in recent years. This is partly in response to the possible implications associated with the new requirements for elevated flares established in the new refinery flare rule. MPGFs also provide many other potential benefits versus conventional elevated flares, however until recently there was no clear procedure for how to permit MPGF's with the EPA. The published AMEL approval framework presents a streamlined procedure for permitting MPGFs in the future. This allows use of MPGFs, and the inherent benefits versus elevated flares, to be more easily achieved.

References:

1. McDaniel, Marc (1983), "Flare Efficiency Study" EPA-600/2-83-0522

2. Pohl, J.H., Payne R., and Lee J. (1984) "Evaluation of Efficiency of Industrial Flares: Test Results" Energy and Environmental Research Corporation, EPA-600/2-84-095

3. Allen, David T. Ph.D. and Torres, Vincent M. (2011), "2010 TCEQ Flare Study Project Final Report", The University of Texas at Austin for Texas Commission on Environmental Quality

4. Federal Register 40CFR Part 60.18 (7-1-06 Edition), "General control device requirements"

5. Federal Register 40CFR Part 63.670 December 1, 2015, "Requirements for flare control devices"

6. Federal Register Volume 81, No 77, Notices, April 21, 2016, "Notice of Final Approval for the Operation of a Pressure-Assisted Multi-Point Ground Flare at Occidental Chemical Corporation"

7. Federal Register 40CFR Parts 60, 61 and 63, Volume 80, No 168, Proposed Rules, August 31, 2015, "Notice of Final Approval for the Operation of a Pressure-Assisted Multi-Point Ground Flare at The Dow Chemical Company and ExxonMobil Chemical Company and Notice of Receipt of Approval Request for the Operation of a Pressure-Assisted Multi-Point Ground Flare at Occidental Chemical Corporation"

AIChE Paper Number: 194d

Challenges in SCR Operation

Robert Alvers
Global Furnace Engineer
Equistar Chemicals, LP

Klavs Beldring
Technical Manager | Air Pollution Control Catalyst & Technology
Haldor Topsoe, Inc.

Prepared for Presentation at the 2017 Spring Meeting
San Antonio, TX, March 27-30, 2017

Challenges in SCR Operation

Authors

Robert Alvers, Equistar Chemicals, LP
Klavs Beldring, Haldor Topsoe, Inc.

Abstract:

Selective Catalytic Reduction (SCR) systems are becoming more common as environmental regulations apply to more and more sources. To achieve desired results, a fundamental understanding of the SCR system and its limitations is needed. Mathematically, most process parameters are measured directly and then used to calculate the values that have permitted limits. Ammonia slip is often a calculated value using a material balance. A stack test is used to verify actual values. Any biases in the system are absorbed into a D-factor. These biases will likely be different in different modes of operation. Physically, the requirement to meet emission limits may lead to a design with operational challenges. This paper will explore the learnings of SCR operation: sensitivity of various inputs on the ammonia slip in different modes of operation, common errors with the calculations, dynamic effects, and operational issues.

Overview

Selective Catalytic Reduction (SCR) systems are common in new ethylene furnace installations and are being considered as possible retrofits in some scenarios to reduce NO_x. Understanding what an SCR does, how an SCR does it, how it is controlled, and how the performance is measured is important to the successful operation of the SCR.

SCR Basics

The goal of an SCR is to reduce nitrogen oxide (NO_x) compounds with ammonia (NH_3) over a catalyst bed to nitrogen and water. NO_x is considered a pollutant and is limited by our environmental permits. The primary reactions are:

1) $4\,NO + O_2 + 4\,NH_3 \rightarrow 4\,N_2 + 6\,H_2O$
2) $NO + NO_2 + 2\,NH_3 \rightarrow 2\,N_2 + 3\,H_2O$
3) $6\,NO_2 + 8\,NH_3 \rightarrow 7\,N_2 + 12\,H_2O$

Source: Haldor Topsoe

The flue gas, containing NO_x, passes through the ammonia injection grid (AIG), where an appropriate amount of ammonia is added to the flue gas stream. To ensure a good ammonia/NO_x distribution the stream is usually mixed via an engineered static mixer or a bank of the convection section. The flue gas then passes through the SCR catalyst bed and the rest of the convection section.

SCR Deactivation

The radiant tubes in cracking furnaces are made of chromium nickel alloys, which when exposed to high temperatures will predominantly release chromium oxyhydroxide ($CrO_2(OH)_2$). The chromium released from the radiant tubes will condense downstream in the SCR catalyst and have a negative impact on the SCR catalyst activity. The general effect of chromium is that each 1,000 ppmw Cr in the catalyst will deactivate it roughly 2.5%. As an example, a catalyst analysis after 5,400 hours of operation showed that about 0.7 %wt Cr had accumulated in the catalyst, which resulted in a 14% drop in activity (at 350°C). After 16,500 hours of operation, it showed an increased chromium uptake. The chromium content was analyzed at three different locations of the catalyst (Table 1).

	Top	Middle	Bottom
Cr, %wt	2	1.6	1.1

Source: Haldor Topsoe

Table 1: Chemical analysis of catalyst sample after 16,500 hours

As the results demonstrate, there is a gradient of chromium uptake in the SCR catalyst. This could indicate that the chromium is deposited as fine aerosols with high diffusivity, which results in a significant overall capture of chromium in the SCR catalyst. The corresponding catalyst deactivation rate was 33% after 16,500 hours (at 350°C).

A comparison of catalyst tests performed with Haldor Topsoe catalyst of similar units is shown in Figure 1.

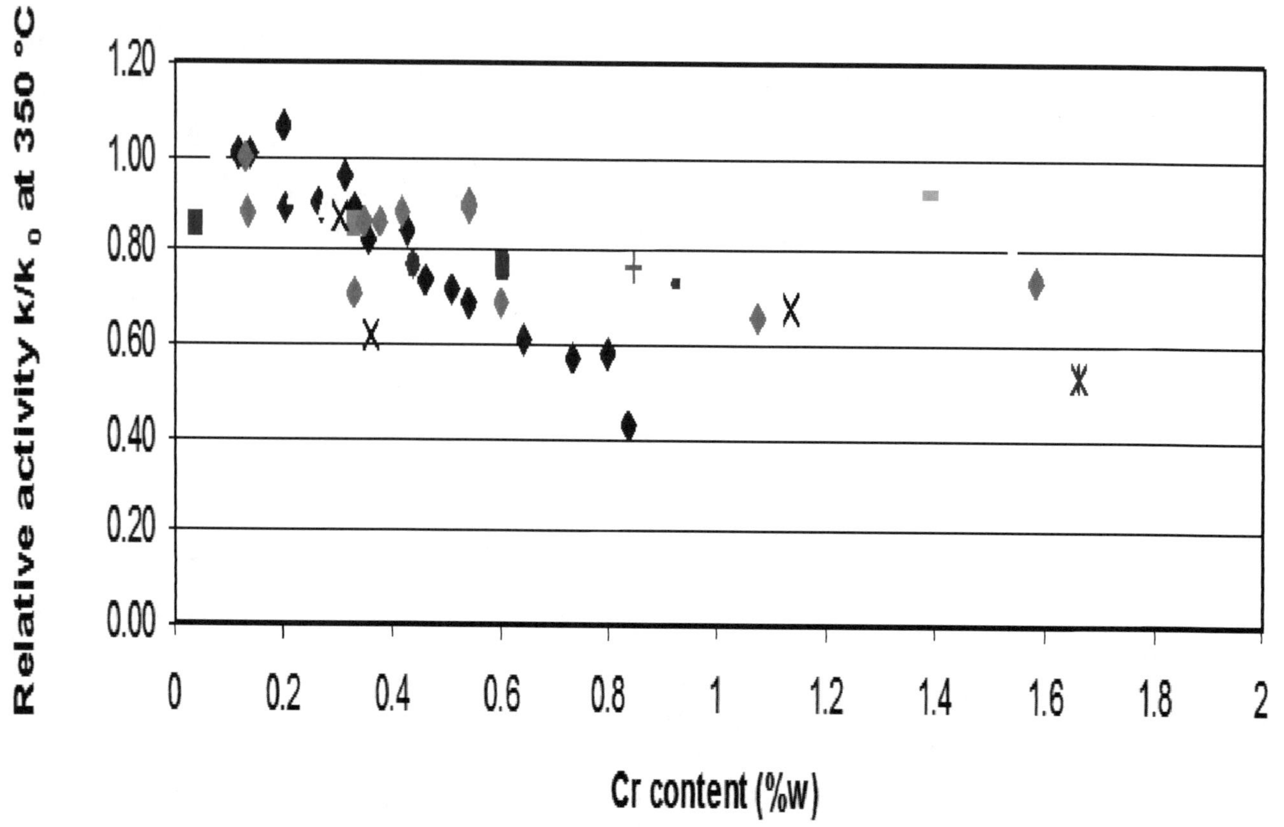

Source: Haldor Topsoe

Figure 1: Catalyst activity relative to fresh catalyst activity at 350°C as a function of chromium accumulation in ethylene cracking furnaces and steam methane reformers on Haldor Topsoe SCR references.

Plotting the two measurements against time, it is clear that there is a relationship between operation hours and catalyst activity (Figure 2). The operation time between plant turnarounds is therefore a significant parameter when designing SCR catalyst systems for ethylene crackers as longer continuous operation requires more catalyst volume. At the same time, the catalyst should be designed for frequent decoking which in some cases increase particulate loading and/or catalyst temperature depending on the unit design – if the furnace decokes to the firebox or not.

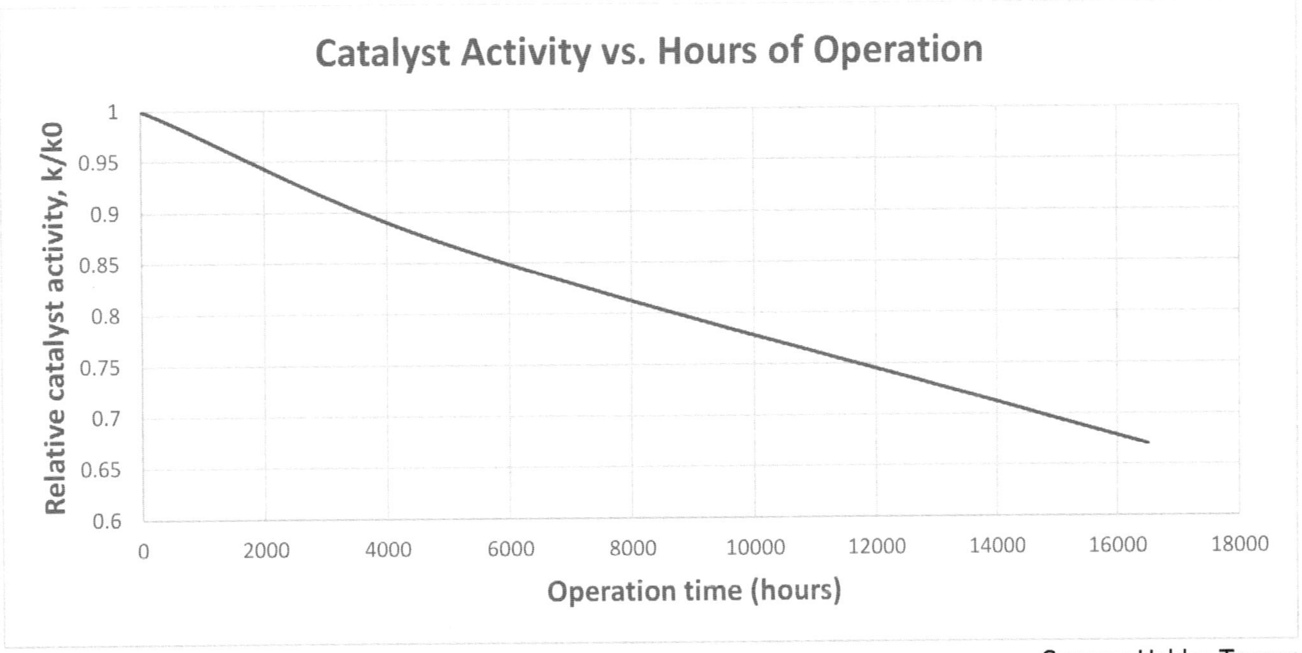

Source: Haldor Topsoe

Figure 2: Relative catalyst activity vs. hours of operation.

As can be seen in Figure 1, there is a correlation between chromium accumulation and catalyst activity, but there is no direct correlation between service hours and catalyst activity, as seen in Figure 3. It is concluded that the chromium uptake is site/application specific.

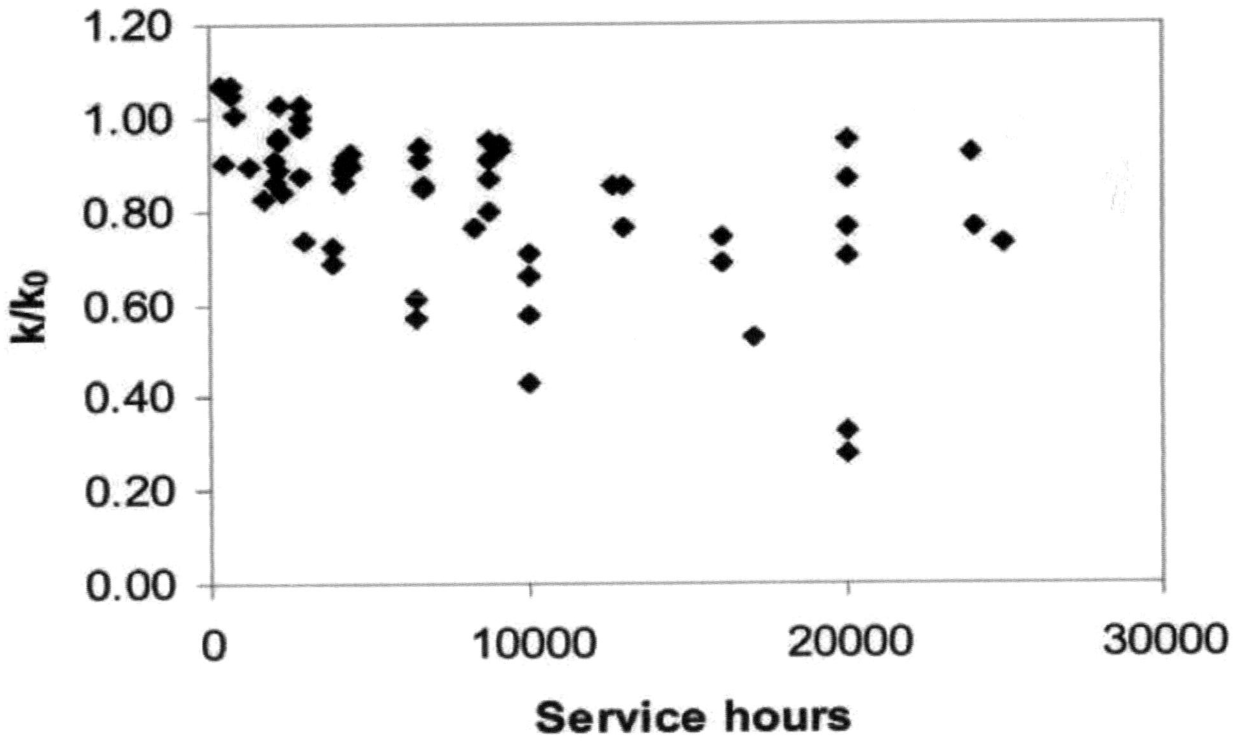

Source: Haldor Topsoe

Figure 3: Relative catalyst activity vs. service hours on similar units

While the measured chromium concentration is only an average over the piece of catalyst sample analyzed, it is also relevant where the chromium is located in the catalyst – especially if one has to predict the performance going forward or for future catalyst designs. Chromium acts as a masking agent and will mainly be found on the catalyst surface but can also diffuse into the catalyst matrix. The diffusion mobility may depend on the actual operating temperature of the SCR catalyst in the convection section because higher temperatures result in higher rates of diffusion. After many hours of operation, the diffusion barrier on the catalyst surface from chromium will reach a level where the mobility of chromium is reduced, whereas the smaller NO_x and NH_3 molecules still can penetrate the catalyst's macro pores and access the active sites in the inner catalyst matrix. At some point, the catalyst activity will enter a steady state where the deactivation has slowed, and the slope of the curve in Figure 2 would approach horizontal.

When looking at the chromium profile though the catalyst wall (Figure 4) for a catalyst sample exposed to 13,000 hours of operation in an ethylene furnace, it shows that chromium mainly accumulates at the catalyst surface with concentrations as high as 9 %wt. The profile is obtained by wavelength dispersive spectroscopy (WDS) in a scanning electron microscope (SEM). It provides quantitative element concentrations. The activity of the catalyst shown below is 76% of fresh catalyst activity (at 350°C). Due to the macro pore system of the catalyst, the surface will not be completely closed, so reactants can still diffuse into the inner catalyst pore system where the SCR reaction can occur. Chromium can continue accumulating on the catalyst surface while increasing the measured surface chromium concentration, but it will have less impact on the catalyst activity.

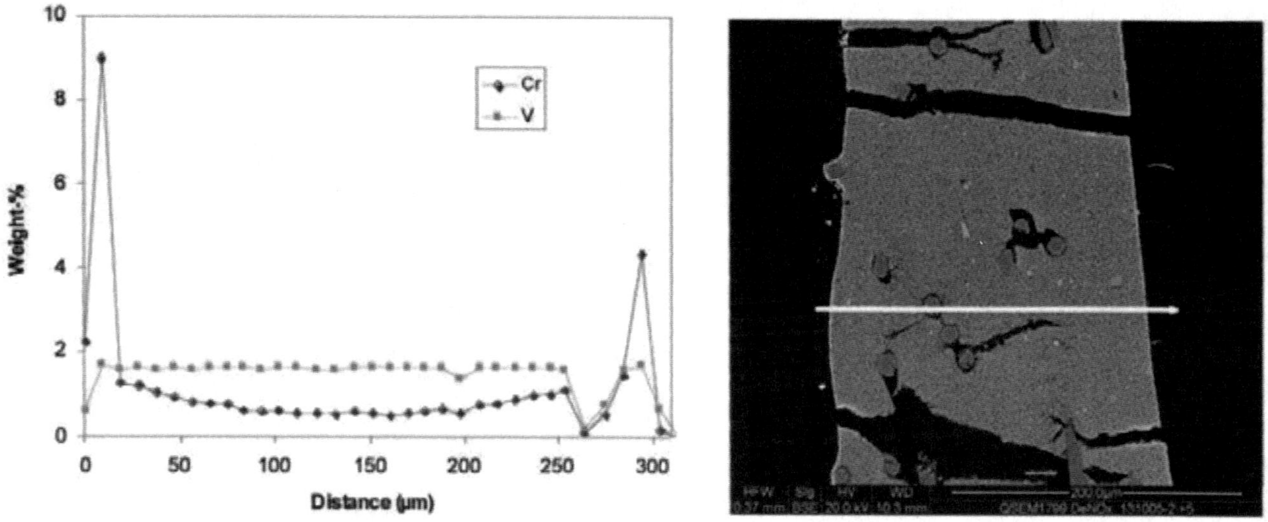

Source: Haldor Topsoe

Figure 4: Vanadium and chromium profile through a catalyst wall.

When designing and operating SCR catalyst systems in ethylene furnaces, it is therefore key to have a thorough understanding of the catalytic behavior and deactivation mechanisms to ensure a trouble free operation during the entire operating period.

Calculation Discussion

The main permitted pollutants of concern in an SCR system are NO_x and NH_3. The purpose of the SCR is to reduce NO_x by reacting it with NH_3. However, NH_3 in the stack is also limited to 10 ppmv by environmental regulations. NO_x concentration into and out of the SCR is measured. The mass flow of NO_x is calculated from the measured SCR outlet NO_x concentration and the calculated flue gas flow. The NH_3 concentration is calculated by material balance. It is assumed that the NO_x is NO, and that it reacts with NH_3 in a 1:1 ratio. The delta NO_x concentration, the flue gas flow rate, and the NH_3 injection rate are measured or calculated. The TCEQ gives the equation for this calculation (Figure: 30 TAC §117.8130(1)):

$$NH_3@O_2 = \left[\left(\frac{a}{b} * 10^6\right) - c\right] * d$$

Where:

- $NH_3@O_2$ = the ammonia concentration (ppmv) at reference oxygen, which is 3 %vol on a dry basis for ethylene furnaces.
- a = ammonia injection rate (lbmol/h) = ammonia injection rate (lb/h) / 17 lb/lbmol.
- b = dry exhaust flow rate (lbmol/h) = dry exhaust flow rate (lb/h) / 29 lb/lbmol.
- c = change in measured NO_x concentration across catalyst (ppmv @ reference O_2).
- d = The D-factor; the ratio of measured to calculated NH_3 slip obtained from stack sampling; this factor accounts for all the biases and systematic errors in the system, as well as, assumptions that are not perfect. It will be accurate at the conditions at which it was determined. The further you get from those conditions, the further the D-factor can drift.

The flue gas flow rate can be derived from a combustion calculation. Inputs are: excess oxygen, fuel gas flow rate, and fuel gas composition. For a single firebox with a convection section, this calculation is straightforward. For a furnace with two fireboxes feeding a convection section, the calculation is straightforward if there is an oxygen measurement of the flue gas going into the SCR. If the available oxygen measurements are in the arch for each firebox, separate calculations must be done for each firebox. The scenario presented in this paper will be the case of two semi-independent fireboxes (each can be fed or decoked separately) flowing into one convection section and SCR. Firebox A is decoking. Firebox B is cracking.

Key questions for calculating the D-factor are: Was the furnace operating at a steady state? Has the data been checked to verify steady state? Did your NO_x analyzer go through a calibration during the stack test?

Consistency of Bases

All the calculations that are done need to have common bases. Does standard temperature and pressure use 60°F or 68°F, 1 bar(g) or 1 atm(g)? Is the measurement or calculation wet or dry? Is it consistent with other parts of the calculation? Are flows and concentrations based on stoichiometric requirements (i.e. 0% excess O_2), at actual excess O_2, or adjusted to 3 %vol excess O_2?

One method to calculate flue gas flow is to use a published fuel specific F-factor. They can be on a wet or dry basis and are usually on a stoichiometric basis. If this flue gas flow is multiplied by the measured stack NO_x at actual O_2, the NO_x mass flow rate will be under predicted.

It is important to know what the instrumentation is measuring. Often, arch O_2 measurements are wet, and stack O_2 measurements are dry. Removing the water will increase the concentration of O_2. If the water is mistakenly removed, mathematically, from an already dry measurement, it will decrease the calculated flue gas rate, which will decrease the NO_x mass flow rate, which will then over predict the NH_3 slip.

Inputs and Sensitivity

The measurements that go into the NO_x and NH_3 slip calculations include: a fuel gas flow to each firebox, an arch O_2 measurement for each firebox, a stack O_2 measurement, fuel gas composition, SCR NO_x inlet and outlet concentrations, and the NH_3 flow rate. Each of these measurements has their own level of accuracy and reliability.

Figure 5 shows the relative sensitivities of several inputs to the NH_3 slip calculation. The underlying values are representative of a two firebox cracking furnace at typical, stable operating conditions, where Firebox A is decoking and Firebox B is cracking.

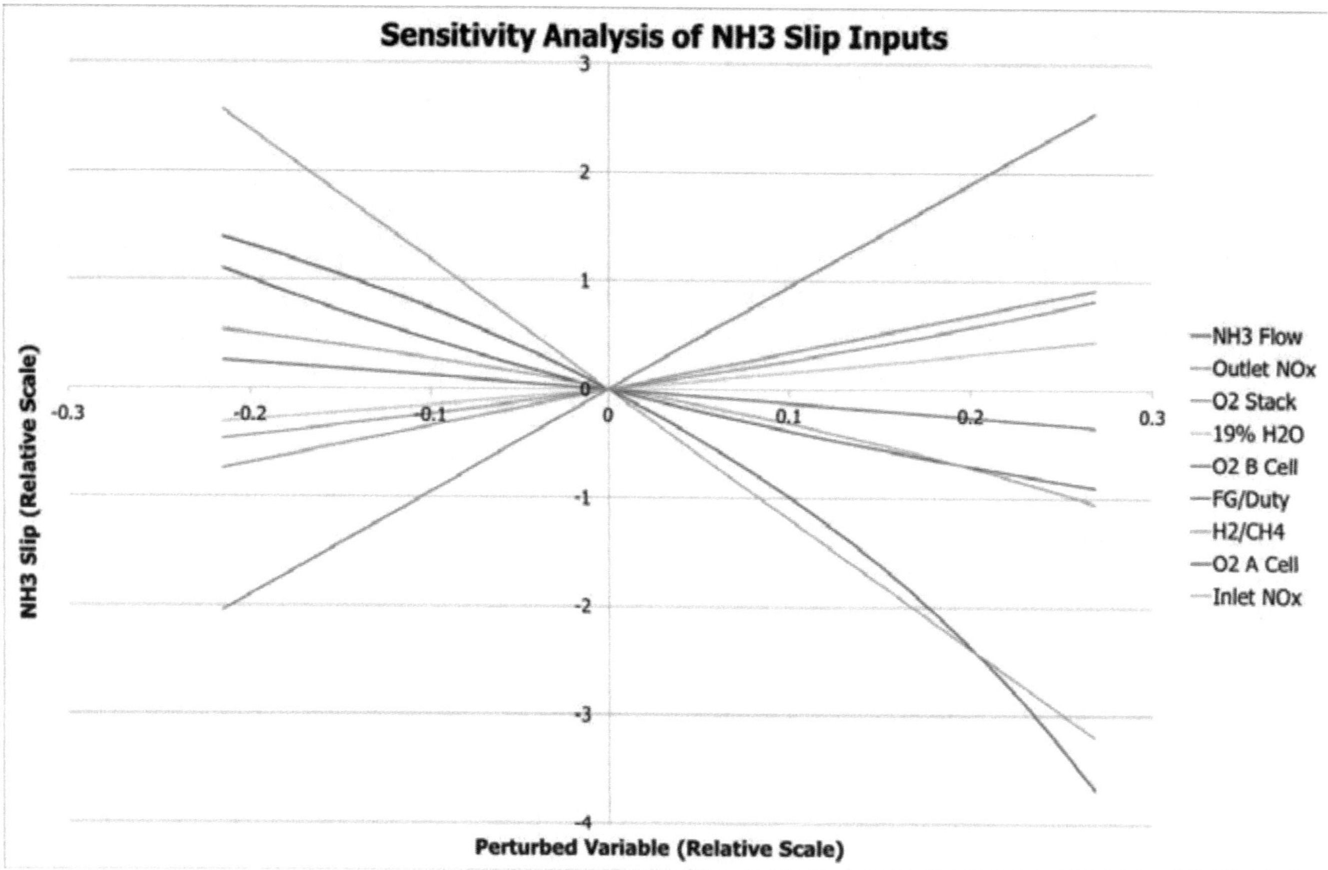

Source: LyondellBasell

Figure 5: Sensitivity analysis of inputs to the NH_3 slip calculation.

For figure 5, each variable was treated as independent, which could represent an instrument drifting. In reality, more than one instrument could be off, and they all have some random error associated with them.

Figure 6 shows how the variability of a function's inputs can synchronize to cause spikes in the function. The size of the spike will vary depending on the number of inputs that become synchronized and the function's sensitivity to each of those inputs.

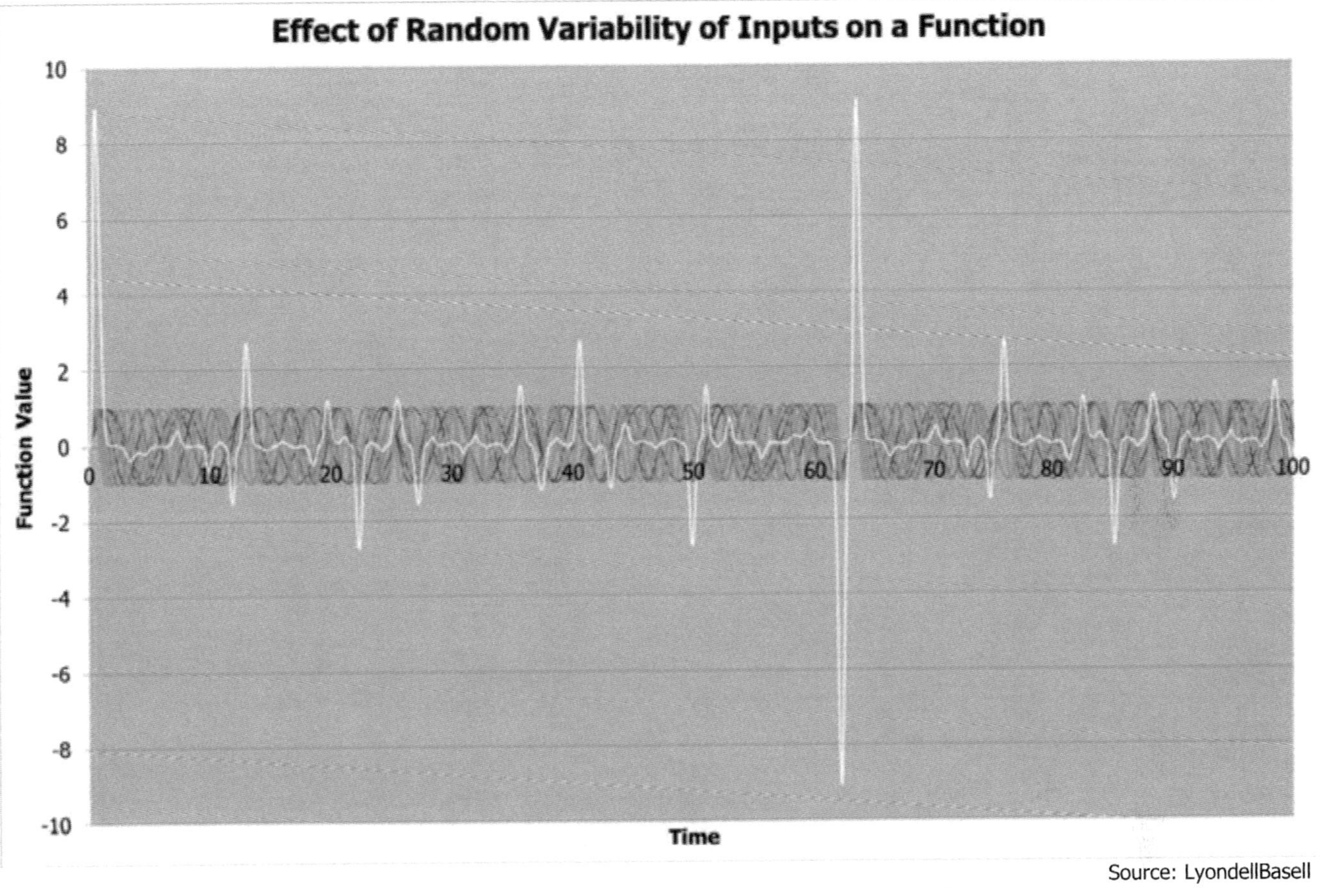

Source: LyondellBasell

Figure 6: The effect of random variability of inputs on a function.

Most of the inputs have a roughly linear effect on NH_3 slip in the ranges where a furnace operates with the exception of higher oxygen levels. Most reporting is required to have a 3 %vol O_2, dry basis. To convert actual O_2 to 3 %vol O_2, an O_2 concentration correction factor is used, which is (20.9-3)/(20.9-Actual O_2). This factor is used to multiply or divide a concentration or a flow rate, respectively.

The sensitivity of O_2 on NH_3 slip displays asymptotic behavior. At typical arch O_2 concentrations during normal cracking, the correction factor is ~0.95. At typical arch O_2 concentrations during decoke, the correction factor is ~2. When the furnace is being ramped up or down, the point at which the SCR NH_3 is turned on varies, but might correspond to a correction factor of 4-8. Random variations in O_2 at higher levels can cause the NH_3 slip to have a large variation. High O_2 levels also have a multiplier effect on other inputs. A 1 ppm change in NO_x at 16.5 vol% O2 becomes a 4 ppm change when converted to a 3 vol% O_2 basis.

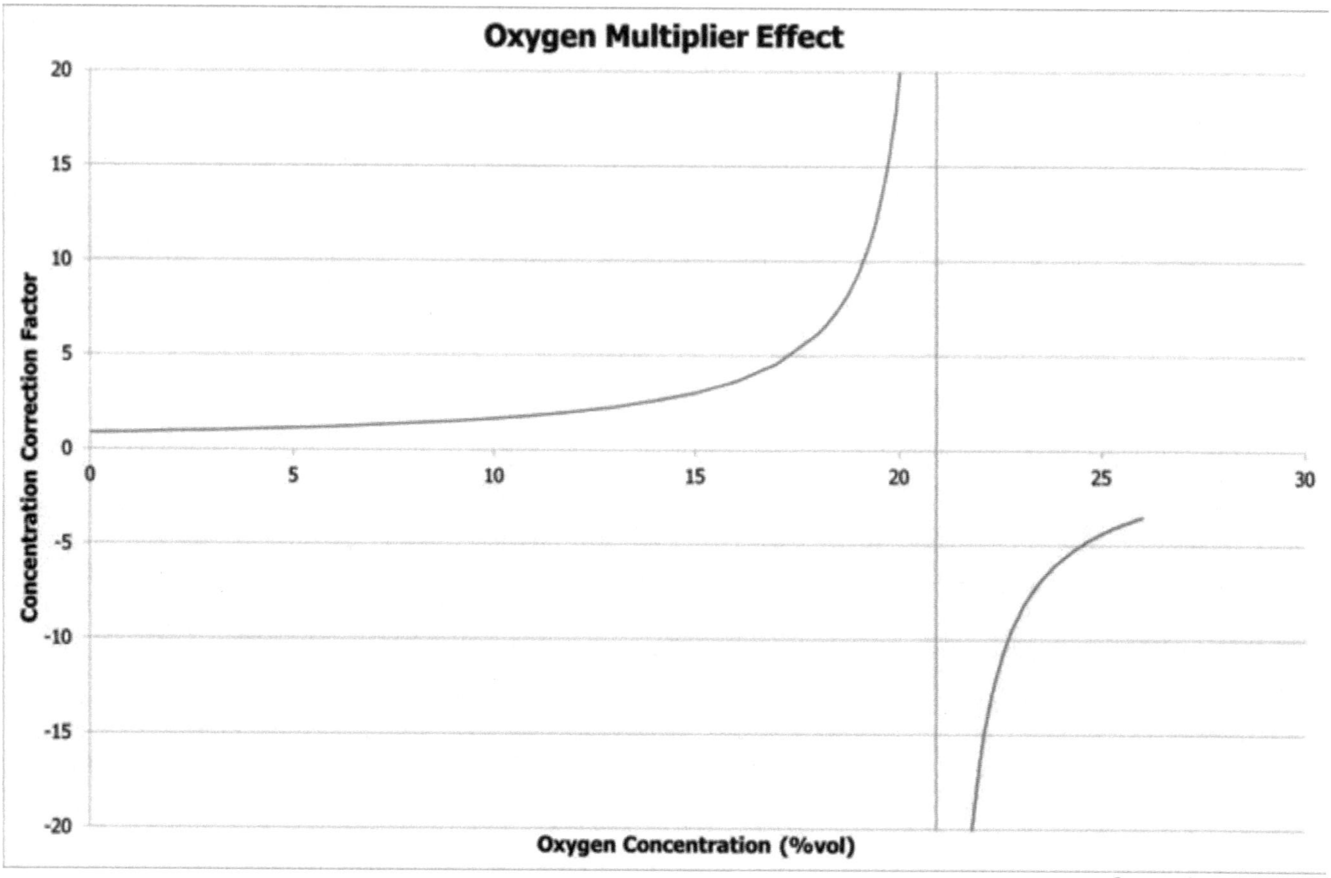

Source: LyondellBasell

Figure 7: Converting flows and concentration to a 3 %vol O_2 basis has a multiplier effect.

Another amplifier of variation in NH_3 slip is the material balance used to calculate NH_3 slip: NH_3 ppmv – NO_x ppmv = NH_3 slip ppmv.

As an exaggerated example:

1,000,001 – 1,000,000 = 1 where 1,000,001 is an exact value, and 1,000,000 has an error of ±0.1%.

The calculation becomes:

1,000,001 – 1,000,000 ±0.1% → 1,000,001 – 1,000,000 ±1,000 = 1 ±1,000

Modes of Operation

The scenario that has been used for discussion in this paper is a two firebox cracking furnace at typical, stable operating conditions, where Firebox A is decoking and Firebox B is cracking.

Normal cracking operation typically has the fewest issues as far as NH_3 slip calculations go. The flue gas O_2 is low. Burners and flow meters are operating at close to design rates. The furnace is operating close to the conditions at which the D-factor was determined. Any biases in the system are accounted for in this mode of operation.

Decoke, whether one firebox or full furnace, operation has higher O_2 and is not operating near the conditions at which the D-factor was determined. If all the system biases combine to be relatively linear or under predict NH_3 slip, an issue will not be noticed.

SU/SD is similar to Decoke but is even further away from the operating conditions at which the D-factor was determined. Additionally, SU/SD is not a steady-state operation. There are some dynamic effects that make an accurate calculation of the NH_3 slip more difficult.

Dynamic Effects

At steady-state, all variables that go into any calculations are constant and do not change. Actually, there are variations due to the limits of control systems and external stressors. When furnace conditions are being changed, the variable of time is introduced. If a change in a variable was to have an instant and complete effect, we would be in a quasi steady-state. A real system has inertia. For example, adding NH_3 will very quickly cause the NH_3 flow meter to read higher, but it will take time for the NH_3 to reach the SCR, to react, and for an effluent sample to reach the NO_x analyzer. This time lag from when NH_3 is increased to when NO_x decreases causes an issue with the NH_3 slip calculation. When the NH_3 increases, the NH_3 slip calculation will immediately show higher NH_3 slip. It will not come back down until the NO_x decreases and is fed back to the NH_3 slip calculation.

These dynamic effects are more or less pronounced depending on the rate of change and the nature of the variable. The measurements that go into the NO_x and NH_3 slip calculations include: a fuel gas flow to each firebox, an arch O_2 measurement for each firebox, a stack O_2 measurement, fuel gas composition, SCR NO_x inlet and outlet concentrations, and the NH_3 flow rate. Flow measurements are quick to show changes, while analyzers are slow.

Recommendations

The material balance calculation for NH_3 can have issues. To minimize these issues, there are three areas on which to focus: the available inputs, which came from the instrumentation design of the furnace, the crafting of the NH_3 slip calculation, and operation of the furnace.

Some inputs have a greater sensitivity on the results. Some inputs have an intrinsic delay. Identify these inputs and design for them. In a two firebox furnace, consider an O_2 measurement directly before the SCR rather than trying to combine a weighted O_2 value from each firebox. Consider a fast loop on an analyzer, or relocate the analyzer to minimize a delay. Select accurate, easily calibrated, instruments.

The calculations need to be setup correctly. Be familiar with the measurement technology being used. Know if an analyzer value is on a wet or dry basis. Question assumptions such as: the fuel gas is only hydrogen and methane, and the water content of the flue gas is a constant 19 %vol. Know the basis for a particular measurement or calculation and make sure the bases align.

How a furnace is operated is the last step to environmental success. Go slow when ramping a furnace to minimize dynamic effects. Know which variables affect the NH_3 slip the most and adjust them to minimize issues. For example, during decoke, keep O_2 as low as possible to minimize the multiplier effect. Tune the NH_3 controller for good, tight control.

One final recommendation is to skip the calculations all together, and install a Tunable Diode Laser (TDL) analyzer for NH_3.

Summary

Selective Catalytic Reduction (SCR) systems are helping reduce NO_x emissions but add complexity to furnace design and operation. A fundamental understanding of the SCR system is needed to understand the inputs, their sensitivities on the outputs, how they should be used in calculations, and how overall furnace operation affects SCR operation. All the inputs and calculations need to have the same bases, whether that is wet or dry, stoichiometric or 3% O_2. Measurements involved in calculations need to be accurate and should be on a critical instrumentation checklist. Dynamic effects introduce a time variable that can be minimized by going slow. Careful consideration of these performance factors will improve SCR performance.

References

None

AIChE Paper Number: 194e

FLAMELESS THERMAL OXIDATION OPERATING EXPERIENCE AT DOW CHEMICAL FREEPORT, TX VCM PLANT

Gene H. Irrgang
Global Product Manager, Thermal Oxidation Technologies
Selas Linde North America

Arturo Lopez
Assoc. EH&S Environmental Director
The Dow Chemical Company

Jon A. Sachs
Global Incineration Technology Manager
Olin Blue Cube Operations LLC

Prepared for Presentation at the 2017 Spring National Meeting
San Antonio, TX, 30 March, 2017

FLAMELESS THERMAL OXIDIZEROPERATING EXPERIENCE AT DOW CHEMICAL FREEPORT, TX VCM PLANT

Gene H. Irrgang
Global Product Manager, Thermal Oxidation Technologies
Selas Linde North America
Arturo Lopez
Assoc. EH&S Environmental Director
The Dow Chemical Company
Jon A. Sachs
Global Incineration Technology Manager
Olin Blue Cube Operations LLC

Abstract:
In 2002 two Flameless Thermal Oxidation systems designed to destroy VOC components and halogenated hydrocarbons started operation at a Dow Chemical's (now Olin Chemical) Oyster Creek VCM plant and their Plant B both in Freeport, TX. The justification for selecting this technology over other common flame based thermal oxidizers was mainly based on the high destruction efficiency of the gaseous hazardous components with low NOx formation required in the Houston non-attainment zone. Since the units were started the on-line availability has exceeded 99% and the cost of operation has proven to be an additional advantage of the technology. This presentation by Olin, Dow and Selas Linde (technology provider) will describe the technology and present operating experience with the FTO units.

Introduction

Faced with the need to reduce air emissions including NO_X, VOC's, and CO from their plants in Freeport, TX, The Dow Chemical Company proactively in the late 1990's to early 2000's evaluated the many incineration technologies. Ultimately they made a decision to install Flameless Thermal Oxidation (FTO) systems. The decision was made based on the demonstrated capabilities of FTO units at other long term operating plants outside Dow to achieve destruction efficiencies exceeding 99.99% for chlorinated compounds while emitting CO and NO_X at less than 5 ppmv. Two separate systems were installed. The first, at their Oyster Creek VCM Plant (now owned by Olin Chemical) had a thermal capacity of nearly 180M Btu/hr and it consisted of two trains of 3 FTO reactors. The other system at Plant B had the capacity of 62M Btu/hr with one, 2 reactor train. In addition to the FTO reactors, each train had conventional waste recovery boilers, HCl acid recovery towers and caustic scrubbers. The initial start-up of the systems occurred in 2002 at Plant B.

Figure 1 Photograph of the Dow Chemical FTO Installation during construction

Within Plant B, a strategy was devised to achieve the desired emissions reductions by replacing older thermal oxidizers that serve various manufacturing processes with newer technologies such as a central vent gas thermal treatment unit to cover the vent component of waste. Taking advantage of an integrated site, these older assets would be retired, and replaced by a central vent gas thermal treatment unit.

The Process Requirements

An evaluation of the various process vent streams by Dow indicated the new central thermal processing unit would need to treat several operational scenarios, with waste vents originating from multiple sources within Plant B. Scenario evaluation set the design flows at a maximum of 6,800 kg/hr to a minimum of zero kg/hr. The vent composition contains varying amounts of the components shown in Table I below.

Table I. Waste Gas Components

Nitrogen	Acetone	Chlorine
Oxygen	Carbon Monoxide	Chloropropanes
Carbon Dioxide	Butane	Chloropropenes
Ethane	Proplyene Oxide	Dichloropropenes
Propane	Dichloropropanes	Trichloropanes
Hydrogen	Ethylene Oxide	Ethylene
Water Vapor	Propylene	Dichloropropanol
Hydrogen Chloride	Other VOCs	

Based on site and business needs, the following requirements had to be met with the new central vent gas thermal treatment unit:

1. Provide on-stream reliability comparable to or better than existing thermal oxidizer technology

2. Operational flexibility to handle the varying vent composition and flow rate

3. Meet new air permit conditions

4. Produce steam

5. Recover aqueous hydrochloric acid

The infrastructure changes associated with a central vent gas treatment unit included the collection of process vents from the major manufacturing facilities within Plant B in vent headers. The vents would then be conveyed by blowers through multiple fume headers to the central thermal treatment facility. Reactive chemicals and stream compatibility studies required the fume headers to be routed through three (3) primary headers, and into three (3) knockout drums to minimize liquid transport to the new thermal oxidizer. Additional safety designs included deflagration arrestors (DFA) mounted in parallel at the connection to each of the new thermal oxidizers. The parallel arrangement of the DFAs was chosen so that vent control could continue in the event one DFA would become plugged with contaminates in the vent line.

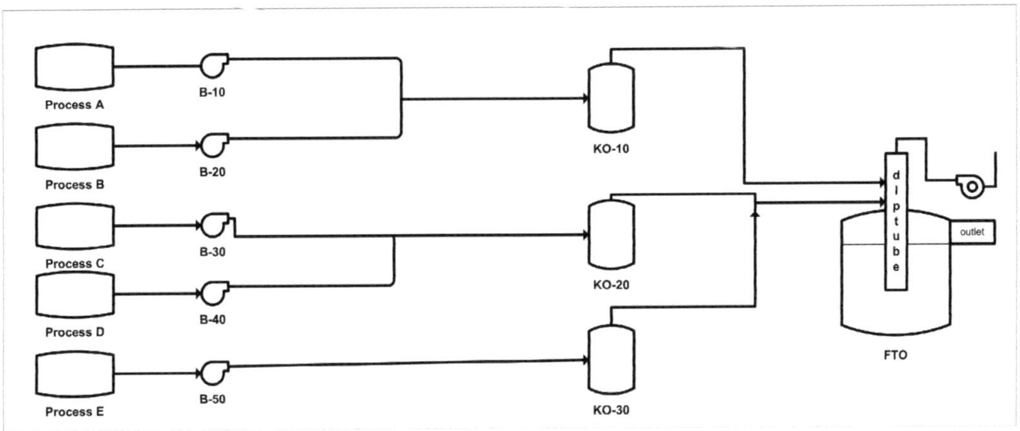

Fig. 2. Simplified Plant B Vent Collection Schematic for transfer of vents to proposed central vent gas thermal treatment unit.

Although the current thermal oxidizers that serviced each source in Plant B were operating within their respective requirements, Dow Chemical was interested in designing for a further reduction in emissions, in anticipation of future permitting and environmental needs. A plant-wide inventory of emissions was conducted, and a challenging goal of significant reductions was put forth. The requirements for the new Plant B central vent gas thermal treatment unit were issued as follows:

- VOC destruction efficiency >99.99%
- Maximum CO emissions 12.5 mg/Nm^3
- Maximum NO_x emissions 4.53 MTPY
- Maximum HCL emissions 4.08 MTPY
- Maximum CL2 emissions 2.04 MTPY

Waste heat recovery producing saturated steam and the production hydrochloric acid was added to the specifications.

Technology Description

Flameless Thermal Oxidation (FTO) was originally developed as a super-efficient energy recovery technology. Initial development was conducted at the Lawrence Livermore Laboratories, under the auspices of the United States Department of Energy, in conjunction with the oil shale project. When that project was aborted, further refinement was supported by private funding and with the expertise of individuals originally associated with the project, the technology was commercialized in the early 1990's. FTO technology has repeatedly demonstrated an organic waste destruction efficiency of >99.99% while producing virtually undetectable NOx and CO emissions. This unique performance assures regulatory compliance with a high degree of reliability.

How the FTO works

The flameless thermal oxidation process thermally reacts waste gases with air to convert organic compounds to their oxidized state, and release heat without producing a flame. This is accomplished by heating the compounds above their auto ignition temperature under controlled conditions in a specially designed reactor that absorbs and dissipates the heat of reaction. A stable oxidation zone is maintained at a precise temperature and residence time sufficient to compensate for variations in waste flow and composition.

The reactor in Figure 3. is made up of a refractory lined, cylindrical vessel (material of construction varies with application) with a central inlet diptube. The vessel is approximately half full of randomly packed inert ceramic media that forms a uniform matrix below and around the diptube.

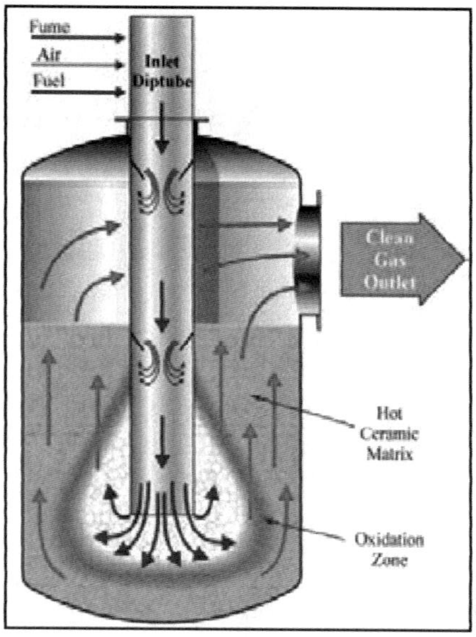

Figure 3. FTO Vessel Schematic

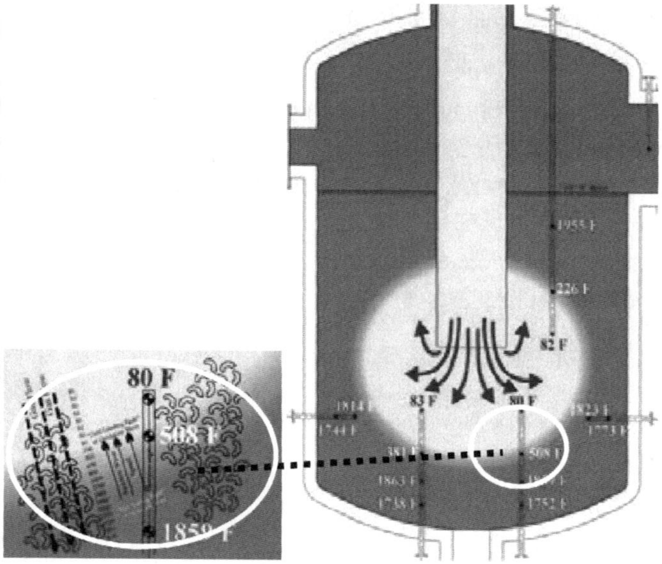

Figure 4. Reaction wave within the FTO

The volume of the chamber and the amount of ceramic matrix above the oxidation zone is designed to ensure a retention time of 2 to 4 seconds, at rated capacity. The system operating turndown typically ranges from 100% to 33%. Higher turndown of up to 15% or 8:1 is available with an advanced diptube design (patent pending). The instantaneous waste flow turndown during any operating range is 100% for maximum process flexibility.

To ensure the reaction temperature in the flameless oxidizer is controlled uniformly throughout the oxidation reaction, the process vents, combustion/dilution air, and auxiliary fuel are premixed to a 99.9% Coefficient of Variability prior to entering the reaction zone. This eliminates localized pockets of reactants that could produce temperatures higher than 1285°C, which could produce unwanted NOx formation. Air is added to provide sufficient oxygen for the oxidation reaction if the process vents themselves do not contain the correct amount of air and to ensure the mixture does not exceed 85% of the LFL. The mixture flows through a cylindrical diptube located in the center of the flameless oxidizer vessel. This diptube is imbedded in ceramic packing. The shape of the packing is selected based on the optimization of pressure drop and void fraction. The composition of the packing is selected for the operating temperature and reaction chemistry; both developed in a joint approach.

Inserted through the vessel shell, and into this packing are multiple thermowells, each containing thermocouples. The length and placement of the thermowells is based on the size of the reaction zone and the modes of operation. The thermowells are constructed of high nickel alloys used in high temperature applications. The thermocouples are used to measure the temperature of the reaction for control purposes.

At start-up, the ceramic packing is preheated to a temperature that is sufficiently above the auto ignition temperature of the fuel and air mixture utilizing a conventional Low NOx gas burner. From a cold start approximately 32 hours is required to reach temperatures set by NFPA recommendations. Hot restarts can be done within shorter periods without the need to reuse the burner. When this temperature is achieved, the burner is shut down and the system begins to establish the profile of the oxidation reaction within the media. The oxidation reaction in the flameless oxidizer is the same as in a flame-based system (except that the reaction is occurring at a lower temperature). The reaction is exothermic, and produces the normal oxidation products such as CO_2, H_2O, N_2, O_2, HCL, and CL_2. As the reaction takes places, it moves in the direction of the incoming reactants at a velocity that is equivalent to the flame speed of the given composition (and is influenced by such factors as temperature, pressure, and fuel to air ratio). Such reaction velocities are typically in the 0.75 ft/s to 2 ft/s range. By controlling the air flow, the reactants are fed at the same velocity as the reaction velocity, and the reaction is balanced as shown in the simplified Figure 4. above.

In the flameless oxidizer the heat that is liberated during this process is transferred to the ceramic packing, and re-radiated back to the incoming reactants. While the packing provides a radiant surface, it also decreases the Coefficient of Variability of the reactants as they flow through its labyrinth, and provides a surface area for anchoring the reaction in a controlled manner. The entire flameless reaction takes place in approximately 12-inch-wide wave. This is the linear distance in the packing in which the reactants are raised from the incoming temperature (typically 80°F) to the final temperature (approximately 1800°F). The reaction wave forms a symmetrical bubble within the ceramic packing that surrounds the dip tube. One hundred percent of the incoming mixture must pass through the reaction wave before exiting the oxidizer vessel. The shape of this bubble is influenced by multiple design characteristics of the flameless oxidizer. The packing outside of the bubble is at the resultant reaction temperature and acts as a "thermal flywheel" absorbing and giving of heat dampening variations resulting in greater operational reliability.

The Dow Chemical FTO Units

In 2016 the Oyster Creek VCM facility was purchased by Olin Chemical along with the FTO system however Dow Chemical retained ownership and operation of the Plant B FTO system.

The Plant B FTO was commissioned in the spring of 2002. Several months were needed to verify control settings and provide extensive operator training. When the vent collection system was brought on line November 2002, process vents were introduced into the FTO system. Performance test were conducted. A test protocol was developed to meet both the requirements for a technology performance evaluation and legal requirements.

The FTO was tested with a chlorinated VOC spike consisting of 68 kg/hr methyl chloride, a Class 1 POHC, methylene chloride, and propane. Pollutants measured were: HX (HCl) and X2 (Cl2), (total hydrogen halides and halogens), total Organic Hazardous Air Pollutants (OHAP's), total VOC's, Methyl Chloride, Methylene Chloride, and Propane for VOC DRE. The test was performed on March 2003. The test consisted of two one-hour runs. The test results are reported in Table II below.

Table II. Plant B FTO emission test results from performance testing on March 26, 2003

SAMPLE TYPE	TEST METHODS	SAMPLING TIME (MINUTES)	ACTUAL EMISSION RATE Run 1	ACTUAL EMISSION RATE Run 2
Total Hydrogen Halides & Halogens (kg/hr)	EPA M26A	60	0.086	0.621
HCl (kg/hr)	EPA M26A	60	0.063	0.081
Cl_2 (kg/hr)	EPA M26A	60	0.022	0.539
Total VOC's * (kg/hr)	EPA M0040	60	0.002	0.002
VOC DRE (Methyl Chloride)	EPA M0040	60	99.999	99.999
Total VOC DRE	EPA M0040	60	99.999	99.999

A second test was performed to demonstrate compliance requirements with the Texas Commission on Environmental Quality (TCEQ) air permit. This test, conducted in April 2003, represented worst-case currently obtainable conditions by operating the oxidizers at a minimum temperature of about 938°C, and operating chlorine scrubber at low pH and recycle flow. Pollutants measured were HCl and Cl2, Total VOC's including VOC DRE, Nitrogen Oxides (NO_x), Carbon Monoxide (CO), Sulfur Dioxide (SO2) and PM (Particulate Matter).

VOC sampling was conducted on the 2 inlet vents lines to the FTO in compliance with the EPA Method 1 location requirements. The number and location of the sampling and traverse points were determined according to the same method. Method 1 required a total of 4 traverse points. The test results are reported in Table III.

Table III. Plant B FTO emission test results from performance testing on April 14, 2003.

SAMPLE TYPE	TEST METHODS	SAMPLING TIME (MIN) (per/run)	ACTUAL EMISSION RATE*
HCl (kg/hr)	EPA M26A	60	< 0.031
Cl_2 (kg/hr)	EPA M26A	60	< 0.022
Total VOC's (kg/hr)	EPA Modified M0040	60	< 0.0005
VOC DRE (%)	EPA Modified Method 0040 and Method 18	60	99.9998
NO_x (kg/hr)	EPA M7E	60	0.09
CO (kg/hr)	EPA M10	60	< 0.031
SO_2 (kg/hr)	EPA M6/8	60	< 0.0005
PM (kg/hr)	EPA M5	60	0.548

* The actual emission rate reported is based on the average of three sampling runs.

Operational Flexibility

Since start-up, the FTO has managed periods of steady vent flows and process upset conditions. Initially the unit was operated to accommodate the worst case design vent flows. The design flow is set by the constant air volume set point, while the oxidation zone is maintained by support fuel. A high air constant volume set point during periods of low vent flow results in high support fuel requirements. After a year of operation Dow performed statistical analysis on process variations and revealed the opportunity to turn down the unit 30% to conserve on support fuel while still being able to respond to process variation.

The FTO is able to handle short duration process excursions in excess of design. One particular case involved a process upset at one of the sending plants where an excessive amount of high energy vent was sent to the Plant B FTO unit. The fuel gas ramped down rapidly to its minimum flow set-point in response to temperature control. The event occurred for less than 10 minutes. During this time the reaction zone shrunk until the event was over, but the unit did not trip on high temperatures as might have been expected.

In another case the FTO oxidizers received a surge of low energy vent flows. This condition was thought to expand the reaction zone beyond the control limits. In this case it demonstrated the FTOs response to the excessive fume flow with a surprise. While the reaction zone did expand, the thermal inertia of the large ceramic beds tempered the dynamic changes of the FTOs while the excessive fume flow was steadily being ramped-down at the process sources.

Long Term Operational and Maintenance Experience

The two systems have now been operating for nearly fifteen years. Dow continues to follow the original control settings from start-up. The on line availability of the FTO reactors is estimated to exceed 99%. The only maintenance points are the thermocouples and there is sufficient redundancy to allow for uninterrupted operation. The readings from periodic stack sampling show that NO_X emissions are less than 2 ppmv, CO less than 5 ppmv, and the Destruction Efficiency for VOC's and all hydrocarbon species exceeds 99.99% (all on corrected basis).

One concern has been the mechanical reliability of the units. In particular Dow had estimated the usable life of the ceramic packing would be 3 to 5 years. A ten-year inspection was performed which included removal and replacement of the ceramic media. The actual finding was that the media on whole was still usable and did not need replacement. There was a little build-up of polymer on the media but it was not enough to affect the operation of the units. The internal insulation was still intact and was not replaced. The diptube showed no deterioration. Thermowells that could not be replaced during operation due to bed shifting were replaced along with the thermocouples.

Technology Improvements

Through the ensuing years since start-up the technology provider (Selas Linde North America) has kept abreast of the operation experience. Based on that information they have continued to develop the technology to add reliability and safety. For instance, an improved diptube design has been submitted for patent that increases turn down and prevents flashback that can occur when sudden changes in composition results in a change in the LFL within the diptube. Additionally, control schemes have been instituted to reduce reliance on the bed thermocouples to more assessable instrumentation.

Summary

Dow considers the installation of the FTO systems to be a success. The 15-year experience with operation and maintenance has exceeded the expectations at the time of installation. The emissions from the unit still meet or exceed the current and future expected permit levels. On whole there is high confidence that the FTO's will continue to allow Dow and Olin to carry out successful plant operation in an environmentally acceptable manner.

AIChE Paper Number: 194f

The Impact of Discharge Criteria on Spent Caustic Treatment Options

William Celenza

Senior Technical Advisor – Environmental

KBR, Inc.

Prepared for Presentation at the 2017 Spring National Meeting
San Antonio, Texas, March 30, 2017

The Impact of Discharge Criteria on Spent Caustic Treatment Options

William Celenza
Senior Technical Advisor – Environmental
KBR, Inc.

Abstract:

Sodium hydroxide, in caustic solutions, is used in the hydrocarbon processing industries to extract and neutralize gases such as hydrogen sulfide and carbon dioxide along with organic acids from hydrocarbon streams. The composition of the produced spent caustic depends on the process generating the waste and on the source of the hydrocarbon stream being washed. Spent caustic is generally classified into three categories: Sulfidic, Naphthenic, and Cresylic. Although all spent caustic streams are environmentally hazardous and require special handling; the sulfidic spent caustics generated at oil refineries and olefins plants are typically easier to manage. Olefins plant sulfidic spent caustic contains hydrogen sulfide along with some mercaptans, phenols, benzene and emulsified oils. The toxic properties of spent caustic need to be mitigated for additional downstream treatment in a conventional wastewater treatment plant; however, the mitigation must not only lead to an effluent acceptable for biological treatment but also to a level that will enable the downstream treatment scheme to be able to produce an acceptable discharge. Thus discharge limits for "conventional" pollutants such as biochemical oxygen demand (BOD), total suspended solids (TSS), pH, and oil and grease (O&G); for "toxic" pollutants such as the priority pollutants; and for "nonconventional" criteria such as whole effluent toxicity all need to be considered. Furthermore, sulfidic spent caustics contain high concentrations of sodium sulfide and sodium bisulfide that when biologically converted to thiosulfates are disruptive to the removal of other organics, in addition to lowering pH.

Even though pretreatment systems at a minimum convert sulfides to thiosulfates to prepare for biological treatment, the thiosulfates still contribute high organic loads, which may be difficult to treat. Thus a reduction of sulfidic spent caustic COD to 1,000 mg/L (Alnaizy 2008 and Hawari 2015) may be required to readily allow for compliance with discharge limits. This is considered a higher end value as influent to a downstream biotreatment scheme. Assuming a pretreated sulfidic spent caustic BOD/COD ratio of 0.4, the minimum suggested for biological treatment, the BOD would be reduced to 400 mg/L by pretreatment. If discharge criteria include an average monthly limit of BOD of 10 mg/L, then even assuming a dilution of the pretreated sulfidic spent caustic with other wastewater streams to a mixed BOD level of 100 mg/L would require a 90% minimum average removal. Thus choosing a pretreatment option should consider more than the minimum acceptable influent to a biologically based treatment plant. This paper discusses sulfidic spent caustic pretreatment options with consideration to optimizing the overall treatment scheme for a typical olefins plant. Working back from existing discharge limits faced by several olefins plants, and considering typical biological treatment efficiencies for these plant wastewater streams, the capabilities of different pretreatment technologies will be compared.

Spent Caustic

Olefins, unsaturated synthetic hydrocarbons, such as ethylene, propylene, and butene products are the largest volume materials produced in the petrochemical industry. In practice, Olefins in the US are mainly manufactured by thermally cracking hydrocarbons which contain a quantity of sulfur. As sulfur is not desired in the products and not destroyed in the cracking process, the resulting cracked gas needs to be scrubbed in a caustic wash tower to remove acid gases. The purge from the caustic scrubbing tower constitutes the spent caustic that includes sulfides, carbonates, mercaptans, phenols, and other soluble or emulsified organic compounds as listed in Table 1. These sulfidic spent caustics which are produced as a liquid stream are usually considered nonrecoverable. Table 1 includes a "typical" range for common parameters of olefin plant sulfidic spent caustic, and values outside of the listed range have been reported typically extending above the higher end.

Table 1 Typical Olefin Plant Sulfidic Spent Caustic Composition (Grover 1993 and Gudideni 2011)

Parameter	Formula	Weight Percent
Sulfates (1)	SO_4^{-2}	6 -- 8
Carbonates (2)	CO_3^{-2}	3 – 4.5
Sulfides (3)	S^{-2}	1.5 – 3.5
Free Sodium Hydroxide	NaOH	1.0 -- 2.5
Mercaptans (4)	RSH	0 – 0.11
Sodium	NaCl	0 – 0.05
Oil and Polymers	C_xH_y	0 – 0.04
Phenol	C_6H_6O	0 – 0.025
Benzene (5)	C_6H_6	0 – 0.01
Water	H_2O	Balance
Chemical Oxygen Demand (COD) (6)	NA	0.8 – 12
Biological Oxygen Demand (BOD)	NA	0.1 – 1.5
Specific Gravity	NA	1.1

pH	NA	>10 S.U.
Conductivity	NA	1120 Micro s/cm
Temperature (7)	NA	100 to 130 °F

1. Sulfates may be present as sulfuric acid (H2SO4), sodium sulfate, and organic sulfates.
2. Carbonate results from the absorption of CO2 gas in the caustic tower. Small amounts of CO2 are produced in the cracking furnaces through steam carbon reforming reactions, and feedstocks typically contain varying amounts of CO2. Carbonate also serves as a pH buffer and helps regulate the pH.
3. The sulfide ion may be present in spent caustic in several forms, including hydrogen sulfide, sodium sulfide, and organic sulfides. However, in alkaline caustic solution over a pH of 9, all the sulfides can be assumed to exist as bisulfide ion HS^-.
4. R represents an alkane, alkene, or other carbon-containing moiety
5. The benzene concentration in spent caustic may be a high enough concentration to cause the entire stream to be regulated under the NESHAP for benzene waste operations. Therefore, in some cases either benzene must be removed from the spent caustic prior to treatment or any off-gas must be treated before discharge to the atmosphere.
6. Sulfides make up the bulk of the spent caustic COD content. Also the COD is soluble, since the suspended solids do not exert much COD. Analytical tests by (Moussavi 2016) indicated that solids are metal oxide / hydroxide resulting from water hardness or corrosion products of pipelines and fittings. Some solids may also have entered the system by local wind and particulates with silt and sand origin.
7. Spent Caustic leaves the stripper typically at a temperature in this range.

The exact composition of the spent caustic will depend on a number of factors including: feedstock type, where and how the type of feedstock was derived, furnace operating conditions such as temperature, residence time, steam / hydrocarbon ratio, and wash tower operation. Steam cracking is performed both on liquefied petroleum gases (LPGs) such as ethane, propane, and/or butane, as well as on liquid feedstocks such as naphtha, gas oil, and other petroleum condensates containing paraffin and naphtha. Acid gases are typically scrubbed on a once through basis or in combination with a regenerative chemical. Regenerative pre-scrubbing, before a final sodium hydroxide treatment, can be used for high sulfur feedstocks to reduce the caustic consumption. Also, raw spent caustic may be stripped with fuel gas in a hydrocarbon stripper to prevent wash oils and polymer from accumulating in the caustic wash tower. Table 2 highlights the sulfur compositions of several feedstocks from different sources.

Table 2 Typical Sulfur Composition of Feedstocks Used for Olefin Production

Component	Naphtha	Kerosine	Gas Oil	Gas Oil	Naphtha	Bio-Ethanol
Origin	Kuwait	North Sea	Kuwait,	Tia Juana	Khafji	USA

			North Sea			
Sulfur content wt%	0.04	0.03	0.1	0.18	0.02	<0.001
Reference	Totten 2003	Totten 2003	Totten 2003	Speight 2000	Speight 2000	USEPA Tier 3 Gasoline Sulfur Regulations (80.1610)

Discharge Limitations

Included in Table 3 is a summary of the discharge criteria for the main process outfall of a Gulf Coast Olefin Facility with a nameplate capacity of 890 MM lbs/yr (400 kta) Ethylene. The sanitary wastes, and all of the non-industrial area storm water are managed separately from this outfall. However, all site cooling tower blowdown and water treatment effluents are managed through this outfall.

Table 3 Evolution of Permitted Wastewater Discharge Criteria for a Gulf Coast Olefin Facility

Criteria	Prior to 1971	1971	1985 (2)	1999	2005
Temp ^oF	Before secondary treatment facilities, only primary treatment for removal of oil then collection of wastewater by a surface drainage ditch to a common	100 max	-----	-----	-----
Flow, MGD		-----	-----	Report	Report
pH (1)		6 to 8.5	6 to 9	6 to 9	6 to 9
COD, lb/day		-----	700 / 1800	-----	-----
BOD, lb/day		85% reduction	100 / 250	100 / 250	100 / 250
Oil, lb/day		No floating oil	30 / 75	No floating solids or visible foam	No floating solids or visible foam
TSS, lb/day		No harmful effects	80 / 175	200 / 600	200 / 600
Ammonia, lb/day		No toxic effects	55 / 170	-----	-----
Sulfide, lb/day		-----	0.5 / 1.6	-----	-----

Phenol, lb/day	holding pond before discharge	No taste or odor effects	0.7 / 2.0	0.03 / 0.06	0.04 / 0.07
Dissolved O2, mg/l		50% of saturation, not less than 4 mg/l	-----	-----	-----
Dioxin (3)		------	-----	20 ppq	-----
VOAs and Base Neutral / Acid Compounds i.e. Benzene lb/day		------	-----	Per EPA listings 0.08 / 0.31	Per EPA listings 0.08 / 0.31
Whole Effluent (Chronic) Toxicity		------	-----	Report	Report
Total Max Daily Loads (TMDLs) – lb/day Total Copper Total Mercury Bromoform Tetrachloroethane PCB-1254 Hexachlorobenzene Hexachlorobutadiene		------	-----	------	 0.1 0.001 4 0.2 0.000001 0.00003 0.012

1. pH range excursions not to exceed 60 minutes per event, with monthly total accumulated time of all excursions to be less than 450 minutes.
2. (USEPA 1985), calculated values per applicable "Effluent Limitation Guidelines" Cracking Subcategory Section 419.22. Not actual permitted values.
3. Dioxin required for any facility managing natural gas pipelines.

Resultant Treatment System Design Requirements

Spent caustic solutions have an inorganic component (i.e. sulfides) and an organic component that both contribute to Biological Oxygen Demand (BOD) and Chemical Oxygen Demand (COD). Sulfide is toxic to microorganisms and known to inhibit several enzymatic systems, thus pretreatment is essential to minimize its impact on a biotreatment system. Pre-treament by neutralization and off-gassing or detoxification by oxidation is often applied to spent caustic.

Spent caustic is also considered a high strength wastewater because of biodegradability criteria, such as an untreated BOD / COD ratio of 0.1 to 0.2, predicting slow biodegradability. A low ratio indicates the level of available organic matter to be degraded by biomass is low, with a ratio of 0.5 or above for untreated wastewater considered readily biodegradable. This slow biodegradability is caused by inhibitory or toxic effects from sulfides, thiosulfates, large molecules, and a high content of non-biodegradable compounds. A ratio less than 0.5 for untreated wastewater is another indicator that physical or chemical treatment is recommended before biological treatment. Note the 1985 column values in Table 3 are derived from the "Guide for the Application of Effluent Limitation Guidelines" for the Petroleum Refining Point Source Category" Subpart B, Cracking Subcategory (USEPA 1985). These EPA guidance values anticipate a "final treated effluent" BOD to COD ratio of 0.14 for the overall process wastewater flow from an Olefins facility in this category. A "treated effluent" BOD to COD ratio near 0.1 indicates a stabilized secondary, or near complete biologically treated, effluent. Thus the expectation is that BOD made available through pre-treatment could readily be removed during biotreatment from a properly designed and operated system.

Sulfides present in the spent caustic can readily be detoxified by oxidation to thiosulfate. Thiosulfate is the main product of chemical sulfide oxidation under alkaline conditions. Sulfate is the sole end product of sulfide oxidation provided that a complete conversion takes place. Oxidation of thiosulfate to sulfate is the rate determining step for overall conversion to sulfate. Sulfate is the preferred end of oxidation since sulfate does not exert any oxygen demand and is thus BOD and COD neutral.

Although biological oxidation of thiosulfates is possible in conventional biotreatment facilities, a 1,000 ppm load of thiosulfates can generate up to 575 ppm of BOD and 625 ppm of COD (Stephenson 1988 and Millano 1985). Also, protons formed during sulfate formation will result in pH neutralization. Thus, thiosulfate bio-oxidation triggers low pH conditions which impact the microbial population favoring autotrophic microbes using sulfur as an O_2 source at the expense of the heterotrophic microbes that contribute to organic BOD removal (Stephenson 1988). As such thiosulfates are usually not readily stabilized to the sulfate form in a biotreatment unit's aeration system, which can lead to a significant level of residual BOD and COD. Also, since the biological oxidation of thiosulfates can result in a reduced ability to remove the other "organic" BOD due to low pH conditions; an aeration basin would be required to include pH control. Finally, thiosulfates have been observed to impact the ability of sludge to settle.

A reduction of sulfidic spent caustic COD to 1,000 mg/L, (Alnaizy 2008 and Hawari 2015) which is considered a higher end value as influent to a downstream biologically based treatment scheme, may be sufficient to readily allow for compliance with discharge limits. A case study on the treatment of ethylene plant spent caustic by Hawari et al also targeted COD and sulfide concentrations at 1,000 mg/l and 2 mg/l, respectively, as pre-treatment requirements for biotreatment. At these concentrations the authors indicated that treated spent caustic can be handled by conventional biotreatment.

The following analysis for the Gulf Coast Olefin Facility from Table 3 presents that a value near this limit may indeed be required.

Back-calculating from the BOD permit limit in Table 3 would provide the maximum BOD loading attributable to the pre-treated spent caustic effluent, as shown in Table 4. This BOD loading would be the maximum attributable to the Wet Air Oxidation (WAO) effluent such that the existing site Wastewater Treatment Plant (WWTP) operation would be expected to routinely meet the State Permit limit.

Table 4 Back-Calculating Max BOD loading into a conventional WWTP to meet discharge limits

Parameter	Permit - Monthly Average Limit		Basis of 90% removal across WWTP	Parameter Loading from non-spent caustic sources	Resultant Parameter Loading available to pre-treated spent caustic
BOD	Lbs/ day	103	1,030	721	309 (b)
	mg/l	11 (a)	114	80	2,576 (c)

(a)	Calculated based on a 750 gpm maximum continuous flow at outfall of WWTP. Note the average continuous flow is anticipated to be around 700 gpm. The Permit limits are listed only in lbs/day.
(b)	By differential: Total Influent BOD to WWTP - BOD loading attributed to non-spent caustic sources
(c)	Calculated based on a 10 gpm spent caustic flowrate.

Thus pre-treated spent caustic effluent would be able to exert up to 2,500 mg/l of BOD without adversely impacting the biotreatment plant's ability to routinely meet the BOD discharge criteria.

Table 2 below shows that a minimum 87.5% conversion of sulfide in the feed to sulfate would be required to routinely meet the WWTP Permit BOD limit.

Table 5 Determination of maximum BOD loading from Pre-treated Spent Caustic

	Sulfide loading in spent caustic influent to pre-treatment at 3 wt. %	Assuming 100% Conversion of sulfide in spent	Assuming 87.5% of Thiosulfate is further converted	BOD Loading attributable to Thiosulfate in pre-

	Sulfide		caustic to Thiosulfate	to Sulfate	treated effluent to biotreatment
BOD	Lbs/day	3,944 Sulfide (a)	3,995 Thio-sulfate	499 Remaining thiosulfate in pre-treated effluent	300 BOD in effluent
	mol /h	2.1	1.05 (b)	0.13	
	mg/l			4,158	2,495 (b)
(a)	Calculated based on a 10 gpm spent caustic flowrate or 131,472 lbs/day				
(b)	0.6 mg/l BOD per 1 mg/l of sulfur as thiosulfate (Stephenson 1998 and Millano 1985).				

This balance is sensitive to several input parameters, such as the BOD reduction across the WWTP's biotreatment unit, as well as the sulfide level in the raw spent caustic. For instance, as listed in Table 6, a 95% BOD removal across the WWTP biotreatment unit would allow up to 1,300 lb/day of BOD to be contributed from the pre-treated spent caustic. Thus the spent caustic pre-treatment system would then need to only achieve up to 45% conversion to sulfate, or equivalent. An example of an Ethylene Facility WWTP with 95% BOD removal capability is the Membrane Bioreactor Technology (MBR) wastewater treatment and reuse scheme at the ExxonMobil Singapore Chemical Plant (Cashion, 2008; and Qin, 2008).

Table 6 Sensitivity analysis to spent caustic pre-treatment technology design requirements

	Parameter	Change	Resultant spent caustic BOD load allowed to biotreatment	Resultant pre-treated spent caustic conversion to sulfate
1.	BOD removal across WWTP biotreatment	From 90% to 95%	1,300 lb/day	45%
2.	Sulfide concentration in raw spent caustic	From 3% to 1.5%	300 lb/day	75%
3.	Both 1. and 2.	Both 1. and 2.	1,300 lb/day	5%

This analysis does not directly consider the other sulfur compounds in raw spent caustic; however, criteria for selecting pre-treatment technologies can also include the ability to manage mercaptans, phenol, and hydrocarbon compounds. Additional concerns regarding odors, toxicity, and worker exposure exist with these compounds also.

Sulfidic spent caustic generated from an olefin plant is typically managed by one of the following methods:

1. Physical Treatment
    - Deep neutralization to a pH of 4 or lower, coupled with venting offgas to emission control followed by high dilution before biotreatment.
    - Neutralization followed by Steam Stripping, coupled with venting offgas to emission control followed by dilution before biotreatment.
    - Chemical precipitation followed by dilution before biotreatment. Depending on the pH adjustment required, if less than 8 then venting offgas to emission control would be required.
2. Chemical Oxidation
    - Chemical Oxidation followed by dilution before biotreatment. Depending on the pH adjustment required, if less than 8 then venting offgas to emission control would be required.
    - Neutralization followed by Fenton Oxidation, coupled with venting offgas to emission control followed by dilution before biotreatment.
3. Wet Air Oxidation
    - Low or Medium Temperature WAO, total oxidation of sulfide to thiosulfate, with partial oxidation of thiosulfate to sulfate, followed by dilution before biotreatment. Venting offgas to emission control.
    - High Temperature WAO, total oxidation of sulfide to sulfate, followed by dilution before biotreatment. Venting offgas to emission control may still be required.
4. Thermal Oxidation / Incineration

These technologies are presented in more detail in the following table:

Table 7 Comparison of Different Pre-treatment Methods

Pre-Treatment Method	Pros	Cons
Physical Treatment followed by biotreatment		
Deep Neutralization	Very low CAPEX Dissolved sulfides converted to H2S gas phase and vented	Dilution of high residual sulfides and mercaptans loading required to avoid disruption of biotreatment High BOD and COD since organics unaffected Large equalization and off-spec storage systems Enhanced monitoring of biotreatment system Acidification accelerates formation of oil polymer by-products Acid Gas handling - Vent gas emission control due to offgassing of H2S, mercaptans Strong odor and health concerns from residual sulfur compounds
Neutralization followed by Steam Stripping	Steam stripping removes additional sulfides and mercaptans after neutralization Low CAPEX	High BOD and COD since most organics unaffected Acidification accelerates formation of oil polymer by-products Acid Gas handling - Vent gas emission control due to offgassing of H2S, mercaptans Odor and health concerns from residual sulfur compounds Steam needs

Chemical Precipitation	Removal of sulfides Removal of emulsified oils and TSS Low CAPEX Low BOD and COD	High OPEX due to chemicals consumption Large chemical sludge generation Management of corrosive and chlorine compounds If pH adjusted to below 8, acid Gas handling - Vent gas emission control due to offgassing of H2S, mercaptans Odor and health concerns from residual sulfur compounds
Chemical Oxidation followed by biotreatment		
Chemical Oxidation by hydrogen peroxide	Oxidation of sulfides Low CAPEX Low BOD and COD	High OPEX due to chemical reagent consumption If pH adjusted to below 8, acid Gas handling - Vent gas emission control due to offgassing of H2S, mercaptans Potential odor and health concerns from residual sulfur compounds
Neutralization followed by Fenton Oxidation	Dissolved sulfides converted to H2S gas phase and vented Oxidation of residual sulfides and organics Low CAPEX Low BOD and COD	High OPEX due to chemical reagents consumption Acid Gas handling - Vent gas emission control due to offgassing of H2S, mercaptans Potential odor and health concerns from residual sulfur compounds Ferrous-containing sludge

		management.
Wet Air Oxidation followed by biotreatment		
Low Pressure 4 – 7 barg and 75 – 100°C Reaction residence time 4 to 6 hrs.	Conversion of sulfide to thiosulfate Further conversion of half the thiosulfate to sulfate Plant Air may meet air supply needs	High CAPEX Only half the thiosulfate gets further converted to sulfate so high BOD and COD and large effluent dilution still required Mercaptans not converted or removed Vent gas emission control due to H2S, mercaptans, organic acids, and methanol Organics not removed Oil polymer by-products not removed Foaming potential
Medium Pressure 7 – 10 barg and 100 – 120 °C Reaction residence time 2 to 4 hrs.	Conversion of sulfide to thiosulfate Further conversion of up to 70% of the thiosulfate to sulfate Partial oxidation of organics	High CAPEX Only up to 70% of the thiosulfate gets further converted to sulfate so high BOD and COD and effluent dilution required Mercaptans not converted or removed Vent gas emission control due to H2S, mercaptans, organic acids, and methanol Residual Organics Oil polymer by-products not removed

		Foaming potential MP Steam needs
High Pressure 14 – 30 barg and 175 – 250 °C Reaction residence time 1 hr.	Conversion of 99% of sulfide and mercaptan to sulfate Low BOD and COD Organics mostly removed Oil polymer by-products partially oxidized Vent gas free of H2S and mercaptans	Very high CAPEX and OPEX Residual oil polymer by-products HP Steam needs Neutral brine effluent still requires dilution Vent gas emission control may still be required.
Thermal Oxidation / Incineration		
Vertical Spray Incinerator 0.2 barg Combustion Temp 800 to 1,000°C Residence time 1.5 to 2 sec	Complete oxidation of sulfides to sulfates Complete oxidation of organics to CO2 and H2O Inorganics converted to molten salt Very low COD	High CAPEX, very high OPEX as auxiliary fuel needed NOx, SOx, and PM10 emissions Injection nozzle cleaning monthly

The base case analysis required over 85% conversion of spent caustic sulfide to sulfate. High temperature WAO technology can convert over 90% spent caustic sulfide to sulfate in one step. See Table 8 for a typical treatment specification for a high temperature WAO. Even if the raw spent caustic sulfide loading was only 1.5% by wt., the required conversion would be around 75% which is near the maximum typically achievable from a medium temperature WAO. For instance, in typical medium temperature WAO effluents, half to two-thirds of the COD is attributable to thiosulfate COD. However, even a low temperature WAO which still converts all of the sulfide to thiosulfate also converts about half of the sulfide to sulfate. Mercaptans and disulfides would not be removed from the liquid; and the vent gas would contain hydrogen sulfide, mercaptans, and low-molecular weight compounds such as acetaldehyde, acetone, acetic acid, and methanol. Thus the vent gas would need to be routed to a control device such as a flare, boiler, or furnace. Only a thermal oxidizer / Incinerator would oxidize most of these compounds.

Table 8 Typical parameters used in specifying a high temperature WAO unit

Effluent	Symbol	Guarantee (mg/l)
Sodium Sulfide	Na_2S	1 max
Sodium Thiosulfate	$Na_2S_2O_3$	1,000 max
Mercaptans	NaRS	<1
Disulfide	S_2	<1
Phenol	C_6H_6O	10 max
Oil & Polymer	CxHy	100 max
Chemical Oxygen Demand	COD	1,500 max

Note that 1.5% by wt. of sulfide is the lower end of the range for sulfidic spent caustic streams generated by an olefins facility, and would likely require a facility to have selected a light feedstock such as an LPG component. However, a lower sulfur content and better BOD removal capability in a biotreatment system may favor the use of a physical and / or chemical pre-treatment step; considering the reagent operating costs are proportional to the constituent loading in the spent caustic. In any case, considerations for "nonconventional" criteria such as whole effluent toxicity should also be included when analyzing pre-treatment choices. The direct cause(s) of observed toxicity can be difficult to determine. Thus a pre-treatment option that provides wider constituent destruction, such as a WAO or thermal oxidation, may provide more tangible benefits than were measured in the past. And the reduction of COD in pre-treatment to 1,000 mg/l may help provide operating margin to ensure compliance with all discharge criteria.

Concluding Thoughts

Current environmental standards have a significant impact on spent caustic treatment system design. Spent caustic can contain high concentrations of toxic constituents that make achieving required effluent qualities through conventional biological treatment systems difficult even after neutralization and dilution. All operations along the path of managing the sulfur in the feedstock, from feedstock preparation, to caustic wash tower and/or pre-scrubbing tower design/operation, to site ultimate discharge options should be reviewed to complete a spent caustic treatment analysis.

In summary, a reduction of sulfidic spent caustic COD to 1,000 mg/L, (Alnaizy, 2008; and Hawari, 2015) which is considered a higher end value as influent to a downstream biologically based treatment scheme, should be sufficient to allow for compliance with most discharge limits.

As the analysis in this paper confirms for a Gulf Coast Olefin Facility, pre-treatment of raw sulfidic spent caustic to a COD between 1,000 mg/l to 10,000 mg/l would be required to enable the required downstream treatment through a conventional biotreatment plant.

References:

Alnaizy, Raafat (2008) "Economic Analysis for Wet Oxidation Processes for the Treatment of Mixed Refinery Spent Caustic," *Environmental Progress* V27 No 3 October, pp. 295-301.

Cashion, Bryan S., Robert J. Wenta, Michele Migliavacca, Mike Howdeshell, Guihe Tao, Kiran A. Kekre, and Jianjun Qin (2008), "Evaluation of Treatment of An Ethylene Plant Waste Water with a Membrane Bioreactor," AIChE Spring Meeting, Ethylene Producers Conference - Tuesday, April 8.

Ghanta, Madhav, Darryl Fahey, and Bala Subramaniam (2014), "Environmental impacts of ethylene production from diverse feedstocks and energy sources," *Appl Petrochem Res* 4, pp. 167-179.

Grover, Rajiv and Hassan M. Gomaa (1993), "Commercially Proven Technologies for the Management of Spent Caustic Generated From Olefin Plants," AIChE Spring Meeting, Ethylene Producers Conference - March.

Gudideni, Veerabhadraiah (2011), N. Mallika, Sanjeev Jindal, "Spent caustic management: Remediation review," Hydrocarbon Processing, Vol. 90, Issue 11

Hawari, Alaa; Hasanat Ramadan; Ibrahim Abu-Reesh; Mabrouk Ouederni (2015). "A comparative study of the treatment of ethylene plant spent caustic by neutralization and classical advanced oxidation," *Journal of Environmental Management* 151, pp. 105-112.

Millano, Elsie F. (1985), "Effect of Thiosulfate on 5-Day Biochemical Oxygen Demand," *Journal (Water Pollution Control Federation)*, Vol. 57, No. 5 (May), pp. 428-431.

Moussavi, Mohsen and Sina Matavos-Aramyan (2016), "Chelate-modified fenton treatment of sulfidic spent caustic," *Korean Journal of Chemical Engineering* 33(8), pp. 2384-2391.

Qin, Jian-Jun, Maung Htun Oo, Guihe Tao, Kiran Arun Kekre, Bryan Cashion and Robert Wenta (2008), "Pilot Study on Treatment of Wastewater from an Ethylene Plant with Membrane Bioreactor Technology," *The Open Chemical Engineering Journal*, 2, pp. 119-124.

Speight, James G. (2000), The Desulfurization of Heavy Oils and Residua, 2nd Edition. Marcel Dekker, Table 7.16

Stephenson, Ralph L., James B. Blackburn (1998), The Industrial Wastewater Systems Handbook, CRC Press LLC, Chapter 11.

Totten, George E., Steven R. Westbrook, Rajesh J. Shah (2003), Fuels and Lubricants Handbook: Technology, Properties, Performance, and Testing. ASTM Manual Series: MNL37WCD.

USEPA, Tier 3 Gasoline Sulfur Regulations, Regulation of Fuels and Fuel Additives: 40 CFR Chapter I, Part 80, Subpart O, Section 80.1610.

USEPA (1985) "Guide for the Application of Effluent Limitation Guidelines for the Petroleum Refining Point Source Category, Subpart B, Cracking Subcategory." 40 CFR Chapter I, Subchapter N, Part 419, Subpart D, Section 419.22. June.

478183 Scarcity of Emission Reduction Credits (ERCs) and Potential Effects on Expansions in Petrochemical and Refinery Sector

Abstract:

Emission Reduction Credits (ERCs) are required for companies with new projects or expansions that exceed certain emission thresholds in nonattainment areas in the country. Depending on the area of the country and source of ERCs, ERCs may never expire, expire after 5 years, or expire after 10 years. This can create and has created significant volatility in prices and supply in certain counties and parishes across the country. The Gulf Coast and Northeast are two areas of the US experiencing a shortage of ERCs and has seen a significant rise in ERC prices due to a reduction of supply. For example, Houston-Galveston-Brazoira (HGB) has seen VOC ERC prices increase from $5,000/ton to a high of $300,000 ton in less than three years. Baton Rouge, Louisiana ERCs have increased over 800% in VOC ERCs over a three period as well. ERCs can be a fatal flaw for a project. Some companies have moved the location of new projects due to shortage of ERCs and other companies have significantly moved up the timing of the procurement of ERCs. This trend may only continue given the forecasted tight ERC supply going forward in certain nonattainment areas.

On October 1st, 2015, EPA announced the lower of the NAAQS standards from 75 ppb to 70 ppb. This is expected to significantly increase the number of counties and parishes that will be classified as nonattainment. This will increase the projects that will require ERCs for their new projects and expansions. Numerous companies that have operated in attainment areas will have to adjust to permitting and offsetting their future emissions which will increase cost and complexity to their expansions.

In response to the shortage of VOC ERCs, many companies and industries are pushing for the ability to use NOx ERCCs to offset VOC emissions. This has occurred in Texas, and Louisiana has just passed a new regulation and states in the Northeast are being approached as well.

The ability for companies to understand the offsetting rules, supply of ERCs, demand of ERCs, procurement and contracting of ERCs along with ERC prices will be critical for companies. I plan to present price charts and discuss supply and demand.

AUTHOR INDEX

AUTHOR INDEX

Institute of Physics
Dirac House, Temple Back
Bristol BS1 6BE UK